PEARSON BACCALAUREATE

STANDARD LEVEL

Mathematics

DEVELOPED SPECIFICALLY FOR THE
IB DIPLOMA

TIM GARRY • IBRAHIM WAZIR
TOK Chapter by Richard Sims

PEARSON

Pearson Education Limited is a company incorporated in England and Wales, having its registered office at Edinburgh Gate, Harlow, Essex, CM20 2JE. Registered company number: 872828.

www.pearsonbaccalaureate.com

Pearson is a registered trademark of Pearson Education Limited.

Text © Pearson Education Limited 2008

First published 2008

15 14 13 12 11
IMP 10 9 8 7 6 5 4 3

ISBN 978 0 435 99 444 0

Copyright notice

All rights reserved. No part of this publication may be reproduced in any form or by any means (including photocopying or storing it in any medium by electronic means and whether or not transiently or incidentally to some other use of this publication) without the written permission of the copyright owner, except in accordance with the provisions of the Copyright, Designs and Patents Act 1988 or under the terms of a licence issued by the Copyright Licensing Agency, Saffron House, 6–10 Kirby Street, London EC1N 8TS (www.cla.co.uk). Applications for the copyright owner's written permission should be addressed to the publisher.

Copyright © 2008 Pearson Education, Inc. or its affiliates. All Rights Reserved. This publication is protected by copyright, and permission should be obtained from the publisher prior to any prohibited reproduction, storage in a retrieval system, or transmission in any form or by any means, electronic, mechanical, photocopying, recording, or likewise. For information regarding permissions, write to Pearson Curriculum Group Rights & Permissions, One Lake Street, Upper Saddle River, New Jersey 07458.

Edited by Maggie Rumble and Gwen Burns
Designed by Tony Richardson
Typeset by Tech-Set Limited
Original illustrations © Pearson Education Ltd 2008
Illustrated by Tech-Set Limited
Cover design by Tony Richardson
Cover photo © Alamy / Marco Regalia Illustration
Indexed by Martin Brooks
Printed in China (GCC/03)

Websites

There are links to an online appendix in this book. In order to ensure that the links are up to date, that the links work, and that the sites are not inadvertently linked to sites that could be considered offensive, we have made the links available on the Pearson website at www.pearsonhotlinks.com. When you access the site, the express code is 4235P.

Acknowledgements

We would like to thank Jane Mann for her support, encouragement, and effectiveness in bringing this project to life. We would also like to thank our editors Maggie Rumble and Gwen Burns for their excellent work.

I dedicate my work to the most important woman in my life: Lody, your untiring love and support inspires me each and every day. I am truly very fortunate to have such a wonderful, patient and devoted wife. To my children Rania, Lydia and Marwan, whose enthusiasm and encouragement spurred me on and kept me motivated. And most importantly, to my grandson Marco, who was born while I was writing this book and who inspired many of the math problems that are included here.

Special thanks go to Lydia for proofreading my chapters. Even though she specializes in English and not Math like her father, she always took the time to read the definitions and problems carefully to make sure the average student would understand them as well. So last but not least, I dedicate this to all the students who had not discovered their love for Math until opening this book.

Ibrahim Wazir

I dedicate the book to my wonderful family: Val, Bethany, Neil and Rhona. My heartfelt thanks to you for tolerating the time and energy I stole away (even in the garage) to work on the book. *Amor est vitae essentia.*

Also special thanks to my friend Marty Kehoe for his helpful comments on the early drafts of some chapters in the book.

Tim Garry

The authors and publisher would like to thank Ric Sims for writing the TOK chapter. The publishers would also like to thank David Harris for his professional guidance, Nicholas Georgiou for checking the answers, Texas Instruments for providing the TI-Smart View program, and the following individuals and organisations for permission to reproduce photographs:

© Alamy/Classic Image p.317; © Alamy/Huntstock, Inc. p.263; © Corbis Sygma/Reuters Raymond p.119; © Science Photo Library/American Institute of Physics p.564; © Science Photo Library/George Bernard p.122; © Science Photo Library/Jean-Loup Charmet p.558; © Science Photo Library/Royal Astronomical Society p.141; © Science Photo Library/Science Source p.439; © Science Photo Library/Science, Industry and Business Library/New York Public Library p.251; © Science Photo Library/Sheila Terry p.33; © Shutterstock/Andres p.315; © Shutterstock/Gary Blakeley p.459; © Shutterstock/Janna Golovacheva p. 490; © Shutterstock/Thomas Hruschka p.390; © Shutterstock/Kasia p.495; © Shutterstock/Troy Kellog p.373; © Shutterstock/Ryby p.567; © Shutterstock/Carlos E Santa Maria p.250; © Shutterstock/David H Seymour p.570.

Cover photograph of Mandelbrot reproduced with kind permission of © Alamy/Marco Regalia Illustration.

The publisher would like to thank the International Baccalaureate Organization for permission to reproduce its intellectual property. This material has been developed independently by the publisher and the content is in no way connected with nor endorsed by the International Baccalaureate Organization.

Every effort has been made to contact copyright holders of material reproduced in this book. Any omissions will be rectified in subsequent printings if notice is given to the publishers.

Contents

Introduction — vii

1 Fundamentals — 1
- 1.1 The real numbers — 1
- 1.2 Roots and radicals (surds) — 7
- 1.3 Exponents (indices) — 9
- 1.4 Scientific notation (standard form) — 12
- 1.5 Algebraic expressions — 14
- 1.6 Equations and formulae — 22

2 Functions and Equations — 32
- 2.1 Relations and functions — 32
- 2.2 Composition of functions — 41
- 2.3 Inverse functions — 45
- 2.4 Transformations of functions — 53
- 2.5 Quadratic functions — 65

3 Sequences and Series — 78
- 3.1 Sequences — 78
- 3.2 Arithmetic sequences — 80
- 3.3 Geometric sequences — 82
- 3.4 Series — 88
- 3.5 The binomial theorem — 96

4 Exponential and Logarithmic Functions — 108
- 4.1 Exponential functions — 108
- 4.2 Exponential growth and decay — 113
- 4.3 The number e — 117
- 4.4 Logarithmic functions — 121
- 4.5 Exponential and logarithmic equations — 130

5 Matrix Algebra — 140
- 5.1 Basic definitions — 141
- 5.2 Matrix operations — 143
- 5.3 Applications to systems — 149

6 Trigonometric Functions and Equations — 160
- 6.1 Angles, circles, arcs and sectors — 161
- 6.2 The unit circle and trigonometric functions — 168
- 6.3 Graphs of trigonometric functions — 177
- 6.4 Solving trigonometric equations and trigonometric identities — 189

7 Triangle Trigonometry — 204
- 7.1 Right triangles and trigonometric functions — 204
- 7.2 Trigonometric functions of any angle — 214
- 7.3 The law of sines — 222
- 7.4 The law of cosines — 229
- 7.5 Applications — 235

8 Vectors I — 250
- 8.1 Vectors as displacements in the plane — 251
- 8.2 Vector operations — 254
- 8.3 Unit vectors and direction angles — 259
- 8.4 Scalar product of two vectors — 266

9 Statistics — 273
- 9.1 Graphical tools — 275
- 9.2 Measures of central tendency — 284
- 9.3 Measures of variability — 288

10 Probability — 315
- 10.1 Randomness — 315
- 10.2 Basic definitions — 318
- 10.3 Probability assignments — 322
- 10.4 Operations with events — 328

11 Differential Calculus I: Fundamentals — 347
- 11.1 Limits of functions — 348
- 11.2 The derivative of a function: definition and basic rules — 352
- 11.3 Maxima and minima – first and second derivatives — 367
- 11.4 Tangents and normals — 382

12 Vectors II — 390
- 12.1 Vectors from a geometric viewpoint — 391
- 12.2 Scalar (dot) product — 398
- 12.3 Equations of lines — 402

13 Differential Calculus II: Further Techniques and Applications — 420
- 13.1 Derivatives of trigonometric, exponential and logarithmic functions — 420
- 13.2 The chain rule — 431
- 13.3 The product and quotient rules — 439
- 13.4 Optimization — 447
- 13.5 Summary of differentiation rules and applications — 455

Contents

14 Integral Calculus		463
14.1	Anti-derivative	463
14.2	Area and definite integral	470
14.3	Areas	477
14.4	Volumes with integrals	482
14.5	Modelling linear motion	485

15 Probability Distributions		495
15.1	Random variables	495
15.2	The binomial distribution	508
15.3	The normal distribution	515

16 Internal Assessment – Portfolio Tasks	534
17 Sample Examination Papers	549
18 Theory of Knowledge	558
Answers	576
Index	611

Introduction

Welcome to your new course! This book is designed to act as a comprehensive course book, covering all the material you will need while studying IB Mathematics Standard Level. It will also help you to prepare for your examinations in a thorough and methodical way.

Content

As you will see when you look at the table of contents all of the course topics are fully covered, though some are split over different chapters in order to group the information as logically as possible. Within each chapter, there are numbered exercises for you to practise and apply the knowledge that you have gained. They will also help you to assess your progress. There are also many examples showing you how to apply the principles you have learned.

Example 9

Find the sine, cosine and tangent of the obtuse angle that measures 150°.

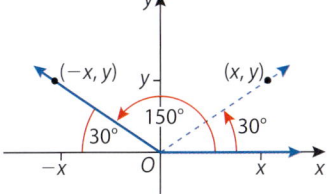

Solution

The terminal side of the angle forms a 30° angle with the x-axis. The sine values for 150° and 30° will be exactly the same, and the cosine and tangent values will be the same but of opposite sign. We know that
$\sin 30° = \frac{1}{2}$, $\cos 30° = \frac{\sqrt{3}}{2}$ and $\tan 30° = \frac{\sqrt{3}}{3}$.
Therefore, $\sin 150° = \frac{1}{2}$, $\cos 150° = -\frac{\sqrt{3}}{2}$ and $\tan 150° = -\frac{\sqrt{3}}{3}$.

At the end of each chapter, there are practice questions, some of which are taken from past exam papers. Towards the end of the book, just before the index, you will find pages with numerical answers to all the exercises and practice questions that have been included. The answers are grouped into Exercises and Practice questions for each chapter.

After the core chapters, you will find a Portfolio chapter, which contains 'problem solvers' intended to prepare you for successfully completing a set of portfolio tasks for the internal assessment component of your course.

Following this is a chapter of sample examination papers.

Finally, you will find a Theory of Knowledge chapter, which should stimulate wider research and the consideration of moral and ethical issues in the field of mathematics.

Information boxes

Throughout the book you will see a number of coloured boxes interspersed through each chapter. Each of these boxes provides different information and stimulus as follows.

> **Assessment statements**
> 3.6 Solution of triangles.
> The cosine rule: $c^2 = a^2 + b^2 - 2ab\cos C$.
> The sine rule: $\dfrac{a}{\sin A} = \dfrac{b}{\sin B} = \dfrac{c}{\sin C}$.
> Area of a triangle as $\frac{1}{2}ab\sin C$.

You will find a box like this at the start of each section in each chapter. They are the numbered objectives for the section you are about to read and they set out what content and aspects of learning are covered in that section.

> Radioactive carbon (carbon-14 or C-14), produced when nitrogen-14 is bombarded by cosmic rays in the atmosphere, drifts down to Earth and is absorbed from the air by plants. Animals eat the plants and take C-14 into their bodies. Humans in turn take C-14 into their bodies by eating both plants and animals. When a living organism dies, it stops absorbing C-14, and the C-14 that is already in the object begins to decay at a slow but steady rate, reverting to nitrogen-14. The half-life of C-14 is 5730 years. Half of the original amount of C-14 in the organic matter will have disintegrated after 5730 years; half of the remaining C-14 will have been lost after another 5730 years, and so forth. By measuring the ratio of C-14 to N-14, archaeologists are able to date organic materials. However, after about 50 000 years, the amount of C-14 remaining will be so small that the organic material cannot be dated reliably.

These boxes contain interesting information which will add to your wider knowledge but which does not fit within the main body of the text.

> The process of 'breaking-up' the vector into its components, as we did in the example, is called **resolving** the vector into its components. Notice that the process of resolving a vector is not unique. That is, you can resolve a vector into several pairs of directions.

These facts are drawn out of the main text and are highlighted. This makes them useful for quick reference and they also enable you to identify the core learning points within a section.

- **Hint:** Notice here that P(B or C) is *not* the sum of P(B) and P(C) because B and C are not disjoint.

These boxes can be found alongside questions, exercises and worked examples and they provide insight into how to answer a question in order to achieve the highest marks in an examination. They also identify common pitfalls when answering such questions and suggest approaches that examiners like to see.

Now you are ready to start. Good luck with your studies!

1 Fundamentals

Introduction

Mathematics is an exciting field of study, concerned with structure, patterns and ideas. To fully appreciate and understand these core aspects of mathematics, you need to be confident and skilled in the rules and language of algebra. Although you have encountered some, perhaps most or all, of the material in this chapter in a previous mathematics course, the aim of this chapter is to ensure that you are familiar with fundamental terminology, notation and algebraic techniques.

1.1 The real numbers

The most fundamental building blocks in mathematics are numbers and the operations that can be performed on them. Algebra, like arithmetic, involves performing operations such as addition, subtraction, multiplication and division on numbers. In arithmetic, we are performing operations on known, specific, numbers (e.g. $5 + 3 = 8$). However, in algebra we often deal with operations on unknown numbers represented by variables – usually symbolized by a letter $\left(\text{e.g. } \frac{a+b}{c} = \frac{a}{c} + \frac{b}{c}\right)$. The use of variables gives us the power to write general statements indicating relationships between numbers. But what types of numbers can variables represent? All of the mathematics in this course involves the real numbers and subsets of the real numbers.

A real number is any number that can be represented by a point on the real number line (Figure 1.1). Each point on the real number line corresponds to one and only one real number, and each real number corresponds to one and only one point on the real number line. This kind of relationship is called a **one-to-one correspondence**. The number associated with a point on the real number line is called the **coordinate** of the point.

> The word *algebra* comes from the 9th-century Arabic book *Hisâb al-Jabr w'al-Muqabala*, written by al-Khowarizmi. The title refers to transposing and combining terms, two processes used in solving equations. In Latin translations, the title was shortened to *Aljabr*, from which we get the word *algebra*. The author's name made its way into the English language in the form of the word *algorithm*.

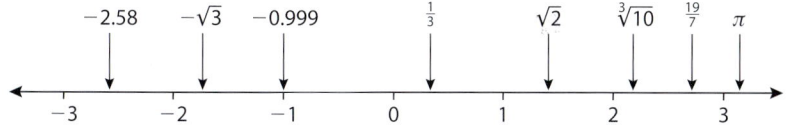

◀ **Figure 1.1** The real number line.

Subsets of the real numbers

The set of real numbers $\mathbb{R}$ contains some important subsets with which you should be familiar.

When we first learn to count, we use the numbers 1, 2, 3, … . These numbers form the set of *counting* numbers or **positive integers** $\mathbb{Z}^+$.

1 Fundamentals

● **Hint:** Do not be confused if you see other textbooks indicate that the set ℕ (usually referred to as the natural numbers) does *not* include zero – and is defined as ℕ = {1, 2, 3, …}. There is disagreement among mathematicians whether zero should be considered a natural number – i.e. reflecting how we *naturally* count. We normally do not start counting at zero. However, zero does represent a counting concept in that it is the absence of any objects in a set. Therefore, some mathematicians (and the IB mathematics curriculum) define the set ℕ as the positive integers *and* zero.

Adding zero to the positive integers (0, 1, 2, 3, …) forms the set referred to as the set ℕ in IB notation.

The set of **integers** consists of the counting numbers with their corresponding negative values and zero (… −3, −2, −1, 0, 1, 2, 3, …) and is denoted by ℤ (from the German word *Zahl* for number).

We construct the **rational numbers** ℚ by taking ratios of integers. Thus, a real number is rational if it can be written as the ratio $\frac{p}{q}$ of any two integers, where $q \neq 0$. The decimal representation of a rational number either repeats or terminates. For example, $\frac{5}{7} = 0.714\,285\,714\,285\ldots = 0.\overline{714\,285}$ (the block of six digits repeats) or $\frac{3}{8} = 0.375$ (the decimal terminates at 5, or, alternatively, has a repeating zero after the 5).

A real number that cannot be written as the ratio of two integers, such as π and $\sqrt{2}$, is called **irrational**. Irrational numbers have infinite non-repeating decimal representations. For example, $\sqrt{2} \approx 1.414\,213\,5623\ldots$ and $\pi \approx 3.141\,592\,653\,59\ldots$. There is no special symbol for the set of irrational numbers.

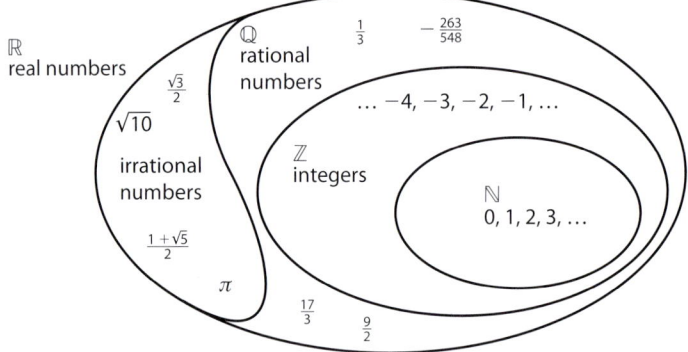

Figure 1.2 A Venn diagram representing the relationships between the different subsets of the real numbers. The rational numbers combined with the irrational numbers make up the entire set of real numbers.

Table 1.1 A summary of the subsets of the real numbers ℝ and their symbols.

Positive integers	$\mathbb{Z}^+ = \{1, 2, 3, \ldots\}$
Positive integers and zero	$\mathbb{N} = \{0, 1, 2, 3, \ldots\}$
Integers	$\mathbb{Z} = \{\ldots -3, -2, -1, 0, 1, 2, 3, \ldots\}$
Rational numbers	$\mathbb{Q}$ = any number that can be written as the ratio $\frac{p}{q}$ of any two integers, where $q \neq 0$

Sets and intervals

If every element of a set C is also an element of a set D, then C is a **subset** of set D, and is written symbolically as $C \subseteq D$. If two sets are equal (i.e. they have identical elements), they satisfy the definition of a subset and each would be a subset of the other. For example, if $C = \{2, 4, 6\}$ and $D = \{2, 4, 6\}$, then $C \subseteq D$ and $D \subseteq C$. What is more common is that a subset is a set that is contained in a larger set and does not contain at least one element of the larger set. Such a subset is called a **proper subset** and is denoted with the symbol $\subset$. For example, if $D = \{2, 4, 6\}$ and $E = \{2, 4\}$, then E is a proper subset of D and is written $E \subset D$. All of the subsets of the real numbers discussed earlier in this section are proper subsets of the real numbers, for example, $\mathbb{N} \subset \mathbb{R}$ and $\mathbb{Z} \subset \mathbb{R}$.

The symbol ∈ indicates that a number, or a number assigned to a variable, belongs to (is an element of) a set. We can write $6 \in \mathbb{Z}$, which is read '6 is an element of the set of integers'. Some sets can be described by listing their elements within brackets. For example, the set A that contains all of the integers between -2 and 2 inclusive can be written as $A = \{-2, -1, 0, 1, 2\}$. We can also use set-builder notation to indicate that the elements of set A are the values that can be assigned to a particular variable. For example, the notation $A = \{x | -2, -1, 0, 1, 2\}$ or $A = \{x \in \mathbb{Z} | -2 \leq x \leq 2\}$ indicates that 'A is the set of all x such that x is an integer greater than or equal to -2 and less than or equal to positive 2'. Set-builder notation is particularly useful for representing sets for which it would be difficult or impossible to list all of the elements. For example, to indicate the set of positive integers n greater than 5, we could write $\{n \in \mathbb{Z} | n > 5\}$ or $\{n | n > 5, n \in \mathbb{Z}\}$.

The **intersection** of A and B (Figure 1.3), denoted by $A \cap B$ and read 'A intersection B', is the set of all elements that are in both set A *and* set B. The **union** of two sets A and B (Figure 1.4), denoted by $A \cup B$ and read 'A union B', is the set of all elements that are in set A *or* in set B (or in both). The set that contains no elements is called the **empty set** (or null set) and is denoted by $\emptyset$.

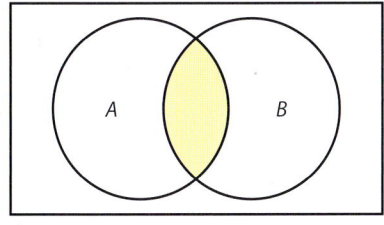

Figure 1.3 Intersection of sets A and B.
$A \cap B$

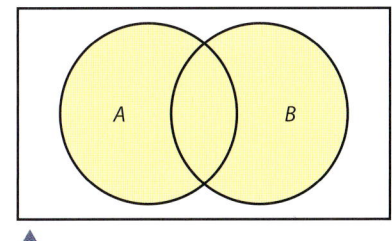

Figure 1.4 Union of sets A and B.
$A \cup B$

• **Hint:** Unless indicated otherwise, if interval notation is used, we assume that it indicates a subset of the real numbers. For example, the expression $x \in [-3, 3]$ is read 'x is any real number between -3 and 3 inclusive.'

Some subsets of the real numbers are a portion, or an **interval**, of the real number line and correspond geometrically to a line segment or a ray. They can be represented either by an inequality or by interval notation. For example, the set of all real numbers x between 2 and 5, including 2 and 5, can be expressed by the inequality $2 \leq x \leq 5$ or by the interval notation $x \in [2, 5]$. This is an example of a **closed interval** (i.e. both endpoints are included in the set) and corresponds to the line segment with endpoints of $x = 2$ and $x = 5$.

An example of an **open interval** is $-3 < x < 1$ or $x \in \,]-3, 1[$, where both endpoints are *not* included in the set. This set corresponds to a line segment with 'open dots' on the endpoints indicating they are excluded.

If an interval, such as $-4 \leq x < 2$ or $x \in [-4, 2[$, includes one endpoint but not the other, it is referred to as a **half-open** interval.

3

1 Fundamentals

• **Hint:** The symbols ∞ (positive infinity) and −∞ (negative infinity) do not represent real numbers. They are simply symbols used to indicate that an interval extends indefinitely in the positive or negative direction.

The three examples of intervals on the real number line given above are all considered **bounded** intervals in that they are line segments with two endpoints (regardless of whether included or excluded). The set of all real numbers greater than 2 is an open interval because the one endpoint is excluded and can be expressed by the inequality $x > 2$, or $x \in (2, \infty)$. This is also an example of an **unbounded** interval and corresponds to a portion of the real number line that is a ray.

Table 1.2 The nine possible types of intervals – both bounded and unbounded. For all of the examples given, we assume that $a < b$.

Interval notation	Inequality	Interval type	Graph
$x \in [a, b]$	$a \leq x \leq b$	closed bounded	
$x \in \,]a, b[$	$a < x < b$	open bounded	
$x \in [a, b[$	$a \leq x < b$	half-open bounded	
$x \in \,]a, b]$	$a < x \leq b$	half-open bounded	
$x \in [a, \infty[$	$x \geq a$	half-open unbounded	
$x \in \,]a, \infty[$	$x > a$	open unbounded	
$x \in \,]-\infty, b]$	$x \leq b$	half-open unbounded	
$x \in \,]-\infty, b[$	$x < b$	open unbounded	
$x \in \,]-\infty, \infty[$	real number line		

Absolute value (modulus)

The **absolute value** (or modulus) of a number a, denoted by $|a|$, is the distance from a to 0 on the real number line. Since a distance must be positive or zero, the absolute value of a number is never negative. Note that if a is a negative number then $-a$ will be positive.

> **Definition of absolute value**
> If a is a real number, the **absolute value** of a is
> $$|a| = \begin{cases} a & \text{if } a \geq 0 \\ -a & \text{if } a < 0 \end{cases}$$

Here are four useful properties of absolute value:

Given that a and b are real numbers, then:

1. $|a| \geq 0$ 2. $|-a| = |a|$ 3. $|ab| = |a||b|$ 4. $\left|\dfrac{a}{b}\right| = \dfrac{|a|}{|b|}, b \neq 0$

Absolute value is used to define the distance between two numbers on the real number line.

> **Distance between two points on the real number line**
> Given that a and b are real numbers, the distance between the points with coordinates a and b on the real number line is $|b-a|$, which is equivalent to $|a-b|$.

Absolute value expressions can appear in inequalities, as shown in the table below.

Inequality	Equivalent form	Graph
$|x| \leq a$	$-a \leq x \leq a$	
$|x| < a$	$-a < x < a$	
$|x| \geq a$	$x \leq -a$ or $x \geq a$	
$|x| > a$	$x < -a$ or $x > a$	

Table 1.3 Properties of absolute value inequalities.

Properties of real numbers

There are four arithmetic operations with real numbers: addition, multiplication, subtraction and division. Since subtraction can be written as addition ($a - b = a + (-b)$), and division can be written as multiplication $\left(\frac{a}{b} = a\left(\frac{1}{b}\right), b \neq 0\right)$, then the properties of the real numbers are defined in terms of addition and multiplication only. In these definitions, $-a$ is the **additive inverse** (or opposite) of a, and $\frac{1}{a}$ is the **multiplicative inverse** (or reciprocal) of a.

Table 1.4 Properties of real numbers.

Property	Rule	Example
commutative property of addition:	$a + b = b + a$	$2x^3 + y = y + 2x^3$
commutative property of multiplication:	$ab = ba$	$(x - 2)3x^2 = 3x^2(x - 2)$
associative property of addition:	$(a + b) + c = a + (b + c)$	$(1 + x) - 5x = 1 + (x - 5x)$
associative property of multiplication:	$(ab)c = a(bc)$	$(3x \cdot 5y)\left(\frac{1}{y}\right) = (3x)\left(5y \cdot \frac{1}{y}\right)$
distributive property:	$a(b + c) = ab + ac$	$x^2(x - 2) = x^2 \cdot x + x^2(-2)$
additive identity property:	$a + 0 = a$	$4y + 0 = 4y$
multiplicative identity property:	$1 \cdot a = a$	$\frac{2}{3} = 1 \cdot \frac{2}{3} = \frac{4}{4} \cdot \frac{2}{3} = \frac{8}{12}$
additive inverse property:	$a + (-a) = 0$	$6y^2 + (-6y^2) = 0$
multiplicative inverse property:	$a \cdot \frac{1}{a} = 1, a \neq 0$	$(y - 3)\left(\frac{1}{y - 3}\right) = 1$

Note: These properties can be applied in either direction as shown in the 'rules' above.

Fundamentals

Exercise 1.1

In questions 1–6, plot the two real numbers on the real number line, and then find the distance between their coordinates.

1 $5; \frac{3}{4}$ **2** $-2; -11$ **3** $13.4; 6$
4 $7; -\frac{5}{3}$ **5** $-3\pi; \frac{2\pi}{3}$ **6** $-\frac{5}{6}; -\frac{9}{4}$

In questions 7–12, write an inequality to represent the given interval and state whether the interval is closed, open or half-open. Also, state whether the interval is bounded or unbounded.

7 $[-5, 3]$ **8** $]-10, -2]$ **9** $[1, \infty[$
10 $]-\infty, 4[$ **11** $[0, 2\pi[$ **12** $[a, b]$

In questions 13–18, use interval notation to represent the subset of real numbers that is indicated by the inequality.

13 $x > 6$ **14** $x \leq -8$ **15** $2 < x < 9$
16 $0 \leq x < 12$ **17** $x > -5$ **18** $-3 \leq x \leq 3$

In questions 19–22, use inequality and interval notation to represent the given subset of real numbers.

19 x is at least 6.
20 x is greater than or equal to 4 and less than 10.
21 x is negative.
22 x is any positive number less than 25.

In questions 23–28, state the indicated set given that $A = \{1, 2, 3, 4, 5, 6, 7, 8\}$, $B = \{1, 3, 5, 7, 9\}$ and $C = \{2, 4, 6\}$.

23 $A \cap B$ **24** $A \cup B$ **25** $B \cap C$
26 $A \cup C$ **27** $A \cap C$ **28** $A \cup B \cup C$

In questions 29–32, use the symbol $\subset$ to write a correct statement involving the two sets.

29 $\mathbb{Z}$ and $\mathbb{R}$ **30** $\mathbb{N}$ and $\mathbb{Q}$ **31** $\mathbb{Z}$ and $\mathbb{N}$ **32** $\mathbb{Q}$ and $\mathbb{Z}$

In questions 33–36, express the inequality, or inequalities, using absolute value.

33 $-6 < x < 6$ **34** $x \leq -4$ or $x \geq 4$
35 $-\pi \leq x \leq \pi$ **36** $x < -1$ or $x > 1$

In questions 37–42, evaluate each absolute value expression.

37 $|-13|$ **38** $|7-11|$ **39** $-5|-5|$
40 $|-3| - |-8|$ **41** $|\sqrt{3} - 3|$ **42** $\frac{-1}{|-1|}$

In questions 43–46, find all values of x that make the equation true.

43 $|x| = 5$ **44** $|x - 3| = 4$
45 $|6 - x| = 10$ **46** $|3x + 5| = 1$

1.2 Roots and radicals (surds)

Roots

If a number can be expressed as the product of two equal factors, that factor is called the **square root** of the number. For example, 7 is the square root of 49 because $7 \times 7 = 49$. Now, 49 is also equal to -7×-7, so -7 is also a square root of 49. Every positive real number will have two real number square roots – one positive and one negative. However, there are many instances where we want only the positive square root. The symbol $\sqrt{}$ (sometimes called the root or radical symbol) indicates only the positive square root – often referred to as the **principal square root**. In words, the square roots of 16 are 4 and -4; but, symbolically, $\sqrt{16} = 4$. The negative square root of 16 is written as $-\sqrt{16}$, and when both square roots are wanted we write $\pm\sqrt{16}$.

When a number can be expressed as the product of three equal factors, then that factor is called the **cube root** of the number. For example, -4 is the cube root of -64 because $(-4)(-4)(-4) = -64$. This is written symbolically as $\sqrt[3]{-64} = -4$.

In general, if a number a can be expressed as the product of n equal factors then that factor is called the ***n*th root** of a and is written as $\sqrt[n]{a}$. n is called the **index** and if no index is written it is assumed to be a 2, thereby indicating a square root. If n is an even number (e.g. square root, fourth root, etc.) then the **principal *n*th root** is positive. For example, since $(-2)(-2)(-2)(-2) = 16$, then -2 is a fourth root of 16. However, the principal fourth root of 16, written $\sqrt[4]{16}$, is equal to $+2$.

Radicals (surds)

Some roots are rational and some are irrational. Consider the two right triangles in Figure 1.5. By applying Pythagoras' theorem, we find the length of the hypotenuse for triangle A to be exactly 5 (an integer and rational number) and the hypotenuse for triangle B to be exactly $\sqrt{80}$ (an irrational number). An irrational root – e.g. $\sqrt{80}, \sqrt{3}, \sqrt{10}, \sqrt[3]{4}$ – is called a **radical** or **surd**. The only way to express irrational roots exactly is in radical, or surd, form.

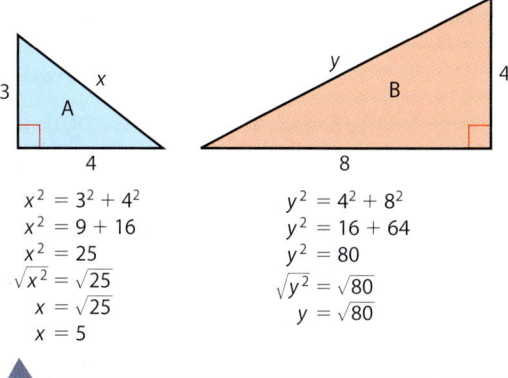

$x^2 = 3^2 + 4^2$
$x^2 = 9 + 16$
$x^2 = 25$
$\sqrt{x^2} = \sqrt{25}$
$x = \sqrt{25}$
$x = 5$

$y^2 = 4^2 + 8^2$
$y^2 = 16 + 64$
$y^2 = 80$
$\sqrt{y^2} = \sqrt{80}$
$y = \sqrt{80}$

Figure 1.5

It is not immediately obvious that the following expressions are all equivalent.

$$\sqrt{80}, \; 2\sqrt{20}, \; \frac{16\sqrt{5}}{\sqrt{16}}, \; 2\sqrt{2}\sqrt{10}, \; \frac{10\sqrt{8}}{\sqrt{10}}, \; 4\sqrt{5}, \; 5\sqrt{\frac{16}{5}}$$

Square roots occur frequently in several of the topics in this course, so it will be useful for us to be able to simplify radicals and recognise equivalent radicals. Two useful rules for manipulating expressions with radicals are given below.

● **Hint:** The solution for the hypotenuse of triangle A in Figure 1.5 involves the equation $x^2 = 25$. Because x represents a length that must be positive, we want only the positive square root when taking the square root of both sides of the equation – i.e. $\sqrt{25}$. However, if there were no constraints on the value of x, we must remember that a positive number will have two square roots and we would write $\sqrt{x^2} = \pm\sqrt{25} \Rightarrow x = \pm 5$.

1 Fundamentals

> **Simplifying radicals**
> For $a \geq 0$, $b \geq 0$ and $n \in \mathbb{Z}^+$, the following rules can be applied:
>
> **1** $\sqrt[n]{a} \times \sqrt[n]{b} = \sqrt[n]{ab}$ $\qquad$ **2** $\dfrac{\sqrt[n]{a}}{\sqrt[n]{b}} = \sqrt[n]{\dfrac{a}{b}}$
>
> Note: Each rule can be applied in either direction.

Example 1

Simplify each of the radicals.

a) $\sqrt{5} \times \sqrt{5}$ $\qquad$ b) $\sqrt{2} \times \sqrt{18}$ $\qquad$ c) $\dfrac{\sqrt{48}}{\sqrt{3}}$ $\qquad$ d) $\sqrt[3]{6} \times \sqrt[3]{36}$

Solution

a) $\sqrt{5} \times \sqrt{5} = \sqrt{5 \cdot 5} = \sqrt{25} = 5$

Note: A special case of the rule $\sqrt[n]{a} \times \sqrt[n]{b} = \sqrt[n]{ab}$ when $n = 2$ is $\sqrt{a} \times \sqrt{a} = a$.

b) $\sqrt{2} \times \sqrt{18} = \sqrt{2 \cdot 18} = \sqrt{36} = 6$

c) $\dfrac{\sqrt{48}}{\sqrt{3}} = \sqrt{\dfrac{48}{3}} = \sqrt{16} = 4$

d) $\sqrt[3]{6} \times \sqrt[3]{36} = \sqrt[3]{6 \cdot 36} = \sqrt[3]{216} = 6$

The radical $\sqrt{24}$ can be simplified because one of the factors of 24 is 4, and the square root of 4 is rational (i.e. 4 is a perfect square).

$$\sqrt{24} = \sqrt{4 \cdot 6} = \sqrt{4}\sqrt{6} = 2\sqrt{6}$$

Rewriting 24 as the product of 3 and 8 (rather than 4 and 6) would not help simplify $\sqrt{24}$ because neither 3 nor 8 are perfect squares.

Example 2

Express each in terms of the simplest possible radical.

a) $\sqrt{18}$ $\qquad$ b) $\sqrt{80}$ $\qquad$ c) $\sqrt{\dfrac{3}{25}}$ $\qquad$ d) $\sqrt{1000}$

Solution

a) $\sqrt{18} = \sqrt{9 \cdot 2} = \sqrt{9}\sqrt{2} = 3\sqrt{2}$

b) $\sqrt{80} = \sqrt{16 \cdot 5} = \sqrt{16}\sqrt{5} = 4\sqrt{5}$

Note: 4 is a factor of 80 and is a perfect square, but 16 is the largest factor that is a perfect square.

c) $\sqrt{\dfrac{3}{25}} = \dfrac{\sqrt{3}}{\sqrt{25}} = \dfrac{\sqrt{3}}{5}$

d) $\sqrt{1000} = \sqrt{100 \cdot 10} = \sqrt{100}\sqrt{10} = 10\sqrt{10}$

We prefer not to have radicals in the denominator of a fraction. Recall, from Example 1a), the special case of the rule $\sqrt[n]{a} \times \sqrt[n]{b} = \sqrt[n]{ab}$ when $n = 2$ is $\sqrt{a} \times \sqrt{a} = a$. The process of eliminating irrational numbers from the denominator is called **rationalising the denominator**.

Example 3

Rationalise the denominator of each expression.   a) $\dfrac{2}{\sqrt{3}}$ b) $\dfrac{\sqrt{7}}{4\sqrt{10}}$

Solution

a) $\dfrac{2}{\sqrt{3}} = \dfrac{2}{\sqrt{3}} \cdot \dfrac{\sqrt{3}}{\sqrt{3}} = \dfrac{2\sqrt{3}}{3}$

b) $\dfrac{\sqrt{7}}{4\sqrt{10}} = \dfrac{\sqrt{7}}{4\sqrt{10}} \cdot \dfrac{\sqrt{10}}{\sqrt{10}} = \dfrac{\sqrt{70}}{4 \cdot 10} = \dfrac{\sqrt{70}}{40}$

Exercise 1.2

In questions 1–9, express each in terms of the simplest possible radical.

1. $\sqrt{8}$
2. $\dfrac{\sqrt{28}}{\sqrt{7}}$
3. $\sqrt{3} \times \sqrt{12}$
4. $\sqrt[3]{9} \times \sqrt[3]{3}$
5. $\dfrac{\sqrt[4]{64}}{\sqrt[4]{4}}$
6. $\sqrt{\dfrac{15}{20}}$
7. $\sqrt{50}$
8. $\sqrt{63}$
9. $\sqrt{288}$

In questions 10–13, completely simplify the expression.

10. $7\sqrt{2} - 3\sqrt{2}$
11. $\sqrt{12} + 8\sqrt{3}$
12. $\sqrt{300} + 5\sqrt{2} - \sqrt{72}$
13. $\sqrt{75} + 2\sqrt{24} - \sqrt{48}$

In questions 14–19, rationalise the denominator, simplifying if possible.

14. $\dfrac{1}{\sqrt{2}}$
15. $\dfrac{3}{\sqrt{5}}$
16. $\dfrac{2\sqrt{3}}{\sqrt{7}}$
17. $\dfrac{1}{\sqrt{27}}$
18. $\dfrac{8}{3\sqrt{2}}$
19. $\dfrac{\sqrt{12}}{\sqrt{18}}$

1.3 Exponents (indices)

Repeated multiplication of identical numbers can be written more efficiently by using exponential notation.

> **Exponential notation**
> If a is any real number ($a \in \mathbb{R}$) and n is a positive integer ($n \in \mathbb{Z}^+$), then
> $$a^n = \underbrace{a \cdot a \cdot a \cdot \ldots \cdot a}_{n \text{ factors}}$$
> where n is the **exponent**, a is the **base** and a^n is called the **nth power** of a.
> Note: n is also called the **power** or **index** (plural: indices).

Integer exponents

We now state seven laws of integer exponents (or indices) that you will have learned in a previous mathematics course. Familiarity with these rules is essential for work throughout this course.

Let a and b be real numbers ($a, b \in \mathbb{R}$) and let m and n be positive integers ($m, n \in \mathbb{Z}^+$). Assume that all denominators and bases are not equal to zero. All of the laws can be applied in either direction.

1 Fundamentals

Table 1.5 Laws of exponents (indices) for integer exponents.

- **Hint:** It is important to recognise the difference between exponential expressions such as $(-3)^2$ and -3^2. In the expression $(-3)^2$, the parentheses make it clear that -3 is the base being raised to the power of 2. However, in -3^2 the negative sign is not considered to be a part of the base with the expression being the same as $-(3)^2$ so that 3 is the base being raised to the power of 2. Hence, $(-3)^2 = 9$ and $-3^2 = -9$.

	Property	Example	Description
1.	$b^m b^n = b^{m+n}$	$x^2 x^5 = x^7$	multiplying like bases
2.	$\dfrac{b^m}{b^n} = b^{m-n}$	$\dfrac{2w^7}{3w^2} = \dfrac{2w^5}{3}$	dividing like bases
3.	$(b^m)^n = b^{mn}$	$(3^x)^2 = 3^{2x} = (3^2)^x = 9^x$	a power raised to a power
4.	$(ab)^n = a^n b^n$	$(4k)^3 = 4^3 k^3 = 64 k^3$	the power of a product
5.	$\left(\dfrac{a}{b}\right)^n = \dfrac{a^n}{b^n}$	$\left(\dfrac{y}{3}\right)^2 = \dfrac{y^2}{3^2} = \dfrac{y^2}{9}$	the power of a quotient
6.	$a^0 = 1$	$(t^2 + 5)^0 = 1$	definition of a zero exponent
7.	$a^{-n} = \dfrac{1}{a^n}$	$2^{-3} = \dfrac{1}{2^3} = \dfrac{1}{8}$	definition of a negative exponent

The last two laws of exponents listed above – the definition of a zero exponent and the definition of a negative exponent – are often assumed without proper explanation. The definition of a^n as repeated multiplication, i.e. n factors of a, is easily understood when n is a positive integer. So how do we formulate appropriate definitions for a^n when n is negative or zero? These definitions will have to be compatible with the laws for positive integer exponents. If the law stating $b^m b^n = b^{m+n}$ is to hold for a zero exponent, then $b^n b^0 = b^{n+0} = b^n$. Since the number 1 is the identity element for multiplication (multiplicative identity property) then $b^n \cdot 1 = b^n$. Therefore, we must define b^0 as the number 1. If the law $b^m b^n = b^{m+n}$ is to also hold for negative integer exponents, then $b^n b^{-n} = b^{n-n} = b^0 = 1$. Since the product of b^n and b^{-n} is 1, they must be reciprocals (multiplicative inverse property). Therefore, we must define b^{-n} as $\dfrac{1}{b^n}$.

Rational exponents

We know that $4^3 = 4 \times 4 \times 4$ and $4^0 = 1$ and $4^{-2} = \dfrac{1}{4^2} = \dfrac{1}{4 \times 4}$, but what meaning are we to give to $4^{\frac{1}{2}}$? In order to carry out algebraic operations with expressions having exponents that are rational numbers, it will be very helpful if they follow the laws established for integer exponents. From the law $b^m b^n = b^{m+n}$, it must follow that $4^{\frac{1}{2}} \times 4^{\frac{1}{2}} = 4^{\frac{1}{2} + \frac{1}{2}} = 4^1$. Likewise, from the law $(b^m)^n = b^{mn}$, it follows that $(4^{\frac{1}{2}})^2 = 4^{\frac{1}{2} \cdot 2} = 4^1$. Therefore, we need to define $4^{\frac{1}{2}}$ as the square root of 4 or, more precisely, as the principal (positive) square root of 4, that is, $\sqrt{4}$. We are now ready to use radicals to define a rational exponent of the form $\dfrac{1}{n}$, where n is a positive integer. If the rule $(b^m)^n = b^{mn}$ is to apply when $m = \dfrac{1}{n}$, it must follow that $(b^{\frac{1}{n}})^n = b^{\frac{n}{n}} = b^1$. This means that the nth power of $b^{\frac{1}{n}}$ is b and, from the discussion of nth roots in Section 1.2, we define $b^{\frac{1}{n}}$ as the principal nth root of b.

> **Definition of $b^{\frac{1}{n}}$**
>
> If $n \in \mathbb{Z}^+$, then $b^{\frac{1}{n}}$ is the principal nth root of b. Using a radical, this means
> $$b^{\frac{1}{n}} = \sqrt[n]{b}$$

This definition allows us to evaluate exponential expressions such as the following:

$$36^{\frac{1}{2}} = \sqrt{36} = 6; \quad (-27)^{\frac{1}{3}} = \sqrt[3]{-27} = -3; \quad \left(\frac{1}{81}\right)^{\frac{1}{4}} = \sqrt[4]{\frac{1}{81}} = \frac{1}{3}$$

Now we can apply the definition of $b^{\frac{1}{n}}$ and the rule $(b^m)^n = b^{mn}$ to develop a rule for expressions with exponents not just of the form $\frac{1}{n}$ but of the more general form $\frac{m}{n}$.

$$b^{\frac{m}{n}} = b^{m \cdot \frac{1}{n}} = (b^m)^{\frac{1}{n}} = \sqrt[n]{b^m}; \text{ or, equivalently, } b^{\frac{m}{n}} = b^{\frac{1}{n} \cdot m} = (b^{\frac{1}{n}})^m = (\sqrt[n]{b})^m$$

This will allow us to evaluate exponential expressions such as $9^{\frac{3}{2}}, (-8)^{\frac{5}{3}}$ and $64^{\frac{5}{6}}$.

> **Definition of rational exponents**
> If m and n are positive integers with no common factors, then
> $$b^{\frac{m}{n}} = \sqrt[n]{b^m} \text{ or } (\sqrt[n]{b})^m$$
> If n is an even number, we must have $b \geq 0$.

The numerator of a rational exponent indicates the power to which the base of the exponential expression is raised, and the denominator indicates the root to be taken. With this definition for rational exponents, we can conclude that the laws of exponents, stated for integer exponents in Section 1.3, also hold true for rational exponents.

Example 4

Evaluate and/or simplify each of the following exponential expressions.
a) $(2xy^2)^3$
b) $2(xy^2)^3$
c) $(-2)^{-3}$

d) $(a-2)^0$
e) $(3^3)^{\frac{1}{2}} \cdot 9^{\frac{3}{4}}$
f) $\dfrac{a^{-2}b^4}{a^{-5}b^5}$

g) $(-32)^{-\frac{4}{5}}$
h) $8^{\frac{2}{3}}$
i) $\left(\frac{1}{2}x^2y\right)^3(x^3y^{-2})^{-1}$

j) $\dfrac{\sqrt{a+b}}{a+b}$
k) $\dfrac{(x+y)^2}{(x+y)^{-2}}$

Solution

a) $(2xy^2)^3 = 2^3x^3(y^2)^3 = 8x^3y^6$

b) $2(xy^2)^3 = 2x^3(y^2)^3 = 2x^3y^6$

c) $(-2)^{-3} = \dfrac{1}{(-2)^3} = -\dfrac{1}{8}$

d) $(a-2)^0 = 1$

e) $(3^3)^{\frac{1}{2}} \cdot 9^{\frac{3}{4}} = 3^{\frac{3}{2}}(3^2)^{\frac{3}{4}} = 3^{\frac{3}{2}} \cdot 3^{\frac{3}{2}} = 3^{\frac{6}{2}} = 3^3 = 27$

f) $\dfrac{a^{-2}b^4}{a^{-5}b^5} = \dfrac{a^{-2-(-5)}}{b^{5-4}} = \dfrac{a^3}{b}$

g) $(-32)^{-\frac{4}{5}} = [-2^5]^{-\frac{4}{5}} = (-2)^{-4} = \dfrac{1}{(-2)^4} = \dfrac{1}{16}$

h) $8^{\frac{2}{3}} = \sqrt[3]{8^2} = \sqrt[3]{64} = 4$ or $8^{\frac{2}{3}} = (\sqrt[3]{8})^2 = (2)^2 = 4$ or $8^{\frac{2}{3}} = (2^3)^{\frac{2}{3}} = 2^2 = 4$

i) $\left(\frac{1}{2}x^2y\right)^3(x^3y^{-2})^{-1} = \left(\dfrac{x^6y^3}{8}\right)(x^{-3}y^2) = \dfrac{x^{6-3}y^{3+2}}{8} = \dfrac{x^3y^5}{8}$

j) $\dfrac{\sqrt{a+b}}{a+b} = \dfrac{(a+b)^{\frac{1}{2}}}{(a+b)^1} = \dfrac{1}{(a+b)^{1-\frac{1}{2}}} = \dfrac{1}{(a+b)^{\frac{1}{2}}} = \dfrac{1}{\sqrt{a+b}}$

Note: Avoid an error here. $\sqrt[n]{a+b} \neq \sqrt[n]{a} + \sqrt[n]{b}$. Also, $\sqrt{a+b} \neq \sqrt{a} + \sqrt{b}$ and $\sqrt{a^2+b^2} \neq a+b$.

k) $\dfrac{(x+y)^2}{(x+y)^{-2}} = (x+y)^{2-(-2)} = (x+y)^4$

Note: Avoid an error here. $(x+y)^n \neq x^n + y^n$.

Although $(x+y)^4 = x^4 + 4x^3y + 6x^2y^2 + 4xy^3 + y^4$, expanding is not generally 'simplifying'.

Exercise 1.3

In questions 1–6, simplify (without your GDC) each expression to a single integer.

1 $16^{\frac{1}{4}}$ **2** $9^{\frac{3}{2}}$ **3** $64^{\frac{2}{3}}$
4 $8^{\frac{4}{3}}$ **5** $32^{\frac{3}{5}}$ **6** $(\sqrt{2})^6$

In questions 7–9, simplify each expression (without your GDC) to a quotient of two integers.

7 $\left(\dfrac{8}{27}\right)^{\frac{2}{3}}$ **8** $\left(\dfrac{9}{16}\right)^{\frac{1}{2}}$ **9** $\left(\dfrac{25}{4}\right)^{\frac{3}{2}}$

In questions 10–13, evaluate (without your GDC) each expression.

10 $(-3)^{-2}$ **11** $(13)^0$ **12** $\dfrac{4 \cdot 3^{-2}}{2^{-2} \cdot 3^{-1}}$ **13** $\left(-\dfrac{3}{4}\right)^{-3}$

In questions 14–28, simplify each exponential expression (leave only positive exponents).

14 $3(-ab^2)^2$ **15** $3(-ab^2)^3$ **16** $(-3ab^2)^2$
17 $5x^3y^{-2} \cdot 2x^2y^5$ **18** $\dfrac{32w^2}{24w^3}$ **19** $\dfrac{6m^3n^{-2}}{8m^{-3}n^2}$
20 $\left(\dfrac{1}{2}m^2n^{-2}\right)^3$ **21** $3^{2m} \cdot 3^n$ **22** $\dfrac{x^{-1}y^5}{xy^3}$
23 $\dfrac{4a^3b^5}{(2a^2b)^4}$ **24** $\dfrac{(\sqrt[3]{x})(\sqrt[3]{x^4})}{\sqrt[3]{x^2}}$ **25** $\dfrac{12(a+b)^3}{9(a+b)}$
26 $\dfrac{(x+4y)^{\frac{1}{2}}}{2(x+4y)^{-1}}$ **27** $\dfrac{p^2+q^2}{\sqrt{p^2+q^2}}$ **28** $4^{3n} \cdot 2^{2m}$

1.4 Scientific notation (standard form)

Exponents provide an efficient way of writing and calculating with very large or very small numbers. The need for this is especially great in science. For example, a light year (the distance that light travels in one year) is 9 460 730 472 581 kilometres, and the mass of a single water molecule is 0.000 000 000 000 000 000 000 0056 grams. It is far more convenient and useful to write such numbers in **scientific notation** (also called **standard form**).

> **Definition of scientific notation**
> A positive number N is written in scientific notation if it is expressed in the form:
> $$N = a \times 10^k, \text{ where } 1 \leq a < 10 \text{ and } k \text{ is an integer}$$

In scientific notation, a light year is about 9.46×10^{12} kilometres. This expression is determined by observing that when a number is multiplied by 10^k and k is **positive**, the decimal point will move k places to the **right**. Therefore, $9.46 \times 10^{12} = 9\,460\,000\,000\,000$ (12 decimal places). Knowing that when a number is multiplied by 10^k and k is **negative** the decimal point will move k places to the **left** helps us to express the mass of a water molecule as 5.6×10^{-24} grams. This expression is equivalent to $0.\underbrace{000\,000\,000\,000\,000\,000\,0056}_{24 \text{ decimal places}}$.

Scientific notation is also a very convenient way of indicating the number of **significant figures** (digits) to which a number has been approximated. A light year expressed to an accuracy of 13 significant figures is $9\,460\,730\,472\,581$ kilometres. However, many calculations will not require such a high degree of accuracy. For a certain calculation it may be more appropriate to have a light year approximated to 4 significant figures, which could be written as $9\,461\,000\,000\,000$ kilometres, or more efficiently and clearly in scientific notation as 9.461×10^{12} kilometres.

Not only is scientific notation conveniently compact, it also allows a quick comparison of the magnitude of two numbers without the need to count zeros. Moreover, it enables us to use the laws of exponents to simplify otherwise unwieldy calculations.

Example 5

Use scientific notation to calculate each of the following.
a) $64\,000 \times 2\,500\,000\,000$
b) $\dfrac{0.000\,000\,78}{0.000\,000\,0012}$
c) $\sqrt[3]{27\,000\,000\,000}$

Solution

a) $64\,000 \times 2\,5000\,000\,000 = (6.4 \times 10^4)(2.5 \times 10^9)$
$= 6.4 \times 2.5 \times 10^4 \times 10^9$
$= 16 \times 10^{4+9}$
$= 1.6 \times 10^1 \times 10^{13} = 1.6 \times 10^{14}$

b) $\dfrac{0.000\,000\,78}{0.000\,000\,0012} = \dfrac{7.8 \times 10^{-7}}{1.2 \times 10^{-9}} = \dfrac{7.8}{1.2} \times \dfrac{10^{-7}}{10^{-9}} = 6.5 \times 10^{-7-(-9)}$
$= 6.5 \times 10^2$ or 650

c) $\sqrt[3]{27\,000\,000\,000} = (2.7 \times 10^{10})^{\frac{1}{3}} = (27 \times 10^9)^{\frac{1}{3}} = (27)^{\frac{1}{3}}(10^9)^{\frac{1}{3}}$
$= 3 \times 10^3$ or 3000

Your GDC will automatically express numbers in scientific notation when a large or small number exceeds its display range. For example, if you use your GDC to compute 2 raised to the 64th power, the display (depending on the GDC model) will show the approximation

$$1.844674407\text{E}19 \text{ or } 1.844674407 \text{ } 19$$

The final digits indicate the power of 10, and we interpret the result as $1.844\,674\,408 \times 10^{19}$. ($2^{64}$ is exactly $18\,446\,744\,073\,709\,551\,616$.)

Exercise 1.4

In questions 1–8, write each number in scientific notation, rounding to 3 significant figures.

1. 253.8
2. 0.007 81
3. 7 405 239
4. 0.000 001 0448
5. 4.9812
6. 0.001 991
7. Land area of Earth: 148 940 000 square kilometres
8. Relative density of hydrogen: 0.000 0899 grams per cm³

In questions 9–12, write each number in ordinary decimal notation.

9. 2.7×10^{-3}
10. 5×10^7
11. 9.035×10^{-8}
12. 4.18×10^{12}

In questions 13–16, use scientific notation and the laws of exponents to perform the indicated operations. Give the result in scientific notation rounded to 2 significant figures.

13. $(2.5 \times 10^{-3})(10 \times 10^5)$
14. $\dfrac{3.2 \times 10^6}{1.6 \times 10^2}$
15. $\dfrac{(1 \times 10^{-3})(3.28 \times 10^6)}{4 \times 10^7}$
16. $(2 \times 10^3)^4(3.5 \times 10^5)$

1.5 Algebraic expressions

Examples of algebraic expressions are:

$$5a^3b^2 \qquad 2x^2 + 7x - 8 \qquad \frac{y^3 - 1}{y + 1} \qquad \frac{(bx + c)^3}{2 - \sqrt{a}}$$

Algebraic expressions are formed by combining variables and constants using addition, subtraction, multiplication, division, exponents and radicals.

Polynomials

An algebraic expression that has only non-negative powers of one or more variable and contains no variable in a denominator is called a **polynomial**.

> **Definition of a polynomial in the variable x**
> Given $a_0, a_1, a_2, \ldots, a_n \in \mathbb{R}$ $a_n \neq 0$ and $n \in \mathbb{Z}^+$, a **polynomial in x** is a sum of distinct **terms** in the form
> $$a_n x^n + a_{n-1} x^{n-1} + \ldots + a_1 x + a_0$$
> where $a_1, a_2, \ldots, a_n$ are the **coefficients**, a_0 is the **constant term** and n (the highest exponent) is the **degree** of the polynomial.

● **Hint:** Polynomials with one, two and three terms are called **monomials**, **binomials** and **trinomials**, respectively. A polynomial of degree one is called linear; degree two is called quadratic; degree three is cubic; and degree four is quartic. Quadratic equations and functions are covered in Chapter 2.

Polynomials are added or subtracted using the properties of real numbers that were discussed in Section 1.1. We do this by combining like terms – terms containing the same variable(s) raised to the same power(s) – and applying the distributive property.

For example,
$2x^2y + 6x^2 - 7x^2y = 2x^2y - 7x^2y + 6x^2$ rearranging terms so the like terms are together

$\qquad\qquad\qquad\qquad = (2 - 7)x^2y + 6x^2$ applying distributive property: $ab + ac = (b + c)a$

$\qquad\qquad\qquad\qquad = -5x^2y + 6x^2$ no like terms remain, so polynomial is simplified

Expanding and factorizing polynomials

We apply the distributive property in the other direction, i.e. $a(b + c) = ab + ac$, in order to multiply polynomials. For example,

$(2x - 3)(x + 5) = 2x(x + 5) - 3(x + 5)$

$\qquad\qquad\qquad = 2x^2 + 10x - 3x - 15$ collecting like terms $10x$ and $-3x$

$\qquad\qquad\qquad = 2x^2 + 7x - 15$ terms written in descending order of the exponents

The process of multiplying polynomials is often referred to as **expanding**. Especially in the case of a polynomial being raised to a power, the number of terms in the resulting polynomial, after applying the distributive property and combining like terms, has increased (expanded) compared to the original number of terms. For example,

$(x + 3)^2 = (x + 3)(x + 3)$ squaring a first degree (linear) binomial

$\qquad\quad\; = x(x + 3) + 3(x + 3)$

$\qquad\quad\; = x^2 + 3x + 3x + 9$

$\qquad\quad\; = x^2 + 6x + 9$ the result is a second degree (quadratic) trinomial

and,

$(x + 1)^3 = (x + 1)(x + 1)(x + 1)$ cubing a first degree binomial

$\qquad\quad\; = (x + 1)(x^2 + x + x + 1)$

$\qquad\quad\; = x(x^2 + 2x + 1) + 1(x^2 + 2x + 1)$

$\qquad\quad\; = x^3 + 2x^2 + x + x^2 + 2x + 1$

$\qquad\quad\; = x^3 + 3x^2 + 3x + 1$ the result is a third degree (cubic) polynomial with four terms

A pair of binomials of the form $a + b$ and $a - b$ are called **conjugates**. In most instances, the product of two binomials produces a trinomial. However, the product of a pair of conjugates produces a binomial such that both terms are squares and the second term is negative – referred to as a **difference of two squares**. For example,

$(x + 5)(x - 5) = x(x - 5) + 5(x - 5)$ multiplying two conjugates

$\qquad\qquad\quad\; = x^2 - 5x + 5x - 25$

$\qquad\qquad\quad\; = x^2 - 25$ $x^2 - 25$ is a difference of two squares

1 Fundamentals

The inverse (or undoing) of multiplication (expansion) is factorization. If it is helpful for us to rewrite a polynomial as a product, then we need to factorize it – i.e. apply the distributive property in the *reverse* direction. The previous four examples can be used to illustrate equivalent pairs of factorized and expanded polynomials.

Factorized		Expanded
$(2x - 3)(x + 5)$	$=$	$2x^2 + 7x - 15$
$(x + 3)^2$	$=$	$x^2 + 6x + 9$
$(x + 1)^3$	$=$	$x^3 + 3x^2 + 3x + 1$
$(x + 5)(x - 5)$	$=$	$x^2 - 25$

Certain polynomial expansions (products) and factorizations occur so frequently you should be able to quickly recognize and apply them. Here is a list of some of the more common ones. You can verify these identities by performing the multiplication.

Common polynomial expansion and factorization patterns

Expanding →

$$(x + a)(x + b) = x^2 + (a + b)x + ab$$
$$(ax + b)(cx + d) = acx^2 + (ad + bc)x + bd$$
$$(a + b)(a - b) = a^2 - b^2$$
$$(a + b)^2 = a^2 + 2ab + b^2$$
$$(a - b)^2 = a^2 - 2ab + b^2$$
$$(a + b)^3 = a^3 + 3a^2b + 3ab^2 + b^3$$
$$(a - b)^3 = a^3 - 3a^2b + 3ab^2 - b^3$$

← Factorizing

These identities are useful patterns into which we can substitute any number or algebraic expression for *a*, *b* or *x*. This allows us to efficiently find products and powers of polynomials and also to factorize many polynomials.

Example 6

Find each product.
a) $(x + 2)(x - 7)$ b) $(3x - 4)(4x + 1)$ c) $(6x + y)(6x - y)$
d) $(4h - 5)^2$ e) $(x^2 + 2)^3$ f) $(3 + 2\sqrt{5})(3 - 2\sqrt{5})$

Solution

a) This product fits the pattern $(x + a)(x + b) = x^2 + (a + b)x + ab$.
$$(x + 2)(x - 7) = x^2 + (2 - 7)x + (2)(-7) = x^2 - 5x - 14$$

b) This product fits the pattern $(ax + b)(cx + d) = acx^2 + (ad + bc)x + bd$.
$$(3x - 4)(4x + 1) = 12x^2 + (3 - 16)x - 4 = 12x^2 - 13x - 4$$

• **Hint:** You should be able to perform the middle step 'mentally' without writing it.

c) This fits the pattern $(a + b)(a - b) = a^2 - b^2$ where the result is a difference of two squares.
$$(5x^3 + 3y)(5x^3 - 3y) = (5x^3)^2 - (3y)^2 = 25x^6 - 9y^2$$

d) This fits the pattern $(a - b)^2 = a^2 - 2ab + b^2$.
$$(4h - 5)^2 = (4h)^2 - 2(4h)(5) + (5)^2 = 16h^2 - 40h + 25$$

e) This fits the pattern $(a + b)^3 = a^3 + 3a^2b + 3ab^2 + b^3$.
$$(x^2 + 2)^3 = (x^2)^3 + 3(x^2)^2(2) + 3(x^2)(2)^2 + (2)^3 = x^6 + 6x^4 + 12x^2 + 8$$

f) The pair of expressions being multiplied do not have a variable but they are conjugates, so they fit the pattern $(a + b)(a - b) = a^2 - b^2$.
$$(3 + 2\sqrt{5})(3 - 2\sqrt{5}) = (3)^2 - (2\sqrt{5})^2 = 9 - (4 \cdot 5) = 9 - 20 = -11$$

Note: The result of multiplying two **irrational** conjugates is a single **rational** number. We will make use of this result to simplify certain fractions.

Example 7

Completely factorize the following expressions.
a) $2x^2 - 14x + 24$
b) $2x^2 + x - 15$
c) $4x^6 - 9$
d) $3y^3 + 24y^2 + 48y$
e) $(x + 3)^2 - y^2$
f) $5x^3y + 20xy^3$

Solution

a) $2x^2 - 14x + 24$
$\quad = 2(x^2 - 7x + 12)$ factor out the greatest common factor

$\quad = 2[x^2 + (-3 - 4)x + (-3)(-4)]$ fits the pattern
$\quad\quad\quad\quad\quad\quad\quad\quad\quad\quad\quad\quad (x + a)(x + b) = x^2 + (a + b)x + ab$

$\quad = 2(x - 3)(x - 4)$ 'trial and error' to find
$\quad\quad\quad\quad\quad\quad\quad\quad -3 - 4 = -7$ and $(-3)(-4) = 12$

b) The terms have no common factor and the leading coefficient is not equal to one. This factorization requires a logical 'trial and error' approach. There are eight possible factorizations.
$(2x - 1)(x + 15)$ $(2x - 3)(x + 5)$ $(2x - 5)(x + 3)$ $(2x - 15)(x + 1)$
$(2x + 1)(x - 15)$ $(2x + 3)(x - 5)$ $(2x + 5)(x - 3)$ $(2x + 15)(x - 1)$

Testing the middle term in each, you find that the correct factorization is $2x^2 + x - 15 = (2x - 5)(x + 3)$.

c) This binomial can be written as the difference of two squares,
$4x^6 - 9 = (2x^3)^2 - (3)^2$, fitting the pattern $a^2 - b^2 = (a + b)(a - b)$.
Therefore, $4x^6 - 9 = (2x^3 + 3)(2x^3 - 3)$.

d) $3y^3 + 24y^2 + 48y = 3y(y^2 + 8y + 16)$ factor out the greatest common factor

$ = 3y(y^2 + 2 \cdot 4y + 4^2)$ fits the pattern $a^2 + 2ab + b^2 = (a+b)^2$

$ = 3y(y+4)^2$

e) Fits the difference of two squares pattern: $a^2 - b^2 = (a+b)(a-b)$ with $a = x+3$ and $b = y$.
Therefore, $(x+3)^2 - y^2 = [(x+3) + y][(x+3) - y]$
$ = (x+y+3)(x-y+3)$

f) $5x^3y + 20xy^3 = 5xy(x^2 + 4y^2)$: although both of the terms x^2 and $4y^2$ are perfect squares, the expression $x^2 + 4y^2$ is not a *difference* of squares and, hence, it cannot be factorized. The sum of two squares, $a^2 + b^2$, cannot be factorized.

Guidelines for factoring polynomials
1 Factor out the greatest common factor, if one exists.
2 Determine if the polynomial, or any factors, fit any of the special polynomial patterns – and factor accordingly.
3 Any quadratic trinomial of the form $ax^2 + bx + c$ will require a logical trial and error approach, if it factorizes.

Most polynomials cannot be factored into a product of polynomials with integer coefficients. In fact, factoring is often difficult, even when possible, for polynomials with degree 3 or higher. Nevertheless, factorizing is a powerful algebraic technique that can be applied in many situations.

Algebraic fractions

An **algebraic fraction** (or rational expression) is a quotient of two algebraic expressions or two polynomials. Given a certain algebraic fraction, we must assume that the variable can only have values such that the denominator is not zero. For example, for the algebraic fraction $\frac{x+3}{x^2-4}$, x cannot be 2 or -2. Most of the algebraic fractions that we will encounter will have numerators and denominators that are polynomials.

Simplifying algebraic fractions

When trying to simplify algebraic fractions, we need to completely factor the numerator and denominator and cancel any common factors.

Example 8

Simplify each algebraic fraction.

a) $\dfrac{2a^2 - 2ab}{6ab - 6b^2}$ b) $\dfrac{1-x^2}{x^2+x-2}$ c) $\dfrac{(x+h)^2 - x^2}{h}$

Solution

a) $\dfrac{2a^2 - 2ab}{6ab - 6b^2} = \dfrac{2a(a-b)}{6b(a-b)} = \dfrac{\overset{1}{\cancel{2}}a}{\underset{3}{\cancel{6}}b} = \dfrac{a}{3b}$

b) $\dfrac{1 - x^2}{x^2 + x - 2} = \dfrac{(1 - x)(1 + x)}{(x - 1)(x + 2)} = \dfrac{-(-1 + x)(1 + x)}{(x - 1)(x + 2)} = \dfrac{-\cancel{(x-1)}(x + 1)}{\cancel{(x-1)}(x + 2)}$

$= -\dfrac{x + 1}{x + 2}$ or $\dfrac{-x - 1}{x + 2}$

c) $\dfrac{(x + h)^2 - x^2}{h} = \dfrac{x^2 + 2hx + h^2 - x^2}{h} = \dfrac{2hx + h^2}{h} = \dfrac{\cancel{h}(2x + h)}{\cancel{h}} = 2x + h$

Adding and subtracting algebraic fractions

Before any fractions – numerical or algebraic – can be added or subtracted they must be expressed with the same denominator, preferably the least common denominator. Then the numerators can be added or subtracted according to the rule: $\dfrac{a}{b} + \dfrac{c}{d} = \dfrac{ad}{bd} + \dfrac{bc}{bd} = \dfrac{ad + bc}{bd}$.

Example 9

Perform the indicated operation and simplify.

a) $x - \dfrac{1}{x}$
b) $\dfrac{2}{a + b} + \dfrac{3}{a - b}$
c) $\dfrac{2}{x + 2} - \dfrac{x - 4}{2x^2 + x - 6}$

Solution

a) $x - \dfrac{1}{x} = \dfrac{x}{1} - \dfrac{1}{x} = \dfrac{x^2}{x} - \dfrac{1}{x} = \dfrac{x^2 - 1}{x}$ or $\dfrac{(x + 1)(x - 1)}{x}$

b) $\dfrac{2}{a + b} + \dfrac{3}{a - b} = \dfrac{2}{a + b} \cdot \dfrac{a - b}{a - b} + \dfrac{3}{a - b} \cdot \dfrac{a + b}{a + b} = \dfrac{2(a - b) + 3(a + b)}{(a + b)(a - b)}$

$= \dfrac{2a - 2b + 3a + 3b}{a^2 - b^2} = \dfrac{5a + b}{a^2 - b^2}$

c) $\dfrac{2}{x + 2} - \dfrac{x - 4}{2x^2 + x - 6} = \dfrac{2}{x + 2} - \dfrac{x - 4}{(2x - 3)(x + 2)}$

$= \dfrac{2}{x + 2} \cdot \dfrac{2x - 3}{2x - 3} - \dfrac{x - 4}{(2x - 3)(x + 2)}$

$= \dfrac{2(2x - 3) - (x - 4)}{(2x - 3)(x + 2)}$

$= \dfrac{4x - 6 - x + 4}{(2x - 3)(x + 2)}$

$= \dfrac{3x - 2}{(2x - 3)(x + 2)}$ or $\dfrac{3x - 2}{2x^2 + x - 6}$

• **Hint:** Although it is true that $\dfrac{a + b}{c} = \dfrac{a}{c} + \dfrac{b}{c}$, be careful to avoid an error here: $\dfrac{a}{b + c} \neq \dfrac{a}{b} + \dfrac{a}{c}$. Also, be sure to only cancel common *factors* between numerator and denominator. It is true that $\dfrac{ac}{bc} = \dfrac{a}{b}$ (with the common factor of c cancelling) because $\dfrac{ac}{bc} = \dfrac{a}{b} \cdot \dfrac{c}{c} = \dfrac{a}{b} \cdot 1 = \dfrac{a}{b}$; but, in general, it is not true that $\dfrac{a + c}{b + c} = \dfrac{a}{b}$. c is not a common factor of the numerator and denominator.

Simplifying a compound fraction

Fractional expressions with fractions in the numerator or denominator, or both, are usually referred to as compound fractions. A compound fraction is best simplified by first simplifying both its numerator and denominator into single fractions, and then multiplying the numerator and denominator by the reciprocal of the denominator, i.e. $\dfrac{\frac{a}{b}}{\frac{c}{d}} = \dfrac{\frac{a}{b} \cdot \frac{d}{c}}{\frac{c}{d} \cdot \frac{d}{c}} = \dfrac{\frac{ad}{bc}}{1} = \dfrac{ad}{bc}$; thereby expressing the compound fraction as a single fraction.

Fundamentals

Example 10

Simplify each compound fraction.

a) $\dfrac{\frac{1}{x+h} - \frac{1}{x}}{h}$ b) $\dfrac{\frac{a}{b} + 1}{1 - \frac{a}{b}}$ c) $\dfrac{x(1-2x)^{-\frac{3}{2}} + (1-2x)^{-\frac{1}{2}}}{1-x}$

Solution

a) $\dfrac{\frac{1}{x+h} - \frac{1}{x}}{h} = \dfrac{\frac{x}{x(x+h)} - \frac{x+h}{x(x+h)}}{\frac{h}{1}} = \dfrac{\frac{x-(x+h)}{x(x+h)}}{\frac{h}{1}} = \dfrac{x-x-h}{x(x+h)} \cdot \dfrac{1}{h}$

$= \dfrac{-\cancel{h}}{x(x+h)} \cdot \dfrac{1}{\cancel{h}} = -\dfrac{1}{x(x+h)}$

b) $\dfrac{\frac{a}{b}+1}{1-\frac{a}{b}} = \dfrac{\frac{a}{b}+\frac{b}{b}}{\frac{b}{b}-\frac{a}{b}} = \dfrac{\frac{a+b}{b}}{\frac{b-a}{b}} = \dfrac{a+b}{\cancel{b}} \cdot \dfrac{\cancel{b}}{b-a} = \dfrac{a+b}{b-a}$

- **Hint:** Factor out the power of $1 - 2x$ with the *smallest* exponent.

c) $\dfrac{x(1-2x)^{-\frac{3}{2}} + (1-2x)^{-\frac{1}{2}}}{1-x} = \dfrac{(1-2x)^{-\frac{3}{2}}[x + (1-2x)^1]}{1-x}$

$= \dfrac{(1-2x)^{-\frac{3}{2}}[x + 1 - 2x]}{1-x}$

$= \dfrac{(1-2x)^{-\frac{3}{2}}\cancel{(1-x)}}{\cancel{1-x}}$

$= \dfrac{1}{(1-2x)^{\frac{3}{2}}}$

With rules for rational exponents and radicals we can do the following, but it's not any *simpler*...

$\dfrac{1}{(1-2x)^{\frac{3}{2}}} = \dfrac{1}{\sqrt{3x-2)^3}} = \dfrac{1}{\sqrt{(3x-2)^2}\sqrt{3x-2}} = \dfrac{1}{|3x-2|\sqrt{3x-2}}$

Rationalizing the denominator

Recall Example 3 from Section 1.2 where we rationalized the denominator of the numerical fractions $\dfrac{2}{\sqrt{3}}$ and $\dfrac{\sqrt{7}}{4\sqrt{10}}$. Also recall from earlier in this section that expressions of the form $a+b$ and $a-b$ are called conjugates and their product is $a^2 - b^2$ (difference of two squares). If a fraction has an irrational denominator of the form $a + b\sqrt{c}$, we can change it to a rational expression ('rationalize') by multiplying numerator and denominator by its conjugate $a - b\sqrt{c}$, given that $(a + b\sqrt{c})(a - b\sqrt{c}) = a^2 - (b\sqrt{c})^2 = a^2 - b^2c$.

Example 11

Rationalize the denominator of each fractional expression.

a) $\dfrac{2}{1+\sqrt{5}}$ b) $\dfrac{1}{\sqrt{x}+1}$

Solution

a) $\dfrac{2}{1+\sqrt{5}} = \dfrac{2}{1+\sqrt{5}} \cdot \dfrac{1-\sqrt{5}}{1-\sqrt{5}} = \dfrac{2(1-\sqrt{5})}{1-(\sqrt{5})^2} = \dfrac{2(1-\sqrt{5})}{1-5} = \dfrac{\cancel{2}(1-\sqrt{5})}{-\cancel{4}_2}$

$= \dfrac{-(1-\sqrt{5})}{2} = \dfrac{-1+\sqrt{5}}{2}$

b) $\dfrac{1}{\sqrt{x}+1} = \dfrac{1}{\sqrt{x}+1} \cdot \dfrac{\sqrt{x}-1}{\sqrt{x}-1} = \dfrac{\sqrt{x}-1}{(\sqrt{x})^2 - 1^2} = \dfrac{\sqrt{x}-1}{x-1}$

Exercise 1.5

In questions 1–12, expand and simplify.

1. $(n + 4)(n - 5)$
2. $(2y - 3)(5y + 3)$
3. $(x + 7)(x - 7)$
4. $(5m + 2)^2$
5. $(x - 1)^3$
6. $(1 + \sqrt{a})(1 - \sqrt{a})$
7. $(a + b)(a - b + 1)$
8. $[(2x + 3) + y][(2x + 3) - y]$
9. $(a + b)^3$
10. $(ax + b)^2$
11. $(1 + \sqrt{5})(1 - \sqrt{5})$
12. $(2x - 1)(2x^2 - 3x + 5)$

In questions 13–30, completely factorize the expression.

13. $12x^2 - 48$
14. $x^3 - 6x^2$
15. $x^2 + x - 12$
16. $7 - 6m - m^2$
17. $x^2 - 10x + 16$
18. $y^2 + 7y + 6$
19. $3n^2 - 21n + 30$
20. $2x^3 + 20x^2 + 18x$
21. $a^2 - 16$
22. $3y^2 - 14y - 5$
23. $25n^4 - 4$
24. $ax^2 + 6ax + 9a$
25. $2n(m + 1)^2 - (m + 1)^2$
26. $x^4 - 1$
27. $9 - (y - 3)^2$
28. $4y^4 - 10y^3 - 96y^2$
29. $4x^2 - 20x + 25$
30. $(2x + 3)^{-2} + 2x(2x + 3)^{-3}$

In questions 31–36, simplify the algebraic fraction.

31. $\dfrac{x + 4}{x^2 + 5x + 4}$
32. $\dfrac{3n - 3}{6n^2 - 6n}$
33. $\dfrac{a^2 - b^2}{5a - 5b}$
34. $\dfrac{x^2 + 4x + 4}{x + 2}$
35. $\dfrac{2a - 5}{5 - 2a}$
36. $\dfrac{(2x + h)^2 - 4x^2}{h}$

In questions 37–46, perform the indicated operation and simplify.

37. $\dfrac{x}{5} - \dfrac{x - 1}{3}$
38. $\dfrac{1}{a} - \dfrac{1}{b}$
39. $\dfrac{2}{2x - 1} - 4$
40. $\dfrac{x}{x + 3} + \dfrac{1}{x}$
41. $\dfrac{1}{x + y} + \dfrac{1}{x - y}$
42. $\dfrac{3}{x - 2} + \dfrac{5}{2 - x}$
43. $\dfrac{2x - 6}{x} \cdot \dfrac{3x}{x - 3}$
44. $\dfrac{3}{y + 2} + \dfrac{5}{y^2 - 3y - 10}$
45. $\dfrac{a + b}{b} \cdot \dfrac{1}{a^2 - b^2}$
46. $\dfrac{3x^2 - 3}{6x} \cdot \dfrac{5x^2}{1 - x}$

In questions 47–50, rationalize the denominator of each fractional expression.

47. $\dfrac{1}{3 - \sqrt{2}}$
48. $\dfrac{5}{2 + \sqrt{3}}$
49. $\dfrac{2\sqrt{2} + \sqrt{3}}{2\sqrt{2} - \sqrt{3}}$
50. $\dfrac{1}{\sqrt{5} + 7}$

1.6 Equations and formulae

Equations, identities and formulae

We will encounter a wide variety of equations in this course. Essentially an equation is a statement equating two algebraic expressions that may be true or false depending upon what value(s) is/are substituted for the variable(s). The value(s) of the variable(s) that make the equation true are called the **solutions** or **roots** of the equation. All of the solutions to an equation comprise the **solution set** of the equation. An equation that is true for all possible values of the variable is called an **identity**. All of the common polynomial expansion and factorization patterns shown in Section 1.5 are identities. For example, $(a + b)^2 = a^2 + 2ab + b^2$ is true for all values of a and b. The following are also examples of identities.

$$3(x - 5) = 2(x + 3) + x - 21 \qquad (x + y)^2 - 2xy = x^2 + y^2$$

Many equations are often referred to as a **formula** (plural: formulae) and typically contain more than one variable and, often, other symbols that represent specific constants or **parameters** (constants that may change in value but do not alter the properties of the expression). Formulae with which you are familiar include:

$$A = \pi r^2, \, d = rt, \, d = \sqrt{(x_1 - x_2)^2 + (y_1 - y_2)^2} \text{ and } V = \tfrac{4}{3}\pi r^3$$

Whereas most equations that we will encounter will have numerical solutions, we can solve a formula for a certain variable in terms of other variables – sometimes referred to as changing the subject of a formula.

Example 12

Solve for the indicated variable in each formula.
a) $a^2 + b^2 = c^2 \quad$ solve for b
b) $T = 2\pi\sqrt{\dfrac{l}{g}} \quad$ solve for l

Solution

a) $a^2 + b^2 = c^2 \Rightarrow b^2 = c^2 - a^2 \Rightarrow b = \pm\sqrt{c^2 - a^2}$
  If b is a length then $b = \sqrt{c^2 - a^2}$.

b) $T = 2\pi\sqrt{\dfrac{l}{g}} \Rightarrow \sqrt{\dfrac{l}{g}} = \dfrac{T}{2\pi} \Rightarrow \dfrac{l}{g} = \dfrac{T^2}{4\pi^2} \Rightarrow l = \dfrac{T^2 g}{4\pi^2}$

The graph of an equation

Two important characteristics of any equation are the number of variables (unknowns) and the type of algebraic expressions it contains (e.g. polynomials, rational expressions, trigonometric, exponential, etc.). Nearly all of the equations in this course will have either one or two variables, and in this introductory chapter we will discuss only equations with algebraic expressions that are polynomials. Solutions for equations with a single variable will consist of individual numbers that can be *graphed* as points on a number line. The **graph** of an equation is a visual representation of the

---

One of the most famous equations in the history of mathematics, $x^n + y^n = z^n$, is associated with Pierre Fermat (1601–1665), a French lawyer and amateur mathematician. Writing in the margin of a French translation of *Arithmetica*, Fermat conjectured that the equation $x^n + y^n = z^n$ ($x, y, z, n \in \mathbb{Z}$) has no non-zero solutions for the variables x, y and z when the parameter n is greater than two. When $n = 2$, the equation is equivalent to Pythagoras' theorem for which there are an infinite number of integer solutions – Pythagorean triples, such as $3^2 + 4^2 = 5^2$ and $5^2 + 12^2 = 13^2$, and their multiples. Fermat claimed to have a proof for his conjecture but that he could not fit it in the margin. All the other margin conjectures in Fermat's copy of *Arithmetica* were proven by the start of the 19th century, but this one remained unproven for over 350 years, until the English mathematician Andrew Wiles proved it in 1994.

equation's solution set. For example, the solution set of the one-variable equation containing quadratic and linear polynomials $x^2 = 2x + 8$ is $x \in \{-2, 4\}$. The graph of this one-variable equation is depicted (Figure 1.6) on a one-dimensional coordinate system, i.e. the real number line.

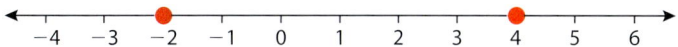

Figure 1.6 The solution set.

The solution set of a two-variable equation will be an **ordered pair** of numbers. An ordered pair corresponds to a location indicated by a point on a two-dimensional coordinate system, i.e. a **coordinate plane**. For example, the solution set of the two-variable **quadratic equation** $y = x^2$ will be an infinite set of ordered pairs (x, y) that satisfy the equation. (Quadratic equations will be covered in detail in Chapter 2.)

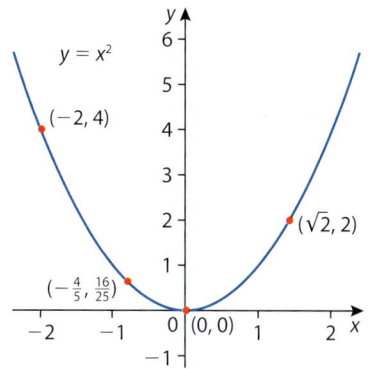

Figure 1.7 Four ordered pairs in the solution set of $y = x^2$ are graphed in red. The graph of all the ordered pairs in the solution set form a curve, as shown in blue.

Equations of lines

A one-variable **linear equation** in x can always be written in the form $ax + b = 0$, $a \neq 0$, and it will have exactly one solution, $x = -\frac{b}{a}$. An example of a two-variable **linear equation in x and y** is $x - 2y = 2$. The graph of this equation's solution set (an infinite set of ordered pairs) is a **line**. (See Figure 1.8.)

The **slope** m, or **gradient**, of a non-vertical line is defined by the formula $m = \frac{y_2 - y_1}{x_2 - x_1} = \frac{\text{vertical change}}{\text{horizontal change}}$. Because division by zero is undefined, the slope of a vertical line is undefined. Using the two points $(1, -\frac{1}{2})$ and $(4, 1)$, we compute the slope of the line with equation $x - 2y = 2$ to be $m = \frac{1 - (-\frac{1}{2})}{4 - 1} = \frac{\frac{3}{2}}{\frac{3}{1}} = \frac{1}{2}$.

If we solve for y, we can rewrite the equation in the form $y = \frac{1}{2}x - 1$. Note that the coefficient of x is the slope of the line and the constant term is the y-coordinate of the point at which the line intersects the y-axis, i.e. the y-intercept. There are several forms in which to write linear equations whose graphs are lines.

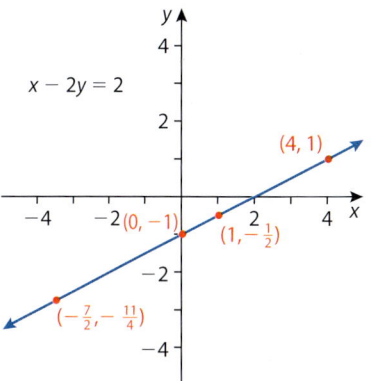

Figure 1.8 The graph of $x - 2y = 2$.

Form	Equation	Characteristics
general form	$ax + by + c = 0$	every line has an equation in this form if both a and $b \neq 0$
slope-intercept form	$y = mx + c$	m is the slope; $(0, c)$ is the y-intercept
point-slope form	$y - y_1 = m(x - x_1)$	m is the slope; (x_1, y_1) is a known point on the line
horizontal line	$y = c$	slope is zero; $(0, c)$ is the y-intercept
vertical line	$x = c$	slope is undefined; unless line is y-axis, no y-intercept

Table 1.6 Forms for equations of lines.

23

1 Fundamentals

Most problems involving equations and graphs fall into two categories: (1) given an equation, determine its graph; and (2) given a graph, or some information about it, find its equation. For lines, the first type of problem is often best solved by using the slope-intercept form. However, for the second type of problem, the point-slope form is usually most useful.

Example 13

Without using a GDC, sketch the line that is the graph of each of the following linear equations, written here in general form.
a) $5x + 3y - 6 = 0$
b) $y - 4 = 0$
c) $x + 3 = 0$

Solution

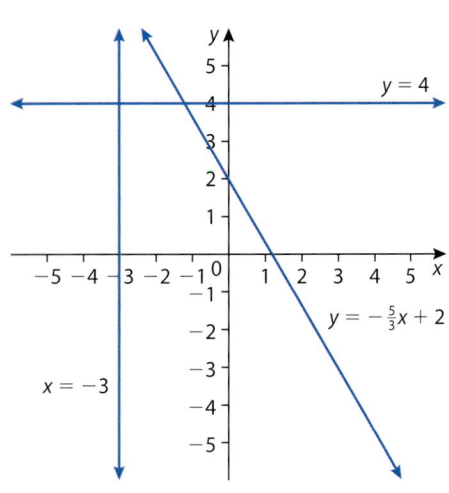

a) Solve for y to write the equation in slope-intercept form.
$5x + 3y - 6 = 0 \Rightarrow 3y = -5x + 6 \Rightarrow y = -\frac{5}{3}x + 2$. The line has a y-intercept of $(0, 2)$ and a slope of $-\frac{5}{3}$.

b) The equation $y - 4 = 0$ is equivalent to $y = 4$, whose graph is a horizontal line with a y-intercept of $(0, 4)$.

c) The equation $x + 3 = 0$ is equivalent to $x = -3$, whose graph is a vertical line with no y-intercept; but, it has an x-intercept of $(-3, 0)$.

Example 14

a) Find the equation of the line that passes through the point $(3, 31)$ and has a slope of 12. Write the equation in slope-intercept form.

b) Find the linear equation in C and F knowing that when $C = 10$ then $F = 50$, and when $C = 100$ then $F = 212$. Solve for F in terms of C.

Solution

a) Substitute into the point-slope form $y - y_1 = m(x - x_1)$; $x_1 = 3$, $y_1 = 31$ and $m = 12$
$y - y_1 = m(x - x_1) \Rightarrow y - 31 = 12(x - 3) \Rightarrow y = 12x - 36 + 31 \Rightarrow y = 12x - 5$

b) The two points, ordered pairs (C, F), that are known to be on the line are $(10, 50)$ and $(100, 212)$. The variable C corresponds to the variable x and F corresponds to y in the definitions and forms stated above. The slope of the line is $m = \dfrac{F_2 - F_1}{C_2 - C_1} = \dfrac{212 - 50}{100 - 10} = \dfrac{162}{90} = \dfrac{9}{5}$. Choose one of the points on the line, say $(10, 50)$, and substitute it and the slope into the point-slope form.

$F - F_1 = m(C - C_1) \Rightarrow F - 50 = \frac{9}{5}(C - 10) \Rightarrow F = \frac{9}{5}C - 18 + 50 \Rightarrow F = \frac{9}{5}C + 32$

The slope of a line is a convenient tool for determining whether two lines are parallel or perpendicular.

The two lines graphed in Figure 1.9 suggest the following property: Two distinct non-vertical lines are **parallel** if, and only if, their slopes are equal, $m_1 = m_2$.

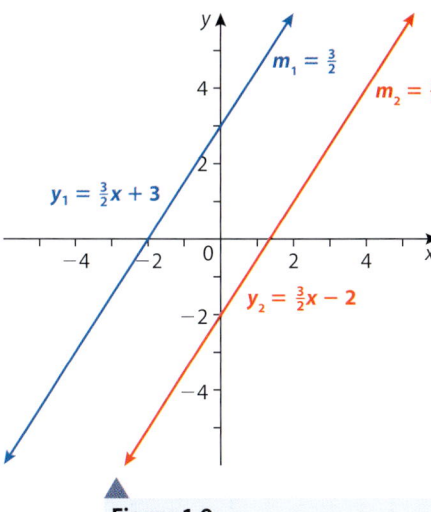

Figure 1.9

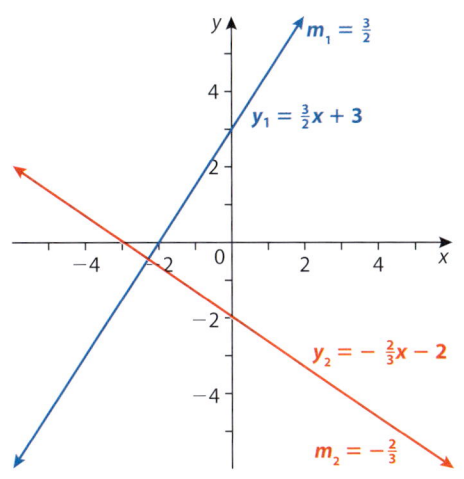

Figure 1.10

The two lines graphed in Figure 1.10 suggest another property: Two non-vertical lines are **perpendicular** if, and only if, their slopes are negative reciprocals – that is, $m_1 = -\frac{1}{m_2}$, which is equivalent to $m_1 \cdot m_2 = -1$.

Distances and midpoints

Recall from Section 1.1 that absolute value (modulus) is used to define the **distance** (always positive) between two points on the real number line.

The distance between the points A and B on the real number line is $|B - A|$, which is equivalent to $|A - B|$.

The points A and B are the endpoints of a line segment that is denoted with the notation $[AB]$ and the length of the line segment is denoted AB. In Figure 1.11, the distance between A and B is $AB = |4 - (-2)| = |-2 - 4| = 6$.

Figure 1.11

The distance between two general points (x_1, y_1) and (x_2, y_2) on a coordinate plane can be found using the definition for distance on a number line and Pythagoras' theorem. For the points (x_1, y_1) and (x_2, y_2), the horizontal distance between them is $|x_1 - x_2|$ and the vertical distance is $|y_1 - y_2|$. As illustrated in Figure 1.12, these distances are the lengths of two legs of a right-angled triangle whose hypotenuse is the distance between the points. If d represents the distance between (x_1, y_1) and (x_2, y_2), then by Pythagoras' theorem $d^2 = |x_1 - x_2|^2 + |y_1 - y_2|^2$. Because the square of any number is positive, the absolute value is not necessary, giving us the **distance formula** for two-dimensional coordinates.

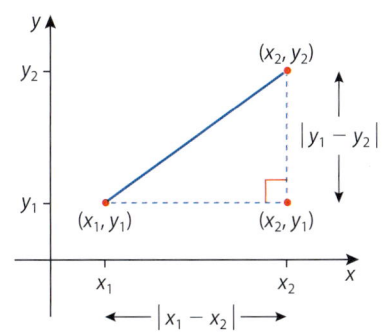

Figure 1.12

1 Fundamentals

> **The distance formula**
> The distance d between the two points (x_1, y_1) and (x_2, y_2) in the coordinate plane is
> $$d = \sqrt{(x_1 - x_2)^2 + (y_1 - y_2)^2}$$

The coordinates of the **midpoint** of a line segment are the average values of the corresponding coordinates of the two endpoints.

> **The midpoint formula**
> The midpoint of the line segment joining the points (x_1, y_1) and (x_2, y_2) in the coordinate plane is
> $$\left(\frac{x_1 + x_2}{2}, \frac{y_1 + y_2}{2}\right)$$

Example 15

a) Show that the points $P(1, 2)$, $Q(3, 1)$ and $R(4, 8)$ are the vertices of a right-angled triangle.

b) Find the midpoint of the hypotenuse.

Solution

a) The three points are plotted and the line segments joining them are drawn in Figure 1.13. Applying the distance formula, we can find the exact lengths of the three sides of the triangle.

$PQ = \sqrt{(1-3)^2 + (2-1)^2} = \sqrt{4+1} = \sqrt{5}$

$QR = \sqrt{(3-4)^2 + (1-8)^2} = \sqrt{1+49} = \sqrt{50}$

$PR = \sqrt{(1-4)^2 + (2-8)^2} = \sqrt{9+36} = \sqrt{45}$

$PQ^2 + PR^2 = QR^2$ because $(\sqrt{5})^2 + (\sqrt{45})^2 = 5 + 45 = 50 = (\sqrt{50})^2$. The lengths of the three sides of the triangle satisfy Pythagoras' theorem, confirming that the triangle is a right-angled triangle.

b) QR is the hypotenuse. Let the midpoint of QR be point M. Using the midpoint formula, $M = \left(\frac{3+4}{2}\right), \left(\frac{1+8}{2}\right) = \left(\frac{7}{2}, \frac{9}{2}\right)$. This point is plotted in Figure 1.13.

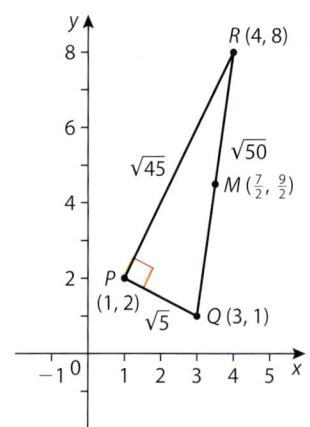

Figure 1.13

Example 16

Find x so that the distance between the points $(1, 2)$ and $(x, -10)$ is 13.

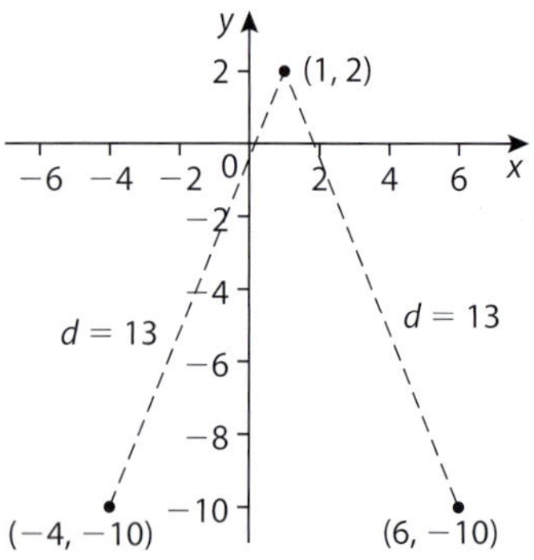

Solution

$d = 13 = \sqrt{(x-1)^2 + (-10-2)^2} \Rightarrow 13^2 = (x-1)^2 + (-12)^2$

$\Rightarrow 169 = x^2 - 2x + 1 + 144 \Rightarrow x^2 - 2x - 24 = 0$

$\Rightarrow (x-6)(x+4) = 0 \Rightarrow x - 6 = 0$ or $x + 4 = 0$

$\Rightarrow x = 6$ or $x = -4$

Figure 1.14 The graph shows the two different points that are both a distance of 13 from $(1, 2)$.

Simultaneous equations

Many problems that we solve with algebraic techniques involve sets of equations with several variables, rather than just a single equation with one or two variables. Such a set of equations is called a set of **simultaneous equations** because we find the values for the variables that solve all of the equations simultaneously. In this section, we consider only the simplest set of simultaneous equations – a pair of linear equations in two variables. We will take a brief look at three methods for solving simultaneous linear equations. They are:

1. Graphical method
2. Elimination method
3. Substitution method

Although we will only look at pairs of linear equations in this section, it is worthwhile mentioning that the graphical and substitution methods are effective for solving sets of equations where not all of the equations are linear, e.g. one linear and one quadratic equation.

Graphical method

The graph of each equation in a system of two linear equations in two unknowns is a line. The graphical interpretation of the solution of a pair of simultaneous linear equations corresponds to determining what point, or points, lies on both lines. Two lines in a coordinate plane can only relate to one another in one of three ways: (1) intersect at exactly one point, (2) intersect at all points on each line (i.e. the lines are identical), or (3) the two lines do not intersect (i.e. the lines are parallel). These three possibilities are illustrated in Figure 1.15.

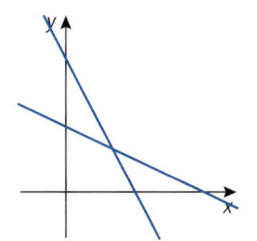

◀ Figure 1.15

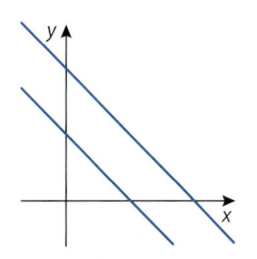

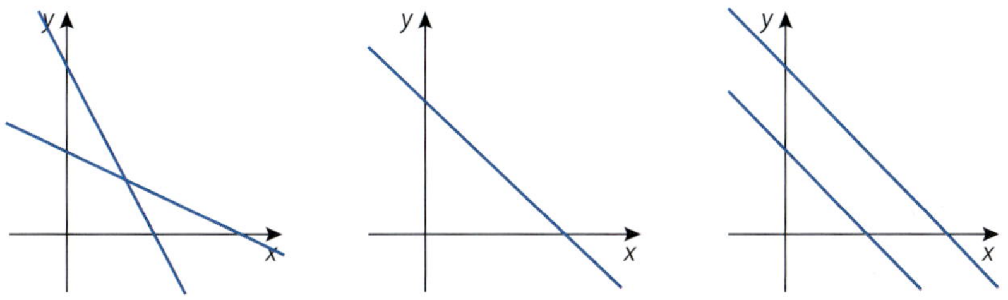

Intersect at exactly one point; exactly one solution

Identical – coincident lines; infinite solutions

Never intersect – parallel lines; no solution

Although a graphical approach to solving simultaneous linear equations provides a helpful visual picture of the number and location of solutions, it can be tedious and inaccurate if done by hand. The graphical method is far more efficient and accurate when performed on a graphical display calculator (GDC).

Example 17

Use the graphical features of a GDC to solve each pair of simultaneous equations.

a) $2x + 3y = 6$
   $2x - y = -10$

b) $7x - 5y = 20$
   $3x + y = 2$

Fundamentals

Solution

a) First, we will rewrite each equation in slope-intercept form, i.e. $y = mx + c$. This is a necessity if we use our GDC, and is also very useful for graphing by hand (manual).

$$2x + 3y = 6 \Rightarrow 3y = -2x + 6 \Rightarrow y = -\frac{2}{3}x + 2 \text{ and } 2x - y = -10 \Rightarrow y = 2x + 10$$

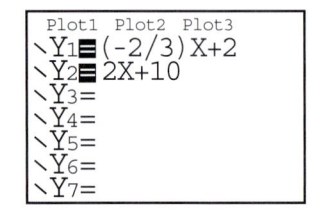

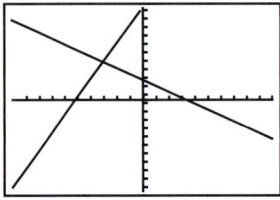

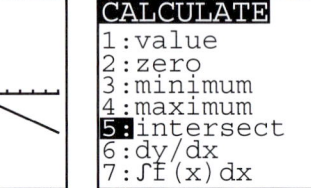

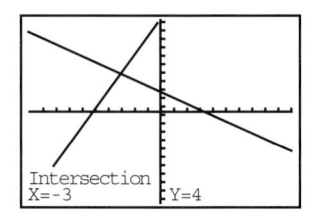

The intersection point and solution to the simultaneous equations is $x = -3$ and $y = 4$, or $(-3, 4)$. If we manually graphed the two linear equations in a) very carefully using graph paper, we may have been able to determine the exact coordinates of the intersection point. However, using a graphical method without a GDC to solve the simultaneous equations in b) would only allow us to crudely approximate the solution.

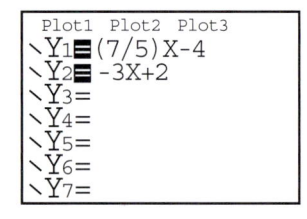

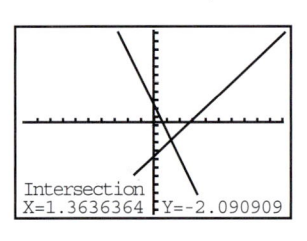

b) $7x - 5y = 20 \Rightarrow 5y = 7x - 20 \Rightarrow y = \frac{7}{5}x - 4$ and
$3x + y = 2 \Rightarrow y = -3x + 2$

The solution to the simultaneous equations is $x = \frac{15}{11}$ and $y = -\frac{23}{11}$, or $\left(\frac{15}{11}, -\frac{23}{11}\right)$.

The full power and efficiency of the GDC is used in this example to find the exact solution.

Elimination method

To solve a system using the **elimination method**, we try to combine the two linear equations using sums or differences in order to eliminate one of the variables. Before combining the equations, we need to multiply one or both of the equations by a suitable constant to produce coefficients for one of the variables that are equal (then subtract the equations), or that differ only in sign (then add the equations).

Example 18

Use the elimination method to solve each pair of simultaneous equations.

a) $5x + 3y = 9$
$2x - 4y = 14$

b) $x - 2y = 3$
$2x - 4y = 5$

Solution

a) We can obtain coefficients for y that differ only in sign by multiplying the first equation by 4 and the second equation by 3. Then we add the equations to eliminate the variable y.

$$\begin{aligned} 5x + 3y &= 9 \rightarrow 20x + 12y = 36 \\ 2x - 4y &= 14 \rightarrow \underline{6x - 12y = 42} \\ & \qquad\qquad\quad 26x \qquad\quad = 78 \\ & \qquad\qquad\qquad\qquad x = \frac{78}{26} \\ & \qquad\qquad\qquad\qquad x = 3 \end{aligned}$$

By substituting the value of 3 for x in either of the original equations we can solve for y.

$$5x + 3y = 9 \Rightarrow 5(3) + 3y = 9 \Rightarrow 3y = -6 \Rightarrow y = -2$$

The solution is $(3, -2)$.

b) To obtain coefficients for x that are equal, we multiply the first equation by 2 and then subtract the equations to eliminate the variable x.

$$\begin{aligned} x - 2y &= 7 \rightarrow 2x - 4y = 14 \\ 2x - 4y &= 5 \rightarrow \underline{2x - 4y = 5} \\ & \qquad\qquad\qquad 0 = 9 \end{aligned}$$

Because it is not possible for 0 to equal 9, there is no solution. The lines that are the graphs of the two equations are parallel. To confirm this we can rewrite each of the equations in the form $y = mx + c$.

$$x - 2y = 7 \Rightarrow 2y = x - 7 \Rightarrow y = \tfrac{1}{2}x - \tfrac{7}{2} \text{ and}$$
$$2x - 4y = 5 \Rightarrow 4y = 2x - 5 \Rightarrow y = \tfrac{1}{2}x - \tfrac{5}{2}$$

Both equations have a slope of $\tfrac{1}{2}$, but different y-intercepts. Therefore, the lines are parallel. This confirms that this pair of simultaneous equations has no solution.

Substitution method

The algebraic method that can be applied effectively to the widest variety of simultaneous equations, including non-linear equations, is the **substitution method**. Using this method, we choose one of the equations and solve for one of the variables in terms of the other variable. We then substitute this expression into the other equation to produce an equation with only one variable, which we can solve directly.

Example 19

Use the substitution method to solve each pair of simultaneous equations.

a) $3x - y = -9$
   $6x + 2y = 2$

b) $-2x + 6y = 4$
   $3x - 9y = -6$

Solution

a) Solve for y in the top equation, $3x - y = -9 \Rightarrow y = 3x + 9$, and substitute $3x + 9$ in for y in the bottom equation:
$6x + 2(3x + 9) = 2 \Rightarrow 6x + 6x + 18 = 2 \Rightarrow 12x = -16 \Rightarrow x = -\frac{16}{12} = -\frac{4}{3}$.
Now substitute $-\frac{4}{3}$ for x in either equation to solve for y.
$3\left(-\frac{4}{3}\right) - y = -9 \Rightarrow y = -4 + 9 \Rightarrow y = 5$.
The solution is $x = -\frac{4}{3}, y = 5$, or $\left(-\frac{4}{3}, 5\right)$.

b) Solve for x in the top equation,
$-2x + 6y = 4 \Rightarrow 2x = 6y - 4 \Rightarrow x = 3y - 2$, and substitute $3y - 2$ in for x in the bottom equation:
$3(3y - 2) - 9y = -6 \Rightarrow 9y - 6 - 9y = -6 \Rightarrow 0 = 0$.
The resulting equation $0 = 0$ is true for any values of x and y. The two equations are equivalent, and their graphs will produce identical lines – i.e. coincident lines. Therefore, the solution set consists of all points (x, y) lying on the line $-2x + 6y = 4$ $\left(\text{or } y = \frac{1}{3}x + \frac{2}{3}\right)$.

Exercise 1.6

In questions 1–8, solve for the indicated variable in each formula.

1. $m(h - x) = n$ solve for x
2. $v = \sqrt{ab - t}$ solve for a
3. $A = \frac{h}{2}(b_1 + b_2)$ solve for b_1
4. $A = \frac{1}{2}r^2\theta$ solve for r
5. $\frac{f}{g} = \frac{h}{k}$ solve for k
6. $at = x - bt$ solve for t
7. $V = \frac{1}{3}\pi r^3 h$ solve for r
8. $F = \dfrac{g}{m_1 k + m_2 k}$ solve for k

In questions 9–12, find the equation of the line that passes through the two given points. Write the line in slope-intercept form ($y = mx + c$), if possible.

9. $(-9, 1)$ and $(3, -7)$
10. $(3, -4)$ and $(10, -4)$
11. $(-12, -9)$ and $(4, 11)$
12. $\left(\frac{7}{3}, -\frac{1}{2}\right)$ and $\left(\frac{7}{3}, \frac{5}{2}\right)$

13. Find the equation of the line that passes through the point $(7, -17)$ and is parallel to the line with equation $4x + y - 3 = 0$. Write the line in slope-intercept form ($y = mx + c$).

14. Find the equation of the line that passes through the point $\left(-5, \frac{11}{2}\right)$ and is perpendicular to the line with equation $2x - 5y - 35 = 0$. Write the line in slope-intercept form ($y = mx + c$).

In questions 15–18, a) find the exact distance between the points, and b) find the midpoint of the line segment joining the two points.

15. $(-4, 10)$ and $(4, -5)$
16. $(-1, 2)$ and $(5, 4)$
17. $\left(\frac{1}{2}, 1\right)$ and $\left(-\frac{5}{2}, \frac{4}{3}\right)$
18. $(12, 2)$ and $(-10, 9)$

In questions 19 and 20, find the value(s) of k so that the distance between the points is 5.

19. $(5, -1)$ and $(k, 2)$
20. $(-2, -7)$ and $(1, k)$

In questions 21–23, show that the given points form the vertices of the indicated polygon.

21. Right-angled triangle: $(4, 0), (2, 1)$ and $(-1, -5)$
22. Isosceles triangle: $(1, -3), (3, 2)$ and $(-2, 4)$
23. Parallelogram: $(0, 1), (3, 7), (4, 4)$ and $(1, -2)$

In questions 24–29, use the elimination method to solve each pair of simultaneous equations.

24 $x + 3y = 8$
   $x - 2y = 3$

25 $x - 6y = 1$
   $3x + 2y = 13$

26 $6x + 3y = 6$
   $5x + 4y = -1$

27 $x + 3y = -1$
   $x - 2y = 7$

28 $8x - 12y = 4$
   $-2x + 3y = 2$

29 $5x + 7y = 9$
   $-11x - 5y = 1$

In questions 30–35, use the substitution method to solve each pair of simultaneous equations.

30 $2x + y = 1$
   $3x + 2y = 3$

31 $3x - 2y = 7$
   $5x - y = -7$

32 $2x + 8y = -6$
   $-5x - 20y = 15$

33 $\dfrac{x}{5} + \dfrac{y}{2} = 8$
   $x + y = 20$

34 $2x - y = -2$
   $4x + y = 5$

35 $0.4x + 0.3y = 1$
   $0.25x + 0.1y = -0.25$

In questions 36–38, solve the pair of simultaneous equations using any method – elimination, substitution or the graphical features of your GDC.

36 $3x + 2y = 9$
   $7x + 11y = 2$

37 $3.62x - 5.88y = -10.11$
   $0.08x - 0.02y = 0.92$

38 $2x - 3y = 4$
   $5x + 2y = 1$

Functions and Equations

Assessment statements

2.1 Concept of function $f: x \mapsto f(x)$; domain, range, image (value). Composite functions $(f \circ g)$; identity function. Inverse function f^{-1}.

2.2 The graph of a function; its equation $y = f(x)$.
Function graphing skills: use of a GDC to graph a variety of functions; investigation of key features of graphs.
Solutions of equations graphically.

2.3 Transformations of graphs: translations; stretches; reflections in the axes. The graph of $y = f^{-1}(x)$ as the reflection in the line $y = x$ of the graph $y = f(x)$.

2.4 The reciprocal function $x \mapsto \frac{1}{x}$, $x \neq 0$: its graph; its self-inverse nature.

2.5 The quadratic function $x \mapsto ax^2 + bx + c$: its graph, y-intercept $(0, c)$, axis of symmetry $x = -\frac{b}{2a}$.
The form $x \mapsto a(x - h)^2 + k$: vertex (h, k).
The form $x \mapsto a(x - p)(x - q)$: x-intercepts $(p, 0)$ and $(q, 0)$.

2.6 The solution of $ax^2 + bx + c = 0$, $a \neq 0$.
The quadratic formula. Use of the discriminant $\Delta = b^2 - 4ac$.

Introduction

This chapter looks at functions and considers how they can be used in describing physical phenomena. We also investigate composite and inverse functions, and transformations such as translations, stretches and reflections. Quadratic functions are treated graphically and algebraically.

2.1 Relations and functions

Relations

There are different scales for measuring temperature. Two of the more commonly used are the Celsius scale and the Fahrenheit scale. A temperature recorded in one scale can be converted to a value in the other scale, based on the fact that there is a constant relationship between the two sets of numbers in each scale. If the variable C represents degrees Celsius and the variable F represents degrees Fahrenheit, this relationship can be expressed by the following equation that converts Celsius to Fahrenheit: $F = \frac{9}{5}C + 32$.

> Most countries, except the United States, use the Celsius scale, invented by the Swedish scientist Anders Celsius (1701–1744). The United States uses the earlier Fahrenheit scale, invented by the Dutch scientist Gabriel Daniel Fahrenheit (1686–1736). A citizen of the USA travelling to other parts of the world will need to convert from degrees Celsius to degrees Fahrenheit.

Many mathematical relationships concern how two sets of numbers relate to one another – and often the best way to express this is with an algebraic equation in two variables. If it's not too difficult, we find it useful to express one variable in terms of the other. For example, in the previous equation, *F* is written in terms of *C* – making *C* the **independent variable** and *F* the **dependent variable**. Since *F* is written in terms of *C*, it is easiest for you to first substitute in a value for *C*, and then evaluate the expression to determine the value of *F*. In other words, the value of *F* is *dependent* upon the value of *C*, which is chosen *independently* of *F*.

A **relation** is a rule that determines how a value of the independent variable corresponds – or is **mapped** – to a value of the dependent variable. A temperature of 30 degrees Celsius corresponds to 86 degrees Fahrenheit.

$$F = \tfrac{9}{5}(30) + 32 = 54 + 32 = 86$$

Along with equations, other useful ways of representing a relation include a graph of the equation on a **Cartesian coordinate system** (also called a **rectangular coordinate system**), a **table**, a set of **ordered pairs**, or a **mapping**. These are illustrated below for the equation $F = \tfrac{9}{5}C + 32$.

René Descartes

The Cartesian coordinate system is named in honour of the French mathematician and philosopher René Descartes (1596-1650). Descartes stimulated a revolution in the study of mathematics by merging its two major fields – algebra and geometry. With his coordinate system utilizing ordered pairs (*Cartesian coordinates*) of real numbers, geometric concepts could be formulated analytically and algebraic concepts (e.g. relationships between two variables) could be viewed graphically. Descartes initiated something that is very helpful to all students of mathematics – that is, considering mathematical concepts from multiple perspectives: graphical (visual) and analytical (algebraic).

Graph

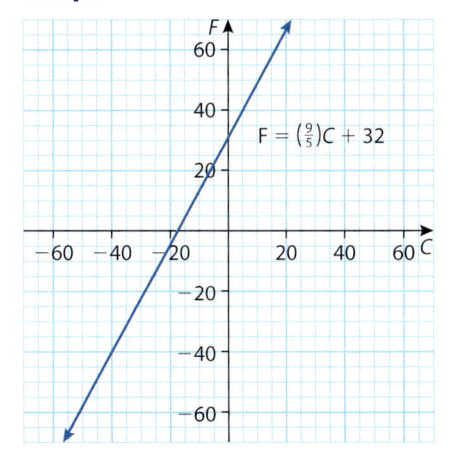

Table

Celsius (C)	Fahrenheit (F)
−40	−40
−30	−22
−20	−4
−10	14
0	32
10	50
20	68
30	86
40	104

Ordered pairs

The graph of the equation $F = \tfrac{9}{5}C + 32$ is a line consisting of an infinite set of ordered pairs (*C*, *F*) – each is a solution of the equation. The following set includes some of the ordered pairs on the line:
$\{(-30, -22), (0, 32), (20, 68), (40, 104)\}$.

Mapping

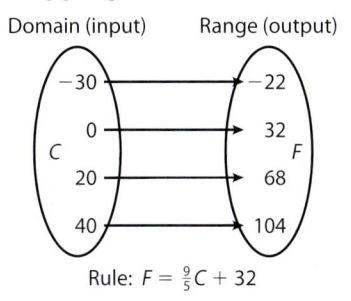

Domain (input) → Range (output)

−30 → −22
0 → 32
20 → 68
40 → 104

Rule: $F = \tfrac{9}{5}C + 32$

● **Hint:** The coordinate system for the graph of an equation has the independent variable on the horizontal axis and the dependent variable on the vertical axis.

The largest possible set of values for the independent variable (the **input** set) is called the **domain** – and the set of resulting values for the dependent variable (the **output** set) is called the **range**. In the context of a mapping, each value in the domain is mapped to its **image** in the range.

2 Functions and Equations

Functions

If the relation is such that each number (or **element**) in the domain produces one and only one number in the range, the relation is called a **function**. Common sense tells us that each numerical temperature in degrees Celsius (C) will convert (or correspond) to only one temperature in degrees Fahrenheit (F). Therefore, the relation given by the equation $F = \frac{9}{5}C + 32$ is a function – any chosen value of C corresponds to exactly one value of F. The idea that a function is a rule that assigns to each number in the domain a unique number in the range is formally defined below.

> **Definition of a function**
> A **function** is a correspondence (**mapping**) between two sets X and Y in which each element of set X corresponds to (maps to) exactly one element of set Y. The **domain** is set X (**independent variable**) and the **range** is set Y (**dependent variable**).

Not only are functions important in the study of mathematics and science, we encounter and use them routinely – often in the form of tables. Examples include height and weight charts, income tax tables, loan payment schedules, and time and temperature charts. The importance of functions in mathematics is evident from the many functions that are installed on your GDC.

For example, the keys labelled SIN, x^{-1}, LN, $\sqrt{}$ each represent a function, because for each input (entry) there is only one output (answer). The calculator screen image shows that for the function $y = \ln x$, the input of $x = 10$ has only one output of $y \approx 2.302\,585\,093$.

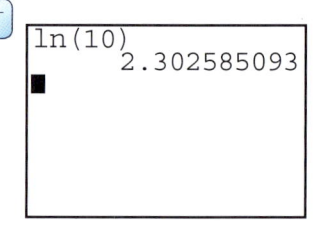

```
ln(10)
          2.302585093
```

For many physical phenomena, we observe that one quantity depends on another. For example, the boiling point of water depends on elevation above sea level; the time for a pendulum to swing through one cycle (its period) depends on the length of the pendulum; and the area of a circle depends on its radius. The word **function** is used to describe this dependence of one quantity on another – i.e. how the value of an independent variable determines the value of a dependent variable.
- Boiling point is a function of elevation (elevation determines boiling point).
- The period of a pendulum is a function of its length (length determines period).
- The area of a circle is a function of its radius (radius determines area).

Example 1

a) Express the volume V of a cube as a function of the length e of each edge.
b) Express the volume V of a cube as a function of its surface area S.

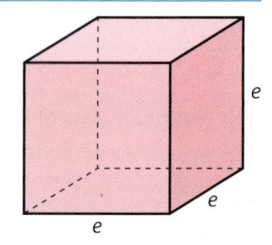

34

Solution

a) V as a function of e is $V = e^3$.

b) The surface area of the cube consists of six squares each with an area of e^2. Hence, the surface area is $6e^2$; that is, $S = 6e^2$. We need to write V in terms of S. We can do this by first expressing e in terms of S, and then substituting this expression in for e in the equation $V = e^3$.

$$S = 6e^2 \Rightarrow e^2 = \frac{S}{6} \Rightarrow e = \sqrt{\frac{S}{6}}.$$

Substituting,

$$V = \left(\sqrt{\frac{S}{6}}\right)^3 = \frac{(S^{\frac{1}{2}})^3}{(6^{\frac{1}{2}})^3} = \frac{S^{\frac{3}{2}}}{6^{\frac{3}{2}}} = \frac{S^1 \cdot S^{\frac{1}{2}}}{6^1 \cdot 6^{\frac{1}{2}}} = \frac{S}{6}\sqrt{\frac{S}{6}}$$

V as a function of S is $V = \frac{S}{6}\sqrt{\frac{S}{6}}$.

Domain and range of a function

The domain of a function may be stated explicitly, or it may be implied by the expression that defines the function. If not explicitly stated, the domain of a function is the set of all real numbers for which the expression is defined as a real number. For example, if a certain value of x is substituted into the algebraic expression defining a function and it causes division by zero or the square root of a negative number (both undefined in the real numbers) to occur, that value of x cannot be in the domain. The domain of a function may also be implied by the physical context or limitations that exist. Usually the range of a function is not given explicitly and is determined by analyzing the output of the function for all values of the input. The range of a function is often more difficult to find than the domain, and analyzing the graph of a function is very helpful in determining it. A combination of algebraic and graphical analysis is very useful in determining the domain and range of a function.

Example 2

Find the domain of each of the following functions.
a) $\{(-6, -3), (-1, 0), (2, 3), (3, 0), (5, 4)\}$
b) Area of a circle: $A = \pi r^2$
c) $y = \frac{1}{x}$
d) $y = \sqrt{x}$

Solution

a) The function consists of a set of ordered pairs. The domain of the function consists of all first coordinates of the ordered pairs. Therefore, the domain is the set $\{-6, -1, 2, 3, 5\}$.

b) The physical context tells you that a circle cannot have a negative radius. You can only choose values for the radius (r) that are greater than zero. Therefore, the domain is the set of all real numbers such that $r > 0$.

c) The value of $x = 0$ cannot be included in the domain because division by zero is not defined for real numbers. Therefore, the domain is the set of all real numbers except zero ($x \neq 0$).

d) Any negative values of x cannot be in the domain because the square root of a negative number is not a real number. Therefore, the domain is all real numbers such that $x \geq 0$.

Determining if a relation is a function

Some relations are not functions – and because of the mathematical significance of functions it is important for us to be able to determine when a relation is, or is not, a function. It follows from the definition of a function that a relation for which a value of the domain (x) corresponds to (or determines) *more than one* value in the range (y) is *not* a function. Any two points (ordered pairs (x, y)) on a *vertical* line have the same x-coordinate. Although a trivial case, it is useful to recognize that the equation for a vertical line, $x = 2$ for example (see Figure 2.1), is a relation but *not* a function. The points with coordinates $(2, -3)$, $(2, 0)$ and $(2, 4)$ are all solutions to the equation $x = 2$. The number two is the only element in the domain of $x = 2$ but it is mapped to *more than one* value in the range ($-3, 0$ and 4, for example). It follows that if a vertical line intersects the graph of a relation at more than one point, then a value in the domain (x) corresponds to more than one value in the range (y) and, hence, the relation is *not* a function. This argument provides an alternative definition of a function and also a convenient visual test to determine whether or not the graph of a relation represents a function.

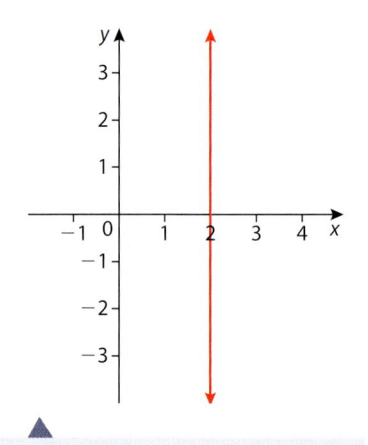

Figure 2.1

Alternative definition of a function
A **function** is a relation in which no two different ordered pairs have the same first coordinate.

Vertical line test for functions
A vertical line intersects the graph of a function at no more than one point.

As the graph in Figure 2.2 clearly shows, a vertical line will intersect the graph of $y = x^2$ at no more than one point – therefore, the relation $y = x^2$ is a function.

Figure 2.2

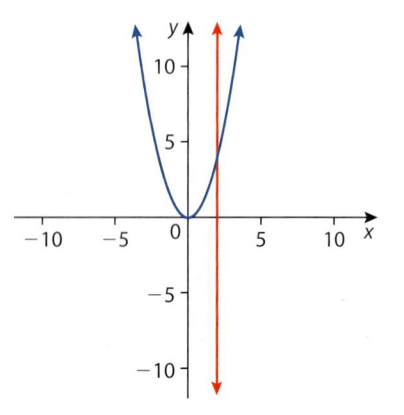

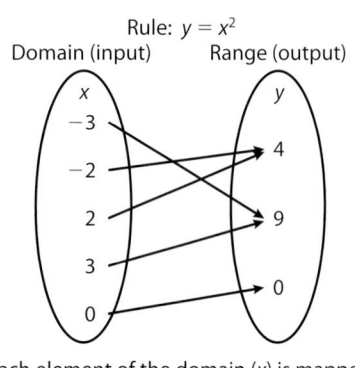

Rule: $y = x^2$
Domain (input)   Range (output)

Each element of the domain (x) is mapped to exactly one element of the range (y).

In contrast, the graph of the equation $y^2 = x$ is a 'sideways' parabola that can clearly be intersected more than once by a vertical line (see Figure 2.3). There are at least two ordered pairs having the same x-coordinate but different y-coordinates (for example, $(9, 3)$ and $(9, -3)$). Therefore, the relation $y^2 = x$ fails the vertical line test indicating that it does *not* represent a function.

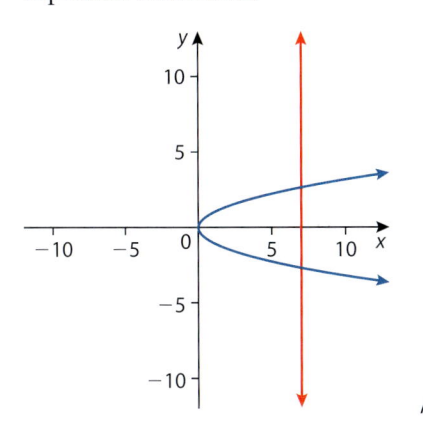

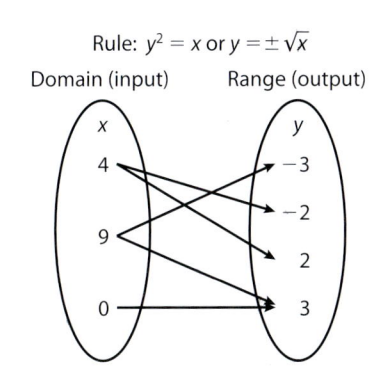

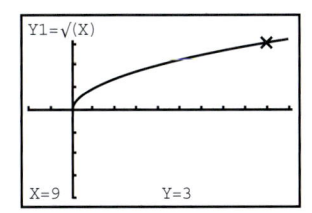

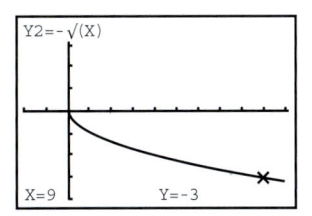

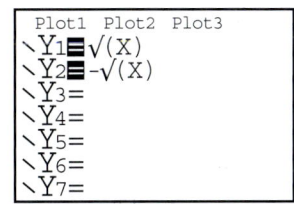

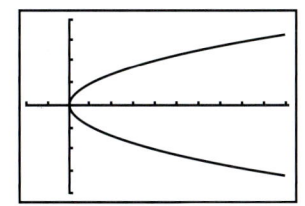

Figure 2.3

At least one element of the domain (x) is mapped to more than one element of the range (y).

- **Hint:** To graph the equation $y^2 = x$ on your GDC, you need to solve for y in terms of x. The result is two separate equations: $y = \sqrt{x}$ and $y = -\sqrt{x}$ (or $y = \pm\sqrt{x}$). Each is one-half of the 'sideways' parabola. Although each represents a function (vertical line test), the combination of the two is a complete graph of $y^2 = x$ that clearly does not satisfy either definition of a function.

Example 3

What is the domain and range for the function $y = x^2$?

Solution

- *Algebraic analysis*: Squaring any real number produces another real number. Therefore, the domain of $y = x^2$ is the set of all real numbers ($\mathbb{R}$). What about the range? Since the square of any positive or negative number will be positive and the square of zero is zero, the range is the set of all real numbers greater than or equal to zero.
- *Graphical analysis*: For the domain, focus on the x-axis and *horizontally* scan the graph from $-\infty$ to $+\infty$. There are no 'gaps' or blank regions in the graph and the parabola will continue to get 'wider' as x goes to either $-\infty$ or $+\infty$. Therefore, the domain is all real numbers. For the range, focus on the y-axis and *vertically* scan from $-\infty$ or $+\infty$. The parabola will continue 'higher' as y goes to $+\infty$, but the graph does not go below the x-axis. The parabola has no points with negative y-coordinates. Therefore, the range is the set of real numbers greater than or equal to zero. See Figure 2.4.

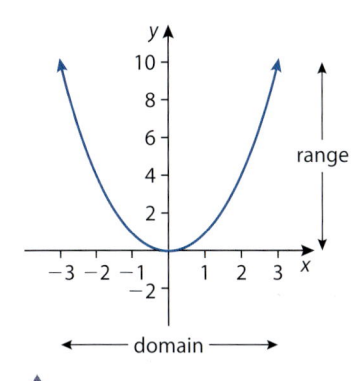

Figure 2.4

2 Functions and Equations

Table 2.1 Different ways of expressing the domain and range of $y = x^2$.

Description in words	Interval notation (both formats)
domain is any real number	domain is $\{x : x \in \mathbb{R}\}$ or domain is $x \in \,]-\infty, \infty[$
range is any real number greater than or equal to zero	range is $\{y : y \geq 0\}$ or range is $y \in [0, \infty[$

- **Hint:** The infinity symbol ∞ does *not* represent a number. When ∞ or $-\infty$ is used in interval notation, it is being used as a convenient notational device to indicate that an interval has no endpoint in a certain direction.

- **Hint:** When asked to determine the domain and range of a function, it is wise for you to conduct both algebraic and graphical analysis – and not rely too much on either approach. For graphical analysis of a function, producing a *comprehensive graph* on your GDC is essential – and an essential skill for this course.

Function notation

It is common practice to assign a name to a function – usually a single letter with f, g and h being the most common. Given that the domain (independent) variable is x and the range (dependent) variable is y, the symbol $f(x)$, read 'f of x', denotes the unique value of y that is generated by the value of x. This **function notation** was devised by the famous Swiss mathematician Leonhard Euler (1707–1783). Another notation – sometimes referred to as **mapping notation** – is based on the idea that the function f is the rule that maps x to $f(x)$ and is written $f : x \mapsto f(x)$. For each value of x in the domain, the corresponding unique value of y in the range is called the **function value** at x, or the **image** of x under f. The image of x may be written as $f(x)$ or as y. For example, for the function $f(x) = x^2$: '$f(3) = 9$'; or 'if $x = 3$ then $y = 9$'.

Table 2.2 Function notation.

Notation	Description in words
$f(x) = x^2$	'the function f, in terms of x, is x^2'; or, simply, 'f of x is x^2'
$f : x \mapsto x^2$	'the function f maps x to x^2'
$f(3) = 9$	'the value of the function f when $x = 3$ is 9'; or, simply, 'f of 3 equals 9'
$f : 3 \mapsto 9$	'the image of 3 under the function f is 9'

Example 4

Find the domain and range of the function $h : x \mapsto \dfrac{1}{x - 2}$.

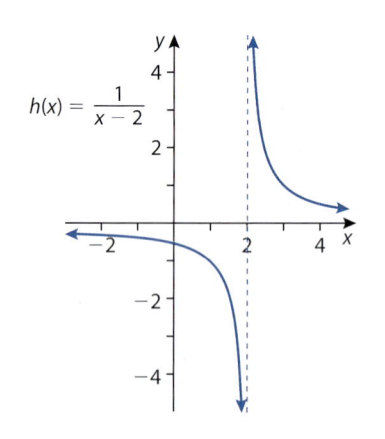

Solution

- *Algebraic analysis*: The function produces a real number for all x, except for $x = 2$ when division by zero occurs. Hence, $x = 2$ is the only real number not in the domain. Since the numerator of $\dfrac{1}{x - 2}$ can never be zero, the value of y cannot be zero. Hence, $y = 0$ is the only real number not in the range.
- *Graphical analysis*: A horizontal scan shows a 'gap' at $x = 2$ dividing the graph of the equation into two branches that both continue indefinitely, with no other 'gaps' as $x \to \pm \infty$. Both branches are **asymptotic** (approach but do not intersect) to the vertical line $x = 2$. This line is a **vertical asymptote** and is drawn as a dashed line (it is *not* part of the graph of the equation). A vertical scan reveals a 'gap' at $y = 0$ (x-axis)

$h(x) = \dfrac{1}{x - 2}$

with both branches of the graph continuing indefinitely, with no other 'gaps' as $y \to \pm \infty$. Both branches are also asymptotic to the x-axis. The x-axis is a **horizontal asymptote**.

Both approaches confirm the following for $h: x \mapsto \dfrac{1}{x-2}$:

The domain is $\{x: x \in \mathbb{R}, x \neq 2\}$ or $x \in \,]-\infty, 2[\,\cup\,]2, \infty[$

The range is $\{y: y \in \mathbb{R}, y \neq 0\}$ or $y \in \,]-\infty, 0[\,\cup\,]0, \infty[$

Example 5

Consider the function $g(x) = \sqrt{x+4}$.

a) Find: (i) $g(7)$
          (ii) $g(32)$
          (iii) $g(-4)$

b) Find the values of x for which g is undefined.

c) State the domain and range of g.

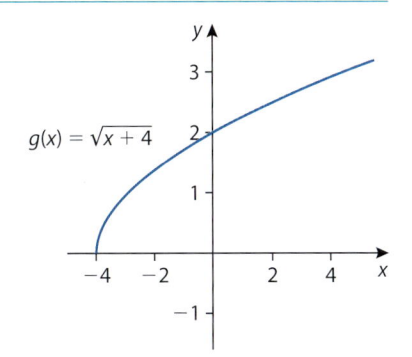

Solution

a) (i) $g(7) = \sqrt{7+4} = \sqrt{11} \approx 3.32$ (3 significant figures)
   (ii) $g(32) = \sqrt{32+4} = \sqrt{36} = 6$
   (iii) $g(-4) = \sqrt{-4+4} = \sqrt{0} = 0$

b) $g(x)$ will be undefined (square root of a negative) when $x + 4 < 0$.
   $x + 4 < 0 \Rightarrow x < -4$. Therefore, $g(x)$ is undefined when $x < -4$.

c) It follows from the result in b) that the domain of g is $\{x: x \geq -4\}$. The symbol $\sqrt{}$ stands for the **principal square root** that, by definition, can only give a result that is positive or zero. Therefore, the range of g is $\{y: y \geq 0\}$. The domain and range are confirmed by analyzing the graph of the function.

Example 6

Find the domain and range of the function $f(x) = \dfrac{1}{\sqrt{9-x^2}}$.

Solution

The graph of $y = \dfrac{1}{\sqrt{9-x^2}}$ on a GDC, shown right, agrees with algebraic analysis indicating that the expression $\dfrac{1}{\sqrt{9-x^2}}$ will be positive for all x, and is defined only for $-3 < x < 3$. Further analysis and tracing the graph reveals that $f(x)$ has a minimum at $\left(0, \dfrac{1}{3}\right)$. The graph on the GDC is misleading in that it appears to show that the function has a maximum value (y) of approximately 2.803 7849 (see screen image next page). Can this be correct? A lack of algebraic thinking and over-reliance on your GDC could easily lead to a mistake. The graph abruptly stops its curve upwards because of low screen resolution.

• **Hint:** As Example 6 illustrates, it is dangerous to completely trust graphs produced on a GDC without also doing some algebraic thinking. It is important to mentally check that the graph shown is comprehensive (shows all important features of the graph), and that the graph agrees with algebraic analysis of the function – e.g. where should the function be zero, positive, negative, undefined, increasing/decreasing without bound, etc.

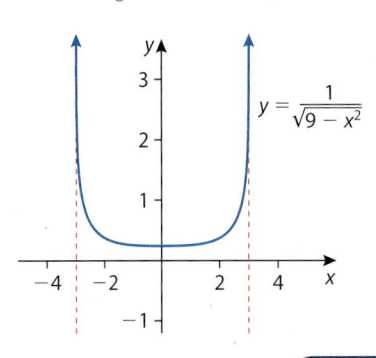

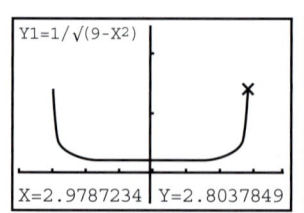

2 Functions and Equations

Function values should get quite large for values of x a little less than 3, because the value of $\sqrt{9-x^2}$ will be small, making the fraction $\dfrac{1}{\sqrt{9-x^2}}$ large. Using your GDC to make a table for $f(x)$, or evaluating the function for values of x very close to -3 or 3, confirms that as x approaches -3 or 3, y increases without bound, i.e. y goes to $+\infty$. Hence, $f(x)$ has vertical asymptotes of $x = -3$ and $x = 3$. This combination of graphical and algebraic analysis leads to the conclusion that the domain of $f(x)$ is $\{x: -3 < x < 3\}$, and the range of $f(x)$ is $\{y: y \geq \tfrac{1}{3}\}$.

Exercise 2.1

For each equation 1–9, a) match it with its graph (choices are labelled A to L), and b) state whether or not the equation represents a function – with a justification. Assume that x is the independent variable and y is the dependent variable.

1 $y = 2x$
2 $y = -3$
3 $x - y = 2$
4 $x^2 + y^2 = 4$
5 $y = 2 - x$
6 $y = x^2 + 2$
7 $y^3 = x$
8 $y = \dfrac{2}{x}$
9 $x^2 + y = 2$

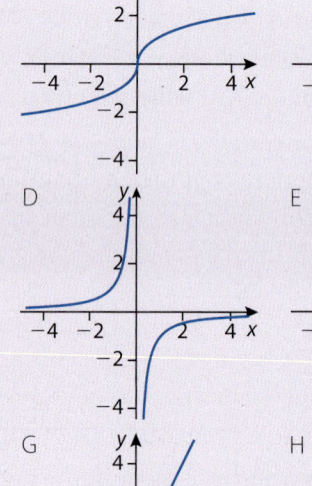

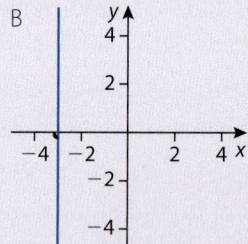

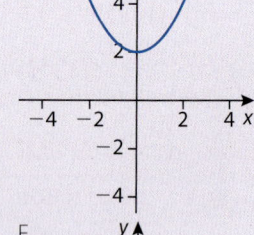

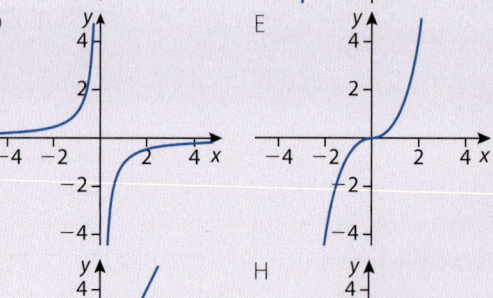

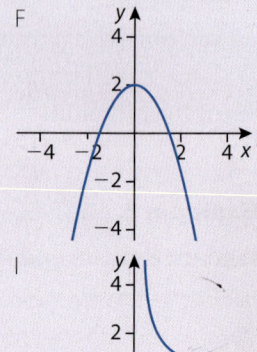

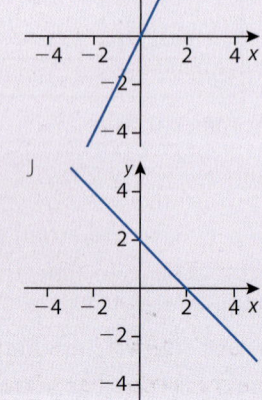

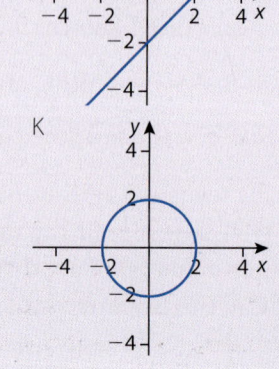

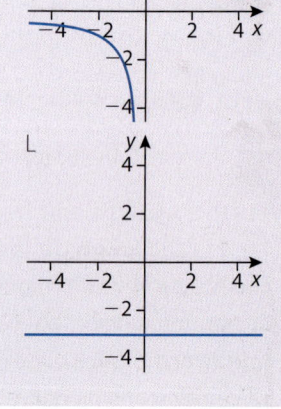

10 Express the area, A, of a circle as a function of its circumference, C.

11 Express the area, A, of an equilateral triangle as a function of the length, ℓ, of each of its sides.

In questions 12–17, find the domain of the function.

12 $f(x) = \frac{2}{5}x - 7$

13 $h(x) = x^2 - 4$

14 $g(t) = \sqrt{3 - t}$

15 $h(t) = \sqrt[3]{t}$

16 Volume of a sphere: $V = \frac{4}{3}\pi r^3$

17 $g(k) = \frac{6}{k^2 - 9}$

18 Do all linear equations represent a function? Explain.

19 Find the domain and range of the function f defined as $f: x \mapsto \frac{1}{x - 5}$.

20 Consider the function $h(x) = \sqrt{x - 4}$.
    a) Find: (i) $h(21)$ (ii) $h(53)$ (iii) $h(4)$
    b) Find the values of x for which h is undefined.
    c) State the domain and range of h.
    d) Sketch a comprehensive graph of the function.

21 Find the domain and range of the function f defined as $f(x) = \frac{1}{\sqrt{x^2 - 9}}$, and sketch a comprehensive graph of the function clearly indicating any intercepts or asymptotes.

2.2 Composition of functions

Composite functions

Consider the function in Example 5 in the previous section, $f(x) = \sqrt{x + 4}$. When you evaluate $f(x)$ for a certain value of x in the domain (for example, $x = 5$) it is necessary for you to perform computations in two separate steps in a certain order.

$f(5) = \sqrt{5 + 4} \Rightarrow f(5) = \sqrt{9}$ Step 1: compute the sum of $5 + 4$
$\Rightarrow f(5) = 3$ Step 2: compute the square root of 9

Given that the function has two separate evaluation 'steps', $f(x)$ can be seen as a combination of two 'simpler' functions that are performed in a specified order. According to how $f(x)$ is evaluated (as shown above), the simpler function to be performed first is the rule of 'adding 4' and the second is the rule of 'taking the square root'. If $h(x) = x + 4$ and $g(x) = \sqrt{x}$, we can create (compose) the function $f(x)$ from a combination of $h(x)$ and $g(x)$ as follows:

$f(x) = g(h(x))$
$\quad\quad = g(x + 4)$ Step 1: substitute $x + 4$ for $h(x)$, making $x + 4$ the argument of $g(x)$
$\quad\quad = \sqrt{x + 4}$ Step 2: apply the function $g(x)$ on the argument $x + 4$

We obtain the rule $\sqrt{x + 4}$ by first applying the rule $x + 4$ and then applying the rule $\sqrt{x}$. A function that is obtained from 'simpler' functions by applying one after another in this way is called a **composite function**. In the example above, $f(x) = \sqrt{x + 4}$ is the **composition** of $h(x) = x + 4$

> From the explanation on how f is the composition (or composite) of g and h, you can see why a composite function is sometimes referred to as a 'function of a function'. Also, note that in the notation $g(h(x))$ the function h that is applied first is written 'inside', and the function g that is applied second is written 'outside'.

Functions and Equations

followed by $g(x) = \sqrt{x}$. In other words, f is obtained by substituting h into g, and can be denoted in function notation by $g(h(x))$ – read 'g of h of x'.

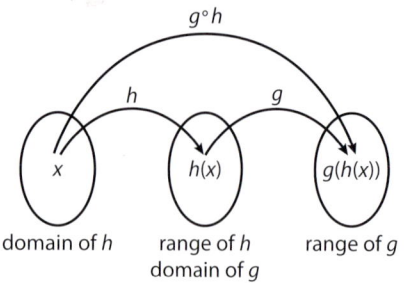

Figure 2.5

We start with a number x in the domain of h and find its image $h(x)$. If this number $h(x)$ is in the domain of g, we then compute the value of $g(h(x))$. The resulting composite function is denoted as $(g \circ h)(x)$. See mapping illustration in Figure 2.5.

• **Hint:** The notations $(g \circ h)(x)$ and $g(h(x))$ are both commonly used to denote a composite function where h is applied first and then followed by applying g. Since we are reading this from left to right, it is easy to apply the functions in the incorrect order. It may be helpful to read $g \circ h$ as 'g following h', or as 'g composed with h' to emphasize the order in which the functions are applied. Also, in either notation, $(g \circ h)(x)$ or $g(h(x))$, the function applied first is closest to the variable x.

Definition of the composition of two functions

The composition of two functions, g and h, such that h is applied first and g second is given by

$$(g \circ h)(x) = g(h(x))$$

The domain of the composite function $g \circ h$ is the set of all x in the domain of h such that $h(x)$ is in the domain of g.

Example 7

If $f(x) = 3x$ and $g(x) = 2x - 6$, find:

a) $(f \circ g)(5)$ b) Express $(f \circ g)(x)$ as a single function rule (expression).

c) $(g \circ f)(5)$ d) Express $(g \circ f)(x)$ as a single function rule (expression).

e) $(g \circ g)(5)$ f) Express $(g \circ g)(x)$ as a single function rule (expression).

Solution

a) $(f \circ g)(5) = f(g(5)) = f(2 \cdot 5 - 6) = f(4) = 3 \cdot 4 = 12$

b) $(f \circ g)(x) = f(g(x)) = f(2x - 6) = 3(2x - 6) = 6x - 18$
Therefore, $(f \circ g)(x) = 6x - 18$.
Check with result from a): $(f \circ g)(5) = 6 \cdot 5 - 18 = 30 - 18 = 12$

c) $(g \circ f)(5) = g(f(5)) = g(3 \cdot 5) = g(15) = 2 \cdot 15 - 6 = 24$

d) $(g \circ f)(x) = g(f(x)) = g(3x) = 2(3x) - 6 = 6x - 6$
Therefore, $(g \circ f)(x) = 6x - 6$.
Check with result from c): $(g \circ f)(5) = 6 \cdot 5 - 6 = 30 - 6 = 24$

e) $(g \circ g)(5) = g(g(5)) = g(2 \cdot 5 - 6) = g(4) = 2 \cdot 4 - 6 = 2$

f) $(g \circ g)(x) = g(g(x)) = g(2x - 6) = 2(2x - 6) - 6 = 4x - 18$
Therefore, $(g \circ g)(x) = 4x - 18$.
Check with result from e): $(g \circ g)(5) = 4 \cdot 5 - 18 = 20 - 18 = 2$

It is important to notice that in parts b) and d) in Example 7, $f \circ g$ is *not* equal to $g \circ f$. At the start of this section, it was shown how the two functions $h(x) = x + 4$ and $g(x) = \sqrt{x}$ could be combined into the composite function $(g \circ h)(x)$ to create the single function $f(x) = \sqrt{x + 4}$. However, the composite function $(h \circ g)(x)$ – the functions applied in reverse order – creates a different function: $(h \circ g)(x) = h(g(x)) = h(\sqrt{x}) = \sqrt{x} + 4$. Since $\sqrt{x} + 4 \neq \sqrt{x + 4}$, then again $f \circ g$ is *not* equal to $g \circ f$. Is it always true that $f \circ g \neq g \circ f$? The next example will answer that question.

Example 8

Given $f: x \mapsto 3x - 6$ and $g: x \mapsto \frac{1}{3}x + 2$, find the following:
a) $(f \circ g)(x)$ b) $(g \circ f)(x)$

Solution
a) $(f \circ g)(x) = f(g(x)) = f\left(\frac{1}{3}x + 2\right) = 3\left(\frac{1}{3}x + 2\right) - 6 = x + 6 - 6 = x$
b) $(g \circ f)(x) = g(f(x)) = g(3x - 6) = \frac{1}{3}(3x - 6) + 2 = x - 2 + 2 = x$

Example 8 shows that it is possible for $f \circ g$ to be equal to $g \circ f$. We will learn in the next section that this occurs in some cases where there is a 'special' relationship between the pair of functions. However, in general, $f \circ g \neq g \circ f$.

Decomposing composite functions

In Examples 7 and 8, we created a single function by forming the composition of two functions. As we did with the function $f(x) = \sqrt{x + 4}$ at the start of this section, it is also important for you to be able to identify two functions that *make up* a composite function, in other words, for you to *decompose* a function into two simpler functions. When you are doing this it is very useful to think of the function which is applied first as the 'inside' function, and the function that is applied second as the 'outside' function. In the function $f(x) = \sqrt{x + 4}$, the 'inside' function is $h(x) = x + 4$ and the 'outside' function is $g(x) = \sqrt{x}$.

• **Hint:** Decomposing composite functions – identifying the component functions that form a composite function – is an important skill when working with certain functions in the topic of calculus. For the composite function $f(x) = (g \circ h)(x)$, g and h are the component functions.

Example 9

Each of the following functions is a composite function of the form $(f \circ g)(x)$. For each, find the two component functions f and g.
a) $h: x \mapsto \dfrac{1}{x + 3}$ b) $k: x \mapsto 2^{4x + 1}$ c) $p(x) = \sqrt[3]{x^2 - 4}$

Solution
a) If you were to evaluate the function $h(x)$ for a certain x in the domain, you would first evaluate the expression $x + 3$, and then evaluate the expression $\frac{1}{x}$. Hence, the 'inside' function (applied first) is $y = x + 3$, and the 'outside' function (applied second) is $y = \frac{1}{x}$. Then, with $g(x) = x + 3$ and $f(x) = \frac{1}{x}$, it follows that $h: x \mapsto (f \circ g)(x)$.

b) Evaluating $k(x)$ requires you to first evaluate the expression $4x + 1$, and then evaluate the expression 2^x. Hence, the 'inside' function is $y = 4x + 1$, and the 'outside' function is $y = 2^x$. Then, with $g(x) = 4x + 1$ and $f(x) = 2^x$, it follows that $k: x \mapsto (f \circ g)(x)$.

c) Evaluating $p(x)$ requires you to perform three separate evaluation 'steps': (1) squaring a number, (2) subtracting four, and then (3) taking the cube root. Hence, it is possible to decompose $p(x)$ into three component functions: if $h(x) = x^2$, $g(x) = x - 4$ and $f(x) = \sqrt[3]{x}$, then $p(x) = (f \circ g \circ h)(x) = f(g(h(x)))$. However, for our purposes it is best to decompose the composite function into only two component functions: if $g(x) = x^2 - 4$ and $f(x) = \sqrt[3]{x}$, then $p: x \mapsto (f \circ g)(x) = f(g(x))$.

Finding the domain of a composition of functions

Referring back to Figure 2.5 (shown again here as Figure 2.6), it is important to note that in order for a value of x to be in the domain of the composite function $g \circ h$, two conditions must be met:

(1) x must be in the domain of h, and (2) $h(x)$ must be in the domain of g.

Likewise, it is also worth noting that $g(h(x))$ is in the range of $g \circ h$ only if x is in the domain of $g \circ h$. The next example illustrates these points – and also that, in general, the domains of $g \circ h$ and $h \circ g$ are not the same.

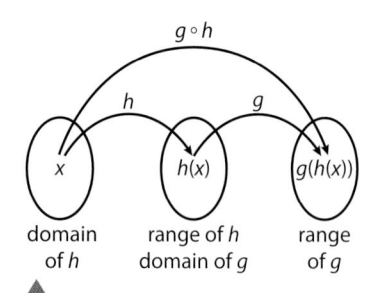

Figure 2.6

Example 10

Let $g(x) = x^2 - 4$ and $h(x) = \sqrt{x}$. Find:

a) $(g \circ h)(x)$ and its domain and range
b) $(h \circ g)(x)$ and its domain and range.

Solution

Firstly, establish the domain and range for both g and h. For $g(x) = x^2 - 4$, the domain is $x \in \mathbb{R}$ and the range is $y \geq -4$. For $h(x) = \sqrt{x}$, the domain is $x \geq 0$ and the range is $y \geq 0$.

a) $(g \circ h)(x) = g(h(x))$
$= g(\sqrt{x})$ To be in the domain of $g \circ h$, $\sqrt{x}$ must be defined for $x \Rightarrow x \geq 0$.
$= (\sqrt{x})^2 - 4$ Therefore, the domain of $g \circ h$ is $x \geq 0$.
$= x - 4$ Since $x \geq 0$, the range for $y = x - 4$ is $y \geq -4$.

Therefore, $(g \circ h)(x) = x - 4$, and its domain is $x \geq 0$, and its range is $y \geq -4$.

b) $(h \circ g)(x) = h(g(x))$ $g(x) = x^2 - 4$ must be in the domain of h
$x^2 - 4 \geq 0 \Rightarrow x^2 \geq 4$
$= h(x^2 - 4)$ Therefore, the domain of $h \circ g$ is $x \leq -2$ or $x \geq 2$
$= \sqrt{x^2 - 4}$ and, with $x \leq -2$ or $x \geq 2$, the range for $y = \sqrt{x^2 - 4}$ is $y \geq 0$.

Therefore, $(h \circ g)(x) = \sqrt{x^2 - 4}$, and its domain is $x \leq -2$ or $x \geq 2$, and its range is $y \geq 0$.

Exercise 2.2

1. Let $f(x) = 2x$ and $g(x) = \dfrac{1}{x-3}, x \neq 0$.
   a) Find the value of (i) $(f \circ g)(5)$ and (ii) $(g \circ f)(5)$.
   b) Find the function rule (expression) for (i) $(f \circ g)(x)$ and (ii) $(g \circ f)(x)$.

2. Let $f: x \mapsto 2x - 3$ and $g: x \mapsto 2 - x^2$.
   In a)–f), evaluate:
   a) $(f \circ g)(0)$ b) $(g \circ f)(0)$ c) $(f \circ f)(4)$
   d) $(g \circ g)(-3)$ e) $(f \circ g)(-1)$ f) $(g \circ f)(-3)$

   In g)–j), find the expression:
   g) $(f \circ g)(x)$ h) $(g \circ f)(x)$ i) $(f \circ f)(x)$ j) $(g \circ g)(x)$

For each pair of functions in questions 3–7, find $(f \circ g)(x)$ and $(g \circ f)(x)$ and state the domain for each.

3. $f(x) = 4x - 1, g(x) = 2 + 3x$
4. $f(x) = x^2 + 1, g(x) = -2x$
5. $f(x) = \sqrt{x+1}, g(x) = 1 + x^2$
6. $f(x) = \dfrac{2}{x+4}, g(x) = x - 1$
7. $f(x) = 3x + 5, g(x) = \dfrac{x-5}{3}$

8. Let $g(x) = \sqrt{x-1}$ and $h(x) = 10 - x^2$. Find:
   a) $(g \circ h)(x)$ and its domain and range
   b) $(h \circ g)(x)$ and its domain and range

In questions 9–14, determine functions g and h so that $f(x) = g(h(x))$.

9. $f(x) = (x+3)^2$
10. $f(x) = \sqrt{x-5}$
11. $f(x) = 7 - \sqrt{x}$
12. $f(x) = \dfrac{1}{x+3}$
13. $f(x) = 10^{x+1}$
14. $f(x) = \sqrt[3]{x-9}$

In questions 15–18, find the domain for a) the function f, b) the function g, and c) the composite function $f \circ g$.

15. $f(x) = \sqrt{x}, g(x) = x^2 + 1$
16. $f(x) = \dfrac{1}{x}, g(x) = x + 3$
17. $f(x) = \dfrac{3}{x^2 - 1}, g(x) = x + 1$
18. $f(x) = 2x + 3, g(x) = \dfrac{x}{2}$

2.3 Inverse functions

Pairs of inverse functions

Let's look again at the function at the start of this chapter – the formula that converts degrees Celsius (C) to degrees Fahrenheit (F): $F = \dfrac{9}{5}C + 32$. If we rearrange the function so that C is the independent variable (i.e. C is expressed in terms of F), we get a different formula that does the reverse, or *inverse* process, and converts F to C. Writing C in terms of F (solving for C) gives: $C = \dfrac{5}{9}(F - 32)$ or $C = \dfrac{5}{9}F - \dfrac{160}{9}$. This new formula could be useful for people travelling to the USA. These two conversion formulae, $F = \dfrac{9}{5}C + 32$ and $C = \dfrac{5}{9}F - \dfrac{160}{9}$, are both linear functions. As mentioned

2 Functions and Equations

Hint: Writing a function using x and y for the independent and dependent variables, such that y is expressed in terms of x, is a good idea because this is the format in which you must enter it on your GDC in order to have the GDC display a graph or table for the function.

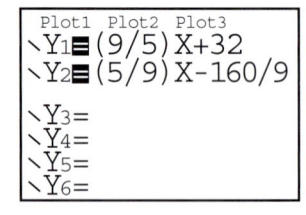

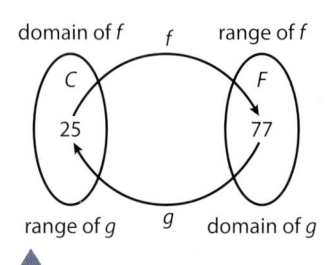

```
Plot1  Plot2  Plot3
\Y1■(9/5)X+32
\Y2■(5/9)X-160/9
\Y3=
\Y4=
\Y5=
\Y6=
```

Figure 2.7

previously, it is typical for the independent variable (domain) of a function to be x and the dependent variable (range) to be y. Let's assign the name f to the function converting C to F, and the name g to the function converting F to C.

converting C to F: $\quad y = \dfrac{9}{5}x + 32 \quad \Rightarrow \quad f(x) = \dfrac{9}{5}x + 32$

converting C to F: $\quad y = \dfrac{5}{9}x - \dfrac{160}{9} \quad \Rightarrow \quad g(x) = \dfrac{5}{9}x - \dfrac{160}{9}$

The two functions, f and g, have a 'special' relationship in that they 'undo' each other.

To illustrate, function f converts $25\,°C$ to $77\,°F$ $\left[f(25) = \dfrac{9}{5}(25) + 32 = 45 + 32 = 77\right]$, and then function g can 'undo' this by converting $77\,°F$ back to $25\,°C$ $\left[g(77) = \dfrac{5}{9}(77) - \dfrac{160}{9} = \dfrac{385 - 160}{9} = \dfrac{225}{9} = 25\right]$. Because function g has this reverse (inverse) effect on function f, we call function g the *inverse* of function f. Function f has the same inverse effect on function g [$g(77) = 25$ and then $f(25) = 77$], making f the inverse function of g. The functions f and g are inverses of each other – they are a *pair of inverse functions*.

In Figure 2.7, the mapping diagram for the functions f and g illustrates the inverse relationship for a pair of inverse functions where the domain of one is the range for the other.

The composition of two inverse functions

The mapping diagram (Figure 2.7) and the numerical examples in the previous paragraph indicate that if function f is applied to a number in its domain (e.g. 25) giving a result in the range of f (e.g. 77) and then function g is applied to this result, the final result (e.g. 25) is the same number first chosen from the domain of f. This process and result can be expressed symbolically as: $(g \circ f)(x) = x$ or $g(f(x)) = x$. The composition of two inverse functions maps any value x back to itself – i.e. one function 'undoing' the other. It must also follow that $(f \circ g) = x$. Let's verify these results for the pair of inverse functions f and g.

$(g \circ f)(x) = g\left(\dfrac{9}{5}x + 32\right) = \dfrac{5}{9}\left(\dfrac{9}{5}x + 32\right) - \dfrac{160}{9} = x + \dfrac{160}{9} - \dfrac{160}{9} = x$

$f(g(x)) = f\left(\dfrac{5}{9}x - \dfrac{160}{9}\right) = \dfrac{9}{5}\left(\dfrac{5}{9}x - \dfrac{160}{9}\right) + 32 = x - \dfrac{160}{5} + 32$

$\qquad\qquad = x - 32 + 32 = x$

Examples 7 and 8 in the previous section on composite functions explored whether $f \circ g = g \circ f$. Example 7 provided a counter-example showing it is not a true statement. However, Example 8 showed a pair of functions for which $(f \circ g)(x) = (g \circ f)(x) = x$; the same result that we just obtained for the pair of inverse functions that convert between C and F. The two functions in Example 8, $f: x \mapsto 3x - 6$ and $g: x \mapsto \dfrac{1}{3}x + 2$, are also a pair of inverse functions.

You are already familiar with pairs of **inverse operations**. Addition and subtraction are inverse operations. For example, the rule of 'adding six' $(x + 6)$ and the rule of 'subtracting six' $(x - 6)$ undo each other. Accordingly, the functions $f(x) = x + 6$ and $g(x) = x - 6$ are a pair of inverse functions. Multiplication and division are also inverse operations.

Definition of the inverse of a function

If f and g are two functions such that $(f \circ g)(x) = x$ for every x in the domain of g and $(g \circ f)(x) = x$ for every x in the domain of f, the function g is the *inverse* of the function f. The notation to indicate the function that is the 'inverse of function f' is f^{-1}. Therefore,

$$(f \circ f^{-1})(x) = x \text{ and } (f^{-1} \circ f)(x) = x$$

The domain of f must be equal to the range of f^{-1}, and the range of f must be equal to the domain of f^{-1}.

Figure 2.8 shows a mapping diagram for a pair of inverse functions.

Finding the inverse of a function

Example 11

Given the linear function $f(x) = 4x - 8$, find its inverse function $f^{-1}(x)$ and verify the result by showing that $(f \circ f^{-1})(x) = x$ and $(f^{-1} \circ f)(x) = x$.

Solution

Recall that the way we found the inverse of the function converting C to F, $F = \frac{9}{5}C + 32$, was by making the independent variable the dependent variable and vice versa. Essentially what we are doing is switching the domain (x) and range (y), since the domain of f becomes the range of f^{-1} and the range of f becomes the domain of f^{-1}, as stated in the definition of the inverse of a function, and depicted in Figure 2.8. Also, recall that $y = f(x)$.

$f(x) = 4x - 8$
$y = 4x - 8$ write $y = f(x)$
$x = 4y - 8$ interchange x and y (i.e. switch the domain and range)
$4y = x + 8$ solve for y (dependent variable) in terms of x (independent variable)
$y = \frac{1}{4}x + 2$

$f^{-1}(x) = \frac{1}{4}x + 2$ resulting equation is $y = f^{-1}(x)$

Verify that f and f^{-1} are inverses by showing that $f(f^{-1}(x)) = x$ and $f^{-1}(f(x)) = x$.

$f\left(\frac{1}{4}x + 2\right) = 4\left(\frac{1}{4}x + 2\right) - 8 = x + 8 - 8 = x$

$f^{-1}(4x - 8) = \frac{1}{4}(4x - 8) + 2 = x - 2 + 2 = x$

This confirms that $y = 4x - 8$ and $y = \frac{1}{4}x + 2$ are inverses of each other.

The method of interchanging x and y to find the inverse function also gives us a way for obtaining the graph of f^{-1} from the graph of f. Given the reversing effect that a pair of inverse functions have on each other, if $f(a) = b$ then $f^{-1}(b) = a$. Hence, if the ordered pair (a, b) is a point on the graph of $y = f(x)$, the 'reversed' ordered pair (b, a) must be on the graph of $y = f^{-1}(x)$. Figure 2.9 shows that the point (b, a) can be found by reflecting the point (a, b) about the line $y = x$.

As Figure 2.10 illustrates, the following is true.

Graphical symmetry of inverse functions

The graph of f^{-1} is a reflection of the graph of f about the line $y = x$.

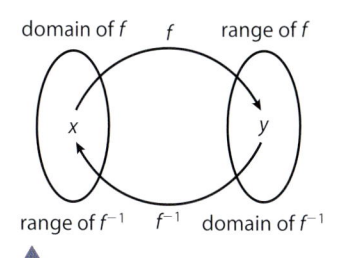

Figure 2.8 $f(x) = y$ and $f^{-1}(y) = x$.

It follows from the definition that if g is the inverse of f, it must also be true that f is the inverse of g.

• **Hint:** Do not mistake the -1 in the notation f^{-1} for an exponent. It is *not* an exponent. f^{-1} does *not* denote the reciprocal of $f(x)$. If a superscript of -1 is applied to the name of a function – as in $f^{-1}(x)$ or $\sin^{-1}(x)$ – it denotes the function that is the inverse of the named function (e.g. $f(x)$ or $\sin(x)$). If a superscript of -1 is applied to an expression, as in 7^{-1} or $(2x + 5)^{-1}$ or $(f(x))^{-1}$, it is an exponent and denotes the reciprocal of the expression. For example, the reciprocal of $f(x)$ is $(f(x))^{-1} = \frac{1}{f(x)}$.

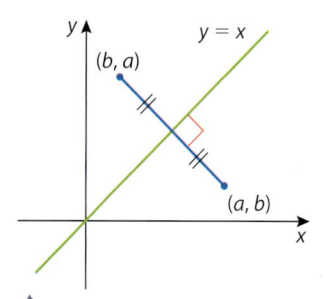

Figure 2.9

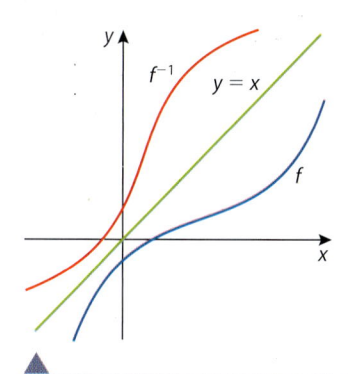

Figure 2.10

47

2 Functions and Equations

The identity function

We have repeatedly demonstrated the fact, and it is formally stated in the definition of the inverse of a function, that the composite function which has a pair of inverse functions as its components is always the linear function $y = x$. That is, $(f \circ f^{-1})(x) = x$ or $(f^{-1} \circ f)(x) = x$. Let's label the function $y = x$ with the name I. Along with the fact that $I(x) = (f \circ f^{-1})(x) = (f^{-1} \circ f)(x) = x$, the function $I(x)$ has other interesting properties. It is obvious that the line $y = x$ is reflected back to itself when reflected about the line $y = x$. Hence, from the graphical symmetry of inverse functions, the function $I(x)$ is its own inverse; that is, $I(x) = I^{-1}(x)$. Most interestingly, $I(x)$ behaves in composite functions just like the number one behaves for real numbers and multiplication. The number one is the **identity element** for multiplication. For any function f, it is true that $f \circ I = f$ and $I \circ f = f$. For this reason, we call the function $f(x) = x$, or $I(x) = x$, the **identity function**.

> When $f(x) = f^{-1}(x)$, the function f is said to be **self-inverse**. The fact that the function $f(x) = x$ is self-inverse should make you wonder if there are any other functions with the same property. Knowing that inverses are symmetric about the line $y = x$, we only need to find a function whose graph has $y = x$ as a line of symmetry.

The existence of an inverse function

Is it possible for the inverse of a function not to be a function? Recall that the definition of a function (Section 2.1) says that a function is a relation such that a certain value x in the domain produces only one value y in the range. The vertical line test for functions followed from this definition.

Example 12

Find the inverse of the function $g(x) = x^2 + 2$ with domain $x \in \mathbb{R}$.

Solution

Following the method used in Example 11:
$$g(x) = x^2 + 2$$
$$y = x^2 + 2$$
$$x = y^2 + 2$$
$$y^2 = x - 2$$
$$y = \pm\sqrt{x - 2}$$

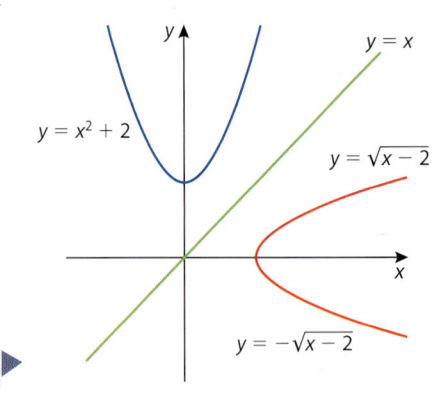

Figure 2.11

Certainly the graphs of $y = x^2 + 2$ and $y = \pm\sqrt{x - 2}$ are reflections about the line $y = x$ (see Figure 2.11). However, the graph of $y = \pm\sqrt{x - 2}$ does not pass the vertical line test. $y = \pm\sqrt{x - 2}$ is the inverse of $g(x) = x^2 + 2$, but it is only a relation and *not* a function.

The inverse of $g(x)$ will be a function only if $g(x)$ is a one-to-one function; that is, a function such that no two elements in the domain (x) of g correspond to the same element in the range (y). The graph of a one-to-one function must pass both a vertical line test and a horizontal line test.

The function $f(x) = x^2$ with domain $x \in \mathbb{R}$ (Figure 2.12) is *not* a one-to-one function. Hence, its inverse is *not* a function. There are two different values of x that correspond to the same value of y; for example, $x = 2$ and $x = -2$ both get mapped to $y = 4$. Hence, f does *not* pass the horizontal line test.

Figure 2.12

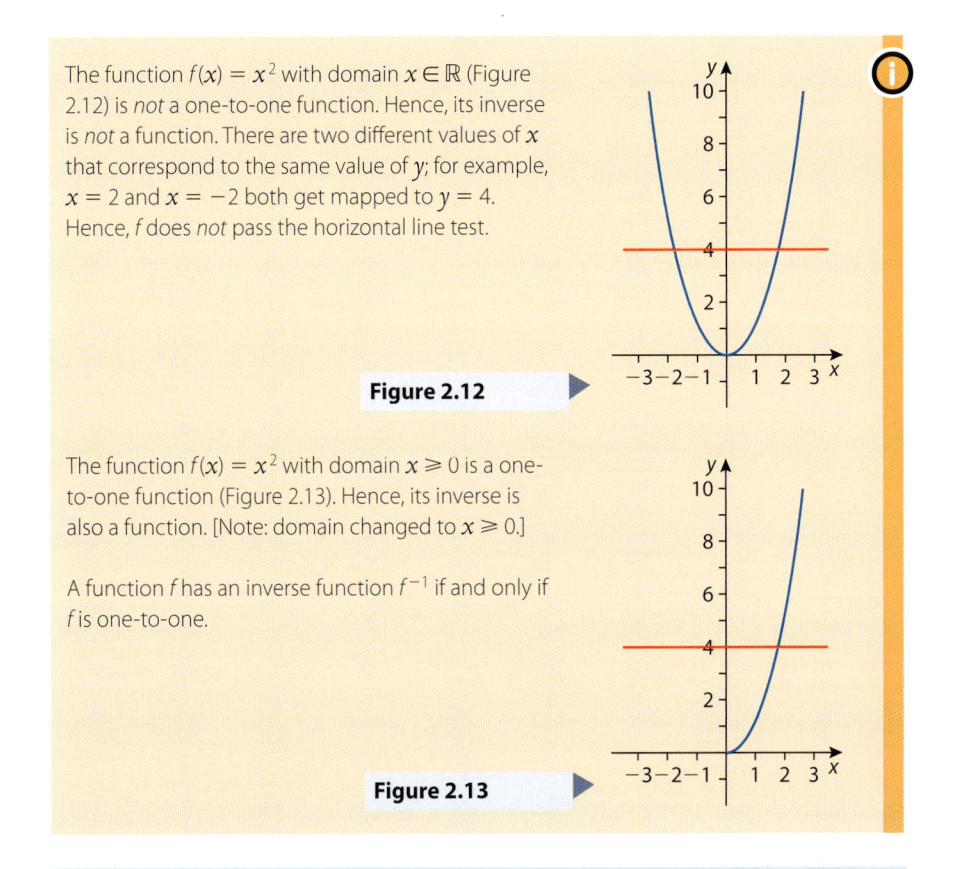

The function $f(x) = x^2$ with domain $x \geq 0$ is a one-to-one function (Figure 2.13). Hence, its inverse is also a function. [Note: domain changed to $x \geq 0$.]

A function f has an inverse function f^{-1} if and only if f is one-to-one.

Figure 2.13

Definition of a one-to-one function
A function is one-to-one if each element y in the range is the image of exactly one element x in the domain. No horizontal line can pass through the graph of a one-to-one function at more than one point (horizontal line test).

Referring back to Example 12, you now understand that the function $g(x) = x^2 + 2$ with domain $x \in \mathbb{R}$ does not have an inverse function $g^{-1}(x)$. However, if the domain is changed so that $g(x)$ is one-to-one, then $g^{-1}(x)$ exists. There is not only one way to change the domain of a function in order to make it one-to-one.

Example 13

Given $g(x) = x^2 + 2$ such that $x \geq 0$, find $g^{-1}(x)$ and state its domain.

Solution

Given that the domain is $x \geq 0$, the range for $g(x)$ will be $y \geq 0$. Since the domain and range are switched for the inverse, for $g^{-1}(x)$ the domain is $x \geq 2$ and the range is $y \geq 2$. Given the working in Example 12, it follows that $g^{-1}(x) = \sqrt{x - 2}$ with domain $x \geq 2$.

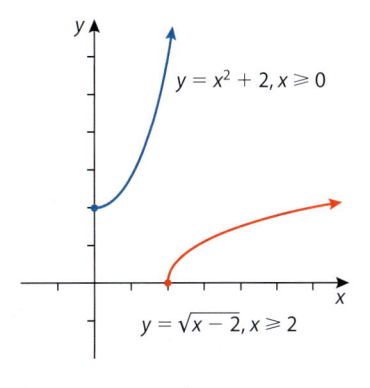

Functions and Equations

Example 14

Given $g(x) = x^2 + 2$ such that $x \leq -1$, find $g^{-1}(x)$ and state its domain.

Solution

Given that the domain is $x \leq -1$, the range for $g(x)$ will be $y \geq 3$. Since the domain and range are switched for the inverse, for $g^{-1}(x)$ the domain is $x \geq 3$ and the range is $y \leq -1$. Given the working in Example 12, it follows that $g^{-1}(x) = -\sqrt{x - 2}$ with domain $x \geq 3$.

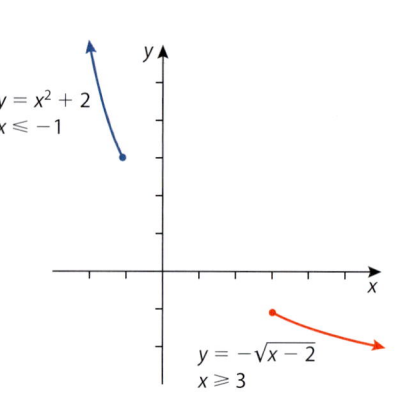

• **Hint:** For the Mathematics Standard Level course, if an inverse function is to be found, the given function will be defined with a domain that ensures it is one-to-one.

Finding the inverse of a function

To find the inverse of a function f, use the following steps:
1. Confirm that f is one-to-one (although, for this course, you can assume this).
2. Replace $f(x)$ with y.
3. Interchange x and y.
4. Solve for y.
5. Replace y with $f^{-1}(x)$.
6. The domain of f^{-1} is equal to the range of f, and the range of f^{-1} is equal to the domain of f.

Example 15

Consider the function $f: x \mapsto \sqrt{x + 3}, x \geq -3$.

a) Determine the inverse function f^{-1}.

b) What is the domain of f^{-1}?

Solution

a) Following the steps for finding the inverse of a function gives:

$y = \sqrt{x + 3}$ replace $f(x)$ with y

$x = \sqrt{y + 3}$ interchange x and y

$x^2 = y + 3$ solve for y (squaring both sides here)

$y = x^2 - 3$ solved for y

$f^{-1}: x \mapsto x^2 - 3$ replace y with $f^{-1}(x)$

b) The domain explicitly defined for f is $x \geq -3$ and since the $\sqrt{}$ symbol stands for the principal square root (positive), then the range of f is all positive real numbers, i.e. $y \geq 0$. The domain of f^{-1} is equal to the range of f; therefore, the domain of f^{-1} is $x \geq 0$.

Graphing $y = \sqrt{x + 3}$ and $y = x^2 - 3$ from Example 15 on your GDC visually confirms these results. Note that since the calculator would have automatically assumed that the domain is $x \in \mathbb{R}$, the domain for the equation $y = x^2 - 3$ has been changed to $x \geq 0$. In order to show that f and f^{-1} are reflections about the line $y = x$, the line $y = x$ has been graphed and a viewing window has been selected to ensure that the scales are equal on each axis. Using the trace feature of your GDC, you can explore a characteristic of inverse functions – that is, if some point (a, b) is on the graph of f, the point (b, a) must be on the graph of f^{-1}.

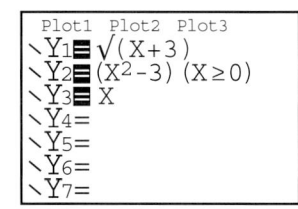

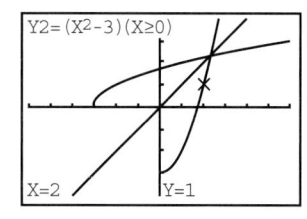

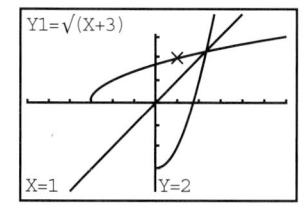

Example 16

Consider the function $f(x) = 2(x + 4)$ and $g(x) = \dfrac{1 - x}{3}$.

a) Find g^{-1} and state its domain and range.
b) Solve the equation $(f \circ g^{-1})(x) = 2$.

Solution

a) $\quad y = \dfrac{1 - x}{3} \quad$ replace $f(x)$ with y

$\quad\quad x = \dfrac{1 - y}{3} \quad$ interchange x and y

$\quad\quad 3x = 1 - y \quad$ solve for y

$\quad\quad y = -3x + 1 \quad$ solved for y

$\quad\quad g^{-1}(x) = -3x + 1 \quad$ replace y with $g^{-1}(x)$

g is a linear function and its domain is $x \in \mathbb{R}$ and its range is $y \in \mathbb{R}$; therefore, for g^{-1} the domain is $x \in \mathbb{R}$ and range is $y \in \mathbb{R}$.

b) $\quad (f \circ g^{-1})(x) = f(g^{-1}(x)) = f(-3x + 1) = 2$

$\quad\quad 2[(-3x + 1) + 4] = 2$

$\quad\quad\quad -6x + 2 + 8 = 2$

$\quad\quad\quad\quad\quad\quad -6x = -8$

$\quad\quad\quad\quad\quad\quad\quad x = \tfrac{4}{3}$

Example 17

Given $f(x) = x^2 - 6x$, find the inverse $f^{-1}(x)$ and state its domain.

Solution

The graph of $f(x) = x^2 - 6x$, $x \in \mathbb{R}$, is a parabola with a vertex at $(3, -9)$. It is not a one-to-one function. There are many ways to restrict the domain of f to make it one-to-one. The choices that have the domain as large as possible are $x \geq 3$ or $x \leq 3$. Let's change the domain of f to $x \geq 3$.

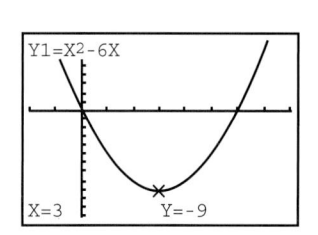

2 Functions and Equations

$$y = x^2 - 6x \quad \text{replace } f(x) \text{ with } y$$
$$x = y^2 - 6y \quad \text{interchange } x \text{ and } y$$
$$y^2 - 6y + 9 = x + 9 \quad \text{add 9 to both sides}$$
(See pg 67 for explanation of method)
$$(y - 3)^2 = x + 9 \quad \text{substituting } (y - 3)^2 \text{ for } y^2 - 6y + 9$$
$$y - 3 = \pm\sqrt{x + 9}$$
$$y = 3 + \sqrt{x + 9} \quad + \text{ rather than } \pm \text{ because range of } f^{-1} \text{ is } x \geq 3$$
(domain of f)

In order for $\sqrt{x + 9}$ to be a real number then $x \geq -9$.

Therefore, $f^{-1}(x) = 3 + \sqrt{x + 9}$ and the domain of f^{-1} is $x \geq -9$.

The inverse relationship between $f(x) = x^2 - 6x$ and $f^{-1}(x) = 3 + \sqrt{x + 9}$ is confirmed graphically in Figure 2.14.

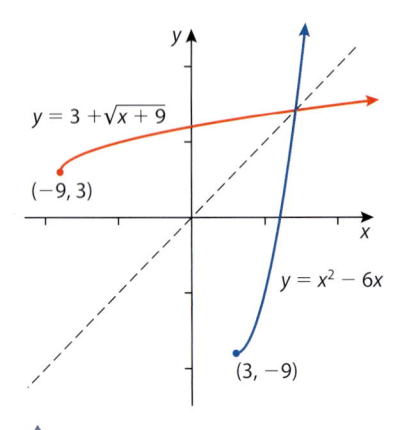

Figure 2.14

Exercise 2.3

In questions 1–4, assume that f is a one-to-one function.

1 a) If $f(2) = -5$, what is $f^{-1}(-5)$?   b) If $f^{-1}(6) = 10$, what is $f(10)$?

2 a) If $f(-1) = 13$, what is $f^{-1}(13)$?   b) If $f^{-1}(b) = a$, what is $f(a)$?

3 If $g(x) = 3x - 7$, what is $g^{-1}(5)$?

4 If $h(x) = x^2 - 8x$, with $x \geq 4$, what is $h^{-1}(-12)$?

In questions 5–12, show a) algebraically and b) graphically that f and g are inverse functions by verifying that $(f \circ g)(x) = x$ and $(g \circ f)(x) = x$, and by sketching the graphs of f and g on the same set of axes, with equal scales on the x- and y-axes. Use your GDC to assist in making your sketches on paper.

5 $f: x \mapsto x + 6; g: x \mapsto x - 6$ **6** $f: x \mapsto 4x; g: x \mapsto \frac{x}{4}$

7 $f: x \mapsto 3x + 9; g: x \mapsto \frac{1}{3}x - 3$ **8** $f: x \mapsto \frac{1}{x}; g: x \mapsto \frac{1}{x}$

9 $f: x \mapsto x^2 - 2, x \geq 0; g: x \mapsto \sqrt{x + 2}, x \geq -2$

10 $f: x \mapsto x^3; g: x \mapsto \sqrt[3]{x}$ **11** $f: x \mapsto \frac{1}{1 + x}; g: x \mapsto \frac{1 - x}{x}$

12 $f: x \mapsto (6 - x)^{\frac{1}{2}}; g: x \mapsto 6 - x^2, x \geq 0$

In questions 13–20, find the inverse function f^{-1} and state its domain.

13 $f(x) = 2x - 3$ **14** $f(x) = \frac{x + 7}{4}$

15 $f(x) = \sqrt{x}$ **16** $f(x) = \frac{1}{x + 2}$

17 $f(x) = 4 - x^2, x \geq 0$ **18** $f(x) = \sqrt{x - 5}$

19 $f(x) = ax + b, a \neq 0$ **20** $f(x) = x^2 + 2x, x \geq -1$

In questions 21–28, use the functions $g(x) = x + 3$ and $h(x) = 2x - 4$ to find the indicated value or the indicated function.

21 $(g^{-1} \circ h^{-1})(5)$ **22** $(h^{-1} \circ g^{-1})(9)$ **23** $(g^{-1} \circ g^{-1})(2)$

24 $(h^{-1} \circ h^{-1})(2)$ **25** $g^{-1} \circ h^{-1}$ **26** $h^{-1} \circ g^{-1}$

27 $(g \circ h)^{-1}$ **28** $(h \circ g)^{-1}$

29 The function in question 8, $f(x) = \frac{1}{x}$, is its own inverse (self-inverse). Show that any function in the form $f(x) = \frac{a}{x + b} - b, a \neq 0$, is its own inverse.

2.4 Transformations of functions

Even when you use your GDC to sketch the graph of a function, it is helpful to know what to expect in terms of the location and shape of the graph – and even more so if you're not allowed to use your GDC for a particular question. In this section, we look at how certain changes to the equation of a function can affect, or **transform**, the location and shape of its graph. We will investigate three different types of **transformations** of functions that include how the graph of a function can be **translated**, **reflected** and **stretched** (or shrunk). This will give us a better understanding of how to efficiently sketch and visualize many different functions.

Graphs of common functions

It is important for you to be familiar with the location and shape of a certain set of common functions. For example, from your previous knowledge about linear equations, you can determine the location of the linear function $f(x) = ax + b$. You know that the graph of this function is a line whose slope is a and whose y-intercept is $(0, b)$.

The eight graphs in Figure 2.15 represent some of the most commonly used functions in algebra. You should be familiar with the characteristics of the graphs of these common functions. This will help you predict and analyze the graphs of more complicated functions that are derived from applying one or more transformations to these simple functions. There are other important basic functions with which you should be familiar – for example, exponential, logarithmic and trigonometric functions – but we will encounter these in later chapters.

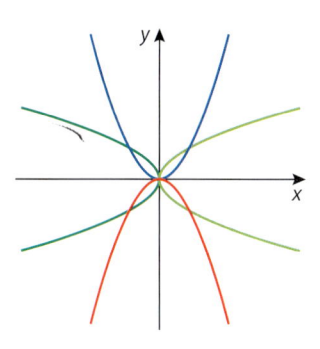

• **Hint:** When analyzing the graph of a function, it is often convenient to express a function in the form $y = f(x)$. As we have done throughout this chapter, we often refer to a function such as $f(x) = x^2$ by the equation $y = x^2$.

Figure 2.15 Graphs of common functions.

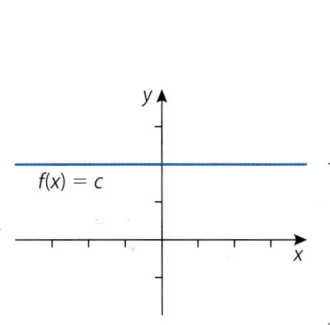

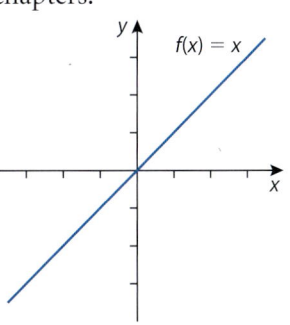

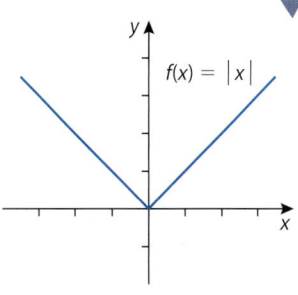

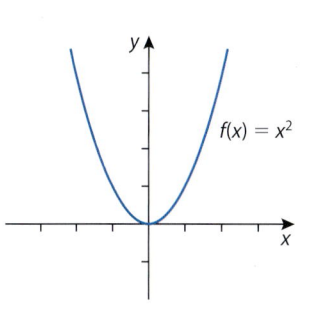

a) Constant function $f(x) = c$

b) Identity function $f(x) = x$

c) Absolute value function $f(x) = |x|$

d) Squaring function $f(x) = x^2$

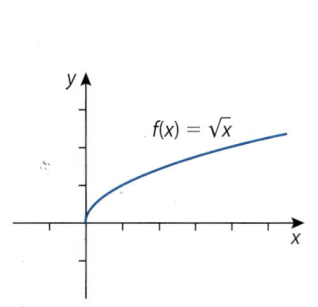

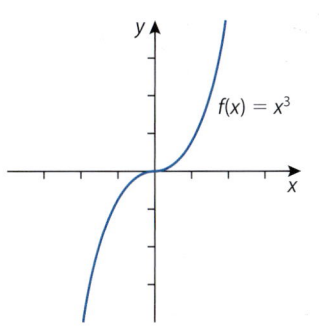

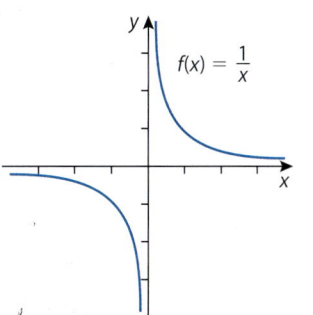

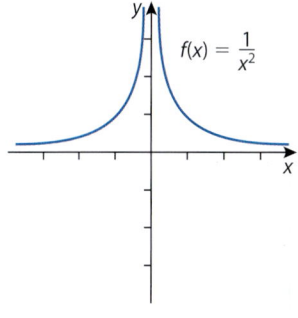

e) Square root function $f(x) = \sqrt{x}$

f) Cubing function $f(x) = x^3$

g) Reciprocal function $f(x) = \frac{1}{x}$

h) Inverse square function $f(x) = \frac{1}{x^2}$

2 Functions and Equations

Hint: The word *inverse* can have different meanings in mathematics depending on the context. In Section 2.3 of this chapter, *inverse* is used to describe operations or functions that undo each other. However, 'inverse' is sometimes used to denote the **multiplicative inverse** (or **reciprocal**) of a number or function. This is how it is used in the names for the functions shown in g) and h) of Figure 2.15. The function in g) is sometimes called the **reciprocal function**.

We will see that many functions have graphs that are a transformation (translation, reflection or stretch), or a combination of transformations, of one of these common functions.

Vertical and horizontal translations

Use your GDC to graph each of the following three functions: $f(x) = x^2$, $g(x) = x^2 + 3$ and $h(x) = x^2 - 2$. How do the graphs of g and h compare with the graph of f that is one of the common functions displayed in Figure 2.15? The graphs of g and h both appear to have the same shape – it's only the location, or position, that has changed compared to f. Although the curves (parabolas) appear to be getting closer together, their vertical separation at every value of x is constant.

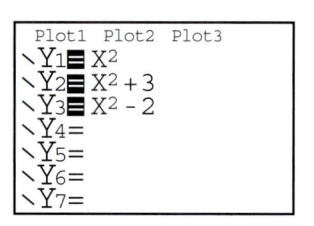

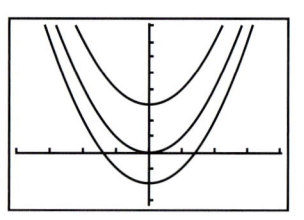

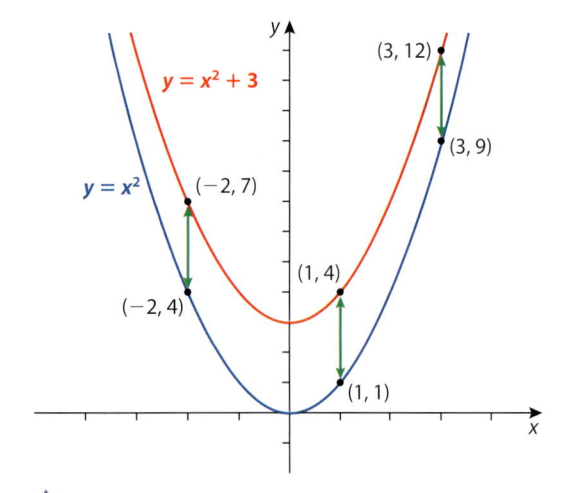

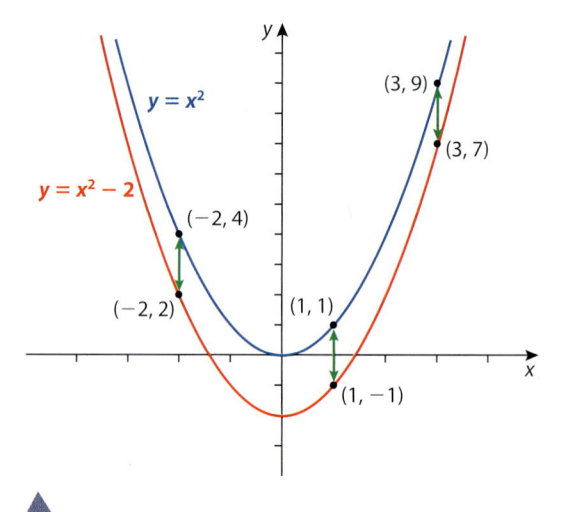

Figure 2.16

Figure 2.17

As Figures 2.16 and 2.17 clearly show, you can obtain the graph of $g(x) = x^2 + 3$ by translating (shifting) the graph of $f(x) = x^2$ *up* three units, and you can obtain the graph of $h(x) = x^2 - 2$ by translating the graph of $f(x) = x^2$ *down* two units.

> **Vertical translations of a function**
> Given $k > 0$, then:
> I. The graph of $y = f(x) + k$ is obtained by translating *up* k units the graph of $y = f(x)$.
> II. The graph of $y = f(x) - k$ is obtained by translating *down* k units the graph of $y = f(x)$.

Change function g to $g(x) = (x + 3)^2$ and change function h to $h(x) = (x - 2)^2$. Graph these two functions along with the 'parent' function

$f(x) = x^2$ on your GDC. This time we observe that functions g and h can be obtained by a horizontal translation of f.

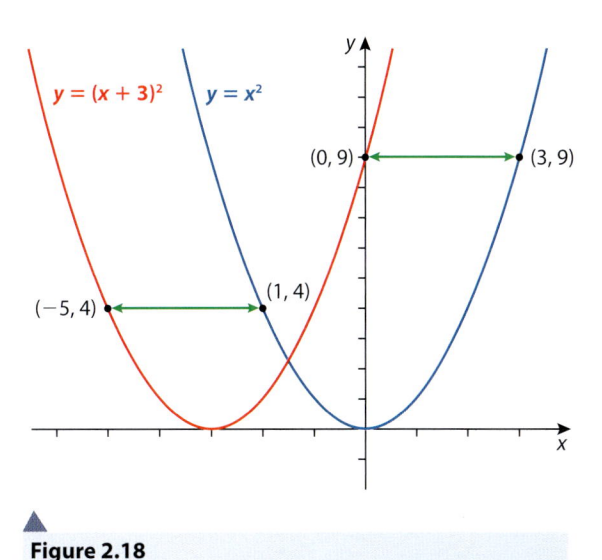

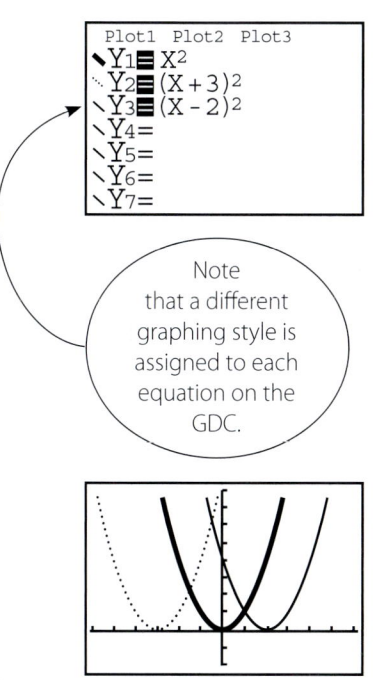

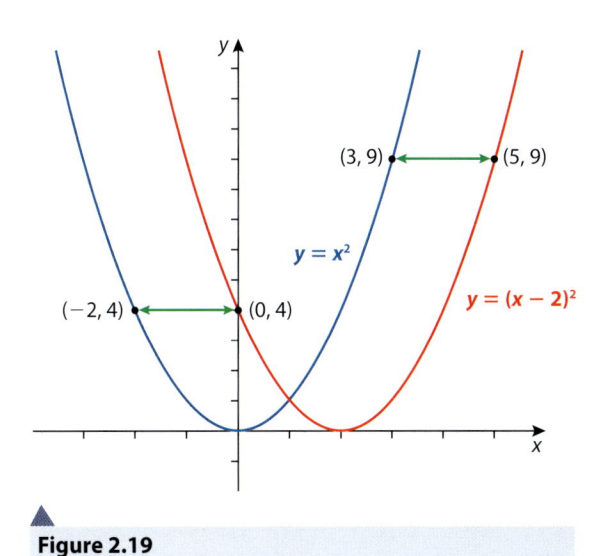

Figure 2.18

Note that a different graphing style is assigned to each equation on the GDC.

Figure 2.19

As Figures 2.18 and 2.19 clearly show, you can obtain the graph of $g(x) = (x + 3)^2$ by translating the graph of $f(x) = x^2$ three units to the *left*, and you can obtain the graph of $h(x) = (x - 2)^2$ by translating the graph of $f(x) = x^2$ two units to the *right*.

Horizontal translations of a function

Given $h > 0$, then:

I. The graph of $y = f(x - h)$ is obtained by translating the graph of $y = f(x)$ h units to the *right*.

II. The graph of $y = f(x + h)$ is obtained by translating the graph of $y = f(x)$ h units to the *left*.

2 Functions and Equations

• **Hint:** A common error is caused by confusion about the direction of a horizontal translation since $f(x)$ is translated *left* if a *positive* number is added *inside* the argument of the function – e.g. $g(x) = (x + 3)^2$ is obtained by translating $f(x) = x^2$ three units *left*. You are in the habit of associating *positive* with movement to the *right* (as on the x-axis) instead of *left*. Whereas $f(x)$ is translated *up* if a *positive* number is added *outside* the function – e.g. $g(x) = x^2 + 3$ is obtained by translating $f(x) = x^2$ three units *up*. This agrees with the convention that a *positive* number is associated with an *upward* movement (as on the y-axis). An alternative (and more consistent) approach to vertical and horizontal translations is to think of what number is being added directly to the x- or y-coordinate. For example, the equation for the graph obtained by translating the graph of $y = x^2$ three units up is $y = x^2 + 3$, which can also be written as $y - 3 = x^2$. In this form, negative three is added to the y-coordinate (vertical coordinate), which causes a vertical translation in the *upward* (or positive) direction. Likewise, the equation for the graph obtained by translating the graph of $y = x^2$ two units to the right is $y = (x - 2)^2$. Negative two is added to the x-coordinate (horizontal coordinate), which causes a horizontal translation to the right (or positive direction). There is consistency between vertical and horizontal translations. Assuming that movement up or to the right is considered positive, and that movement down or to the left is negative, then the direction for either type of translation is opposite to the sign ($\pm$) of the number being added to the vertical (y) or horizontal (x) coordinate. In fact, what is actually being translated is the y-axis or the x-axis. For example, the graph of $y - 3 = x^2$ can also be obtained by not changing the graph of $y = x^2$ but instead translating the y-axis three units down – which creates exactly the same effect as translating the graph of $y = x^2$ three units up.

Example 18

The diagrams show how the graph of $y = \sqrt{x}$ is transformed to the graph of $y = f(x)$ in three steps. For each diagram, a) and b), give the equation of the curve.

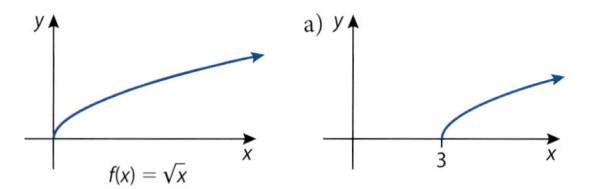

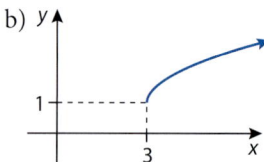

> Note that in Example 18, if the transformations had been performed in reverse order – that is, the vertical translation followed by the horizontal translation – it would produce the same final graph (in part b)) with the same equation. In other words, when applying both a vertical and horizontal translation on a function it does not make any difference which order they are applied (i.e. they are commutative). However, as we will see further on in the chapter, it *can* make a difference to how other sequences of transformations are applied. In general, transformations are *not* commutative.

Solution

To obtain graph a), the graph of $y = \sqrt{x}$ is translated three units to the right. To produce the equation of the translated graph, -3 is added *inside* the argument of the function $y = \sqrt{x}$. Therefore, the equation of the curve graphed in a) is $y = \sqrt{x - 3}$.

To obtain graph b), the graph of $y = \sqrt{x - 3}$ is translated up one unit. To produce the equation of the translated graph, $+1$ is added *outside* the function. Therefore, the equation of the curve graphed in b) is $y = \sqrt{x - 3} + 1$ (or $y = 1 + \sqrt{x - 3}$).

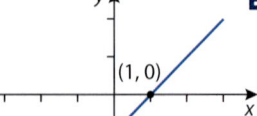

Example 19

Write the equation of the absolute value function whose graph is shown on the left.

Solution

The graph shown is exactly the same shape as the graph of the equation $y = |x|$ but in a different position. Given that the vertex is $(-2, -3)$, it is clear that this graph can be obtained by translating $y = |x|$ two units left and then three units down. When we move $y = |x|$ two units left we get the graph of $y = |x + 2|$. Moving the graph of $y = |x + 2|$ three units down gives us the graph of $y = |x + 2| - 3$. Therefore, the equation of the graph shown is $y = |x + 2| - 3$. (Note: The two translations applied in reverse order produce the same result.)

Reflections

Use your GDC to graph the two functions $f(x) = x^2$ and $g(x) = -x^2$. The graph of $g(x) = -x^2$ is a reflection in the x-axis of $f(x) = x^2$. This certainly makes sense because g is formed by multiplying f by -1, causing the y-coordinate of each point on the graph of $y = -x^2$ to be the negative of the y-coordinate of the point on the graph of $y = x^2$ with the same x-coordinate.

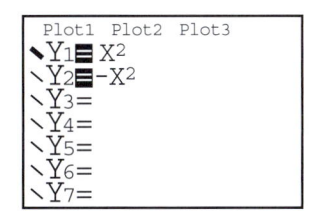

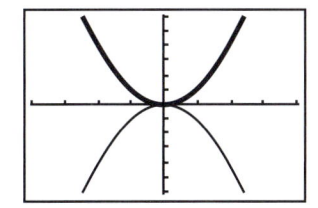

Figures 2.20 and 2.21 illustrate that the graph of $y = -f(x)$ is obtained by reflecting the graph of $y = f(x)$ in the x-axis.

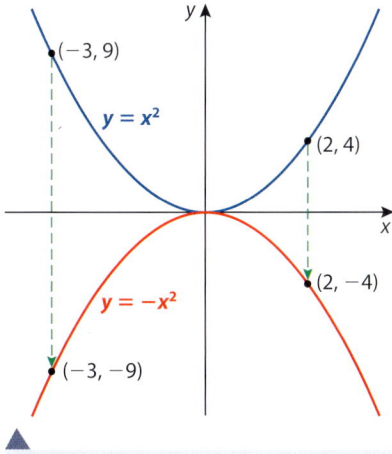

Figure 2.20

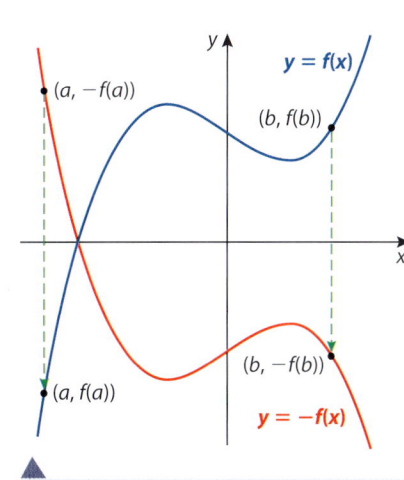

Figure 2.21

• **Hint:** The expression $-x^2$ is potentially ambiguous. It is accepted to be equivalent to $-(x)^2$. It is *not* equivalent to $(-x)^2$. For example, if you enter the expression -3^2 into your GDC, it gives a result of -9, *not* $+9$. In other words, the expression -3^2 is consistently interpreted as 3^2 being multiplied by -1. The same as $-x^2$ is interpreted as x^2 being multiplied by -1.

Graph the functions $f(x) = \sqrt{x - 2}$ and $g(x) = \sqrt{-x - 2}$. Previously, with $f(x) = x^2$ and $g(x) = -x^2$, g was formed by multiplying the entire function f by -1. However, for $f(x) = \sqrt{x - 2}$ and $g(x) = \sqrt{-x - 2}$, g is formed by multiplying the variable x by -1. In this case, the graph of $g(x) = \sqrt{-x - 2}$ is a reflection in the y-axis of $f(x) = \sqrt{x - 2}$. This makes sense if you recognize that the y-coordinate on the graph of $y = \sqrt{-x}$ will be the same as the y-coordinate on the graph of $y = \sqrt{x}$, if the value substituted for x in $y = \sqrt{-x}$ is the opposite of the value of x in $y = \sqrt{x}$. For example, if $x = 9$ then $y = \sqrt{9} = 3$; and, if $x = -9$ then $y = \sqrt{-(-9)} = \sqrt{9} = 3$. Opposite values of x in the two functions produce the same y-coordinate for each.

Functions and Equations

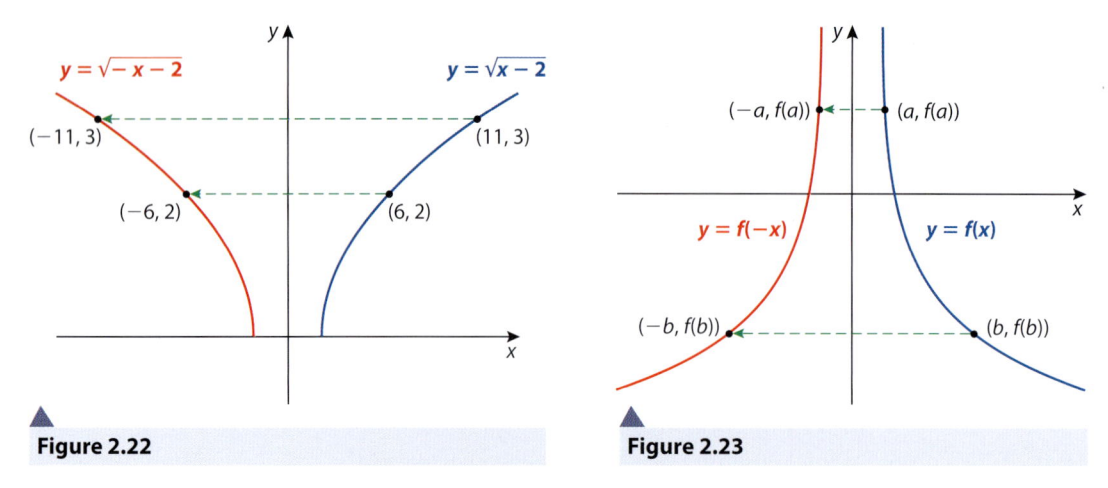

Figure 2.22

Figure 2.23

Figures 2.22 and 2.23 illustrate that the graph of $y = f(-x)$ is obtained by reflecting the graph of $y = f(x)$ in the y-axis.

> **Reflections of a function in the coordinate axes**
> I. The graph of $y = -f(x)$ is obtained by reflecting the graph of $y = f(x)$ in the x-axis.
> II. The graph of $y = f(-x)$ is obtained by reflecting the graph of $y = f(x)$ in the y-axis.

Example 20

For $g(x) = 2x^3 - 6x^2 + 3$, find:
a) the function $h(x)$ that is the reflection of $g(x)$ in the x-axis
b) the function $p(x)$ that is the reflection of $g(x)$ in the y-axis.

Solution

a) Knowing that $y = -f(x)$ is the reflection of $y = f(x)$ in the x-axis, then $h(x) = -g(x) = -(2x^3 - 6x^2 + 3) \Rightarrow h(x) = -2x^3 + 6x^2 - 3$ will be the reflection of $g(x)$ in the x-axis. We can verify the result on the GDC – graphing the original equation $y = 2x^3 - 6x^2 + 3$ in bold style.

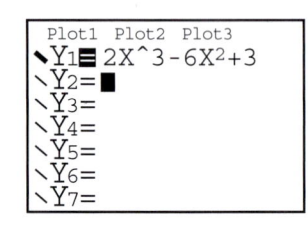

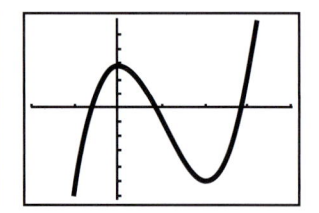

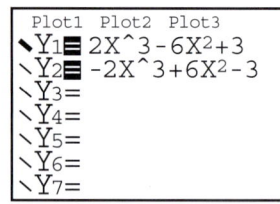

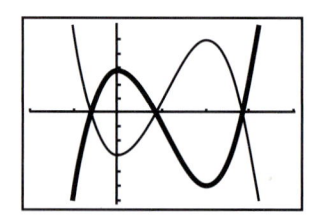

b) Knowing that $y = f(-x)$ is the reflection of $y = f(x)$ in the y-axis, we need to substitute $-x$ for x in $y = g(x)$. Thus,
$p(x) = g(-x) = 2(-x)^3 - 6(-x)^2 + 3 \Rightarrow p(x) = -2x^3 - 6x + 3$ will be the reflection of $g(x)$ in the y-axis. Again, we can verify the result on the GDC – graphing the original equation $y = 2x^3 - 6x^2 + 3$ in bold style.

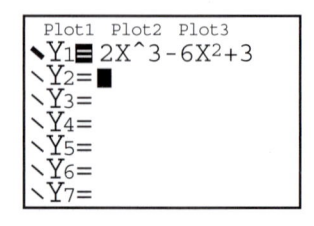

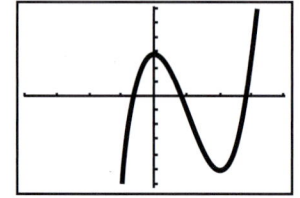

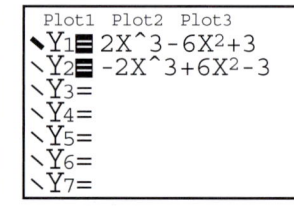

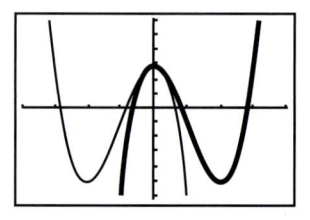

Non-rigid transformations: stretching and shrinking

Horizontal and vertical translations, and reflections in the *x*- and *y*-axes are called **rigid transformations** because the shape of the graph does not change – only its position is changed. **Non-rigid transformations** cause the shape of the original graph to change. The non-rigid transformations that we will study cause the shape of a graph to *stretch* or *shrink* in either the vertical or horizontal direction.

Vertical stretch or shrink

Graph the following three functions: $f(x) = x^2$, $g(x) = 3x^2$ and $h(x) = \frac{1}{3}x^2$. How do the graphs of *g* and *h* compare to the graph of *f*? Clearly, the shape of the graphs of *g* and *h* is not the same as the graph of *f*. Multiplying the function *f* by a positive number greater than one, or less than one, has distorted the shape of the graph. For a certain value of *x*, the *y*-coordinate of $y = 3x^2$ is three times the *y*-coordinate of $y = x^2$. Therefore, the graph of $y = 3x^2$ can be obtained by *vertically stretching* the graph of $y = x^2$ by a factor of 3 (**scale factor 3**). Likewise, the graph of $y = \frac{1}{3}x^2$ can be obtained by *vertically shrinking* the graph of $y = x^2$ by **scale factor** $\frac{1}{3}$.

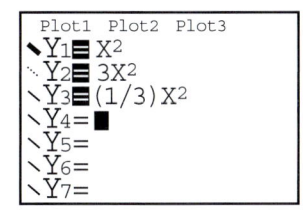

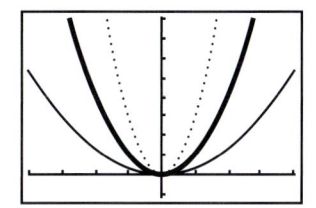

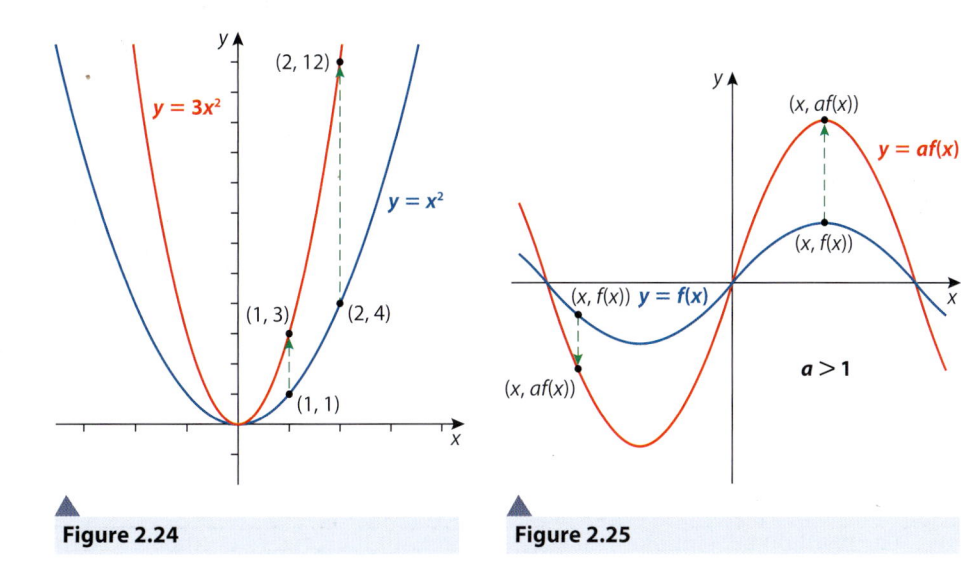

Figures 2.24 and 2.25 illustrate how multiplying a function by a positive number, *a*, *greater than one* causes a transformation by which the function *stretches* vertically by scale factor *a*. A point (x, y) on the graph of $y = f(x)$ is transformed to the point (x, ay) on the graph of $y = af(x)$.

Figure 2.24

Figure 2.25

2 Functions and Equations

Figures 2.26 and 2.27 illustrate how multiplying a function by a positive number, *a*, *greater than zero and less than one* causes the function to *shrink* vertically by scale factor *a*. A point (x, y) on the graph of $y = f(x)$ is transformed to the point (x, ay) on the graph of $y = af(x)$.

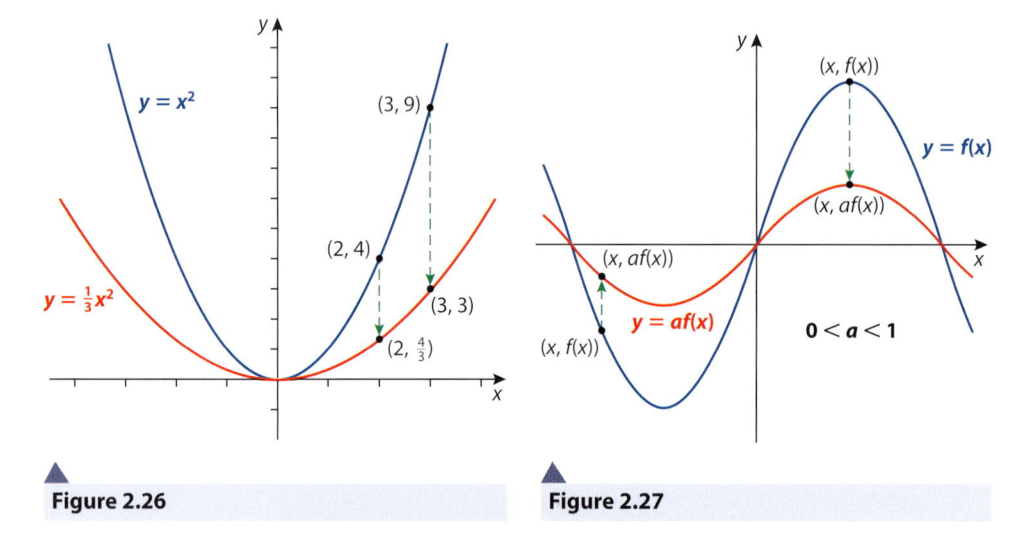

Figure 2.26

Figure 2.27

> **Vertical stretching and shrinking of functions**
> I. If $a > 1$, the graph of $y = af(x)$ is obtained by *vertically stretching* the graph of $y = f(x)$.
> II. If $0 < a < 1$, the graph of $y = af(x)$ is obtained by *vertically shrinking* the graph of $y = f(x)$.

Horizontal stretch or shrink

Let's investigate how the graph of $y = f(ax)$ is obtained from the graph of $y = f(x)$. Given $f(x) = x^2 - 4x$, find another function, $g(x)$, such that $g(x) = f(2x)$. We substitute $2x$ for x in the function f, giving $g(x) = (2x)^2 - 4(2x)$. For the purposes of our investigation, let's leave $g(x)$ in this form. On your GDC, graph these two functions, $f(x) = x^2 - 4x$ and $g(x) = (2x)^2 - 4(2x)$, using the indicated viewing window and graphing *f* in bold style.

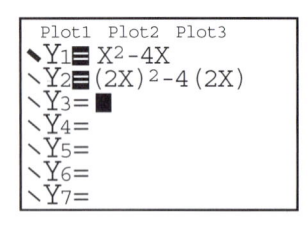

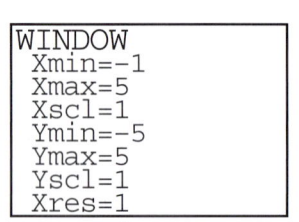

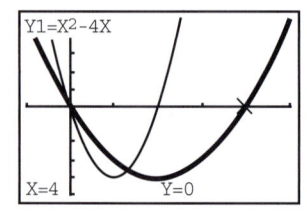

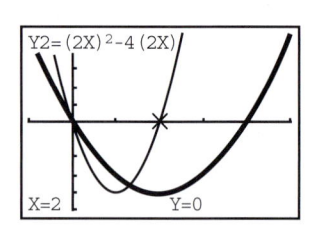

Comparing the graphs of the two equations, we see that $y = g(x)$ is *not* a translation or a reflection of $y = f(x)$. It is similar to the *shrinking* effect that occurs for $y = af(x)$ when $0 < a < 1$, except, instead of a vertical shrinking, the graph of $y = g(x) = f(2x)$ is obtained by *horizontally* shrinking the graph of $y = f(x)$. Given that it is a shrinking – rather than a stretching – the scale factor must be less than one. Consider the point $(4, 0)$ on the graph of $y = f(x)$. The point on the graph of $y = g(x) = f(2x)$ with the same *y*-coordinate and on the same side of the parabola is (2, 0). The x-coordinate of the point on $y = f(2x)$ is the x-coordinate of the point on $y = f(x)$ multiplied by $\frac{1}{2}$. Use your GDC to confirm this for other pairs of corresponding points on $y = x^2 - 4x$ and $y = (2x)^2 - 4(2x)$ that have the same y-coordinate. The graph of $y = f(2x)$ can be obtained by *horizontally shrinking* the graph of $y = f(x)$ by scale factor $\frac{1}{2}$. This makes sense because if $f(2x_2) = (2x_2)^2 - 4(2x_2)$ and $f(x_1) = x_1^2 - 4x_1$ are to produce the same y-value then $2x_2 = x_1$; and, thus, $x_2 = \frac{1}{2}x_1$. Figures 2.28 and 2.29 illustrate how multiplying the x-variable of a function by a positive number, a, *greater than one* causes the function to *shrink* horizontally by scale factor $\frac{1}{a}$. A point (x, y) on the graph of $y = f(x)$ is transformed to the point $\left(\frac{1}{a}x, y\right)$ on the graph of $y = f(ax)$.

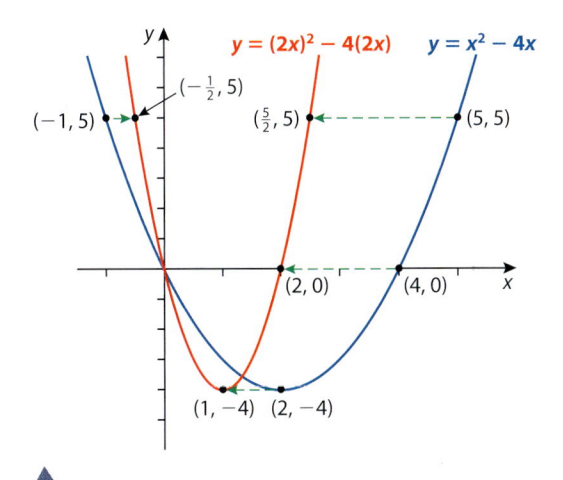

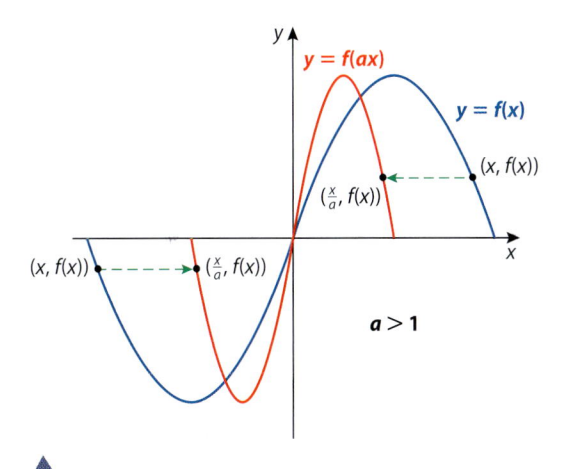

Figure 2.28

Figure 2.29

If $0 < a < 1$, the graph of the function $y = f(ax)$ is obtained by a *horizontal stretching* of the graph of $y = f(x)$ – rather than a shrinking – because the scale factor $\frac{1}{a}$ will be a value greater than 1 if $0 < a < 1$. Now, letting $a = \frac{1}{2}$ and, again using the function $f(x) = x^2 - 4x$, find $g(x)$, such that $g(x) = f(\frac{1}{2}x)$. We substitute $\frac{x}{2}$ for x in f, giving $g(x) = \left(\frac{x}{2}\right)^2 - 4\left(\frac{x}{2}\right)$. On your GDC, graph the functions f and g using the indicated viewing window with f in bold.

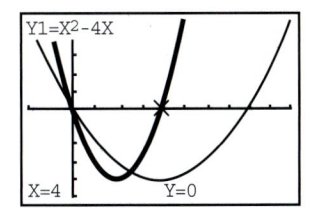

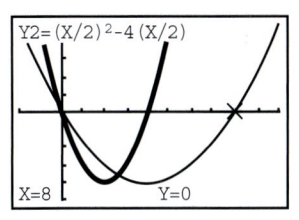

The graph of $y = \left(\frac{x}{2}\right)^2 - 4\left(\frac{x}{2}\right)$ is a horizontal stretching of the graph of $y = x^2 - 4x$ by scale factor $\frac{1}{a} = \frac{1}{\frac{1}{2}} = 2$. For example, the point (4, 0) on $y = f(x)$ has been moved horizontally to the point (8, 0) on $y = g(x) = f\left(\frac{x}{2}\right)$.

Functions and Equations

Figures 2.30 and 2.31 illustrate how multiplying the *x*-variable of a function by a positive number, *a*, *greater than zero and less than one* causes the function to *stretch* horizontally by scale factor $\frac{1}{a}$. A point (x, y) on the graph of $y = f(x)$ is transformed to the point $\left(\frac{1}{a}x, y\right)$ on the graph of $y = f(ax)$.

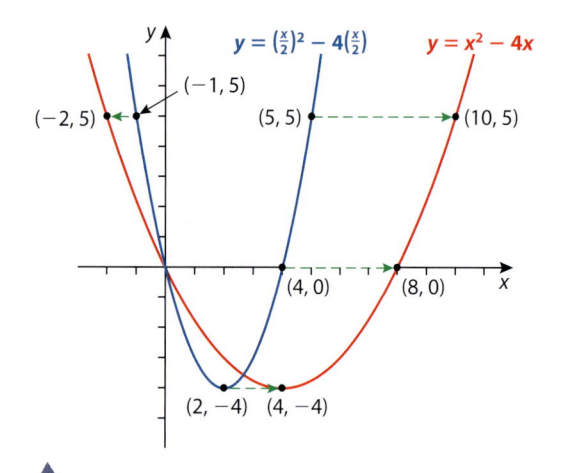

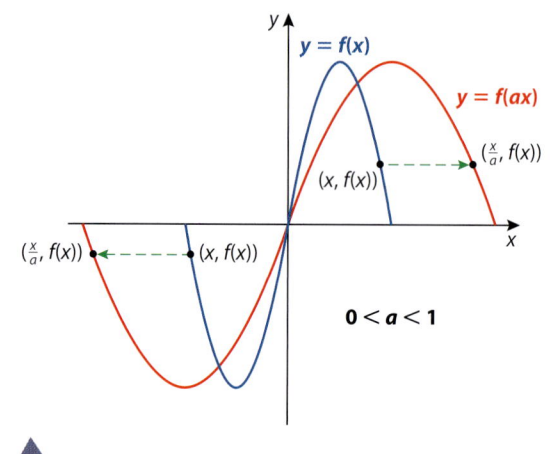

Figure 2.30

Figure 2.31

> **Horizontal stretching and shrinking of functions**
> I. If $a > 1$, the graph of $y = f(ax)$ is obtained by *horizontally shrinking* the graph of $y = f(x)$.
> II. If $0 < a < 1$, the graph of $y = f(ax)$ is obtained by *horizontally stretching* the graph of $y = f(x)$.

Example 21

The graph of $y = f(x)$ is shown. Sketch the graph of each of the following two functions.

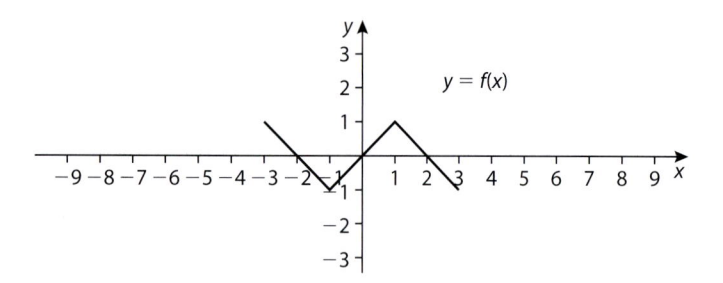

a) $y = 3f(x)$
b) $y = \frac{1}{3}f(x)$
c) $y = f(3x)$
d) $y = f\left(\frac{1}{3}x\right)$

Solution

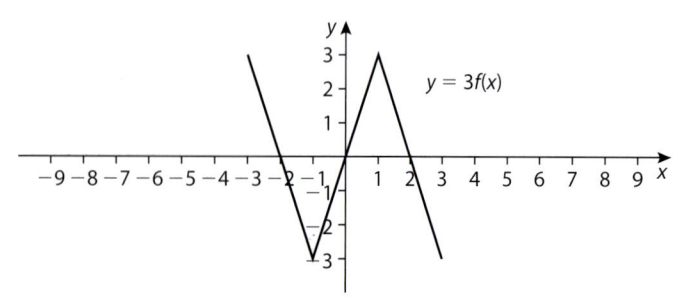

a) The graph of $y = 3f(x)$ is obtained by vertically stretching the graph of $y = f(x)$ by scale factor 3.

b) The graph of $y = \frac{1}{3}f(x)$ is obtained by vertically shrinking the graph of $y = f(x)$ by scale factor $\frac{1}{3}$.

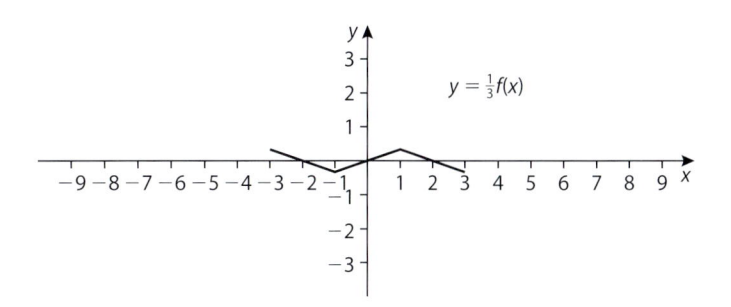

c) The graph of $y = f(3x)$ is obtained by horizontally shrinking the graph of $y = f(x)$ by scale factor $\frac{1}{3}$.

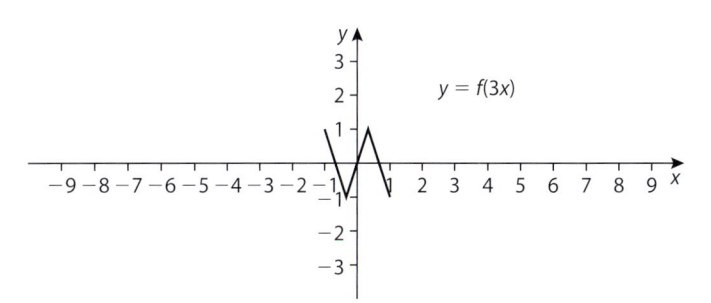

d) The graph of $y = f\left(\frac{1}{3}x\right)$ is obtained by horizontally stretching the graph of $y = f(x)$ by scale factor 3.

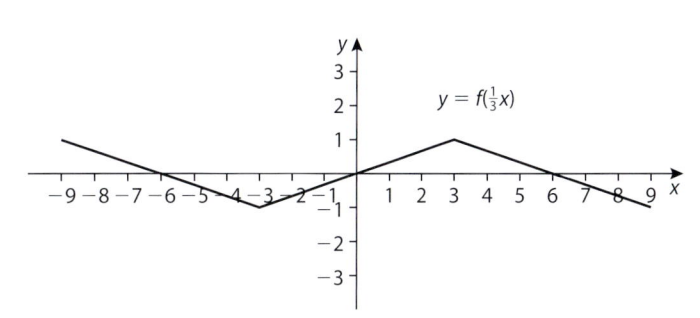

Example 22

Describe the sequence of transformations performed on the graph of $y = x^2$ to obtain the graph of $y = 4x^2 - 3$.

Solution

Step 1: Start with the graph of $y = x^2$.

Step 2: Vertically stretch $y = x^2$ by scale factor 4.

Step 3: Vertically translate $y = 4x^2$ three units down.

Step 1:

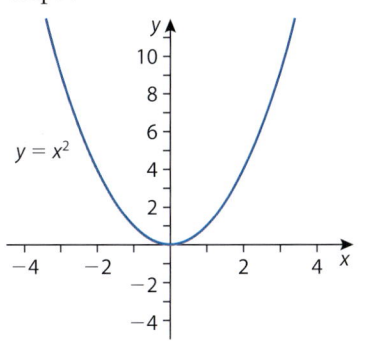

Step 2:

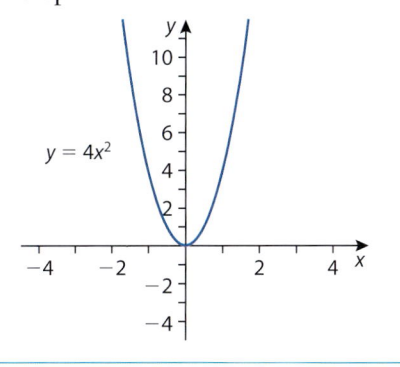

Step 3:

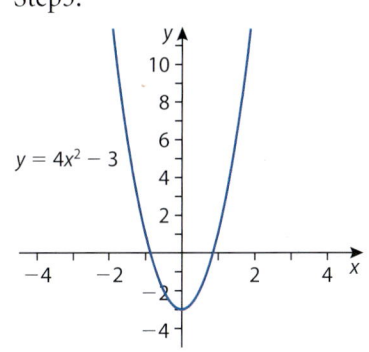

Note that in Example 22, a vertical stretch followed by a vertical translation does not produce the same graph if the two transformations are performed in reverse order. A vertical translation followed by a vertical stretch would generate the following sequence of equations:

Step 1: $y = x^2$ Step 2: $y = x^2 - 3$ Step 3: $y = 4(x^2 - 3) = 4x^2 - 12$

This final equation is not the same as $y = 4x^2 - 3$.

When combining two or more transformations, the order in which they are performed can make a difference. In general, when a sequence of transformations includes a vertical/horizontal stretch or shrink, or a reflection through the x-axis, the order may make a difference.

Summary of transformations on the graphs of functions
Assume that a, h and k are positive real numbers.

Transformed function	Transformation performed on $y = f(x)$
$y = f(x) + k$	vertical translation k units up
$y = f(x) - k$	vertical translation k units down
$y = f(x - h)$	horizontal translation h units right
$y = f(x + h)$	horizontal translation h units left
$y = -f(x)$	reflection in the x-axis
$y = f(-x)$	reflection in the y-axis
$y = af(x)$	vertical stretch ($a > 1$) or shrink ($0 < a < 1$)
$y = f(ax)$	horizontal stretch ($0 < a < 1$) or shrink ($a > 1$)

Exercise 2.4

In questions 1–14, sketch the graph of f, without a GDC or by plotting points, by using your knowledge of some of the basic functions shown in Figure 2.15.

1 $f: x \mapsto x^2 - 6$
2 $f: x \mapsto (x - 6)^2$
3 $f: x \mapsto |x| + 4$
4 $f: x \mapsto |x + 4|$
5 $f: x \mapsto 5 + \sqrt{x - 2}$
6 $f: x \mapsto \dfrac{1}{x - 3}$
7 $f: x \mapsto \dfrac{1}{(x + 5)^2} + 2$
8 $f: x \mapsto -x^3 - 4$
9 $f: x \mapsto -|x - 1| + 6$
10 $f: x \mapsto \sqrt{-x + 3}$
11 $f: x \mapsto 3\sqrt{x}$
12 $f: x \mapsto \tfrac{1}{2}x^2$
13 $f: x \mapsto \left(\tfrac{1}{2}x\right)^2$
14 $f: x \mapsto (-x)^3$

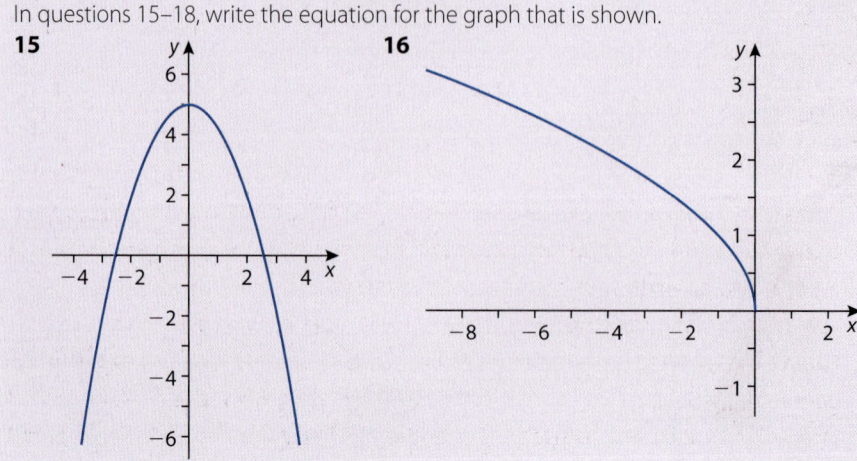

In questions 15–18, write the equation for the graph that is shown.

15

16

17

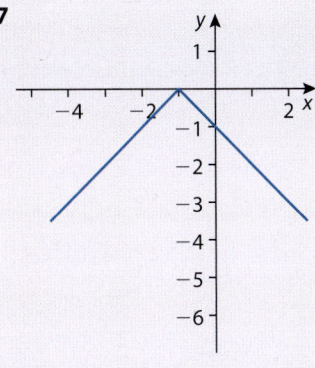

18 Vertical and horizontal asymptotes shown:

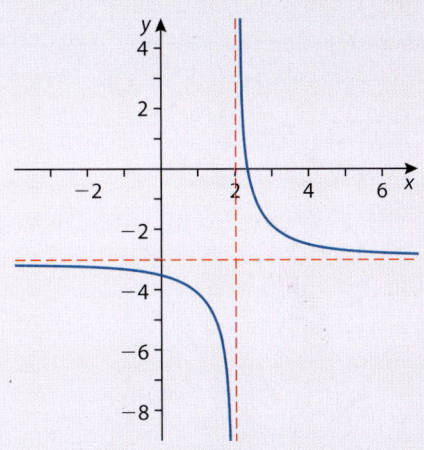

19 The graph of f is given. Sketch the graphs of the following functions.
a) $y = f(x) - 3$
b) $y = f(x - 3)$
c) $y = 2f(x)$
d) $y = f(2x)$
e) $y = -f(x)$
f) $y = f(-x)$
g) $y = 2f(x) + 4$

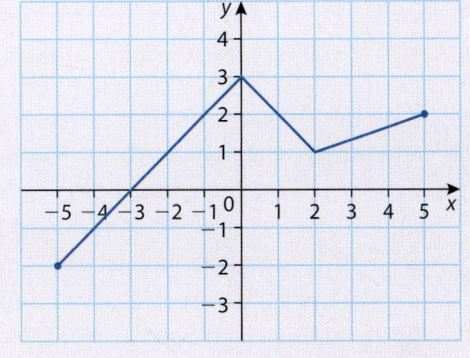

In questions 20–23, specify a sequence of transformations to perform on the graph of $y = x^2$ to obtain the graph of the given function.

20 $g: x \mapsto (x - 3)^2 + 5$

21 $h: x \mapsto -x^2 + 2$

22 $p: x \mapsto \frac{1}{2}(x + 4)^2$

23 $f: x \mapsto [3(x - 1)]^2 - 6$

In questions 24–26, a) express the quadratic function in the form $f(x) = a(x - h)^2 + k$, and b) state the coordinates of the vertex of the parabola with equation $y = f(x)$.

24 $f(x) = x^2 + 6x + 2$

25 $f(x) = x^2 - 2x + 4$

26 $f(x) = 4x^2 - 4x - 1$

2.5 Quadratic functions

A **linear function** is a polynomial function of degree one that can be written in the general form $f(x) = ax + b$, where $a \neq 0$. The **degree** of a polynomial written in terms of x refers to the largest exponent for x in any terms of the polynomial. In this section, we will consider **quadratic functions** that are second degree polynomial functions, often written in the general form $f(x) = ax^2 + bx + c$. Examples of quadratic functions, such as $f(x) = x^2 + 2$ (where $a = 1$, $b = 0$ and $c = 2$) and $f(x) = x^2 - 4x$ (where $a = 1$, $b = -4$ and $c = 0$), appeared earlier in this chapter.

> The word *quadratic* comes from the Latin word *quadratus* that means four-sided, to make square, or simply a square. *Numerus quadratus* means a square number. Before modern algebraic notation was developed in the 17th and 18th centuries, the geometric figure of a square was used to indicate a number multiplying itself. Hence, raising a number to the power of two (in modern notation) is commonly referred to as the operation of squaring. *Quadratic* then came to be associated with a polynomial of degree two rather than being associated with the number four, as the prefix *quad* often indicates (e.g. quadruple).

65

2 Functions and Equations

Definition of a quadratic function
If a, b and c are real numbers, and $a \neq 0$, the function $f(x) = ax^2 + bx + c$ is a **quadratic function**. The graph of f is the graph of the equation $y = ax^2 + bx + c$ and is called a **parabola**.

Figure 2.32

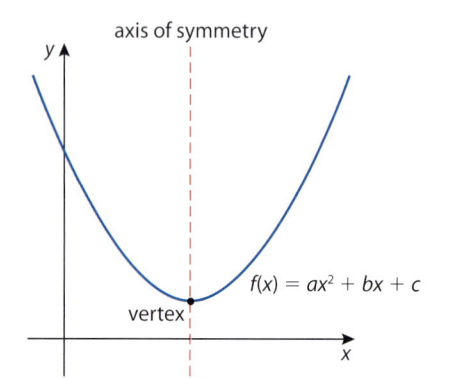

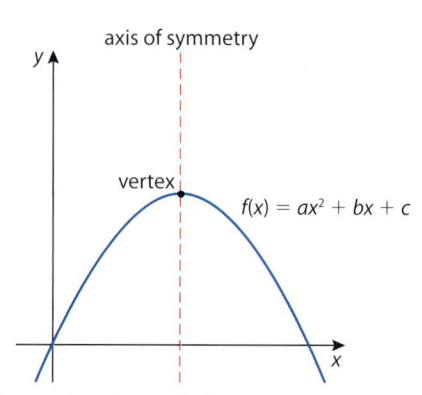

If $a > 0$ then the parabola opens upward

If $a < 0$ then the parabola opens downward

Figure 2.33

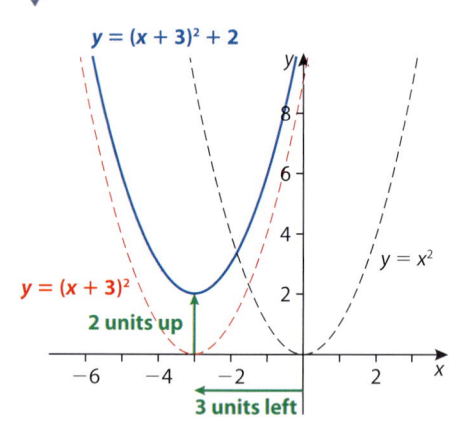

$y = (x + 3)^2 + 2$

$y = (x + 3)^2$

2 units up

3 units left

axis of symmetry $x = -3$

$y = (x + 3)^2 + 2$

vertex $(-3, 2)$

Figure 2.34

Each parabola is symmetric about a vertical line called its **axis of symmetry**. The axis of symmetry passes through a point on the parabola called the **vertex** of the parabola, as shown in Figure 2.32. If the leading coefficient, a, of the quadratic function $f(x) = ax^2 + bx + c$ is positive, the parabola opens upward (concave up) – and the y-coordinate of the vertex will be a **minimum value** for the function. If the leading coefficient, a, of $f(x) = ax^2 + bx + c$ is negative, the parabola opens downward (concave down) – and the y-coordinate of the vertex will be a **maximum value** for the function.

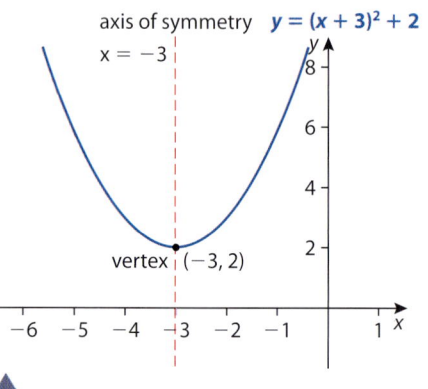

The graph of $f(x) = a(x - h)^2 + k$

From the previous section, we know that the graph of the equation $y = (x + 3)^2 + 2$ can be obtained by translating $y = x^2$ three units to the left and two units up. Being familiar with the shape and position of the graph of $y = x^2$, and knowing the two translations that transform $y = x^2$ to $y = (x + 3)^2 + 2$, we can easily visualize and/or sketch the graph of $y = (x + 3)^2 + 2$ (see Figure 2.33). We can also determine the axis of symmetry and the vertex of the graph. Figure 2.34 shows that the graph of $y = (x + 3)^2 + 2$ has an axis of symmetry of $x = -3$ and a vertex at $(-3, 2)$. The equation $y = (x + 3)^2 + 2$ can also be written as $y = x^2 + 6x + 11$. Because we can easily identify the vertex of the parabola when the equation is written as $y = (x + 3)^2 + 2$, we often refer to this as the **vertex form** of the quadratic equation, and $y = x^2 + 6x + 11$ as the **general form**.

- **Hint:** $f(x) = a(x - h)^2 + k$ is sometimes referred to as the **standard form** of a quadratic function.

Vertex form of a quadratic function
If a quadratic function is written in the form $f(x) = a(x - h)^2 + k$, with $a \neq 0$, the graph of f has an axis of symmetry of $x = h$ and a vertex at (h, k).

Completing the square

For visualizing and sketching purposes, it is helpful to have a quadratic function written in vertex form. How do we rewrite a quadratic function written in the form $f(x) = ax^2 + bx + c$ (general form) into the form $f(x) = a(x - h)^2 + k$ (vertex form)? We use the technique of **completing the square**.

For any real number p, the quadratic expression $x^2 + px + \left(\frac{p}{2}\right)^2$ is the square of $\left(x + \frac{p}{2}\right)$. Convince yourself of this by expanding $\left(x + \frac{p}{2}\right)^2$. The technique of *completing the square* is essentially the process of adding a constant to a quadratic expression to make it the square of a binomial. If the coefficient of the quadratic term (x^2) is a positive one, the coefficient of the linear term is p, and the constant term is $\left(\frac{p}{2}\right)^2$, then $x^2 + px + \left(\frac{p}{2}\right)^2 = \left(x + \frac{p}{2}\right)^2$ and the square is completed.

Remember that the coefficient of the quadratic term (leading coefficient) must be equal to positive one before completing the square.

Example 23

Find the equation of the axis of symmetry and the coordinates of the vertex of the graph of $f(x) = x^2 - 8x + 18$ by rewriting the function in the form $f(x) = a(x - h)^2 + k$.

Solution

To complete the square and get the quadratic expression $x^2 - 8x + 18$ in the form $x^2 + px + \left(\frac{p}{2}\right)^2$, the constant term needs to be $\left(\frac{-8}{2}\right)^2 = 16$. We need to add 16, but also subtract 16, so that we are adding zero overall and, hence, not changing the original expression.

$f(x) = x^2 - 8x + 16 - 16 + 18$ actually adding zero ($-16 + 16$) to the right side

$f(x) = x^2 - 8x + 16 + 2$ $x^2 - 8x + 16$ fits the pattern $x^2 + px + \left(\frac{p}{2}\right)^2$ with $p = -8$

$f(x) = (x - 4)^2 + 2$ $x^2 - 8x + 16 = (x - 4)^2$

The axis of symmetry of the graph of f is the vertical line $x = 4$ and the vertex is at $(4, 2)$. See Figure 2.35.

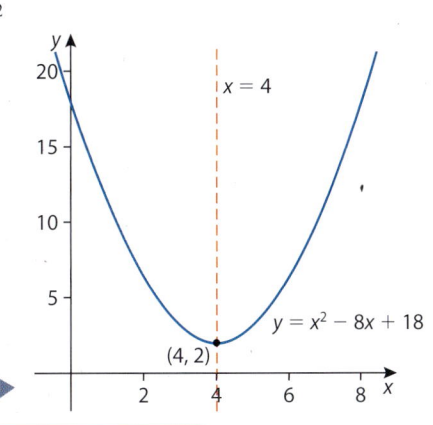

Figure 2.35

Functions and Equations

Example 24

For the function $g: x \mapsto -2x^2 - 12x + 7$,
a) find the axis of symmetry and the vertex of the graph
b) indicate the transformations that can be applied to $y = x^2$ to obtain the graph
c) find the minimum or maximum value.

Solution

a) $g: x \mapsto -2\left(x^2 + 6x - \dfrac{7}{2}\right)$ factorize so that the coefficient of the quadratic term is $+1$

$g: x \mapsto -2\left(x^2 + 6x + 9 - 9 - \dfrac{7}{2}\right)$ $p = 6 \Rightarrow \left(\dfrac{p}{2}\right)^2 = 9$; hence, add $+9 - 9$ (zero)

$g: x \mapsto -2\left[(x+3)^2 - \dfrac{18}{2} - \dfrac{7}{2}\right]$ $x^2 + 6x + 9 = (x+3)^2$

$g: x \mapsto -2\left[(x+3)^2 - \dfrac{25}{2}\right]$

$g: x \mapsto -2(x+3)^2 + 25$ multiply through by -2 to remove outer brackets

$g: x \mapsto -2(x-(-3))^2 + 25$ express in vertex form: $g: x \mapsto a(x-h)^2 + k$

The axis of symmetry of the graph of g is the vertical line $x = -3$ and the vertex is at $(-3, 25)$. See Figure 2.36.

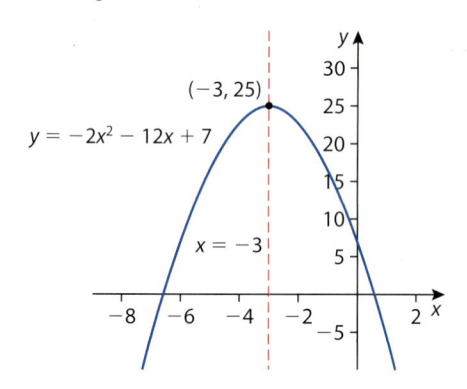

Figure 2.36

b) Since $g: x \mapsto -2x^2 - 12x + 7 = -2(x+3)^2 + 25$, the graph of g can be obtained by applying the following transformations (in the order given) on the graph of $y = x^2$: horizontal translation of 3 units left; reflection in the x-axis (parabola opening down); vertical stretch of factor 2; and a vertical translation of 25 units up.

c) The parabola opens down because the leading coefficient is negative. Therefore, g has a maximum and no minimum value. The maximum value is 25 (y-coordinate of vertex) at $x = -3$.

The technique of completing the square can be used to derive the quadratic formula. The following example derives a general expression for the axis of symmetry and vertex of a quadratic function in the general form $f(x) = ax^2 + bx + c$ by completing the square.

Example 25

Find the axis of symmetry and the vertex for the general quadratic function $f(x) = ax^2 + bx + c$.

Solution

$f(x) = a\left(x^2 + \dfrac{b}{a}x + \dfrac{c}{a}\right)$ factorize so that the coefficient of the x^2 term is $+1$

$f(x) = a\left[x^2 + \dfrac{b}{a}x + \left(\dfrac{b}{2a}\right)^2 - \left(\dfrac{b}{2a}\right)^2 + \dfrac{c}{a}\right]$ $p = \dfrac{b}{a} \Rightarrow \left(\dfrac{p}{2}\right)^2 = \left(\dfrac{b}{2a}\right)^2$

$f(x) = a\left[\left(x + \dfrac{b}{2a}\right)^2 - \dfrac{b^2}{4a^2} + \dfrac{c}{a}\right]$ $x^2 + \dfrac{b}{a}x + \left(\dfrac{b}{2a}\right)^2 = \left(x + \dfrac{b}{2a}\right)^2$

$f(x) = a\left(x + \dfrac{b}{2a}\right)^2 - \dfrac{b^2}{4a} + c$ multiply through by a

$f(x) = a\left(x - \left(-\dfrac{b}{2a}\right)\right)^2 + c - \dfrac{b^2}{4a}$ express in vertex form: $f(x) = a(x - h)^2 + k$

This result leads to the following generalization.

> **Symmetry and vertex of $f(x) = ax^2 + bx + c$**
> For the graph of the quadratic function $f(x) = ax^2 + bx + c$, the axis of symmetry is the vertical line with the equation $x = -\dfrac{b}{2a}$ and the vertex has coordinates $\left(-\dfrac{b}{2a}, c - \dfrac{b^2}{4a}\right)$.

Check the results for Example 24 using the formulae for the axis of symmetry and vertex. For the function $g: x \mapsto -2x^2 - 12x + 7$:

$x = -\dfrac{b}{2a} = -\dfrac{-12}{2(-2)} = -3 \Rightarrow$ axis of symmetry is the vertical line $x = -3$

$c - \dfrac{b^2}{4a} = 7 - \dfrac{(-12)^2}{4(-2)} = \dfrac{56}{8} + \dfrac{144}{8} = 25 \Rightarrow$ vertex has coordinates $(-3, 25)$

These results agree with the results from Example 24.

Zeros of a quadratic function

A specific value for x is a **zero** (or **root**) of a quadratic function $f(x) = ax^2 + bx + c$ if it is a solution to the equation $ax^2 + bx + c = 0$. For this course, we are only concerned with values of x that are real numbers. The x-coordinate of any point(s) where f crosses the x-axis (y-coordinate is zero) is a zero of the function. A quadratic function can have no, one or two real zeros as Table 2.3 illustrates. Finding the zeros of a quadratic function requires you to solve quadratic equations of the form $ax^2 + bx + c = 0$. Although $a \neq 0$, it is possible for b or c to be equal to zero. There are five general methods for solving quadratic equations as outlined in Table 2.3.

2 Functions and Equations

Square root **Examples**	If $a^2 = c$ and $c > 0$, then $a = \pm\sqrt{c}$. $x^2 - 25 = 0$ $\qquad$ $(x + 2)^2 = 15$ $x^2 = 25$ $\qquad\qquad$ $x + 2 = \pm\sqrt{15}$ $x = \pm\sqrt{5}$ $\qquad\quad$ $x = -2 \pm \sqrt{15}$
Factorizing **Examples**	If $ab = 0$, then $a = 0$ or $b = 0$. $x^2 + 3x - 10 = 0$ $\qquad$ $x^2 - 7x = 0$ $(x + 5)(x - 2) = 0$ $\qquad$ $x(x - 7) = 0$ $x = -5$ or $x = 2$ $\qquad\;$ $x = 0$ or $x = 7$
Completing the square	If $x^2 + px + q = 0$, then $x^2 + px + \left(\dfrac{p}{2}\right)^2 = -q + \left(\dfrac{p}{2}\right)^2$ which leads to $\left(x + \dfrac{p}{2}\right)^2 = -q + \dfrac{p^2}{4}$ ... and then the square root of both sides (as above).
Example	$x^2 - 8x + 5 = 0$ $x^2 - 8x + 16 = -5 + 16$ $(x - 4)^2 = 11$ $x - 4 = \pm\sqrt{11}$ $x = 4 \pm \sqrt{11}$
Quadratic formula	If $ax^2 + bx + c = 0$, then $x = \dfrac{-b \pm \sqrt{b^2 - 4ac}}{2a}$.
Example	$2x^2 - 3x - 4 = 0$ $x = \dfrac{-(-3) \pm \sqrt{(-3)^2 - 4(2)(-4)}}{2(2)}$ $x = \dfrac{3 \pm \sqrt{41}}{4}$
Graphing	Graph the equation $y = ax^2 + bx + c$ on your GDC. Use the calculating features of your GDC to determine the x-coordinates of the point(s) where the parabola intersects the x-axis.
Example	$2x^2 - 5x - 7 = 0$ $\quad$ GDC calculations reveal that the zeros are at $x = \dfrac{7}{2}$ and $x = -1$

Table 2.3 Methods for solving quadratic equations.

The quadratic formula and the discriminant

The expression $b^2 - 4ac$ in the quadratic formula has special significance because you need to take the positive and negative square root of $b^2 - 4ac$ when using the quadratic formula. Hence, whether $b^2 - 4ac$ (often labelled Δ; read 'delta') is positive, negative or zero will determine the number of real solutions for the quadratic equation $ax^2 + bx + c = 0$, and, consequently, also the number of times the graph of $f(x) = ax^2 + bx + c$ intersects the x-axis ($y = 0$).

For the quadratic function $f(x) = ax^2 + bx + c$, $a \neq 0$:

If $\Delta = b^2 - 4ac > 0$, f has two distinct real solutions, and the graph of f intersects the x-axis twice.

If $\Delta = b^2 - 4ac = 0$, f has one real solution (a double root), and the graph of f intersects the x-axis once (i.e. it is tangent to the x-axis).

If $\Delta = b^2 - 4ac < 0$, f has no real solutions, and the graph of f does not intersect the x-axis.

Example 26

Use the discriminant to determine how many real solutions each equation has. Visually confirm the result by graphing the corresponding quadratic function for each equation on your GDC.
a) $x^2 + 3x - 1 = 0$ b) $4x^2 - 12x + 9 = 0$ c) $2x^2 - 5x + 6 = 0$

Solution

a) The discriminant is $\Delta = 3^2 - 4(1)(-1) = 13 > 0$. Therefore, the equation has two distinct real zeros. This result is confirmed by the graph of the quadratic function $y = x^2 + 3x - 1$ which clearly shows it intersecting the x-axis twice as shown in GDC image on the right.

b) The discriminant is $\Delta = (-12)^2 - 4(4)(9) = 0$. Therefore, the equation has one real zero. The graph on the GDC of $y = 4x^2 - 12x + 9$ appears to intersect the x-axis at only one point. We can be more confident with this conclusion by investigating further – for example, tracing or looking at a table of values on the GDC as shown in GDC images below.

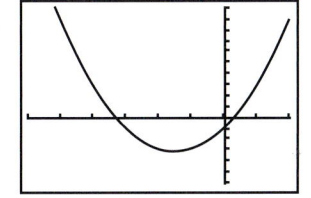

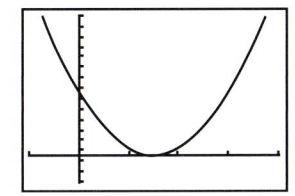

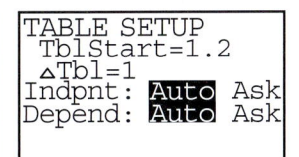

c) The discriminant is $\Delta = (-5)^2 - 4(2)(6) = -23 < 0$. Therefore, the equation has no real zeros. This result is confirmed by the graph of the quadratic function $y = 2x^2 - 5x + 6$ which clearly shows that the graph does not intersect the x-axis as shown in GDC image on the right.

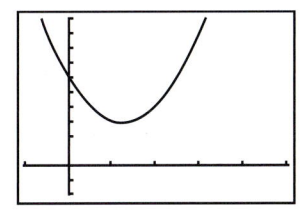

Example 27

For $4x^2 + 4kx + 9 = 0$, determine the value(s) of k so that the equation has a) one real zero, b) two distinct real zeros, and c) no real zeros.

Solution

a) For one real zero: $\Delta = (4k)^2 - 4(4)(9) = 0 \Rightarrow 16k^2 - 144 = 0$
   $\Rightarrow 16k^2 = 144 \Rightarrow k^2 = 9 \Rightarrow k = \pm 3$

b) For two distinct real zeros: $\Delta = (4k)^2 - 4(4)(9) > 0$
   $\Rightarrow 16k^2 > 144 \Rightarrow k^2 > 9 \Rightarrow k < -3$ or $k > 3$

c) For no real zeros: $\Delta = (4k)^2 - 4(4)(9) < 0 \Rightarrow 16k^2 < 144$
   $\Rightarrow k^2 < 9 \Rightarrow k > -3$ and $k < 3 \Rightarrow -3 < k < 3$

Functions and Equations

The graph of $f(x) = a(x - p)(x - q)$

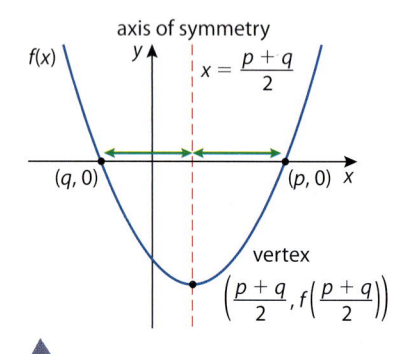

If a quadratic function is written in the form $f(x) = a(x - p)(x - q)$ then we can easily identify the x-intercepts of the graph of f. Consider that $f(p) = a(p - p)(p - q) = a(0)(p - q) = 0$ and that $f(q) = a(q - p)(q - q) = a(q - p)(0) = 0$. Therefore, the quadratic function $f(x) = a(x - p)(x - q)$ will intersect the x-axis at the points $(p, 0)$ and $(q, 0)$. We need to factorize in order to rewrite a quadratic function in the form $f(x) = ax^2 + bx + c$ to the form $f(x) = a(x - p)(x - q)$. Hence, $f(x) = a(x - p)(x - q)$ can be referred to as the **factorized** form of a quadratic function. Recalling the symmetric nature of a parabola, it is clear that the x-intercepts $(p, 0)$ and $(q, 0)$ will be equidistant from the axis of symmetry (see Figure 2.37). As a result, the equation of the axis of symmetry and the x-coordinate of the vertex of the parabola can be found from finding the average of p and q.

Figure 2.37

Factorized form of a quadratic function

If a quadratic function is written in the form $f(x) = a(x - p)(x - q)$, with $a \neq 0$, the graph of f has x-intercepts at $(p, 0)$ and $(q, 0)$, an axis of symmetry with equation $x = \frac{p+q}{2}$, and a vertex at $\left(\frac{p+q}{2}, f\left(\frac{p+q}{2}\right)\right)$.

Example 28

Find the equation of each quadratic function from the graph in the form $f(x) = a(x - p)(x - q)$ and also in the form $f(x) = ax^2 + bx + c$.

a)

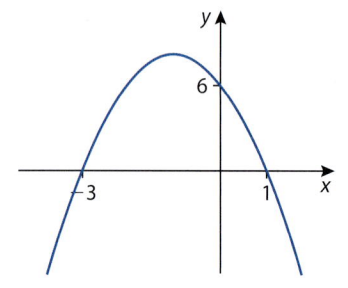

b)

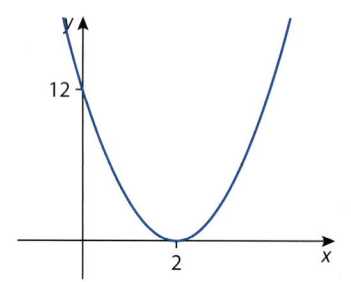

Solution

a) Since the x-intercepts are -3 and 1 then $y = a(x + 3)(x - 1)$. The y-intercept is 6, so when $x = 0$, $y = 6$. Hence,
$6 = a(0 + 3)(0 - 1) = -3a \Rightarrow a = -2$ ($a < 0$ agrees with the fact that the parabola is opening down). The function is $f(x) = -2(x + 3)(x - 1)$, and expanding to remove brackets reveals that the function can also be written as $f(x) = -2x^2 - 4x + 6$.

b) The function has one x-intercept at 2 (double root), so $p = q = 2$ and $y = a(x - 2)(x - 2) = a(x - 2)^2$. The y-intercept is 12, so when $x = 0$, $y = 12$. Hence, $12 = a(0 - 2)^2 = 4a \Rightarrow a = 3$ ($a > 0$ agrees with the parabola opening up). The function is $f(x) = 3(x - 2)^2$. Expanding reveals that the function can also be written as $f(x) = 3x^2 - 12x + 12$.

Example 29

The graph of a quadratic function intersects the *x*-axis at the points $(-6, 0)$ and $(-2, 0)$ and also passes through the point $(2, 16)$. a) Write the function in the form $f(x) = a(x - p)(x - q)$. b) Find the vertex of the parabola. c) Write the function in the form $f(x) = a(x - h)^2 + k$.

Solution

a) The *x*-intercepts of -6 and -2 gives $f(x) = a(x + 6)(x + 2)$. Since f passes through $(2, 16)$, then $f(2) = 16 \Rightarrow f(2) = a(2 + 6)(2 + 2) = 16$ $\Rightarrow 32a = 16 \Rightarrow a = \frac{1}{2}$. Therefore, $f(x) = \frac{1}{2}(x + 6)(x + 2)$.

b) The *x*-coordinate of the vertex is the average of the *x*-intercepts. $x = \frac{-6 - 2}{2} = -4$, so the *y*-coordinate of the vertex is $y = f(-4) = \frac{1}{2}(-4 + 6)(-4 + 2) = -2$. Hence, the vertex is $(-4, -2)$.

c) In vertex form, the quadratic function is $f(x) = \frac{1}{2}(x + 4)^2 - 2$.

Exercise 2.5

For each of the quadratic functions *f* in questions 1–5, find the following:
a) the axis of symmetry and the vertex, by algebraic methods
b) the transformation(s) that can be applied to $y = x^2$ to obtain the graph of $y = f(x)$
c) the minimum or maximum value of *f*.

Check your results using your GDC.

1 $f: x \mapsto x^2 - 10x + 32$ **2** $f: x \mapsto x^2 + 6x + 8$ **3** $f: x \mapsto -2x^2 - 4x + 10$
4 $f: x \mapsto 4x^2 - 4x + 9$ **5** $f: x \mapsto \frac{1}{2}x^2 + 7x + 26$

In questions 6–13, solve the quadratic equation using factorization.

6 $x^2 + 2x - 8 = 0$ **7** $x^2 = 3x + 10$
8 $6x^2 - 9x = 0$ **9** $6 + 5x = x^2$
10 $x^2 + 9 = 6x$ **11** $3x^2 + 11x - 4 = 0$
12 $3x^2 + 18 = 15x$ **13** $9x - 2 = 4x^2$

In questions 14–19, use the method of completing the square to solve the quadratic equation.

14 $x^2 + 4x - 3 = 0$ **15** $x^2 - 4x - 5 = 0$
16 $x^2 - 2x + 3 = 0$ **17** $2x^2 + 16x + 6 = 0$
18 $x^2 + 2x - 8 = 0$ **19** $-2x^2 + 4x + 9 = 0$

20 Let $f(x) = x^2 - 4x - 1$. a) Use the quadratic formula to find the zeros of the function. b) Use the zeros to find the equation for the axis of symmetry of the parabola. c) Find the minimum or maximum value of *f*.

In questions 21–24, determine the number of real solutions to each equation.

21 $x^2 + 3x + 2 = 0$ **22** $2x^2 - 3x + 2 = 0$
23 $x^2 - 1 = 0$ **24** $2x^2 - \frac{9}{4}x + 1 = 0$

25 Find the value(s) of *p* for which the equation $2x^2 + px + 1 = 0$ has one real solution.

26 Find the value(s) of k for which the equation $x^2 + 4x + k = 0$ has two distinct real solutions.

27 The equation $x^2 - 4kx + 4 = 0$ has two distinct real solutions. Find the set of all possible values of k.

28 Find all possible values of m so that the graph of the function $g: x \mapsto mx^2 + 6x + m$ does not touch the x-axis.

Practice questions

1 Let $f: x \mapsto \sqrt{x-3}$ and $g: x \mapsto x^2 + 2x$. The function $(f \circ g)(x)$ is defined for all $x \in \mathbb{R}$ **except** for the interval $]a, b[$.
   a) Calculate the values of a and b.
   b) Find the range of $f \circ g$.

2 Two functions g and h are defined as $g(x) = 2x - 7$ and $h(x) = 3(2 - x)$. Find: **a)** $g^{-1}(3)$ **b)** $(h \circ g)(6)$

3 Consider the functions $f(x) = 5x - 2$ and $g(x) = \dfrac{4-x}{3}$.
   a) Find g^{-1}.
   b) Solve the equation $(f \circ g^{-1})(x) = 8$.

4 The functions g and h are defined by $g: x \mapsto x - 3$ and $h: x \mapsto 2x$.
   a) Find an expression for $(g \circ h)(x)$.
   b) Show that $g^{-1}(14) + h^{-1}(14) = 24$.

5 The function f is defined by $f(x) = x^2 + 8x + 11$, for $x \geq -4$.
   a) Write $f(x)$ in the form $(x - h)^2 + k$.
   b) Find the inverse function f^{-1}.
   c) State the domain of f^{-1}.

6 The diagram right shows the graph of $y = f(x)$. It has maximum and minimum points at $(0, 0)$ and $(1, -1)$, respectively.
   a) Copy the diagram and, on the same diagram, draw the graph of $y = f(x + 1) - \frac{1}{2}$.
   b) What are the coordinates of the minimum and maximum points of $y = f(x + 1) - \frac{1}{2}$?

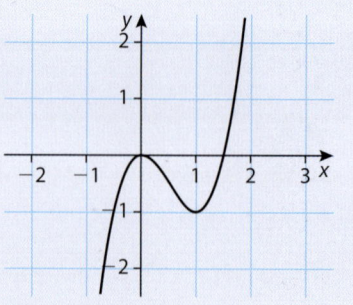

7 The diagram shows parts of the graphs of $y = x^2$ and $y = -\frac{1}{2}(x + 5)^2 + 3$.

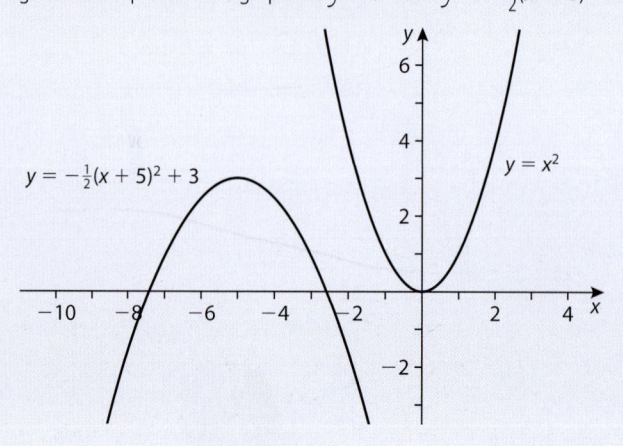

The graph of $y = x^2$ may be transformed into the graph of $y = -\frac{1}{2}(x + 5)^2 + 3$ by these transformations.

> A reflection in the line $y = 0$, followed by
> a vertical stretch by scale factor k, followed by
> a horizontal translation of p units, followed by
> a vertical translation of q units.

Write down the value of
a) k b) p c) q.

8 The function f is defined by $f(x) = \dfrac{4}{\sqrt{16 - x^2}}$, for $-4 < x < 4$.
   a) Without using a GDC, sketch the graph of f.
   b) Write down the equation of each vertical asymptote.
   c) Write down the range of the function f.

9 Let $g: x \mapsto \dfrac{1}{x}, x \neq 0$.
   a) Without using a GDC, sketch the graph of g.

The graph of g is transformed to the graph of h by a translation of 4 units to the left and 2 units down.

   b) Find an expression for the function h.
   c) (i) Find the x- and y-intercepts of h.
      (ii) Write down the equations of the asymptotes of h.
      (iii) Sketch the graph of h.

10 Consider $f(x) = \sqrt{x + 3}$.
   a) Find:
      (i) $f(8)$
      (ii) $f(46)$
      (iii) $f(-3)$
   b) Find the values of x for which f is undefined.
   c) Let $g: x \mapsto x^2 - 5$. Find $(g \circ f)(x)$.

11 Let $g(x) = \dfrac{x - 8}{2}$ and $h(x) = x^2 - 1$.
   a) Find $g^{-1}(-2)$.
   b) Find an expression for $(g^{-1} \circ h)(x)$.
   c) Solve $(g^{-1} \circ h)(x) = 22$.

12 Given the functions $f: x \mapsto 3x - 1$ and $g: x \mapsto \dfrac{4}{x}$, find the following:
   a) f^{-1} b) $f \circ g$ c) $(f \circ g)^{-1}$ d) $g \circ g$

13 The quadratic function f is defined by $f(x) = 2x^2 + 8x + 17$.
   a) Write f in the form $f(x) = 2(x - h)^2 + k$.
   b) The graph of f is translated 5 units in the positive x-direction and 2 units in the positive y-direction. Find the function g for the translated graph, giving your answer in the form $g(x) = 2(x - h)^2 + k$.

14 Let $g(x) = 3x^2 - 6x - 4$.
  a) Express $g(x)$ in the form $g(x) = 3(x - h)^2 + k$.
  b) Write down the vertex of the graph of g.
  c) Write down the equation of the axis of symmetry of the graph of g.
  d) Find the y-intercept of the graph of g.
  e) The x-intercepts of g can be written as $\dfrac{p \pm \sqrt{q}}{r}$, where $p, q, r \in \mathbb{Z}$. Find the value of p, q and r.

15 a) The diagram shows part of the graph of the function $h(x) = \dfrac{a}{x - b}$. The curve passes through the point $A(-4, -8)$. The vertical line (MN) is an asymptote. Find the value of: **(i)** a **(ii)** b.

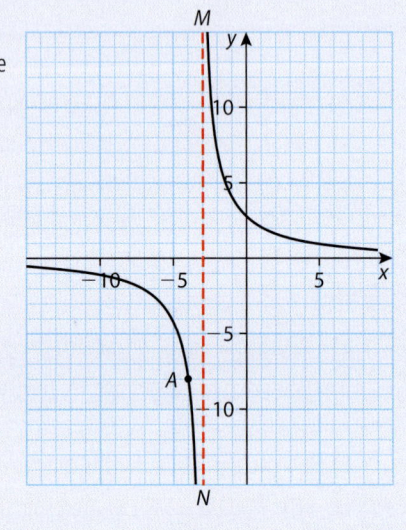

b) The graph of $h(x)$ is transformed as shown in the diagram right. The point A is transformed to $A'(-4, 8)$. Give a full geometric description of the transformation.

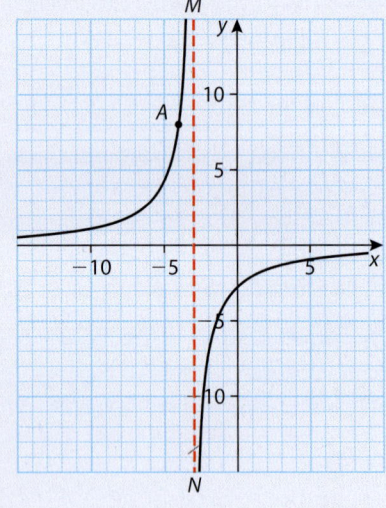

16 The graph of $y = f(x)$ is shown in the diagram.

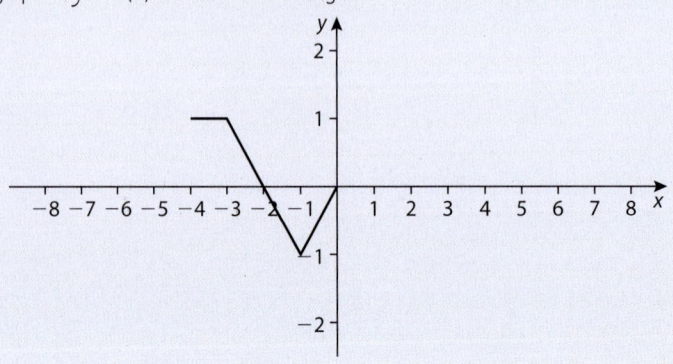

a) Make two copies of the coordinate system as shown in the diagram but without the graph of $y = f(x)$. On the first diagram sketch a graph of $y = 2f(x)$, and on the second diagram sketch a graph of $y = f(x - 4)$.

b) The point $A(-3, 1)$ is on the graph of $y = f(x)$. The point A' is the corresponding point on the graph of $y = -f(x) - 1$. Find the coordinates of A'.

17 The diagram represents the graph of the function $f(x) = (x - p)(x - q)$.

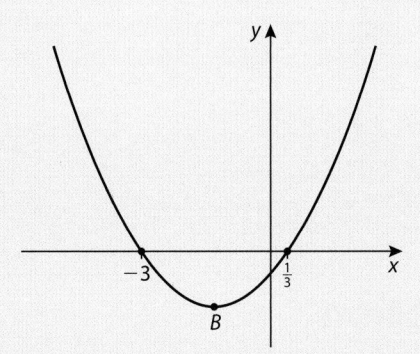

a) Write down the values of p and q.
b) The function has a minimum value at the point B. Find the x-coordinate of B.
c) Write the expression for $f(x)$ in the form $ax^2 + bx + c$.

18 The diagram shows the parabola $y = (5 + x)(2 - x)$. The points A and C are the x-intercepts and the point B is the maximum point. Find the coordinates of A, B and C.

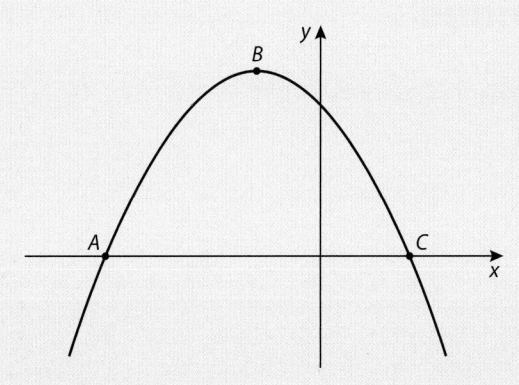

Sequences and Series

Assessment statements
1.1 Arithmetic sequences and series; sum of finite arithmetic sequences; geometric sequences and series; sum of finite and infinite geometric series.
Sigma notation.
1.3 The binomial theorem: expansion of $(a + b)^n$, $n \in \mathbb{N}$.

Introduction

The heights of consecutive bounds of a ball, compound interest and Fibonacci numbers are only a few of the applications of sequences and series that you have seen in previous courses. In this chapter, you will review these concepts, consolidate your understanding and take them one step further.

3.1 Sequences

Take the following pattern as an example:

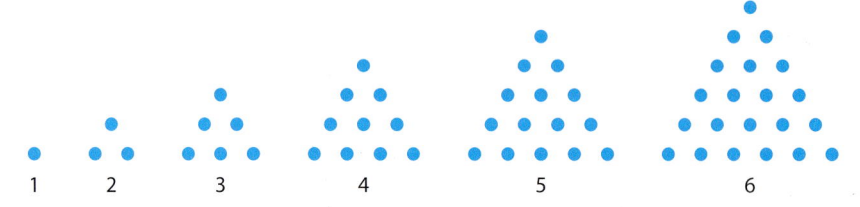

The first figure represents 1 dot, the second represents 3 dots, etc. This pattern can also be described differently. For example, in function notation:

$$f(1) = 1, f(2) = 3, f(3) = 6, \text{ etc., where the domain is } \mathbb{Z}^+$$

Here are some more examples of sequences:
1. 6, 12, 18, 24, 30
2. 3, 9, 27,..., 3^k, ...
3. $\left\{\dfrac{1}{i^2}; i = 1, 2, 3, ..., 10\right\}$
4. $\{b_1, b_2, ..., b_n, ...\}$, sometimes used with an abbreviation $\{b_n\}$

The first and third sequences are **finite** and the second and fourth are **infinite**. Notice that, in the second and third sequences, we were able to define a rule that yields the nth number in the sequence (called the nth term) as a function of n, the term's number. In this sense, a sequence is a **function** that assigns a **unique** number (a_n) to each positive integer n.

Example 1

Find the first five terms and the 50th term of the sequence $\{b_n\}$ such that $b_n = 2 - \dfrac{1}{n^2}$.

Solution

Since we know an *explicit* expression for the nth term as a *function* of its number n, we only need to find the value of that function for the required terms:

$b_1 = 2 - \dfrac{1}{1^2} = 1;\ b_2 = 2 - \dfrac{1}{2^2} = 1\dfrac{3}{4};\ b_3 = 2 - \dfrac{1}{3^2} = 1\dfrac{8}{9};\ b_4 = 2 - \dfrac{1}{4^2} = 1\dfrac{15}{16};$

$b_5 = 2 - \dfrac{1}{5^2} = 1\dfrac{24}{25};\ \text{and}\ b_{50} = 2 - \dfrac{1}{50^2} = 1\dfrac{2499}{2500}.$

So, informally, **a sequence is an ordered set of real numbers**. That is, there is a first number, a second, and so forth. The notation used for such sets is shown above. The way we defined the function in Example 1 is called the **explicit** definition of a sequence. There are other ways to define sequences, one of which is the **recursive** definition. The following example will show you how this is used.

Example 2

Find the first five terms and the 20th term of the sequence $\{b_n\}$ such that $b_1 = 5$ and $b_n = 2(b_{n-1} + 3)$.

● **Hint:** This can easily be done using a GDC.

Solution

The defining formula for this sequence is recursive. It allows us to find the nth term b_n if we know the preceding term b_{n-1}. Thus, we can find the second term from the first, the third from the second, and so on. Since we know the first term, $b_1 = 5$, we can calculate the rest:

$b_2 = 2(\mathbf{b_1} + 3) = 2(\mathbf{5} + 3) = 16$
$b_3 = 2(\mathbf{b_2} + 3) = 2(\mathbf{16} + 3) = 38$
$b_4 = 2(\mathbf{b_3} + 3) = 2(\mathbf{38} + 3) = 82$
$b_5 = 2(\mathbf{b_4} + 3) = 2(\mathbf{82} + 3) = 170$

Thus, the first five terms of this sequence are 5, 16, 38, 82, 170. However, to find the 20th term, we must first find all 19 preceding terms. This is one of the drawbacks of the recursive definition, unless we can change the definition into explicit form.

```
Plot1 Plot2 Plot3
 nMin=1
\U(n)■2(u(n−1)+3
)
 U(nMin)■5■
\V(n)=
 V(nMin)=
\W(n)=
```

```
U(5)
              170
U(20)
          5767162
```

However, you need to understand that not all sequences have formulae, either recursive or explicit. Some sequences are given only by listing their terms. Among the many kinds of sequences that there are, two types are of interest to us: arithmetic and geometric sequences.

3 Sequences and Series

Exercise 3.1

Find the first five terms and the 50th term of each infinite sequence defined in questions 1–8.

1. $a_n = 2n - 3$
2. $b_n = 2 \times 3^{n-1}$
3. $u_n = (-1)^{n-1} \dfrac{2n}{n^2 + 2}$
4. $a_n = n^{n-1}$
5. $a_n = 2a_{n-1} + 5$ and $a_1 = 3$
6. $u_{n+1} = \dfrac{3}{2u_n + 1}$ and $u_1 = 0$
7. $b_n = 3 \cdot b_{n-1}$ and $b_1 = 2$
8. $a_n = a_{n-1} + 2$ and $a_1 = -1$

3.2 Arithmetic sequences

Examine the following sequences and the most likely recursive formula for each of them.

$7, 14, 21, 28, 35, 42, \ldots$ $a_1 = 7$ and $a_n = a_{n-1} + 7$, for $n > 1$
$2, 11, 20, 29, 38, 47, \ldots$ $a_1 = 2$ and $a_n = a_{n-1} + 9$, for $n > 1$
$48, 39, 30, 21, 12, 3, -6, \ldots$ $a_1 = 48$ and $a_n = a_{n-1} - 9$, for $n > 1$

Note that in each case above, every term is formed by adding a constant number to the preceding term. Sequences formed in this manner are called **arithmetic sequences**.

> **Definition of an arithmetic sequence**
> A sequence $a_1, a_2, a_3, \ldots$ is an **arithmetic sequence** if there is a constant d for which
> $$a_n = a_{n-1} + d$$
> for all integers $n > 1$. d is called the **common difference** of the sequence, and $d = a_n - a_{n-1}$ for all integers $n > 1$.

So, for the sequences above, 7 is the common difference for the first, 9 is the common difference for the second and -9 is the common difference for the third.

This description gives us the recursive definition of the arithmetic sequence. It is possible, however, to find the explicit definition of the sequence.

Applying the recursive definition repeatedly will enable you to see the expression we are seeking:

$$a_2 = a_1 + d;\ a_3 = a_2 + d = a_1 + d + d = a_1 + 2d;$$
$$a_4 = a_3 + d = a_1 + 2d + d = a_1 + 3d; \ldots$$

So, as you see, you can get to the nth term by adding d to a_1, $(n - 1)$ times, and therefore:

> **nth term of an arithmetic sequence**
> The general (nth) term of an arithmetic sequence, a_n, with first term a_1 and common difference d, may be expressed explicitly as
> $$a_n = a_1 + (n - 1)d$$

This result is useful in finding any term of the sequence without knowing all the previous terms.

Note: The arithmetic sequence can be looked at as a linear function as explained in the introduction to this chapter, i.e. for every increase of one unit in n, the value of the term will increase by d units. As the first term is a_1, the point $(1, a_1)$ belongs to this function. The constant increase d can be considered to be the gradient (slope) of this linear model; hence, the nth term, the dependent variable in this case, can be found by using the *point-slope* form of the equation of a line:

$$y - y_1 = m(x - x_1)$$
$$a_n - a_1 = d(n - 1) \Leftrightarrow a_n = a_1 + (n - 1)d$$

This agrees with our definition of an arithmetic sequence.

Example 3

Find the nth and the 50th terms of the sequence 2, 11, 20, 29, 38, 47, …

Solution

This is an arithmetic sequence whose first term is 2 and common difference is 9. Therefore,

$$a_n = a_1 + (n - 1)d = 2 + (n - 1) \times 9 = 9n - 7$$
$$\Rightarrow a_{50} = 9 \times 50 - 7 = 443$$

Example 4

Find the recursive and the explicit forms of the definition of the following sequence, then calculate the value of the 25th term.

$$13, 8, 3, -2, \ldots$$

Solution

This is clearly an arithmetic sequence, since we observe that -5 is the common difference.

Recursive definition: $a_1 = 13$
$$a_n = a_{n-1} - 5$$

Explicit definition: $a_n = 13 - 5(n - 1) = 18 - 5n$, and
$$a_{25} = 18 - 5 \times 25 = -107$$

Example 5

Find a definition for the arithmetic sequence whose first term is 5 and fifth term is 11.

Solution

Since the fifth term is given, using the explicit form, we have
$$a_5 = a_1 + (5 - 1)d \Rightarrow 11 = 5 + 4d \Rightarrow d = \tfrac{3}{2}$$

This leads to the general term,
$$a_n = 5 + \tfrac{3}{2}(n - 1), \text{ or, equivalently, the recursive form}$$
$$\begin{cases} a_1 = 5 \\ a_n = a_{n-1} + \tfrac{3}{2}, n > 1 \end{cases}$$

3 Sequences and Series

● **Hint:** Definition: In a finite arithmetic sequence $a_1, a_2, a_3, \ldots, a_k$, the terms $a_2, a_3, \ldots, a_{k-1}$ are called **arithmetic means** between a_1 and a_k.

Example 6

Insert four arithmetic means between 3 and 7.

Solution

Since there are four means between 3 and 7, the problem can be reduced to a situation similar to Example 5 by considering the first term to be 3 and the sixth term to be 7. The rest is left as an exercise for you!

Exercise 3.2

1. Insert four arithmetic means between 3 and 7.

2. Say whether each given sequence is an arithmetic sequence. If yes, find the common difference and the 50th term; if not, say why not.
   a) $a_n = 2n - 3$
   b) $b_n = n + 2$
   c) $c_n = c_{n-1} + 2$, and $c_1 = -1$
   d) $u_n = 3u_{n-1} + 2$
   e) $2, 5, 7, 12, 19, \ldots$
   f) $2, -5, -12, -19, \ldots$

For each arithmetic sequence in questions 3–8, find:
   a) the 8th term
   b) an explicit formula for the nth term
   c) a recursive formula for the nth term.

3. $-2, 2, 6, 10, \ldots$
4. $29, 25, 21, 17, \ldots$
5. $-6, 3, 12, 21, \ldots$
6. $10.07, 9.95, 9.83, 9.71, \ldots$
7. $100, 97, 94, 91, \ldots$
8. $2, \frac{3}{4}, -\frac{1}{2}, -\frac{7}{4}, \ldots$

9. Find five arithmetic means between 13 and -23.

10. Find three arithmetic means between 299 and 300.

11. In an arithmetic sequence, $a_5 = 6$ and $a_{14} = 42$. Find an explicit formula for the nth term of this sequence.

12. In an arithmetic sequence, $a_3 = -40$ and $a_9 = -18$. Find an explicit formula for the nth term of this sequence.

3.3 Geometric sequences

Examine the following sequences and the most likely recursive formula for each of them.

$7, 14, 28, 56, 112, 224, \ldots$ $a_1 = 7$ and $a_n = a_{n-1} \times 2$, for $n > 1$
$2, 18, 162, 1458, 13\,122, \ldots$ $a_1 = 2$ and $a_n = a_{n-1} \times 9$, for $n > 1$
$48, -24, 12, -6, 3, -1.5, \ldots$ $a_1 = 48$ and $a_n = a_{n-1} \times -0.5$, for $n > 1$

Note that in each case above, every term is formed by multiplying a constant number with the preceding term. Sequences formed in this manner are called **geometric sequences**.

Definition of a geometric sequence

A sequence $a_1, a_2, a_3 \ldots$ is a **geometric sequence** if there is a constant r for which
$$a_n = a_{n-1} \times r$$
for all integers $n > 1$. r is called the **common ratio** of the sequence, and $r = a_n \div a_{n-1}$ for all integers $n > 1$.

Thus, for the sequences above, 2 is the common ratio for the first, 9 is the common ratio for the second and -0.5 is the common ratio for the third.

This description gives us the recursive definition of the geometric sequence. It is possible, however, to find the explicit definition of the sequence.

Applying the recursive definition repeatedly will enable you to see the expression we are seeking:

$$a_2 = a_1 \times r;\ a_3 = a_2 \times r = a_1 \times r \times r = a_1 \times r^2;$$
$$a_4 = a_3 \times r = a_1 \times r^2 \times r = a_1 \times r^3;\ \ldots$$

So, as you see, you can get to the nth term by multiplying a_1 with r, $(n-1)$ times, and therefore:

> **nth term of geometric sequence**
> The general (nth) term of a geometric sequence, a_n, with common ratio r and first term a_1, may be expressed explicitly as
> $$a_n = a_1 \times r^{(n-1)}$$

This result is useful in finding any term of the sequence without knowing all the previous terms.

Example 7

a) Find the geometric sequence with $a_1 = 2$ and $r = 3$.

b) Describe the sequence $3, -12, 48, -192, 768, \ldots$

c) Describe the sequence $1, \frac{1}{2}, \frac{1}{4}, \frac{1}{8}, \ldots$

d) Graph the sequence $a_n = \frac{1}{4} \cdot 3^{n-1}$

Solution

a) The geometric sequence is $2, 6, 18, 54, \ldots, 2 \times 3^{n-1}$. Notice that the ratio of a term to the preceding term is 3.

b) This is a geometric sequence with $a_1 = 3$ and $r = -4$. The nth term is $a_n = 3 \times (-4)^{n-1}$. Notice that, when the common ratio is negative, the terms of the sequence alternate in sign.

c) The nth term of this sequence is $a_n = 1 \cdot \left(\frac{1}{2}\right)^{n-1}$. Notice that the ratio of any two consecutive terms is $\frac{1}{2}$. Also, notice that the terms decrease in value.

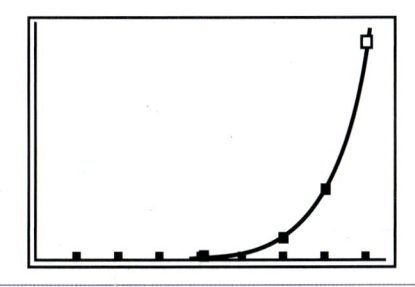

d) The graph of the geometric sequence is shown on the left. Notice that the points lie on the graph of the function $y = \frac{1}{4} \cdot 3^{x-1}$.

3 Sequences and Series

Example 8

At 8:00 a.m., 1000 mg of medicine is administered to a patient. At the end of each hour, the concentration of medicine is 60% of the amount present at the beginning of the hour.

a) What portion of the medicine remains in the patient's body at noon if no additional medication has been given?

b) If a second dosage of 1000 mg is administered at 10:00 a.m., what is the total concentration of the medication in the patient's body at noon?

Solution

a) We use the geometric model, as there is a constant multiple by the end of each hour. Hence, the concentration at the end of any hour after administering the medicine is given by:

$$a_n = a_1 \times r^{(n-1)}, \text{ where } n \text{ is the number of hours}$$

Thus, at noon $n = 5$, and $a_5 = 1000 \times 0.6^{(5-1)} = 129.6$.

b) For the second dosage, the amount of medicine at noon corresponds to $n = 3$, and $a_3 = 1000 \times 0.6^{(3-1)} = 360$.

So, the concentration of medicine is $129.6 + 360 = 489.6$ mg.

Compound interest

See also Section 4.2.

Interest compounded annually

When we borrow money we pay interest, and when we invest money we receive interest. Suppose an amount of €1000 is put into a savings account that bears an annual interest of 6%. How much money will we have in the bank at the end of four years?

It is important to note that the 6% interest is given annually and is added to the savings account, so that in the following year it will also earn interest, and so on.

Time in years	Amount in the account
0	1000
1	$1000 + 1000 \times 0.06 = 1000(1 + 0.06)$
2	$1000(1 + 0.06) + (1000(1 + 0.06)) \times 0.06 = 1000(1 + 0.06)(1 + 0.06) = 1000(1 + 0.06)^2$
3	$1000(1 + 0.06)^2 + (1000(1 + 0.06)^2) \times 0.06 = 1000(1 + 0.06)^2 (1 + 0.06) = 1000(1 + 0.06)^3$
4	$1000(1 + 0.06)^3 + (1000(1 + 0.06)^3) \times 0.06 = 1000(1 + 0.06)^3 (1 + 0.06) = \mathbf{1000(1 + 0.06)^4}$

Table 3.1 Compound interest.

This appears to be a geometric sequence with five terms. You will notice that the number of terms is five, as both the beginning and the end of the first year are counted. (Initial value, when time = 0, is the first term.)

In general, if a **principal** of P euros is invested in an account that yields an interest rate r (expressed as a decimal) annually, and this interest is added at the end of the year, every year, to the principal, then we can use the geometric sequence formula to calculate the **future value** A, which is accumulated after t years.

If we repeat the steps above, with

$A_0 = P =$ initial amount
$r =$ annual interest rate
$t =$ number of years

it becomes easier to develop the formula:

Time in years	Amount in the account
0	$A_0 = P$
1	$A_1 = P + Pr = P(1 + r)$
2	$A_2 = A_1(1 + r) = P(1 + r)^2$
⋮	
t	$A_t = P(1 + r)^t$

◀ **Table 3.2** Compound interest formula.

Notice that since we are counting from 0 to t, we have $t + 1$ terms, and hence using the geometric sequence formula,

$$a_n = a_1 \times r^{(n-1)} \Rightarrow A_t = A_0 \times (1 + r)^t$$

Interest compounded *n* times per year

Suppose that the principal P is invested as before but the interest is paid n times per year. Then $\frac{r}{n}$ is the interest paid every compounding period. Since every year we have n periods, for t years, we have nt periods. The amount A in the account after t years is

$$A = P\left(1 + \frac{r}{n}\right)^{nt}$$

Example 9

€1000 is invested in an account paying compound interest at a rate of 6%. Calculate the amount of money in the account after 10 years if

a) the compounding is annual
b) the compounding is quarterly
c) the compounding is monthly.

Solution

a) The amount after 10 years is
$A = 1000(1 + 0.06)^{10} = €1790.85$.

b) The amount after 10 years quarterly compounding is
$A = 1000\left(1 + \frac{0.06}{4}\right)^{40} = €1814.02$.

c) The amount after 10 years monthly compounding is
$A = 1000\left(1 + \frac{0.06}{12}\right)^{120} = €1819.40$.

Sequences and Series

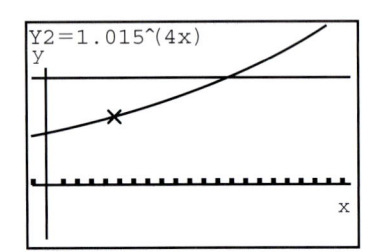

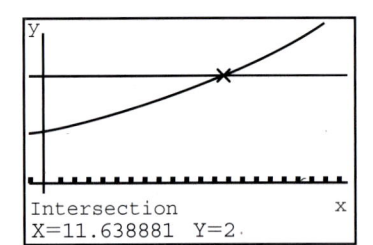

Example 10

You invested €1000 at 6% compounded quarterly. How long will it take this investment to increase to €2000?

Solution

Let $P = 1000$, $r = 0.06$, $n = 4$ and $A = 2000$ in the compound interest formula:
$$A = P\left(1 + \frac{r}{n}\right)^{nt}$$
Then solve for t:
$$2000 = 1000\left(1 + \frac{0.06}{4}\right)^{4t} \Rightarrow 2 = 1.015^{4t}$$
Using a GDC, we can graph the functions $y = 2$ and $y = 1.015^{4t}$ and then find the intersection between their graphs.

As you can see, it will take the €1000 investment 11.64 years to double to €2000. This translates into approximately 47 quarters.

You can check your work to see that this is accurate by using the compound interest formula:
$$A = 1000\left(1 + \frac{0.06}{4}\right)^{47} = €2013.28$$
In the next chapter you will learn how to solve the problem algebraically.

Example 11

You want to invest €1000. What interest rate is required to make this investment grow to €2000 in 10 years if interest is compounded quarterly?

Solution

Let $P = 1000$, $n = 4$, $t = 10$ and $A = 2000$ in the compound interest formula:
$$A = P\left(1 + \frac{r}{n}\right)^{nt}$$
Now solve for r:
$$2000 = 1000\left(1 + \frac{r}{4}\right)^{40} \Rightarrow 2 = \left(1 + \frac{r}{4}\right)^{40} \Rightarrow 1 + \frac{r}{4} = \sqrt[40]{2} \Rightarrow r = 4(\sqrt[40]{2} - 1)$$
$$= 0.0699$$

So, at a rate of 7% compounded quarterly, the €1000 investment will grow to at least €2000 in 10 years.

You can check to see whether your work is accurate by using the compound interest formula:
$$A = 1000\left(1 + \frac{0.07}{4}\right)^{40} = €2001.60$$

Population growth

The same formulae can be applied when dealing with population growth.

Example 12

The city of Baden in Lower Austria grows at an annual rate of 0.35%. The population of Baden in 1981 was 23 140. What is the estimate of the population of this city for 2011?

Solution

This situation can be modelled by a geometric sequence whose first term is 23 140 and whose common ratio is 1.0035. Since we count the population of 1981 among the terms, the number of terms is 31.

2011 is equivalent to the 31st term in this sequence. The estimated population for Baden is, therefore,

$$\text{Population }(2011) = a_{31} = 23\,140(1.0035)^{30} = 25\,697$$

Note: In Chapter 4, more realistic population growth models will be explored and more efficient methods will be developed, as well as the ability to calculate interest that is continuously compounded.

Exercise 3.3

1. Insert four geometric means between 3 and 96.

2. Determine whether the sequence in each question is arithmetic, geometric or neither. Find the common difference for the arithmetic ones and the common ratio for the geometric ones. Find the common difference or ratio, and the 10th term for each arithmetic or geometric one as appropriate.
   a) $a_n = 3n - 3$
   b) $b_n = 2^{n+2}$
   c) $c_n = 2c_{n-1} - 2$, and $c_1 = -1$
   d) $u_n = 3u_{n-1}$ and $u_1 = 4$
   e) 2, 5, 12.5, 31.25, 78.125 …
   f) 2, −5, 12.5, −31.25, 78.125 …
   g) 2, 2.75, 3.5, 4.25, 5, …
   h) 18, −12, 8, $-\frac{16}{3}$, $\frac{32}{9}$, …

For each geometric sequence in questions 3–8, find
   a) the 8th term
   b) an explicit formula for the *n*th term
   c) a recursive formula for the *n*th term.

3. $-2, 3, -\frac{9}{2}, \frac{27}{4}, \ldots$
4. $35, 25, \frac{125}{7}, \frac{625}{49}, \ldots$
5. $-6, -3, -\frac{3}{2}, -\frac{3}{4}, \ldots$
6. 9.5, 19, 38, 76, …
7. 100, 95, 90.25, …
8. $2, \frac{3}{4}, \frac{9}{32}, \frac{27}{256}, \ldots$

9. Find three geometric means between 7 and 4375.

10. Find a geometric *mean* between 16 and 81. • **Hint:** This is also called the **mean proportional**.

11. The first term of a geometric sequence is 24 and the fourth term is 3. Find the fifth term and an expression for the *n*th term.

12. The common ratio in a geometric sequence is $\frac{2}{7}$ and the fourth term is $\frac{14}{3}$. Find the third term.

13. Which term of the geometric sequence 6, 18, 54, … is 118 098?

14. The fourth term and the seventh term of a geometric sequence are 18 and $\frac{729}{8}$. Is $\frac{59049}{128}$ a term of this sequence? If so, which term is it?

15. Jim put €1500 into a savings account that pays 4% interest compounded semi-annually. How much will his account hold 10 years later if he does not make any additional investments in this account?

16. At her daughter Jane's birth, Charlotte set aside £500 into a savings account. The interest she earned was 4% compounded quarterly. How much money will Jane have on her 16th birthday?

17. How much money should you invest now if you wish to have an amount of €4000 in your account after 6 years if interest is compounded quarterly at an annual rate of 5%?

18. In 2007, the population of Switzerland (in thousands) was estimated to be 7554. How large would the Swiss population be in 2012 if it grows at a rate of 0.5% annually?

• **Hint:** Definition: In a finite geometric sequence $a_1, a_2, a_3, \ldots, a_k$, the terms $a_2, a_3, \ldots a_{k-1}$ are called **geometric means** between a_1 and a_k.

3.4 Series

The word 'series' in common language implies much the same thing as 'sequence'. But in mathematics when we talk of a series, we are referring in particular to sums of terms in a sequence, e.g. for a sequence of values a_n, the corresponding series is the sequence of S_n with

$$S_n = a_1 + a_2 + \ldots + a_{n-1} + a_n$$

If the terms are in an arithmetic sequence, we call the sum an **arithmetic series**.

Sigma notation

Most of the series we consider in mathematics are **infinite** series. This name is used to emphasize the fact that the series contain infinitely many terms. Any sum in the series S_k will be called a partial sum and is given by

$$S_k = a_1 + a_2 + \ldots + a_{k-1} + a_k$$

For convenience, this partial sum is written using the sigma notation:

$$S_k = \sum_{i=1}^{i=k} a_i = a_1 + a_2 + \ldots + a_{k-1} + a_k$$

Sigma notation is a concise and convenient way to represent long sums. Here, the symbol Σ is the Greek capital letter *sigma* that refers to the initial letter of the word 'sum'. So, the expression $\sum_{i=1}^{i=k} a_i$ means the sum of all the terms a_i, where i takes the values from 1 to k. We can also write $\sum_{i=m}^{n} a_i$ to mean the sum of the terms a_i, where i takes the values from m to n. In such a sum, m is called the lower limit and n the upper limit.

Example 13

Write out what is meant by:

a) $\sum_{i=1}^{5} i^4$ b) $\sum_{r=3}^{7} 3^r$ c) $\sum_{j=1}^{n} x_j p(x_j)$

Solution

a) $\sum_{i=1}^{5} i^4 = 1^4 + 2^4 + 3^4 + 4^4 + 5^4$

b) $\sum_{r=3}^{7} 3^r = 3^3 + 3^4 + 3^5 + 3^6 + 3^7$

c) $\sum_{j=1}^{n} x_j p(x_j) = x_1 p(x_1) + x_2 p(x_2) + \ldots + x_n p(x_n)$

Example 14

Evaluate $\displaystyle\sum_{n=0}^{5} 2^n$

Solution

$$\sum_{n=0}^{5} 2^n = 2^0 + 2^1 + 2^2 + 2^3 + 2^4 + 2^5 = 63$$

Example 15

Write the sum $\frac{1}{2} - \frac{2}{3} + \frac{3}{4} - \frac{4}{5} + \ldots + \frac{99}{100}$ in sigma notation.

Solution

We notice that each term's numerator and denominator are consecutive integers, so they take on the absolute value of $\dfrac{k}{k+1}$ or any equivalent form.

We also notice that the signs of the terms alternate and that we have 99 terms. To take care of the sign, we use some power of (-1) that will start with a positive value. If we use $(-1)^k$, the first term will be negative, so we can use $(-1)^{k+1}$ instead. We can, therefore, write the sum as

$$(-1)^{1+1}\frac{1}{2} + (-1)^{2+1}\frac{2}{3} + (-1)^{3+1}\frac{3}{4} + \ldots + (-1)^{99+1}\frac{99}{100} = \sum_{k=1}^{99}(-1)^{k+1}\frac{k}{k+1}$$

Properties of the sigma notation

There are a number of useful results that we can obtain when we use sigma notation.

1 For example, suppose we had a sum of constant terms

   $$\sum_{i=1}^{5} 2$$

   What does this mean? If we write this out in full, we get

   $$\sum_{i=1}^{5} 2 = 2 + 2 + 2 + 2 + 2 = 5 \times 2 = 10.$$

   In general, if we sum a constant n times then we can write

   $$\sum_{i=1}^{n} k = k + k + \ldots + k = n \times k = nk.$$

2 Suppose we have the sum of a constant times i. What does this give us? For example,

   $$\sum_{i=1}^{5} 5i = 5 \times 1 + 5 \times 2 + 5 \times 3 + 5 \times 4 + 5 \times 5 = 5 \times (1 + 2 + 3 + 4 + 5) = 75.$$

   However, this can also be interpreted as follows

   $$\sum_{i=1}^{5} 5i = 5 \times 1 + 5 \times 2 + 5 \times 3 + 5 \times 4 + 5 \times 5 = 5 \times (1 + 2 + 3 + 4 + 5) = 5\sum_{i=1}^{5} i$$

which implies that
$$\sum_{i=1}^{5} 5i = 5\sum_{i=1}^{5} i$$

In general, we can say
$$\sum_{i=1}^{n} ki = k \times 1 + k \times 2 + \ldots + k \times n$$
$$= k \times (1 + 2 + \ldots + n)$$
$$= k\sum_{i=1}^{n} i$$

3 Suppose that we need to consider the summation of two different functions, such as
$$\sum_{k=1}^{n}(k^2 + k^3) = (1^2 + 1^3) + (2^2 + 2^3) + \ldots + n^2 + n^3$$
$$= (1^2 + 2^2 + \ldots + n^2) + (1^3 + 2^3 + \ldots + n^3)$$
$$= \sum_{k=1}^{n}(k^2) + \sum_{k=1}^{n}(k^3)$$

In general,
$$\sum_{k=1}^{n}(f(k)) + g(k)) = \sum_{k=1}^{n} f(k) + \sum_{k=1}^{n} g(k)$$

Arithmetic series

In arithmetic series, we are concerned with adding the terms of arithmetic sequences. It is very helpful to be able to find an easy expression for the partial sums of this series.

Let us start with an example:

Find the partial sum for the first 50 terms of the series

$$3 + 8 + 13 + 18 + \ldots$$

We express S_{50} in two different ways:

$$S_{50} = 3 + 8 + 13 + \ldots + 248, \text{ and}$$
$$S_{50} = 248 + 243 + 238 + \ldots + 3$$
$$\overline{2S_{50} = 251 + 251 + 251 + \ldots + 251}$$

There are 50 terms in this sum, and hence
$$2S_{50} = 50 \times 251 \Rightarrow S_{50} = \frac{50}{2}(251) = 6275.$$

This reasoning can be extended to any arithmetic series in order to develop a formula for the nth partial sum S_n.

Let $\{a_n\}$ be an arithmetic sequence with first term a_1 and a common difference d. We can construct the series in two ways: Forward, by adding d to a_1 repeatedly, and backwards by subtracting d from a_n repeatedly. We get the following two expressions for the sum:

$S_n = a_1 + a_2 + a_3 + \ldots + a_n = a_1 + (a_1 + d) + (a_1 + 2d) + \ldots + (a_1 + (n-1)d)$
and
$S_n = a_n + a_{n-1} + a_{n-2} + \ldots + a_1 = a_n + (a_n - d) + (a_n - 2d) + \ldots + (a_n - (n-1)d)$

By adding, term by term vertically, we get

$$S_n = a_1 + (a_1 + d) + (a_1 + 2d) + \ldots + (a_1 + (n-1)d)$$
$$S_n = a_n + (a_n - d) + (a_n - 2d) + \ldots + (a_n - (n-1)d)$$
$$2S_n = (a_1 + a_n) + (a_1 + a_n) + (a_1 + a_n) + \ldots + (a_1 + a_n)$$

Since we have n terms, we can reduce the expression above to

$$2S_n = n(a_1 + a_n), \quad \text{which can be reduced to}$$
$$S_n = \frac{n}{2}(a_1 + a_n), \quad \text{which in turn can be changed to give an interesting perspective of the sum,}$$
$$\text{i.e. } S_n = n\left(\frac{a_1 + a_n}{2}\right) \text{ is } n \text{ times the average of the first and last terms!}$$

If we substitute $a_1 + (n-1)d$ for a_n then we arrive at an alternative formula for the sum:

$$S_n = \frac{n}{2}(a_1 + a_1 + (n-1)d) = \frac{n}{2}(2a_1 + (n-1)d)$$

Example 16

Find the partial sum for the first 50 terms of the series

$$3 + 8 + 13 + 18 + \ldots$$

Solution

Using the second formula for the sum, we get

$$S_{50} = \frac{50}{2}(2 \times 3 + (50-1)5) = 25 \times 251 = 6275.$$

Using the first formula requires that we know the nth term. So, $a_{50} = 3 + 49 \times 5 = 248$, which now can be used:

$$S_{50} = 25(3 + 248) = 6275.$$

Geometric series

As is the case with arithmetic series, it is often desirable to find a general expression for the nth partial sum of a geometric series.

Let us start with an example:

Find the partial sum for the first 20 terms of the series

$$3 + 6 + 12 + 24 + \ldots$$

We express S_{20} in two different ways and subtract them:

$$S_{20} = 3 + 6 + 12 + \ldots + 1\,572\,864$$
$$2S_{20} = 6 + 12 + \ldots + 1\,572\,864 + 3\,145\,728$$
$$-S_{20} = 3 - 3\,145\,728$$
$$\Rightarrow S_{20} = 3\,145\,725$$

This reasoning can be extended to any geometric series in order to develop a formula for the nth partial sum S_n.

Sequences and Series

Let $\{a_n\}$ be a geometric sequence with first term a_1 and a common ratio $r \neq 1$. We can construct the series in two ways as before and using the definition of the geometric sequence, i.e. $a_n = a_{n-1} \times r$, then

$$S_n = a_1 + a_2 + a_3 + \ldots + a_{n-1} + a_n, \text{ and}$$
$$rS_n = ra_1 + ra_2 + ra_3 + \ldots + ra_{n-1} + ra_n$$
$$= a_2 + a_3 + \ldots + a_{n-1} + a_n + ra_n$$

Now, we subtract the first and last expressions to get

$$S_n - rS_n = a_1 - ra_n \Rightarrow S_n(1-r) = a_1 - ra_n \Rightarrow S_n = \frac{a_1 - ra_n}{1-r}; r \neq 1.$$

This expression, however, requires that r, a_1, as well as a_n be known in order to find the sum. However, using the nth term expression developed earlier, we can simplify this sum formula to

$$S_n = \frac{a_1 - ra_n}{1-r} = \frac{a_1 - ra_1 r^{n-1}}{1-r} = \frac{a_1(1-r^n)}{1-r}; r \neq 1.$$

Example 17

Find the partial sum for the first 20 terms of the series $3 + 6 + 12 + 24 + \ldots$ in the opening example for this section.

Solution

$$S_{20} = \frac{3(1 - 2^{20})}{1-2} = \frac{3(1 - 1\,048\,576)}{-1} = 3\,145\,725$$

Infinite geometric series

Consider the series

$$\sum_{k=1}^{n} 2\left(\tfrac{1}{2}\right)^{k-1} = 2 + 1 + \tfrac{1}{2} + \tfrac{1}{4} + \tfrac{1}{8} + \ldots$$

Consider also finding the partial sums for 10, 20 and 100 terms. The sums we are looking for are the partial sums of a geometric series. So,

$$\sum_{k=1}^{10} 2\left(\tfrac{1}{2}\right)^{k-1} = 2 \times \frac{1 - \left(\tfrac{1}{2}\right)^{10}}{1 - \tfrac{1}{2}} \approx 3.996$$

$$\sum_{k=1}^{20} 2\left(\tfrac{1}{2}\right)^{k-1} = 2 \times \frac{1 - \left(\tfrac{1}{2}\right)^{20}}{1 - \tfrac{1}{2}} \approx 3.999\,996$$

$$\sum_{k=1}^{100} 2\left(\tfrac{1}{2}\right)^{k-1} = 2 \times \frac{1 - \left(\tfrac{1}{2}\right)^{100}}{1 - \tfrac{1}{2}} \approx 4$$

As the number of terms increases, the partial sum appears to be approaching the number 4. This is no coincidence. In the language of limits,

$$\lim_{n \to \infty} \sum_{k=1}^{n} 2\left(\tfrac{1}{2}\right)^{k-1} = \lim_{n \to \infty} 2 \times \frac{1 - \left(\tfrac{1}{2}\right)^k}{1 - \tfrac{1}{2}} = 2 \times \frac{1 - 0}{\tfrac{1}{2}} = 4, \text{ since } \lim_{n \to \infty} \left(\tfrac{1}{2}\right)^n = 0.$$

This type of problem allows us to extend the usual concept of a 'sum' of a **finite** number of terms to make sense of sums in which an **infinite** number of terms is involved. Such series are called **infinite series**.

One thing to be made clear about infinite series is that they are not true sums! The associative property of addition of real numbers allows us to extend the definition of the sum of two numbers, such as $a + b$, to three or four or n numbers, but not to an infinite number of numbers. For example, you can add any specific number of 5s together and get a real number, but if you add an *infinite* number of 5s together, you cannot get a real number! The remarkable thing about infinite series is that, in some cases, such as the example above, the sequence of partial sums (which are true sums) approach a finite limit L. The limit in our example is 4. This we write as

$$\lim_{n \to \infty} \sum_{k=1}^{n} a_k = \lim_{n \to \infty} (a_1 + a_2 + \ldots + a_n) = L.$$

We say that the series **converges** to L, and it is convenient to define L as the **sum of the infinite series**. We use the notation

$$\sum_{k=1}^{\infty} a_k = \lim_{n \to \infty} \sum_{k=1}^{n} a_k = L.$$

We can, therefore, write the limit above as

$$\sum_{k=1}^{\infty} 2\left(\tfrac{1}{2}\right)^{k-1} = \lim_{n \to \infty} \sum_{k=1}^{n} 2\left(\tfrac{1}{2}\right)^{k-1} = 4.$$

If the series does not have a limit, it **diverges** and does not have a sum.

We are now ready to develop a general rule for **infinite geometric series**. As you know, the sum of the geometric series is given by

$$S_n = \frac{a_1 - ra_n}{1 - r} = \frac{a_1 - ra_1 r^{n-1}}{1 - r} = \frac{a_1(1 - r^n)}{1 - r}; r \neq 1.$$

If $|r| < 1$, then $\lim_{n \to \infty} r^n = 0$ and

$$S_n = S = \lim_{n \to \infty} \frac{a_1(1 - r^n)}{1 - r} = \frac{a_1}{1 - r}.$$

We will call this **the sum of the infinite geometric series**. In all other cases the series diverges. The proof is left as an exercise.

$$\sum_{k=1}^{\infty} 2\left(\tfrac{1}{2}\right)^{k-1} = \frac{2}{1 - \tfrac{1}{2}} = 4, \text{ as already shown.}$$

Example 18

A rational number is a number that can be expressed as a quotient of two integers. Show that $0.\overline{6} = 0.666\ldots$ is a rational number.

Solution

$$0.\overline{6} = 0.666\ldots = 0.6 + 0.06 + 0.006 + 0.0006 + \ldots$$
$$= \tfrac{6}{10} + \tfrac{6}{10} \cdot \tfrac{1}{10} + \tfrac{6}{10} \cdot \left(\tfrac{1}{10}\right)^2 + \tfrac{6}{10} \cdot \left(\tfrac{1}{10}\right)^3 + \ldots$$

This is an infinite geometric series with $a_1 = \tfrac{6}{10}$ and $r = \tfrac{1}{10}$; therefore,

$$0.\overline{6} = \frac{\tfrac{6}{10}}{1 - \tfrac{1}{10}} = \tfrac{6}{10} \cdot \tfrac{10}{9} = \tfrac{2}{3}$$

Example 19

If a ball has elasticity such that it bounces up 80% of its previous height, find the total vertical distances travelled down and up by this ball when it is dropped from an altitude of 3 metres. Ignore friction and air resistance.

Solution

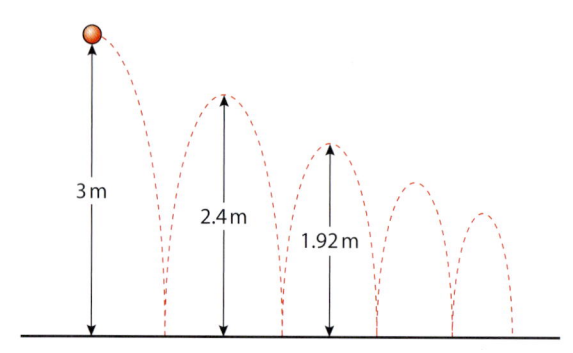

After the ball is dropped the initial 3 m, it bounces up and down a distance of 2.4 m. Each bounce after the first bounce, the ball travels 0.8 times the previous height twice – once upwards and once downwards. So, the total vertical distance is given by

$$h = 3 + 2(2.4 + (2.4 \times 0.8) + (2.4 \times 0.8^2) + \ldots) = 3 + 2 \times l$$

The amount in parenthesis is an infinite geometric series with $a_1 = 2.4$ and $r = 0.8$. The value of that quantity is

$$l = \frac{2.4}{1 - 0.8} = 12.$$

Hence, the total distance required is

$$h = 3 + 2(12) = 27 \text{ m}.$$

Applications of series to compound interest calculations (optional)

Annuities

An **annuity** is a sequence of equal periodic payments. If you are saving money by depositing the same amount at the end of each compounding period, the annuity is called **ordinary annuity**. Using geometric series you can calculate the **future value** (FV) of this annuity, which is the amount of money you have after making the last payment.

You invest €1000 at the end of each year for 10 years at a fixed annual interest rate of 6%. See table below.

Table 3.3 Calculating the future value.

Year	Amount invested	Future value
10	1000	1000
9	1000	$1000(1 + 0.06)$
8	1000	$1000(1 + 0.06)^2$
⋮		
1	1000	$1000(1 + 0.06)^9$

The future value of this investment is the sum of all the entries in the last column, so it is

$$FV = 1000 + 1000(1 + 0.06) + 1000(1 + 0.06)^2 + \ldots + 1000(1 + 0.06)^9$$

This sum is a partial sum of a geometric series with $n = 10$ and $r = 1 + 0.06$. Hence,

$$FV = \frac{1000(1 - (1 + 0.06)^{10})}{1 - (1 + 0.06)} = \frac{1000(1 - (1 + 0.06)^{10})}{-0.06} = 13\,180.79.$$

This result can also be produced with a GDC, as shown.

We can generalize the previous formula in the same manner. Let the periodic payment be R and the periodic interest rate be i, i.e. $i = \frac{r}{n}$. Let the number of periodic payments be m.

```
Plot1 Plot2 Plot3
nMin=1
∴U(n) =U(n−1) * (1+
0.06)
 U(nMin) =1000
∴V(n) =
 V(nMin) =
∴W(n) =
```

```
sum(seq(u(n),n,1,
10)
            13180.79494
```

Period	Amount invested	Future value
m	R	R
m − 1	R	R(1 + i)
m − 2	R	R(1 + i)²
⋮		
1	R	R(1 + i)^(m − 1)

◀ **Table 3.4** Calculating the future value — formula.

The future value of this investment is the sum of all the entries in the last column, so it is

$$FV = R + R(1 + i) + R(1 + i)^2 + \ldots + R(1 + i)^{m-1}$$

This sum is a partial sum of a geometric series with m terms and $r = 1 + i$. Hence,

$$FV = \frac{R(1 - (1 + i)^m)}{1 - (1 + i)} = \frac{R(1 - (1 + i)^m)}{-i} = R\left(\frac{(1 + i)^m - 1}{i}\right)$$

Note: If the payment is made at the beginning of the period rather than at the end, the annuity is called **annuity due** and the future value after m periods will be slightly different. The table for this situation is given below.

Period	Amount invested	Future value
m	R	R(1 + i)
m − 1	R	R(1 + i)²
m − 2	R	R(1 + i)³
⋮		
1	R	R(1 + i)^m

◀ **Table 3.5** Calculating the future value (annuity due).

The future value of this investment is the sum of all the entries in the last column, so it is

$$FV = R(1 + i) + R(1 + i)^2 + \ldots + R(1 + i)^{m-1} + R(1 + i)^m$$

This sum is a partial sum of a geometric series with m terms and $r = 1 + i$. Hence,

$$FV = \frac{R(1 + i(1 - (1 + i)^m)}{1 - (1 + i)} = \frac{R(1 + i - (1 + i)^{m+1})}{-i} = R\left(\frac{(1 + i)^{m+1} - 1}{i} - 1\right)$$

If the previous investment is made at the beginning of the year rather than at the end, then in 10 years we have

$$FV = R\left(\frac{(1 + i)^{m+1} - 1}{i} - 1\right) = 1000\left(\frac{(1 + 0.06)^{10+1} - 1}{0.006} - 1\right) = 13\,971.64.$$

Exercise 3.4

1. Find the sum of the arithmetic series $11 + 17 + \ldots + 365$.

2. Find the sum:
$$2 - 3 + \frac{9}{2} - \frac{27}{4} + \ldots - \frac{177\,147}{1024}$$

3. Evaluate $\sum_{k=0}^{13} (2 - 0.3k)$.

4. Evaluate $2 - \frac{4}{5} + \frac{8}{25} - \frac{16}{125} + \ldots$

5. Evaluate $\frac{1}{3} + \frac{\sqrt{3}}{12} + \frac{1}{16} + \frac{\sqrt{3}}{64} + \frac{3}{256} + \ldots$

6. Express each repeating decimal as a fraction:
   a) $0.\overline{52}$ b) $0.4\overline{53}$ c) $3.01\overline{37}$

7. At the beginning of every month, Maggie invests £150 in an account that pays 6% annual rate. How much money will there be in the account after six years?

3.5 The binomial theorem

In this section, you will learn about a sequence of numbers called Pascal's triangle and work with one of its applications, the binomial theorem.

A binomial is a polynomial with two terms. For example, $x + y$ is a binomial. In principle, it is easy to raise $x + y$ to any power; but raising it to high powers would be tedious. In this book, we will find a formula that gives the expansion of $(x + y)^n$ for any positive integer n, but we will leave the proof for higher level courses.

Let us look at some special cases of the expansion of $(x + y)^n$:

$(x+y)^0 = 1$
$(x+y)^1 = x + y$
$(x+y)^2 = x^2 + 2xy + y^2$
$(x+y)^3 = x^3 + 3x^2y + 3xy^2 + y^3$

$(x+y)^4 = x^4 + 4x^3y + 6x^2y^2 + 4xy^3 + y^4$
$(x+y)^5 = x^5 + 5x^4y + 10x^3y^2 + 10x^2y^3 + 5xy^4 + y^5$
$(x+y)^6 = x^6 + 6x^5y + 15x^4y^2 + 20x^3y^3 + 15x^2y^4 + 6xy^5 + y^6$

There are several things that you will have noticed after looking at the expansion:
- There are $n+1$ terms in the expansion of $(x+y)^n$.
- The degree of each term is n.
- The powers on x begin with n and decrease to 0.
- The powers on y begin with 0 and increase to n.
- The coefficients are symmetric.

For instance, notice how the exponents of x and y behave in the expansion of $(x+y)^5$.

The exponents of x decrease:
$(x+y)^5 = x^{\boxed{5}} + 5x^{\boxed{4}}y + 10x^{\boxed{3}}y^2 + 10x^{\boxed{2}}y^3 + 5x^{\boxed{1}}y^4 + x^{\boxed{0}}y^5$

The exponents of y increase:
$(x+y)^5 = x^5y^{\boxed{0}} + 5x^4y^{\boxed{1}} + 10x^3y^{\boxed{2}} + 10x^2y^{\boxed{3}} + 5xy^{\boxed{4}} + y^{\boxed{5}}$

Using this pattern, we can now proceed to expand any binomial raised to power n: $(x+y)^n$. For example, leaving a blank for the missing coefficients, the expansion for $(x+y)^7$ can be written as

$(x+y)^7$
$= \Box x^7 + \Box x^6y + \Box x^5y^2 + \Box x^4y^3 + \Box x^3y^4 + \Box x^2y^5 + \Box xy^6 + \Box y^7$

To finish the expansion we need to determine these coefficients. In order to see the pattern, let us look at the coefficients of the expansion we started the section with.

$(x+y)^0$	1								row 0
$(x+y)^1$	1	1							row 1
$(x+y)^2$	1	2	1						row 2
$(x+y)^3$	1	3	3	1					row 3
$(x+y)^4$	1	4	6	4	1				row 4
$(x+y)^5$	1	5	10	10	5	1			row 5
$(x+y)^6$	1	6	15	20	15	6	1		row 6
	0 column	1 column	2 column	3 column	4 column	5 column	6 column		

A triangle like the one above is known as Pascal's triangle. Notice how the first and **second** terms in row **3** give you the **second** term in row **4**; the third and **fourth** terms in row **3** give you the **fourth** term of row **4**; the second and **third** terms in row **5** give you the **third** term in row **6**; and the fifth and **sixth** terms in row **5** give you the **sixth** term in row **6**, and so on. So now we can state the key property of Pascal's triangle.

3 Sequences and Series

> **Pascal's triangle**
> Every entry in a row is the sum of the term directly above it and the entry diagonally above and to the left of it. When there is no entry, the value is considered zero.

Pascal's triangle was known to Persian and Chinese mathematicians in the 13th century.

Take the last entry in row 5, for example; there is no entry directly above it, so its value is $0 + 1 = 1$.

From this property it is easy to find all the terms in any row of Pascal's triangle from the row above it. So, for the expansion of $(x + y)^7$, the terms are found from row 6 as follows:

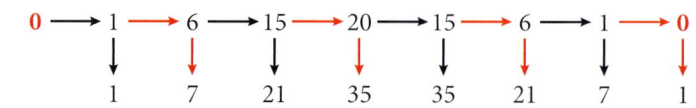

$$0 \to 1 \to 6 \to 15 \to 20 \to 15 \to 6 \to 1 \to 0$$
$$\downarrow \quad \downarrow \quad \downarrow \quad \downarrow \quad \downarrow \quad \downarrow \quad \downarrow \quad \downarrow$$
$$1 \quad 7 \quad 21 \quad 35 \quad 35 \quad 21 \quad 7 \quad 1$$

So, $(x + y)^7 = x^7 + \boxed{7}x^6y + \boxed{21}x^5y^2 + \boxed{35}x^4y^3 + \boxed{35}x^3y^4 + \boxed{21}x^2y^5 + \boxed{7}xy^6 + y^7$.

Note: Several sources use a slightly different arrangement for Pascal's triangle. The common usage considers the triangle as isosceles and uses the principle that every two entries add up to give the entry diagonally below them, as shown in the following diagram.

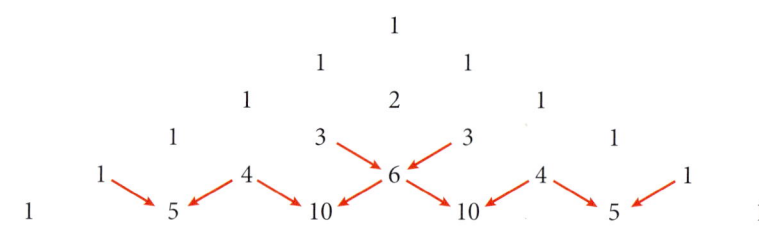

$$1$$
$$1 \quad 1$$
$$1 \quad 2 \quad 1$$
$$1 \quad 3 \quad 3 \quad 1$$
$$1 \quad 4 \quad 6 \quad 4 \quad 1$$
$$1 \quad 5 \quad 10 \quad 10 \quad 5 \quad 1$$

Example 20

Use Pascal's triangle to expand $(2k - 3)^5$.

Solution

We can find the expansion above by replacing x by $2k$ and y by -3 in the binomial expansion of $(x + y)^5$.

Using the fifth row of Pascal's triangle for the coefficients will give us the following:

$\mathbf{1}(2k)^5 + \mathbf{5}(2k)^4(-3) + \mathbf{10}(2k)^3(-3)^2 + \mathbf{10}(2k)^2(-3)^3 + \mathbf{5}(2k)(-3)^4 + \mathbf{1}(-3)^5 = 32k^5 - 240k^4 + 720k^3 - 1080k^2 + 810k - 243$.

Pascal's triangle is an easy and useful tool in finding the coefficients of the binomial expansion for relatively small values of n. It is not very efficient doing that for large values of n. Imagine you want to evaluate $(x + y)^{20}$. Using Pascal's triangle, you will need the terms in the 19th row and the 18th row and so on. This makes the process tedious and not practical.

Luckily, we have a formula that can find the coefficients of any Pascal's triangle row. This formula is the binomial formula, whose proof is beyond the scope of this book. Every entry in Pascal's triangle is denoted by $\binom{n}{r}$, which is also known as the binomial coefficient.

The proof that Pascal's entry and the binomial coefficient are the same can be found by visiting www.heinemann.co.uk/hotlinks and entering the express code 4235P, then clicking on weblink 1.

In $\binom{n}{r}$, n is the row number and r is the column number. To understand the binomial coefficient, we need to understand what the factorial notation means.

> **Factorial notation**
> The product of the first n positive integers is denoted by $n!$ and is called **n factorial**:
> $$n! = 1 \times 2 \times 3 \times 4 \ldots (n-2) \times (n-1) \times n$$
> We also define $0! = 1$.

This definition of the factorial makes many formulae involving the multiplication of consecutive positive integers shorter and easier to write. That includes the binomial coefficient.

> **The binomial coefficient**
> With n and r as non-negative integers such that $n \geq r$, the binomial coefficient $\binom{n}{r}$ is defined by
> $$\binom{n}{r} = \frac{n!}{r!(n-r)!}$$

Note: The GDC uses $_nC_r$ to represent $\binom{n}{r}$.

Example 21

Find the value of a) $\binom{7}{3}$ b) $\binom{7}{4}$ c) $\binom{7}{0}$ d) $\binom{7}{7}$

Solution

a) $\binom{7}{3} = \frac{7!}{3!(7-3)!} = \frac{7!}{3!4!} = \frac{1 \cdot 2 \cdot 3 \cdot 4 \cdot 5 \cdot 6 \cdot 7}{(1 \cdot 2 \cdot 3)(1 \cdot 2 \cdot 3 \cdot 4)} = \frac{5 \cdot 6 \cdot 7}{1 \cdot 2 \cdot 3} = 35$

b) $\binom{7}{4} = \frac{7!}{3!(7-4)!} = \frac{7!}{4!3!} = \frac{1 \cdot 2 \cdot 3 \cdot 4 \cdot 5 \cdot 6 \cdot 7}{(1 \cdot 2 \cdot 3 \cdot 4)(1 \cdot 2 \cdot 3)} = \frac{5 \cdot 6 \cdot 7}{1 \cdot 2 \cdot 3} = 35$

c) $\binom{7}{0} = \frac{7!}{0!(7-0)!} = \frac{7!}{0!7!} = \frac{1}{1} = 1$

d) $\binom{7}{7} = \frac{7!}{7!(7-7)!} = \frac{7!}{7!0!} = \frac{1}{1} = 1$

• **Hint:** Your calculator can do the tedious work of evaluating the binomial coefficient. If you have a TI, the binomial coefficient appears as $_nC_r$, which is another notation frequently used in mathematical literature.

```
7 nCr 3
            35
7 nCr 4
            35
7 nCr 0
             1
```

Although the binomial coefficient $\binom{n}{r}$ appears as a fraction, all its results where n and r are non-negative integers are positive integers. Also, notice the **symmetry** of the coefficient in the previous examples. This is a property that you are asked to prove in the exercises:

$$\binom{n}{r} = \binom{n}{n-r}$$

Example 22

Calculate the following:

$\binom{6}{0}$, $\binom{6}{1}$, $\binom{6}{2}$, $\binom{6}{3}$, $\binom{6}{4}$, $\binom{6}{5}$, $\binom{6}{6}$

Solution

$\binom{6}{0} = 1$, $\binom{6}{1} = 6$, $\binom{6}{2} = 15$, $\binom{6}{3} = 20$, $\binom{6}{4} = 15$, $\binom{6}{5} = 6$, $\binom{6}{6} = 1$

The values we calculated above are precisely the entries in the sixth row of Pascal's triangle.

We can write Pascal's triangle in the following manner:

$$\binom{0}{0}$$
$$\binom{1}{0} \quad \binom{1}{1}$$
$$\binom{2}{0} \quad \binom{2}{1} \quad \binom{2}{2}$$
$$\binom{3}{0} \quad \binom{3}{1} \quad \binom{3}{2} \quad \binom{3}{3}$$
$$\cdots \quad \cdots \quad \cdots \quad \cdots$$
$$\binom{n}{0} \quad \binom{n}{1} \quad \cdots \quad \cdots \quad \cdots \quad \binom{n}{n}$$

Example 23

Calculate $\binom{n}{r-1} + \binom{n}{r}$.

• **Hint:** You will be able to provide reasons for the steps after you do the exercises!

Solution

$$\binom{n}{r-1} + \binom{n}{r} = \frac{n!}{(r-1)!(n-r+1)!} + \frac{n!}{r!(n-r)!}$$

$$= \frac{n! \cdot r}{r \cdot (r-1)!(n-r+1)!} + \frac{n! \cdot (n-r+1)}{r!(n-r)! \cdot (n-r+1)}$$

$$= \frac{n! \cdot r}{r!(n-r+1)!} + \frac{n! \cdot (n-r+1)}{r!(n-r+1)!}$$

$$= \frac{n! \cdot r + n! \cdot (n-r+1)}{r!(n-r+1)!} = \frac{n!(r+n-r+1)}{r!(n-r+1)!}$$

$$= \frac{n!(n+1)}{r!(n-r+1)!} = \frac{(n+1)!}{r!(n+1-r)!} = \binom{n+1}{r}$$

If we read the result above carefully, it says that the sum of the terms in the nth row $(r-1)$th and rth columns is equal to the entry in the $(n+1)$th row and rth column. That is, the two entries on the left are adjacent entries in the nth row of Pascal's triangle and the entry on the right is the entry in the $(n+1)$th row directly below the rightmost entry. This is precisely the principle behind Pascal's triangle!

Using the binomial theorem

We are now prepared to state the binomial theorem.

$$(x+y)^n = \binom{n}{0}x^n + \binom{n}{1}x^{n-1}y + \binom{n}{2}x^{n-2}y^2 + \binom{n}{3}x^{n-3}y^3 + \ldots + \binom{n}{n-1}xy^{n-1} + \binom{n}{n}y^n$$

In a compact form, we can use sigma notation to express the theorem as follows:

$$(x+y)^n = \sum_{i=0}^{n} \binom{n}{i} x^{n-i} y^i$$

Example 24

Use the binomial theorem to expand $(x + y)^7$.

Solution

$$(x + y)^7 = \binom{7}{0}x^7 + \binom{7}{1}x^{7-1}y + \binom{7}{2}x^{7-2}y^2 + \binom{7}{3}x^{7-3}y^3 + \binom{7}{4}x^{7-4}y^4$$
$$+ \binom{7}{5}x^{7-5}y^5 + \binom{7}{6}xy^6 + \binom{7}{7}y^7$$
$$= x^7 + 7x^6y + 21x^5y^2 + 35x^4y^3 + 35x^3y^4 + 21x^2y^5 + 7xy^6 + y^7$$

Example 25

Find the expansion for $(2k - 3)^5$.

Solution

$$(2k - 3)^5 = \binom{5}{0}(2k)^5 + \binom{5}{1}(2k)^4(-3) + \binom{5}{2}(2k)^3(-3)^2 + \binom{5}{3}(2k)^2(-3)^3$$
$$+ \binom{5}{4}(2k)(-3)^4 + \binom{5}{5}(-3)^5$$
$$= 32k^5 - 240k^4 + 720k^3 - 1080k^2 + 810k - 243$$

Example 26

Find the term containing a^3 in the expansion $(2a - 3b)^9$.

Solution

To find the term, we do not need to expand the whole expression.

Since $(x + y)^n = \sum_{i=0}^{n} \binom{n}{i} x^{n-i} y^i$, the term containing a^3 is the term where $n - i = 3$, i.e. when $i = 6$. So, the required term is

$$\binom{9}{6}(2a)^{9-6}(-3b)^6 = 84 \cdot 8a^3 \cdot 729b^6 = 489\,888 a^3 b^6.$$

Example 27

Find the term independent of x in $\left(2x^2 - \dfrac{3}{x}\right)^9$.

Solution

To get such a term, we need the power of the first term to be equal to the power of the second term, i.e. $2k = 9 - k \Rightarrow k = 3$, so the term is

$$\binom{9}{6}(2x^2)^3\left(-\dfrac{3}{x}\right)^6 = 84 \cdot 8x^6 \cdot \dfrac{729}{x^6} = 489\,888.$$

3 Sequences and Series

Example 28

Find the coefficient of b^6 in the expansion of $\left(2b^2 - \dfrac{1}{b}\right)^{12}$.

Solution

The general term is

$$\binom{12}{i}(2b^2)^{12-i}\left(-\dfrac{1}{b}\right)^i = \binom{12}{i}(2)^{12-i}(b^2)^{12-i}\left(-\dfrac{1}{b}\right)^i$$

$$= \binom{12}{i}(2)^{12-i}b^{24-2i}b^{-i}(-1)^i = \binom{12}{i}(2)^{12-i}b^{24-3i}(-1)^i$$

$24 - 3i = 6 \Rightarrow i = 6$. So, the coefficient in question is $\binom{12}{6}(2)^6(-1)^6 = 59\,136$.

Exercise 3.5

1 Use Pascal's triangle to expand each binomial.
  a) $(x + 2y)^5$
  b) $(a - b)^4$
  c) $(x - 3)^6$
  d) $(2 - x^3)^4$
  e) $(x - 3b)^7$
  f) $\left(2n + \dfrac{1}{n^2}\right)^6$
  g) $\left(\dfrac{3}{x} - 2\sqrt{x}\right)^4$

2 Evaluate each expression.
  a) $\binom{8}{3}$
  b) $\binom{18}{5} - \binom{18}{13}$
  c) $\binom{7}{4}\binom{7}{3}$
  d) $\binom{5}{0} + \binom{5}{1} + \binom{5}{2} + \binom{5}{3} + \binom{5}{4} + \binom{5}{5}$
  e) $\binom{6}{0} - \binom{6}{1} + \binom{6}{2} - \binom{6}{3} + \binom{6}{4} - \binom{6}{5} + \binom{6}{6}$

3 Use the binomial theorem to expand each of the following.
  a) $(x + 2y)^7$
  b) $(a - b)^6$
  c) $(x - 3)^5$
  d) $(2 - x^3)^6$
  e) $(x - 3b)^7$
  f) $\left(2n + \dfrac{1}{n^2}\right)^6$
  g) $\left(\dfrac{3}{x} - 2\sqrt{x}\right)^4$
  h) $(1 + \sqrt{5})^4 + (1 - \sqrt{5})^4$
  i) $(\sqrt{3} + 1)^8 - (\sqrt{3} - 1)^8$
  j) $(1 + i)^8$, where $i^2 = -1$
  k) $(\sqrt{2} - i)^6$, where $i^2 = -1$

4 Consider the expression $\left(x - \dfrac{2}{x}\right)^{45}$.
  a) Find the first three terms of this expansion.
  b) Find the constant term if it exists or justify why it does not exist.
  c) Find the last three terms of the expansion.
  d) Find the term containing x^3 if it exists or justify why it does not exist.

5 Prove that $\binom{n}{k} = \binom{n}{n-k}$ for all $n, k \in \mathbb{N}$ and $n \geqslant k$.

6 Prove that for any positive integer n,
$$\binom{n}{1} + \binom{n}{2} + \ldots + \binom{n}{n-1} + \binom{n}{n} = 2^n - 1 \quad \bullet \text{ Hint: } 2^n = (1+1)^n$$

7 Consider all $n, k \in \mathbb{N}$ and $n \geqslant k$.
  a) Verify that $k! = k(k-1)!$
  b) Verify that $(n - k + 1)! = (n - k + 1)(n - k)!$
  c) Justify the steps given in the proof of $\binom{n}{r-1} + \binom{n}{r} = \binom{n+1}{r}$ in the examples.

8 Find the value of the expression:
$$\binom{6}{0}\left(\frac{1}{3}\right)^6 + \binom{6}{1}\left(\frac{1}{3}\right)^5\left(\frac{2}{3}\right) + \binom{6}{2}\left(\frac{1}{3}\right)^4\left(\frac{2}{3}\right)^2 + \ldots + \binom{6}{6}\left(\frac{2}{3}\right)^6$$

9 Find the value of the expression:
$$\binom{8}{0}\left(\frac{2}{5}\right)^8 + \binom{8}{1}\left(\frac{2}{5}\right)^7\left(\frac{3}{5}\right) + \binom{8}{2}\left(\frac{2}{5}\right)^6\left(\frac{3}{5}\right)^2 + \ldots + \binom{8}{8}\left(\frac{3}{5}\right)^8$$

10 Find the value of the expression:
$$\binom{n}{0}\left(\frac{1}{7}\right)^n + \binom{n}{1}\left(\frac{1}{7}\right)^{n-1}\left(\frac{6}{7}\right) + \binom{n}{2}\left(\frac{1}{7}\right)^{n-2}\left(\frac{6}{7}\right)^2 + \ldots + \binom{n}{n}\left(\frac{6}{7}\right)^n$$

Practice questions

Find the first five terms of each infinite sequence defined in questions 1–6.

1 $s(n) = 2n - 3$

2 $g(k) = 2^k - 3$

3 $f(n) = 3 \times 2^{-n}$

4 $\begin{cases} a_1 = 5 \\ a_n = a_{n-1} + 3; \text{ for } n > 1 \end{cases}$

5 $a_n = (-1)^n(2^k) + 3$

6 $\begin{cases} b_1 = 3 \\ b_n = b_{n-1} + 2n; \text{ for } n \geq 2 \end{cases}$

Determine whether each sequence in questions 7–12 is arithmetic, geometric or neither. Find the common difference for the arithmetic ones and the common ratio for the geometric ones.

7 52, 55, 58, 61, ...

8 −1, 3, −9, 27, −81, ...

9 0.1, 0.2, 0.4, 0.8, 1.6, 3.2, ...

10 3, 6, 12, 18, 21, 27, ...

11 6, 14, 20, 28, 34, ...

12 2.4, 3.7, 5, 6.3, 7.6, ...

For each arithmetic or geometric sequence in questions 13–23, find
  a) the 8th term
  b) an explicit formula for the *n*th term
  c) a recursive formula for the *n*th term.

13 −3, 2, 7, 12, ...

14 19, 15, 11, 7, ...

15 −8, 3, 14, 25, ...

16 10.05, 9.95, 9.85, 9.75, ...

17 100, 99, 98, 97, ...

18 $2, \frac{1}{2}, -1, -\frac{5}{2}, \ldots$

19 3, 6, 12, 24, ...

20 4, 12, 36, 108, ...

21 5, −5, 5, −5, ...

22 3, −6, 12, −24, ...

23 972, −324, 108, −361, ...

24 Find five arithmetic means between 15 and −21.

25 Find three arithmetic means between 99 and 100.

26 In an arithmetic sequence, $a_3 = 11$ and $a_{12} = 47$. Find an explicit formula for the nth term of this sequence.

27 In an arithmetic sequence, $a_7 = -48$ and $a_{13} = -10$. Find an explicit formula for the nth term of this sequence.

28 Find four geometric means between 7 and 1701.

29 Find a geometric mean between 9 and 64.   • **Hint:** This is also called the **mean proportional.**

30 The first term of a geometric sequence is 24 and the third term is 6. Find the fourth term and an expression for the nth term.

31 The common ratio in a geometric sequence is $\frac{3}{7}$ and the fourth term is $\frac{14}{3}$. Find the third term.

32 Which term of the geometric sequence 7, 21, 63, ... is 137 781?

33 The third term and the sixth term of a geometric sequence are 18 and $\frac{243}{4}$. Is $\frac{19\,683}{64}$ a term of this sequence? If so, which term is it?

34 Tim put €2500 into a savings account that pays 4% interest compounded semi-annually. How much will his account hold 10 years later if he does not make any additional investments in this account?

35 At her son William's birth, Jane set aside £1000 into a savings account. The interest she earned was 6% compounded quarterly. How much money will William have on his 18th birthday?

36 How much money should you invest now if you wish to have an amount of €3000 in your account after six years if interest is compounded quarterly at an annual rate of 6%?

37 Find the sum of the arithmetic series $13 + 19 + \ldots + 367$.

38 Find the sum of
$$2 - \frac{4}{3} + \frac{8}{9} - \frac{16}{27} + \ldots - \frac{4096}{177\,147}$$

39 Evaluate $\sum_{k=0}^{11}(3 + 0.2k)$.

40 Evaluate $2 - \frac{4}{3} + \frac{8}{9} - \frac{16}{27} + \ldots$

41 Evaluate $\frac{1}{2} + \frac{\sqrt{2}}{2\sqrt{3}} + \frac{1}{3} + \frac{\sqrt{2}}{3\sqrt{3}} + \frac{2}{9} + \ldots$

42 Express each repeating decimal as a fraction:
  a) $0.\overline{7}$ b) $0.3\overline{45}$ c) $3.21\overline{29}$

43 Find the coefficient of x^6 in the expansion of $(2x - 3)^9$.

44 Find the coefficient of $x^3 b^4$ in $(ax + b)^7$.

45 Find the constant term of $\left(\frac{2}{z^2} - z\right)^{15}$.

46 Expand $(3n - 2m)^5$.

47 Find the coefficient of r^{10} in $(4 + 3r^2)^9$.

48 In an arithmetic sequence, the first term is 4, the fourth term is 19 and the nth term is 99. Find the common difference and the number of terms n.

49 Two students, Nick and Charlotte, decide to start preparing for their IB exams 15 weeks ahead of the exams. Nick starts by studying for 12 hours in the first week and plans to increase the amount by 2 hours per week. Charlotte starts with 12 hours in the first week and decides to increase her time by 10% every week.
   a) How many hours did each student study in week 5?
   b) How many hours in total does each student study for the 15 weeks?
   c) In which week will Charlotte exceed 40 hours per week?
   d) In which week does Charlotte catch up with Nick in the number of hours spent on studying per week?

50 Two diet schemes are available for relatively overweight people to lose weight. Plan A promises the patient an initial weight loss of 1000 g the first month, with a steady loss of an additional 80 g every month after the first. So, the second month the patient will lose 1080 g and so on for a maximum duration of 12 months.

Plan B starts with a weight loss of 1000 g the first month and an increase in weight loss by 6% more every following month.
   a) Write down the amount of grams lost under Plan B in the second and third months.
   b) Find the weight lost in the 12th month for each plan.
   c) Find the total weight loss during a 12-month period under
      (i) Plan A   (ii) Plan B.

51 Planning on buying your first car in 10 years, you start a savings plan where you invest €500 at the beginning of the year for 10 years. Your investment scheme offers a fixed rate of 6% per year compounded annually.

Calculate, giving your answers to the nearest euro (€),
   (a) how much the first €500 is worth at the end of 10 years
   (b) the total value your investment will give you at the end of the 10 years.

52 The first three terms of an arithmetic sequence are 6, 9.5, 13.
   a) What is the 40th term of the sequence?
   b) What is the sum of the first 103 terms of the sequence?

53 A marathon runner plans her training programme for a 20 km race. On the first day she plans to run 2 km, and then she wants to increase her distance by 500 m on each subsequent training day.
   a) On which day of her training does she first run a distance of 20 km?
   b) By the time she manages to run the 20 km distance, what is the total distance she would have run for the whole training programme?

54 In the nation of Telefonica, cellular phones were first introduced in the year 2000. During the first year, the number of people who bought a cellular phone was 1600. In 2001, the number of new participants was 2400, and in 2002 the new participants numbered 3600.
   a) You notice that the trend is a geometric sequence; find the common ratio.

Assuming that the trend continues,
   b) how many participants will join in 2012?
   c) in what year would the number of new participants first exceed 50 000?

Between 2000 and 2002, the total number of participants reaches 7600.
   d) What is the total number of participants between 2000 and 2012?

During this period, the total adult population of Telefonica remains at approximately 800 000.
   e) Use this information to suggest a reason why this trend in growth would not continue.

55 In an arithmetic sequence, the fist term is 25, the fourth term is 13 and the nth term is $-11\,995$. Find the common difference d and the number of terms n.

56 The midpoints M, N, P, Q of the sides of a square of side 1 cm are joined to form a new square.

  a) Show that the side of the second square $MNPQ$ is $\frac{\sqrt{2}}{2}$.

  b) Find the area of square $MNPQ$.

  A new third square $RSTU$ is constructed in the same manner.

  c) (i) Find the area of the third square just constructed.

  (ii) Show that the areas of the squares are in a geometric sequence and find its common ratio.

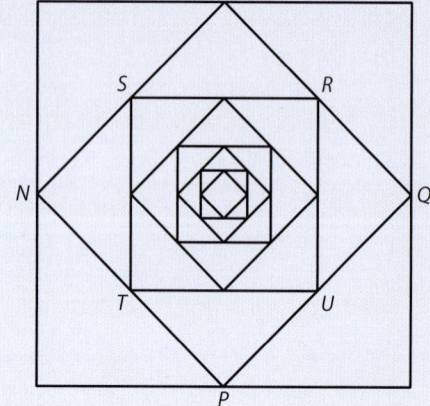

  The procedure continues indefinitely.

  d) (i) Find the area of the tenth square.

  (ii) Find the sum of the areas of all the squares.

57 Tim is a dedicated swimmer. He goes swimming once every week. He starts the first week of the year by swimming 200 metres. Each week after that he swims 20 m more than the previous week. He does that all year long (52 weeks).

  a) How far does he swim in the final week?

  b) How far does he swim altogether?

58 The diagram below shows three iterations of constructing squares in the following manner: A square of side 3 units is given, then it is divided into nine smaller squares as shown and the middle square is shaded. Each of the unshaded squares is in turn divided into nine squares and the process is repeated. The area of the first shaded square is 1 unit.

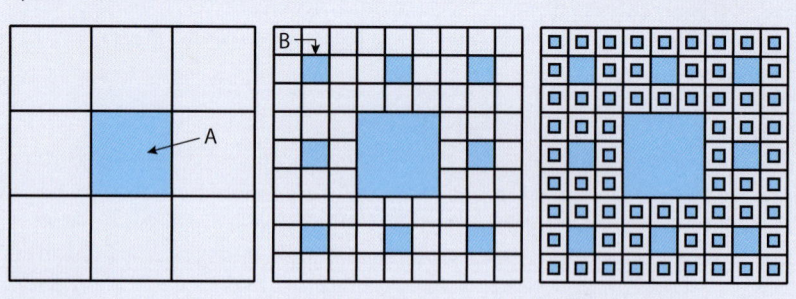

  a) Find the area of each of the squares A and B.

  b) Find the area of any small square in the third diagram.

  c) Find the area of the shaded regions in the second and third iterations.

  d) If the process is continued indefinitely, find the area left unshaded.

59 The table below shows four series of numbers. One series is an arithmetic one, one is a converging geometric series, one is a diverging geometric series and the fourth is neither geometric nor arithmetic.

Series		Type of series
(i)	$2 + 22 + 222 + 2222 + \ldots$	
(ii)	$2 + \frac{4}{3} + \frac{8}{9} + \frac{16}{27} + \ldots$	
(iii)	$0.8 + 0.78 + 0.76 + 0.74 + \ldots$	
(iv)	$2 + \frac{8}{3} + \frac{32}{9} + \frac{128}{27} + \ldots$	

a) Complete the table by stating the type of each series.
b) Find the sum of the infinite geometric series above.

60 Two IT companies offer 'apparently' similar salary schemes for their new appointees. Kell offers a starting salary of €18 000 per year and then an annual increase of €400 every year after the first. YBO offers a starting salary of €17 000 per year and an annual increase of 7% for the rest of the years after the first.
  a) (i) Write down the salary paid during the second and third years for each company.
  (ii) Calculate the total amount that an employee working for 10 years will accumulate in each company.
  (iii) Calculate the salary paid during the tenth year for each company.
  b) Tim works at Kell and Merijayne works at YBO.
  (i) When would Merijayne start earning more than Tim?
  (ii) What is the minimum number of years that Merijayne requires so that her total earnings exceed Tim's total earnings?

61 A theatre has 24 rows of seats. There are 16 seats in the first row and each successive row increases by 2 seats, 1 on each side.
  a) Calculate the number of seats in the 24th row.
  b) Calculate the number of seats in the whole theatre.

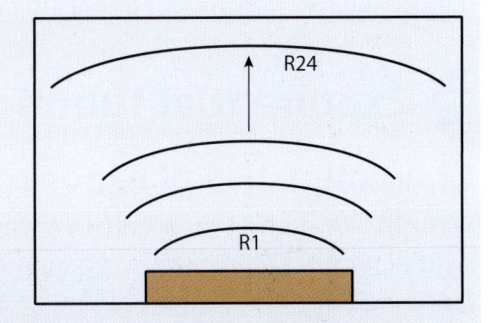

62 The amount of €7000 is invested at 5.25% annual compound interest.
  a) Write down an expression for the value of this investment after *t* full years.
  b) Calculate the minimum number of years required for this amount to become €10 000.
  c) For the same number of years as in part b), would an investment of the same amount be better if it were at a 5% rate compounded quarterly?

63 With S_n denoting the sum of the first *n* terms of an arithmetic sequence, we are given that $S_1 = 9$ and $S_2 = 20$.
  a) Find the second term.
  b) Calculate the common difference of the sequence.
  c) Find the fourth term.

4 Exponential and Logarithmic Functions

Assessment statements
1.2 Exponents and logarithms.
    Laws of exponents; laws of logarithms. Change of base.
2.7 The function $x \mapsto a^x$, $a > 0$.
    The inverse function $x \mapsto \log_a x$, $x > 0$.
    Graphs of $y = a^x$ and $y = \log_a x$.
    Solutions of $a^x = b$ using logarithms.
2.8 The exponential function $x \mapsto e^x$.
    The logarithmic function $x \mapsto \ln x$, $x > 0$.

Introduction

A variety of functions have already been considered in this text (see Figure 2.15 in Section 2.4): polynomial functions (e.g. linear, quadratic and cubic functions), functions with radicals (e.g. square root function), rational functions (e.g. inverse and inverse square functions) and the absolute value functions. This chapter examines two very important and useful functions: the exponential function and its inverse function, the logarithmic function.

4.1 Exponential functions

Characteristics of exponential functions

We begin our study of exponential functions by comparing two algebraic expressions that represent two seemingly similar but very different functions. The two expressions x^2 and 2^x are similar in that they both contain a **base** and an **exponent** (or power). In x^2, the base is the variable x and the exponent is the constant 2. In 2^x, the base is the constant 2 and the exponent is the variable x. However, x^2 and 2^x are examples of two different types of functions.

• **Hint:** Another word for exponent is **index** (plural: **indices**).

The quadratic function $y = x^2$ is in the form 'variable base$^{\text{constant power}}$', where the base is a variable and the exponent is an integer greater than or equal to zero (non-negative integer). Any function in this form is called a **polynomial** (or **power**) **function**.

The function $y = 2^x$ is in the form 'constant base$^{\text{variable power}}$', where the base is a positive real number (not equal to one) and the exponent is a variable. Any function in this form is called an **exponential function**.

To illustrate a fundamental difference between exponential functions and power functions, consider the function values for $y = x^2$ and $y = 2^x$ when

x is an integer from 0 to 10. Both a table and a graph (Figure 4.1) showing these results display clearly how the values for the exponential function eventually increase at a significantly faster rate than the power function.

x	$y = x^2$	$y = 2^x$
0	0	1
1	1	2
2	4	4
3	9	8
4	16	16
5	25	32
6	36	64
7	49	128
8	64	256
9	81	512
10	100	1024

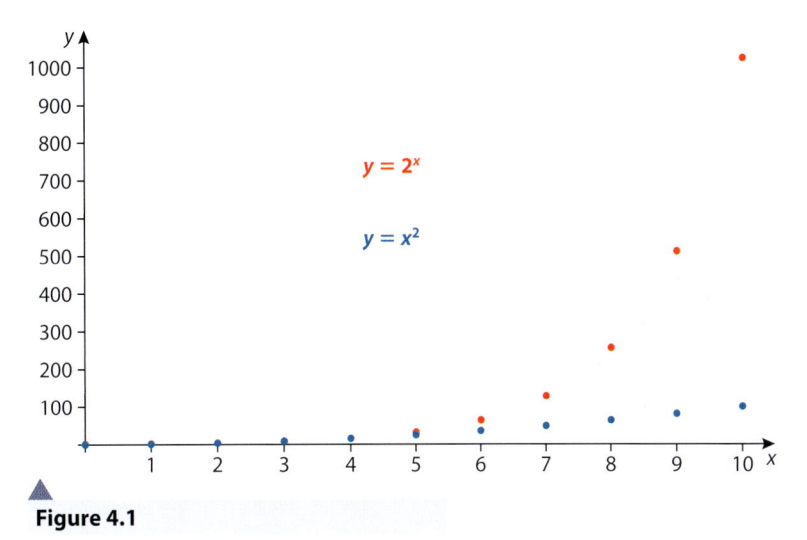

Figure 4.1

Another important point to make is that polynomial, or power, functions can easily be defined (and computed) for any real number. For any power function $y = x^n$, where n is any positive integer, y is found by simply taking x and repeatedly multiplying it n times. Hence, x can be any real number. For example, for the power function $y = x^3$, if $x = \pi$, then $y = \pi^3 \approx$ 31.006 276 68…. Since a power function like $y = x^3$ is defined for all real numbers, we can graph it as a continuous curve so that every real number is the x-coordinate of some point on the curve. What about the exponential function $y = 2^x$? Can we compute a value for y for any real number x? Before we try, let's first consider x being any rational number and recall the following laws of exponents (indices) that were covered in Section 1.3.

Laws of exponents

For $b > 0$ and $m, n \in \mathbb{Q}$ (rational numbers):

$$b^m \cdot b^n = b^{m+n} \qquad \frac{b^m}{b^n} = b^{m-n} \qquad (b^m)^n = b^{mn} \qquad b^0 = 1 \qquad b^{-m} = \frac{1}{b^m}$$

Also, in Section 1.3, we covered the definition of a rational exponent.

Rational exponent

For $b > 0$ and $m, n \in \mathbb{Z}$ (integers):

$$b^{\frac{m}{n}} = \sqrt[n]{b^m} = (\sqrt[n]{b})^m$$

From these established facts, we are able to compute b^x ($b > 0$) when x is any rational number. For example, $b^{4.7} = b^{\frac{47}{10}}$ represents the 10th root of b raised to the 47th power i.e. $\sqrt[10]{b^{47}}$. Now, we would like to define b^x when x is any real number such as π or $\sqrt{2}$. We know that π has a non-terminating, non-repeating decimal representation that begins $\pi = 3.141\ 5\ 92\ 653\ 589\ 793\ \ldots$. Consider the sequence of numbers

$$b^3,\ b^{3.1},\ b^{3.14},\ b^{3.141},\ b^{3.1415},\ b^{3.14159},\ \ldots$$

> To demonstrate just how quickly $y = 2^x$ increases, consider what would happen if you were able to repeatedly fold a piece of paper in half 50 times. A typical piece of paper is about five thousandths of a centimetre thick. Each time you fold the piece of paper the thickness of the paper doubles, so after 50 folds the thickness of the folded paper is the height of a stack of 2^{50} pieces of paper. The thickness of the paper after being folded 50 times would be $2^{50} \times$ 0.005 cm – which is more than 56 million kilometres (nearly 35 million miles)! Compare that with the height of a stack of 50^2 pieces of paper that would be a meagre $12\frac{1}{2}$ cm – only 0.000 125 km.

4 Exponential and Logarithmic Functions

Every term in this sequence is defined because each has a rational exponent. Although it is beyond the scope of this text, it can be proved that each number in the sequence gets closer and closer to a certain real number – defined as b^π. Similarly, we can define other irrational exponents, thus proving that the laws of exponents hold for all real exponents. Figure 4.2 shows a sequence of exponential expressions approaching the value of 2^π.

x	2^x (12 s.f.)
3	8.000 000 000 00
3.1	8.574 187 700 29
3.14	8.815 240 927 01
3.141	8.821 353 304 55
3.1415	8.824 411 082 48
3.141 59	8.824 961 595 06
3.141 592	8.824 973 829 06
3.141 5926	8.824 977 499 27
3.141 592 65	8.824 977 805 12

Your GDC will give an approximate value for 2^π to at least 10 significant figures, as shown below.

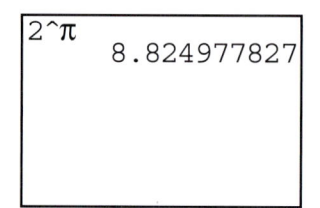

```
2^π
         8.824977827
```

Figure 4.2

Graphs of exponential functions

Using this definition of irrational powers, we can now construct a complete graph of any exponential function $f(x) = b^x$ such that b is a number greater than zero and x is any real number.

Example 1

Graph each exponential function by plotting points.

a) $f(x) = 3^x$
b) $g(x) = \left(\frac{1}{3}\right)^x$

Solution

We can easily compute values for each function for integral values of x from -3 to 3. Knowing that exponential functions are defined for all real numbers – not just integers – we can sketch a smooth curve in Figure 4.3, filling in between the ordered pairs shown in the table.

x	$f(x) = 3^x$	$g(x) = \left(\frac{1}{3}\right)^x$
-3	$\frac{1}{27}$	27
-2	$\frac{1}{9}$	9
-1	$\frac{1}{3}$	3
0	1	1
1	3	$\frac{1}{3}$
2	9	$\frac{1}{9}$
3	27	$\frac{1}{27}$

Figure 4.3

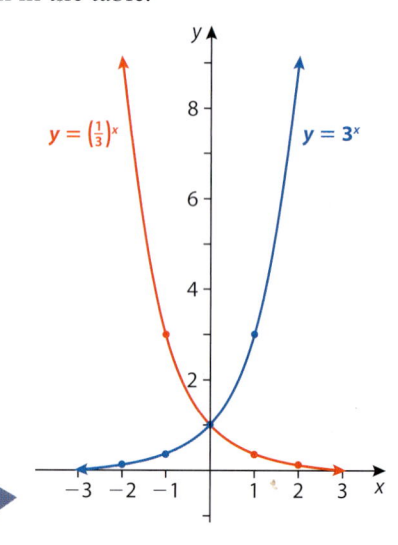

110

Remember that in Section 2.4 we established that the graph of $y = f(-x)$ is obtained by reflecting the graph of $y = f(x)$ in the y-axis. It is clear from the table and the graph in Figure 4.3 that the graph of function g is a reflection of function f about the y-axis. Let's use some laws of exponents to show that $g(x) = f(-x)$.

$$g(x) = \left(\frac{1}{3}\right)^x = \frac{1^x}{3^x} = \frac{1}{3^x} = 3^{-x} = f(-x)$$

It is useful to point out that both of the graphs, $y = 3^x$ and $y = \left(\frac{1}{3}\right)^x$, pass through the point (0, 1) and have a horizontal asymptote of $y = 0$ (x-axis). The same is true for the graph of all exponential functions in the form $y = b^x$ given that $b \neq 1$. If $b = 1$, then $y = 1^x = 1$ and the graph is a horizontal line rather than a constantly increasing or decreasing curve.

Exponential functions

If $b > 0$ and $b \neq 1$, the **exponential function** with base b is the function defined by

$$f(x) = b^x$$

The **domain** of f is the set of real numbers ($x \in \mathbb{R}$) and the **range** of f is the set of positive real numbers ($y > 0$). The graph of f passes through (0, 1), has the x-axis as a **horizontal asymptote**, and, depending on the value of the base of the exponential function b, will either be a continually increasing **exponential growth curve** or a continually decreasing **exponential decay curve**.

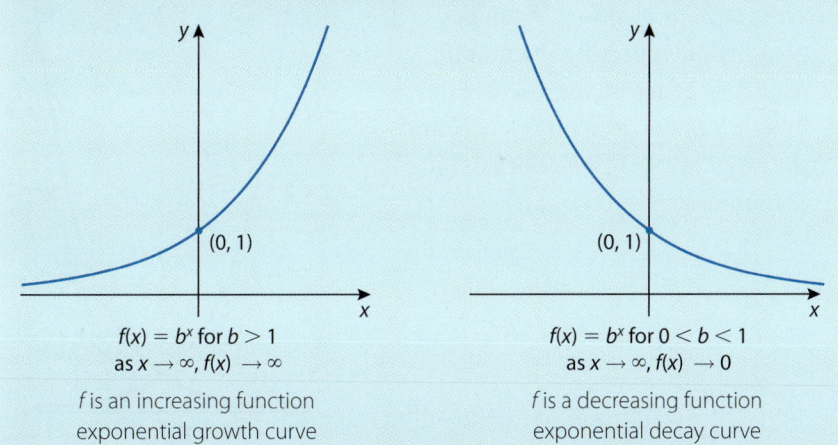

$f(x) = b^x$ for $b > 1$
as $x \to \infty$, $f(x) \to \infty$

f is an increasing function
exponential growth curve

$f(x) = b^x$ for $0 < b < 1$
as $x \to \infty$, $f(x) \to 0$

f is a decreasing function
exponential decay curve

The graphs of all exponential functions will display a characteristic growth or decay curve. As we shall see, many natural phenomena exhibit exponential growth or decay. Also, the graphs of exponential functions behave **asymptotically** for either very large positive values of x (decay curve) or very large negative values of x (growth curve). This means that there will exist a horizontal line that the graph will approach, but not intersect, as either $x \to \infty$ or as $x \to -\infty$.

Transformations of exponential functions

Recalling from Section 2.4 how the graphs of functions are translated and reflected, we can efficiently sketch the graph of many exponential functions.

4 Exponential and Logarithmic Functions

Example 2

Using the graph of $f(x) = 2^x$, sketch the graph of each function. State the domain and range for each function and the equation of its horizontal asymptote.

a) $g(x) = 2^x + 3$
b) $h(x) = 2^{-x}$
c) $p(x) = -2^x$
d) $r(x) = 2^{x-4}$
e) $v(x) = 3(2^x)$

Solution

a) The graph of $g(x) = 2^x + 3$ can be obtained by translating the graph of $f(x) = 2^x$ vertically three units up. For function g, the domain is x is any real number ($x \in \mathbb{R}$) and the range is $y > 3$. The horizontal asymptote for g is $y = 3$.

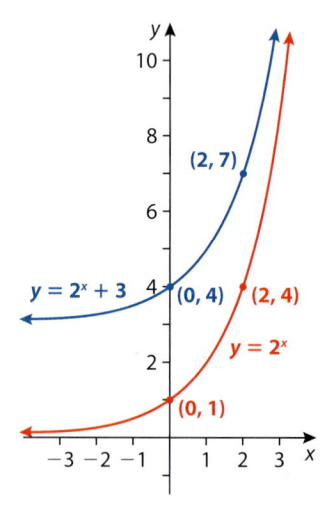

b) The graph of $h(x) = 2^{-x}$ can be obtained by reflecting the graph of $f(x) = 2^x$ across the x-axis. For function h, the domain is $x \in \mathbb{R}$ and the range is $y > 0$. The horizontal asymptote is $y = 0$ (x-axis).

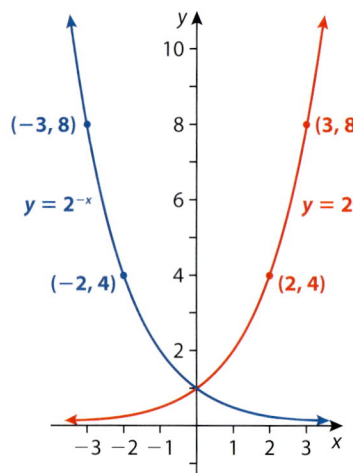

c) The graph of $p(x) = -2^x$ can be obtained by reflecting the graph of $f(x) = 2^x$ across the x-axis. For function p, the domain is $x \in \mathbb{R}$ and the range is $y < 0$. The horizontal asymptote is $y = 0$ (x-axis).

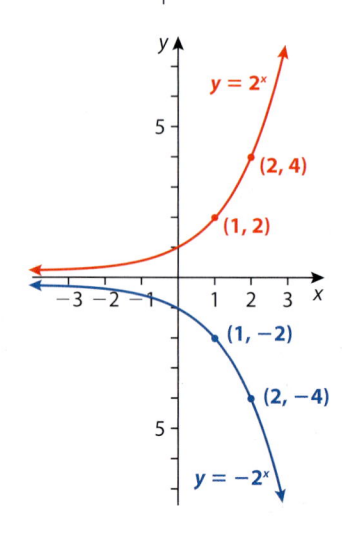

d) The graph of $r(x) = 2^{x-4}$ can be obtained by translating the graph of $f(x) = 2^x$ four units to the right. For function r, the domain is $x \in \mathbb{R}$ and the range is $y > 0$. The horizontal asymptote is $y = 0$ (x-axis).

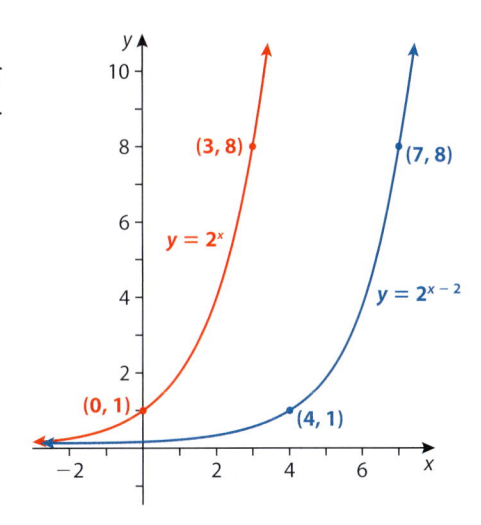

e) The graph of $v(x) = 3(2^x)$ can be obtained by a vertical stretch of the graph of $f(x) = 2^x$ by scale factor 3. For function v, the domain is $x \in \mathbb{R}$ and the range is $y > 0$. The horizontal asymptote is $y = 0$ (x-axis).

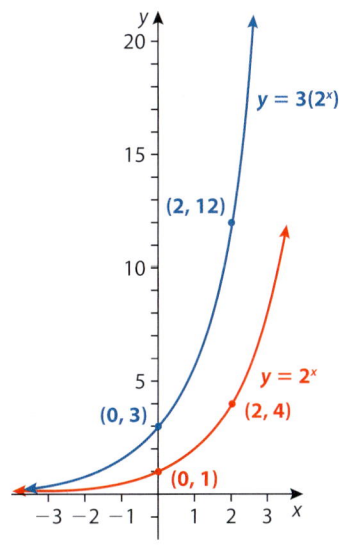

Note that for function p in part c) of Example 2 the horizontal asymptote is an **upper bound** (i.e. no function value is equal to or greater than $y = 0$). Whereas, in parts a), b), d) and e) the horizontal asymptote for each function is a **lower bound** (i.e. no function value is equal to or less than the y-value of the asymptote).

4.2 Exponential growth and decay

Mathematical models of growth and decay

Exponential functions are well suited as a mathematical model for a wide variety of steadily increasing or decreasing phenomena of many kinds, including population growth (or decline), investment of money with compound interest and radioactive decay. Recall from the previous chapter that the formula for finding terms in a geometric sequence (repeated multiplication by common ratio r) is an exponential function. Many instances of growth or decay occur geometrically (repeated multiplication by a growth or decay factor).

4 Exponential and Logarithmic Functions

> **Exponential models**
> Exponential models are equations of the form $A(t) = A_0 b^t$, where $A_0 \neq 0$, $b > 0$ and $b \neq 1$. $A(t)$ is the **amount after time t**. $A(0) = A_0 b^0 = A_0(1) = A_0$, so A_0 is called the **initial amount** or value (often the value at time $(t) = 0$). If $b > 1$, then $A(t)$ is an **exponential growth model**. If $0 < b < 1$, then $A(t)$ is an **exponential decay model**. The value of b, the base of the exponential function, is often called the **growth or decay factor**.

Example 3

A sample count of bacteria in a culture indicates that the number of bacteria is doubling every hour. Given that the estimated count at 15:00 was 12 000 bacteria, find the estimated count three hours earlier at 12:00 and write an exponential growth function for the number of bacteria at any hour t.

Solution

Consider the time at 12:00 to be the starting, or initial, time and label it $t = 0$ hours. Then the time at 15:00 is $t = 3$. The amount at any time t (in hours) will double after an hour so the growth factor, b, is 2. Therefore, $A(t) = A_0(2)^t$. Knowing that $A(3) = 12\,000$, compute A_0: $12\,000 = A_0(2)^3$ $\Rightarrow 12\,000 = 8A_0 \Rightarrow A_0 = 1500$
$A(t) = 1500(2)^t$

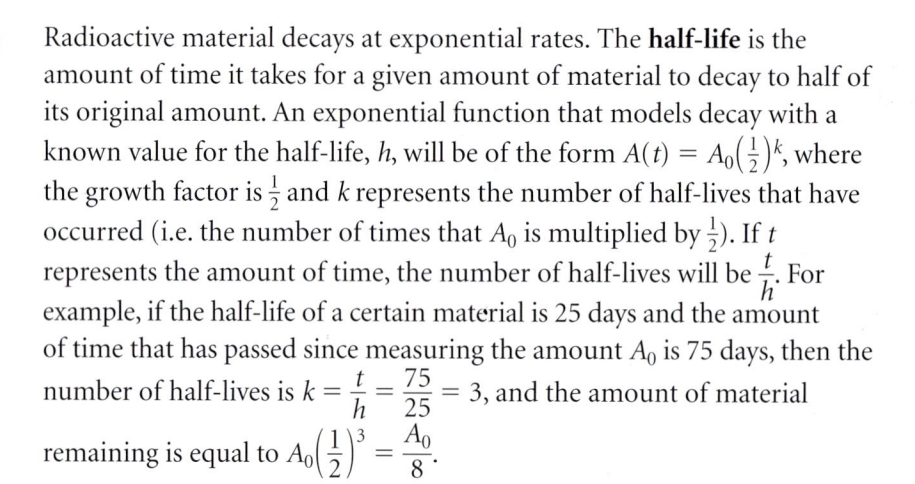

Radioactive material decays at exponential rates. The **half-life** is the amount of time it takes for a given amount of material to decay to half of its original amount. An exponential function that models decay with a known value for the half-life, h, will be of the form $A(t) = A_0\left(\frac{1}{2}\right)^k$, where the growth factor is $\frac{1}{2}$ and k represents the number of half-lives that have occurred (i.e. the number of times that A_0 is multiplied by $\frac{1}{2}$). If t represents the amount of time, the number of half-lives will be $\frac{t}{h}$. For example, if the half-life of a certain material is 25 days and the amount of time that has passed since measuring the amount A_0 is 75 days, then the number of half-lives is $k = \frac{t}{h} = \frac{75}{25} = 3$, and the amount of material remaining is equal to $A_0\left(\frac{1}{2}\right)^3 = \frac{A_0}{8}$.

> **Half-life formula**
> If a certain initial amount, A_0, of material decays with a half-life of h, the amount of material that remains at time t is given by the exponential decay model $A(t) = A_0\left(\frac{1}{2}\right)^{\frac{t}{h}}$. The time units (e.g. seconds, hours, years) for h and t must be the same.

Example 4

The half-life of radioactive carbon-14 is approximately 5730 years. How much of a 10 g sample of carbon-14 remains after 15 000 years?

Solution

The exponential decay model for the carbon-14 is $A(t) = A_0\left(\frac{1}{2}\right)^{\frac{t}{5730}}$. What remains of 10 g after 15 000 years is given by
$A(15\,000) = 10\left(\frac{1}{2}\right)^{\frac{15\,000}{5730}} \approx 1.63$ g.

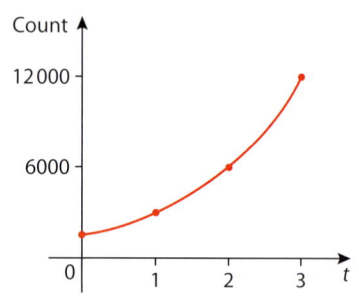

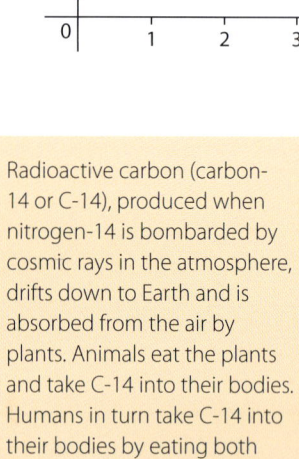

Radioactive carbon (carbon-14 or C-14), produced when nitrogen-14 is bombarded by cosmic rays in the atmosphere, drifts down to Earth and is absorbed from the air by plants. Animals eat the plants and take C-14 into their bodies. Humans in turn take C-14 into their bodies by eating both plants and animals. When a living organism dies, it stops absorbing C-14, and the C-14 that is already in the object begins to decay at a slow but steady rate, reverting to nitrogen-14. The half-life of C-14 is 5730 years. Half of the original amount of C-14 in the organic matter will have disintegrated after 5730 years; half of the remaining C-14 will have been lost after another 5730 years, and so forth. By measuring the ratio of C-14 to N-14, archaeologists are able to date organic materials. However, after about 50 000 years, the amount of C-14 remaining will be so small that the organic material cannot be dated reliably.

Compound interest

Recall from Chapter 3 that exponential functions occur in calculating compound interest. If an initial amount of money P, called the **principal**, is invested at an interest rate r per time period, then after one time period the amount of interest is $P \times r$ and the total amount of money is $A = P + Pr = P(1 + r)$. If the interest is added to the principal, the new principal is $P(1 + r)$, and the total amount after another time period is $A = P(1 + r)(1 + r) = P(1 + r)^2$. In the same way, after a third time period the amount is $A = P(1 + r)^3$. In general, after k periods the total amount is $A = P(1 + r)^k$, an exponential function with growth factor $1 + r$. For example, if the amount of money in a bank account is earning interest at a rate of 6.5% per time period, the growth factor is $1 + 0.065 = 1.065$. Is it possible for r to be negative? Yes, if an amount (not just money) is decreasing. For example, if the population of a town is decreasing by 12% per time period, the decay factor is $1 - 0.12 = 0.88$.

For compound interest, if the annual interest rate is r and interest is compounded (number of times added in) n times per year, then each time period the interest rate is $\frac{r}{n}$, and there are $n \times t$ time periods in t years.

> **Compound interest formula**
> The exponential function for calculating the amount of money after t years, $A(t)$, where P is the initial amount or principal, the annual interest rate is r and the number of times interest is compounded per year is n, is given by
> $$A(t) = P\left(1 + \frac{r}{n}\right)^{nt}$$

Example 5

An initial amount of 1000 euros is deposited into an account earning $5\frac{1}{4}\%$ interest per year. Find the amounts in the account after eight years if interest is compounded annually, semi-annually, quarterly, monthly and daily.

Solution

We use the exponential function associated with compound interest with values of $P = 1000$, $r = 0.0525$ and $t = 8$.

◀ Table 4.1

Compounding	n	Amount after 8 years
Annual	1	$1000\left(1 + \frac{0.0525}{1}\right)^8 = 1505.83$
Semi-annual	2	$1000\left(1 + \frac{0.0525}{2}\right)^{2(8)} = 1513.74$
Quarterly	4	$1000\left(1 + \frac{0.0525}{4}\right)^{4(8)} = 1517.81$
Monthly	12	$1000\left(1 + \frac{0.0525}{12}\right)^{12(8)} = 1520.57$
Daily	365	$1000\left(1 + \frac{0.0525}{365}\right)^{365(8)} = 1521.92$

4 Exponential and Logarithmic Functions

Example 6

A new car is purchased for $22 000. If the value of the car decreases (depreciates) at a rate of approximately 15% per year, what will be the approximate value of the car to the nearest whole dollar in $4\frac{1}{2}$ years?

Solution

The decay rate for the exponential function is $1 - r = 1 - 0.15 = 0.85$. In other words, after each year the car's value is 85% of what it was one year before. We use the exponential decay model $A(t) = A_0 b^t$ with values $A_0 = 22\,000$, $b = 0.85$ and $t = 4.5$.

$$A(4.5) = 22\,000(0.85)^{4.5} \approx 10\,588$$

The value of the car will be approximately $10 588.

Exercise 4.1 and 4.2

For questions 1–3, sketch a graph of the function and state its domain, range, y-intercept and the equation of its horizontal asymptote.

1 $f(x) = 3^{x+4}$
2 $g(x) = -2^x + 8$
3 $h(x) = 4^{-x} - 1$

4 If a general exponential function is written in the form $f(x) = a(b)^{x-c} + d$, state the domain, range, y-intercept and the equation of the horizontal asymptote in terms of the parameters a, b, c and d.

5 Using your GDC and a graph-viewing window with Xmin = −2, Xmax = 2, Ymin = 0 and Ymax = 4, sketch a graph for each exponential equation on the same set of axes.
   a) $y = 2^x$ b) $y = 4^x$ c) $y = 8^x$
   d) $y = 2^{-x}$ e) $y = 4^{-x}$ f) $y = 8^{-x}$

6 Write equations that are equivalent to the equations in 5 d), e) and f) but have an exponent of positive x rather than negative x.

7 If $1 < a < b$, which is steeper: the graph of $y = a^x$ or $y = b^x$?

8 The population of a city triples every 25 years. At time $t = 0$, the population is 100 000. Write a function for the population $P(t)$ as a function of t. What is the population after:
   a) 50 years   b) 70 years   c) 100 years?

9 An experiment involves a colony of bacteria in a solution. It is determined that the number of bacteria doubles approximately every 3 minutes and the initial number of bacteria at the start of the experiment is 10^4. Write a function for the number of bacteria $N(t)$ as a function of t (in minutes). Approximately how many bacteria are there after:
   a) 3 minutes   b) 9 minutes   c) 27 minutes   d) one hour?

10 If $10 000 is invested at an annual interest rate of 11%, compounded quarterly, find the value of the investment after the given number of years.
   a) 5 years   b) 10 years   c) 15 years

11 A sum of $5000 is deposited into an investment account that earns interest at a rate of 9% per year compounded monthly.
   a) Write the function $A(t)$ that computes the value of the investment after t years.
   b) Use your GDC to sketch a graph of $A(t)$ with values of t on the horizontal axis ranging from $t = 0$ years to $t = 25$ years.
   c) Use the graph on your GDC to determine the minimum number of years (to the nearest whole year) for this investment to have a value greater than $20 000.

12 If $10 000 is invested at an annual interest rate of 11% for a period of five years, find the value of the investment for the following compounding periods.
   a) annually   b) monthly   c) daily   d) hourly

13 Imagine a bank account that has the fantastic annual interest rate of 100%. If you deposit $1 into this account, how much will be in the account exactly one year later, for the following compounding periods?
   a) annually   b) monthly   c) daily   d) hourly   e) every minute

14 Each year for the past eight years, the population of deer in a national park increases at a steady rate of 3.2% per year. The present population is approximately 248 000.
   a) What was the approximate number of deer one year ago?
   b) What was the approximate number of deer eight years ago?

15 Radioactive carbon-14 has a half-life of 5730 years. The remains of an animal are found 20 000 years after it died. About what percentage (to 3 significant figures) of the original amount of carbon-14 (when the animal was alive) would you expect to find?

16 Once a certain drug enters the bloodstream of a human patient, it has a half-life of 36 hours. An amount of the drug, A_0, is injected in the bloodstream at 12:00 on Monday. How much of the drug will be in the bloodstream of the patient five days later at 12:00 on Friday?

17 Why are exponential functions of the form $f(x) = b^x$ defined so that $b > 0$?

18 You are offered a highly paid job that lasts for just one month – exactly 30 days. Which of the following payment plans, I or II, would give you the largest salary? How much would you get paid?
   I   One dollar on the first day of the month, two dollars on the second day, three dollars on the third day, and so on (getting paid one dollar more each day) until the end of the 30 days. (You would have a total of $55 after 10 days.)
   II One cent ($0.01) on the first day of the month, two cents ($0.02) on the second day, four cents on the third day, eight cents on the fourth day, and so on (each day getting paid double from the previous day) until the end of the 30 days. (You would have a total of $10.23 after 10 days.)

4.3 The number e

Recalling the definition of an exponential function $f(x) = b^x$, we recognize that any positive number can be used as the base b. Given that our number system is a base 10 system and that a base 2 number system (binary numbers) has useful applications (e.g. computers), it is understandable that exponential functions with base 2 or 10 are commonly used for modelling certain applications. However, the most important base is an irrational number that is denoted with the letter e. The value of e, approximated to 6 significant figures, is 2.71 828. The importance of e will be clearer when we get to calculus topics. The number π – another very useful irrational number – has a natural geometric significance as the ratio of circumference to diameter for any circle. Although not geometric, the number e also occurs in a 'natural' manner. We can see this by revisiting compound interest and considering **continuous change** rather than **incremental change**.

> The 'discovery' of the constant e is attributed to Jakob Bernoulli (1654–1705). He was a member of the famous Bernoulli family of distinguished mathematicians, scientists and philosophers. This included his brother Johann (1667–1748), who made important developments in calculus, and his nephew Daniel (1700–1782), who is most well known for Bernoulli's principle in physics. The constant e is of enormous mathematical significance – and it appears 'naturally' in many mathematical processes. Jakob Bernoulli first observed e when studying sequences of numbers in connection to compound interest problems.

Continuously compounded interest

In the previous section and in Chapter 3, we computed amounts of money resulting from an initial amount (principal) with interest being compounded (added in) at discrete intervals (e.g. yearly, monthly, daily). In the formula that we used, $A(t) = P\left(1 + \frac{r}{n}\right)^{nt}$, n is the number of times that interest is compounded per year. Instead of adding interest only at discrete intervals, let's investigate what happens if we try to add interest continuously – that is, let the value of n increase without bound ($n \to \infty$).

Consider investing just $1 at a very generous annual interest rate of 100%. How much will be in the account at the end of just one year? It depends on how often the interest is compounded. If it is only added at the end of the year ($n = 1$), the account will have $2 at the end of the year. Is it possible to compound the interest more often to get a one-year balance of $2.50 or of $3.00? We use the compound interest formula with $P = \$1$, $r = 1.00$ (100%) and $t = 1$, and compute the amounts for increasing values of n. $A(1) = 1\left(1 + \frac{1}{n}\right)^{n \cdot 1} = \left(1 + \frac{1}{n}\right)^n$. This can be done very efficiently on your GDC by entering the equation $y = \left(1 + \frac{1}{x}\right)^x$ to display a table showing function values of increasing values of x.

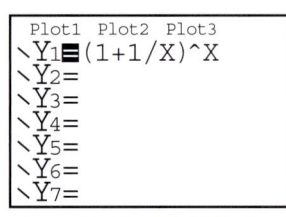

As the number of compounding periods during the year increases, the amount at the end of the year appears to approach a limiting value.

As $n \to \infty$, the quantity of $\left(1 + \frac{1}{n}\right)^n$ approaches the number e. To 13 decimal places, e is approximately 2.718 281 828 4590.

Table 4.2

Compounding	n	$A(1) = \left(1 + \frac{1}{n}\right)^n$
Annual	1	2
Semi-annual	2	2.25
Quarterly	4	2.441 406 25…
Monthly	12	2.613 035 290 22…
Daily	365	2.714 567 482 02…
Hourly	8 760	2.718 126 690 63…
Every minute	525 600	2.718 279 2154…
Every second	31 536 000	2.718 282 472 54…

Leonhard Euler (1701–1783) was the dominant mathematical figure of the 18th century and is one of the most influential and prolific mathematicians of all time. Euler's collected works fill over 70 large volumes. Nearly every branch of mathematics has significant theorems that are attributed to Euler.

Euler proved mathematically that the limit of $\left(1 + \frac{1}{n}\right)^n$ as n goes to infinity is precisely equal to an irrational constant which he labelled e. His mathematical writings were influential not just because of the content and quantity but also because of Euler's insistence on clarity and efficient mathematical notation. Euler introduced many of the common algebraic notations that we use today. Along with the symbol e for the base of natural logarithms (1727), Euler introduced $f(x)$ for a function (1734), i for the square root of negative one (1777), π for pi, Σ for summation (1755), and many others. His introductory algebra text, written originally in German (Euler was Swiss), is still available in English translation. Euler spent most of his working life in Russia and Germany. Switzerland honoured Euler by placing his image on the 10 Swiss franc banknote.

Definition of e

$$e = \lim_{n \to \infty} \left(1 + \frac{1}{n}\right)^n$$

The definition is read as 'e equals the limit of $\left(1 + \frac{1}{n}\right)^n$ as n goes to infinity'.

As the number of compoundings, n, increase without bound, we approach continuous compounding – where interest is being added continuously. In the formula for calculating amounts resulting from compound interest, letting $m = \frac{n}{r}$ produces

$$A(t) = P\left(1 + \frac{r}{n}\right)^{nt} = P\left(1 + \frac{1}{m}\right)^{mrt} = P\left[\left(1 + \frac{1}{m}\right)^m\right]^{rt}$$

Now if $n \to \infty$ and the interest rate r is constant, then $\frac{n}{r} = m \to \infty$. From the limit definition of e, we know that if $m \to \infty$, then $\left(1 + \frac{1}{m}\right)^m \to e$. Therefore, for continuous compounding, it follows that

$$A(t) = P\left[\left(1 + \frac{1}{m}\right)^n\right]^{rt} = P[e]^{rt}.$$

This result is part of the reason that e is the best choice for the base of an exponential function modelling change that occurs continually (e.g. radioactive decay) rather than in discrete intervals.

Continuous compound interest formula

The exponential function for calculating the amount of money after t years, $A(t)$, for interest compounded continuously, where P is the initial amount or principal and r is the annual interest rate, is given by $A(t) = Pe^{rt}$.

Example 7

An initial investment of 1000 euros earns interest at an annual rate of $7\frac{1}{2}\%$. Find the total amount after five years if the interest is compounded
a) quarterly, and b) continuously.

Solution

a) $A(t) = P\left(1 + \frac{r}{n}\right)^{nt} = 1000\left(1 + \frac{0.075}{4}\right)^{4 \cdot 5} = 1449.95$ euros

b) $A(t) = Pe^{rt} = 1000e^{0.075(5)} = 1454.99$ euros

4 Exponential and Logarithmic Functions

The natural exponential function and continuous change

For many applications involving continuous change, the most suitable choice for a mathematical model is an exponential function with a base having the value of e.

> **The natural exponential function**
> The natural exponential function is the function defined as
> $$f(x) = e^x$$
> As with other exponential functions, the domain of the natural exponential function is the set of all real numbers ($x \in \mathbb{R}$), and its range is the set of positive real numbers ($y > 0$). The natural exponential function is often referred to as *the* exponential function.

The formula developed for continuously compounded interest does not apply only to applications involving adding interest to financial accounts. It can be used to model growth or decay of a quantity that is changing *geometrically* (i.e. repeated multiplication by a constant ratio, or growth/decay factor) and the change is continuous, or approaching continuous. Another version of a formula for continuous change, which we will learn more about in calculus, follows.

> **Continuous exponential growth/decay**
> If an initial quantity C (when $t = 0$) grows or decays continuously at a rate r over a certain time period, the amount $A(t)$ after t time periods is given by the function $A(t) = Ce^{rt}$.
> If $r > 0$, the quantity is increasing (growing). If $r < 0$, the quantity is decreasing (decaying).

Example 8

A programme to reduce the number of rabbits has been taking place in a certain Australian city park. At the start of the programme there were 230 rabbits. After t years the number of rabbits, R, is modelled by $R = 230e^{-0.2t}$. How many rabbits are there after three years?

Solution

$R = 230e^{-0.2(3)} \approx 126.2$. There are approximately 126 rabbits after three years of the programme.

> **Exercise 4.3**
>
> 1 Use your GDC to graph the curve $y = \left(1 + \frac{1}{x}\right)^x$ and the horizontal line $y = 2.72$. Use a graph window so that x ranges from 0 to 20 and y ranges from 0 to 3. Describe the behaviour of the graph of $y = \left(1 + \frac{1}{x}\right)^x$. Will it ever intersect the graph of $y = 2.72$? Explain.
>
> 2 Two different banks, Bank A and Bank B, offer accounts with exactly the same annual interest rate of 6.85%. However, the account from Bank A has the interest compounded monthly whereas the account from Bank B compounds the interest continuously. To decide which bank to open an account with, you

calculate the **amount of interest** you would earn after three years from an initial deposit of 500 euros in each bank's account. It is assumed that you make no further deposits and no withdrawals during the three years. How much interest would you earn from each of the accounts? Which bank's account earns more – and how much more?

3 Dina wishes to deposit $1000 into an investment account and then withdraw the total in the account in five years. She has the choice of two different accounts. *Blue Star account*: interest is earned at an annual interest rate of 6.13% compounded weekly (52 weeks in a year). *Red Star account*: interest is earned at an annual interest rate of 5.95% compounded continuously. Which investment account – *Blue Star* **or** *Red Star* – will result in the greatest total at the end of five years? What is the total after five years for this account? How much more is it than the total for the other account?

4 Strontium-90 is a radioactive isotope of strontium. Strontium-90 decays according to the function $A(t) = Ce^{-0.0239t}$, where t is time in years and C is the initial amount of strontium-90 when $t = 0$. If you have 1 kilogram of strontium-90 to start with, how much (approximated to 3 significant figures) will you have after:

a) 1 year?
b) 10 years?
c) 100 years?
d) 250 years?

5 A radioactive substance decays in such a way that the mass (in kilograms) remaining after t days is given by the function $A(t) = 5e^{-0.0347t}$.

a) Find the mass (i.e. initial mass) at time $t = 0$.
b) What **percentage** of the initial mass remains after 10 days?
c) On your GDC and then on paper, draw a graph of the function $A(t)$ for $0 \leq t \leq 50$.
d) Use one of your graphs to approximate, to the nearest whole day, the half-life of the radioactive substance.

6 Which of the given interest rates and compounding periods would provide the better investment?

a) $8\frac{1}{2}$% per year, compounded semi-annually
b) $8\frac{1}{4}$% per year, compounded quarterly
c) 8% per year, compounded continuously

4.4 Logarithmic functions

In Example 7 of the previous section, we used the equation $A(t) = 1000e^{0.075t}$ to compute the amount of money in an account after t years. Now suppose we wish to determine how much time, t, it takes for the initial investment of 1000 euros to double. To find this we need to solve the following equation for t: $2000 = 1000e^{0.075t} \Rightarrow 2 = e^{0.075t}$. The unknown t is in the exponent. At this point in the book, we do not have an algebraic method to solve such an equation, but developing the concept of a **logarithm** will provide us with the means to do so.

4 Exponential and Logarithmic Functions

John Napier (1550–1617) was a Scottish landowner, scholar and mathematician who 'invented' logarithms – a word he coined which derives from two Greek words: *logos* – meaning ratio, and *arithmos* – meaning number. Logarithms made numerical calculations much easier in areas such as astronomy, navigation, engineering and warfare. English mathematician Henry Briggs (1561–1630) came to Scotland to work with Napier and together they perfected logarithms, which included the idea of using the base ten. After Napier died in 1617, Briggs took over the work on logarithms and published a book of tables in 1624. By the second half of the 17th century, the use of logarithms had spread around the world. They became as popular as electronic calculators in our time. The great French mathematician Pierre-Simon Laplace (1749–1827) even suggested that the logarithms of Napier and Briggs doubled the life of astronomers, because it so greatly reduced the labours of calculation. In fact, without the invention of logarithms it is difficult to imagine how Kepler and Newton could have made their great scientific advances. In 1621, an English mathematician and clergyman, William Oughtred (1574–1660) used logarithms as the basis for the invention of the slide rule. The slide rule was a very effective calculation tool that remained in common use for over three hundred years.

The inverse of an exponential function

For $b > 1$, an exponential function with base b is increasing for all x, and for $0 < b < 1$ an exponential function is decreasing for all x. It follows from this that all exponential functions must be one-to-one. Recall from Section 2.3 that a one-to-one function passes both a vertical line test and a horizontal line test. We demonstrated that an inverse function would exist for any one-to-one function. Therefore, an exponential function with base b such that $b > 0$ and $b \neq 1$ will have an inverse function, which is given in the following definition. Also recall from Section 2.3 that the domain of a function $f(x)$ is the range of its inverse function $f^{-1}(x)$, and, similarly, the range of $f(x)$ is the domain of $f^{-1}(x)$. The domain and range are switched around for a function and its inverse.

> **Definition of a logarithmic function**
> For $b > 0$ and $b \neq 1$, the **logarithmic function** $y = \log_b x$ (read as 'logarithm with base b of x') is the inverse of the exponential function with base b.
>
> $$y = \log_b x \text{ if and only if } x = b^y$$
>
> The domain of the logarithmic function $y = \log_b x$ is the set of positive real numbers ($x > 0$) and its range is all real numbers ($y \in \mathbb{R}$).

Logarithmic expressions and equations

When evaluating logarithms, note that *a logarithm is an exponent*. This means that the value of $\log_b x$ is the exponent to which b must be raised to obtain x. For example, $\log_2 8 = 3$ because 2 must be raised to the power of 3 to obtain 8 – that is, $\log_2 8 = 3$ if and only if $2^3 = 8$.

We can use the definition of a logarithmic function to translate a logarithmic equation into an exponential equation and vice versa. When doing this, it is helpful to remember, as the definition stated, that in either form – logarithmic or exponential – the base is the same.

logarithmic equation
exponent
$$y = \log_b(x)$$
base

exponential equation
exponent
$$x = b^y$$
base

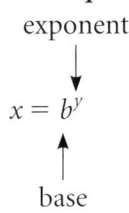

Example 9

Find the value of each of the following logarithms.

a) $\log_7 49$ b) $\log_5(\frac{1}{5})$ c) $\log_6 \sqrt{6}$ d) $\log_4 64$ e) $\log_{10} 0.001$

Solution

For each logarithmic expression in a) to e), we set it equal to y and use the definition of a logarithmic function to obtain an equivalent equation in exponential form. We then solve for y by applying the logical fact that if $b > 0$, $b \neq 1$ and $b^y = b^k$ then $y = k$.

a) Let $y = \log_7 49$ which is equivalent to the exponential equation $7^y = 49$. Since $49 = 7^2$, then $7^y = 7^2$. Therefore, $y = 2 \Rightarrow \log_7 49 = 2$.

b) Let $y = \log_5(\frac{1}{5})$ which is equivalent to the exponential equation $5^y = \frac{1}{5}$. Since $\frac{1}{5} = 5^{-1}$, then $5^y = 5^{-1}$. Therefore, $y = -1 \Rightarrow \log_5(\frac{1}{5}) = -1$.

c) Let $y = \log_6 \sqrt{6}$ which is equivalent to the exponential equation $6^y = \sqrt{6}$. Since $\sqrt{6} = 6^{\frac{1}{2}}$, then $6^y = 6^{\frac{1}{2}}$. Therefore, $y = \frac{1}{2} \Rightarrow \log_6 \sqrt{6} = \frac{1}{2}$.

d) Let $y = \log_4 64$ which is equivalent to the exponential equation $4^y = 64$. Since $64 = 4^3$, then $4^y = 4^3$. Therefore, $y = 3 \Rightarrow \log_4 64 = 3$.

e) Let $y = \log_{10} 0.001$ which is equivalent to the exponential equation $10^y = 0.001$. Since $0.001 = \frac{1}{1000} = \frac{1}{10^3} = 10^{-3}$, then $10^y = 10^{-3}$. Therefore, $y = -3 \Rightarrow \log_{10} 0.001 = -3$.

Example 10

Find the domain of the function $f(x) = \log_2(4 - x^2)$.

Solution

From the definition of a logarithmic function the domain of $y = \log_b x$ is $x > 0$, thus for $f(x)$ it follows that
$4 - x^2 > 0 \Rightarrow (2 + x)(2 - x) > 0 \Rightarrow -2 < x < 2$.
Hence, the domain is $-2 < x < 2$.

Properties of logarithms

As with all functions and their inverses, their graphs are reflections of each other over the line $y = x$. Figure 4.4 illustrates this relationship for exponential and logarithmic functions, and also confirms the domain and range for the logarithmic function stated in the previous definition.

4 Exponential and Logarithmic Functions

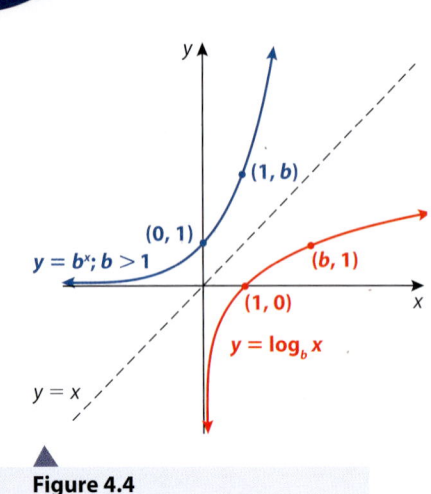

Figure 4.4

Notice that the points $(0, 1)$ and $(1, 0)$ are mirror images of each other over the line $y = x$. This corresponds to the fact that since $b^0 = 1$ then $\log_b 1 = 0$. Another pair of mirror image points, $(1, b)$ and $(b, 1)$, highlight the fact that $\log_b b = 1$.

Notice also that since the x-axis is a horizontal asymptote of $y = b^x$, the y-axis is a vertical asymptote of $y = \log_b x$.

In Section 2.3, we established that a function f and its inverse function f^{-1} satisfy the equations

$$f^{-1}(f(x)) = x \qquad \text{for } x \text{ in the domain of } f$$
$$f(f^{-1}(x)) = x \qquad \text{for } x \text{ in the domain of } f^{-1}$$

When applied to $f(x) = b^x$ and $f^{-1}(x) = \log_b x$, these equations become

$$\log_b(b^x) = x \qquad x \in \mathbb{R}$$
$$b^{\log_b x} = x \qquad x > 0$$

> **Properties of logarithms I**
> For $b > 0$ and $b \neq 1$, the following statements are true:
> 1. $\log_b 1 = 0$ (because $b^0 = 1$)
> 2. $\log_b b = 1$ (because $b^1 = b$)
> 3. $\log_b(b^x) = x$ (because $b^x = b^x$)
> 4. $b^{\log_b x} = x$ (because $\log_b x$ is the power to which b must be raised to get x)

The logarithmic function with base 10 is called the **common logarithmic function**. On calculators and on your GDC, this function is denoted by **log**. The value of the expression $\log_{10} 1000$ is 3 because 10^3 is 1000. Generally, for common logarithms (i.e. base 10) we omit writing the base of 10. Hence, if **log** is written with no base indicated, it is assumed to have a base of 10. For example, $\log 0.01 = -2$.

Common logarithm: $\log_{10} x = \log x$

As with exponential functions, the most widely used logarithmic function – and the other logarithmic function supplied on all calculators – is the logarithmic function with the base of e. This function is known as the **natural logarithmic function** and it is the inverse of the natural exponential function $y = e^x$. The natural logarithmic function is denoted by the symbol **ln**, and the expression $\ln x$ is read as 'the natural logarithm of x'.

Natural logarithm: $\log_e x = \ln x$

Example 11

Evaluate the following expressions:

a) $\log\left(\frac{1}{10}\right)$ b) $\log(\sqrt{10})$ c) $\log 1$ d) $10^{\log 47}$ e) $\log 50$

f) $\ln e$ g) $\ln\left(\frac{1}{e^3}\right)$ h) $\ln 1$ i) $e^{\ln 5}$ j) $\ln 50$

Solution

a) $\log\left(\frac{1}{10}\right) = \log(10^{-1}) = -1$
b) $\log(\sqrt{10}) = \log(10^{\frac{1}{2}}) = \frac{1}{2}$
c) $\log 1 = \log(10^0) = 0$
d) $10^{\log 47} = 47$
e) $\log 50 \approx 1.699$ (using GDC)
f) $\ln e = 1$
g) $\ln\left(\frac{1}{e^3}\right) = \ln(e^{-3}) = -3$
h) $\ln 1 = \ln(e^0) = 0$
i) $e^{\ln 5} = 5$
j) $\ln 50 \approx 3.912$ (using GDC)

Example 12

The diagram shows the graph of the line $y = x$ and two curves. Curve A is the graph of the equation $y = \log x$. Curve B is the reflection of curve A in the line $y = x$.
a) Write the equation for curve B.
b) Write the coordinates of the y-intercept of curve B.

Solution

a) Curve A is the graph of $y = \log x$, the common logarithm with base 10, which could also be written as $y = \log_{10} x$. Curve B is the inverse of $y = \log_{10} x$, since it is the reflection of it in the line $y = x$. Hence, the equation for curve B is the exponential equation $y = 10^x$.

b) The y-intercept occurs when $x = 0$. For curve B, $y = 10^0 = 1$. Therefore, the y-intercept for curve B is $(0, 1)$.

The logarithmic function with base b is the inverse of the exponential function with base b. Therefore, it makes sense that the laws of exponents (Section 1.3) should have corresponding properties involving logarithms. For example, the exponential property $b^0 = 1$ corresponds to the logarithmic property $\log_b 1 = 0$. We will state and prove three further important logarithmic properties that correspond to the following three exponential properties.

1. $b^m \cdot b^n = b^{m+n}$
2. $\dfrac{b^m}{b^n} = b^{m-n}$
3. $(b^m)^n = b^{mn}$

Properties of logarithms II

Given $M > 0$, $N > 0$ and k is any real number, the following properties are true for logarithms with $b > 0$ and $b \neq 1$.

Property	Description
1. $\log_b(MN) = \log_b M + \log_b N$	the log of a product is the sum of the logs of its factors
2. $\log_b\left(\dfrac{M}{N}\right) = \log_b M - \log_b N$	the log of a quotient is the log of the numerator minus the log of the denominator
3. $\log_b(M^k) = k \log_b M$	the log of a number raised to an exponent is the exponent times the log of the number

Any of these properties can be applied in either direction.

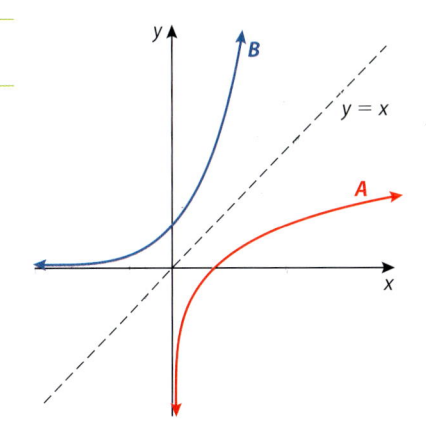

Exponential and Logarithmic Functions

Proofs

Property 1: Let $x = \log_b M$ and $y = \log_b N$.

The corresponding exponential forms of these two equations are
$$b^x = M \text{ and } b^y = N$$
Then, $\log_b(MN) = \log_b(b^x b^y) = \log_b(b^{x+y}) = x + y$.

It's given that $x = \log_b M$ and $y = \log_b N$; hence, $x + y = \log_b M + \log_b N$.

Therefore, $\log_b(MN) = \log_b M + \log_b N$.

Property 2: Again, let $x = \log_b M$ and $y = \log_b N \Rightarrow b^x = M$ and $b^y = N$.

Then, $\log_b\left(\dfrac{M}{N}\right) = \log_b\left(\dfrac{b^x}{b^y}\right) = \log_b(b^{x-y}) = x - y$.

With $x = \log_b M$ and $y = \log_b N$, then $x - y = \log_b M - \log_b N$.

Therefore, $\log_b\left(\dfrac{M}{N}\right) = \log_b M - \log_b N$.

Property 3: Let $x = \log_b M \Rightarrow b^x = M$.

Now, let's take the logarithm of M^k and substitute b^x for M:
$$\log_b(M^k) = \log_b[(b^x)^k] = \log_b(b^{kx}) = kx$$
It's given that $x = \log_b M$; hence, $kx = k\log_b M$.

Therefore, $\log_b(M^k) = k\log_b M$.

• **Hint:** The notation $f(x)$ uses brackets *not* to indicate multiplication but to indicate the argument of the function f. The symbol f is the name of a function, not a variable – it is not multiplying the variable x. Therefore, $f(x + y)$ is NOT equal to $f(x) + f(y)$. Likewise, the symbol **log** is also the name of a function. Therefore, $\log_b(x + y)$ is not equal to $\log_b(x) + \log_b(y)$. Other mistakes to avoid include incorrectly simplifying quotients or powers of logarithms. Specifically, $\dfrac{\log_b x}{\log_b y} \neq \log\left(\dfrac{x}{y}\right)$ and $(\log_b x)^k \neq k(\log_b x)$.

Example 13

Use the properties of logarithms to write each logarithmic expression as a sum, difference, and/or constant multiple of simple logarithms (i.e. logarithms without sums, products, quotients or exponents).

a) $\log_2(8x)$
b) $\ln\left(\dfrac{3}{y}\right)$
c) $\log(\sqrt{7})$
d) $\log_b\left(\dfrac{x^3}{y^2}\right)$
e) $\ln(5e^2)$
f) $\log\left(\dfrac{m+n}{n}\right)$

Solution

a) $\log_2(8x) = \log_2 8 + \log_2 x = 3 + \log_2 x$

b) $\ln\left(\dfrac{3}{y}\right) = \ln 3 - \ln y$

c) $\log(\sqrt{7}) = \log(7^{\frac{1}{2}}) = \frac{1}{2}\log 7$

d) $\log_b\left(\dfrac{x^3}{y^2}\right) = \log_b(x^3) - \log_b(y^2) = 3\log_b x - 2\log_b y$

e) $\ln(5e^2) = \ln 5 + \ln(e^2) = \ln 5 + 2\ln e = \ln 5 + 2(1) = 2 + \ln 5$
   $(2 + \ln 5 \approx 3.609 \text{ using GDC})$

f) $\log\left(\dfrac{m+n}{m}\right) = \log(m+n) - \log m$
   (remember $\log(m + n) \neq \log m + \log n$)

Example 14

Write each expression as the logarithm of a single quantity.
a) $\log 6 + \log x$
b) $\log_2 5 + 2 \log_2 3$
c) $\ln y - \ln 4$
d) $\log_b 12 - \frac{1}{2} \log_b 9$
e) $\log_3 M + \log_3 N - 2 \log_3 P$
f) $\log_2 80 - \log_2 5$

Solution

a) $\log 6 + \log x = \log(6x)$

b) $\log_2 5 + 2 \log_2 3 = \log_2 5 + \log_2(3^2) = \log_2 5 + \log_2 9 = \log_2(5 \cdot 9)$
$= \log_2 45$

c) $\ln y - \ln 4 = \ln\left(\dfrac{y}{4}\right)$

d) $\log_b 12 - \frac{1}{2} \log_b 9 = \log_b 12 - \log_b(9^{\frac{1}{2}}) = \log_b 12 - \log_b(\sqrt{9})$
$= \log_b 12 - \log_b 3 = \log_b\left(\dfrac{12}{3}\right) = \log_b 4$

e) $\log_3 M + \log_3 N - 2 \log_3 P = \log_3(MN) - \log_3(P^2) = \log_3\left(\dfrac{MN}{P^2}\right)$

f) $\log_2 80 - \log_2 5 = \log_2\left(\dfrac{80}{5}\right) = \log_2 16 = 4$ (because $2^4 = 16$)

Change of base

The answer to part f) of Example 14 was $\log_2 16$ which we can compute to be exactly 4 because we know that $2^4 = 16$. The answer to part e) of Example 13 was $2 + \ln 5$ which we approximated to 3.609 using the natural logarithm function key (**ln**) on our GDC. But, what if we wanted to compute an approximate value for $\log_2 45$, the answer to part b) of Example 14? Our GDC can only evaluate common logarithms (base 10) and natural logarithms (base e). To evaluate logarithmic expressions and graph logarithmic functions to other bases we need to apply a **change of base formula.**

> **Change of base formula**
> Let a, b and x be positive real numbers such that $a \neq 1$ and $b \neq 1$. Then $\log_b x$ can be expressed in terms of logarithms to any other base a as follows:
> $$\log_b x = \dfrac{\log_a x}{\log_a b}$$

Proof

$y = \log_b x \Rightarrow b^y = x$ convert from logarithmic form to exponential form

$\log_a x = \log_a(b^y)$ if $b^y = x$, then log of each with same bases must be equal

$\log_a x = y \log_a b$ applying the property $\log_b(M^k) = k \log_b M$

$y = \dfrac{\log_a x}{\log_a b}$ divide both sides by $\log_a b$

$\log_b x = \dfrac{\log_a x}{\log_a b}$ substitute $\log_b x$ for y

Exponential and Logarithmic Functions

To apply the change of base formula, let $a = 10$ or $a = e$. Then the logarithm of any base b can be expressed in terms of either common logarithms or natural logarithms. For example:

$$\log_2 x = \frac{\log x}{\log 2} \quad \text{or} \quad \frac{\ln x}{\ln 2}$$

$$\log_5 x = \frac{\log x}{\log 5} \quad \text{or} \quad \frac{\ln x}{\ln 5}$$

$$\log_2 45 = \frac{\log 45}{\log 2} = \frac{\ln 45}{\ln 2} \approx 5.492 \quad \text{(using GDC)}$$

Example 15

Use the change of base formula and common or natural logarithms to evaluate each logarithmic expression. Start by making a rough mental estimate. Approximate your answer to 4 significant figures.

a) $\log_3 30$

b) $\log_9 6$

Solution

a) The value of $\log_3 30$ is the power to which 3 is raised to obtain 30. Because $3^3 = 27$ and $3^4 = 81$, the value of $\log_3 30$ is between 3 and 4, and will be much closer to 3 than 4 – perhaps around 3.1. Using the change of base formula and common logarithms, we obtain
$\log_3 30 = \frac{\log 30}{\log 3} \approx 3.096$. This agrees well with the mental estimate.

After computing the answer on your GDC, use your GDC to also check it by raising 3 to the answer and confirming that it gives a result of 30.

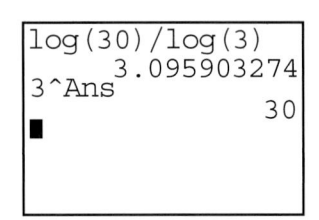

```
log(30)/log(3)
            3.095903274
3^Ans
                     30
```

b) The value of $\log_9 6$ is the power to which 9 is raised to obtain 6. Because $9^{\frac{1}{2}} = \sqrt{9} = 3$ and $9^1 = 9$, the value of $\log_9 6$ is between $\frac{1}{2}$ and 1 – perhaps around 0.75. Using the change of base formula and natural logarithms, we obtain $\log_9 6 = \frac{\ln 6}{\ln 9} \approx 0.815$. This agrees well with the mental estimate.

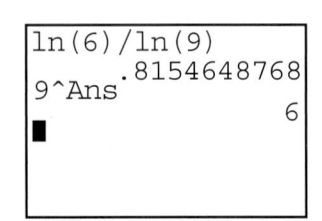

```
ln(6)/ln(9)
            .8154648768
9^Ans
                      6
```

Exercise 4.4

In questions 1–9, express each logarithmic equation as an exponential equation.

1 $\log_2 16 = 4$ **2** $\ln 1 = 0$ **3** $\log 100 = 2$

4 $\log 0.01 = -2$ **5** $\log_7 343 = 3$ **6** $\ln\left(\frac{1}{e}\right) = -1$

7 $\log 50 = y$ **8** $\ln x = 12$ **9** $\ln(x+2) = 3$

In questions 10–18, express each exponential equation as a logarithmic equation.

10 $2^{10} = 1024$ **11** $10^{-4} = 0.0001$ **12** $4^{-\frac{1}{2}} = \frac{1}{2}$

13 $3^4 = 81$ **14** $10^0 = 1$ **15** $e^x = 5$

16 $2^{-3} = 0.125$ **17** $e^4 = y$ **18** $10^{x+1} = y$

In questions 19–34, find the exact value of the expression without using your GDC.

19 $\log_2 64$ **20** $\log_4 64$ **21** $\log_2\left(\frac{1}{8}\right)$ **22** $\log_3(3^5)$

23 $\log_8 1$ **24** $10^{\log 6}$ **25** $\log_3\left(\frac{1}{27}\right)$ **26** $e^{\ln\sqrt{2}}$

27 $\log 1000$ **28** $\ln(\sqrt{e})$ **29** $\ln\left(\frac{1}{e^2}\right)$ **30** $\log 0.001$

31 $\log_4 2$ **32** $3^{\log_3 18}$ **33** $\log_5(\sqrt[3]{5})$ **34** $10^{\log \pi}$

In questions 35–42, use a GDC to evaluate the expression, correct to 4 significant figures.

35 $\log 50$ **36** $\log \sqrt{3}$ **37** $\ln 50$ **38** $\ln \sqrt{3}$

39 $\log 25$ **40** $\log\left(\frac{1+\sqrt{5}}{2}\right)$ **41** $\ln 100$ **42** $\ln(100^3)$

In questions 43–45, find the domain of the logarithmic function.

43 $f(x) = \log(x-2)$ **44** $g(x) = \ln(x^2)$ **45** $h(x) = \log(x) - 2$

For questions 46–49, find the equation of the function that is graphed in the form $f(x) = \log_b x$.

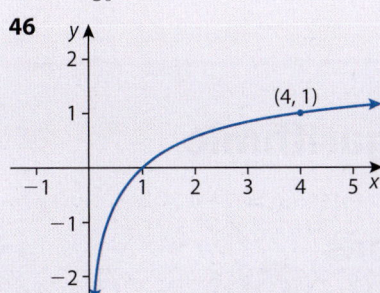

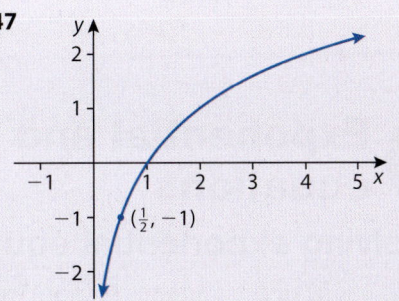

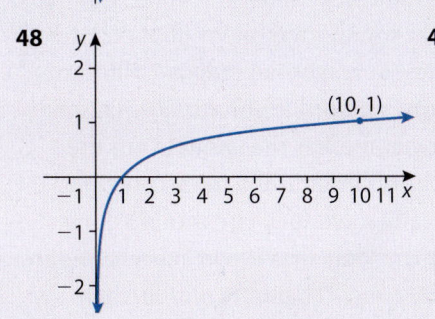

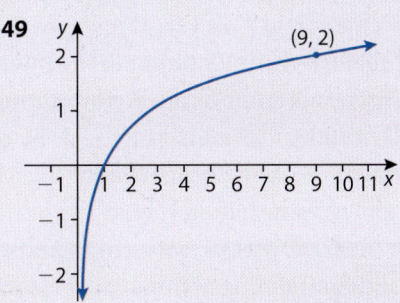

46 Graph passing through $(4, 1)$.

47 Graph passing through $\left(\frac{1}{2}, -1\right)$.

48 Graph passing through $(10, 1)$.

49 Graph passing through $(9, 2)$.

4 Exponential and Logarithmic Functions

In questions 50–55, use properties of logarithms to write each logarithmic expression as a sum, difference and/or constant multiple of simple logarithms (i.e. logarithms without sums, products, quotients or exponents).

50 $\log_2(2m)$

51 $\log\left(\dfrac{9}{x}\right)$

52 $\ln(\sqrt[5]{x})$

53 $\log_3(ab^3)$

54 $\log[10x(1+r)^t]$

55 $\ln\left(\dfrac{m^3}{n}\right)$

In questions 56–61, write each expression as the logarithm of a single quantity.

56 $\log(x^2) + \log\left(\dfrac{1}{x}\right)$

57 $\log_3 9 + 3\log_3 2$

58 $4\ln y - \ln 4$

59 $\log_b 12 - \dfrac{1}{2}\log_b 9$

60 $\log p - \log q - \log r$

61 $2\ln 6 - 1$ • **Hint:** $\ln(?) = 1$

In questions 62–65, use the change of base formula and common or natural logarithms to evaluate each logarithmic expression. Approximate your answer to 3 significant figures.

62 $\log_2 1000$

63 $\log_{\frac{1}{2}} 40$

64 $\log_6 40$

65 $\log_5(0.75)$

In questions 66 and 67, use the change of base formula to evaluate $f(20)$.

66 $f(x) = \log_2 x$

67 $f(x) = \log_5 x$

68 Use the change of base formula to prove the following statement.
$$\log_b a = \dfrac{1}{\log_a b}$$

69 Show that $\log e = \dfrac{1}{\ln 10}$.

70 The relationship between the number of decibels dB (one variable) and the intensity I of a sound (in watts per square metre) is given by the formula $dB = 10\log\left(\dfrac{I}{10^{-16}}\right)$. Use properties of logarithms to write the formula in simpler form. Then find the number of decibels of a sound with an intensity of 10^{-4} watts per square metre.

4.5 Exponential and logarithmic equations

Solving exponential equations

At the start of the previous section, we wanted to find a way to determine how much time t (in years) it would take for an investment of 1000 euros to double, if the investment earns interest at an annual rate of 7.5%. Since the interest is compounded continuously, we need to solve this equation: $2000 = 1000e^{0.075t} \Rightarrow 2 = e^{0.075t}$. The equation has the variable t in the exponent. With the properties of logarithms established in the previous section, we now have a way to algebraically solve such equations. Along with these properties, we need to apply the logic that if two expressions are equal then their logarithms must also be equal. That is, if $m = n$, then $\log_b m = \log_b n$.

Example 16

Solve the equation for the variable t. Give your answer accurate to 3 significant figures.
$$2 = e^{0.075t}$$

Solution

$$2 = e^{0.075t}$$
$\ln 2 = \ln(e^{0.075t})$ take natural logarithm of both sides
$\ln 2 = 0.075t$ apply the property $\log_b(b^x) = x$ and $\ln e = 1$
$$t = \frac{\ln 2}{0.075} \approx 9.24$$

With interest compounding continuously at an annual interest rate of 7.5%, it takes about 9.24 years for the investment to double.

This example serves to illustrate a general strategy for solving exponential equations. To solve an exponential equation, first isolate the exponential expression and take the logarithm of both sides. Then apply a property of logarithms so that the variable is no longer in the exponent and it can be isolated on one side of the equation. By taking the logarithm of both sides of an exponential equation, we are making use of the inverse relationship between exponential and logarithmic functions. Symbolically, this method can be represented as follows – solving for x:

(i) If $b = 10$ or e: $y = b^x \Rightarrow \log_b y = \log_b b^x \Rightarrow \log_b y = x$

(ii) If $b \neq 10$ or e:
$y = b^x \Rightarrow \log_a y = \log_a b^x \Rightarrow \log_a y = x \log_a b \Rightarrow x = \dfrac{\log_a y}{\log_a b}$

Example 17

Solve for x in the equation $3^{x-4} = 24$. Approximate the answer to 3 significant figures.

Solution

$3^{x-4} = 24$
$\log(3^{x-4}) = \log 24$ take common logarithm of both sides
$(x-4)\log 3 = \log 24$ apply the property $\log_b(M^k) = k \log_b M$
$x - 4 = \dfrac{\log 24}{\log 3}$ divide both sides by $\log 3$ $\left[\text{note: } \dfrac{\log 24}{\log 3} \neq \log 8 \right]$
$x = \dfrac{\log 24}{\log 3} + 4$
$x \approx 6.89$ using GDC

• **Hint:** We could have used natural logarithms instead of common logarithms to solve the equation in Example 17. Using the same method but with natural logarithms, we get
$$x = \frac{\ln 24}{\ln 3} + 4 \approx 6.89.$$

Recall Example 10 in Section 3.3 in which we solved an exponential equation graphically, because we did not yet have the tools to solve it algebraically. Let's solve it now using logarithms.

Exponential and Logarithmic Functions

• **Hint:** Be sure to use brackets appropriately when entering the expression $\frac{\ln 2}{4 \ln 1.015}$ on your GDC. Following the rules for order of operations, your GDC will give an incorrect result if entered as shown here.

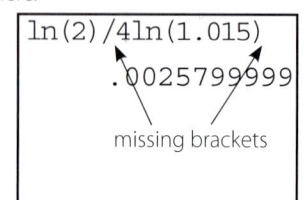

```
ln(2)/4ln(1.015)
           .0025799999
```
missing brackets

Example 18

You invested €1000 at 6% compounded quarterly. How long will it take this investment to increase to €2000?

Solution

Using the compound interest formula from Section 4.2, $A(t) = P\left(1 + \frac{r}{n}\right)^{nt}$, with $P = €1000$, $r = 0.06$ and $n = 4$, we need to solve for t when $A(t) = 2P$.

$$2P = P\left(1 + \frac{0.06}{4}\right)^{4t} \quad \text{substitute } 2P \text{ for } A(t)$$

$$2 = 1.015^{4t} \quad \text{divide both sides by } P$$

$$\ln 2 = \ln(1.015^{4t}) \quad \text{take natural logarithm of both sides}$$

$$\ln 2 = 4t \ln 1.015 \quad \text{apply the property } \log_b(M^k) = k \log_b M$$

$$t = \frac{\ln 2}{4 \ln 1.015}$$

$$t \approx 11.6389 \quad \text{evaluated on GDC}$$

The investment will double in 11.64 years – about 11 years and 8 months.

```
ln(2)/(4ln(1.015
))
           11.63888141
```

Example 19

The bacteria that cause 'strep throat' will grow in number at a rate of about 2.3% per minute. To the nearest whole minute, how long will it take for these bacteria to double in number?

Solution

Let t represent time in minutes and let A_0 represent the number of bacteria at $t = 0$.

Using the exponential growth model from Section 4.2, $A(t) = A_0 b^t$, the growth factor, b, is $1 + 0.023 = 1.023$ giving $A(t) = A_0(1.023)^t$. The same equation would apply to money earning 2.3% annual interest with the money being added (compounded) once per year rather than once per minute. So, our mathematical model assumes that the number of bacteria increase incrementally, with the number increasing by 2.3% at the end of each minute. To find the doubling time, find the value of t so that $A(t) = 2A_0$.

$$2A_0 = A_0(1.023)^t \quad \text{substitute } 2A_0 \text{ for } A(t)$$

$$2 = 1.023^t \quad \text{divide both sides by } A_0$$

$$\ln 2 = \ln(1.023^t) \quad \text{take natural logarithm of both sides}$$

$$\ln 2 = t \ln 1.023 \quad \text{apply the property } \log_b(M^k) = k \log_b M$$

$$t = \frac{\ln 2}{\ln 1.023} \approx 30.482$$

The number of bacteria will double in about 30 minutes.

Alternative solution

What if we assumed continuous growth instead of incremental growth? We apply the continuous exponential growth model from Section 4.3: $A(t) = Ce^{rt}$ with initial amount C and $r = 0.023$.

$$2C = Ce^{0.023t} \qquad \text{substitute } 2C \text{ for } A(t)$$

$$2 = e^{0.023t} \qquad \text{divide both sides by } C$$

$$\ln 2 = \ln(e^{0.023t}) \qquad \text{take natural logarithm of both sides}$$

$$\ln 2 = 0.023t \qquad \text{apply the property } \log_b(b^x) = x$$

$$t = \frac{\ln 2}{0.023} \approx 30.137$$

Continuous growth has a slightly shorter doubling time, but rounded to the nearest minute it also gives an answer of 30 minutes.

Example 20

$1000 is invested in an investment account that earns interest at an annual rate of 10% compounded monthly. Calculate the minimum number of years needed for the amount in the account to exceed $4000.

Solution

We use the exponential function associated with compound interest,

$A(t) = P\left(1 + \frac{r}{n}\right)^{nt}$ with $P = 1000$, $r = 0.1$ and $n = 12$.

$$4000 = 1000\left(1 + \frac{0.1}{12}\right)^{12t} \Rightarrow 4 = (1.008\overline{3})^{12t} \Rightarrow \log 4 = \log[(1.008\overline{3})^{12t}] \Rightarrow$$

$$\log 4 = 12t \log(1.008\overline{3}) \Rightarrow t = \frac{\log 4}{12 \log(1.008\overline{3})} \approx 13.92 \text{ years}$$

The minimum number of years needed for the account to exceed $4000 is 14 years.

Example 21

A 20 g sample of radioactive iodine decays so that the mass remaining after t days is given by the equation $A(t) = 20e^{-0.087t}$, where $A(t)$ is measured in grams. After how many days (to the nearest whole day) is there only 5 g remaining?

Solution

$$5 = 20e^{-0.087t} \Rightarrow \frac{5}{20} = e^{-0.087t} \Rightarrow \ln 0.25 = \ln(e^{-0.087t}) \Rightarrow$$

$$\ln 0.25 = -0.087t \Rightarrow t = \frac{\ln 0.25}{-0.087} \approx 15.93$$

After about 16 days there is only 5 g remaining.

Solving logarithmic equations

A logarithmic equation is an equation where the variable appears within the argument of a logarithm. For example, $\log x = \frac{1}{2}$ or $\ln x = 4$. We can solve both of these logarithmic equations directly by applying the definition of a logarithmic function (Section 4.4):

$$y = \log_b x \text{ if and only if } x = b^y$$

The logarithmic equation $\log x = \frac{1}{2}$ is equivalent to the exponential equation $x = 10^{\frac{1}{2}} = \sqrt{10}$, which leads directly to the solution. Likewise, the equation $\ln x = 4$ is equivalent to $x = e^4 \approx 54.598$. Both of these equations could have been solved by means of another method that makes use of the following two facts:

(i) if $a = b$ then $n^a = n^b$; and (ii) $b^{\log_b x} = x$

To understand (ii) above, remember that a **logarithm is an exponent**. The value of $\log_b x$ is the exponent to which b is raised to give x. And b is being raised to this value; hence, the expression $b^{\log_b x}$ is equivalent to x. Therefore, another method for solving the logarithmic equation $\ln x = 4$ is to **exponentiate** both sides, i.e. use the expressions on either side of the equal sign as exponents for exponential expressions with equal bases. The base needs to be the base of the logarithm.

$$\ln x = 4 \Rightarrow e^{\ln x} = e^4 \Rightarrow x = e^4$$

Example 22

Solve for x: $\log_3(2x - 5) = 2$

Solution

$$\log_3(2x - 5) = 2 \Rightarrow 3^{\log_3(2x - 5)} = 3^2$$
$$2x - 5 = 9$$
$$2x = 14$$
$$x = 7$$

Example 23

Solve for x in terms of k: $\log_2(5x) = 3 + k$

Solution

$\log_2(5x) = 3 + k \Rightarrow 2^{\log_2(5x)} = 2^{3+k}$ exponentiate both sides with base $= 2$
$5x = 2^3 \cdot 2^k$ law of exponents $b^m \cdot b^n = b^{m+n}$ used 'in reverse'

$$x = \frac{8}{5}(2^k)$$

For some logarithmic equations, it is necessary to first apply a property, or properties, of logarithms to simplify combinations of logarithmic expressions before solving.

Example 24

Solve for x: $\log_2 x + \log_2(10 - x) = 4$

Solution

$\log_2 x + \log_2(10 - x) = 4$

$\log_2[x(10 - x)] = 4$ property of logarithms:
$\log_b M + \log_b N = \log_b(MN)$

$10x - x^2 = 2^4$ changing from logarithmic form to exponential form

$x^2 - 10x + 16 = 0$

$(x - 2)(x - 8) = 0$

$x = 2$ or $x = 8$

When solving logarithmic equations, you should be careful to always check if the *original* equation is a true statement when any solutions are substituted in for the variable. For Example 24, both of the solutions $x = 2$ and $x = 8$ produce true statements when substituted into the original equations. Sometimes 'extra' (extraneous) invalid solutions are produced, as illustrated in the next example.

Example 25

Solve for x: $\ln(x - 2) + \ln(2x - 3) = 2 \ln x$

Solution

$\ln(x - 2) + \ln(2x - 3) = 2 \ln x$

$\ln[(x - 2)(2x - 3)] = \ln x^2$ properties of logarithms

$\ln(2x^2 - 7x + 6) = \ln x^2$

$e^{\ln(2x^2 - 7x + 6)} = e^{\ln x^2}$ exponentiate both sides

$2x^2 - 7x + 6 = x^2$

$x^2 - 7x + 6 = 0$

$(x - 6)(x - 1) = 0$ factorize

$x = 6$ or $x = 1$

Substituting these two *possible* solutions indicates that $x = 1$ is not a valid solution. The reason is that if you try to substitute 1 for x into the original equation, we are not able to evaluate the expression $\ln(2x - 3)$ because we can only take the logarithm of a positive number. Therefore, $x = 6$ is the only solution. $x = 1$ is an extraneous solution that is not valid.

Solving, or checking the solutions to, a logarithmic equation on your GDC will help you avoid, or determine, extraneous solutions. To solve Example 25 on your GDC, a useful approach is to first set the equation equal to zero. Then graph the expression (after setting it equal to y) and observe where the graph intersects the x-axis (i.e. $y = 0$).

Graphical solution for Example 25:

$\ln(x-2) + \ln(2x-3) = 2\ln x \Rightarrow \ln(x-2) + \ln(2x-3) - 2\ln x = 0$

Graph the equation $y = \ln(x-2) + \ln(2x-3) - 2\ln x$ on your GDC and find x-intercepts.

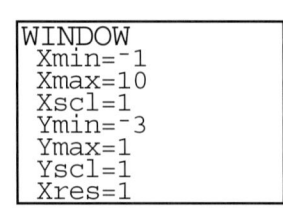

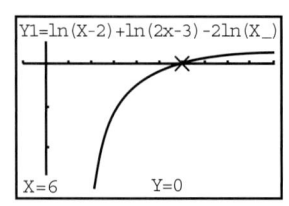

```
Plot1 Plot2 Plot3
\Y1■ ln(X-2)+ln(2
X-3)-2ln(X)
\Y2=
\Y3=
\Y4=
\Y5=
\Y6=
```

```
WINDOW
Xmin=-1
Xmax=10
Xscl=1
Ymin=-3
Ymax=1
Yscl=1
Xres=1
```

```
Y1=ln(X-2)+ln(2x-3)-2ln(X_)



X=6          Y=0
```

The graph only intersects the x-axis at $x = 6$ and not at $x = 1$. Hence, $x = 6$ is the only valid solution and $x = 1$ is an extraneous solution.

Exercise 4.5

In questions 1–12, solve for x in the exponential equation. Give x accurate to 3 significant figures.

1 $10^x = 5$ **2** $4^x = 32$ **3** $8^{x-6} = 60$

4 $2^{x+3} = 100$ **5** $\left(\frac{1}{5}\right)^x = 22$ **6** $e^x = 15$

7 $10^x = e$ **8** $3^{2x-1} = 35$ **9** $2^{x+1} = 3^{x-1}$

10 $2e^{10x} = 19$ **11** $6^{\frac{x}{2}} = 5^{1-x}$ **12** $\left(1 + \frac{0.05}{12}\right)^{12x} = 3$

13 $5000 is invested in an account that pays 7.5% interest per year, compounded quarterly.
  a) Find the amount in the account after three years.
  b) How long will it take for the money in the account to double? Give the answer to the nearest quarter of a year.

14 How long will it take for an investment of €500 to triple in value if the interest is 8.5% per year, compounded continuously. Give the answer in number of years accurate to 3 significant figures.

15 A single bacterium begins a colony in a laboratory dish. If the colony doubles every hour, after how many hours does the colony first have more than one million bacteria?

16 Find the least number of years for an investment to double if interest is compounded annually with the following interest rates.
  a) 3%   b) 6%   c) 9%

17 A new car purchased in 2005 decreases in value by 11% per year. When is the first year that the car is worth less than one-half of its original value?

18 Uranium-235 is a radioactive substance that has a half-life of 2.7×10^5 years.
  a) Find the amount remaining from a 1 g sample after a thousand years.
  b) How long will it take a 1 g sample to decompose until its mass is 700 milligrams (i.e. 0.7 g)? Give the answer in years accurate to 3 significant figures.

19 The stray dog population in a town is growing exponentially with about 18% more stray dogs each year. In 2008, there are 16 stray dogs.
   a) Find the projected population of stray dogs after five years.
   b) When is the first year that the number of stray dogs is greater than 70?

20 Initially a water tank contains one thousand litres of water. At the time $t = 0$ minutes, a tap is opened and water flows out of the tank. The volume, V litres, which remains in the tank after t minutes is given by the following exponential function: $V(t) = 1000(0.925)^t$.
   a) Find the value of V after 10 minutes.
   b) Find how long, to the nearest second, it takes for half of the initial amount of water to flow out of the tank.
   c) The tank is considered 'empty' when only 5% of the water remains. From when the tap is first opened, how many whole minutes have passed before the tank can first be considered empty?

21 The mass m kilograms of a radioactive substance at time t days is given by $m = 5e^{-0.13t}$.
   a) What is the initial mass?
   b) How long does it take for the substance to decay to 0.5 kg? Give the answer in days accurate to 3 significant figures.

In questions 22–32, solve for x in the logarithmic equation. Give exact answers and be sure to check for extraneous solutions.

22 $\log_2(3x - 4) = 4$

23 $\log(x - 4) = 2$

24 $\ln x = -3$

25 $\log_{16} x = \frac{1}{2}$

26 $\log \sqrt{x + 2} = 1$

27 $\ln(x^2) = 16$

28 $\log_2(x^2 + 8) = \log_2 x + \log_2 6$

29 $\log_3(x - 8) + \log_3 x = 2$

30 $\log 7 - \log(4x + 5) + \log(2x - 3) = 0$

31 $\log_3 x + \log_3(x - 2) = 1$

32 $\log x^8 = (\log x)^4$

Practice questions

1 Solve for x in each equation.
   a) $\log_x 16 = 4$
   b) $\log_3 27 = x$
   c) $\log_8 x = -\frac{1}{3}$
   d) $\log(x + 2) + \log(x - 2) = \log 5$

2 Solve for x in each equation.
   a) $4^x = 36$
   b) $5 \times 3^x = 18$
   c) $8^{-x} = \left(\frac{1}{4}\right)^3$
   d) $6^x = 0.25^{2x-1}$

3 Write each expression as the logarithm of a single quantity.
   a) $\log_2 x^2 - \log_2 x + 2 \log_2 3$
   b) $\ln 3 + \frac{1}{2}\ln(x - 4) - \ln x$

4 If $\log_b M = 5.42$ and $\log_b N^2 = 3.78$, find the following:
   a) $\log_b N$
   b) $\log_b\left(\frac{N^4}{\sqrt{M}}\right)$

5 Pablo invested 2000 euros at an annual rate of 6.75%, compounded annually.
   a) Find the value of Pablo's investment after four years. Give your answer to the nearest euro.
   b) How many years will it take for Pablo's investment to double in value?
   c) What should the interest rate be if Pablo's initial investment were to double in value in 10 years?

6 Let $\log P = x$, $\log Q = y$ and $\log R = z$.

Express $\log\left(\dfrac{P}{QR^3}\right)^2$ in terms of x, y and z.

7 $1000 is deposited into a bank account that earns interest at an annual rate of 4% compounded annually. After three years, the annual interest rate is increased to 7% for a further four years.
a) How much money is in the account after the seven years?
b) Find what constant rate of annual interest compounded annually would have given the same amount of money in the seven years. Give your answer as a percentage to 1 decimal place.

8 Express each of the following expressions as simply as possible.
a) $\log_2 5 \times \log_5 2$
b) $\log_4 8$
c) $4^{\log_2 6}$

9 At the start of the year 2000 there were 500 elephants in a game reserve. After t years, the number of elephants E is given by $500(1.032)^t$.
a) Find the number of elephants at the start of 2006.
b) After how many full years will the number of elephants first become greater than 750?

10 The half-life of radioactive radium is 1620 years. What percentage of an initial amount of radioactive radium will remain after 100 years?

11 A car, when purchased new, had an initial value of $25 000. After one year, the car had decreased in value to $22 000.
a) After one year, what percentage of the initial value is the new value of the car?
b) If the car continues to decrease in value at the same annual rate, what is the car's value after six years? Give your answer to the nearest dollar.
c) If the car was purchased in 2002 in which year is the car first worth less than $8000?

12 Consider the function $f: x \mapsto e^{x-2}$.
a) Write down the domain and range of f.
b) Write down the coordinates of any y-intercept, and the equation of any asymptotes for the graph of f.
c) Find f^{-1}.
d) Write down the domain and range of f^{-1}.

13 A population of a certain insect grows at a rate of 6% per month. Initially there are 500 insects.
a) Find the size of the population after four months.
b) Find the size of the population after sixteen months.
c) Let the size of the population after t months be given by the function $f(t) = A_0 b^t$.
Write down
  (i) the value A_0
  (ii) the value of b.
An alternative way of modelling the size of the insect population is given by the function $g(t) = 500 e^{kt}$.
d) By equating $f(t)$ and $g(t)$, find the value of k. Give your answer correct to 5 decimal places.

14 State the domain for each of the following two functions.
   a) $f(x) = \log\left(\dfrac{x}{x-2}\right)$
   b) $g(x) = \log x - \log(x-2)$

   Solve each of the following equations.
   c) $\log\left(\dfrac{x}{x-2}\right) = -2$
   d) $\log x - \log(x-2) = -2$

15 An experiment is designed to study a certain type of bacteria. The number of bacteria after t minutes is given by an exponential function of the form $A(t) = Ce^{kt}$, where C and k are constants. At the start of the experiment (when $t = 0$) there are 5000 bacteria. After 22 minutes, the number of bacteria has increased to 17 000.
   a) Find the exact value of C and an approximate value of k (to 3 significant figures).
   b) How many bacteria does the exponential function predict there will be after one hour?

5 Matrix Algebra

Assessment statements

4.1 Definition of a matrix: the terms 'element', 'row', 'column' and 'order'.
4.2 Algebra of matrices: equality; addition; subtraction; multiplication by a scalar.
Multiplication of matrices.
Identity and zero matrices.
4.3 Determinant of a square matrix.
Calculation of 2×2 and 3×3 determinants.
Inverse of a 2×2 matrix.
Conditions for the existence of the inverse of a matrix.
4.4 Solution of systems of linear equations using inverse matrices (a maximum of three equations in three unknowns).

Introduction

Matrices can be found anywhere and everywhere. If you have ever used a spreadsheet such as Excel or Lotus, or have ever created a table, then you have used a matrix. Matrices make the presentation of data understandable and help make calculations easy to perform. For example, your teacher's grade book may look something like this:

Student	Quiz 1	Quiz 2	Test 1	Test 2	Homework	Grade
Tim	70	80	86	82	95	A
Maher	89	56	80	60	55	C
...	...	...	...	...	...	...

If we want to know Tim's grade on Test 2, we simply follow along the row 'Tim' to the column 'Test 2' and find that he received a score of 82. Take a look at the matrix below about the sale of cameras in a store according to location and type.

	City	Donau	Neubau	Moedling
Nikon	153	98	74	56
Canon	211	120	57	29
Olympus	82	31	12	5
Other	308	242	183	107

If we want to know how many Canon cameras were sold in the Neubau shop, we follow along the row 'Canon' to the column 'Neubau' and find that 57 Canons were sold.

5.1 Basic definitions

What is a matrix?

A matrix is a rectangular array of elements. The elements can be symbolic expressions or numbers.

Matrix [A] is denoted by

$$A = \begin{pmatrix} a_{11} & a_{12} & \cdots & a_{1n} \\ a_{21} & a_{22} & \cdots & a_{2n} \\ \vdots & \vdots & \vdots & \vdots \\ a_{m1} & a_{m2} & \cdots & a_{mn} \end{pmatrix} \begin{matrix} \leftarrow \\ \leftarrow \\ \vdots \\ \leftarrow \end{matrix} \Bigg\} m \text{ rows}$$

$$\underbrace{\uparrow \quad \uparrow \quad \cdots \quad \uparrow}_{n \text{ columns}}$$

Row i of A has n elements and is $(a_{i1} \quad a_{i2} \quad \cdots \quad a_{in})$.

Column j of A has m elements and is $\begin{pmatrix} a_{1j} \\ a_{2j} \\ \vdots \\ a_{mj} \end{pmatrix}$.

The number of rows and columns of the matrix define its size (order). So, a matrix that has m rows and n columns is said to have an $m \times n$ (m by n) size (order). A matrix A with $m \times n$ order (size) is sometimes denoted as $[A]_{m \times n}$ or $[A]_{mn}$ to show that A is a matrix with m rows and n columns. (Some authors use $[a_{ij}]$ to represent a matrix.) The sales matrix has a 4×4 order. When $m = n$, the matrix is said to be a square matrix with order n, so the sales matrix is a square matrix of order 4.

Every entry in the matrix is called an **entry** or **element** of the matrix, and is denoted by a_{ij}, where i is the row number and j is the column number of that element. The ordered pair (i, j) is also called the **address** of the element. So, in the grades matrix example, the entry (2, 4) is 60, the student Maher's grade on Test 2, while (2, 4) in the sales matrix example is 29, Canon's sales in the Moedling shop.

Arthur Cayley (1821–1895)

Arthur Cayley entered Trinity College, Cambridge in 1838. While still an undergraduate, he published three papers in the *Cambridge Mathematical Journal*. Cayley graduated as Senior Wrangler in 1842 and won the first Smith's prize. Winning a fellowship enabled him to teach for four years at Cambridge. He published 28 papers in the *Cambridge Mathematical Journal* during these years. Since a fellowship had limited tenure, Cayley needed to find a profession. He spent 14 years as a lawyer but, although very skilled in his legal specialty, he always considered it as a means to make money so that he could pursue mathematics. During these 14 years as a lawyer he published around 250 mathematical papers.

His published work comprises over 900 papers and notes covering several fields of modern mathematics. The most important aspect of his work was in developing the algebra of matrices.

5 Matrix Algebra

Vectors

A vector is a matrix that has only one row or one column. There are two types of vectors – row vectors and column vectors.

Row vector

If a matrix has one row, it is called a row vector.

$B = (b_1 \; b_2 \; \ldots \; b_m)$ is a row vector with dimension m.

$B = (1 \; 2)$ could represent the position of a point in a plane and is an example of a row vector of dimension 2.

Column vector

If a matrix has one column, it is called a column vector.

$C = \begin{pmatrix} c_1 \\ c_2 \\ \vdots \\ c_n \end{pmatrix}$ is a column vector with dimension n.

$C = \begin{pmatrix} 1 \\ 2 \end{pmatrix}$ again could represent the position of a point in a plane and is an example of a column vector of dimension 2.

As you see, vectors can be represented by row or column matrices.

Submatrix

If some row(s) and/or column(s) of a matrix A are deleted, the remaining matrix is called a submatrix of A.

For example, if we are interested in the sales of the three main types of cameras in the central part of the city, we can represent them with the following *submatrix* of the original matrix:

$$\begin{pmatrix} 153 & 98 \\ 211 & 120 \\ 82 & 31 \end{pmatrix}$$

Zero matrix

A matrix for which all entries are equal to zero ($a_{ij} = 0$ for all i and j).

$(0 \; 0), \begin{pmatrix} 0 & 0 \\ 0 & 0 \end{pmatrix}, \begin{pmatrix} 0 & 0 & 0 \\ 0 & 0 & 0 \end{pmatrix}$ are zero matrices.

Diagonal

A square matrix where all entries except the diagonal entries are zero is called a **diagonal matrix**.

In a square matrix, the entries $a_{11}, a_{22}, \ldots, a_{nn}$ are called the **diagonal elements** of the matrix. Sometimes the diagonal of the matrix is also called the **principal** or **main** of the matrix.

$$\begin{pmatrix} 153 & 0 & 0 & 0 \\ 0 & 120 & 0 & 0 \\ 0 & 0 & 12 & 0 \\ 0 & 0 & 0 & 107 \end{pmatrix}$$

What is the diagonal in our sales matrix? Here, $a_{11} = 153$, $a_{22} = 120$, $a_{33} = 12$ and $a_{44} = 107$.

Triangular matrix

You can use a matrix to present data showing distances between different cities.

	Graz	Salzburg	Innsbruck	Linz
Vienna	191	298	478	185
Graz		282	461	220
Salzburg			188	135
Innsbruck				320

◀ Table 5.1

The data in Table 5.1 can be represented by a triangular matrix (upper triangular in this case).

$$\begin{pmatrix} 191 & 298 & 478 & 185 \\ 0 & 282 & 461 & 220 \\ 0 & 0 & 188 & 135 \\ 0 & 0 & 0 & 320 \end{pmatrix}$$

In a triangular matrix, the entries on one side of its diagonal are all zero.

> **Definition of a triangular matrix**
> A triangular matrix is a square matrix with order n for which $a_{ij} = 0$ when $i > j$ (upper triangular) or, alternatively, when $i < j$ (lower triangular).

5.2 Matrix operations

When are two matrices considered to be equal?

Two matrices A and B are equal if the size of A and B is the same (number of rows and columns are the same for A and B) and $a_{ij} = b_{ij}$ for all i and j.

For example, $\begin{pmatrix} 2 & 3 \\ 5 & 7 \end{pmatrix}$ and $\begin{pmatrix} 2 & x \\ x^2 - 4 & 7 \end{pmatrix}$ can only be equal if $x = 3$ and $x^2 - 4 = 5$, which can only be true if $x = 3$.

How do you add/subtract two matrices?

Two matrices A and B can be added only if they have the *same size*. If C is the sum of the two matrices, then we write

$$C = A + B$$

where $c_{ij} = a_{ij} + b_{ij}$, i.e. we add 'corresponding' terms, one by one.

For example,

$$\begin{pmatrix} 2 & 3 \\ 5 & 7 \end{pmatrix} + \begin{pmatrix} x & y \\ a & b \end{pmatrix} = \begin{pmatrix} 2+x & 3+y \\ 5+a & 7+b \end{pmatrix}$$

Subtraction is done similarly:

$$\begin{pmatrix} 2 & 3 & 1 \\ 5 & 7 & 0 \end{pmatrix} - \begin{pmatrix} x & y & 8 \\ a & b & 2 \end{pmatrix} = \begin{pmatrix} 2-x & 3-y & -7 \\ 5-a & 7-b & -2 \end{pmatrix}$$

The operations of addition and subtraction of matrices obey all rules of addition and subtraction of real numbers. That is,

$A + B = B + A$; $A + (B + C) = (A + B) + C$; $A - (B + C) = A - B - C$.

5 Matrix Algebra

How do we multiply a scalar by a matrix?

A scalar is any object that is not a matrix. The multiplication by a scalar is straightforward. You multiply each term of the matrix by the scalar.

If A is an $m \times n$ matrix, and c is a scalar, the scalar product of c and A is another matrix $B = cA$ such that every entry b_{ij} of B is a multiple of its corresponding A entry, i.e. $b_{ij} = c \times a_{ij}$.

> It is often convenient to rewrite the scalar multiple cA by factoring c out of every entry in the matrix. For instance, in the following example, the scalar $\frac{1}{2}$ has been factored out of the matrix.
>
> $$\begin{pmatrix} \frac{1}{2} & -\frac{3}{2} \\ \frac{5}{2} & \frac{1}{2} \end{pmatrix} = \frac{1}{2}\begin{pmatrix} 1 & -3 \\ 5 & 1 \end{pmatrix}$$

Matrix multiplication

At first glance, the following definition may seem unusual. You will see later, however, that this definition of the product of two matrices has many practical applications.

> **Matrix multiplication**
>
> If $A = (a_{ij})$ is an $m \times n$ matrix and B is an $n \times p$ matrix, the product AB is an $m \times p$ matrix, $AB = (c_{ij})$, where
>
> $$c_{ij} = \sum_{k=1}^{n} a_{ik}b_{kj} = a_{i1}b_{1j} + a_{i2}b_{2j} + \ldots + a_{in}b_{nj}$$
>
> For each $i = 1, 2, \ldots, m$ and $j = 1, 2, \ldots, n$.

This definition means that each entry in the product with an address ij in the product AB is obtained by multiplying the entries in the ith row of A by the *corresponding* entries in the jth column of B and then adding the results. The following shows the process in detail:

$$c_{ij} = \begin{pmatrix} a_{i1} & a_{i2} & \ldots & a_{in} \end{pmatrix} \begin{pmatrix} b_{1j} \\ b_{2j} \\ \vdots \\ b_{nj} \end{pmatrix} = a_{i1}b_{1j} + a_{i2}b_{2j} + \ldots + a_{in}b_{nj}$$

Example 1

Find $C = AB$ if $A = \begin{pmatrix} 3 & -5 & 2 \\ 2 & 1 & 7 \end{pmatrix}$ and $B = \begin{pmatrix} 3 & -2 & 1 & 5 \\ 5 & 8 & -4 & 0 \\ -9 & 10 & 5 & 3 \end{pmatrix}$.

Solution

A is a 2×3 matrix and B is a 3×4 matrix, so the product must be a 2×4 matrix. Every entry in the product is the result of multiplying the entries in the rows of A and columns of B. For example:

$$c_{12} = \sum_{k=1}^{3} a_{1k}b_{k2} = \begin{pmatrix} a_{11} & a_{12} & a_{13} \end{pmatrix} \begin{pmatrix} b_{12} \\ b_{22} \\ b_{32} \end{pmatrix} = \begin{pmatrix} 3 & -5 & 2 \end{pmatrix} \begin{pmatrix} -2 \\ 8 \\ 10 \end{pmatrix}$$

$$= 3 \times (-2) - 5 \times 8 + 2 \times 10 = -26$$

or

$$c_{23} = \sum_{k=1}^{3} a_{2k}b_{k3} = \begin{pmatrix} a_{21} & a_{22} & a_{23} \end{pmatrix} \begin{pmatrix} b_{13} \\ b_{23} \\ b_{33} \end{pmatrix} = \begin{pmatrix} 2 & 1 & 7 \end{pmatrix} \begin{pmatrix} 1 \\ -4 \\ 5 \end{pmatrix}$$

$$= 2 \times 1 + 1 \times (-4) + 7 \times 5 = 33$$

The operation is repeated eight times to get

$$C = AB = \begin{pmatrix} -34 & -26 & 33 & 21 \\ -52 & 74 & 33 & 31 \end{pmatrix}$$

This product can also be found using a GDC.

```
[A] [B]
[[-34  -26  33  21…
 [-52  -74  33  31…
```

For the product of two matrices to be defined, the number of columns in the first matrix should be the same as the number of rows in the second matrix.

$$\begin{array}{ccc} A & B & = AB \\ m \times n & n \times p & m \times p \end{array}$$

equal — order of AB

Examples – matrix multiplication

a) $\begin{pmatrix} 5 & 0 & 3 \\ -2 & 1 & 2 \end{pmatrix} \begin{pmatrix} -2 & 4 \\ 1 & -1 \\ 3 & -2 \end{pmatrix} = \begin{pmatrix} -1 & 14 \\ 11 & -13 \end{pmatrix}$

$\quad\quad 2 \times 3 \quad\quad\quad 3 \times 2 \quad\quad\quad 2 \times 2$

b) $\begin{pmatrix} 4 & -5 \\ 1 & 7 \end{pmatrix} \begin{pmatrix} 1 & 0 \\ 0 & 1 \end{pmatrix} = \begin{pmatrix} 4 & -5 \\ 1 & 7 \end{pmatrix}$

$\quad\quad 2 \times 2 \quad\; 2 \times 2 \quad\;\; 2 \times 2$

c) $\begin{pmatrix} 5 & 0 & 3 \\ -2 & 1 & 2 \\ 2 & 1 & 3 \end{pmatrix} \begin{pmatrix} -\frac{1}{7} & -\frac{3}{7} & \frac{3}{7} \\ -\frac{10}{7} & -\frac{9}{7} & \frac{16}{7} \\ \frac{4}{7} & \frac{5}{7} & -\frac{5}{7} \end{pmatrix} = \begin{pmatrix} 1 & 0 & 0 \\ 0 & 1 & 0 \\ 0 & 0 & 1 \end{pmatrix}$

$\quad\quad 3 \times 3 \quad\quad\quad\quad 3 \times 3 \quad\quad\quad\quad 3 \times 3$

As you see from part b) above, the matrix $\begin{pmatrix} 1 & 0 \\ 0 & 1 \end{pmatrix}$ does not create a new value when it is multiplied by another matrix. This is why it is called the **identity** matrix of order 2.

> **The identity matrix**
> A diagonal matrix where $a_{ij} = 1$ is called the identity matrix of order n.

Examples – identity matrices

a) $\begin{pmatrix} a & b & c \\ d & e & f \\ g & h & i \end{pmatrix} \begin{pmatrix} 1 & 0 & 0 \\ 0 & 1 & 0 \\ 0 & 0 & 1 \end{pmatrix} = \begin{pmatrix} a & b & c \\ d & e & f \\ g & h & i \end{pmatrix}$

b) $\begin{pmatrix} 1 & 0 & 0 \\ 0 & 1 & 0 \\ 0 & 0 & 1 \end{pmatrix} \begin{pmatrix} a & b & c \\ d & e & f \\ g & h & i \end{pmatrix} = \begin{pmatrix} a & b & c \\ d & e & f \\ g & h & i \end{pmatrix}$

c) $\begin{pmatrix} a & b & c & m \\ d & e & f & n \\ g & h & i & p \\ j & k & l & q \end{pmatrix} \begin{pmatrix} 1 & 0 & 0 & 0 \\ 0 & 1 & 0 & 0 \\ 0 & 0 & 1 & 0 \\ 0 & 0 & 0 & 1 \end{pmatrix} = \begin{pmatrix} a & b & c & m \\ d & e & f & n \\ g & h & i & p \\ j & k & l & q \end{pmatrix}$

Sometimes, the identity matrix is denoted by I_n, where n is the order. So, in parts a) and b) above, the identity is I_3, and in c) it is I_4.

Examples – comparing AB with BA

a) $(2 \quad -1 \quad 3) \begin{pmatrix} 2 \\ 5 \\ 4 \end{pmatrix} = (11)$

$\quad 1 \times 3 \quad\quad 3 \times 1 \quad 1 \times 1$

b) $\begin{pmatrix} 2 \\ 5 \\ 4 \end{pmatrix} (2 \quad -1 \quad 3) = \begin{pmatrix} 4 & -2 & 6 \\ 10 & -5 & 15 \\ 8 & -4 & 12 \end{pmatrix}$

$\quad 3 \times 1 \quad\quad 1 \times 3 \quad\quad\quad 3 \times 3$

Notice the difference between the products in parts a) and b). Matrix multiplication, in general, is **not commutative**. It is usually not true that $AB = BA$.

Let $A = \begin{pmatrix} 3 & 6 \\ 5 & 2 \end{pmatrix}$ and $B = \begin{pmatrix} -2 & 3 \\ 1 & 5 \end{pmatrix}$, then $AB = \begin{pmatrix} 3 & 6 \\ 5 & 2 \end{pmatrix} \begin{pmatrix} -2 & 3 \\ 1 & 5 \end{pmatrix} = \begin{pmatrix} 0 & 39 \\ -8 & 25 \end{pmatrix}$

but

$BA = \begin{pmatrix} -2 & 3 \\ 1 & 5 \end{pmatrix} \begin{pmatrix} 3 & 6 \\ 5 & 2 \end{pmatrix} = \begin{pmatrix} 9 & -6 \\ 28 & 16 \end{pmatrix} \Rightarrow AB \neq BA$

However, if we let

$A = \begin{pmatrix} 3 & 6 \\ 5 & 2 \end{pmatrix}$ and $B = \begin{pmatrix} 2 & 6 \\ 5 & 1 \end{pmatrix}$, then $AB = \begin{pmatrix} 3 & 6 \\ 5 & 2 \end{pmatrix} \begin{pmatrix} 2 & 6 \\ 5 & 1 \end{pmatrix} = \begin{pmatrix} 36 & 24 \\ 20 & 32 \end{pmatrix}$ and

$BA = \begin{pmatrix} 2 & 6 \\ 5 & 1 \end{pmatrix} \begin{pmatrix} 3 & 6 \\ 5 & 2 \end{pmatrix} = \begin{pmatrix} 36 & 24 \\ 20 & 32 \end{pmatrix} \Rightarrow AB = BA$

Thus, in general, $AB \neq BA$. However, for some matrices A and B, it may happen that $AB = BA$.

Example 2

Find the average sales in each of the regions (City, Donau, Neubau and Moedling), given the following information.

	City	Donau	Neubau	Moedling
Nikon	153	98	74	56
Canon	211	120	57	29
Olympus	82	31	12	5
Other	308	242	183	107

The average selling price for each make of camera is as follows:
Nikon €1200, Canon €1100, Olympus €900, Other €600

Solution

We set up a matrix multiplication in which the individual camera sales are multiplied by the corresponding price. Since the rows represent the sales of the different makes of camera, create a row matrix of the different prices and perform the multiplication.

$$(1200 \ 1100 \ 900 \ 600) \begin{pmatrix} 153 & 98 & 74 & 56 \\ 211 & 120 & 57 & 29 \\ 82 & 31 & 12 & 5 \\ 308 & 242 & 183 & 107 \end{pmatrix} = (674\,300 \ 422\,700 \ 272\,100 \ 167\,800)$$

So, the regions' sales are:

	City	Donau	Neubau	Moedling
Sales	674 300	422 700	272 100	167 800

Remember that we are multiplying a 1 × 4 matrix with a 4 × 4 matrix and hence we get a 1 × 4 matrix.

Exercise 5.1 and 5.2

1 Consider the following matrices

$$A = \begin{pmatrix} -2 & x \\ y-1 & 3 \end{pmatrix}, B = \begin{pmatrix} x+1 & -3 \\ 4 & y-2 \end{pmatrix}$$

a) Evaluate each of the following
   (i) $A + B$ (ii) $3A - B$.
b) Find x and y such that $A = B$.
c) Find x and y such that $A + B$ is a diagonal matrix.
d) Find AB and BA.

2 Solve for the variables.

a) $\begin{pmatrix} 3 & 0 \\ 4 & 2 \end{pmatrix} \begin{pmatrix} x \\ y \end{pmatrix} = \begin{pmatrix} 6 \\ -12 \end{pmatrix}$ b) $\begin{pmatrix} 2 & p \\ 3 & q \end{pmatrix} \begin{pmatrix} 4 \\ 5 \end{pmatrix} = \begin{pmatrix} 18 \\ -8 \end{pmatrix}$

3 The diagram on the right shows the major highways connecting some European cities: Vienna (V), Munich (M), Frankfurt (F), Stuttgart (S), Zurich (Z), Milano (L) and Paris (P).

a) Write the number of *direct* routes between each pair of cities into a matrix as started below:

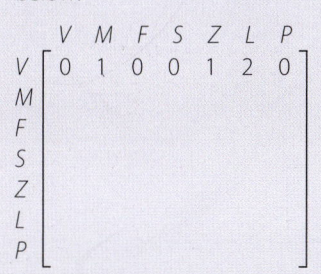

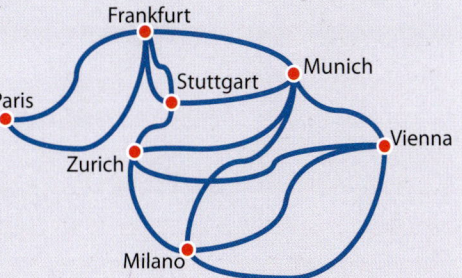

$$\begin{array}{c} \\ V \\ M \\ F \\ S \\ Z \\ L \\ P \end{array} \begin{array}{c} \begin{matrix} V & M & F & S & Z & L & P \end{matrix} \\ \begin{bmatrix} 0 & 1 & 0 & 0 & 1 & 2 & 0 \\ & & & & & & \\ & & & & & & \\ & & & & & & \\ & & & & & & \\ & & & & & & \\ & & & & & & \end{bmatrix} \end{array}$$

b) Multiply the matrix from part a) by itself and interpret what it signifies.

4 Consider the following matrices

$$A = \begin{pmatrix} 2 & 5 & 1 \\ 0 & -3 & 2 \\ 7 & 0 & -1 \end{pmatrix}, B = \begin{pmatrix} m & -2 \\ 3m & -1 \\ 2 & 3 \end{pmatrix}, C = \begin{pmatrix} x-1 & 5 & y \\ 0 & -x & y+1 \\ 2x+y & x-3y & 2y-x \end{pmatrix}$$

a) Find $A + C$. b) Find AB. c) Find BA.
d) Solve for x and y if $A = C$. e) Find $B + C$.
f) Solve for m if $3B + 2\begin{pmatrix} -1 & m^2 \\ -5 & 2 \\ 1 & -1 \end{pmatrix} = \begin{pmatrix} 7 & 12 \\ 17 & 1 \\ 2m+2 & 7 \end{pmatrix}$.

5 Find a, b and c so that the following equation is true:

$$2 \cdot \begin{pmatrix} a-1 & b \\ c+2 & 3 \end{pmatrix} + \begin{pmatrix} 3 & -1 \\ 0 & 5 \end{pmatrix} = \begin{pmatrix} -5 & 5 \\ 8 & c+9 \end{pmatrix}$$

6 Find x and y such that:

$$\begin{pmatrix} 2 & -3 \\ -5 & 7 \end{pmatrix} \begin{pmatrix} x-11 & 1-x \\ -5 & x+2y \end{pmatrix} = \begin{pmatrix} 1 & 0 \\ 0 & 1 \end{pmatrix}$$

7 Find m and n if

$$\begin{pmatrix} m^2-1 & m+2 \\ 5 & -2 \end{pmatrix} = \begin{pmatrix} 3 & n+1 \\ 5 & n-5 \end{pmatrix}.$$

8 There are two supermarkets in your area. Your shopping list consists of 2 kg of tomatoes, 500 g of meat and 3 litres of milk. Prices differ between the different shops, and it is difficult to switch between stores to make certain you are paying the least amount of money. A better strategy is to check and see where you pay less on *average*! The prices of the different items are given below. Which shop should you go to?

Product	Price in shop A	Price in shop B
Tomato	€1.66/kg	€1.58/kg
Meat	€2.55/100 g	€2.6/100 g
Milk	€0.90/litre	€0.95/litre

9 Consider the matrices

$$A = \begin{pmatrix} 2 & 0 \\ -5 & 1 \end{pmatrix}, B = \begin{pmatrix} 3 & -1 \\ 1 & 4 \end{pmatrix} \text{ and } C = \begin{pmatrix} -3 & 5 \\ 2 & 7 \end{pmatrix}.$$

a) Find $A + (B + C)$ and $(A + B) + C$.
b) Make a conjecture about the addition of 2×2 matrices observed in a) above and prove it.
c) Find $A(BC)$ and $(AB)C$.
d) Make a conjecture about the multiplication of 2×2 matrices observed in c) above and prove it.

10 A company stores and sells air conditioning units, electric heaters and humidifiers. Row matrix A represents the number of each unit sold last year, and matrix B represents the profit margin for each unit. Find AB and describe what the product represents.

$$A = (235 \quad 562 \quad 117), B = \begin{pmatrix} €120 \\ €95 \\ €56 \end{pmatrix}$$

11 Find r and s such that the following equation is true: $rA + B = A$, where
$A = \begin{pmatrix} 2 & 3 \\ 5 & 7 \end{pmatrix}$ and $B = \begin{pmatrix} -4 & -6 \\ s-8 & -37 \end{pmatrix}$.

12 Let $A = \begin{pmatrix} 1 & 1 \\ 0 & 1 \end{pmatrix}$.

a) Find:
  (i) A^2 (ii) A^3 (iii) A^4 (iv) A^n

Let $B = \begin{pmatrix} 3 & 3 \\ 0 & 3 \end{pmatrix}$.

b) Find:
  (i) B^2 (ii) B^3 (iii) B^4 (iv) B^n

5.3 Applications to systems

There is a wide range of applications of matrices in solving systems of equations. Recall from your algebra that the equation of a straight line can take the form

$$ax + by = c$$

where a, b and c are constants and x and y are variables. We call this equation a **linear equation in two variables**. Similarly, the equation of a plane in three-dimensional space has the form

$$ax + by + cz = d$$

where a, b, c and d are constants. We call this equation a **linear equation in three variables**.

A **solution** of a linear equation in n variables (in this case two or three) is an ordered set of real numbers (x_0, y_0, z_0) so that the equation in question is satisfied when these values are substituted for the corresponding variables. For example, the equation

$$x + 2y = 4$$

is satisfied when $x = 2$ and $y = 1$. Some other solutions are $x = -4$ and $y = 4$, $x = 0$ and $y = 2$, and $x = -2$ and $y = 3$.

The set of all solutions of a linear equation is its **solution set**, and when this set is found, the equation is said to have been **solved**. To describe the entire solution set we often use a **parametric representation** as illustrated in the following examples.

Example 3

Solve the linear equation $x + 2y = 4$.

Solution

To find the solution set of an equation in two variables, we solve for one variable in terms of the other. For instance, if we solve for x, we obtain

$$x = 4 - 2y.$$

In this form, y is **free**, in the sense that it can take on any real value, while x is not free, since its value depends on that of y. To represent this solution set in general terms, we introduce a third variable, for example, t, called a **parameter**, and by letting $y = t$ we represent the solution set as

$$x = 4 - 2t, y = t, t \text{ is any real number.}$$

Particular solutions can then be obtained by assigning values to the parameter t. For instance, $t = 1$ yields the solution $x = 2$ and $y = 1$, and $t = 3$ yields the solution $x = -2$ and $y = 3$.

Note that the solution set of a linear equation can be represented parametrically in several ways. For instance, in this example, if we solve for y in terms of x, the parametric representation would take the following form

$$x = m, y = 2 - \tfrac{1}{2}m, m \text{ is a real number.}$$

Also, by choosing $m = 2$, one particular solution would be $(x, y) = (2, 1)$, and by choosing $m = -2$, another particular solution would be $(-2, 3)$.

Example 4

Solve the linear equation $3x + 2y - z = 3$.

Solution

Choosing x and y as the *free* variables, we solve for z.

$$z = 3x + 2y - 3$$

Letting $x = p$ and $y = q$, we obtain the parametric representation:

$$x = p, y = q, z = 3x + 2y - 3, p \text{ and } q \text{ any real numbers.}$$

A particular solution $(x, y, z) = (1, 1, 2)$.

Parametric representation is very important when we study vectors and lines later on in the book.

Systems of linear equations – refresher

A system of k equations in n variables is a set of k linear equations in the same n variables. For example,

$$2x + 3y = 3$$
$$x - y = 4$$

is a system of two linear equations in two variables, while

$$x - 2y + 3z = 9$$
$$x - 3y = 4$$

is a system with two equations and three variables, and

$$x - 2y + 3z = 9$$
$$x - 3y = 4$$
$$2x - 5y + 5z = 17$$

is a system with three equations and three variables.

A **solution** of a system of equations is an ordered set of numbers $x_0, y_0, \ldots$ which satisfy every equation in the system. For example, $(3, -1)$ is a solution of

$$2x + 3y = 3$$
$$x - y = 4$$

Both equations in the system are satisfied when $x = 3$ and $y = -1$ are substituted into the equations. On the contrary, $(0, 1)$ is not a solution of the system, even though it satisfies the first equation, as it does not satisfy the second.

As you already know, there are several ways of finding solutions to systems. In this book, we will consider using matrix methods to solve systems of equations.

Taking our example above, notice how we can write the system of equations in matrix form:

$$\begin{cases} 2x + 3y = 3 \\ x - y = 4 \end{cases} \Rightarrow \begin{pmatrix} 2 & 3 \\ 1 & -1 \end{pmatrix} \begin{pmatrix} x \\ y \end{pmatrix} = \begin{pmatrix} 3 \\ 4 \end{pmatrix}$$

The representation of the system of equations in this way enables us to use matrix operations in solving systems. This matrix equation can be written as

$$\begin{pmatrix} 2 & 3 \\ 1 & -1 \end{pmatrix} \begin{pmatrix} x \\ y \end{pmatrix} = \begin{pmatrix} 3 \\ 4 \end{pmatrix} \Rightarrow AX = C$$

where A is the coefficient matrix, X is the variables' matrix and C is the constants' matrix. However, to solve this equation, the inverse of a matrix has to be defined as the solution of the system in the form

$$X = A^{-1}C$$

where A^{-1} is the inverse of the matrix A.

Matrix inverse

To solve the equation $2x = 6$ for x, we need to multiply both sides of the equation by $\frac{1}{2}$:

$\frac{1}{2} \times 2x = \frac{1}{2} \times 6 \Rightarrow x = 3$. This is so, because $\frac{1}{2} \times 2 = 2 \times \frac{1}{2} = 1$.

$\frac{1}{2}$ is called the multiplicative inverse of 2. The inverse of a matrix is defined in a similar manner and plays a similar role in solving a matrix equation, such as $AX = C$.

> **Inverse of a matrix**
> A square matrix B is the inverse of a square matrix A if $AB = BA = I$, where I is the identity matrix.

The notation A^{-1} is used to denote the inverse of a matrix A. Thus, $B = A^{-1}$. Note that only square matrices can have multiplicative inverses.

Example – matrix inverse

$A = \begin{pmatrix} 7 & 5 \\ 4 & 3 \end{pmatrix}$ and $B = \begin{pmatrix} 3 & -5 \\ -4 & 7 \end{pmatrix}$ are multiplicative inverses since

$$AB = \begin{pmatrix} 7 & 5 \\ 4 & 3 \end{pmatrix}\begin{pmatrix} 3 & -5 \\ -4 & 7 \end{pmatrix} = \begin{pmatrix} 21-20 & -35+35 \\ 12-12 & -20+21 \end{pmatrix} = \begin{pmatrix} 1 & 0 \\ 0 & 1 \end{pmatrix}$$

$$BA = \begin{pmatrix} 3 & -5 \\ -4 & 7 \end{pmatrix}\begin{pmatrix} 7 & 5 \\ 4 & 3 \end{pmatrix} = \begin{pmatrix} 21-20 & 15-15 \\ -28+28 & -20+21 \end{pmatrix} = \begin{pmatrix} 1 & 0 \\ 0 & 1 \end{pmatrix}$$

Finding the inverse can also be achieved using a GDC.

```
[A]⁻¹
          [[3  -5]
           [-4  7 ]]
[A]⁻¹[A]
          [[1  0]
           [0  1]]
```

There are a few methods available for finding the inverse of a 2 × 2 matrix. We will be using the following method only, since the other methods are beyond the scope of this textbook.

Let $A = \begin{pmatrix} a & b \\ c & d \end{pmatrix}$ and assume $A^{-1} = \begin{pmatrix} e & f \\ g & h \end{pmatrix}$ and then solve the following matrix equation for e, f, g and h in terms of a, b, c and d.

$$\begin{pmatrix} a & b \\ c & d \end{pmatrix}\begin{pmatrix} e & f \\ g & h \end{pmatrix} = \begin{pmatrix} 1 & 0 \\ 0 & 1 \end{pmatrix} \Rightarrow \begin{pmatrix} ae+bg & af+bh \\ ce+dg & cf+dh \end{pmatrix} = \begin{pmatrix} 1 & 0 \\ 0 & 1 \end{pmatrix}$$

Now we can set up two systems to solve for the required variables, i.e.:

$$\begin{pmatrix} ae+bg & af+bh \\ ce+dg & cf+dh \end{pmatrix} = \begin{pmatrix} 1 & 0 \\ 0 & 1 \end{pmatrix}$$

$\left.\begin{array}{l} ae+bg=1 \\ ce+dg=0 \end{array}\right\} \Rightarrow \left.\begin{array}{l} dae+\mathbf{dbg}=d \\ bce+\mathbf{bdg}=0 \end{array}\right\} \Rightarrow e = \dfrac{d}{ad-bc}, g = \dfrac{-c}{ad-bc}$

$\left.\begin{array}{l} af+bh=0 \\ cf+dh=1 \end{array}\right\} \Rightarrow \left.\begin{array}{l} daf+\mathbf{dbh}=0 \\ bcf+\mathbf{bdh}=b \end{array}\right\} \Rightarrow f = \dfrac{-b}{ad-bc}, h = \dfrac{a}{ad-bc}$

Therefore, $A^{-1} = \begin{pmatrix} \dfrac{d}{ad-bc} & \dfrac{-b}{ad-bc} \\ \dfrac{-c}{ad-bc} & \dfrac{a}{ad-bc} \end{pmatrix}$ or $A^{-1} = \dfrac{1}{ad-bc}\begin{pmatrix} d & -b \\ -c & a \end{pmatrix}$.

Example 5

Find the inverse of $\begin{pmatrix} 4 & 7 \\ 3 & 5 \end{pmatrix}$.

Solution

Here $a = 4$, $b = 7$, $c = 3$ and $d = 5$, so $ad - bc = -1$. Thus,

$$A^{-1} = \frac{1}{ad - bc}\begin{pmatrix} d & -b \\ -c & a \end{pmatrix} = \frac{1}{-1}\begin{pmatrix} 5 & -7 \\ -3 & 4 \end{pmatrix} = \begin{pmatrix} -5 & 7 \\ 3 & -4 \end{pmatrix}.$$

```
[A]
                          [[4 7]
                           [3 5]]
[A]-1
                          [[-5  7]
                           [ 3 -4]]
■
```

The determinant

The number $ad - bc$ is called the **determinant** of the 2×2 matrix $A = \begin{pmatrix} a & b \\ c & d \end{pmatrix}$.

The notation we will use for this number is **det A**, so $\det A = ad - bc$.

The determinant plays an important role in determining whether a matrix has an inverse or not.

If the determinant is zero, i.e. $ad - bc = 0$, the matrix does not have an inverse. If a matrix has no inverse, it is called a **singular matrix**; if it is invertible, it is called **non-singular**.

Example 6

Solve the system of equations.
$$2x + 3y = 3$$
$$x - y = 4$$

Solution

In matrix form, the system can be written as

$$\begin{pmatrix} 2 & 3 \\ 1 & -1 \end{pmatrix}\begin{pmatrix} x \\ y \end{pmatrix} = \begin{pmatrix} 3 \\ 4 \end{pmatrix} \Rightarrow \begin{pmatrix} x \\ y \end{pmatrix} = \begin{pmatrix} 2 & 3 \\ 1 & -1 \end{pmatrix}^{-1}\begin{pmatrix} 3 \\ 4 \end{pmatrix}$$

$$\Rightarrow \begin{pmatrix} x \\ y \end{pmatrix} = \frac{1}{-5}\begin{pmatrix} -1 & -3 \\ -1 & 2 \end{pmatrix}\begin{pmatrix} 3 \\ 4 \end{pmatrix}$$

$$\Rightarrow \begin{pmatrix} x \\ y \end{pmatrix} = \frac{1}{-5}\begin{pmatrix} -15 \\ 5 \end{pmatrix} = \begin{pmatrix} 3 \\ -1 \end{pmatrix}$$

```
[A]-1 [C]
                              [[3 ]
                               [-1]]
■
```

5 Matrix Algebra

Solving systems of equations in three variables follows similar procedures. However, finding the inverse of a 3 × 3 matrix will be delegated to the GDC at this level. As in the case of a 2 × 2 matrix, the existence of an inverse for a 3 × 3 matrix depends on the value of its determinant.

The determinant of a 3 × 3 matrix A can be achieved in one of two ways:

1. $A = \begin{pmatrix} a & b & c \\ d & e & f \\ g & h & i \end{pmatrix} \Rightarrow \det A = a(ei - fh) - b(di - fg) + c(dh - eg)$

   For example, if

   $A = \begin{pmatrix} 5 & 1 & -4 \\ 2 & -3 & -5 \\ 7 & 2 & -6 \end{pmatrix} \Rightarrow \det A = 5(18 + 10) - 1(-12 + 35) - 4(4 + 21) = 17$

   ```
   [A]
           [[5  1  -4]
            [2 -3  -5]
            [7  2  -6]]
   det([A])
                    17
   ```

2. A practical method is to use a 'special' set up as follows:

   $\det A = \begin{vmatrix} a & b & c \\ d & e & f \\ g & h & i \end{vmatrix} \begin{matrix} a & b \\ d & e \\ g & h \end{matrix} = aei + bfg + cdh - gec - hfa - idb$

   This is done by 'copying' the first two columns and adding them to the end of the matrix, multiplying down the main diagonals and adding the products, and then multiplying up the second diagonals and subtracting them from the previous product, as shown. In the example above:

   $\begin{vmatrix} 5 & 1 & -4 \\ 2 & -3 & -5 \\ 7 & 2 & -6 \end{vmatrix} \begin{matrix} 5 & 1 \\ 2 & -3 \\ 7 & 2 \end{matrix}$

   $= 5(-3)(-6) + 1(-5)(7) + (-4) \cdot 2 \cdot 2 - 7(-3)(-4) - 2(-5) \cdot 5 - (-6) \cdot 2 \cdot 1$
   $= 90 - 35 - 16 - 84 + 50 + 12$
   $= 152 - 135$
   $= 17$

   In fact, this arrangement is simply a reordering of the calculations involved in the previous method.

Example 7

Solve the system of equations.
$$5x + y - 4z = 5$$
$$2x - 3y - 5z = 2$$
$$7x + 2y - 6z = 5$$

Solution

We write this system in matrix form:

$$\begin{pmatrix} 5 & 1 & -4 \\ 2 & -3 & -5 \\ 7 & 2 & -6 \end{pmatrix} \begin{pmatrix} x \\ y \\ z \end{pmatrix} = \begin{pmatrix} 5 \\ 2 \\ 5 \end{pmatrix}$$

Since det $A \neq 0$, we can find the solution in the same way we did for the 2 × 2 matrix, i.e.

$$\begin{pmatrix} 5 & 1 & -4 \\ 2 & -3 & -5 \\ 7 & 2 & -6 \end{pmatrix} \begin{pmatrix} x \\ y \\ z \end{pmatrix} = \begin{pmatrix} 5 \\ 2 \\ 5 \end{pmatrix} \Rightarrow \begin{pmatrix} x \\ y \\ z \end{pmatrix} = \begin{pmatrix} 5 & 1 & -4 \\ 2 & -3 & -5 \\ 7 & 2 & -6 \end{pmatrix}^{-1} \begin{pmatrix} 5 \\ 2 \\ 5 \end{pmatrix}$$

Using a GDC:

```
[A]-1 [C]
          [[3 ]
           [-2]
           [2 ]]
```

To check your work, you can store the answer matrix as D and then substitute the values into the system:

$$\begin{pmatrix} 5 & 1 & -4 \\ 2 & -3 & -5 \\ 7 & 2 & -6 \end{pmatrix} \begin{pmatrix} 3 \\ -2 \\ 2 \end{pmatrix} = \begin{pmatrix} 15 - 2 - 8 \\ 6 + 6 - 10 \\ 21 - 4 - 12 \end{pmatrix} = \begin{pmatrix} 5 \\ 2 \\ 5 \end{pmatrix}, \text{ or}$$

```
[A] [D]
          [[5]
           [2]
           [5]]
```

Exercise 5.3

1 Consider the matrix M which satisfies the matrix equation

$$\begin{pmatrix} 3 & 7 \\ -4 & -9 \end{pmatrix} M = \begin{pmatrix} 2 & 1 \\ 3 & 5 \end{pmatrix}.$$

a) Write out the inverse of matrix $\begin{pmatrix} 3 & 7 \\ -4 & -9 \end{pmatrix}$.

b) Hence, write M as a product of two matrices.

c) Evaluate M.

d) Now consider the equation containing the matrix N:

$$N \begin{pmatrix} 3 & 7 \\ -4 & -9 \end{pmatrix} = \begin{pmatrix} 2 & 1 \\ 3 & 5 \end{pmatrix}$$

(i) Write N as a product of two matrices.
(ii) Evaluate N.

e) Write a short paragraph describing your work on this problem.

Matrix Algebra

2 Find the matrix E in the following equation:
$$\begin{pmatrix} 1 & 3 \\ 3 & 4 \end{pmatrix} = \begin{pmatrix} 1 & 0 \\ 3 & 1 \end{pmatrix} E \begin{pmatrix} 1 & 0 \\ 0 & -5 \end{pmatrix}$$

3 a) Prove that the matrix $A = \begin{pmatrix} 2 & -3 & 1 \\ 1 & 1 & -3 \\ 3 & -2 & -3 \end{pmatrix}$ should have an inverse.

b) Write out A^{-1}.

c) Hence, solve the system of equations:
$$\begin{cases} 2x - 3y + z = 4.2 \\ x + y - 3z = -1.1 \\ 3x - 2y - 3z = 2.9 \end{cases}$$

4 Find the inverse for each matrix.

a) $A = \begin{pmatrix} \frac{\sqrt{3}}{2} & -\frac{1}{2} \\ \frac{1}{2} & \frac{\sqrt{3}}{2} \end{pmatrix}$

b) $B = \begin{pmatrix} a & 1 \\ a+2 & \frac{3}{a}+1 \end{pmatrix}$

5 For what values of x is the following matrix singular?
$$A = \begin{pmatrix} x+1 & 3 \\ 3x-1 & x+3 \end{pmatrix}$$

6 Find n such that $\begin{pmatrix} 2 & -1 & 4 \\ 2n & 2 & 0 \\ 2 & 1 & 4n \end{pmatrix}$ is the inverse of $\begin{pmatrix} -2 & -3 & 4 \\ 1 & 2 & -2 \\ 3n & 2 & -5n \end{pmatrix}$.

7 Consider the two matrices $A = \begin{pmatrix} 4 & 2 \\ 0 & -3 \end{pmatrix}$ and $B = \begin{pmatrix} 2 & 1 \\ 3 & 5 \end{pmatrix}$.

a) Find X such that $XA = B$.
b) Find Y such that $AY = B$.
c) Is $X = Y$? Explain.

8 Consider the two matrices
$$P = \begin{pmatrix} 2 & 0 & -1 \\ 3 & 5 & 4 \\ 1 & 0 & -1 \end{pmatrix} \text{ and } Q = \begin{pmatrix} 3 & -1 & 1 \\ 4 & 0 & 0 \\ 1 & 2 & -1 \end{pmatrix}.$$

a) Find PQ and QP.
b) Find $P^{-1}, Q^{-1}, P^{-1}Q^{-1}, Q^{-1}P^{-1}, (PQ)^{-1},$ and $(QP)^{-1}$.
c) Write a few sentences about your observations in parts a) and b).

Practice questions

1 Solve each system of equations using matrix methods.

a) $\begin{cases} x + y + 3z = 3 \\ x + y + 6z = 3 \\ x + 2y + 4z = 7 \end{cases}$

b) $\begin{cases} x + 2y - z = 1 \\ y - 2z = -3 \\ 3x - y + 2z = 6 \end{cases}$

2 For what value(s) of a, if any, is each of the following matrices singular? For all other values, find the inverse.

a) $\begin{pmatrix} a & a \\ -a & a \end{pmatrix}$

b) $\begin{pmatrix} 3 & a \\ 2a & a^2 \end{pmatrix}$

c) $\begin{pmatrix} e^{3a} & e^{2a} \\ e^{-2a} & e^a \end{pmatrix}$

d) $\begin{pmatrix} \sin a & -\cos a \\ \cos a & \sin a \end{pmatrix}$

3 If $A = \begin{pmatrix} 3x & x \\ 2 & x-1 \end{pmatrix}$ and $\det A = 2$, find the possible values of x.

4 Let $A = \begin{pmatrix} k & 3 \\ 3 & -1 \end{pmatrix}$, where $k \in \mathbb{Z}$.

a) Find A^2 in terms of k.

b) If A^2 is equal to $\begin{pmatrix} 13 & 3 \\ 3 & 10 \end{pmatrix}$, find the value of k.

c) Using the value of k found in b), find A^{-1} and hence solve the system of equations:
$2x + 3y = 13$
$3x - y = 3$

5 M and N are two 2×2 matrices. $M = \begin{pmatrix} 2 & 1 \\ 3 & 5 \end{pmatrix}$ and $MN = \begin{pmatrix} 3 & 3 \\ 8 & 8 \end{pmatrix}$. Find N.

6 $I = \begin{pmatrix} 1 & 0 \\ 0 & 1 \end{pmatrix}$ is the identity matrix for 2×2 matrix multiplication. Consider matrix $A = \begin{pmatrix} 4 & -1 \\ 3 & 5 \end{pmatrix}$. Find the possible values of the real number k such that the matrix $(A - kI)$ is singular.

7 Recall that for two matrices A and B to be inverses of each other $AB = I$, where I is the identity matrix.

a) Find the values of m and n so that matrices A and B are inverses of each other:

$A = \begin{pmatrix} 1 & -1 & m \\ 2 & -3 & 2 \\ -1 & 2 & -2 \end{pmatrix}; B = \begin{pmatrix} 2 & 0 & 1 \\ 2 & n & 0 \\ 1 & -1 & -1 \end{pmatrix}$

b) Hence, for the values of m and n found above, solve the system of equations:
$\begin{cases} x - y + z = 4 \\ 2x - 3y + 2z = 9 \\ -x + 2y - 2z = -7 \end{cases}$

8 Find all values of m so that the following system is inconsistent.
$(m-4)x + 3y = 7$
$-2x + (m+1)y = 12$

9 Matrices A and B are given such that $AB = BA$. Find the values of m and n.
$A = \begin{pmatrix} m & 3 \\ 4 & 2 \end{pmatrix}, B = \begin{pmatrix} 2 & n \\ 8 & 4 \end{pmatrix}$

Matrix Algebra

10 Find x such that the matrix $\begin{pmatrix} 1 & -2 & 2x \\ 1+2x & -2 & 0 \\ -1 & 2 & 1 \end{pmatrix}$ is singular.

11 Consider the matrices A, B and C which are given by

$$A = \begin{pmatrix} 2 & 1 \\ 7 & 4 \end{pmatrix}, B = \begin{pmatrix} m & n \\ p & q \end{pmatrix} \text{ and } C = \begin{pmatrix} 2 & 9 \\ -3 & 8 \end{pmatrix}.$$

Find the values of m, n, p and q such that $AB + B = C$.

12 Consider the matrix $M = \begin{pmatrix} 6a+1 & 2 \\ 3a & 1 \end{pmatrix}$, $a \neq 0$.

   a) Find M^{-1}.
   b) Additionally, you are given $a = 1$, $N = \begin{pmatrix} 6 & 5 \\ -2 & 3 \end{pmatrix}$ and $P = \begin{pmatrix} 3 & -1 \\ 2 & 7 \end{pmatrix}$.
   Solve the equation $XM + N = P$ for X.

13 Consider the matrix $M = \begin{pmatrix} 1 & 3 & a \\ 2 & 2 & 1 \\ -a & 2 & -2 \end{pmatrix}$.

   a) Find a such that $\det M = 7$.
   b) Write down the inverse of M for the values of a found above.
   c) Hence, solve the system of equations:
   $$\begin{cases} x + 3y - z = 7 \\ 2x + 2y + z = 35 \\ x + 2y - 2z = 14 \end{cases}$$

14 If $A = \begin{pmatrix} 2x & 3 \\ -4x & x \end{pmatrix}$ and $\det A = 14$, find x.

15 Let $M = \begin{pmatrix} a & 2 \\ 2 & -1 \end{pmatrix}$, where $a \in \mathbb{Z}$.

   a) Find M^2 in terms of a.
   b) If M^2 is equal to $\begin{pmatrix} 5 & -4 \\ -4 & 5 \end{pmatrix}$, find the value of a.

   Using this value of a, find M^{-1} and hence solve the system of equations:
   $$-x + 2y = -3$$
   $$2x - y = 3$$

16 Two matrices are given, where $A = \begin{pmatrix} 5 & 2 \\ 2 & 0 \end{pmatrix}$ and $BA = \begin{pmatrix} 11 & 2 \\ 44 & 8 \end{pmatrix}$. Find B.

17 The matrices A, B and X are given, where

$$A = \begin{pmatrix} 3 & 1 \\ -5 & 6 \end{pmatrix}, B = \begin{pmatrix} 4 & 8 \\ 0 & -3 \end{pmatrix} \text{ and } X = \begin{pmatrix} a & b \\ c & d \end{pmatrix}, \text{ with } a, b, c, d \in \mathbb{R}.$$

Find the values of a, b, c and d such that $AX + X = B$.

18 $A = \begin{pmatrix} 5 & -2 \\ 7 & 1 \end{pmatrix}$ is a 2 × 2 matrix.

   a) Write out A^{-1}.

   b) (i) If $XA + B = B$, where B, C and X are 2 × 2 matrices, express X in terms of A^{-1}, B and C.

   (ii) Find X if $B = \begin{pmatrix} 6 & 7 \\ 5 & -2 \end{pmatrix}$ and $C = \begin{pmatrix} -5 & 0 \\ -8 & 7 \end{pmatrix}$.

19 Given $A = \begin{pmatrix} a & b \\ c & 1 \end{pmatrix}$ and $B = \begin{pmatrix} 1 & 2 \\ d & c \end{pmatrix}$,

   a) write out $A + B$

   b) find AB.

20 a) Write out the inverse of the matrix $A = \begin{pmatrix} 1 & -3 & 1 \\ 2 & 2 & -1 \end{pmatrix}$.

   b) Hence, solve the system of simultaneous equations:

   $x - 3y + z = 1$

   $2x + 2y - z = 2$

   $x - 5y + 3z = 3$

21 The two matrices C and D are given, where

   $C = \begin{pmatrix} -2 & 4 \\ 1 & 7 \end{pmatrix}$ and $D = \begin{pmatrix} 5 & 2 \\ -1 & a \end{pmatrix}$.

   The matrix Q is given such that $3Q = 2C - D$.

   a) Find Q.

   b) Find CD.

   c) Find D^{-1}.

Questions 14–18, 20–21: © International Baccalaureate Organization

6 Trigonometric Functions and Equations

Assessment statements

3.1 The circle: radian measure of angles; length of an arc; area of a sector.
3.2 Definition of $\cos\theta$ and $\sin\theta$ in terms of the unit circle.
Definition of $\tan\theta$ as $\frac{\sin\theta}{\cos\theta}$.
The identity $\cos^2\theta + \sin^2\theta = 1$.
3.3 Double angle formulae: $\sin 2\theta = 2\sin\theta\cos\theta$; $\cos 2\theta = \cos^2\theta - \sin^2\theta$.
3.4 The circular functions $\sin x$, $\cos x$ and $\tan x$: their domains and ranges; their periodic nature; and their graphs.
Composite functions of the form $f(x) = a\sin(b(x + c)) + d$.
3.5 Solution of trigonometric equations in a finite interval.
Equations of the type $a\sin(b(x + c)) = k$.
Equations leading to quadratic equations in, for example, $\sin x$.
Graphical interpretation of the above.

Introduction

The word *trigonometry* comes from two Greek words, *trigonon* and *metron*, meaning 'triangle measurement'. Trigonometry developed out of the use and study of triangles, in surveying, navigation, architecture and astronomy, to find relationships between lengths of sides of triangles and measurement of angles. As a result, trigonometric functions were initially defined as functions of angles – that is, functions with angle measurements as their domains. With the development of calculus in the seventeenth century and the growth of knowledge in the sciences, the application of trigonometric functions grew to include a wide variety of periodic (repetitive) phenomena such as wave motion, vibrating strings, oscillating pendulums, alternating electrical current and biological cycles. These applications of trigonometric functions require their domains to be sets of real numbers without reference to angles or triangles. Hence, trigonometry can be approached from two different perspectives – **functions of angles**, or **functions of real numbers**. The first perspective is the focus of the next chapter where trigonometric functions will be defined in terms of the **ratios of sides of a right triangle**. The second perspective is the focus of this chapter where trigonometric functions will be defined in terms of a real number that is the **length of an arc along the unit circle**. While it is possible to define trigonometric functions in these two different ways, they assign the same value (interpreted as an angle, an arc length, or simply a real number) to a particular real number. Although this chapter will not refer much to triangles, it seems fitting to begin by looking at angles and arc lengths – geometric objects indispensable to the two different ways of viewing trigonometry.

6.1 Angles, circles, arcs and sectors

Angles

An **angle** in a plane is made by rotating a ray about its endpoint, called the **vertex** of the angle. The starting position of the ray is called the **initial side** and the position of the ray after rotation is called the **terminal side** of the angle (Figure 6.1). An angle having its vertex at the origin and its initial side lying on the positive x-axis is said to be in **standard position** (Figure 6.2). A **positive angle** is produced when a ray is rotated in an anticlockwise direction, and a **negative angle** when a ray is rotated in a clockwise direction. Two angles in standard position that have the same terminal sides – regardless of the direction or number of rotations – are called **coterminal angles**. Greek letters are often used to represent angles, and the direction of rotation is indicated by an arc with an arrow at its endpoint. The x- and y-axes divide the coordinate plane into four quadrants (numbered with Roman numerals). Figure 6.3 shows a positive angle α (alpha) and a negative angle β (beta) that are coterminal in quadrant III.

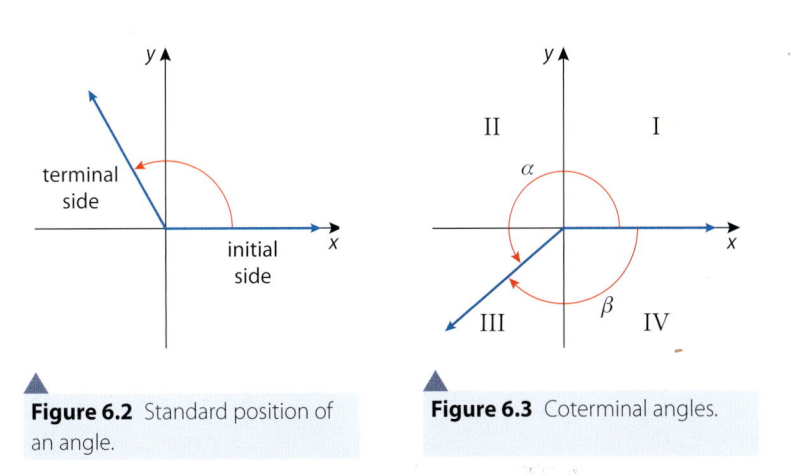

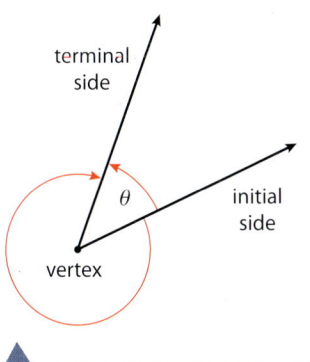

Figure 6.1

Figure 6.2 Standard position of an angle.

Figure 6.3 Coterminal angles.

Measuring angles: degree measure and radian measure

Perhaps the most natural unit for measuring large angles is the **revolution**. For example, most cars have an instrument (a tachometer) that indicates the number of revolutions per minute (rpm) at which the engine is operating. However, to measure smaller angles, we need a smaller unit. A common unit for measuring angles is the **degree**, of which there are 360 in one revolution. Hence, the unit of one degree (1°) is defined to be 1/360 of one anticlockwise revolution about the vertex.

6 Trigonometric Functions and Equations

> The convention of having 360 degrees in one revolution can be traced back around 4000 years to ancient Babylonian civilizations. The number system most widely used today is a base 10, or **decimal**, system. Babylonian mathematics used a base 60, or **sexagesimal**, number system. Although 60 may seem to be an awkward number to have as a base, it does have certain advantages. It is the smallest number that has 2, 3, 4, 5 and 6 as factors – and it also has factors of 10, 12, 15, 20 and 30. But why 360 degrees? We're not certain but it may have to do with the Babylonians assigning 60 divisions to each angle in an equilateral triangle and exactly six equilateral triangles can be arranged around a single point. That makes $6 \times 60 = 360$ equal divisions in one full revolution. There are few numbers as small as 360 that have so many different factors. This makes the degree a useful unit for dividing one revolution into an equal number of parts. 120 degrees is $\frac{1}{3}$ of a revolution, 90 degrees is $\frac{1}{4}$ of a revolution, 60 degrees is $\frac{1}{6}$, 45 degrees is $\frac{1}{8}$, and so on.

There is another method of measuring angles that is more natural. Instead of dividing a full revolution into an arbitrary number of equal divisions (e.g. 360), consider an angle that has its vertex at the centre of a circle (a **central angle**) and subtends (or intercepts) a part of the circle, called an **arc of the circle**. Figure 6.4 shows three circles with radii of different lengths ($r_1 < r_2 < r_3$) and the same central angle θ (theta) subtending (intercepting) the arc lengths s_1, s_2 and s_3. Regardless of the size of the circle (i.e. length of the radius), the ratio of arc length (s) to radius (r) for a given circle will be constant. For the angle θ in Figure 6.4, $\frac{s_1}{r_1} = \frac{s_2}{r_2} = \frac{s_3}{r_3}$. Because this ratio is an arc length divided by another length (radius), it is just an ordinary real number and has no units.

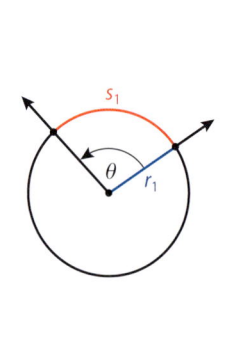

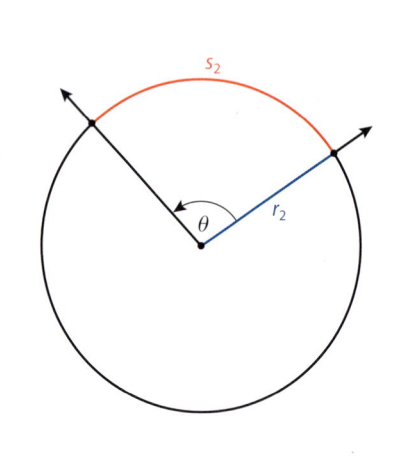

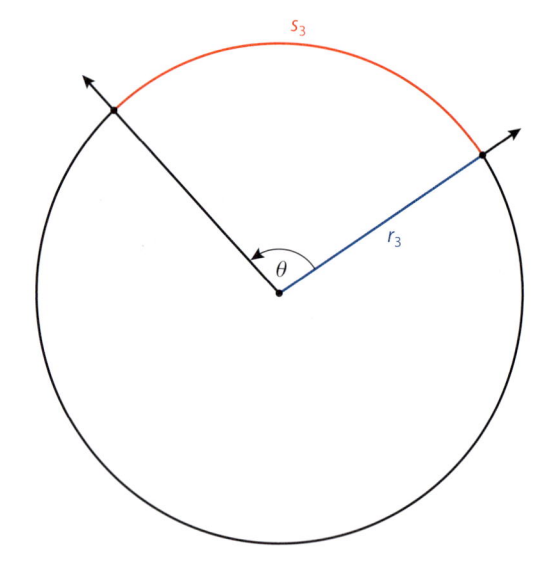

Figure 6.4 Different circles with the same central angle θ subtending different arcs, but the ratio of arc length to radius remains constant.

Minor and major arcs
If a central angle is **less** than 180°, the subtended arc is referred to as a **minor arc**. If a central angle is **greater** than 180°, the subtended arc is referred to as a **major arc**.

The ratio $\frac{s}{r}$ indicates how many radius lengths, r, fit into the length of the arc s. For example, if $\frac{s}{r} = 2$, the length of s is equal to two radius lengths. This accounts for the name **radian** and leads to the following definition.

Radian measure
One **radian** is the measure of a central angle θ of a circle that subtends an arc s of the circle that is exactly the same length as the radius r of the circle. That is, when $\theta = 1$ radian, arc length = radius.

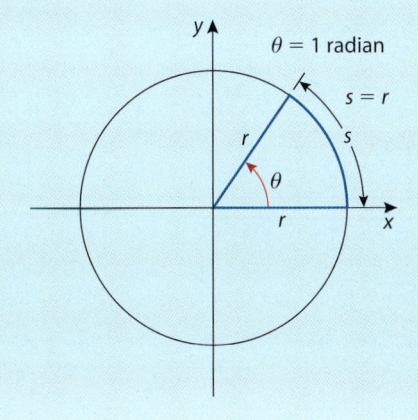

> When the measure of an angle is, for example, 5 radians, the word 'radians' does not indicate units (as when writing centimetres, seconds or degrees) but indicates the *method* of angle measurement. If the measure of an angle is in units of degrees, we must indicate this by word or symbol. For example, $\theta = 5$ degrees or $\theta = 5°$. However, when radian measure is used it is customary to write no units or symbol. For example, a central angle θ that subtends an arc equal to five radius lengths (radians) is simply given as $\theta = 5$.

The unit circle

When an angle is measured in radians it makes sense to draw it, or visualize it, so that it is in standard position. It follows that the angle will be a central angle of a circle whose centre is at the origin, as shown above. As Figure 6.4 illustrated, it makes no difference what size circle is used. The most practical circle to use is the circle with a radius of one unit so the radian measure of an angle will simply be equal to the length of the subtended arc.

$$\text{Radian measure: } \theta = \frac{s}{r} \qquad \text{If } r = 1, \text{ then } \theta = \frac{s}{1} = s.$$

The circle with a radius of one unit and centre at the origin $(0, 0)$ is called the **unit circle** (Figure 6.5). The equation for the unit circle is $x^2 + y^2 = 1$. Because the circumference of a circle with radius r is $2\pi r$, a central angle of one full anticlockwise revolution (360°) subtends an arc on the unit circle equal to 2π units. Hence, if an angle has a degree measure of 360°, its radian measure is exactly 2π. It follows that an angle of 180° has a radian measure of exactly π. This fact can be used to convert between degree measure and radian measure, and vice versa.

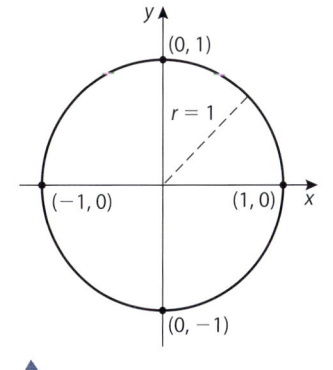

Figure 6.5 The unit circle.

Conversion between degrees and radians
Because $180° = \pi$ radians, $1° = \frac{\pi}{180}$ radians, and 1 radian $= \frac{180°}{\pi}$. An angle with a radian measure of 1 has a degree measure of approximately 57.3° (to 3 significant figures).

Example 1

The angles of 30° and 45°, and their multiples, are often encountered in trigonometry. Convert 30° and 45° to radian measure and sketch the corresponding arc on the unit circle. Use these results to convert 60° and 90° to radian measure.

163

6 Trigonometric Functions and Equations

• **Hint:** It is very helpful to be able to quickly recall the results from Example 1:
$30° = \frac{\pi}{6}$, $45° = \frac{\pi}{4}$, $60° = \frac{\pi}{3}$ and $90° = \frac{\pi}{2}$. Of course, not all angles are multiples of 30° or 45° when expressed in degrees, and not all angles are multiples of $\frac{\pi}{6}$ or $\frac{\pi}{4}$ when expressed in radians. However, these 'special' angles often appear in problems and applications. Knowing these four facts can help you to quickly convert mentally between degrees and radians for many common angles. For example, to convert 225° to radians, apply the fact that $225° = 5(45°)$. Since $45° = \frac{\pi}{4}$, then $225° = 5(45°) = 5\left(\frac{\pi}{4}\right) = \frac{5\pi}{4}$. And another example, convert $\frac{11\pi}{6}$ to degrees: $\frac{11\pi}{6} = 11\left(\frac{\pi}{6}\right) = 11(30°) = 330°$.

Solution

(Note that the 'degree' units cancel.)

$$30° = 30°\left(\frac{\pi}{180°}\right) = \frac{30°}{180°}\pi = \frac{\pi}{6}$$

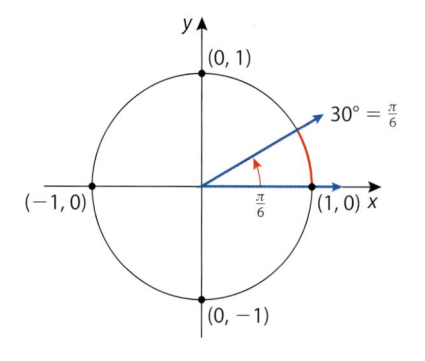

$$45° = 45°\left(\frac{\pi}{180°}\right) = \frac{45°}{180°}\pi = \frac{\pi}{4}$$

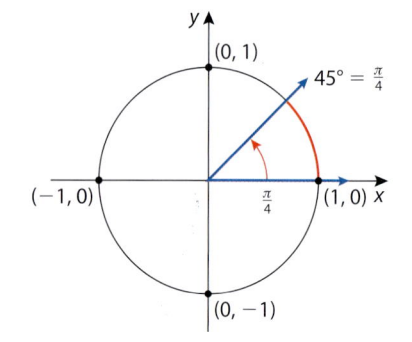

Since $60° = 2(30°)$ and $30° = \frac{\pi}{6}$, then $60° = 2\left(\frac{\pi}{6}\right) = \frac{\pi}{3}$. Similarly, $90° = 2(45°)$ and $45° = \frac{\pi}{4}$, so $90° = 2\left(\frac{\pi}{4}\right) = \frac{\pi}{2}$.

Example 2

a) Convert the following radian measures to degrees. Express exactly, if possible. Otherwise, express accurate to 3 significant figures.
   (i) $\frac{4\pi}{3}$ (ii) $-\frac{3\pi}{2}$ (iii) 5   (iv) 1.38

b) Convert the following degree measures to radians. Express exactly, if possible. Otherwise, express accurate to 3 significant figures.
   (i) 135°   (ii) −150°   (iii) 175°   (iv) 10°

• **Hint:** All GDCs will have a degree mode and a radian mode. Before doing any calculations with angles on your GDC, be certain that the mode setting for angle measurement is set correctly. Although you may be more familiar with degree measure, as you progress further in mathematics – and especially in calculus – radian measure is far more useful.

Solution

a) (i) $\frac{4\pi}{3} = 4\left(\frac{\pi}{3}\right) = 4(60°) = 240°$

   (ii) $-\frac{3\pi}{2} = -\frac{3}{2}(\pi) = -\frac{3}{2}(180°) = -270°$

   (iii) $5\left(\frac{180°}{\pi}\right) \approx 286.479° \approx 286°$

   (iv) $1.38\left(\frac{180°}{\pi}\right) \approx 79.068° \approx 79.1°$

b) (i) $135° = 3(45°) = 3\left(\dfrac{\pi}{4}\right) = \dfrac{3\pi}{4}$

(ii) $-150° = -5(30°) = -5\left(\dfrac{\pi}{6}\right) = -\dfrac{5\pi}{6}$

(iii) $175°\left(\dfrac{\pi}{180°}\right) \approx 3.0543 \approx 3.05$

(iv) $10°\left(\dfrac{\pi}{180°}\right) \approx 0.174\,53 \approx 0.175$

Because 2π is approximately 6.28 (3 significant figures), there are a little more than six radius lengths in one revolution, as shown in Figure 6.6.

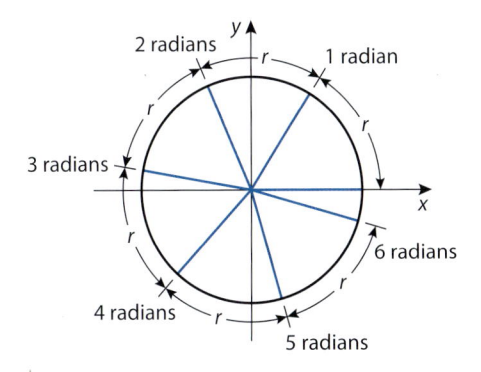

◀ Figure 6.6

Arc length

For any angle θ, its radian measure is given by $\theta = \dfrac{s}{r}$. Simple rearrangement of this formula leads to another formula for computing arc length.

> **Arc length**
> For a circle of radius r, a central angle θ subtends an arc of the circle of length s given by
> $$s = r\theta$$
> where θ is in radian measure.

Example 3

A circle has a radius of 10 cm. Find the length of the arc of the circle subtended by a central angle of 150°.

Solution

To use the formula $s = r\theta$, we must first convert 150° to radian measure.

$$150° = 150°\left(\dfrac{\pi}{180°}\right) = \dfrac{150\pi}{180} = \dfrac{5\pi}{6}$$

Given that the radius, r, is 10 cm, substituting into the formula gives

$$s = r\theta \Rightarrow s = 10\left(\dfrac{5\pi}{6}\right) = \dfrac{25\pi}{3} \approx 26.179\,94$$

The length of the arc is approximately 26.18 cm (4 significant figures).

Note that the units of the product $r\theta$ are the same as the units of r because in radian measure θ has no units.

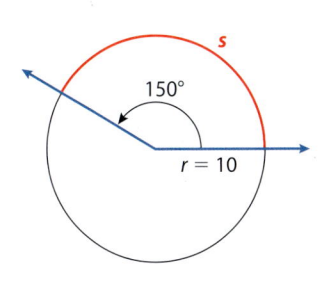

6 Trigonometric Functions and Equations

Example 4

The diagram shows a circle of centre O with radius $r = 6$ cm. Angle AOB subtends the minor arc AB such that the length of the arc is 10 cm. Find the measure of angle AOB in degrees to 3 significant figures.

Solution

From the arc length formula, $s = r\theta$, we can state that $\theta = \frac{s}{r}$. Remember that the result for θ will be in radian measure. Therefore, angle $AOB = \frac{10}{6} = \frac{5}{3}$ or $1.\overline{6}$ radians. Now, we convert to degrees: $\frac{5}{3}\left(\frac{180°}{\pi}\right) \approx 95.492\,97°$. The degree measure of angle AOB is approximately $95.5°$.

Geometry of a circle

It is helpful to recall some fundamental properties of a circle (Figure 6.7).

Figure 6.7

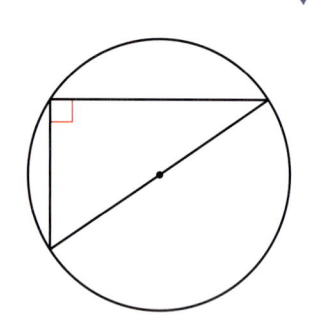

The angle inscribed in a semicircle is a right angle.

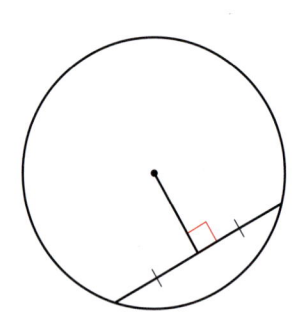

The line segment from the centre perpendicular to a chord also bisects the chord.

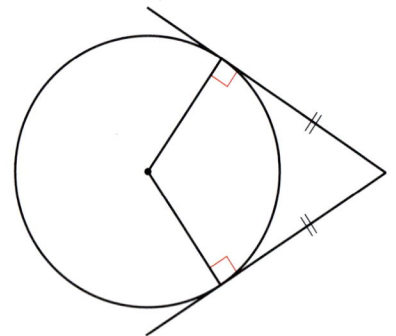

A tangent to a circle is perpendicular to the radius drawn to the point of tangency.

If two tangents share an external point, the distances from the external point to the point of tangency are equal.

Sector of a circle

A **sector of a circle** is the region bounded by an arc of the circle and the two sides of a central angle (Figure 6.8). The ratio of the area of a sector to the area of the circle (πr^2) is equal to the ratio of the length of the subtended arc to the circumference of the circle ($2\pi r$). If s is the arc length and A is the area of the sector, we can write the following proportion: $\frac{A}{\pi r^2} = \frac{s}{2\pi r}$. Solving for A gives $A = \frac{\pi r^2 s}{2\pi r} = \frac{1}{2}rs$. From the formula for arc length we have $s = r\theta$, with θ the radian measure of the central angle. Substituting $r\theta$ for s gives the area of a sector to be $A = \frac{1}{2}rs = \frac{1}{2}r(r\theta) = \frac{1}{2}r^2\theta$. This result makes sense because, if the sector is the entire circle, $\theta = 2\pi$ and area $A = \frac{1}{2}r^2\theta = \frac{1}{2}r^2(2\pi) = \pi r^2$, which is the formula for the area of a circle.

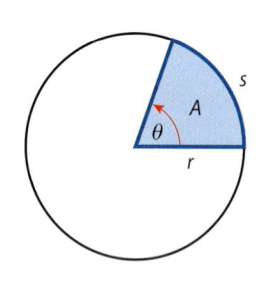

Figure 6.8 Sector of a circle.

Area of a sector

In a circle of radius r, the area of a sector with a central angle θ measured in radians is
$$A = \frac{1}{2}r^2\theta$$

Example 5

A circle of radius 9 cm has a sector whose central angle has radian measure $\frac{2\pi}{3}$. Find the exact values of the following: a) the length of the arc subtended by the central angle, and b) the area of the sector.

Solution

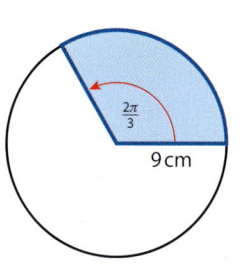

a) $s = r\theta \Rightarrow s = 9\left(\frac{2\pi}{3}\right) = 6\pi$

The length of the arc is exactly 6π cm.

b) $A = \frac{1}{2}r^2\theta \Rightarrow A = \frac{1}{2}(9)^2\left(\frac{2\pi}{3}\right) = 27\pi$

The area of the sector is exactly 27π cm^2.

• **Hint:** The formula for arc length, $s = r\theta$, and the formula for area of a sector, $A = \frac{1}{2}r^2\theta$, are true only when θ is in radians.

Exercise 6.1

In questions 1–9, find the exact radian measure of the angle given in degree measure.

1 60° **2** 150° **3** −270°
4 36° **5** 135° **6** 50°
7 −45° **8** 400° **9** −480°

In questions 10–18, find the degree measure of the angle given in radian measure. If possible, express exactly. Otherwise, express accurate to 3 significant figures.

10 $\frac{3\pi}{4}$ **11** $-\frac{7\pi}{2}$ **12** 2
13 $\frac{7\pi}{6}$ **14** −2.5   **15** $\frac{5\pi}{3}$
16 $\frac{\pi}{12}$ **17** 1.57   **18** $\frac{8\pi}{3}$

In questions 19–24, the measure of an angle in standard position is given. Find two angles – one positive and one negative – that are coterminal with the given angle. If no units are given, assume the angle is in radian measure.

19 30°   **20** $\frac{3\pi}{2}$ **21** 175°
22 $-\frac{\pi}{6}$ **23** $\frac{5\pi}{3}$ **24** 3.25

In questions 25 and 26, find the length of the arc s in the figure.

25 120°, $r = 6$ cm

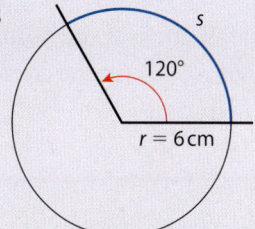

26 70°, $r = 12$ cm

167

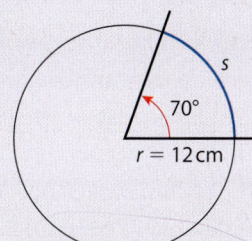

27 Find the angle θ in the figure in both radian measure and degree measure.

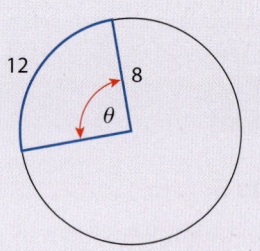

28 Find the radius r of the circle in the figure.

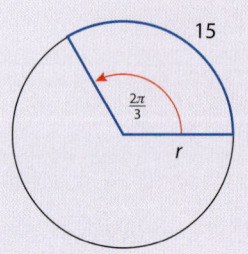

In questions 29 and 30, find the area of the sector in each figure.

29

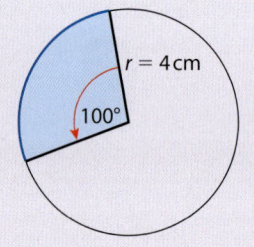

30

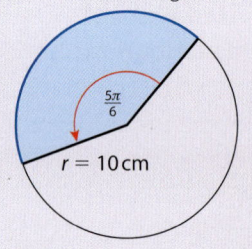

31 An arc of length 60 cm subtends a central angle α in a circle of radius 20 cm. Find the measure of α in both degrees and radians, approximate to 3 significant figures.

32 Find the length of an arc that subtends a central angle with radian measure of 2 in a circle of radius 16 cm.

33 The area of a sector of a circle with a central angle of 60° is 24 cm². Find the radius of the circle.

6.2 The unit circle and trigonometric functions

Several important functions can be described by mapping the coordinates of points on the real number line onto the points of the unit circle. Recall from the previous section that the unit circle has its centre at (0, 0), it has a radius of one unit and its equation is $x^2 + y^2 = 1$.

A wrapping function: the real number line and the unit circle

Suppose that the real number line is tangent to the unit circle at the point (1, 0) – and that zero on the number line matches with (1, 0) on the circle, as shown in Figure 6.9. Because of the properties of circles, the real number line in this position will be perpendicular to the x-axis. The scales on the number line and the x- and y-axes need to be the same. Imagine that the real number line is flexible like a string and can wrap around the circle, with zero on the number line remaining fixed to the point (1, 0) on the unit circle. When the top portion of the string moves along the circle, the wrapping is anticlockwise ($t > 0$), and when the bottom portion of the string moves along the circle, the wrapping is clockwise ($t < 0$). As the string wraps around the unit circle, each real number t on the string is mapped onto a point (x, y) on the circle. Hence, the real number line from 0 to t makes an arc of length t starting on the circle at $(1, 0)$ and ending at the point (x, y) on the circle. For example, since the circumference of the unit circle is 2π, the number $t = 2\pi$ will be wrapped anticlockwise around the circle to the point $(1, 0)$. Similarly, the number $t = \pi$ will be wrapped anticlockwise halfway around the circle to the point $(-1, 0)$ on the circle. And the number $t = -\frac{\pi}{2}$ will be wrapped clockwise one-quarter of the way around the circle to the point $(0, -1)$ on the circle. Note that each number t on the real number line is mapped (corresponds) to *exactly one* point on the unit circle, thereby satisfying the definition of a function (Section 2.1) – consequently this mapping is called a **wrapping function**.

Before we leave our mental picture of the string (representing the real number line) wrapping around the unit circle, consider any pair of points on the string that are exactly 2π units from each other. Let these two points represent the real numbers t_1 and $t_1 + 2\pi$. Because the circumference of the unit circle is 2π, these two numbers will be mapped to the same point on the unit circle. Furthermore, consider the infinite number of points whose distance from t_1 is any integer multiple of 2π, i.e. $t_1 + k \cdot 2\pi, k \in \mathbb{Z}$, and again all of these numbers will be mapped to the same point on the unit circle. Consequently, the wrapping function is not a one-to-one function as defined in Section 2.3. Output for the function (points on the unit circle) are unchanged by the addition of any integer multiple of 2π to any input value (a real number). Functions that behave in such a repetitive (or cyclic) manner are called **periodic**.

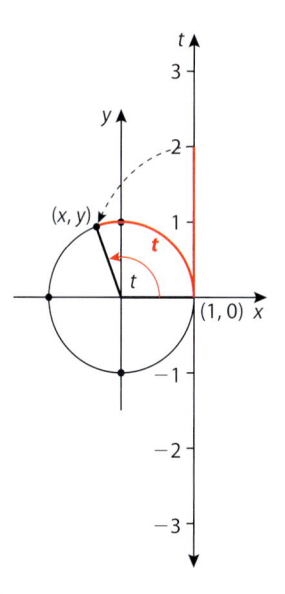

Figure 6.9

Definition of a periodic function
A function f such that $f(x) = f(x + p)$ is a **periodic function**. If p is the least positive constant for which $f(x) = f(x + p)$ is true, p is called the **period** of the function.

Trigonometric functions

From our discussions about functions in Chapter 2, it is customary for a function to have a domain (input) and range (output) that are sets having individual numbers as elements. We use the individual coordinates x and y of the points on the unit circle to define a certain set of functions called **trigonometric functions**. For this course, we define three trigonometric functions: the **sine** function, the **cosine** function and the **tangent** function. The names of these functions are often abbreviated in writing (but not speaking) as **sin**, **cos** and **tan**, respectively. When the real number t is wrapped to a point (x, y) on the unit circle, the value of the y-coordinate is assigned to the sine function; the x-coordinate is assigned to the cosine function; and the ratio of the two coordinates $\frac{y}{x}$ is assigned to the tangent function.

> We are surrounded by periodic functions. A few examples include: the average daily temperature variation during the year; sunrise and the day of the year; animal populations over many years; the height of tides and the position of the Moon; and your height above ground when riding a Ferris wheel and the rotation of the wheel.

169

6 Trigonometric Functions and Equations

> **The trigonometric functions: sine, cosine and tangent**
> Let t be any real number and (x, y) a point on the unit circle to which t is mapped. Then the function definitions are:
> $$\sin t = y \qquad \cos t = x \qquad \tan t = \frac{\sin t}{\cos t} = \frac{y}{x}, x \neq 0$$
>
>
>
> On the unit circle: $x = \cos t, y = \sin t$.
> Signs of the trigonometric functions depend on the quadrant where the arc t terminates.

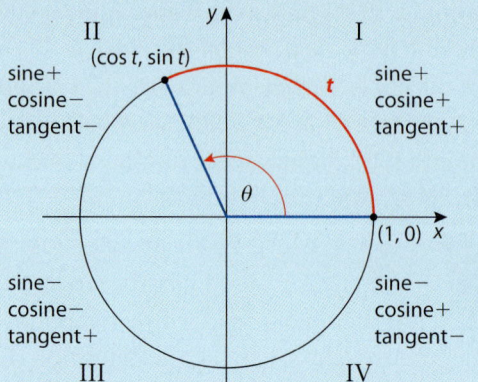

• **Hint:** When sine, cosine and tangent are defined as circular functions based on the unit circle, radian measure is used. The values for the domain of the sine and cosine functions are real numbers that are arc lengths on the unit circle. As we know from the previous section, the arc length on the unit circle subtends an angle in standard position, whose radian measure is equivalent to the arc length (see definition box above).

Because the definitions for the sine, cosine and tangent functions given here do not refer to triangles or angles, but rather to a real number representing an arc length on the unit circle, the name **circular functions** is also given to them. In fact, from this chapter's perspective that these functions are *functions of real numbers* rather than *functions of angles*, 'circular' is a more appropriate adjective than 'trigonometric'. Nevertheless, trigonometric is the more common label and will be used throughout the book.

Let's use the definitions for these three trigonometric, or circular, functions to evaluate them for some 'easy' values of t.

Example 6

Evaluate the sine, cosine and tangent functions for the following values of t.

a) $t = 0$
b) $t = \dfrac{\pi}{2}$
c) $t = \pi$
d) $t = \dfrac{3\pi}{2}$
e) $t = 2\pi$

Solution

Evaluating the sin, cos and tan functions for any value of t involves finding the coordinates of the point on the unit circle where the arc of length t will 'wrap to' (or terminate), starting at the point $(1, 0)$. It is useful to remember that an arc of length π is equal to one-half of the circumference of the unit circle. All of the values for t in this example are positive, so the arc length will wrap along the unit circle in an anticlockwise direction.

a) An arc of length $t = 0$ has no length so it 'terminates' at the point $(1, 0)$. Therefore, by definition
$$\sin 0 = y = 0$$
$$\cos 0 = x = 1$$
$$\tan 0 = \frac{y}{x} = \frac{0}{1} = 0$$

b) An arc of length $t = \frac{\pi}{2}$ is equivalent to one-quarter of the circumference of the unit circle (Figure 6.10), so it terminates at the point $(0, 1)$. By definition:

$\sin \frac{\pi}{2} = y = 1$

$\cos \frac{\pi}{2} = x = 0$

$\tan \frac{\pi}{2} = \frac{y}{x} = \frac{1}{0}$ which is undefined

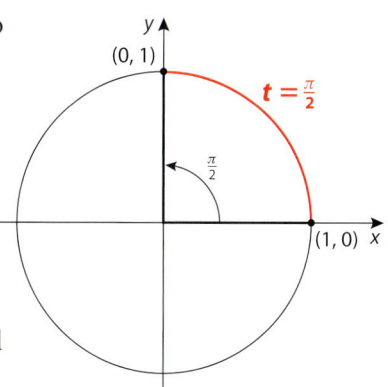

◀ **Figure 6.10**

c) An arc of length $t = \pi$ is equivalent to one-half of the circumference of the unit circle (Figure 6.11), so it terminates at the point $(-1, 0)$. By definition:

$\sin \pi = y = 0$

$\cos \pi = x = -1$

$\tan \pi = \frac{y}{x} = \frac{0}{-1} = 0$

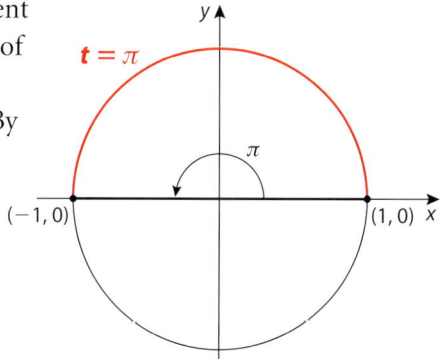

◀ **Figure 6.11**

d) An arc of length $t = \frac{3\pi}{2}$ is equivalent to three-quarters of the circumference of the unit circle (Figure 6.12), so it terminates at the point $(0, -1)$.
By definition:

$\sin \frac{3\pi}{2} = y = -1$

$\cos \frac{3\pi}{2} = x = 0$

$\tan \frac{3\pi}{2} = \frac{y}{x} = \frac{-1}{0}$ which is undefined

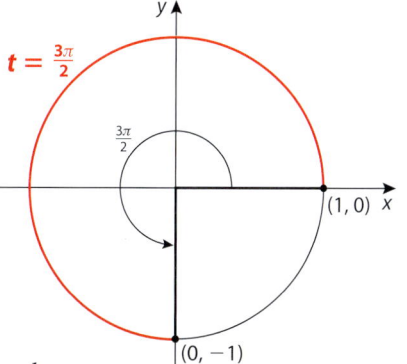

◀ **Figure 6.12**

e) An arc of length $t = 2\pi$ is equivalent to the circumference of the unit circle (Figure 6.13), so it terminates at the point $(1, 0)$. By definition:

$\sin 2\pi = y = 0$

$\cos 2\pi = x = 1$

$\tan 2\pi = \frac{y}{x} = \frac{0}{1} = 0$

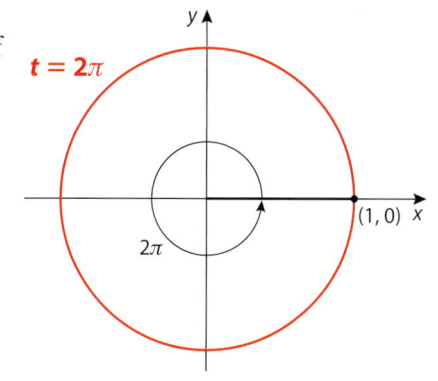

◀ **Figure 6.13**

6 Trigonometric Functions and Equations

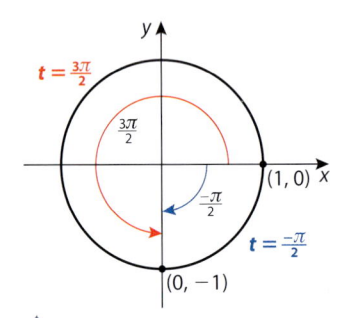

Figure 6.14

Domain and range of trigonometric functions

Because every real number t corresponds to exactly one point on the unit circle, the domain for both the sine function and the cosine function is the set of all real numbers. From Example 6, parts b) and d), where the value of $\tan t$ is undefined, it is clear that the domain for the tangent function is not all real numbers. Given the definitions $\tan t = \frac{y}{x}$, $x \neq 0$, and $\cos t = x$, it is clear that any value of t that corresponds to a point on the unit circle with an x-coordinate equal to zero cannot be in the domain of the tangent function (division by zero is undefined). From Example 6, we can see that $\cos t = 0$ for $t = \frac{\pi}{2}$, $t = \frac{3\pi}{2}$ and then for $t = \frac{5\pi}{2}$, and for $t = \frac{7\pi}{2}$, and so on. What about negative values for t (arc lengths wrapped in a clockwise direction)? Clearly an arc length of $t = -\frac{\pi}{2}$ will terminate at $(0, -1)$, the same as when $t = \frac{3\pi}{2}$, as shown in Figure 6.14. And $\cos t = 0$ also for $t = -\frac{3\pi}{2}$, $t = -\frac{5\pi}{2}$, and so on. Therefore, the domain of the tangent function is all real numbers but *not* including the infinite set of numbers generated by adding any integer multiple of π to $\frac{\pi}{2}$.

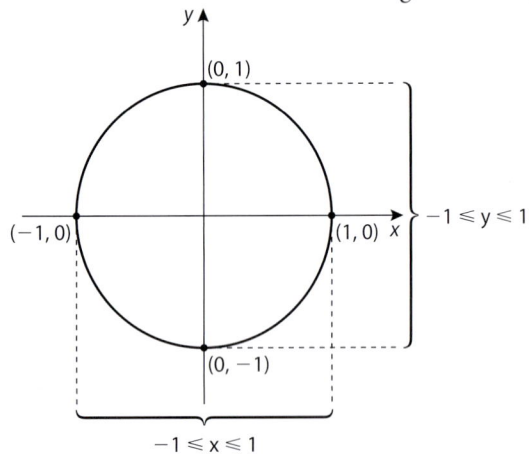

To determine the range of the sine and cosine functions, consider the unit circle shown in Figure 6.15. Because $\sin t = y$ and $\cos t = x$ and (x, y) is on the unit circle, we can see that $-1 \leq y \leq 1$ and $-1 \leq x \leq 1$. Therefore, $-1 \leq \sin t \leq 1$ and $-1 \leq \cos t \leq 1$. The range for the tangent function will not be bounded as for sine and cosine. As t approaches values where $x = \cos t = 0$, the value of $\frac{y}{x} = \tan t$ will become very large – either negative or positive, depending on which quadrant t is in. Therefore, $-\infty < \tan t < \infty$; or, in other words, $\tan t$ can be any real number.

Figure 6.15

Domain and range of sine, cosine and tangent functions
$f(t) = \sin t$ domain: $\{t : t \in \mathbb{R}\}$ range: $-1 \leq f(t) \leq 1$
$f(t) = \cos t$ domain: $\{t : t \in \mathbb{R}\}$ range: $-1 \leq f(t) \leq 1$
$f(t) = \tan t$ domain: $\left\{t : t \in \mathbb{R}, t \neq \frac{\pi}{2} + k\pi, k \in \mathbb{Z}\right\}$ range: $f(t) \in \mathbb{R}$

From our previous discussion of periodic functions, we can conclude that all three of these trigonometric functions are periodic. Given that the sine and cosine functions are generated directly from the wrapping function, the period of each of these functions is 2π. That is,

$$\sin t = \sin(t + k \cdot 2\pi), k \in \mathbb{Z} \text{ and } \cos t = \cos(t + k \cdot 2\pi), k \in \mathbb{Z}$$

Initial evidence from Example 6 indicates that the period of the tangent function is π. That is,

$$\tan t = \tan(t + k \cdot \pi), k \in \mathbb{Z}$$

We will establish these results graphically in the next section. Also note that since these functions are periodic then they are not one-to-one functions.

Evaluating trigonometric functions

In Example 6, the unit circle was divided into four equal arcs corresponding to t values of $0, \frac{\pi}{2}, \pi, \frac{3\pi}{2}$ and 2π. Let's evaluate the sine, cosine and tangent functions for further values of t that would correspond to dividing the unit circle into eight equal arcs. Let's also make use of the symmetry of the unit circle. That is, any points on the unit circle which are reflections about the x-axis will have the same x-coordinate (same value of cosine), and any points on the unit circle which are reflections about the y-axis will have the same y-coordinate (same value of sine), as shown in Figure 6.16.

Example 7

Evaluate the sine, cosine and tangent functions for $t = \frac{\pi}{4}$, and then use that result to evaluate the same functions for $t = \frac{3\pi}{4}, t = \frac{5\pi}{4}$ and $t = \frac{7\pi}{4}$.

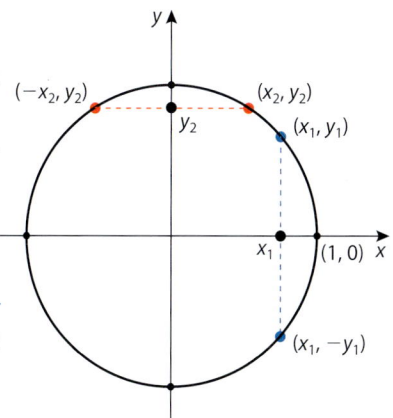

Figure 6.16

Solution

When an arc of length $t = \frac{\pi}{4}$ is wrapped along the unit circle starting at $(1, 0)$, it will terminate at a point (x_1, y_1) in quadrant I that is equidistant from $(1, 0)$ and $(0, 1)$. Since the line $y = x$ is a line of symmetry for the unit circle, (x_1, y_1) is on this line. Hence, the point (x_1, y_1) is the point of intersection of the unit circle $x^2 + y^2 = 1$ with the line $y = x$. Let's find the coordinates of the intersection point by solving this pair of simultaneous equations by substituting x for y into the equation $x^2 + y^2 = 1$.

$$x^2 + y^2 = 1 \Rightarrow x^2 + x^2 = 1 \Rightarrow 2x^2 = 1 \Rightarrow x^2 = \tfrac{1}{2} \Rightarrow x = \pm\sqrt{\tfrac{1}{2}} = \pm\tfrac{1}{\sqrt{2}}$$

Rationalising the denominator gives $x = \pm\frac{\sqrt{2}}{2}$ and, since the point is in the first quadrant, $x = \frac{\sqrt{2}}{2}$. Given that the point is on the line $y = x$ then $y = \frac{\sqrt{2}}{2}$. Therefore, the arc of length $t = \frac{\pi}{4}$ will terminate at the point $\left(\frac{\sqrt{2}}{2}, \frac{\sqrt{2}}{2}\right)$ on the unit circle. Using the symmetry of the unit circle, we can also determine the points on the unit circle where arcs of length $t = \frac{3\pi}{4}, t = \frac{5\pi}{4}$ and $t = \frac{7\pi}{4}$ terminate. These arcs and the coordinates of their terminal points are given in Figure 6.17.

Using the coordinates of these points, we can now evaluate the trigonometric functions for $t = \frac{\pi}{4}, \frac{3\pi}{4}, \frac{5\pi}{4}$ and $\frac{7\pi}{4}$. By definition:

$$t = \frac{\pi}{4}: \quad \sin\frac{\pi}{4} = y = \frac{\sqrt{2}}{2} \qquad \cos\frac{\pi}{4} = x = \frac{\sqrt{2}}{2} \qquad \tan\frac{\pi}{4} = \frac{y}{x} = \frac{\frac{\sqrt{2}}{2}}{\frac{\sqrt{2}}{2}} = 1$$

$$t = \frac{3\pi}{4}: \quad \sin\frac{3\pi}{4} = y = \frac{\sqrt{2}}{2} \qquad \cos\frac{3\pi}{4} = x = -\frac{\sqrt{2}}{2} \qquad \tan\frac{3\pi}{4} = \frac{y}{x} = \frac{\frac{\sqrt{2}}{2}}{-\frac{\sqrt{2}}{2}} = -1$$

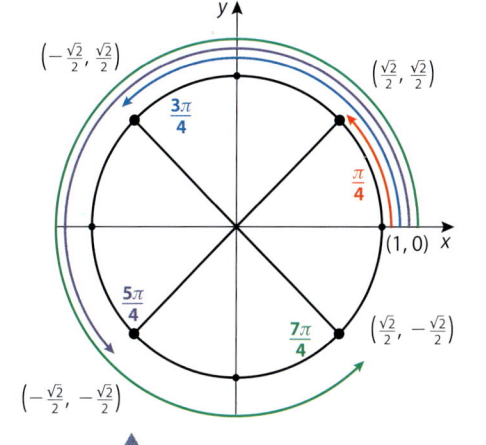

Figure 6.17

6 Trigonometric Functions and Equations

$$t = \frac{5\pi}{4}: \sin\frac{5\pi}{4} = y = -\frac{\sqrt{2}}{2} \quad \cos\frac{5\pi}{4} = x = -\frac{\sqrt{2}}{2} \quad \tan\frac{5\pi}{4} = \frac{y}{x} = \frac{-\frac{\sqrt{2}}{2}}{-\frac{\sqrt{2}}{2}} = 1$$

$$t = \frac{7\pi}{4}: \sin\frac{7\pi}{4} = y = -\frac{\sqrt{2}}{2} \quad \cos\frac{7\pi}{4} = x = \frac{\sqrt{2}}{2} \quad \tan\frac{7\pi}{4} = \frac{y}{x} = \frac{-\frac{\sqrt{2}}{2}}{\frac{\sqrt{2}}{2}} = -1$$

We can use a method similar to that of Example 7 to find the point on the unit circle where an arc of length $t = \frac{\pi}{6}$ terminates in the first quadrant. Then we can again apply symmetry about the line $y = x$ and the y- and x-axes to find points on the circle corresponding to arcs whose lengths are multiples of $\frac{\pi}{6}$, e.g. $\frac{2\pi}{6} = \frac{\pi}{3}$, $\frac{4\pi}{6} = \frac{2\pi}{3}$, etc. Arcs whose lengths are multiples of $\frac{\pi}{4}$ and $\frac{\pi}{6}$ correspond to eight equally spaced points and twelve equally spaced points, respectively, around the unit circle, as shown in Figures 6.18 and 6.19. The coordinates of these points give us the sine, cosine and tangent values for common values of t.

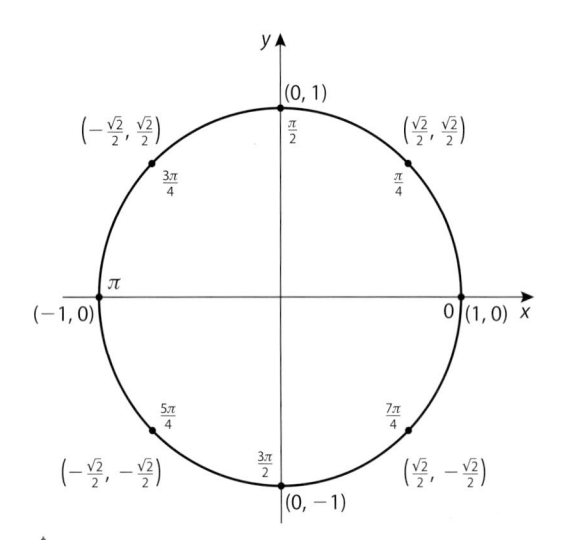

Figure 6.18 Arc lengths that are multiples of $\frac{\pi}{4}$ divide the unit circle into eight equally spaced points.

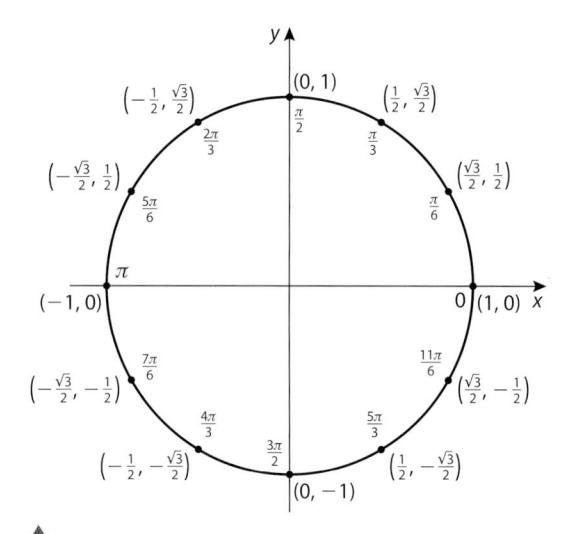

Figure 6.19 Arc lengths that are multiples of $\frac{\pi}{6}$ divide the unit circle into twelve equally spaced points.

You will find it very helpful to know from memory the exact values of sine and cosine for numbers that are multiples of $\frac{\pi}{6}$ and $\frac{\pi}{4}$. Use the unit circle diagrams shown in Figures 6.18 and 6.19 as a guide to help you do this and to visualize the location of the terminal points of different arc lengths. With the symmetry of the unit circle and a point's location in the coordinate plane telling us the sign of x and y (see definition box page 170), we only need to remember the sine and cosine of common values of t in the first quadrant and on the positive x- and y-axes. These are organized in Table 6.1.

t	$\sin t$	$\cos t$	$\tan t$
0	0	1	0
$\dfrac{\pi}{6}$	$\dfrac{1}{2}$	$\dfrac{\sqrt{3}}{2}$	$\dfrac{\sqrt{3}}{3}$
$\dfrac{\pi}{4}$	$\dfrac{\sqrt{2}}{2}$	$\dfrac{\sqrt{2}}{2}$	1
$\dfrac{\pi}{3}$	$\dfrac{\sqrt{3}}{2}$	$\dfrac{1}{2}$	$\sqrt{3}$
$\dfrac{\pi}{2}$	1	0	undefined

Table 6.1 The sine, cosine and tangent of common values of t.

If t is not a multiple of one of these common values, the values of the trigonometric functions for that number can be found using your GDC.

Example 8

Find the following function values. Find the exact value, if possible. Otherwise, find the approximate value accurate to 3 significant figures.

a) $\sin\dfrac{2\pi}{3}$ b) $\cos\dfrac{5\pi}{4}$ c) $\tan\dfrac{11\pi}{6}$ d) $\sin\dfrac{13\pi}{6}$ e) $\cos 3.75$

Solution

a) The terminal point for $\dfrac{2\pi}{3}$ is in the second quadrant and is the reflection in the y-axis of the terminal point for $\dfrac{\pi}{3}$, whose y-coordinate is $\dfrac{\sqrt{3}}{2}$. Therefore, $\sin\dfrac{2\pi}{3} = \dfrac{\sqrt{3}}{2}$.

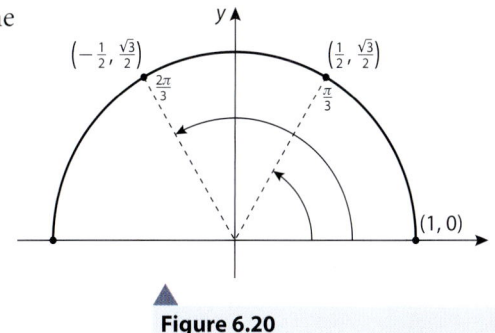

Figure 6.20

b) $\dfrac{5\pi}{4}$ is in the third quadrant. Hence, its x-coordinate and cosine must be negative. All of the odd multiples of $\dfrac{\pi}{4}$ have terminal points with x- and y-coordinates of $\pm\dfrac{\sqrt{2}}{2}$. Therefore, $\cos\dfrac{5\pi}{4} = -\dfrac{\sqrt{2}}{2}$.

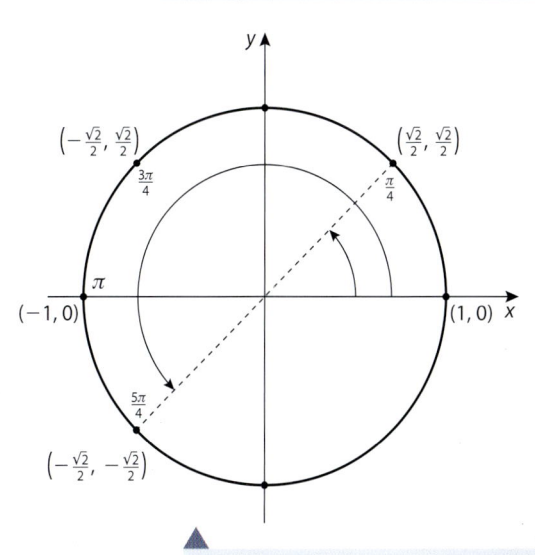

Figure 6.21

6 Trigonometric Functions and Equations

c) $\dfrac{11\pi}{6}$ is in the fourth quadrant, so its tangent will be negative. Its terminal point is the reflection in the x-axis of the terminal point for $\dfrac{\pi}{6}$, whose coordinates are $\left(\dfrac{\sqrt{3}}{2}, \dfrac{1}{2}\right)$. Therefore,

$$\tan\dfrac{11\pi}{6} = \dfrac{y}{x} = \dfrac{-\dfrac{1}{2}}{\dfrac{\sqrt{3}}{2}} = -\dfrac{1}{\sqrt{3}} = -\dfrac{\sqrt{3}}{3}.$$

d) $\dfrac{13\pi}{6}$ is more than one revolution. Because $\dfrac{13\pi}{6} = \dfrac{\pi}{6} + 2\pi$ and the period of the sine function is 2π [i.e. $\sin t = \sin(t + k \cdot 2\pi)$, $k \in \mathbb{Z}$] then $\sin\dfrac{13\pi}{6} = \sin\dfrac{\pi}{6} = \dfrac{1}{2}$.

e) An arc of length 3.75 will have its terminal point in the third quadrant since $\pi \approx 3.14$ and $\dfrac{3\pi}{2} \approx 4.71$, meaning $\pi < 3.75 < \dfrac{3\pi}{2}$. Hence, $\cos 3.75$ must be negative. To evaluate $\cos 3.75$ you must use your GDC. Be certain that it is set to radian mode. To an accuracy of 3 significant figures, $\cos 3.75 \approx -0.821$.

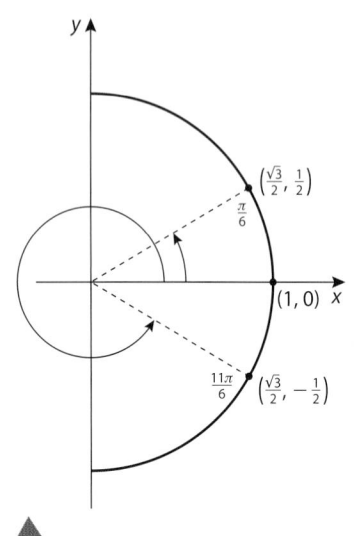

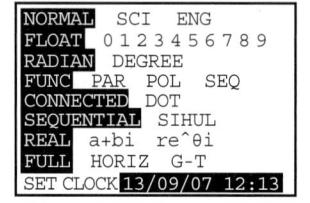

Figure 6.22

Exercise 6.2

In questions 1–9, t is the length of an arc on the unit circle starting from (1, 0). a) State the quadrant in which the terminal point of the arc lies. b) Find the coordinates of the terminal point (x, y) on the unit circle. Give exact values for x and y, if possible. Otherwise, approximate values to 3 significant figures.

1. $t = \dfrac{\pi}{6}$ 2. $t = \dfrac{5\pi}{3}$ 3. $t = \dfrac{7\pi}{4}$

4. $t = \dfrac{3\pi}{2}$ 5. $t = 2$ 6. $t = -\dfrac{\pi}{4}$

7. $t = -1$ 8. $t = -\dfrac{5\pi}{4}$ 9. $t = 3.52$

In questions 10–18, state the exact value (if possible) of the sine, cosine and tangent of the given real number.

10. $\dfrac{\pi}{3}$ 11. $\dfrac{5\pi}{6}$ 12. $-\dfrac{3\pi}{4}$

13. $\dfrac{\pi}{2}$ 14. $-\dfrac{4\pi}{3}$ 15. 3π

16. $\dfrac{3\pi}{2}$ 17. $-\dfrac{7\pi}{6}$ 18. $t = 1.25\pi$

In questions 19–22, use the periodic properties of the sine and cosine functions to find the exact value of $\sin x$ and $\cos x$.

19. $x = \dfrac{13\pi}{6}$ 20. $x = \dfrac{10\pi}{3}$

21. $x = \dfrac{15\pi}{4}$ 22. $x = \dfrac{17\pi}{6}$

6.3 Graphs of trigonometric functions

The graph of a function provides a useful visual image of its behaviour. For example, from the previous section we know that trigonometric functions are periodic, i.e. their values repeat in a regular manner. The graphs of the trigonometric functions should provide a picture of this periodic behaviour. In this section, we will graph the sine, cosine and tangent functions and transformations of the sine and cosine functions.

Graphs of the sine and cosine functions

Since the period of the sine function is 2π, we know that two values of t (domain) that differ by 2π (e.g. $\frac{\pi}{6}$ and $\frac{13\pi}{6}$ in Example 8) will produce the same value for y (range). This means that any portion of the graph of $y = \sin t$ with a t-interval of length 2π (called one **period** or **cycle** of the graph) will repeat. Remember that the domain of the sine function is all real numbers, so one period of the graph of $y = \sin t$ will repeat indefinitely in the positive and negative direction. Therefore, in order to construct a complete graph of $y = \sin t$, we need to graph just one period of the function, that is, from $t = 0$ to $t = 2\pi$, and then repeat the pattern in both directions.

We know from the previous section that $\sin t$ is the y-coordinate of the terminal point on the unit circle corresponding to the real number t (Figure 6.23). In order to generate one period of the graph of $y = \sin t$, we need to record the y-coordinates of a point on the unit circle and the corresponding value of t as the point travels anticlockwise one revolution, starting from the point (1, 0). These values are then plotted on a graph with t on the horizontal axis and y (i.e. $\sin t$) on the vertical axis. Figure 6.24 illustrates this process in a sequence of diagrams.

```
sin(2.53)
        .5741721484
sin(2.53+2π)
        .5741721484
sin(2.53+4π)
        .5741721484
```

The period of $y = \sin x$ is 2π.

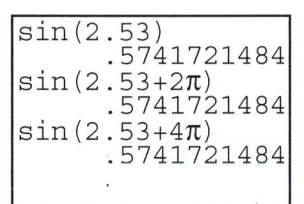

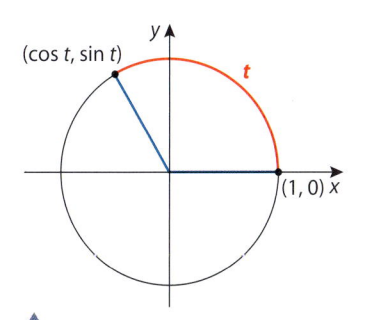

Figure 6.23

Figure 6.24 Graph of the sine function for $0 \leq t \leq 2\pi$ generated from a point travelling along the unit circle.

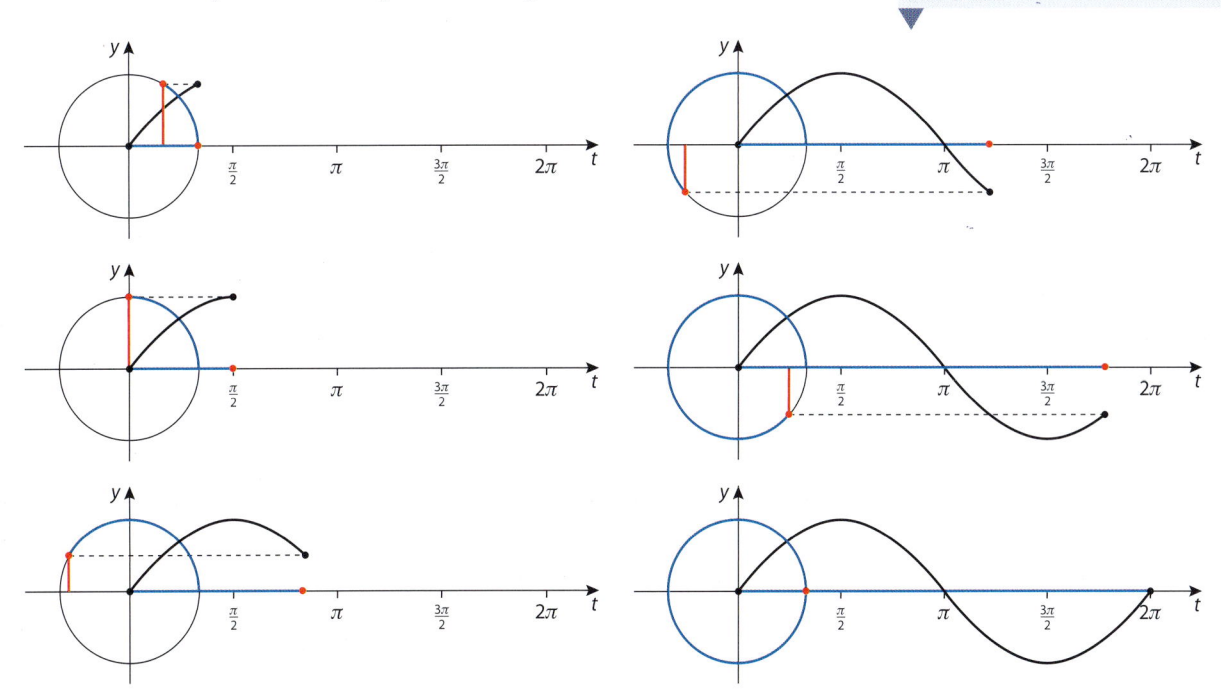

6 Trigonometric Functions and Equations

As the point (cos t, sin t) travels along the unit circle, the x-coordinate (i.e. cos t) goes through the same cycle of values as the y-coordinate (sin t) does. The only difference is that the x-coordinate begins at a different value in the cycle – when $t = 0$, $y = 0$, but $x = 1$. The result is that the graph of $y = \cos t$ is the exact same shape as $y = \sin t$ but it has been shifted to the left $\frac{\pi}{2}$ units. The graph of $y = \cos t$ for $0 \leq t \leq 2\pi$ is shown in Figure 6.25.

Figure 6.25

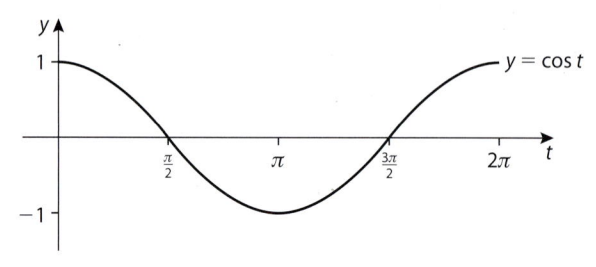

The convention is to use the letter x to denote the variable in the domain of the function. Hence, we will use the letter x rather than t and write the trigonometric functions as $y = \sin x$, $y = \cos x$ and $y = \tan x$.

Because the period for both the sine function and cosine function is 2π, to graph $y = \sin x$ and $y = \cos x$ for wider intervals of x we simply need to repeat the shape of the graph that we generated from the unit circle for $0 \leq x \leq 2\pi$ (Figures 6.24 and 6.25). Figure 6.26 shows the graphs of $y = \sin x$ and $y = \cos x$ for $-4\pi \leq x \leq 4\pi$.

Figure 6.26

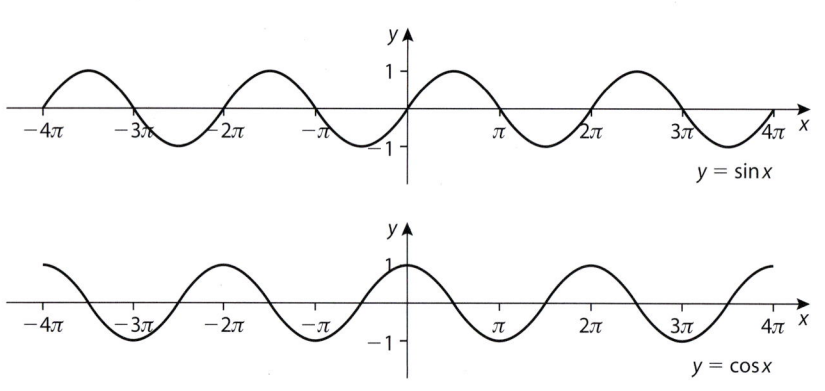

Aside from their periodic behaviour, these graphs reveal further properties of the graphs of $y = \sin x$ and $y = \cos x$. Note that the sine function has a maximum value of $y = 1$ for all $x = \frac{\pi}{2} + k \cdot 2\pi$, $k \in \mathbb{Z}$, and has a minimum value of $y = -1$ for all $x = -\frac{\pi}{2} + k \cdot 2\pi$, $k \in \mathbb{Z}$. The cosine function has a maximum value of $y = 1$ for all $x = k \cdot 2\pi$, $k \in \mathbb{Z}$, and has a minimum value of $y = -1$ for all $x = \pi + k \cdot 2\pi$, $k \in \mathbb{Z}$. This also confirms – as established in the previous section – that both functions have a domain of all real numbers and a range of $-1 \leq y \leq 1$.

Closer inspection of the graphs, in Figure 6.26, shows that the graph of $y = \sin x$ has rotational symmetry about the origin – that is, it can be rotated one-half of a revolution about (0, 0) and it remains the same. This graph symmetry can be expressed with the identity: $\sin(-x) = -\sin x$. For example, $\sin\left(-\frac{\pi}{6}\right) = -\frac{1}{2}$ and $-\left[\sin\left(\frac{\pi}{6}\right)\right] = -\left[\frac{1}{2}\right] = -\frac{1}{2}$. A function that is symmetric about the origin is called an **odd function**. The graph of $y = \cos x$ has line symmetry in the y-axis – that is, it can be reflected in the line $x = 0$ and it remains the same. This graph symmetry can be expressed with the identity: $\cos(-x) = \cos x$. For example, $\cos\left(-\frac{\pi}{6}\right) = \frac{\sqrt{3}}{2}$ and $\cos\frac{\pi}{6} = \frac{\sqrt{3}}{2}$. A function that is symmetric about the y-axis is called an **even function**.

> **Odd and even functions**
>
> A function is **odd** if, for each x in the domain of f, $f(-x) = -f(x)$.
>
> The graph of an odd function is symmetric with respect to the origin (rotational symmetry).
>
> A function is **even** if, for each x in the domain of f, $f(-x) = f(x)$.
>
> The graph of an even function is symmetric with respect to the y-axis (line symmetry).

Graphs of transformations of the sine and cosine functions

In Section 2.4, we learned how to transform the graph of a function by horizontal and vertical translations, by reflections in the coordinate axes, and by stretching and shrinking – both horizontal and vertical. The following is a review of these transformations.

> **Review of transformations of graphs of functions**
>
> Assume that a, b, c and d are real numbers.
>
> | To obtain the graph of: | From the graph of $y = f(x)$: |
> |---|---|
> | $y = f(x) + d$ | Translate d units up for $d > 0$, d units down for $d < 0$. |
> | $y = f(x + c)$ | Translate c units left for $c > 0$, c units right for $c < 0$. |
> | $y = -f(x)$ | Reflect in the x-axis. |
> | $y = af(x)$ | Vertical stretch ($a > 1$) or shrink ($0 < a < 1$) of factor a. |
> | $y = f(-x)$ | Reflect in the y-axis. |
> | $y = f(bx)$ | Horizontal stretch ($0 < b < 1$) or shrink ($b > 1$) of factor $\frac{1}{b}$. |

In this section, we will look at the composition of sine and cosine functions of the form

$$f(x) = a\sin[b(x + c)] + d \quad \text{and} \quad f(x) = a\cos[b(x + c)] + d$$

Example 9

Sketch the graph of each function on the interval $-\pi \leqslant x \leqslant 3\pi$.

a) $f(x) = 2\cos x$

b) $g(x) = \cos x + 3$

c) $h(x) = 2\cos x + 3$

d) $p(x) = \frac{1}{2}\sin x - 2$

Trigonometric Functions and Equations

Solution

a) Since $a = 2$, the graph of $y = 2\cos x$ is obtained by vertically stretching the graph of $y = \cos x$ by a factor of 2.

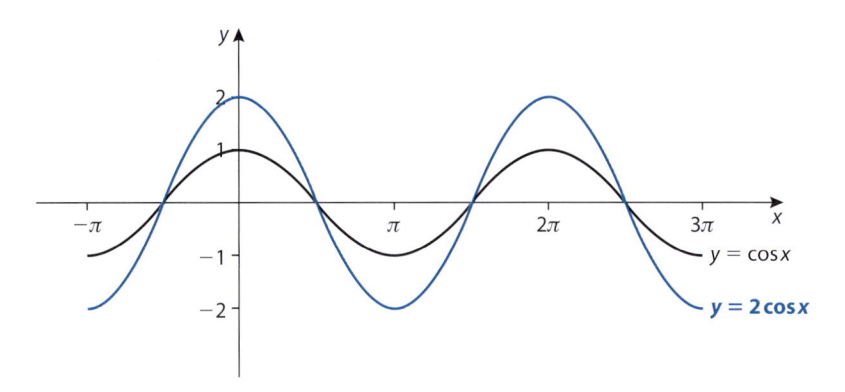

b) Since $d = 3$, the graph of $y = \cos x + 3$ is obtained by translating the graph of $y = \cos x$ three units up.

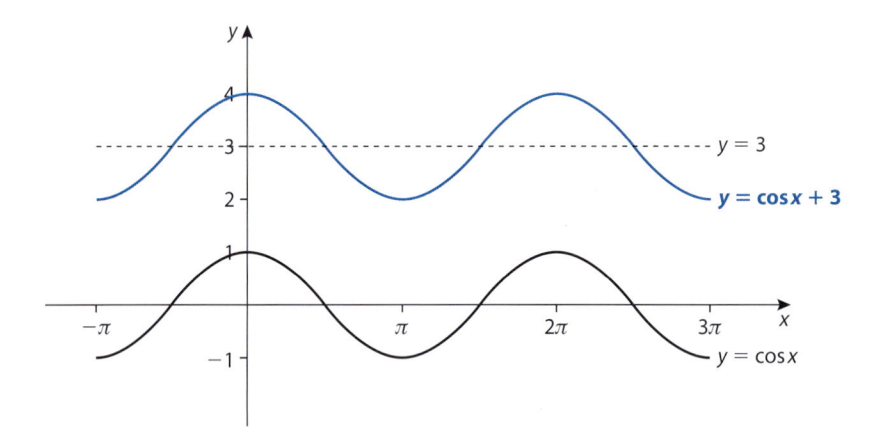

c) We can obtain the graph of $y = 2\cos x + 3$ by combining both of the transformations to the graph of $y = \cos x$ performed in parts a) and b) – namely, a vertical stretch of factor 2 and a translation 3 units up.

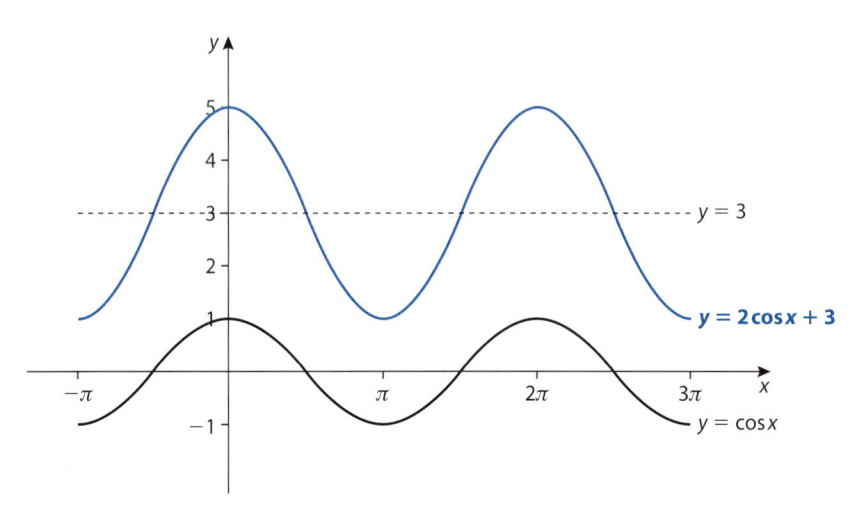

d) The graph of $y = \frac{1}{2}\sin x - 2$ can be obtained by vertically shrinking the graph of $y = \sin x$ by a factor of $\frac{1}{2}$ and then translating it down 2 units.

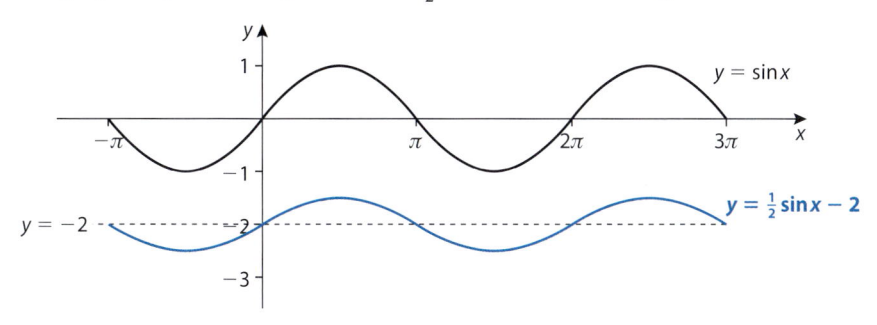

In part a), the graph of $y = 2\cos x$ has many of the same properties as the graph of $y = \cos x$: same period, and the maximum and minimum values occur at the same values of x. However, the graph ranges between -2 and 2 instead of -1 and 1. This difference is best described by referring to the **amplitude** of each graph. The amplitude of $y = \cos x$ is 1 and the amplitude of $y = 2\cos x$ is 2. The amplitude of a sine or cosine graph is not always equal to its maximum value. In part b), the amplitude of $y = \cos x + 3$ is 1; in part c), the amplitude of $y = 2\cos x + 3$ is 2; and the amplitude of $y = \frac{1}{2}\sin x - 2$ is $\frac{1}{2}$. For all three of these, the graphs oscillate about the horizontal line $y = d$. How *high* and *low* the graph oscillates with respect to the mid-line, $y = d$, is the graph's amplitude. With respect to the general form $y = af(x)$, changing the amplitude is equivalent to a vertical stretching or shrinking. Thus, we can give a more precise definition of amplitude in terms of the parameter a.

> **Amplitude of the graph of sine and cosine functions**
> The graphs of $f(x) = a\sin[b(x + c)] + d$ and $f(x) = a\cos[b(x + c)] + d$ have an **amplitude** equal to $|a|$.

Example 10

Waves are produced in a long tank of water. The depth of the water, d metres, at t seconds, at a fixed location in the tank, is modelled by the function $d(t) = M\cos\left(\frac{\pi}{2}t\right) + K$, where M and K are positive constants. On the right is the graph of $d(t)$ for $0 \leqslant t \leqslant 12$ indicating that the point $(2, 5.1)$ is a minimum and the point $(8, 9.7)$ is a maximum.

a) Find the value of K and the value of M.

b) After $t = 0$, find the first time when the depth of the water is 9.7 metres.

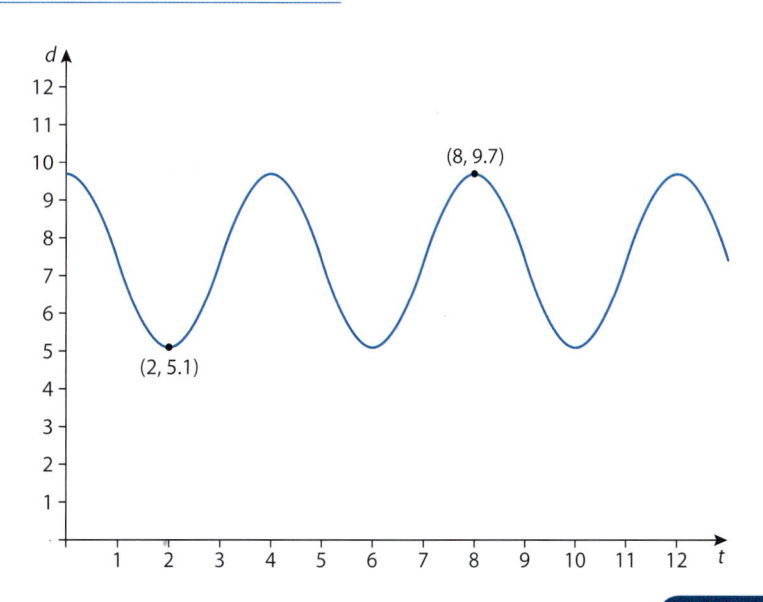

Solution

a) The constant K is equivalent to the constant d in the general form of a cosine function: $f(x) = a\cos[b(x+c)] + d$. To find the value of K and the equation of the horizontal mid-line, $y = K$, find the average of the function's maximum and minimum value: $K = \dfrac{9.7 + 5.1}{2} = 7.4$.

The constant M is equivalent to the constant a whose absolute value is the amplitude. The amplitude is the difference between the function's maximum value and the mid-line: $|M| = 9.7 - 7.4 = 2.3$. Thus, $M = 2.3$ or $M = -2.3$. Try $M = 2.3$ by evaluating the function at one of the known values:

$d(2) = 2.3\cos\left(\dfrac{\pi}{2}(2)\right) + 7.4 = 2.3\cos\pi + 7.4 = 2.3(-1) + 7.4 = 5.1$.

This agrees with the point $(2, 5.1)$ on the graph. Therefore, $M = 2.3$.

b) Maximum values of the function ($d(8) = 9.7$) occur at values of t that differ by a value equal to the period. From the graph, we can see that the difference in t values from the minimum $(2, 5.1)$ to the maximum $(8, 9.7)$ is equivalent to one-and-a-half periods. Therefore, the period is 4 and the first time after $t = 0$ at which $d = 9.7$ is $t = 4$.

All four of the functions in Example 9 had the same period of 2π, but the function in Example 10 had a period of 4. Because $y = \sin x$ completes one period from $x = 0$ to $x = 2\pi$, it follows that $y = \sin bx$ completes one period from $bx = 0$ to $bx = 2\pi$. This implies that $y = \sin bx$ completes one period from $x = 0$ to $x = \dfrac{2\pi}{b}$. This agrees with the period for the function $d(t) = 2.3\cos\left(\dfrac{\pi}{2}t\right) + 7.4$ in Example 10: period $= \dfrac{2\pi}{b} = \dfrac{2\pi}{\frac{\pi}{2}} = \dfrac{2\pi}{1} \cdot \dfrac{2}{\pi} = 4$.

Note that the change in amplitude and vertical translation had no effect on the period. We should also expect that a horizontal translation of a sine or cosine curve should not affect the period. The next example looks at a function that is horizontally translated (shifted) and has a period different from 2π.

Example 11

Sketch the function $f(x) = \sin\left(2x + \dfrac{2\pi}{3}\right)$.

Solution

To determine how to transform the graph of $y = \sin x$ to obtain the graph of $y = \sin\left(2x + \dfrac{2\pi}{3}\right)$, we need to make sure the function is written in the form $f(x) = a\sin[b(x+c)] + d$. Clearly, $a = 1$ and $d = 0$, but we will need to factorize a 2 from the expression $2x + \dfrac{2\pi}{3}$ to get $f(x) = \sin\left[2\left(x + \dfrac{\pi}{3}\right)\right]$.

According to our general transformations from Chapter 2, we expect that the graph of f is obtained by first translating the graph of $y = \sin x$ to the left $\dfrac{\pi}{3}$ units and then a horizontal shrinking by factor $\dfrac{1}{2}$ (see Section 2.4).

The graphs below illustrate the two-stage sequence of transforming $y = \sin x$ to $y = \sin\left[2\left(x + \frac{\pi}{3}\right)\right]$.

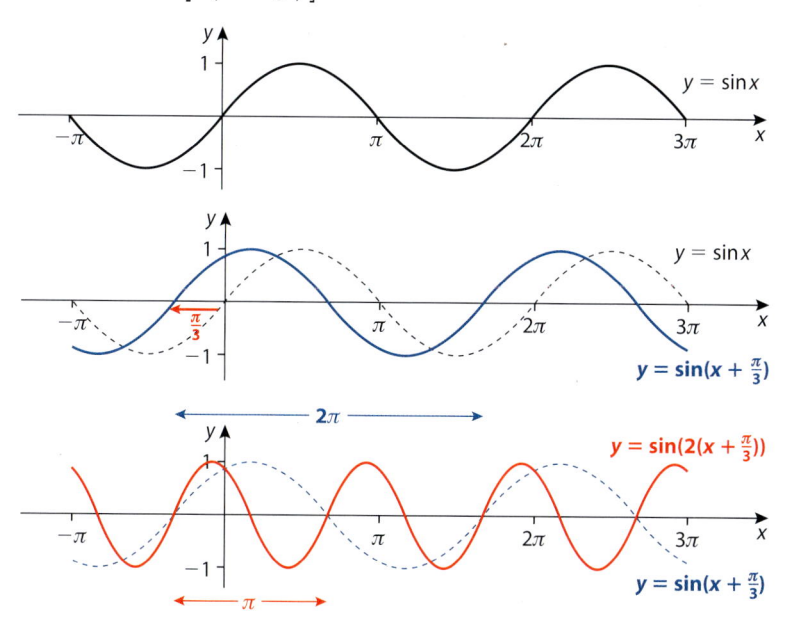

Note: A horizontal translation of a sine or cosine curve is often referred to as a **phase shift**. The equations $y = \sin\left(x + \frac{\pi}{3}\right)$ and $y = \sin\left[2\left(x + \frac{\pi}{3}\right)\right]$ both underwent a phase shift of $-\frac{\pi}{3}$.

> **Period and horizontal translation (phase shift) of sine and cosine functions**
>
> Given that b is a positive real number, $y = a\sin[b(x + c)] + d$ and $y = a\cos[b(x + c)] + d$ have a **period** of $\frac{2\pi}{b}$ and a horizontal translation (**phase shift**) of $-c$.

Example 12

The graph of a function in the form $y = a\cos bx$ is given in the diagram right.

a) Write down the value of a.

b) Calculate the value of b.

Solution

a) The amplitude of the graph is 14. Therefore, $a = 14$.

b) From inspecting the graph we can see that the period is $\frac{\pi}{4}$.

Period $= \frac{2\pi}{b} = \frac{\pi}{4}$

$b\pi = 8\pi \Rightarrow b = 8$.

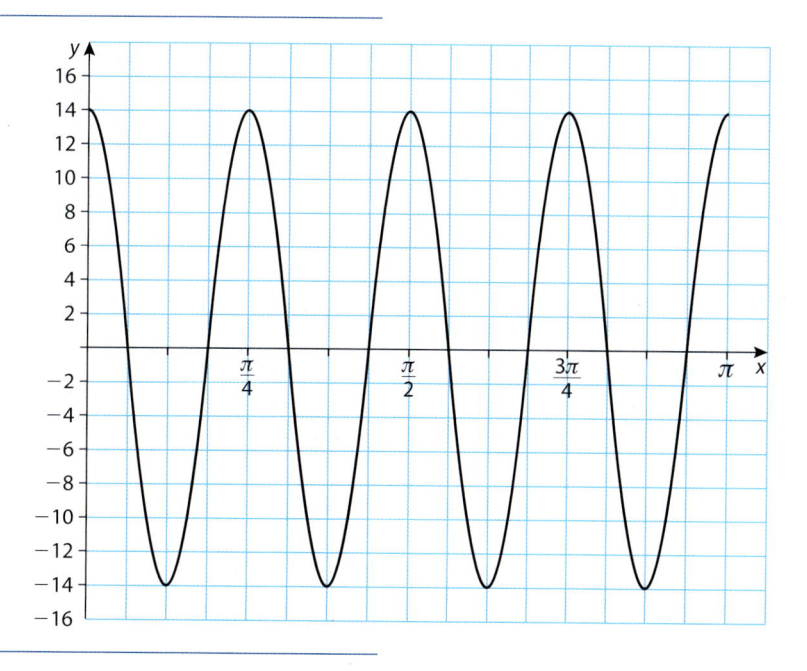

Example 13

For the function $f(x) = 2\cos\left(\dfrac{x}{2}\right) - \dfrac{3}{2}$:

a) Sketch the function for the interval $-\pi \leq x \leq 5\pi$. Write down its amplitude and period.

b) Determine the domain and range for $f(x)$.

c) Write $f(x)$ as a trigonometric function in terms of sine rather than cosine.

Solution

a) $a = 2 \Rightarrow$ amplitude $= 2$; $b = \dfrac{1}{2} \Rightarrow$ period $= \dfrac{2\pi}{\frac{1}{2}} = 4\pi$. To obtain the graph of $y = 2\cos\left(\dfrac{x}{2}\right) - \dfrac{3}{2}$, we perform the following transformations on $y = \cos x$: (i) a horizontal stretch by factor $\dfrac{1}{\frac{1}{2}} = 2$, (ii) a vertical stretch by factor 2, and (iii) a vertical translation down $\dfrac{3}{2}$ units.

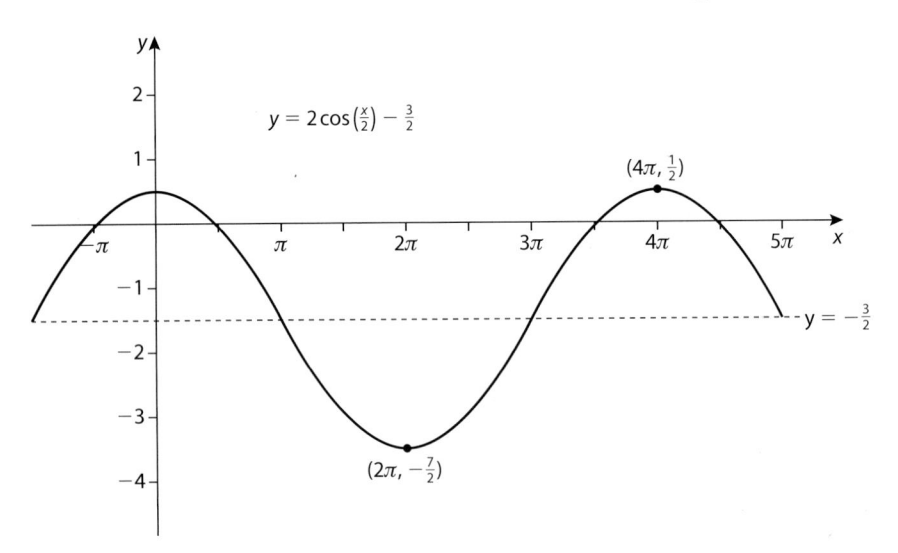

b) The domain is all real numbers. The function will reach a maximum value of $d + a = -\dfrac{3}{2} + 2 = \dfrac{1}{2}$, and a minimum value of $d - a = -\dfrac{3}{2} - 2 = -\dfrac{7}{2}$.

Hence, the range is $-\dfrac{7}{2} \leq y \leq \dfrac{1}{2}$.

c) The graph of $y = \cos x$ can be obtained by translating the graph of $y = \sin x$ to the left $\dfrac{\pi}{2}$ units. Thus, $\cos x = \sin\left(x + \dfrac{\pi}{2}\right)$, or, in other words, any cosine function can be written as a sine function with a phase shift $= -\dfrac{\pi}{2}$. Therefore, $f(x) = 2\cos\left(\dfrac{x}{2}\right) - \dfrac{3}{2} = 2\sin\left(\dfrac{x}{2} + \dfrac{\pi}{2}\right) - \dfrac{3}{2}$.

Horizontal translation (phase shift) identities

The following are true for all values of x:
$$\cos x = \sin\left(x + \frac{\pi}{2}\right) \qquad \sin x = \cos\left(x - \frac{\pi}{2}\right)$$
$$\cos x = \sin\left(\frac{\pi}{2} - x\right) \qquad \sin x = \cos\left(\frac{\pi}{2} - x\right)$$

The identity $\cos x = \sin\left(x + \frac{\pi}{2}\right)$ is equivalent to the identity $\cos x = \sin\left(\frac{\pi}{2} - x\right)$ because $\sin\left(\frac{\pi}{2} - x\right) = \sin\left[-\left(x - \frac{\pi}{2}\right)\right]$ and the graph of $y = \sin\left[-\left(x - \frac{\pi}{2}\right)\right]$ can be obtained by first translating $y = \sin x$ to the right $\frac{\pi}{2}$ units, and then reflecting the graph in the y-axis. This produces the same graph as $y = \cos x$. This can be confirmed nicely on your GDC as shown.
Therefore, $\cos x = \sin\left(\frac{\pi}{2} - x\right)$. In fact, it is also true that $\sin x = \cos\left(\frac{\pi}{2} - x\right)$. Clearly, $x + \left(\frac{\pi}{2} - x\right) = \frac{\pi}{2}$. If the domain ($x$) values were being treated as angles, then x and $\frac{\pi}{2} - x$ would be complementary angles.
This is why cosine is considered the co-function of sine. Two trigonometric functions f and g are co-functions if the following are true for all x: $f(x) = g\left(\frac{\pi}{2} - x\right)$ and $f\left(\frac{\pi}{2} - x\right) = g(x)$.

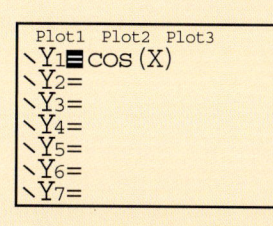

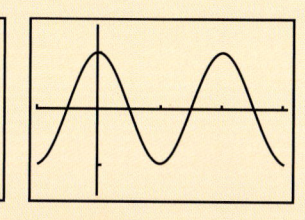

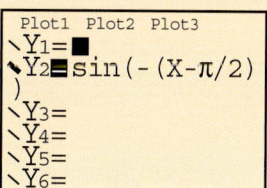

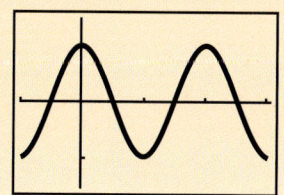

Graph of the tangent function

From work done earlier in this chapter, we expect that the behaviour of the tangent function will be significantly different from that of the sine and cosine functions. In Section 6.2, we concluded that the function $f(x) = \tan x$ has a domain of all real numbers such that $x \neq \frac{\pi}{2} + k\pi$, $k \in \mathbb{Z}$, and that its range is all real numbers. Also, the results for Example 6 in Section 6.2 led us to speculate that the period of the tangent function is π. This makes sense since the identity $\tan x = \frac{\sin x}{\cos x}$ informs us that $\tan x$ will be zero whenever $\sin x = 0$, which occurs at values of x that differ by π (visualize arcs on the unit circle whose terminal points are either $(1, 0)$ or $(-1, 0)$). The values of x for which $\cos x = 0$ cause $\tan x$ to be undefined ('gaps' in the domain) also differ by π (the points $(0, 1)$ or $(0, -1)$ on the unit circle). As x approaches these values where $\cos x = 0$, the value of $\tan x$ will become very large – either very large negative or very large positive.
Thus, the graph of $y = \tan x$ has vertical asymptotes at $x = \frac{\pi}{2} + k\pi$, $k \in \mathbb{Z}$. Consequently, the graphical behaviour of the tangent function will not be a wave pattern such as that produced by the sine and cosine functions, but rather a series of separate curves that repeat every π units. Figure 6.27 shows the graph of $y = \tan x$ for $-2\pi \leq x \leq 2\pi$.

6 Trigonometric Functions and Equations

Figure 6.27

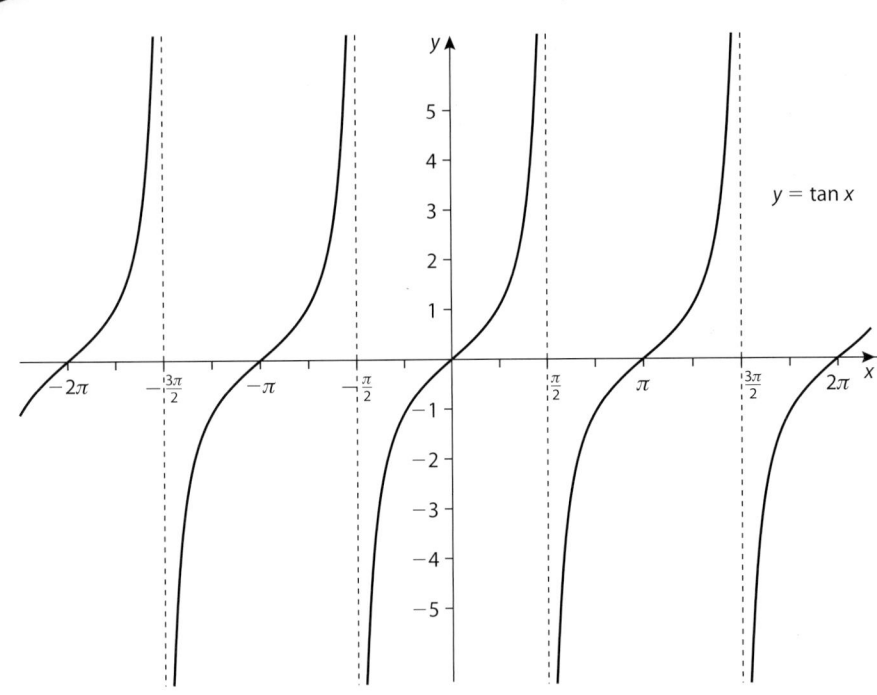

The graph gives clear confirmation that the period of the tangent function is π, that is, $\tan x = \tan(x + k \cdot \pi)$, $k \in \mathbb{Z}$.

The graph of $y = \tan x$ has rotational symmetry about the origin – that is, it can be rotated one-half of a revolution about $(0, 0)$ and it remains the same. Hence, like the sine function, tangent is an odd function and $\tan(-x) = -\tan x$.

Although the graph of $y = \tan x$ can undergo a vertical stretch or shrink, it is meaningless to consider its amplitude since the tangent function has no maximum or minimum value. However, other transformations can affect the period of the tangent function.

Example 14

Sketch each function.

a) $f(x) = \tan 2x$
b) $g(x) = \tan\left[2\left(x - \dfrac{\pi}{4}\right)\right]$

Solution

a) An equation in the form $y = f(bx)$ indicates a horizontal shrinking of $f(x)$ by a factor of $\dfrac{1}{b}$. Hence, the period of $y = \tan 2x$ is $\dfrac{1}{2} \cdot \pi = \dfrac{\pi}{2}$.

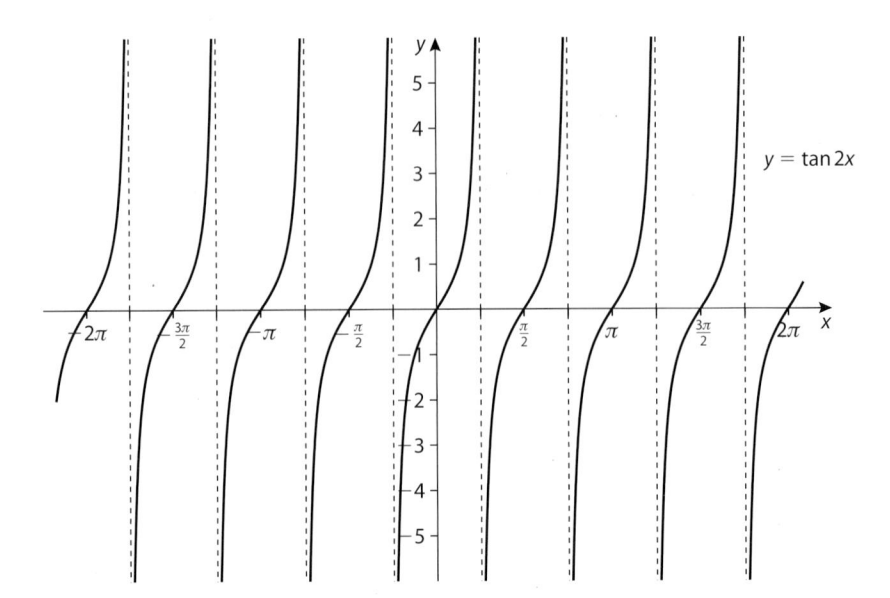

186

b) The graph of $y = \tan\left[2\left(x - \frac{\pi}{4}\right)\right]$ is obtained by first translating the graph of $y = \tan x$ to the right $\frac{\pi}{4}$ units, and then a horizontal shrinking by a factor of $\frac{1}{2}$. As for $f(x) = \tan 2x$ in part a), the period of $g(x) = \tan\left[2\left(x - \frac{\pi}{4}\right)\right]$ is $\frac{\pi}{2}$.

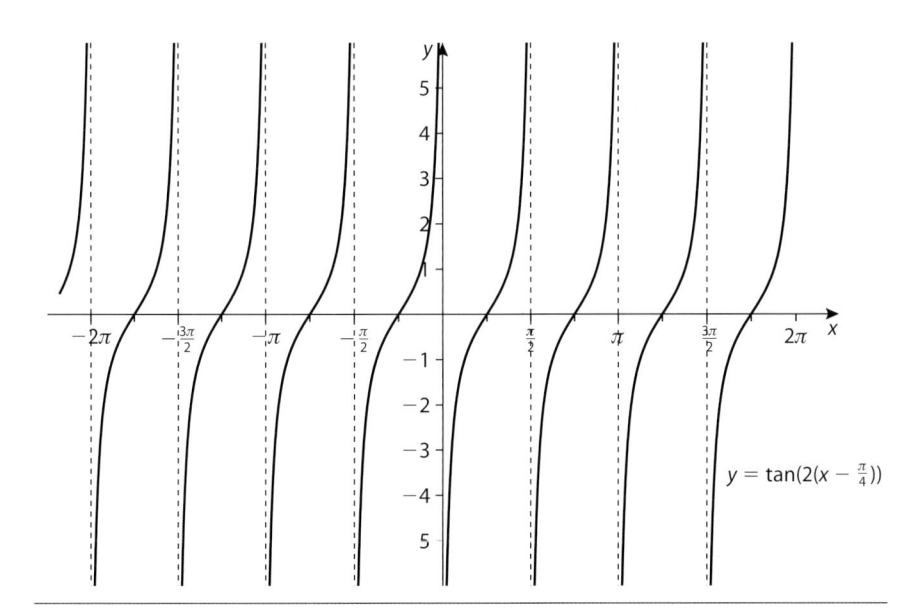

Exercise 6.3

In questions 1–9, without using your GDC, sketch a graph of each equation on the interval $-\pi \leq x \leq 3\pi$.

1 $y = 2\sin x$

2 $y = \cos x - 2$

3 $y = \frac{1}{2}\cos x$

4 $y = \sin\left(x - \frac{\pi}{2}\right)$

5 $y = \cos(2x)$

6 $y = 1 + \tan x$

7 $y = \sin\left(\frac{x}{2}\right)$

8 $y = \tan\left(x + \frac{\pi}{2}\right)$

9 $y = \cos\left(2x - \frac{\pi}{4}\right)$

For each function in questions 10–12:
a) Sketch the function for the interval $-\pi \leq x \leq 5\pi$. Write down its amplitude and period.
b) Determine the domain and range for $f(x)$.

10 $f(x) = \frac{1}{2}\cos x - 3$

11 $g(x) = 3\sin(3x) - \frac{1}{2}$

12 $g(x) = 1.2\sin\left(\frac{x}{2}\right) + 4.3$

6 Trigonometric Functions and Equations

In questions 13–15, a graph for the interval $0 \leq x \leq 12$ is given for a trigonometric equation that can be written in the form $y = A\sin\left(\frac{\pi}{4}x\right) + B$. Two points – one a minimum and the other a maximum – are indicated on the graph. Find the value of A and B for each.

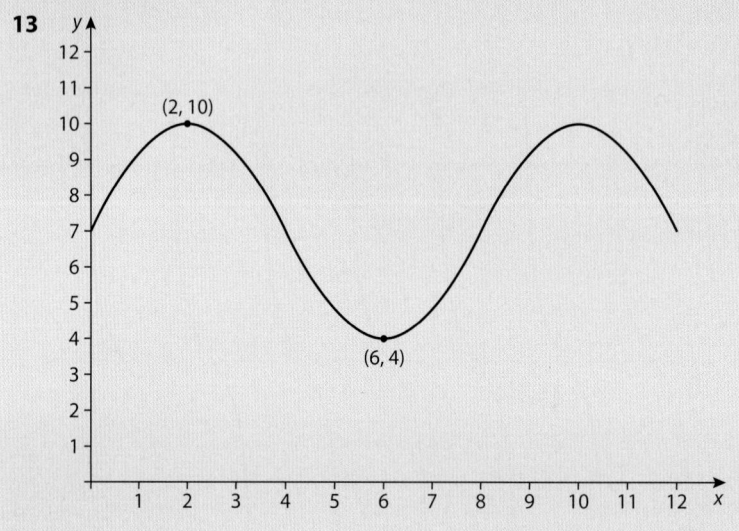

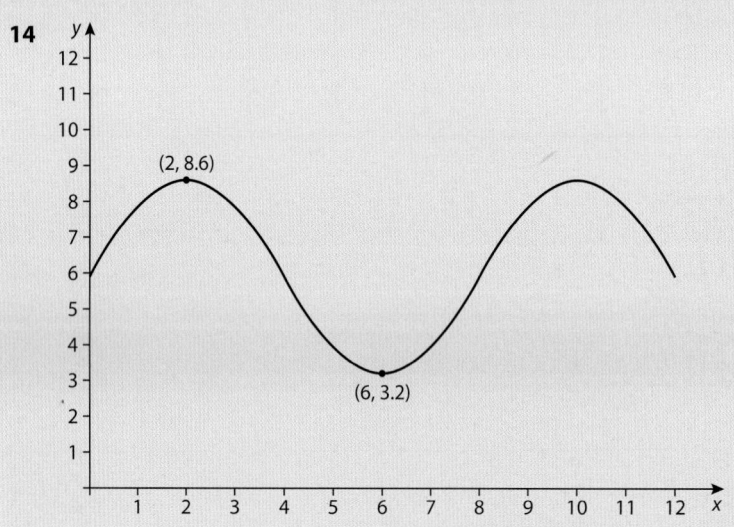

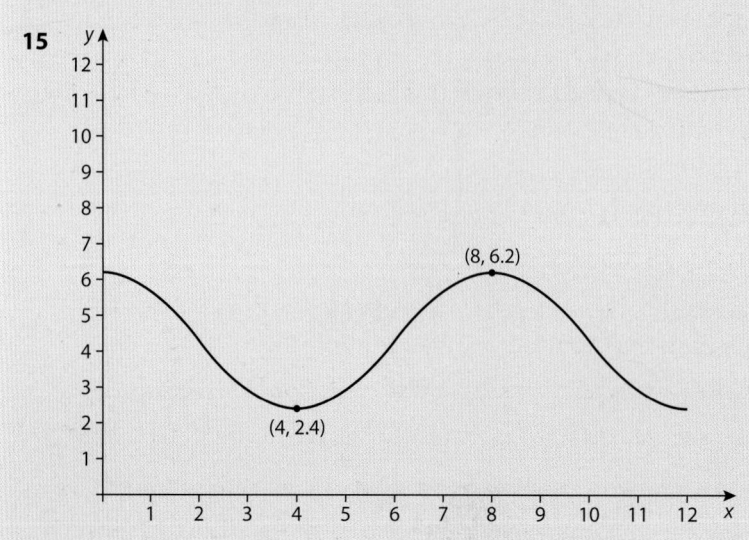

13 Points: $(2, 10)$ maximum, $(6, 4)$ minimum.

14 Points: $(2, 8.6)$ maximum, $(6, 3.2)$ minimum.

15 Points: $(4, 2.4)$ minimum, $(8, 6.2)$ maximum.

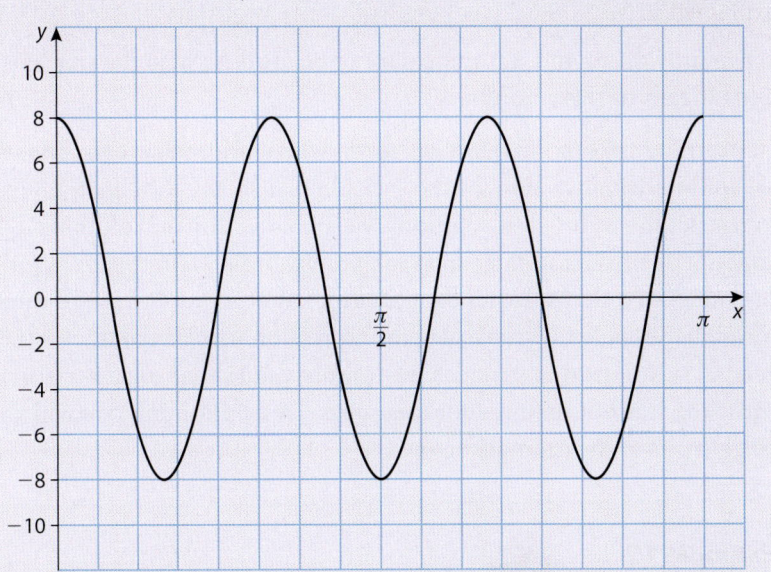

16 The graph of a function in the form $y = p \cos qx$ is given in the diagram below.
  a) Write down the value of p.
  b) Calculate the value of q.

6.4 Solving trigonometric equations and trigonometric identities

The primary focus of this section is to examine methods for solving equations that contain the sine, cosine and tangent functions. For example, the following are **trigonometric equations**:

$\sin x = \frac{1}{2}$ $3\cos x = 5\sin x$ $\tan x = \frac{\sin x}{\cos x}$ $1 + \sin x = 3\cos^2 x$ $\sin^2 x + \cos^2 x = 1$

The equations $\tan x = \frac{\sin x}{\cos x}$ and $\sin^2 x + \cos^2 x = 1$ are examples of special equations called **identities**. As we learned in Section 1.6, an identity is an equation that is true for all possible values of the variable. The other equations are true for only certain values of x. Identities can be helpful in solving trigonometric equations by allowing us to simplify some trigonometric expressions. Equations that contain trigonometric functions often can be solved using the same graphical and algebraic methods that solve other equations.

The unit circle and exact solutions to trigonometric equations

When you are asked to solve a trigonometric equation, there are two important questions you need to consider:
1. Is it possible, or required, to express any solution(s) exactly?
2. For what interval of the variable (usually x) are all solutions to be found?

With regard to the first question, exact solutions are only attainable, in most cases, if they are an integer multiple of $\frac{\pi}{6}$ or $\frac{\pi}{4}$. The variable for which we are trying to solve in trigonometric equations is a real number that can be interpreted as the length of an arc on the unit circle. As explained in Section 6.2, arc lengths that are multiples of $\frac{\pi}{6}$ or $\frac{\pi}{4}$ commonly occur and it is important to be familiar with the sine, cosine and tangent of these numbers.

Concerning the second question, for most trigonometric equations there are infinitely many values of the variable that satisfy the equation. In order to restrict the number of solutions, we are asked for the solutions to be contained within a suitable interval. For example, we may search for all the values of x that solve an equation such that $0 \leq x \leq 2\pi$. Although it is certainly possible to write a general expression using a parameter (i.e. the general solution) that specifies the infinite values that solve a trigonometric equation, it is not required for this course. A solution interval will always be given, as in the example below.

Example 15

Find the exact solution(s) to the equation $\sin x = \frac{1}{2}$ for $0 \leq x \leq 2\pi$.

Solution

Recalling the definition of the sine function, this equation can be interpreted as asking for the length, x, of arcs along the unit circle that have a terminal point with a y-coordinate equal to $\frac{1}{2}$. We know, from Section 6.2, that arc lengths of $\frac{\pi}{6}$ and $\frac{5\pi}{6}$ have terminal points with y-coordinates of $\frac{1}{2}$. There are clearly an infinite number of arcs – both positive and negative – that will terminate at the same points which can be written as $x = \frac{\pi}{6} + k \cdot 2\pi$ and $x = \frac{5\pi}{6} + k \cdot 2\pi$, $k \in \mathbb{Z}$. However, we are only asked for the solutions in the interval $0 \leq x \leq 2\pi$. Therefore, $x = \frac{\pi}{6}$ or $x = \frac{5\pi}{6}$.

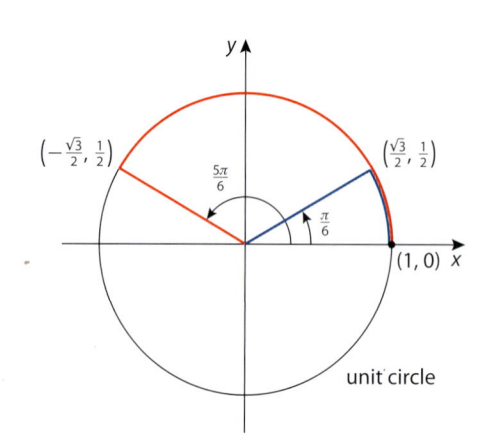

Another way to see that the equation $\sin x = \frac{1}{2}$ has infinitely many solutions is to graph the equations $y = \sin x$ and $y = \frac{1}{2}$ (Figure 6.28) and search for intersection points, i.e. where the two equations are equal.

◀ **Figure 6.28**

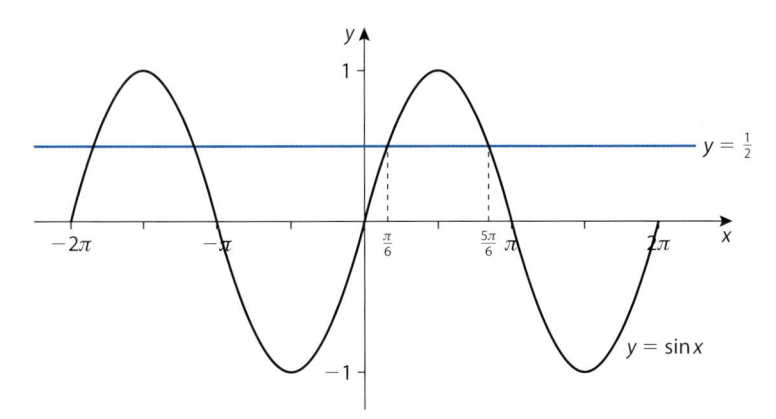

The graphs of the two equations will intersect repeatedly as they extend indefinitely in both directions.

Your GDC can be a very effective tool for searching for solutions graphically. However, it can be limited when exact solutions are requested. The sequence of GDC images below show a graphical solution for the equation in Example 15.

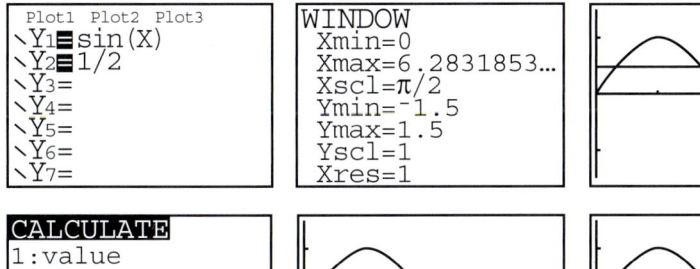

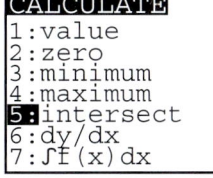

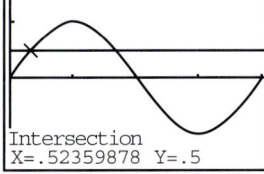

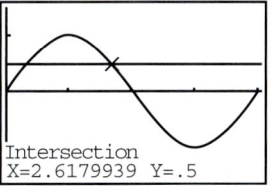

The GDC gives two solutions in the interval $0 \leq x \leq 2\pi$ as $x = 0.523\,598\,78$ and $x = 2.617\,9939$. These values are approximations (to 8 significant figures) of two irrational numbers: $x = \dfrac{\pi}{6}$ and $x = \dfrac{5\pi}{6}$. Therefore, if you wish, or need, to find exact solutions, you will need to remember the trigonometric function values for the multiples of $\dfrac{\pi}{6}$ and $\dfrac{\pi}{4}$ (see Figures 6.18 and 6.19 in Section 6.2).

Example 16

Find the exact solution(s) to the equation $\tan(x) + 1 = 0$ for $-\pi \leq x < \pi$.

Solution

It's important to note that the solution interval is different than for Example 15. The possible values of x include negative values (from 0 to $-\pi$) and positive values (from 0 to π). With respect to the unit circle, the solutions will correspond to points in any of the quadrants (as for Example 15) but points in quadrants III and IV will correspond to arcs rotating clockwise (negative direction). Solutions to this equation are values of x such

● **Hint:** The expression $\tan x + 1$ is not equivalent to $\tan(x + 1)$. In the first expression, x alone is the argument of the function, and in the second expression, $x + 1$ is the argument of the function. It is a good habit to use brackets to make it absolutely clear what is, or is not, the argument of a function. For example, there is no ambiguity if $\tan x + 1$ is written as $\tan(x) + 1$, or as $1 + \tan x$.

191

6 Trigonometric Functions and Equations

that $\tan x = -1$. Given $\tan x = \frac{\sin x}{\cos x}$ and since $\sin x$ and $\cos x$ correspond to the y-coordinate and x-coordinate, respectively, on the unit circle, then any solutions will be in quadrants II and IV where the x- and y-coordinates have opposite signs. The arcs terminating midway in the quadrants will terminate at points having opposite values for x and y. Therefore, as shown in the figure, the solutions are exactly $x = -\frac{\pi}{4}$ or $x = \frac{3\pi}{4}$.

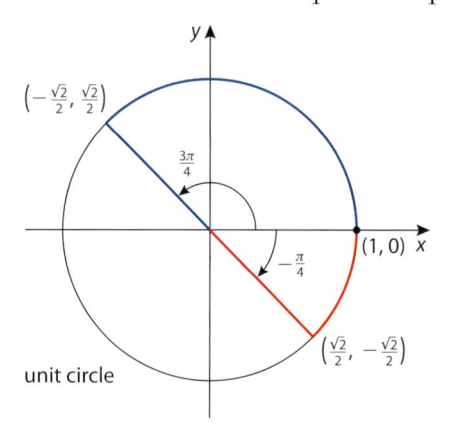

It is possible to arrive at exact answers that are not multiples of $\frac{\pi}{6}$ or $\frac{\pi}{4}$, as the next example illustrates.

Example 17

Find the exact solution(s) to the equation $\cos^2\left(x - \frac{\pi}{3}\right) = \frac{1}{2}$ for $0 \leq x \leq 2\pi$.

Solution

The expression $\cos^2\left(x - \frac{\pi}{3}\right)$ can also be written as $\left[\cos\left(x - \frac{\pi}{3}\right)\right]^2$. The first step is to take the square root of both sides – remembering that every positive number has two square roots – which gives $\cos\left(x - \frac{\pi}{3}\right) = \pm\sqrt{\frac{1}{2}} = \pm\frac{1}{\sqrt{2}} = \pm\frac{\sqrt{2}}{2}$. All of the odd integer multiples of $\frac{\pi}{4}$ $\left(\ldots -\frac{3\pi}{4}, -\frac{\pi}{4}, \frac{\pi}{4}, \frac{3\pi}{4}, \ldots\right)$ have a cosine equal to either $\frac{\sqrt{2}}{2}$ or $-\frac{\sqrt{2}}{2}$. That is, $x - \frac{\pi}{3} = \frac{\pi}{4} + k \cdot \frac{\pi}{2}$. Now, solve for x.
$x = \frac{\pi}{4} + \frac{\pi}{3} + k \cdot \frac{\pi}{2} = \frac{7\pi}{12} + k \cdot \frac{6\pi}{12}$. The last step is to substitute in different integer values for k to generate all the possible values for x so that $0 \leq x \leq 2\pi$.
When $k = 0$: $x = \frac{7\pi}{12}$; when $k = 1$: $x = \frac{7\pi}{12} + \frac{6\pi}{12} = \frac{13\pi}{12}$;
when $k = 2$: $x = \frac{7\pi}{12} + \frac{12\pi}{12} = \frac{19\pi}{12}$;
when $k = 3$: $x = \frac{7\pi}{12} + \frac{18\pi}{12} = \frac{25\pi}{12}$; ... however, $\frac{25\pi}{12} > 2\pi$ ...but
when $k = -1$: $x = \frac{7\pi}{12} - \frac{6\pi}{12} = \frac{\pi}{12}$. There are four exact solutions in the interval $0 \leq x \leq 2\pi$ and they are: $x = \frac{\pi}{12}, \frac{7\pi}{12}, \frac{13\pi}{12}$ or $\frac{19\pi}{12}$.

• **Hint:** Check the solutions to trigonometric equations with your GDC. The sequence of GDC images here verifies that $x = \frac{\pi}{12}$ is the first solution to the equation in Example 17.

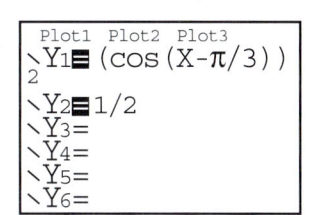

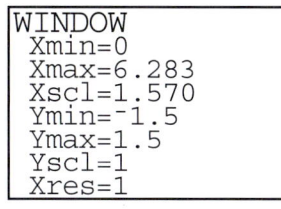

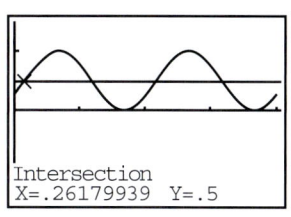

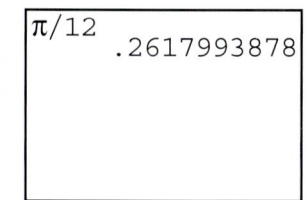

When entering the equation $y = \cos^2\left(x - \frac{\pi}{3}\right)$ into your GDC (as shown in the first GDC image), you will have to enter it in the form $y = \left[\cos\left(x - \frac{\pi}{3}\right)\right]^2$. Be aware that $\cos^2\left(x - \frac{\pi}{3}\right)$ is not equivalent to $\cos\left(x - \frac{\pi}{3}\right)^2$. The expression $\cos\left(x - \frac{\pi}{3}\right)^2$ indicates that the quantity $x - \frac{\pi}{3}$ is squared first and then the cosine of the resulting value is found. However, the expression $y = \cos^2\left(x - \frac{\pi}{3}\right)$ indicates that the cosine of $x - \frac{\pi}{3}$ is found first and then that value is squared.

Graphical solutions to trigonometric equations

If exact solutions are not required then a graphical solution using your GDC is a very effective way to find approximate solutions to trigonometric equations. Unless instructed to do otherwise, you should give approximate solutions to an accuracy of 3 significant figures.

Let's solve the equation in Example 16 again. If the instructions do not explicitly ask for exact solutions, approximate solutions are acceptable.

Example 18

Find the solution(s) to the equation $\tan(x) + 1 = 0$ for $-\pi \leq x < \pi$.

Solution

Graph the equation $y = \tan(x) + 1$ and find all of its zeros (x-intercepts) in the interval $-\pi \leq x < \pi$.

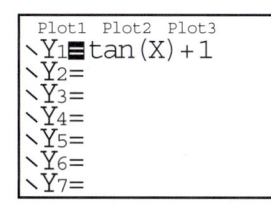

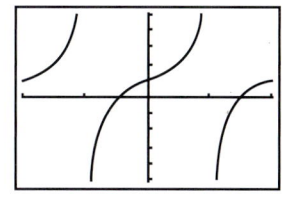

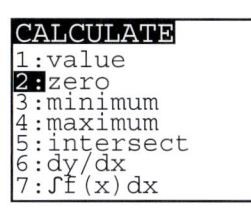

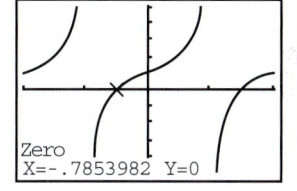

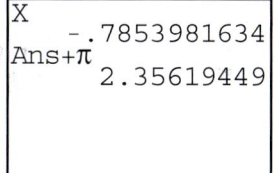

This sequence of GDC images indicates an approximate solution $x \approx -0.785$ between 0 and $-\pi$. Since we know that the period of $y = \tan x + 1$ is π (same as for $y = \tan x$), we can simply add π to this first solution to find the one between 0 and π, as shown in the final GDC image. Therefore, two solutions for x in the interval $-\pi \leq x < \pi$ are $x \approx -0.785$ and $x \approx 2.36$ (accuracy to 3 significant figures).

6 Trigonometric Functions and Equations

A graphical approach is effective and appropriate when it is not possible, or very difficult, to find exact solutions.

Example 19

The peak height, h metres, of ocean waves during a storm is given by the equation $h = 9 + 4\sin\left(\dfrac{t}{2}\right)$, where t is the number of hours after midnight.

A tsunami alarm is triggered when the peak height goes above 12.5 metres. Find the value of t when the alarm first sounds.

Solution

Graph the equations $y = 9 + 4\sin\left(\dfrac{x}{2}\right)$ and $y = 12.5$ and find the first point of intersection for $x > 0$.

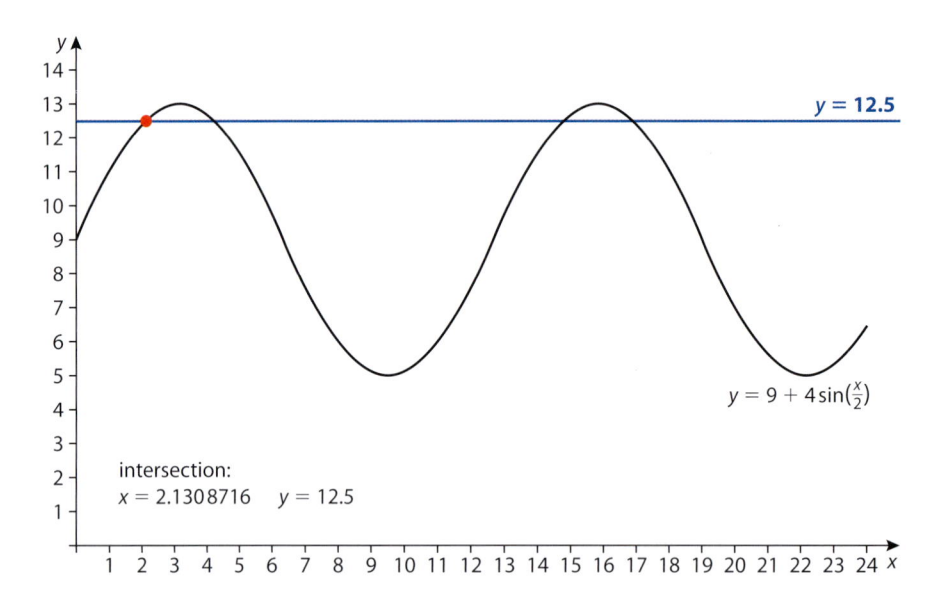

intersection: $x = 2.1308716 \quad y = 12.5$

Using the Intersect command on the GDC indicates that the first point of intersection has an x-coordinate of approximately 2.13. Therefore, the alarm will first sound when $t \approx 2.13$ hours.

Analytic solutions to trigonometric equations

In this section, we will see how general algebraic techniques and trigonometric identities can be applied to solve trigonometric equations. An analytical approach requires you to devise a solution strategy utilizing algebraic methods that you have applied to other types of equations – such as quadratic equations. Often, trigonometric equations that demand an analytic approach will result in exact solutions, but not always. Although our approach for equations in this section focuses on algebraic techniques, it is important to use graphical methods to support or confirm our analytical solutions.

Example 20

Solve $2\sin^2 x - \sin x = 0$ for $-\pi \leq x \leq \pi$.

Solution

We can factorize and apply the rule that if $a \cdot b = 0$ then either $a = 0$ or $b = 0$.

$2\sin^2 x - \sin x = 0 \Rightarrow \sin x(2\sin x - 1) = 0 \Rightarrow \sin x = 0$ or $\sin x = \frac{1}{2}$

For $\sin x = 0$: $x = -\pi, 0, \pi$; for $\sin x = \frac{1}{2}$: $x = \frac{\pi}{6}, \frac{5\pi}{6}$.

Therefore, $x = -\pi, 0, \frac{\pi}{6}, \frac{5\pi}{6}, \pi$.

● **Hint:** Although exact answers were not demanded in Example 20, given our knowledge of the unit circle and familiarity with the sine of common values (i.e. multiples of $\frac{\pi}{6}$ and $\frac{\pi}{4}$), we are able to give exact answers without any difficulty. It would have been acceptable to give approximate solutions, but it is worth recognizing that this would have required considerable more effort than providing exact solutions. Entering and graphing the equation $y = 2\sin^2 x - \sin x$ on your GDC (see GDC images) would not be the most efficient or appropriate solution method, but, if sufficient time is available, it is an effective way to confirm your exact solutions.

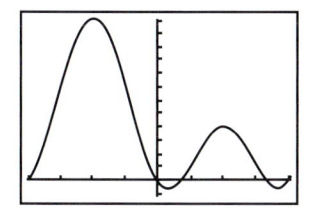

The next example illustrates how the application of a trigonometric identity can be helpful to rewrite an equation in a way that allows us to solve it algebraically.

Example 21

Solve $3\sin x + \tan x = 0$ for $0 \leq x \leq 2\pi$.

Solution

Since the structure of this equation is such that an expression is set equal to zero, it would be nice to be able to use the same algebraic technique as the previous example – that is, factorize and solve for when each factor is zero. However, it is not possible to factorize the expression $3\sin x + \tan x$, and rewriting the equation as $3\sin x = -\tan x$ does not help. Are there any expressions in the equation for which we can substitute an equivalent expression that will make the equation accessible to an algebraic solution? We do not have any equivalent expressions for $\sin x$, but we do have an identity for $\tan x$. From the definition of $\tan x$, we know that $\tan x = \frac{\sin x}{\cos x}$. Let's see what happens when we substitute $\frac{\sin x}{\cos x}$ for $\tan x$.

$3\sin x + \tan x = 0 \Rightarrow 3\sin x + \frac{\sin x}{\cos x} = 0$

Now, multiply both sides by $\cos x$ while recognising that $\cos x \neq 0$ ($x \neq \frac{\pi}{2} + k \cdot \pi, k \in \mathbb{Z}$).

$3\sin x + \frac{\sin x}{\cos x} = 0 \Rightarrow 3\sin x \cos x + \sin x = 0 \Rightarrow \sin x(3\cos x + 1) = 0 \Rightarrow$

$\sin x = 0$ or $\cos x = -\frac{1}{3}$

For $\sin x = 0$: $x = 0, \pi, 2\pi$.

We know that $(1, 0)$ and $(-1, 0)$ are the points on the unit circle that correspond to $\sin x = 0$ giving the three exact solutions above. Although

6 Trigonometric Functions and Equations

we know that the points on the unit circle that correspond to $\cos x = -\frac{1}{3}$ will be in the second and third quadrants, we do not know their exact coordinates. So, we will need to use our GDC to find approximate solutions to $\cos x = -\frac{1}{3}$ for $0 \leq x \leq 2\pi$.

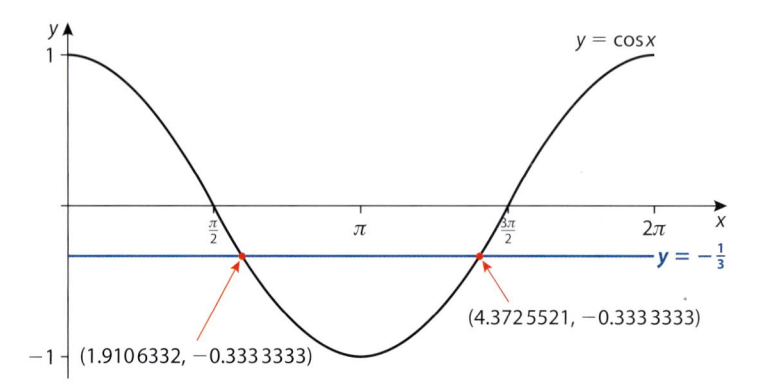

• **Hint:** A strategy that often proves fruitful is to try and rewrite a trigonometric equation in terms of just one trigonometric function. If that is not possible, try and rewrite it in terms of only the sine and cosine functions. This strategy was used in Example 21.

Thus, for $\cos x = -\frac{1}{3}$: $x \approx 1.91$ or $x \approx 4.37$ (three significant figures).

Therefore, the full solution set for the equation is $x = 0, \pi, 2\pi; x \approx 1.91, 4.37$.

Trigonometric identities

As Example 21 illustrated, sometimes an analytical method for solving a trigonometric equation relies on a trigonometric identity providing a useful substitution. There are a few trigonometric identities, other than $\tan x = \frac{\sin x}{\cos x}$, required for this course which can be used to help simplify trigonometric expressions and solve equations.

At the start of this section, it was stated that the equation $\sin^2 x + \cos^2 x = 1$ is an identity; that is, it's true for all possible values of x. Let's prove that this is the case.

Recall from Section 6.1 that the equation for the unit circle is $x^2 + y^2 = 1$. That is, the coordinates (x, y) of any point on the circle will satisfy the equation $x^2 + y^2 = 1$. Also, in Section 6.2, we learned that the sine and cosine functions are defined in terms of the coordinates of the terminal point of an arc on the unit circle starting at $(1, 0)$, as shown in Figure 6.29. If t is any real number that is the length of an arc on the unit circle that terminates at (x, y), then $x = \cos t$ and $y = \sin t$. Substituting directly into the equation for the circle gives $\sin^2 t + \cos^2 t = 1$. As mentioned in Section 6.3, the convention is to use x to denote the domain variable rather than t. Therefore, the equation $\sin^2 x + \cos^2 x = 1$ is true for any real number x.

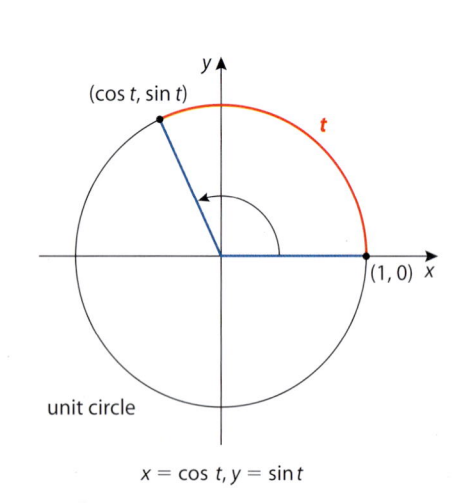

◀ Figure 6.29

The Pythagorean identities for sine and cosine
The following equations are true for all real numbers x:
$$\sin^2 x + \cos^2 x = 1 \qquad \sin^2 x = 1 - \cos^2 x \qquad \cos^2 x = 1 - \sin^2 x$$

Another useful set of trigonometric identities are referred to as the **double angle identities** because they are equations involving $\sin 2x$ and $\cos 2x$. As discussed back in Section 6.1, the argument of a trigonometric function (x in $\sin x$, θ in $\cos \theta$) can be interpreted as an angle (in degrees or radians), or as just a real number. Even though these identities are called double *angle* identities they apply for either interpretation.

> The identity $\sin^2 x + \cos^2 x = 1$ is often referred to as a Pythagorean identity because, as we will see in the other chapter on trigonometry, $\sin x$ and $\cos x$ can represent the legs of a right-angled triangle with a hypotenuse equal to one. Substituting into the Pythagorean theorem gives $\sin^2 x + \cos^2 x = 1$.

Double angle identities for sine and cosine
The following equations are true for all real numbers x:
$$\sin 2x = 2 \sin x \cos x$$
$$\cos 2x = \begin{cases} \cos^2 x - \sin^2 x \\ 2\cos^2 x - 1 \\ 1 - 2\sin^2 x \end{cases}$$

It is quite easy to verify the double angle identities by means of graphical analysis on your GDC.

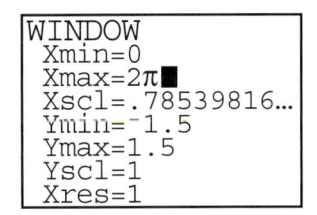

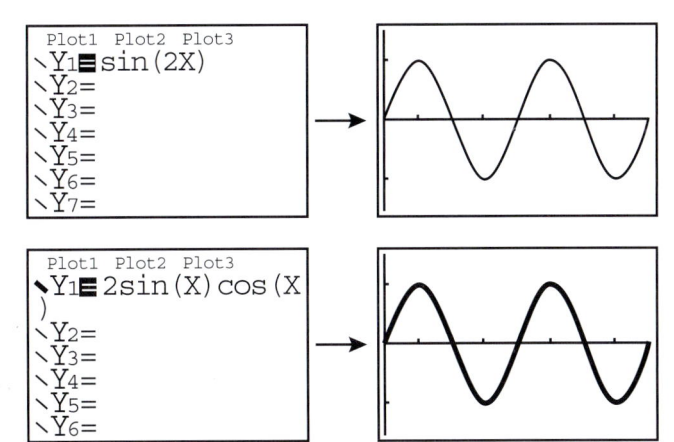

```
WINDOW
Xmin=0
Xmax=2π
Xscl=.78539816…
Ymin=-1.5
Ymax=1.5
Yscl=1
Xres=1
```

```
Plot1 Plot2 Plot3
\Y1■sin(2X)
\Y2=
\Y3=
\Y4=
\Y5=
\Y6=
\Y7=
```

```
Plot1 Plot2 Plot3
\Y1■2sin(X)cos(X)
\Y2=
\Y3=
\Y4=
\Y5=
\Y6=
```

The GDC screen images shown here illustrate that $\sin 2x$ is equivalent to $2 \sin x \cos x$. Use your GDC to verify that $\cos 2x$ is equivalent to $\cos^2 x - \sin^2 x$. Once the identity $\cos 2x = \cos^2 x - \sin^2 x$ is established we can use one of the Pythagorean identities to rewrite it in terms of only sine or cosine; thus, establishing the other two double angle identities for cosine.

$$\cos 2x = \cos^2 x - \sin^2 x$$
$$\cos 2x = \cos^2 x - (1 - \cos^2 x) \qquad \text{substitute } 1 - \cos^2 x \text{ for } \sin^2 x$$
$$\cos 2x = 2\cos^2 x - 1$$

Similar steps can be performed to show that $\cos 2x = 1 - 2\sin^2 x$. Now let's see how these identities can help us with algebraic solutions of trigonometric equations.

Trigonometric Functions and Equations

Example 22

Solve the equation $\cos 2x + \cos x = 0$ for $0 \leq x \leq 2\pi$.

Solution

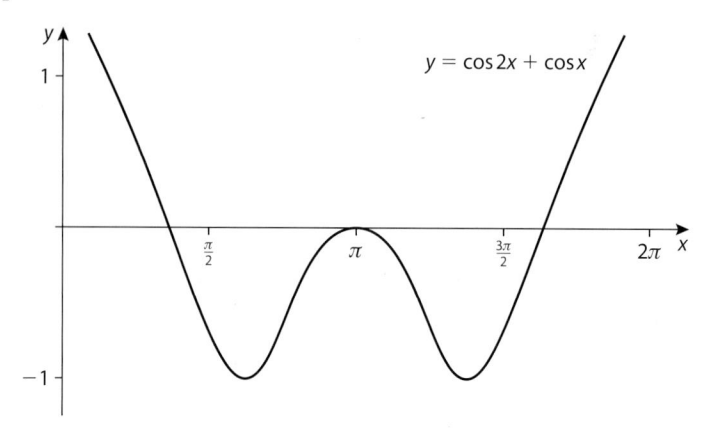

Taking an initial look at the graph of $y = \cos 2x + \cos x$ suggests that there are possibly three solutions in the interval $x \in [0, 2\pi]$. Although the expression $\cos 2x + \cos x$ contains terms with only the cosine function, it is not possible to perform any algebraic operations on them because they have different arguments. In order to solve algebraically, we need both cosine functions to have arguments of x (rather than $2x$). There are three different double angle identities for $\cos 2x$. It is best to have the equation in terms of one trigonometric function, so we choose to substitute $2\cos^2 x - 1$ for $\cos 2x$.

$\cos 2x + \cos x = 0 \Rightarrow 2\cos^2 x - 1 + \cos x = 0 \Rightarrow 2\cos^2 x + \cos x - 1 = 0$

$(2\cos x - 1)(\cos x + 1) = 0 \Rightarrow \cos x = \dfrac{1}{2}$ or $\cos x = -1$

For $\cos x = \dfrac{1}{2}$: $x = \dfrac{\pi}{3}, \dfrac{5\pi}{3}$; for $\cos x = -1$: $x = \pi$.

Therefore, all of the solutions in the interval $0 \leq x \leq 2\pi$ are: $x = \dfrac{\pi}{3}, \pi, \dfrac{5\pi}{3}$.

Example 23

a) Express $2\cos^2 x + \sin x$ in terms of $\sin x$ only.
b) Solve the equation $2\cos^2 x + \sin x = -1$ for x in the interval $0 \leq x \leq 2\pi$, expressing your answer(s) exactly.

Solution

a) $2\cos^2 x + \sin x = 2(1 - \sin^2 x) + \sin x$ using Pythagorean identity
$\cos^2 x = 1 - \sin^2 x$
$= 2 - 2\sin^2 x + \sin x$

b) $2\cos^2 x + \sin x = -1$
$2 - 2\sin^2 x + \sin x = -1$ substitute result from a)
$2\sin^2 x - \sin x - 3 = 0$ [alternatively: let $\sin x = y$, then $2y^2 - y - 3 = 0$]
$(2\sin x - 3)(\sin x + 1) = 0$ factorize [alt: $(2y - 3)(y + 1) = 0$]
$\sin x = \dfrac{3}{2}$ or $\sin x = -1$ [alt: $y = \dfrac{3}{2}$ or $y = -1 \Rightarrow \sin x = \dfrac{3}{2}$ or $\sin x = -1$]

For $\sin x = \frac{3}{2}$: no solution because $\frac{3}{2}$ is not in the range of the sine function.

For $\sin x = -1$: $x = \frac{3\pi}{2}$. Therefore, only one solution in $0 \leq x \leq 2\pi$: $x = \frac{3\pi}{2}$.

Use your GDC to check this result by rewriting $2\cos^2 x + \sin x = -1$ as $2\cos^2 x + \sin x + 1 = 0$ and then graph $y = 2\cos^2 x + \sin x + 1$; confirming a single zero at $x = \frac{3\pi}{2}$ in the interval $x \in [0, 2\pi]$.

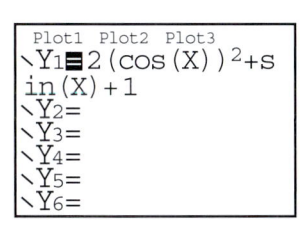

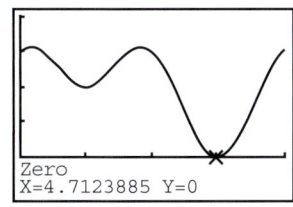

```
Plot1  Plot2  Plot3
\Y1■2(cos(X))²+s
in(X)+1
\Y2=
\Y3=
\Y4=
\Y5=
\Y6=
```

```
WINDOW
Xmin=0
Xmax=6.2831853...
Xscl=π/2
Ymin=-1
Ymax=4
Yscl=1
Xres=1
```

```
X
          4.712388457
3π/2
          4.71238898
```

```
Zero
X=4.7123885  Y=0
```

Example 24

Solve the equation $2\sin 2x = 3\cos x$ for $0 \leq x \leq \pi$.

Solution

$$2\sin 2x = 3\cos x$$
$$2(2\sin x \cos x) = 3\cos x \quad \text{using double angle identity for sine}$$
$$4\sin x \cos x = 3\cos x \quad \text{do } not \text{ divide by } \cos x \text{ as solution(s) may be eliminated}$$
$$4\sin x \cos x - 3\cos x = 0 \quad \text{set equal to zero to prepare for solving by factorization}$$
$$\cos x(4\sin x - 3) = 0 \quad \text{factorize}$$
$$\cos x = 0 \text{ or } \sin x = \frac{3}{4}$$

For $\cos x = 0$: $x = \frac{\pi}{2}$.

For $\sin x = \frac{3}{4}$: $x \approx 0.848$ or 2.29.

Approximate solutions found using Intersect command on GDC. All solutions in the interval $0 \leq x \leq \pi$ are: $x = \frac{\pi}{2}$ and $x \approx 0.848, 2.29$.

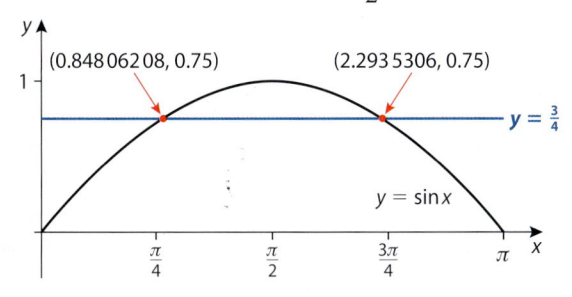

6 Trigonometric Functions and Equations

The final example illustrates how trigonometric identities can be applied to find exact values for trigonometric expressions.

Example 25

Given that $\cos x = \frac{1}{4}$ and $0 < x < \frac{\pi}{2}$, find the *exact* values of

a) $\sin x$ b) $\sin 2x$.

Solution

a) Given $0 < x < \frac{\pi}{2}$ it follows that $\sin x > 0$, because the arc with length x will terminate in the first quadrant. The Pythagorean identity is useful when relating $\sin x$ and $\cos x$.

$$\sin^2 x = 1 - \cos^2 x \Rightarrow \sin x = \sqrt{1 - \cos^2 x} \Rightarrow \sin x = \sqrt{1 - \left(\frac{1}{4}\right)^2}$$
$$= \sqrt{\frac{15}{16}} = \frac{\sqrt{15}}{4}.$$

b) $\sin 2x = 2\sin x \cos x = 2\left(\frac{\sqrt{15}}{4}\right)\left(\frac{1}{4}\right) = \frac{\sqrt{15}}{8}$

Summary of fundamental trigonometric identities

Definition of tangent function: $\tan x = \frac{\sin x}{\cos x}$

Odd/even function identities: $\sin(-x) = -\sin x$ $\cos(-x) = \cos x$
                                            $\tan(-x) = -\tan x$

Co-function identities: $\sin\left(\frac{\pi}{2} - x\right) = \cos x$ $\cos\left(\frac{\pi}{2} - x\right) = \sin x$

Pythagorean identities: $\sin^2 x + \cos^2 x = 1$
                                 $\sin^2 x = 1 - \cos^2 x$ $\cos^2 x = 1 - \sin^2 x$

Double angle identities: $\sin 2x = 2\sin x \cos x$
                               $\cos 2x = \cos^2 x - \sin^2 x$
                               $\cos 2x = 2\cos^2 x - 1$ $\cos 2x = 1 - 2\sin^2 x$

Exercise 6.4

In questions 1–10, find the exact solution(s) for $0 \leq x \leq 2\pi$. Verify your solution(s) with your GDC.

1. $\cos x = \frac{1}{2}$
2. $2\sin x + 1 = 0$
3. $1 - \tan x = 0$
4. $\sqrt{3} = 2\sin x$
5. $2\sin^2 x = 1$
6. $4\cos^2 x = 3$
7. $\tan^2 x - 1 = 0$
8. $4\cos^2 x = 1$
9. $\tan x(\tan x + 1) = 0$
10. $\sin x \cos x = 0$

In questions 11–16, use your GDC to find approximate solution(s) for $0 \leq x \leq 2\pi$. Express solutions accurate to 3 significant figures.

11. $\sin x = 0.4$
12. $3\cos x + 1 = 0$
13. $\tan x = 2$
14. $\sin 2x = 0.85$
15. $\cos(x - 1) = -0.38$
16. $3\tan x = 10$

In questions 17–20, given that k is any integer, list all of the possible values for x that are in the specified interval.

17 $x = \frac{\pi}{2} + k \cdot \pi, \ -3\pi \leq x \leq 3\pi$

18 $x = \frac{\pi}{6} + k \cdot 2\pi, \ -2\pi \leq x \leq 2\pi$

19 $x = \frac{7\pi}{12} + k \cdot \pi, \ 0 \leq x \leq 2\pi$

20 $x = \frac{\pi}{4} + k \cdot \frac{\pi}{4}, \ 0 \leq x \leq 2\pi$

In questions 21–24, find the exact solutions for $0 \leq x \leq 2\pi$.

21 $\cos\left(x - \frac{\pi}{6}\right) = -\frac{1}{2}$

22 $\tan(x + \pi) = 1$

23 $\sin 2x = \frac{\sqrt{3}}{2}$

24 $\sin^2\left(x + \frac{\pi}{2}\right) = \frac{3}{4}$

25 The number, N, of empty birds' nests in a park is approximated by the function $N = 74 + 42\sin\left(\frac{\pi}{12}t\right)$, where t is the number of hours after midnight.
Find the value of t when the number of empty nests first equals 90.
Approximate the answer to 1 decimal place.

26 In Edinburgh, the number of hours of daylight on day D is modelled by the function $H = 12 + 7.26\sin\left[\frac{2\pi}{365}(D - 80)\right]$, where D is the number of days after December 31 (e.g. January 1 is $D = 1$, January 2 is $D = 2$, and so on).
Do not use your GDC on part a).
a) Which days of the year have 12 hours of daylight?
b) Which days of the year have about 15 hours of daylight?
c) How many days of the year have more than 17 hours of daylight?

In questions 27–34, solve the equation for the stated solution interval. Find exact solutions, if possible. Otherwise, give solutions to 3 significant figures. Verify solutions with your GDC.

27 $2\cos^2 x + \cos x = 0; \ 0 \leq x \leq 2\pi$

28 $2\sin^2 x - \sin x - 1 = 0; \ 0 \leq x \leq 2\pi$

29 $2\cos x + \sin 2x = 0; \ -\pi \leq x \leq \pi$

30 $2\sin x = \cos 2x; \ -\pi \leq x \leq \pi$

31 $\tan^2 x - \tan x = 2; \ -\frac{\pi}{2} \leq x \leq \frac{\pi}{2}$

32 $\sin^2 x = \cos^2 x; \ 0 \leq x \leq \pi$

33 $2\sin^2 x + 3\cos x - 3 = 0; \ 0 \leq x \leq 2\pi$

34 $2\sin x = 3\cos x; \ 0 \leq x \leq 2\pi$

35 Given that $\sin x = \frac{3}{5}$ and $0 < x < \frac{\pi}{2}$, find the *exact* values of
a) $\cos x$
b) $\cos 2x$
c) $\sin 2x$.

36 Given that $\cos x = -\frac{2}{3}$ and $\frac{\pi}{2} < x < \pi$, find the *exact* values of
a) $\sin x$
b) $\sin 2x$
c) $\cos 2x$.

6 Trigonometric Functions and Equations

Practice questions

1. A toy on an elastic string is attached to the top of a doorway. It is pulled down and released, allowing it to bounce up and down. The length of the elastic string, L centimetres, is modelled by the function $L = 110 + 25\cos(2\pi t)$, where t is time in seconds after release.
   a) Find the length of the elastic string after 2 seconds.
   b) Find the minimum length of the string.
   c) Find the first time after release that the string is 85 cm.
   d) What is the period of the motion?

2. Find the exact solution(s) to the equation $2\sin^2 x - \cos x + 1 = 0$ for $0 \leqslant x \leqslant 2\pi$.

3. The diagram shows a circle of radius 6 cm.
   The perimeter of the shaded sector is 25 cm.
   Find the radian measure of the angle θ.

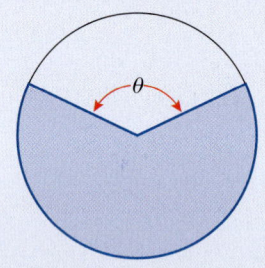

4. Consider the two functions $f(x) = \cos 4x$ and $g(x) = \cos\left(\dfrac{x}{2}\right)$.
   a) Write down: (i) the minimum value of the function f
   (ii) the period of g.
   b) For the equation $f(x) = g(x)$, find the number of solutions in the interval $0 \leqslant x \leqslant \pi$.

5. A reflector is attached to the spoke of a bicycle wheel. As the wheel rolls along the ground, the distance, d centimetres, that the reflector is above the ground after t seconds is modelled by the function
   $$d = p + q\cos\left(\dfrac{2\pi}{m}t\right),$$ where p, q and m are constants.
   The distance d is at a maximum of 64 cm at $t = 0$ seconds and at $t = 0.5$ seconds, and is at a minimum of 6 cm at $t = 0.25$ seconds and at $t = 0.75$ seconds. Write down the value of:
   a) p b) q c) m.

6. Find all solutions to $1 + \sin 3x = \cos(0.25x)$ such that $x \in [0, \pi]$.

7. Find all solutions to both trigonometric equations in the interval $x \in [0, 2\pi]$. Express the solutions exactly.
   a) $2\cos^2 x + 5\cos x + 2 = 0$ b) $\sin 2x - \cos x = 0$

8. The value of x is in the interval $\dfrac{\pi}{2} < x < \pi$ and $\cos^2 x = \dfrac{8}{9}$. Without using your GDC, find the exact values for the following:
   a) $\sin x$ b) $\cos 2x$ c) $\sin 2x$

9. The depth, d metres, of water in a harbour varies with the tides during each day. The first high (maximum) tide after midnight occurs at 5:00 a.m. with a depth of 5.8 m. The first low (minimum) tide occurs at 10:30 a.m. with a depth of 2.6 m.
   a) Find a trigonometric function that models the depth, d, of the water t hours after midnight.
   b) Find the depth of the water at 12 noon.
   c) A large boat needs at least 3.5 m of water to dock in the harbour. During what time interval after 12 noon can the boat dock safely?

10 Solve the equation $\tan^2 x + 2\tan x - 3 = 0$ for $0 \leq x \leq \pi$. Give solutions exactly, if possible. Otherwise, give solutions to 3 significant figures.

11 The following diagram shows a circle of centre O and radius 10 cm. The arc ABC subtends an angle of $\frac{3}{2}$ radians at the centre O.
  a) Find the length of the arc ACB.
  b) Find the area of the shaded region.

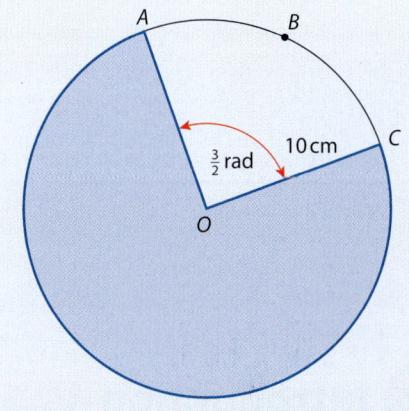

12 Consider the function $f(x) = \frac{5}{2}\cos\left(2x - \frac{\pi}{2}\right)$. For what values of k will the equation $f(x) = k$ have no solutions?

13 A portion of the graph of $y = k + a\sin x$ is shown below. The graph passes through the points $(0, 1)$ and $\left(\frac{3\pi}{2}, 3\right)$. Find the value of k and a.

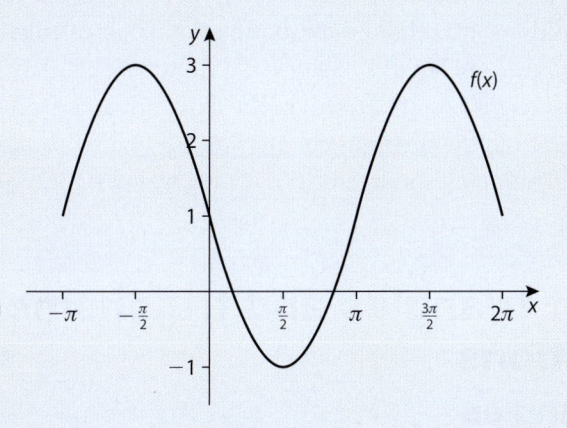

7 Triangle Trigonometry

Assessment statements
3.6 Solution of triangles.
The cosine rule: $c^2 = a^2 + b^2 - 2ab \cos C$.
The sine rule: $\dfrac{a}{\sin A} = \dfrac{b}{\sin B} = \dfrac{c}{\sin C}$.
Area of a triangle as $\tfrac{1}{2} ab \sin C$.

Introduction

In this chapter, we approach trigonometry from a **right triangle** perspective where trigonometric functions will be defined in terms of the **ratios of sides of a right triangle**. Over two thousand years ago, the Greeks developed trigonometry to make helpful calculations for surveying, navigating, building and other practical pursuits. Their calculations were based on the angles and lengths of sides of a right triangle. The modern development of trigonometry, based on the length of an arc on the unit circle, was covered in the previous chapter. We begin a more classical approach by introducing some terminology regarding right triangles.

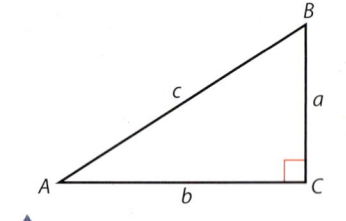

Figure 7.1

- **Hint:** In IB notation, [AC] denotes the line segment connecting points A and C. The notation AC represents the *length* of this line segment. Also, the notation $A\hat{B}C$ denotes the angle with its vertex at point B, with one side of the angle containing the point A and the other side containing point C.

7.1 Right triangles and trigonometric functions

Right triangles

The conventional notation for triangles is to label the three vertices with capital letters, for example A, B and C. The same capital letters can be used to represent the measure of the angles at these vertices. However, we will often use a Greek letter, such as α (alpha), β (beta) or θ (theta) to do so. The corresponding lower-case letters, a, b and c, represent the lengths of the sides opposite the vertices. For example, b represents the length of the side opposite angle B, that is, the line segment AC, or [AC] (Figure 7.1).

In a right triangle, the longest side is opposite the right angle (i.e. measure of 90°) and is called the **hypotenuse**, and the two shorter sides adjacent to the right angle are often called the **legs** (Figure 7.2). Because the sum of the three angles in any triangle in plane geometry is 180°, then the two non-right angles are both **acute angles** (i.e. measure between 0 and 90 degrees). It also follows that the two acute angles in a right triangle are a pair of **complementary angles** (i.e. have a sum of 90°).

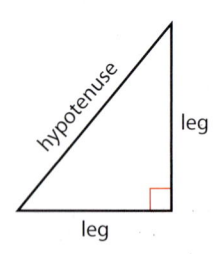

Figure 7.2

Trigonometric functions of an acute angle

We can use properties of similar triangles and the definitions of the sine, cosine and tangent functions from Chapter 6 to define these functions in terms of the sides of a right triangle.

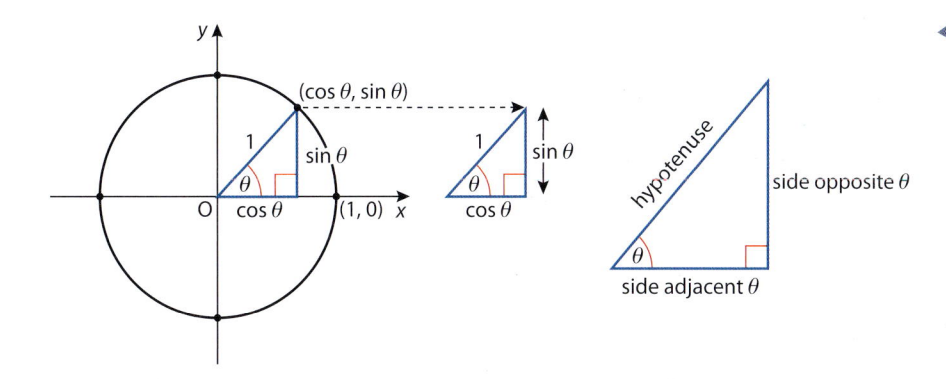

Figure 7.3

The right triangles shown in Figure 7.3 are **similar triangles** because corresponding angles have equal measure – each has a right angle and an acute angle of measure θ. It follows that the ratios of corresponding sides are equal, allowing us to write the following three proportions involving the sine, cosine and tangent of the acute angle θ.

$$\frac{\sin \theta}{1} = \frac{\text{opposite}}{\text{hypotenuse}} \qquad \frac{\cos \theta}{1} = \frac{\text{adjacent}}{\text{hypotenuse}} \qquad \frac{\tan \theta}{1} = \frac{\sin \theta}{\cos \theta} = \frac{\text{opposite}}{\text{adjacent}}$$

The definitions of the trigonometric functions in terms of the sides of a right triangle follow directly from these three equations.

Right triangle definition of the trigonometric functions

Let θ be an **acute angle** of a right triangle, then the sine, cosine and tangent functions of the angle θ are defined as the following ratios in the right triangle:

$$\sin \theta = \frac{\text{side opposite angle } \theta}{\text{hypotenuse}}$$

$$\cos \theta = \frac{\text{side adjacent angle } \theta}{\text{hypotenuse}}$$

$$\tan \theta = \frac{\text{side opposite angle } \theta}{\text{side adjacent angle } \theta}$$

It follows that the sine, cosine and tangent of an acute angle are positive.

It is important to understand that properties of similar triangles are the foundation of right triangle trigonometry. Regardless of the size (i.e. lengths of sides) of a right triangle, so long as the angles do not change, the ratio of any two sides in the right triangle will remain *constant*. All the right triangles in Figure 7.4 have an acute angle with a measure of 30° (thus, the other acute angle is 60°). For each triangle, the ratio of the side opposite the 30° angle to the hypotenuse is exactly $\frac{1}{2}$. In other words, the sine of 30° is always $\frac{1}{2}$. This agrees with results from the previous chapter knowing that an angle of 30° is equivalent to $\frac{\pi}{6}$ in radian measure.

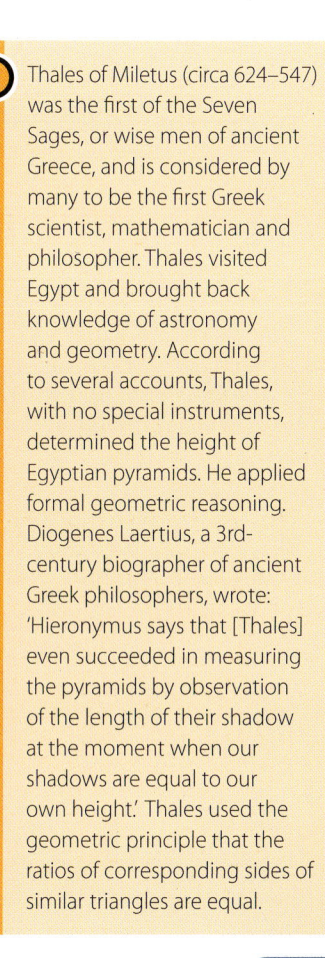

> Thales of Miletus (circa 624–547) was the first of the Seven Sages, or wise men of ancient Greece, and is considered by many to be the first Greek scientist, mathematician and philosopher. Thales visited Egypt and brought back knowledge of astronomy and geometry. According to several accounts, Thales, with no special instruments, determined the height of Egyptian pyramids. He applied formal geometric reasoning. Diogenes Laertius, a 3rd-century biographer of ancient Greek philosophers, wrote: 'Hieronymus says that [Thales] even succeeded in measuring the pyramids by observation of the length of their shadow at the moment when our shadows are equal to our own height.' Thales used the geometric principle that the ratios of corresponding sides of similar triangles are equal.

7 Triangle Trigonometry

Figure 7.4

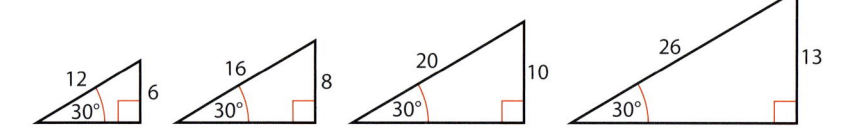

For any right triangle, the sine ratio for 30° is always $\frac{1}{2}$: $\sin 30° = \frac{1}{2}$.

The trigonometric functions of acute angles are not always rational numbers such as $\frac{1}{2}$. We will see in upcoming examples that the sine of 60° is exactly $\frac{\sqrt{3}}{2}$.

Evaluating trigonometric functions for 30°, 45° and 60°

We can use Pythagoras' theorem and properties of triangles to find the exact values for the most common acute angles: 30°, 45° and 60°.

Example 1

Find the values of sin 45°, cos 45° and tan 45°.

Solution

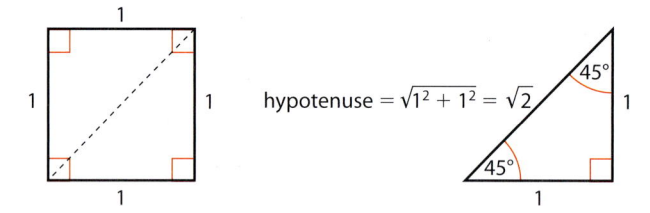

hypotenuse $= \sqrt{1^2 + 1^2} = \sqrt{2}$

Consider a square with each side equal to one unit. Draw a diagonal of the square, forming two isosceles right triangles. From geometry, we know that the diagonal will bisect each of the two right angles forming two isosceles right triangles, each with two acute angles of 45°. The isosceles right triangles have legs of length one unit and, from Pythagoras' theorem, a hypotenuse of exactly $\sqrt{2}$ units. The trigonometric functions are then calculated as follows:

$$\sin 45° = \frac{\text{opposite}}{\text{hypotenuse}} = \frac{1}{\sqrt{2}} = \frac{\sqrt{2}}{2} \quad \text{[multiplying by } \frac{\sqrt{2}}{\sqrt{2}} \text{ to rationalize the denominator]}$$

$$\cos 45° = \frac{\text{adjacent}}{\text{hypotenuse}} = \frac{1}{\sqrt{2}} = \frac{\sqrt{2}}{2}$$

$$\tan 45° = \frac{\text{opposite}}{\text{adjacent}} = \frac{1}{1} = 1$$

Example 2

Find the values of the sine, cosine and tangent functions for 30° and 60°.

Solution

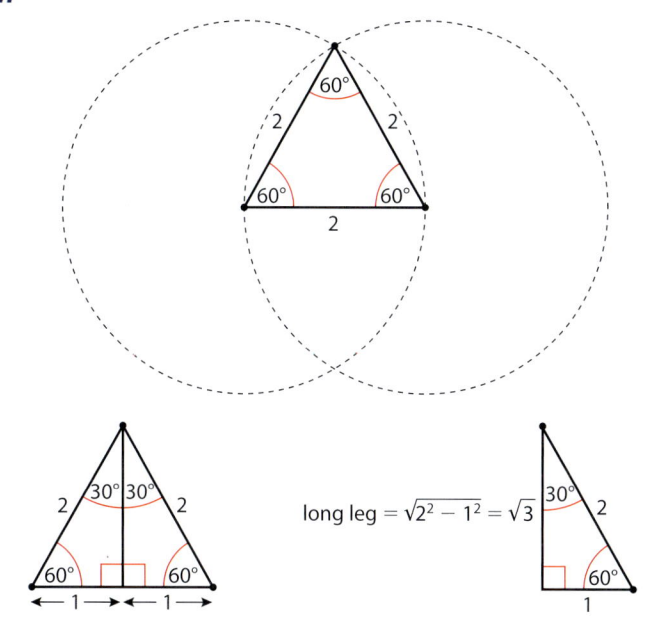

Start with a line segment of length two units. Using each endpoint as a centre and the segment as a radius, construct two circles. The endpoints of the original line segment and the point of intersection of the two circles are the vertices of an equilateral triangle. Each side has a length of two units and the measure of each angle is 60°. From geometry, the altitude drawn from one of the vertices bisects the angle at that vertex and also bisects the opposite side to which it is perpendicular. Two right triangles are formed that have acute angles of 30° and 60°, a hypotenuse of two units, and a short leg of one unit. Using Pythagoras' theorem, the long leg is $\sqrt{3}$ units. The trigonometric functions of 30° and 60° are then calculated as follows:

$\sin 60° = \dfrac{\text{opposite}}{\text{hypotenuse}} = \dfrac{\sqrt{3}}{2}$ $\sin 30° = \dfrac{\text{opposite}}{\text{hypotenuse}} = \dfrac{1}{2}$

$\cos 60° = \dfrac{\text{adjacent}}{\text{hypotenuse}} = \dfrac{1}{2}$ $\cos 30° = \dfrac{\text{adjacent}}{\text{hypotenuse}} = \dfrac{\sqrt{3}}{2}$

$\tan 60° = \dfrac{\text{opposite}}{\text{adjacent}} = \dfrac{\sqrt{3}}{1} = \sqrt{3}$ $\tan 30° = \dfrac{\text{opposite}}{\text{adjacent}} = \dfrac{1}{\sqrt{3}} = \dfrac{\sqrt{3}}{3}$ rationalizing the denominator

The geometric derivation of the values of the sine, cosine and tangent functions for the 'special' acute angles 30°, 45° and 60°, in Examples 1 and 2, agree with the results from the previous chapter. The results for these angles – in both degree and radian measure – are summarised in the box below.

Values of sine, cosine and tangent for common acute angles

$\sin 30° = \sin \dfrac{\pi}{6} = \dfrac{1}{2}$ $\cos 30° = \cos \dfrac{\pi}{6} = \dfrac{\sqrt{3}}{2}$ $\tan 30° = \tan \dfrac{\pi}{6} = \dfrac{\sqrt{3}}{3}$

$\sin 45° = \sin \dfrac{\pi}{4} = \dfrac{\sqrt{2}}{2}$ $\cos 45° = \cos \dfrac{\pi}{4} = \dfrac{\sqrt{2}}{2}$ $\tan 45° = \tan \dfrac{\pi}{4} = 1$

$\sin 60° = \sin \dfrac{\pi}{3} = \dfrac{\sqrt{3}}{2}$ $\cos 60° = \cos \dfrac{\pi}{3} = \dfrac{1}{2}$ $\tan 60° = \tan \dfrac{\pi}{3} = \sqrt{3}$

● **Hint:** It is important that you are able to recall – without a calculator – the exact trigonometric values for these common angles.

7 Triangle Trigonometry

Observe that $\sin 30° = \cos 60° = \frac{1}{2}$, $\sin 60° = \cos 30° = \frac{\sqrt{3}}{2}$ and $\sin 45° = \cos 45° = \frac{\sqrt{2}}{2}$. Complementary angles (sum of 90°) have equal function values for sine and cosine. That is, for all angles x measured in degrees, $\sin x = \cos(90° - x)$ or $\sin(90° - x) = \cos x$. As noted in Chapter 6, it is for this reason that sine and cosine are called co-functions.

Solution of right triangles

Every triangle has three sides and three angles – six different parts. The ancient Greeks knew how to solve for all of the unknown angles and sides in a right triangle given that either the length of two sides, or the length of one side and the measure of one angle, were known. To **solve a right triangle** means to find the measure of any unknown sides or angles. We can accomplish this by applying Pythagoras' theorem and trigonometric functions. We will utilize trigonometric functions in two different ways when solving for missing parts in right triangles – to find the length of a side, and to find the measure of an angle. Solving right triangles using the sine, cosine and tangent functions is essential to finding solutions to problems in fields such as astronomy, navigation, engineering and architecture. In Sections 7.3 and 7.4, we will see how trigonometry can also be used to solve for missing parts in triangles that are not right triangles.

Angles of depression and elevation

An imaginary line segment from an observation point O to a point P (representing the location of an object) is called the **line of sight** of P. If P is above O, the acute angle between the line of sight of P and a horizontal line passing through O is called the **angle of elevation** of P. If P is below O, the angle between the line of sight and the horizontal is called the **angle of depression** of P. This is illustrated in Figure 7.5.

Figure 7.5

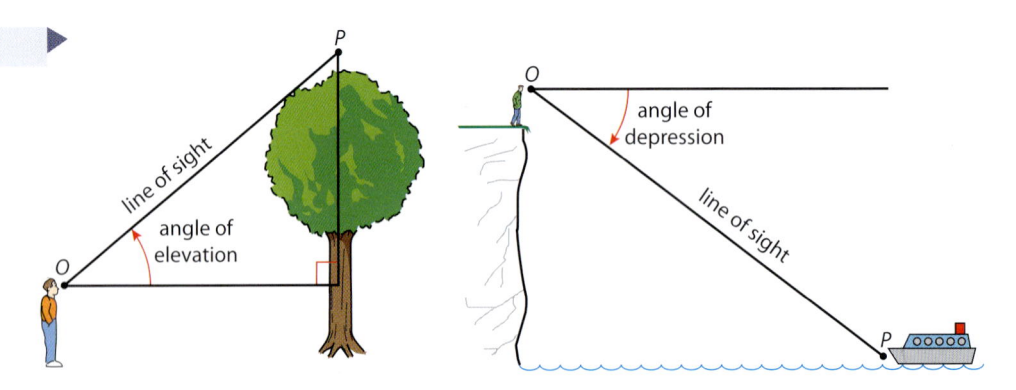

Example 3

Solve triangle ABC given $c = 8.76$ cm and angle $A = 30°$, where the right angle is at C. Give exact answers when possible, otherwise give to an accuracy of 3 significant figures.

Solution

Knowing that the conventional notation is to use a lower-case letter to represent the length of a side opposite the vertex denoted with the corresponding upper-case letter, we sketch triangle ABC indicating the known measurements.

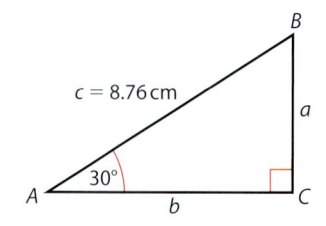

From the definition of sine and cosine functions, we have

$$\sin 30° = \frac{\text{opposite}}{\text{hypotenuse}} = \frac{a}{8.76} \qquad \cos 30° = \frac{\text{adjacent}}{\text{hypotenuse}} = \frac{b}{8.76}$$

$$a = 8.76 \sin 30° \qquad\qquad\qquad b = 8.76 \cos 30°$$

$$a = 8.76\left(\frac{1}{2}\right) = 4.38 \qquad b = 8.76\left(\frac{\sqrt{3}}{2}\right) \approx 7.586\,382\,537 \approx 7.59$$

Therefore, $a = 4.38$ cm, $b \approx 7.59$ cm, and it's clear that angle $B = 60°$.

We can use Pythagoras' theorem to check our results for a and b.

$$a^2 + b^2 = c^2 \Rightarrow \sqrt{a^2 + b^2} = 8.76$$

Be aware that the result for a is exactly 4.38 cm (assuming measurements given for angle A and side c are exact), but the result for b can only be approximated. To reduce error when performing the check, we should use the most accurate value (i.e. most significant figures) for b possible. The most effective way to do this on our GDC is to use results that are stored to several significant figures, as shown in the GDC screen image.

```
8.76(√(3)/2)
            7.586382537
Ans→B
            7.586382537
√(4.38²+B²)
                    8.76
```

Example 4

A scientist involved in forest management wants to measure the height of a tree without climbing it. From a point 34.5 m from the base of a large tree, the scientist determines that the angle of elevation from horizontal ground to the top of the tree is 52.4°. What is the height of the tree, approximated to the nearest tenth of a metre?

Solution

$$\tan 52.4° = \frac{\text{opposite}}{\text{adjacent}} = \frac{h}{34.5} \Rightarrow h = 34.5 \tan 52.4°$$

$$h \approx 34.5(1.2985)$$

$$h \approx 44.799\,16$$

The height of the tree is approximately 44.8 m.

In both Examples 3 and 4, one of the acute angles of a right triangle was given so the third angle is easily determined from the fact that the sum of the angles is 180°. Let's look at how we can use trigonometric functions to solve a right triangle for which the lengths of two of the sides are known, but the measure of both acute angles are unknown.

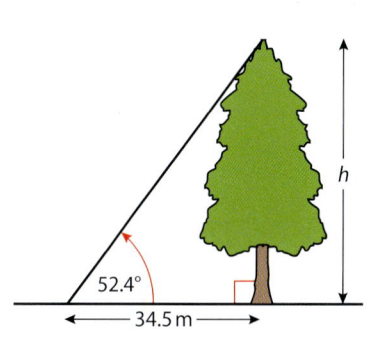

7 Triangle Trigonometry

Example 5

Solve triangle *PQR* given $QR = 9$ cm and $PQ = 12$ cm, where the right angle is at *R*. Give exact answers when possible, otherwise give to an accuracy of 3 s.f.

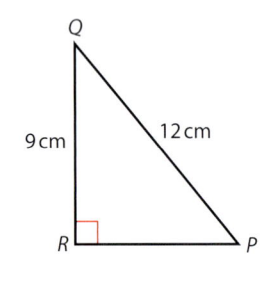

Solution

Using Pythagoras' theorem: $PR = \sqrt{12^2 - 9^2} = \sqrt{63} = 3\sqrt{7} \approx 7.94$.

Both of the acute angles, $\angle P$ and $\angle Q$, are unknown. We know the lengths of all three sides, so it is possible to evaluate any of the trigonometric functions for either of these angles. For example, it is clear that $\sin P = \frac{9}{12} = 0.75$. To determine the acute angle that has a sine ratio of 0.75, we need to perform the **inverse** of the sine function (written as $\sin^{-1}$). We can do this by solving the equation $\sin x = 0.75$ graphically, as we did in Section 6.4. (See GDC screen images at bottom of page.) Using a graphical method is particularly suitable if *x* represents a real number, or perhaps an angle in radian measure, and there is more than one solution for *x*. For triangle *PQR*, there is only one solution for *P* in the equation $\sin P = 0.75$ and it must be between 0° and 90°. Your GDC (in 'degree' mode) can be used to find the acute angle *P*, either graphically or by directly computing the inverse sine of 0.75. Although, as we will realize, there are an infinite number of angles with a sine ratio of 0.75, your GDC is programmed so that the inverse sine ($\sin^{-1}$) computation gives only the one acute angle with a sine ratio of 0.75. The GDC screen images illustrate that having your GDC compute an inverse trigonometric value is the most efficient method for finding an acute angle.

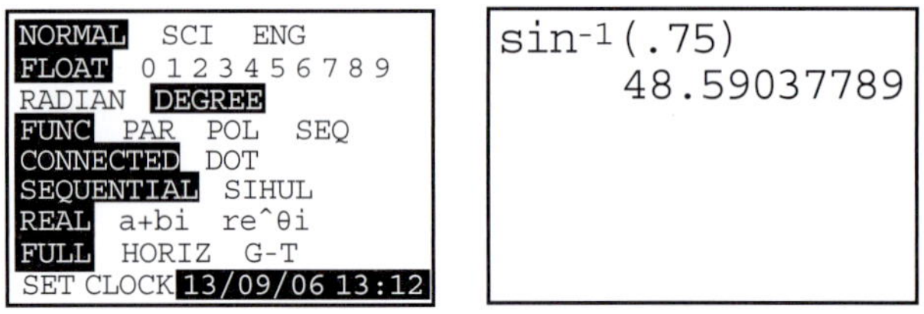

Thus, $\angle P \approx 48.6°$ from which it follows that $\angle Q \approx 90° - 48.6° \approx 41.4°$.

Therefore, the missing parts of triangle *PQR* are $PR \approx 7.94$ cm, $\angle P \approx 48.6°$ and $\angle Q \approx 41.4°$.

Graphical solution:

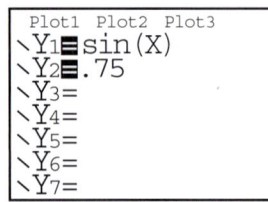

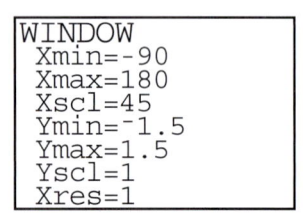

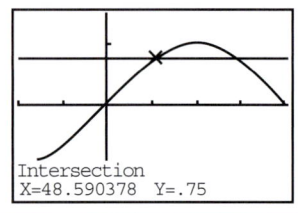

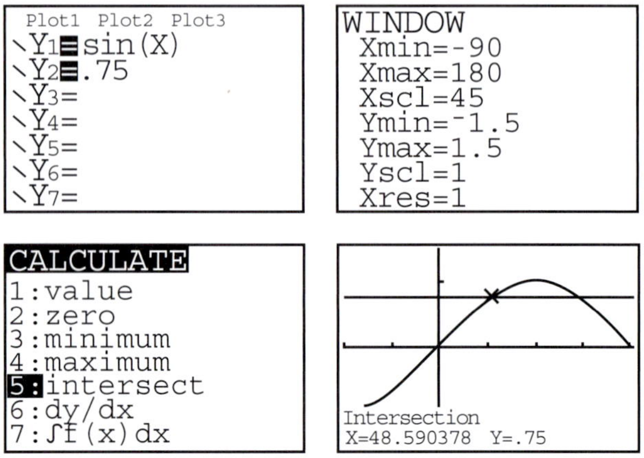

• **Hint:** As mentioned in Section 2.3, the notation for indicating the inverse of a function is a superscript of negative one. For example, the inverse of the cosine function is written as $\cos^{-1}$. The negative one is *not* an exponent, so it does not denote reciprocal. Do not make this error: $\cos^{-1} x \neq \frac{1}{\cos x}$. And as stated in Section 2.3, if $f(a) = b$ then $f^{-1}(b) = a$. For example, for the sine function, if $\sin 60° = \frac{\sqrt{3}}{2}$ then $\sin^{-1}\left(\frac{\sqrt{3}}{2}\right) = 60°$.

Example 6

From the top of a perpendicular cliff 93 m high, the angle of depression of a boat is 26.5°. How far is the boat from the foot of the cliff? Give your answer accurate to 3 s.f.

Solution

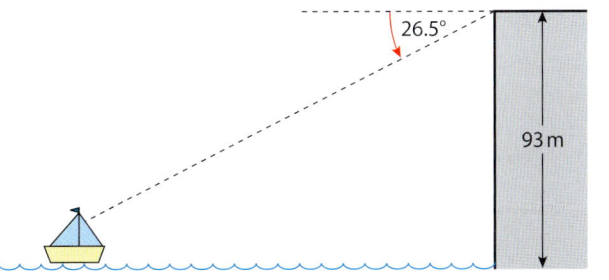

If the angle of depression of the boat from the top of the cliff is 26.5°, the angle of elevation of the top of the cliff from the boat is also 26.5°. Thus, we can use the right triangle below to solve for d.

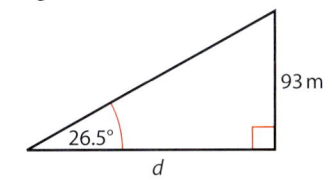

$$\tan 26.5° = \frac{93}{d} \Rightarrow d = \frac{93}{\tan 26.5} \Rightarrow d \approx \frac{93}{0.498\,58} \approx 186.53$$

The boat is approximately 187 m from the foot of the cliff.

Example 7

A man who is 183 cm tall casts a 72 cm long shadow on the horizontal ground. What is the angle of elevation of the sun to the nearest tenth of a degree?

Solution

In the diagram, the angle of elevation of the sun is labelled θ.

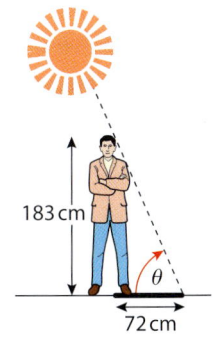

$$\tan \theta = \frac{183}{72}$$

$$\theta = \tan^{-1}\left(\frac{183}{72}\right)$$

$$\theta \approx 68.5°$$

```
tan⁻¹(183/72)
         68.52320902
```

GDC computation

The angle of elevation of the sun is approximately 68.5°.

7 Triangle Trigonometry

Exercise 7.1

In questions 1–6, find the exact value of the trigonometric function for the specified acute angle in triangle ABC.

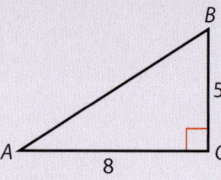

1. $\sin A$
2. $\cos A$
3. $\tan A$
4. $\sin B$
5. $\cos B$
6. $\tan B$

7. Using your GDC, find (accurate to 3 s.f.) the degree measure of $B\hat{A}C$ and $A\hat{B}C$ in right triangle ABC above.

In questions 8–13, sketch a right triangle corresponding to the given trigonometric function of the acute angle θ. Use Pythagoras' theorem to determine the third side, and then find the value of the other two trigonometric functions of θ.

8. $\sin\theta = \dfrac{3}{5}$
9. $\cos\theta = \dfrac{5}{8}$
10. $\tan\theta = 2$
11. $\cos\theta = \dfrac{7}{10}$
12. $\tan\theta = \dfrac{1}{3}$
13. $\cos\theta = \dfrac{\sqrt{7}}{4}$

In questions 14–19, find the exact value of the trigonometric function.

14. $\sin 45°$
15. $\cos\dfrac{\pi}{6}$
16. $\tan 45°$
17. $\sin\dfrac{\pi}{3}$
18. $\tan\dfrac{\pi}{6}$
19. $\cos 60°$

In questions 20–25, find the exact value of θ in degree measure $(0 < \theta < 90°)$ and in radian measure $\left(0 < \theta < \dfrac{\pi}{2}\right)$ without using your GDC.

20. $\cos\theta = \dfrac{1}{2}$
21. $\sin\theta = \dfrac{\sqrt{2}}{2}$
22. $\tan\theta = \sqrt{3}$
23. $\sin\theta = \dfrac{\sqrt{3}}{2}$
24. $\tan\theta = 1$
25. $\cos\theta = \dfrac{\sqrt{3}}{2}$

In questions 26–31, find the approximate value (to 3 s.f.) of θ in degree measure $(0 < \theta < 90°)$ and in radian measure $\left(0 < \theta < \dfrac{\pi}{2}\right)$ by using the inverse key on your GDC.

26. $\sin\theta = 0.7258$
27. $\cos\theta = 0.7258$
28. $\tan\theta = 1.2953$
29. $\cos\theta = 0.1638$
30. $\sin\theta = 0.4721$
31. $\tan\theta = 0.6507$

In questions 32–37, solve for x. Give your answer to 3 s.f.

32. 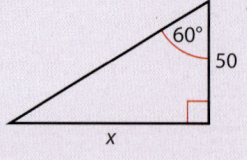
33. 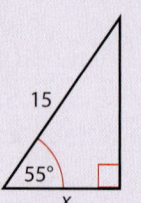
34. 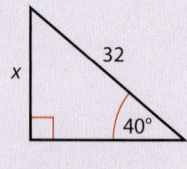
35. 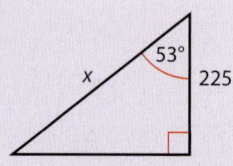
36. 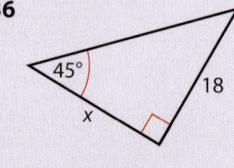
37.

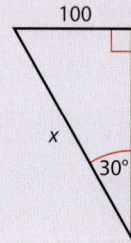

In questions 38 and 39, solve for all of the unknown sides and angles.

38

39

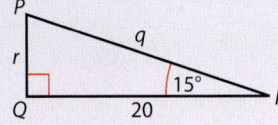

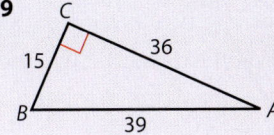

40 The tallest tree in the world is reputed to be a giant redwood named *Hyperion* located in Redwood National Park in California, USA. At a point 41.5 m from the centre of its base and on the same elevation, the angle of elevation of the top of the tree is 70°. How tall is the tree? Give your answer to 3 s.f.

41 The top of the Eiffel Tower in Paris (not including the antenna) is 300 m high. What will be the angle of elevation of the top of the tower from a point on the ground (assumed level) that is 125 m from the centre of the tower's base?

42 A woman, 1.62 m tall, standing 3 m from a street light casts a 2 m long shadow (see diagram). What is the height of the street light?

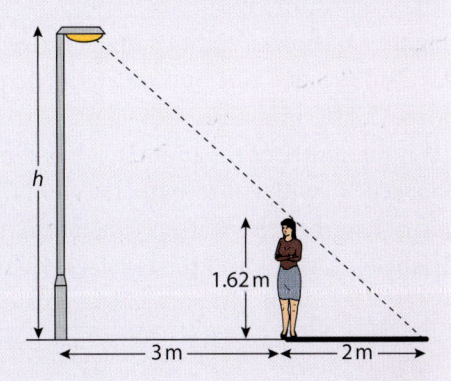

43 A 6 m ladder leaning against the side of a building makes a 72° angle with the ground (see diagram). How far up the side of the house does the ladder reach?

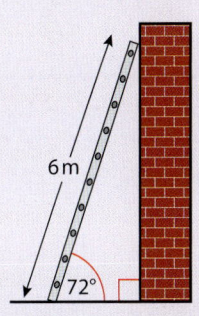

44 An isosceles triangle has sides of length 8 cm, 8 cm and 6 cm (see diagram). Find the angle between the two equal sides.

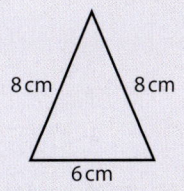

45 From a 50 m observation tower on the shoreline, a coastguard sights a boat in difficulty. The angle of depression of the boat is 5° (see diagram). How far is the boat from the shoreline?

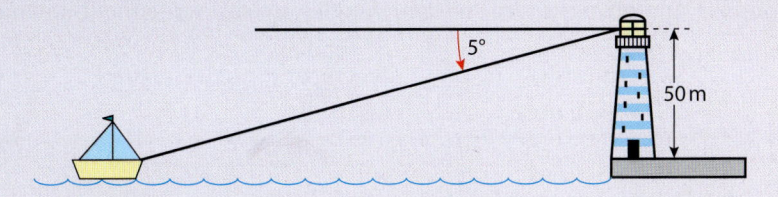

213

7.2 Trigonometric functions of any angle

In this section, we will extend the trigonometric ratios to all angles allowing us to solve problems involving any size angle.

Functions of an angle related to functions of a real number

It is useful to pause for a moment in our consideration of the trigonometric functions as functions of an acute angle in a right triangle, and take a look at how this approach relates to the one taken in Chapter 6, where the trigonometric functions were functions of a real number.

Figure 7.6 shows a right triangle, $\triangle ABC$, where the angle at vertex A is labelled θ. Side BC is opposite to angle θ and side AC is adjacent to angle θ. Place $\triangle ABC$ in the coordinate plane so that angle θ is in standard position (A is the centre of the unit circle) as shown in Figure 7.7. The point labelled B', with coordinates (x, y) on the unit circle, is the point where the arc of length t terminates. Note that $\triangle ABC$ is similar to the smaller right triangle, $\triangle AB'C'$, and the two legs of $\triangle AB'C'$ are x and y (Figure 7.8).

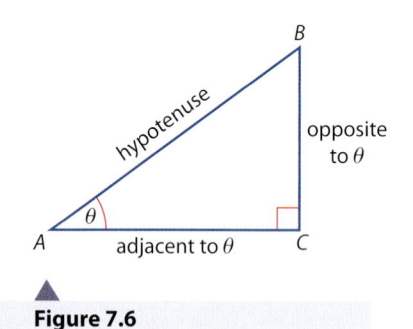

Figure 7.6

Figure 7.7 The radian measure of angle θ is t.

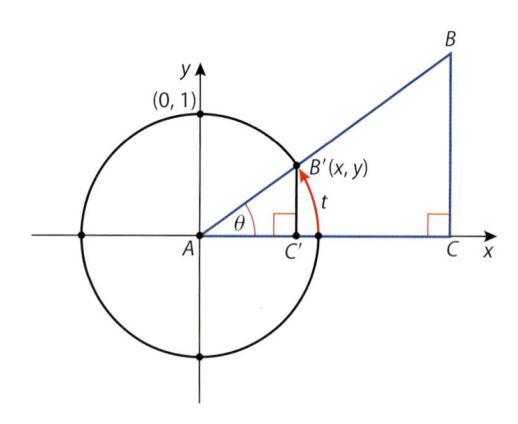

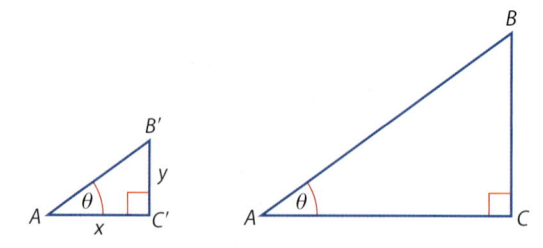

Figure 7.8

From the definitions of the trigonometric functions of an acute angle in Section 7.1 and properties of similar triangles, we can write the following:

$$\sin\theta = \frac{\text{opposite}}{\text{hypotenuse}} = \frac{BC}{AB} = \frac{B'C'}{AB'} = \frac{y}{1} = y$$

$$\cos\theta = \frac{\text{adjacent}}{\text{hypotenuse}} = \frac{AC}{AB} = \frac{AC'}{AB'} = \frac{x}{1} = x$$

From the definitions of the trigonometric functions for the real number t in Section 6.2, we know that $\sin t = y$ and $\cos t = x$. Furthermore, if θ is given in radian measure, then $\theta = t$. Therefore, the trigonometric functions **of the angle** with radian measure θ are precisely the same as the trigonometric functions **of the real number** t. One of the reasons why trigonometric functions are so useful in a range of applications is because they can be applied in these two different ways.

Now let's consider angles other than acute angles.

Defining trigonometric functions for any angle in standard position

Consider the point $P(x, y)$ on the terminal side of an angle θ in standard position (Figure 7.9) such that r is the distance from the origin O to P. If θ is an acute angle then we can construct a right triangle POQ (Figure 7.10) by dropping a perpendicular from P to a point Q on the x-axis, and it follows that:

$$\sin \theta = \frac{y}{r}, \cos \theta = \frac{x}{r} \text{ and } \tan \theta = \frac{y}{x} \, (x \neq 0).$$

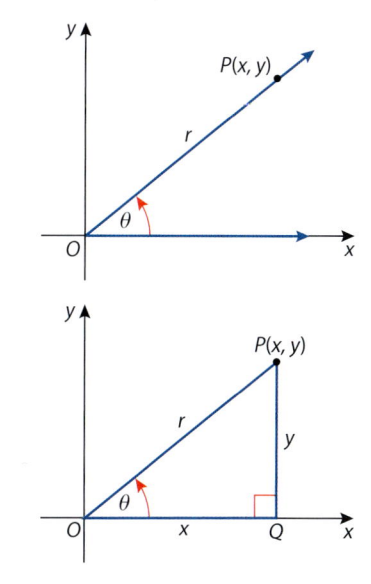

Figure 7.9

Figure 7.10

Extending this to angles other than acute angles allows us to define the trigonometric functions for any angle – positive or negative. It is important to note that the values of the trigonometric ratios do not depend on the choice of the point $P(x, y)$. If $P'(x', y')$ is any other point on the terminal side of angle θ, as in Figure 7.11, then triangles POQ and $P'OQ'$ are similar and the trigonometric ratios for corresponding angles are equal.

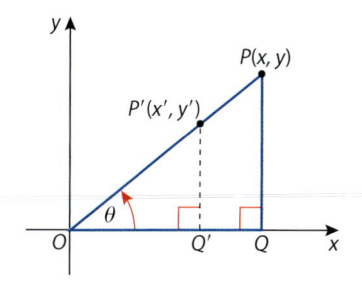

Figure 7.11

7 Triangle Trigonometry

Definition of basic trigonometric functions

Let θ be any angle (in degree or radian measure) in standard position, with (x, y) any point on the terminal side of θ, and $r = \sqrt{x^2 + y^2}$, the distance from the origin to the point (x, y), as shown below.

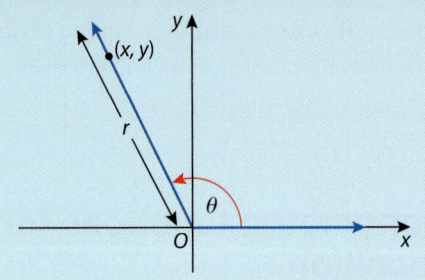

Then the trigonometric functions are defined as follows:

$$\sin \theta = \frac{y}{r} \quad \cos \theta = \frac{x}{r} \quad \tan \theta = \frac{y}{x} \, (x \neq 0)$$

Example 8

Find the sine, cosine and tangent of an angle α that contains the point $(-3, 4)$ on its terminal side when in standard position.

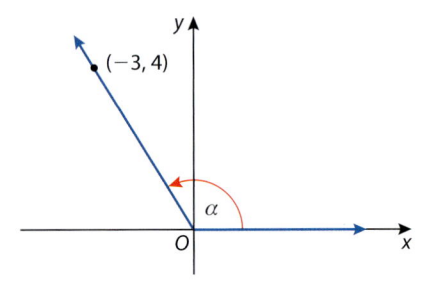

Solution

$$r = \sqrt{x^2 + y^2} = \sqrt{(-3)^2 + 4^2} = \sqrt{25} = 5$$

Then, $\sin \alpha = \dfrac{y}{r} = \dfrac{4}{5}$

$\cos \alpha = \dfrac{x}{r} = \dfrac{-3}{5} = -\dfrac{3}{5}$

$\tan \alpha = \dfrac{y}{x} = \dfrac{4}{-3} = -\dfrac{4}{3}$

Note that for the angle α in Example 8, we can form a right triangle by constructing a line segment from the point $(-3, 4)$ perpendicular to the x-axis, as shown in Figure 7.12. Clearly, $\theta = 180° - \alpha$. Furthermore, the values of the sine, cosine and tangent of the angle θ are the same as that for the angle α, except that the *sign* may be different.

Figure 7.12

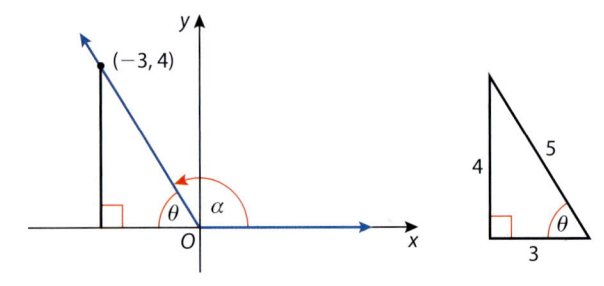

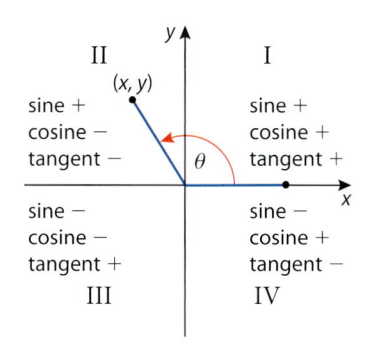

Figure 7.13 Sign of trigonometric function values depends on the quadrant in which the terminal side of the angle lies.

Whether the trigonometric functions are defined in terms of the length of an arc or in terms of an angle, the signs of trigonometric function values are determined by the quadrant in which the arc or angle lies, when in standard position (Figure 7.13).

Example 9

Find the sine, cosine and tangent of the obtuse angle that measures 150°.

Solution

The terminal side of the angle forms a 30° angle with the x-axis. The sine values for 150° and 30° will be exactly the same, and the cosine and tangent values will be the same but of opposite sign. We know that
$\sin 30° = \frac{1}{2}$, $\cos 30° = \frac{\sqrt{3}}{2}$ and $\tan 30° = \frac{\sqrt{3}}{3}$.
Therefore, $\sin 150° = \frac{1}{2}$, $\cos 150° = -\frac{\sqrt{3}}{2}$ and $\tan 150° = -\frac{\sqrt{3}}{3}$.

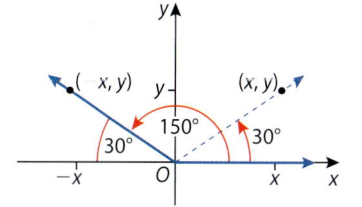

Example 9 illustrates three trigonometric identities for angles whose sum is 180° (i.e. a pair of supplementary angles). The following are true for any acute angle θ:

$\sin(180° - \theta) = \sin \theta$
$\cos(180° - \theta) = -\cos \theta$
$\tan(180° - \theta) = -\tan \theta$

Example 10

Given that $\sin \theta = \frac{5}{13}$ and $90° < \theta < 180°$, find the exact values of $\cos \theta$ and $\tan \theta$.

Solution

θ is an angle in the second quadrant. It follows from the definition $\sin \theta = \frac{y}{r}$ that with θ in standard position there must be a point on the terminal side of the angle that is 13 units from the origin (i.e. $r = 13$) and which has a y-coordinate of 5, as shown in the diagram.

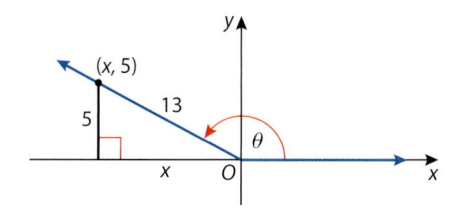

Using Pythagoras' theorem, $|x| = \sqrt{13^2 - 5^2} = \sqrt{144} = 12$. Because θ is in the second quadrant, the x-coordinate of the point must be negative, thus $x = -12$.

Therefore, $\cos \theta = \frac{-12}{13} = -\frac{12}{13}$, and $\tan \theta = \frac{5}{-12} = -\frac{5}{12}$.

Example 11

a) Find the acute angle with the same sine ratio as (i) 135°, and (ii) 117°.
b) Find the acute angle with the same cosine ratio as (i) 300°, and (ii) 342°.

Solution

a) (i) Angles in the first and second quadrants have the same sine ratio. Hence, the identity $\sin(180° - \theta) = \sin \theta$. Since $180° - 135° = 45°$, then $\sin 135° = \sin 45°$.

(ii) Since $180° - 117° = 63°$, then $\sin 117° = \sin 63°$

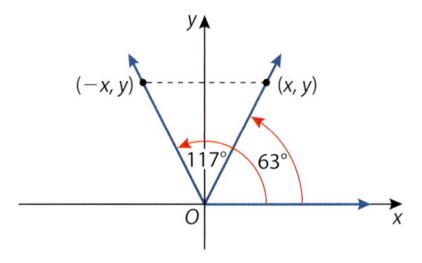

b) (i) Angles in the first and fourth quadrants have the same cosine ratio. Hence, the identity $\cos(360° - \theta) = \cos \theta$. Since $360° - 300° = 60°$, then $\cos 300° = \cos 60°$.

(ii) Since $360° - 342° = 18°$, then $\cos 342° = \cos 18°$.

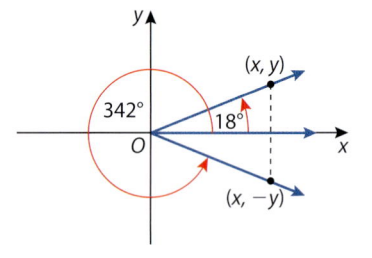

Areas of triangles

You are familiar with the standard formula for the area of a triangle, area = $\frac{1}{2}$ × base × height (or area = $\frac{1}{2}bh$), where the base, b, is a side of the triangle and the height, h, (or altitude) is a line segment perpendicular to the base (or the line containing it) and drawn to the vertex opposite to the base, as shown in Figure 7.14.

Figure 7.14

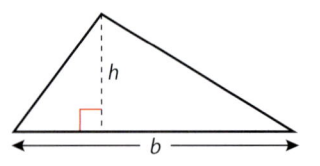

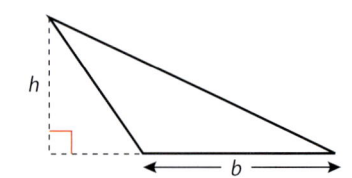

If the lengths of two sides of a triangle and the measure of the angle between these sides (often called the included angle) are known, then the triangle is unique and has a fixed area. Hence, we should be able to calculate the area from just these measurements, i.e. from knowing two sides and the included angle. This calculation is quite straightforward if the triangle is a right triangle (Figure 7.15) and we know the lengths of the two legs on either side of the right angle.

Let's develop a general area formula that will apply to any triangle – right, acute or obtuse. For triangle ABC shown in Figure 7.16, suppose we know the lengths of the two sides a and b and the included angle C. If the length of the height from B is h, the area of the triangle is $\frac{1}{2}bh$. From right triangle trigonometry, we know that $\sin C = \frac{h}{a}$, or $h = a \sin C$. Substituting $a \sin C$ for h, area $= \frac{1}{2}bh = \frac{1}{2}b(a \sin C) = \frac{1}{2}ab \sin C$.

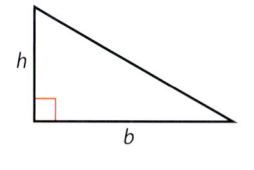

Figure 7.15

Figure 7.16

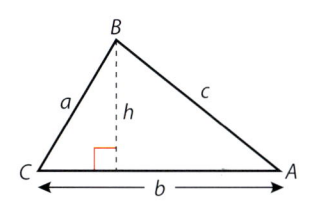

If the angle C is obtuse, then from Figure 7.17 we see that $\sin(180° - C) = \frac{h}{a}$. So, the height is $h = a \sin(180° - C)$. However, $\sin(180° - C) = \sin C$. Thus, $h = a \sin C$ and, again, area $= \frac{1}{2}ab \sin C$.

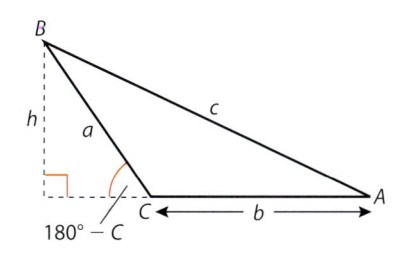

Figure 7.17

Area of a triangle

For a triangle with sides of lengths a and b and included angle C,

$$\text{Area of } \triangle = \tfrac{1}{2}ab \sin C$$

● **Hint:** Note that the procedure for finding the area of a triangle from a pair of sides and the included angle can be performed three different ways. For any triangle labelled in the manner of the triangles in Figures 7.16 and 7.17, its area is expressed by any of the following three expressions.

$$\text{Area of } \triangle = \tfrac{1}{2}ab \sin C$$
$$= \tfrac{1}{2}ac \sin B$$
$$= \tfrac{1}{2}bc \sin A$$

These three equivalent expressions will prove to be helpful for developing an important formula for solving non-right triangles in the next section.

Example 12

Find the area of each triangle. Express the area exactly, or, if not possible, express it accurate to 3 s.f.

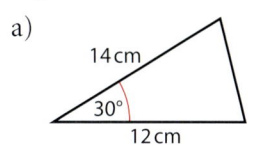

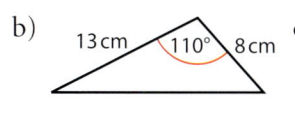

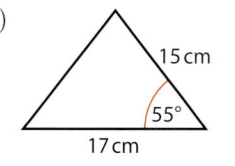

a) 14 cm, 30°, 12 cm

b) 13 cm, 110°, 8 cm

c) 15 cm, 55°, 17 cm

Solution

a) Area $= \frac{1}{2}(12)(14) \sin 30° = 84(0.5) = 42 \text{ cm}^2$

b) Area $= \frac{1}{2}(8)(13) \sin 110° \approx 52(0.939\,69) \approx 48.9 \text{ cm}^2$

c) Area $= \frac{1}{2}(15)(17) \sin 55° \approx 127.5(0.819\,152) \approx 104 \text{ cm}^2$

7 Triangle Trigonometry

Example 13

The circle shown has a radius of 1 cm and the central angle θ subtends an arc of length of $\frac{2\pi}{3}$ cm. Find the area of the shaded region.

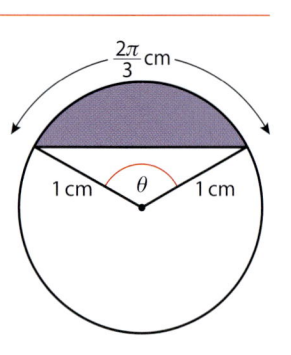

Solution

The formula for the area of a sector is $A = \frac{1}{2}r^2\theta$ (Section 6.1), where θ is the central angle in radian measure. Since the radius of the circle is one, the length of the arc subtended by θ is the same as the radian measure of θ. Thus, area of sector
$= \frac{1}{2}(1)^2\left(\frac{2\pi}{3}\right) = \frac{\pi}{3}$ cm². The area of the triangle formed by the two radii and the chord is equal to
$\frac{1}{2}(1)(1)\sin\left(\frac{2\pi}{3}\right) = \frac{1}{2}\left(\frac{\sqrt{3}}{2}\right) = \frac{\sqrt{3}}{4}$ cm².
$\left[\sin\frac{2\pi}{3} = \sin\left(\pi - \frac{2\pi}{3}\right) = \sin\frac{\pi}{3} = \frac{\sqrt{3}}{2}\right]$

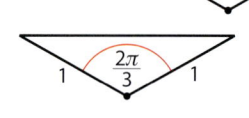

The area of the shaded region is found by subtracting the area of the triangle from the area of the sector. Area $= \frac{\pi}{3} - \frac{\sqrt{3}}{4}$ or $\frac{4\pi - 3\sqrt{3}}{12}$ or approximately 0.614 cm² (3 s.f.).

Example 14

Show that it is possible to construct two different triangles with an area of 35 cm² that have sides measuring 8 cm and 13 cm. For each triangle, find the measure of the (included) angle between the sides of 8 cm and 13 cm to the nearest tenth of a degree.

Solution

We can visualize the two different triangles with equal areas – one with an acute included angle (α) and the other with an obtuse included angle (β).

$$\text{Area} = \tfrac{1}{2}(\text{side})(\text{side})(\text{sine of included angle}) = 35 \text{ cm}^2$$
$$= \tfrac{1}{2}(8)(13)(\sin\alpha) = 35$$
$$52\sin\alpha = 35$$
$$\sin\alpha = \frac{35}{52}$$
$$\alpha = \sin^{-1}\left(\frac{35}{52}\right) \quad \text{recall that the GDC will only give the acute angle}$$
$$\text{with sine ratio of } \frac{35}{52}$$
$$\alpha \approx 42.3° \qquad \text{rounded to the nearest tenth}$$

Knowing that $\sin(180° - \alpha) = \sin\alpha$, the obtuse angle β is equal to $180° - 42.3° = 137.7°$.

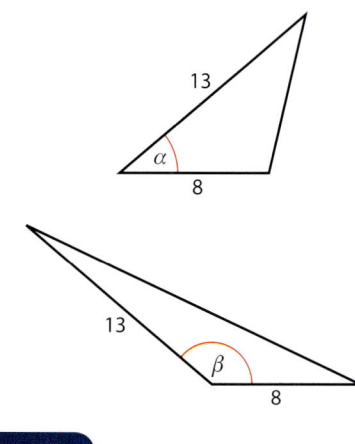

Check this answer by computing on your GDC:
$\frac{1}{2}(8)(13)(\sin 137.7°) \approx 34.997 \approx 35$ cm^2.

Therefore, there are two different triangles with sides 8 cm and 13 cm and area of 35 cm^2 – one with an included angle of 42.3° and the other with an included angle of 137.7°.

Exercise 7.2

In questions 1–6, find the exact value of the sine, cosine and tangent functions of the angle θ.

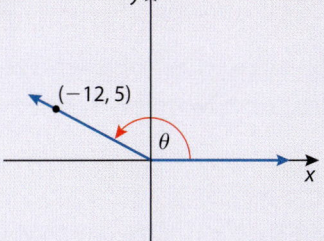

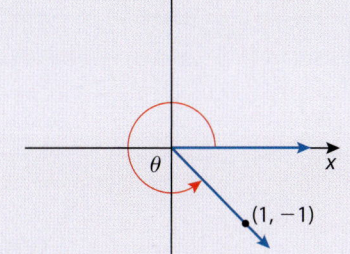

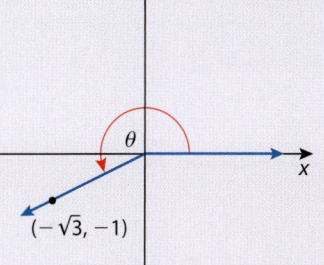

1 (4, 3)

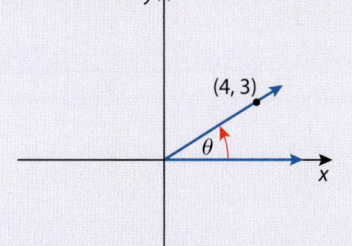

2 (−12, 5)

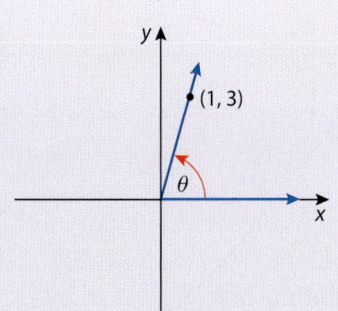

3 (1, −1)

4 (−√3, −1)

5 (1, 3)

6 (−3, −3)

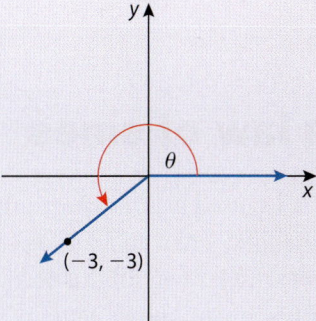

7 By using the symmetry of the unit circle, or otherwise, determine the exact sine, cosine and tangent function values for the following common obtuse angles.
   a) 120° b) 135° c) 150°

8 Evaluate the sine, cosine and tangent of each angle without using your GDC.
   a) 225° b) 330° c) $\frac{7\pi}{6}$
   d) −60° e) 270° f) $\frac{5\pi}{3}$
   g) −120° h) $-\frac{\pi}{4}$ i) π

9. Given that $\cos\theta = \frac{3}{5}$ and $-90° < \theta < 0°$, find the exact values of $\sin\theta$ and $\tan\theta$.

10. Given that $\sin\theta = \frac{8}{17}$ and $90° < \theta < 180°$, find the exact values of $\cos\theta$ and $\tan\theta$.

11. Given that $\tan\theta = -\frac{12}{5}$ and $\sin\theta < 0°$, find the exact values of $\sin\theta$ and $\cos\theta$.

12. Given that $\sin\theta = 0$ and $\cos\theta < 0°$, find the exact values of $\cos\theta$ and $\tan\theta$.

13. a) Find the acute angle with the same sine ratio as (i) 150°, and (ii) 95°.
    b) Find the acute angle with the same cosine ratio as (i) 315°, and (ii) 353°.
    c) Find the acute angle with the same tangent ratio as (i) 240°, and (ii) 200°.

14. Find the area of each triangle. Express the area exactly, or, if not possible, express it accurate to 3 s.f.
    a) triangle with sides 4 and 6, included angle 60°
    b) triangle with side 8, angle 105°, side 23
    c) triangle with side 30, angle 45°, side 90

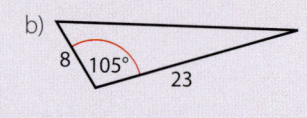

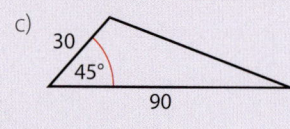

15. A chord AB subtends an angle of 120° at O, the centre of a circle with radius 15 cm. Find the area of a) the sector AOB, and b) the triangle AOB.

16. Find the area of the shaded region (called a *segment*) in each circle.
    a) circle radius 10 cm, angle $\frac{\pi}{3}$
    b) circle radius 12 cm, angle 135°

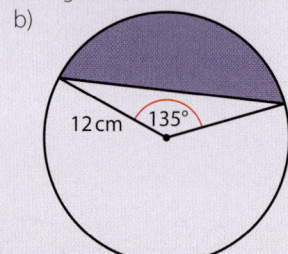

17. Find the area of a parallelogram with two sides of length 15 cm and 10 cm, if the angle between these sides has a measure of 54° (see diagram).

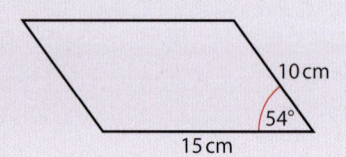

7.3 The law of sines

In Section 7.1 we used techniques from right triangle trigonometry to solve right triangles when an acute angle and one side are known, or when two sides are known. In this section and the next, we will study methods for finding unknown lengths and angles in triangles that are not right triangles. These general methods are effective for solving problems involving any kind of triangle – right, acute or obtuse.

Possible triangles constructed from three given parts

As mentioned in the previous paragraph, we've solved right triangles by either knowing an acute angle and one side, or knowing two sides. Since the triangles also have a right angle, each of those two cases actually involved knowing three different parts of the triangle – either two angles and a side, or two sides and an angle. We need to know at least three parts of a triangle in order to solve for other unknown parts. Different arrangements of the three known parts can be given. Before solving for unknown parts, it is helpful to know whether the three known parts determine a unique triangle, or possibly more than one triangle. The table below summarizes the five different arrangements of three parts and the number of possible triangles for each. You are encouraged to confirm these results on your own with manual or computer generated sketches.

Possible triangles formed with three known parts

Known parts	Number of possible triangles
Three angles (AAA)	Infinite triangles (not possible to solve)
Three sides (SSS) (sum of any two must be greater than the third)	One unique triangle
Two sides and their included angle (SAS)	One unique triangle
Two angles and any side (ASA or AAS)	One unique triangle
Two sides and a non-included angle (SSA)	No triangle, one triangle or two triangles (ambiguous case).

ASA, AAS and SSA can be solved using the **law of sines**, whereas SSS and SAS can be solved using the **law of cosines** (next section).

The law of sines (or sine rule)

In the previous section, we showed that we can write three equivalent expressions for the area of any triangle for which we know two sides and the included angle.

$$\text{Area of } \triangle = \tfrac{1}{2}ab\sin C = \tfrac{1}{2}ac\sin B = \tfrac{1}{2}bc\sin A$$

If each of these expressions is divided by $\tfrac{1}{2}abc$,

$$\frac{\tfrac{1}{2}ab\sin C}{\tfrac{1}{2}abc} = \frac{\tfrac{1}{2}ac\sin B}{\tfrac{1}{2}abc} = \frac{\tfrac{1}{2}bc\sin A}{\tfrac{1}{2}abc}$$

we obtain three equivalent ratios – each containing the sine of an angle divided by the length of the side opposite the angle.

The law of sines

If A, B and C are the angle measures of any triangle and a, b and c are, respectively, the lengths of the sides opposite these angles, then

$$\frac{\sin A}{a} = \frac{\sin B}{b} = \frac{\sin C}{c}$$

Alternatively, the law of sines can also be written as $\dfrac{a}{\sin A} = \dfrac{b}{\sin B} = \dfrac{c}{\sin C}$.

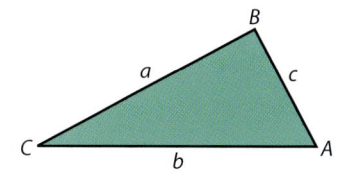

7 Triangle Trigonometry

Solving triangles given two angles and any side (ASA or AAS)

If we know two angles and any side of a triangle, we can use the law of sines to find any of the other angles or sides of the triangle.

Example 15

Find all of the unknown angles and sides of triangle *DEF* shown in the diagram. Approximate all measurements to 1 decimal place.

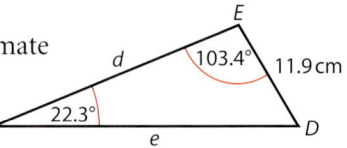

Solution

The third angle of the triangle is
$$D = 180° - E - F = 180° - 103.4° - 22.3° = 54.3°$$
Using the law of sines, we can write the following proportion to solve for the length *e*:
$$\frac{\sin 22.3°}{11.9} = \frac{\sin 103.4°}{e}$$
$$e = \frac{11.9 \sin 103.4°}{\sin 22.3°} \approx 30.507 \text{ cm}$$

We can write another proportion from the law of sines to solve for *d*:
$$\frac{\sin 22.3°}{11.9} = \frac{\sin 54.3°}{d}$$
$$d = \frac{11.9 \sin 54.3°}{\sin 22.3°} \approx 25.467 \text{ cm}$$

Therefore, the other parts of the triangle are $D = 54.3°$, $e \approx 30.5$ cm and $d \approx 25.5$ cm.

• **Hint:** When using your GDC to find angles and lengths with the law of sines (or the law of cosines), remember to store intermediate answers on the GDC for greater accuracy. By not rounding until the final answer, you reduce the amount of round-off error.

Example 16

A tree on a sloping hill casts a shadow 45 m along the side of the hill. The gradient of the hill is $\frac{1}{5}$ (or 20%) and the angle of elevation of the sun is 35°. How tall is the tree to the nearest tenth of a metre?

Solution

α is the angle that the hill makes with the horizontal. Its measure can be found by computing the inverse tangent of $\frac{1}{5}$.
$$\alpha = \tan^{-1}\left(\frac{1}{5}\right) \approx 11.3099°$$

224

The height of the tree is labelled h. The angle of elevation of the sun is the angle between the sun's rays and the horizontal. In the diagram, this angle of elevation is the sum of α and β. Thus, $\beta \approx 35° - 11.3099° \approx 23.6901°$. For the larger right triangle with $\alpha + \beta = 35°$ as one of its acute angles, the other acute angle – and the angle in the obtuse triangle opposite the side of 45 m – must be 55°. Now we can apply the law of sines for the obtuse triangle to solve for h.

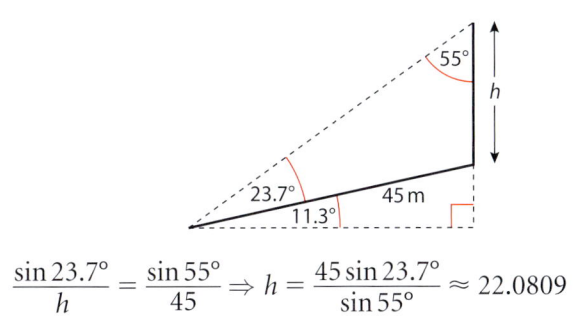

$$\frac{\sin 23.7°}{h} = \frac{\sin 55°}{45} \Rightarrow h = \frac{45 \sin 23.7°}{\sin 55°} \approx 22.0809$$

Therefore, the tree is approximately 22.1 m tall.

Two sides and a non-included angle (SSA) – the ambiguous case

The arrangement where we are given the lengths of two sides of a triangle and the measure of an angle not between those two sides can produce three different results: no triangle, one unique triangle or two different triangles. Let's explore these possibilities with the following example.

Example 17

Find all of the unknown angles and sides of triangle ABC where $a = 35$ cm, $b = 50$ cm and $A = 30°$. Approximate all measurements to 1 decimal place.

Solution

Figure 7.18 shows the three parts we have from which to try and construct a triangle.

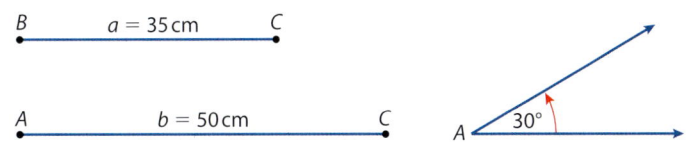

◀ Figure 7.18

We attempt to construct the triangle, as shown in Figure 7.19. We first draw angle A with its initial side (or base line of the triangle) extended. We then measure off the known side $b = AC = 50$. To construct side a (opposite angle A), we take point C as the centre and with radius $a = 35$ we draw an arc of a circle. The points on this arc are all possible positions for vertex B – one of the endpoints of side a, or BC. Point B must be on the base line, so B can be located at any point of intersection of the circular arc and the base line. In this instance, with these particular measurements for the two sides and non-included angle, there are two points of intersection, which we label B_1 and B_2.

7 Triangle Trigonometry

Figure 7.19

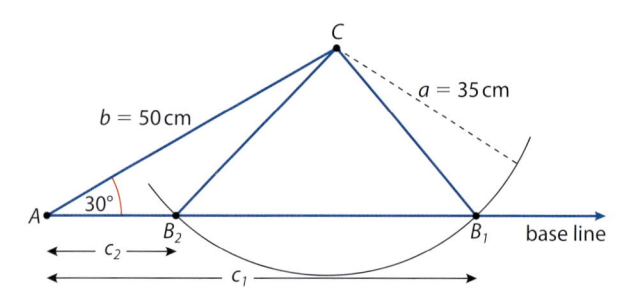

Therefore, we can construct two different triangles, triangle AB_1C (Figure 7.20) and triangle AB_2C (Figure 7.21). Angle B_1 will be acute and angle B_2 will be obtuse. To complete the solution of this problem, we need to solve each of these triangles.

- Solve triangle AB_1C:

Figure 7.20

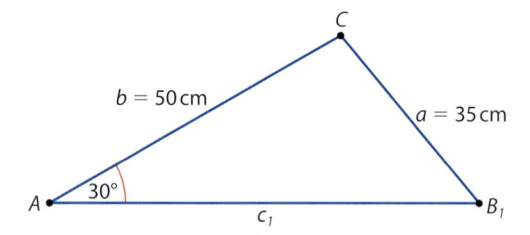

We can solve for acute angle B_1 using the law of sines:

$$\frac{\sin 30°}{35} = \frac{\sin B_1}{50}$$

$$\sin B_1 = \frac{50 \sin 30°}{35} = \frac{50(0.5)}{35}$$

$$B_1 = \sin^{-1}\left(\frac{5}{7}\right) \approx 45.5847°$$

Then, $C \approx 180° - 30° - 45.5847° \approx 104.4153°$.

With another application of the law of sines, we can solve for side c_1:

$$\frac{\sin 30°}{35} = \frac{\sin 104.4153°}{c_1}$$

$$c_1 = \frac{35 \sin 104.4153°}{\sin 30°} \approx \frac{35(0.96852)}{0.5} \approx 67.7964 \text{ cm}$$

Therefore, for triangle AB_1C, $B_1 \approx 45.6°$, $C \approx 104.4°$ and $c_1 \approx 67.8$ cm.

- Solve triangle AB_2C:

Figure 7.21

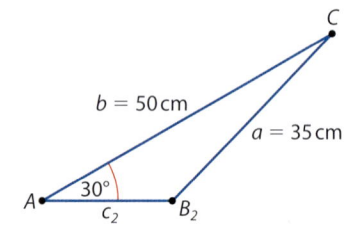

Solving for obtuse angle B_2 using the law of sines gives the same result as above, except we know that $90° < B_2 < 180°$.

We also know that $\sin(180° - \theta) = \sin \theta$.

Thus, $B_2 = 180° - B_1 \approx 180° - 45.5847° \approx 134.4153°$.

Then, $C \approx 180° - 30° - 134.4153° \approx 15.5847°$.

With another application of the law of sines, we can solve for side c_2:
$$\frac{\sin 30°}{35} = \frac{\sin 15.5847°}{c_2}$$
$$c_2 \approx \frac{35 \sin 15.5847°}{\sin 30°} \approx \frac{35(0.26866)}{0.5} \approx 18.8062 \text{ cm}$$

Therefore, for triangle AB_2C, $B_2 \approx 134.4°$, $C \approx 15.6°$ and $c_2 \approx 18.8$ cm.

Now that we have solved this specific example, let's take a more general look and examine all the possible conditions and outcomes for the SSA arrangement. In general, we are given the lengths of two sides – call them a and b – and a non-included angle – for example, angle A that is opposite side a. From these measurements, we can determine the number of different triangles. Figure 7.22 shows the four different possibilities (or cases) when angle A is acute. The number of triangles depends on the length of side a.

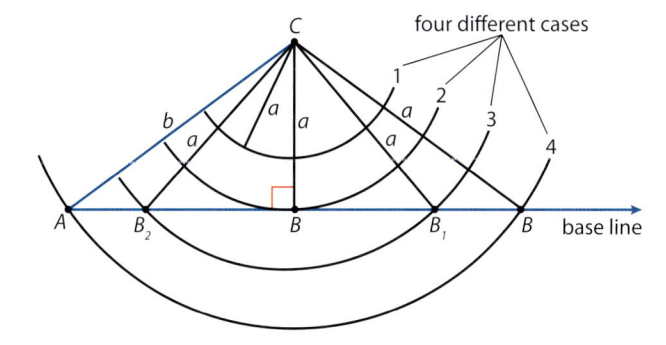

Figure 7.22

In case 2, side a is perpendicular to the base line resulting in a single right triangle, shown in Figure 7.23. In this case, clearly $\sin A = \frac{a}{b}$ and $a = b \sin A$. In case 1, the length of a is shorter than it is in case 2, i.e $b \sin A$. In case 3, which occurred in Example 17, the length of a is longer than $b \sin A$, but less than b. And, in case 4, the length of a is greater than b. These results are summarized in the table below. Because the number of triangles may be none, one or two, depending on the length of a (the side opposite the given angle), the SSA arrangement is called the ambiguous case.

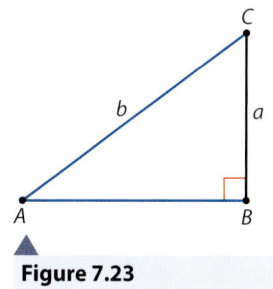

Figure 7.23

The ambiguous case (SSA)
Given the lengths of sides a and b and the fact that the non-included angle A is acute, the following four cases and resulting triangles can occur.

Length of a	Number of triangles
$a < b \sin A$	No triangle
$a = b \sin A$	One right triangle
$b \sin A < a < b$	Two triangles
$a \geq b$	One triangle

7 Triangle Trigonometry

The situation is considerably simpler if angle A is obtuse rather than acute. Figure 7.24 shows that if $a > b$ then there is only one possible triangle, and if $a \leq b$ then no triangle that contains angle A is possible.

Figure 7.24 Angle A is obtuse.

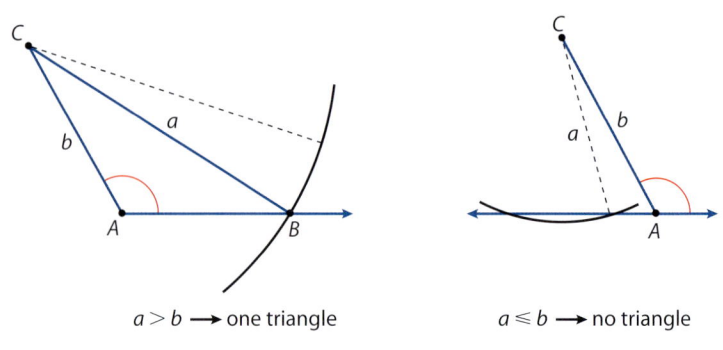

$a > b \longrightarrow$ one triangle

$a \leq b \longrightarrow$ no triangle

Example 18

For triangle ABC, if side $b = 50$ cm and angle $A = 30°$, find the values for the length of side a that will produce: (i) no triangle, (ii) one triangle, (iii) two triangles. This is the same SSA information given in Example 17 with the exception that side a is not fixed at 35 cm, but is allowed to vary.

Solution

Because this is a SSA arrangement and given A is an acute angle, then the number of different triangles that can be constructed is dependent on the length of a. First calculate the value of $b \sin A$:

$$b \sin A = 50 \sin 30° = 50(0.5) = 25 \text{ cm}$$

Thus, if a is exactly 25 cm then triangle ABC is a right triangle, as shown in the figure.

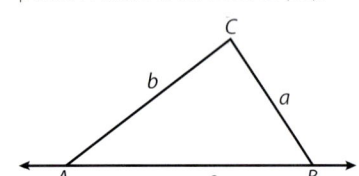

- **Hint:** It is important to be familiar with the notation for line segments and angles commonly used in IB exam questions. For example, the line segment labelled b in the diagram (below) is denoted as [AC] in IB notation. Angle A, the angle between [BA] and [AC], is denoted as $B\hat{A}C$. Also, the line containing points A and B is denoted as (AB).

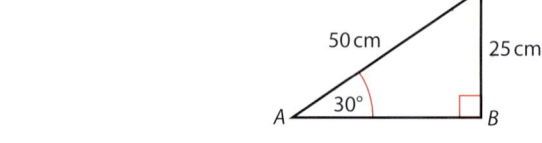

(i) If $a < 25$ cm, there is no triangle.
(ii) If $a = 25$ cm, or $a > 50$ cm, there is one unique triangle.
(iii) If 25 cm $< a < 50$ cm, there are two different possible triangles.

Example 19

The diagrams below show two different triangles both satisfying the conditions: $HK = 18$ cm, $JK = 15$ cm, $J\hat{H}K = 53°$.

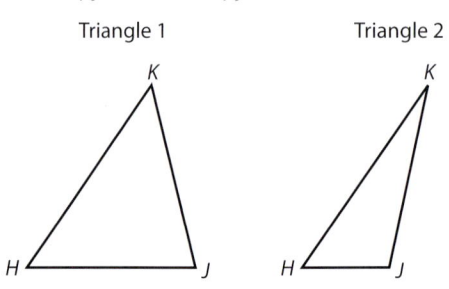

Triangle 1   Triangle 2

a) Calculate the size of $H\hat{J}K$ in Triangle 2.
b) Calculate the area of Triangle 1.

Solution

a) From the law of sines, $\dfrac{\sin(H\hat{J}K)}{18} = \dfrac{\sin 53°}{15} \Rightarrow \sin(H\hat{J}K) = \dfrac{18 \sin 53°}{15}$

$\approx 0.958\,36 \Rightarrow \sin^{-1}(0.95836) \approx 73.408°$

However, $H\hat{J}K > 90° \Rightarrow H\hat{J}K \approx 180° - 73.408° \approx 106.592°$.

Therefore, in Triangle 2 $H\hat{J}K \approx 107°$ (3 s.f.).

b) In Triangle 1, $H\hat{J}K < 90° \Rightarrow H\hat{J}K \approx 73.408°$

$\Rightarrow H\hat{K}J \approx 180° - (73.408° + 53°) \approx 53.592°$

Area $= \tfrac{1}{2}(18)(15) \sin(53.592°) \approx 108.649 \text{ cm}^2$.

Therefore, the area of Triangle 1 is approximately 109 cm^2 (3 s.f.).

7.4 The law of cosines

Two cases remain in our list of different ways to arrange three known parts of a triangle. If three sides of a triangle are known (SSS arrangement), or two sides of a triangle and the angle between them are known (SAS arrangement), then a unique triangle is determined. However, in both of these cases, the law of sines cannot solve the triangle.

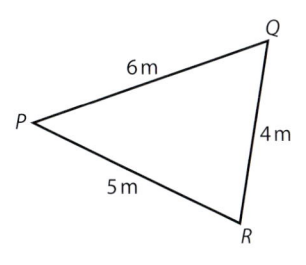

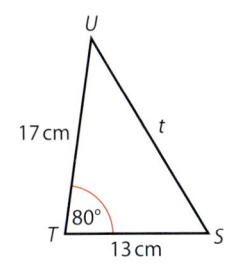

◀ **Figure 7.25**

For example, it is not possible to set up an equation using the law of sines to solve triangle PQR or triangle STU in Figure 7.25.

- Trying to solve $\triangle PQR$: $\dfrac{\sin P}{4} = \dfrac{\sin R}{6} \Rightarrow$ two unknowns; cannot solve for angle P or angle R

- Trying to solve $\triangle STU$: $\dfrac{\sin 80°}{t} = \dfrac{\sin U}{13} \Rightarrow$ two unknowns; cannot solve for angle U or side R

The law of cosines (or cosine rule)

We will need the **law of cosines** to solve triangles with these kinds of arrangements of sides and angles. To derive this law, we need to place a general triangle ABC in the coordinate plane so that one of the vertices is at the origin and one of the sides is on the positive x-axis. Figure 7.26 shows both an acute triangle ABC and an obtuse triangle ABC. In either case, the coordinates of vertex C are $x = b \cos C$ and $y = b \sin C$. Because c is the distance from A to B, then we can use the distance formula to write

7 Triangle Trigonometry

$$c = \sqrt{(b\cos C - a)^2 + (b\sin C - 0)^2}$$ distance between $(b\cos C, b\sin C)$ and $(a, 0)$

$$c^2 = (b\cos C - a)^2 + (b\sin C - 0)^2$$ squaring both sides

$$c^2 = b^2\cos^2 C - 2ab\cos C + a^2 + b^2\sin^2 C$$ expand

$$c^2 = b^2(\cos^2 C + \sin^2 C) - 2ab\cos C + a^2$$ factor out b^2 from two terms

$$c^2 = b^2 - 2ab\cos C + a^2$$ apply trigonometric identity $\cos^2\theta + \sin^2\theta = 1$

$$c^2 = a^2 + b^2 - 2ab\cos C$$ rearrange terms

This equation gives one form of the law of cosines. Two other forms are obtained in a similar manner by having either vertex A or vertex B, rather than C, located at the origin.

Figure 7.26

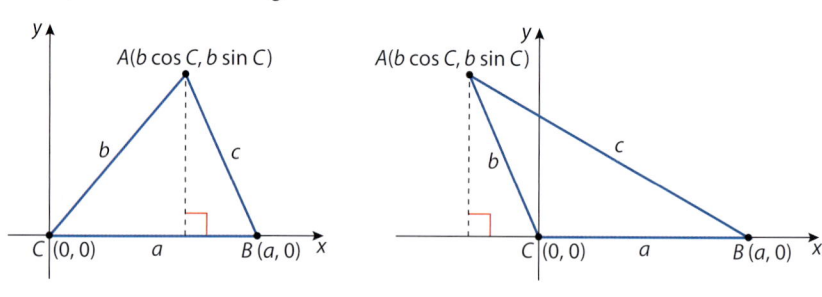

The law of cosines

In any triangle ABC with corresponding sides a, b and c:
$$c^2 = a^2 + b^2 - 2ab\cos C$$
$$b^2 = a^2 + c^2 - 2ac\cos B$$
$$a^2 = b^2 + c^2 - 2bc\cos A$$

It is helpful to understand the underlying pattern of the law of cosines when applying it to solve for parts of triangles. The pattern relies on choosing one particular angle of the triangle and then identifying the two sides that are adjacent to the angle and the one side that is opposite to it. The law of cosines can be used to solve for the chosen angle or the side opposite the chosen angle.

$$c^2 = a^2 + b^2 - 2ab\cos C$$

side opposite the chosen angle → c^2
chosen angle → C
one side adjacent to the chosen angle, other side adjacent to the chosen angle

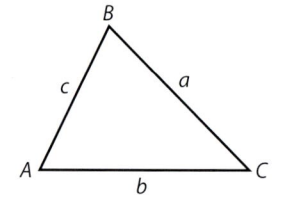

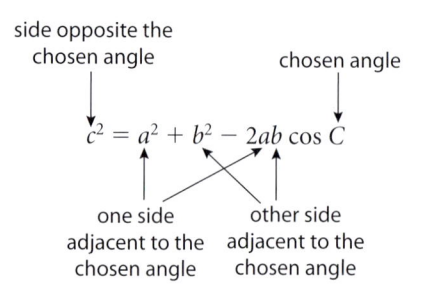

Solving triangles given two sides and the included angle (SAS)

If we know two sides and the included angle, we can use the law of cosines to solve for the side opposite the given angle. Then it is best to solve for one of the two remaining angles using the law of sines.

Example 20

Find all of the unknown angles and sides of triangle STU, one of the triangles shown earlier in Figure 7.25. Approximate all measurements to 1 decimal place.

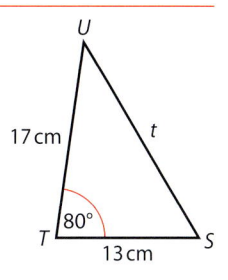

Solution

We first solve for side t, opposite the known angle $S\hat{T}U$, using the law of cosines:

$$t^2 = 13^2 + 17^2 - 2(13)(17)\cos 80°$$

$$t = \sqrt{13^2 + 17^2 - 2(13)(17)\cos 80°}$$

$$t \approx 19.5256$$

Now use the law of sines to solve for one of the other angles, say $T\hat{S}U$:

$$\frac{\sin T\hat{S}U}{17} = \frac{\sin 80°}{19.5256}$$

$$\sin T\hat{S}U = \frac{17 \sin 80°}{19.5256}$$

$$T\hat{S}U = \sin^{-1}\left(\frac{17 \sin 80°}{19.5256}\right)$$

$$T\hat{S}U \approx 59.0288°$$

Then, $S\hat{U}T \approx 180° - (80° + 59.0288°) \approx 40.9712°$.

Therefore, the other parts of the triangle are $t \approx 19.5$ cm, $T\hat{S}U \approx 59.0°$ and $S\hat{U}T \approx 41.0°$.

- **Hint:** As previously mentioned, remember to store intermediate answers on the GDC for greater accuracy. By not rounding until the final answer, you reduce the amount of round-off error. The GDC screen images below show the calculations in the solution for Example 20 above.

```
√(13²+17-2(13)(
17)cos(80))
         19.52556031
Ans→T
         19.52556031
```

```
Ans→T
         19.52556031
sin⁻¹(17sin(80)/T
)
         59.02884098
Ans→S
         59.02884098
```

```
sin⁻¹(17sin(80)/T
)
         59.02884098
Ans→S
         59.02884098
180-(80+S)
         40.97115902
```

You may have noticed that the formula for the law of cosines looks similar to the formula for Pythagoras' theorem. In fact, Pythagoras' theorem can be considered a special case of the law of cosines. When the chosen angle in the law of cosines is 90°, and since $\cos 90° = 0$, the law of cosines becomes Pythagoras' theorem.

If angle $C = 90°$, then
$c^2 = a^2 + b^2 - 2ab\cos C$
$\Rightarrow c^2 = a^2 + b^2 - 2ab\cos 90°$
$\Rightarrow c^2 = a^2 + b^2 - 2ab(0)$
$\Rightarrow c^2 = a^2 + b^2$ or $a^2 + b^2 = c^2$

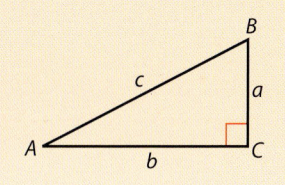

Example 21

A ship travels 50 km due west, then changes its course 18° northward, as shown in the diagram. After travelling 75 km in that direction, how far is the ship from its point of departure? Give your answer to the nearest tenth of a kilometre.

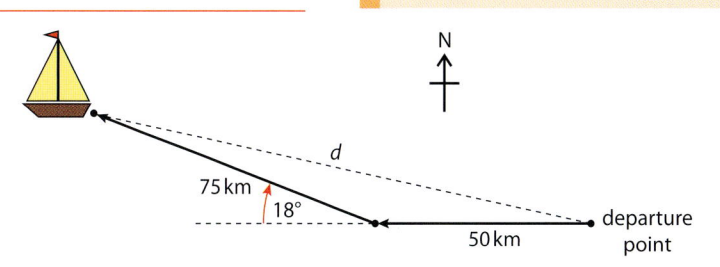

Solution
Let d be the distance from the departure point to the position of the ship. A large obtuse triangle is formed by the three distances of 50 km, 75 km and d km. The angle opposite side d is $180° - 18° = 162°$. Using the law of cosines, we can write the following equation to solve for d:
$$d^2 = 50^2 + 75^2 - 2(50)(75) \cos 162°$$
$$d = \sqrt{50^2 + 75^2 - 2(50)(75) \cos 162°} \approx 123.523$$
Therefore, the ship is approximately 123.5 km from its departure point.

Solving triangles given three sides (SSS)

Given three line segments such that the sum of the lengths of any two is greater than the length of the third, then they will form a unique triangle. Therefore, if we know three sides of a triangle we can solve for the three angle measures. To use the law of cosines to solve for an unknown angle, it is best to first rearrange the formula so that the chosen angle is the subject of the formula.

Solve for angle C in:
$$c^2 = a^2 + b^2 - 2ab \cos C \Rightarrow 2ab \cos C = a^2 + b^2 - c^2 \Rightarrow \cos C = \frac{a^2 + b^2 - c^2}{2ab}$$
Then, $C = \cos^{-1}\left(\frac{a^2 + b^2 - c^2}{2ab}\right)$.

Example 22
Find all of the unknown angles of triangle PQR, the second triangle shown earlier in Figure 7.25. Approximate all measurements to 1 decimal place.

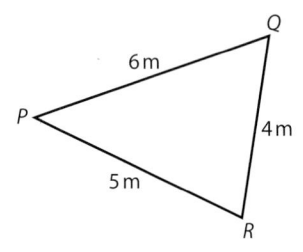

Solution
Note that the smallest angle will be opposite the shortest side. Let's first solve for the smallest angle – thus, writing the law of cosines with chosen angle P:
$$P = \cos^{-1}\left(\frac{5^2 + 6^2 - 4^2}{2(5)(6)}\right) \approx 41.4096°$$
Now that we know the measure of angle P, we have two sides and a non-included angle (SSA), and the law of sines can be used to find the other non-included angle. Consider the sides $QR = 4$, $RP = 5$ and the angle $P \approx 41.4096°$. Substituting into the law of sines, we can solve for angle Q that is opposite RP.
$$\frac{\sin Q}{5} = \frac{\sin 41.4096°}{4}$$

$$\sin Q = \frac{5 \sin 41.4096°}{4}$$

$$Q = \sin^{-1}\left(\frac{5 \sin 41.4096°}{4}\right) \approx 55.7711°$$

Then, $R \approx 180° - (41.4096° + 55.7711°) \approx 82.8192°$.

Therefore, the three angles of triangle PQR are $P \approx 41.4°$, $Q \approx 55.8°$ and $R \approx 82.8°$.

Example 23

A ladder that is 8 m long is leaning against a non-vertical wall that slopes away from the ladder. The foot of the ladder is 3.5 m from the base of the wall, and the distance from the top of the ladder down the wall to the ground is 5.75 m. To the nearest tenth of a degree, what is the acute angle at which the wall is inclined to the horizontal?

Solution

Let's start by drawing a diagram that accurately represents the given information. θ marks the acute angle of inclination of the wall. Its supplement is $F\hat{B}T$. From the law of cosines:

$$\cos F\hat{B}T = \frac{3.5^2 + 5.75^2 - 8^2}{2(3.5)(5.75)}$$

$$F\hat{B}T = \cos^{-1}\left(\frac{3.5^2 + 5.75^2 - 8^2}{2(3.5)(5.75)}\right) \approx 117.664°$$

$$\theta \approx 180° - 117.664° \approx 62.336°$$

Therefore, the angle of inclination of the wall is approximately 62.3°.

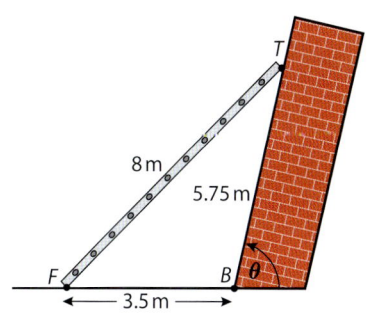

Exercise 7.3 and 7.4

In questions 1–6, state the number of distinct triangles (none, one, two or infinite) that can be constructed with the given measurements. If the answer is one or two triangles, provide a sketch of each triangle.

1. $A\hat{C}B = 30°$, $A\hat{B}C = 50°$ and $B\hat{A}C = 100°$
2. $A\hat{C}B = 30°$, $AC = 12$ cm and $BC = 17$ cm
3. $A\hat{C}B = 30°$, $AB = 7$ cm and $AC = 14$ cm
4. $A\hat{C}B = 47°$, $BC = 20$ cm and $A\hat{B}C = 55°$
5. $B\hat{A}C = 25°$, $AB = 12$ cm and $BC = 7$ cm
6. $AB = 23$ cm, $AC = 19$ cm and $BC = 11$ cm

In questions 7–15, solve the triangle. In other words, find the measurements of all unknown sides and angles. If two triangles are possible, solve for both.

7. $B\hat{A}C = 37°$, $A\hat{B}C = 28°$ and $AC = 14$
8. $A\hat{B}C = 68°$, $A\hat{C}B = 47°$ and $AC = 23$
9. $B\hat{A}C = 18°$, $A\hat{C}B = 51°$ and $AC = 4.7$
10. $A\hat{C}B = 112°$, $A\hat{B}C = 25°$ and $BC = 240$
11. $BC = 68$, $A\hat{C}B = 71°$ and $AC = 59$
12. $BC = 16$, $AC = 14$ and $AB = 12$
13. $BC = 42$, $AC = 37$ and $AB = 26$
14. $BC = 34$, $A\hat{B}C = 43°$ and $AC = 28$
15. $AC = 0.55$, $B\hat{A}C = 62°$ and $BC = 0.51$

16 Find the lengths of the diagonals of a parallelogram whose sides measure 14 cm and 18 cm and which has one angle of 37°.

17 Find the measures of the angles of an isosceles triangle whose sides are 10 cm, 8 cm and 8 cm.

18 A boat is sailing directly towards a cliff. The angle of elevation of a point on the top of the cliff and straight ahead of the boat increases from 10° to 15° as the ship sails a distance of 50 m (see diagram). Find the height of the cliff.

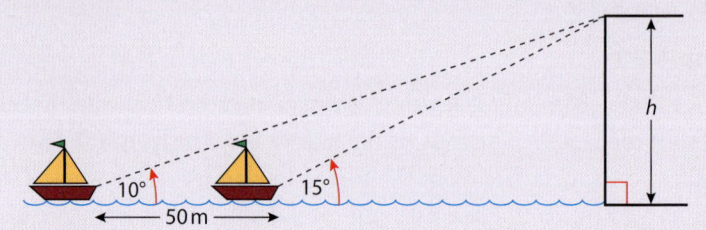

19 Given that for triangle DEF, $E\hat{D}F = 43°$, $DF = 24$ and $FE = 18$, find the two possible measures of $D\hat{F}E$.

20 A tractor drove from a point A directly north for 500 m, and then drove north-east (i.e. bearing of 45°) for 300 m, stopping at point B. What is the distance between points A and B?

21 Find the measure of the smallest angle in the triangle shown.

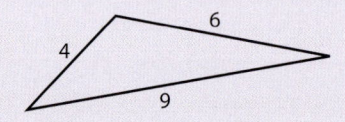

22 Find the area of triangle PQR.

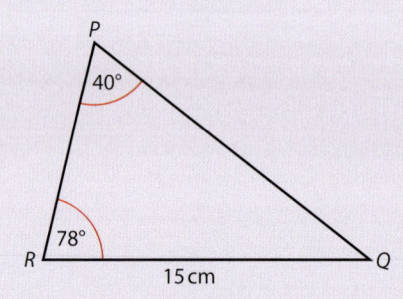

In questions 23 and 24, find a value for the length of BC so that the number of possible triangles is: a) one, b) two and c) none.

23 $B\hat{A}C = 36°$, $AB = 5$

24 $B\hat{A}C = 60°$, $AB = 10$

25 A 50 m vertical pole is to be erected on the side of a sloping hill that makes a 8° angle with the horizontal (see diagram). Find the length of each of the two supporting wires (x and y) that will be anchored 35 m uphill and downhill from the base of the pole.

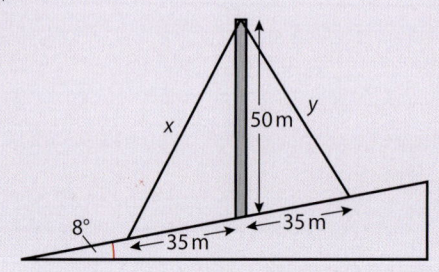

7.5 Applications

There are some additional applications of triangle trigonometry – both right triangles and non-right triangles – that we should take some time to examine.

Equations of lines and angles between two lines

Recall from Section 1.6, the slope m, or gradient, of a non-vertical line is defined as $m = \dfrac{y_2 - y_1}{x_2 - x_1} = \dfrac{\text{vertical change}}{\text{horizontal change}}$.

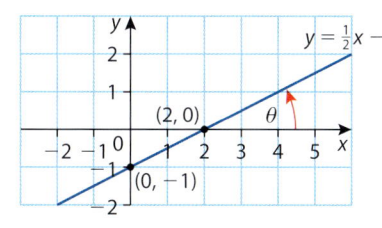

◀ Figure 7.27

The equation of the line shown in Figure 7.27 has a slope $m = \frac{1}{2}$ and a y-intercept of $(0, -1)$. So, the equation of the line is $y = \frac{1}{2}x - 1$. We can find the measure of the acute angle θ between the line and the x-axis by using the tangent function (Figure 7.28).

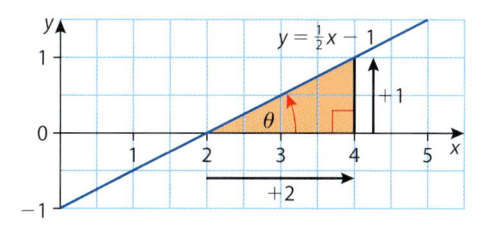

◀ Figure 7.28

$\theta = \tan^{-1}(m) = \tan^{-1}\left(\frac{1}{2}\right) \approx 26.6°$.

Clearly, the slope, m, of this line is equal to $\tan \theta$. If we know the angle between the line and the x-axis, and the y-intercept $(0, c)$, we can write the equation of the line in slope-intercept form $(y = mx + c)$ as $y = (\tan \theta)x + c$.

Before we can generalize for any non-horizontal line, let's look at a line with a negative slope.

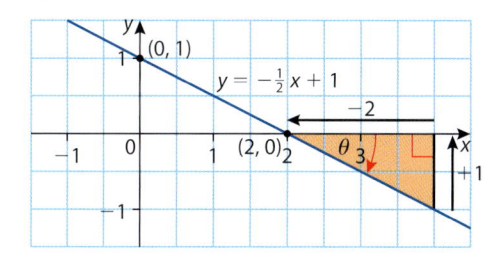

◀ Figure 7.29

The slope of the line is $-\frac{1}{2}$. In order for $\tan \theta$ to be equal to the slope of the line, the angle θ must be the angle that the line makes with the x-axis in the positive direction, as shown in Figure 7.29. In this example,
$\theta = \tan^{-1}(m) = \tan^{-1}\left(-\frac{1}{2}\right) \approx -26.6°$.

Remember, an angle with a negative measure indicates a clockwise rotation from the initial side to the terminal side of the angle.

7 Triangle Trigonometry

> **Equations of lines intersecting the x-axis**
> If a line has a y-intercept of (0, c) and makes an angle of θ with the positive direction of the x-axis, such that $-90° < \theta < 90°$, then the slope (gradient) of the line is $m = \tan\theta$ and the equation of the line is $y = (\tan\theta)x + c$. Note: The angle this line makes with any horizontal line will be θ.

Let's use triangle trigonometry to find the angle between any two intersecting lines – not just for a line intersecting the x-axis. Realize that any pair of intersecting lines that are not perpendicular will have both an acute angle and an obtuse angle between them. When asked for an angle between two lines, the convention is to give the acute angle.

Example 24

Find the acute angle between the lines $y = 3x$ and $y = -x$.

Solution

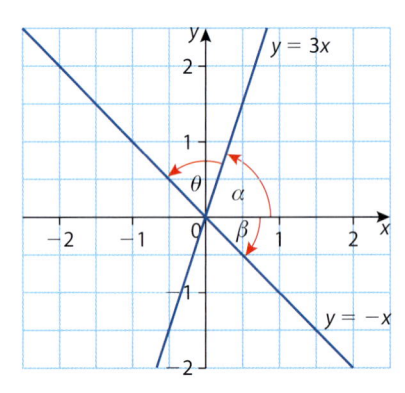

The angle between the line $y = 3x$ and the positive x-axis is α, and the angle between the line $y = -x$ and the positive x-axis is β.

$$\alpha = \tan^{-1}(3) \approx 71.565°$$
$$\beta = \tan^{-1}(-1) = -45°$$

The obtuse angle between the two lines is
$\alpha - \beta \approx 71.565° - (-45°) \approx 116.565°$.

Therefore, the acute angle θ between the two lines is
$\theta = 180° - 116.565° \approx 63.4°$.

Example 25

Find the acute angle between the lines $y = 5x - 2$ and $y = \frac{1}{3}x - 1$.

Solution

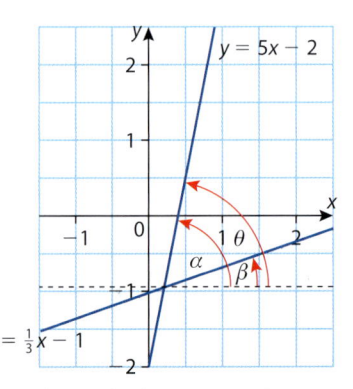

A horizontal line is drawn through the point of intersection.

The angle between $y = 5x - 2$ and this horizontal line is α, and the angle between $y = \frac{1}{3}x - 1$ and this horizontal line is β.

$$\alpha = \tan^{-1}(5) \approx 78.690° \quad \text{and} \quad \beta = \tan^{-1}\left(\frac{1}{3}\right) = 18.435°$$

The acute angle θ between the two lines is
$\theta = \alpha - \beta \approx 78.690° - 18.435° \approx 60.3°$.

We can generalize the procedure for finding the angle between two lines as follows.

> Given two non-vertical lines with equations of $y_1 = m_1x + c_1$ and $y_2 = m_2x + c_2$, the angle between the two lines is $|\tan^{-1}(m_1) - \tan^{-1}(m_2)|$. Note: This angle may be acute or obtuse.

Example 26

a) Find the exact equation of line L_1 that passes through the origin and makes an angle of $-60°$ with the positive direction of the x-axis (or $120°$).
b) The equation of line L_2 is $7x + y + 1 = 0$. Find the acute angle between the lines L_1 and L_2.

Solution

a) The equation of the line is given by $y = (\tan\theta)x$

$$\Rightarrow y = [\tan(-60°)]x = \left[\frac{\sin(-60°)}{\cos(-60°)}\right]x = \left[\frac{\frac{\sqrt{3}}{2}}{-\frac{1}{2}}\right]x = (-\sqrt{3})x$$

Therefore, the equation of L_1 is $y = (-\sqrt{3})x$ or $y = -x\sqrt{3}$.
Note: $\tan(-60°) = \tan 120° = -\sqrt{3}$.

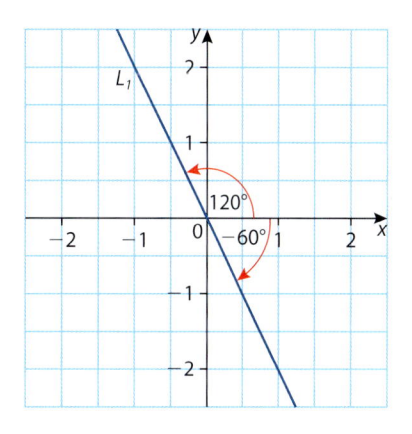

b) $L_2: 7x + y + 1 = 0 \Rightarrow y = -7x - 1$

θ is the acute angle between the lines L_1 and L_2.
$\theta = |\tan^{-1}(m_1) - \tan^{-1}(m_2)| = |\tan^{-1}(-\sqrt{3}) - \tan^{-1}(-7)|$
$\Rightarrow \theta \approx |-60° - (-81.870°)| \approx |-21.87°|$

Therefore, the acute angle between the lines is approximately $21.9°$ (3 s.f.).

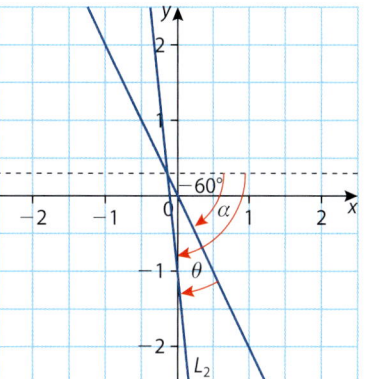

Further applications involving the solution of triangles

Many problems that involve distances and angles are represented by diagrams with multiple triangles – right and otherwise. These diagrams can be confusing and difficult to interpret correctly. In these situations, it is important to carry out a careful analysis of the given information and diagram – this will usually lead to drawing additional diagrams. Often we can extract a triangle, or triangles, for which we have enough information to allow us to solve the triangle(s).

Example 27

Two boats, J and K, are 500 m apart. A lighthouse is on top of a 470 m cliff. The base, B, of the cliff is in line horizontally with $[JK]$. From the top, T, of the lighthouse, the angles of depression of J and K are, respectively, 25° and 40°. Find, correct to the nearest metre, the height, h, of the lighthouse from its base on the clifftop ground to the top T.

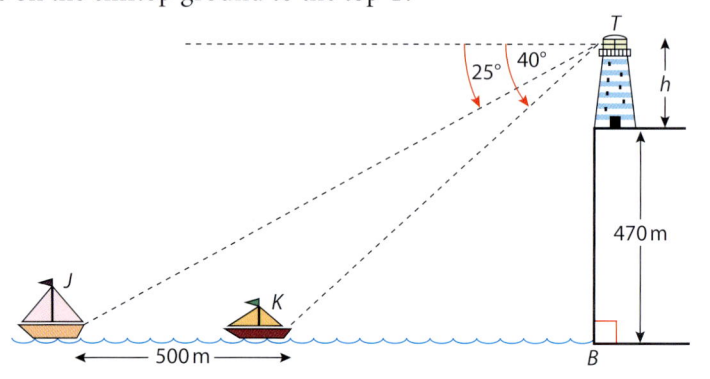

Solution

First, extract obtuse triangle JKT and apply the law of sines to solve for the side KT, which is also the hypotenuse of the right triangle KBT.

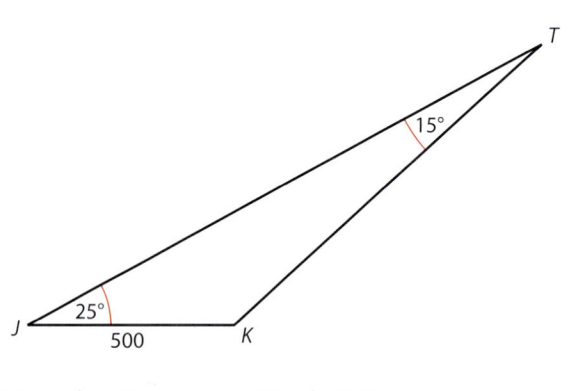

$$\frac{\sin 25°}{KT} = \frac{\sin 15°}{500} \Rightarrow KT = \frac{500 \sin 25°}{\sin 15°} \approx 816.436 \text{ m}$$

We can now use the right triangle KBT to find the side BT – which is equal to the height of the cliff plus the height of the lighthouse.

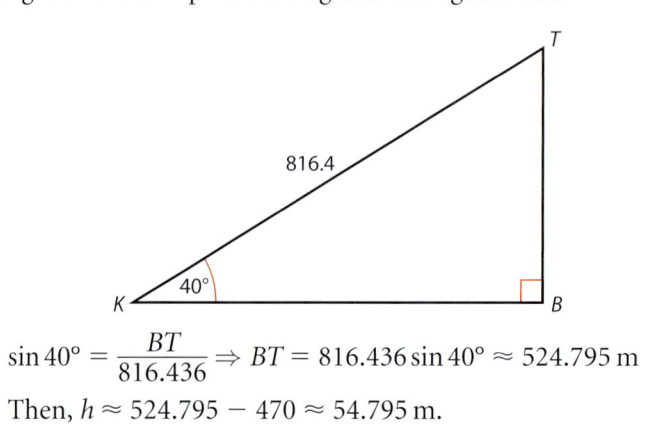

$$\sin 40° = \frac{BT}{816.436} \Rightarrow BT = 816.436 \sin 40° \approx 524.795 \text{ m}$$

Then, $h \approx 524.795 - 470 \approx 54.795$ m.

Therefore, the height of the lighthouse is 54.8 m.

Example 28

As viewed from the surface of the Earth, the angle subtended by the full Moon is 0.5182°. Given that the distance from the Earth's surface to the Moon's surface is approximately 383 500 km, find the radius of the Moon to 3 s.f.

Solution

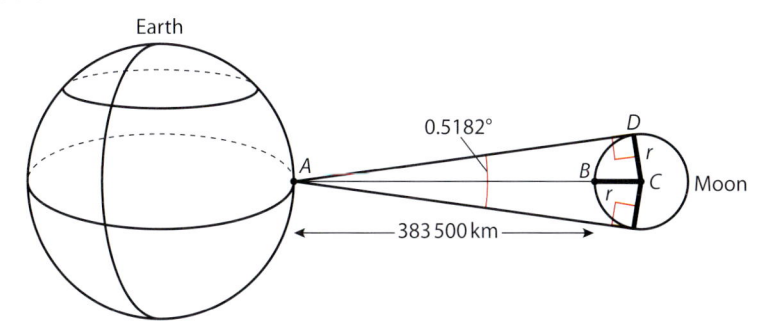

Remember that the radius of a circle drawn to a point of tangency will be perpendicular to the tangent line. This gives us two right triangles in the diagram – each with one acute angle having a measure of $\frac{1}{2}(0.5182°) = 0.2591°$. Extract right triangle ADC from the diagram.

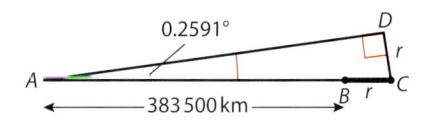

$$\sin(0.2591°) = \frac{r}{383\,500 + r}$$

$r = (383\,500 + r)\sin(0.2591°)$

$r = 383\,500 \sin(0.2591°) + r\sin(0.2591°)$

$r - r\sin(0.2591°) = 383\,500 \sin(0.2591°)$ Collect terms containing r on the left side.

$r(1 - \sin(0.2591°)) = 383\,500 \sin(0.2591°)$ Factor out r from the expression on the left side.

$r = \dfrac{383\,500 \sin(0.2591°)}{1 - \sin(0.2591°)} \approx 1742.12 \text{ km}$

Therefore, the approximate radius of the Moon is 1740 km to 3 s.f.

Example 29

The diagram shows a point P that is 10 km due south of a point D. A straight road PQ is such that the (compass) bearing of Q from P is 45°. A and B are two points on this road which are both 8 km from D. Find the bearing of B from D, approximated to 3 s.f.

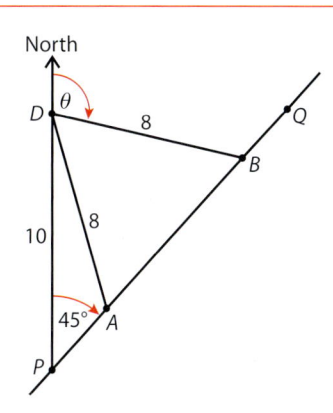

Solution

The angle θ in the diagram is the bearing *of B from D*. A strategy that will lead to finding θ is:

(1) Extract triangle *PDB* and use the law of sines to solve for $D\hat{B}P$.

(2) Triangle *ADB* is isosceles (two sides equal), so $D\hat{A}B = D\hat{B}P$; and since the sum of angles in triangle *ADB* is 180°, we can solve for $A\hat{D}B$.

(3) We can solve for $D\hat{A}P$ because it is supplementary to $D\hat{A}B$, and then we can find the third angle in triangle *APD*.

(4) Since $\theta + A\hat{D}B + A\hat{D}P = 180°$, we can solve for θ.

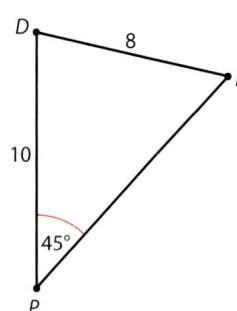

$$\frac{\sin D\hat{B}P}{10} = \frac{\sin 45°}{8}$$

$$\sin D\hat{B}P = \frac{10 \sin 45°}{8}$$

$$D\hat{B}P = \sin^{-1}\left(\frac{10 \sin 45°}{8}\right) \approx 62.11°$$

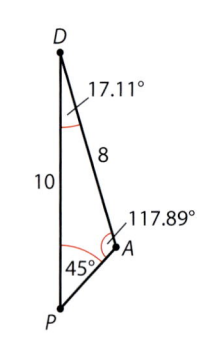

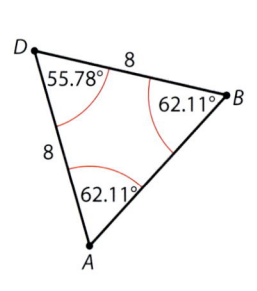

$D\hat{A}B = D\hat{B}P \approx 62.11°$

$A\hat{D}B \approx 180° - 2(62.11°) \approx 55.78°$

$P\hat{A}D \approx 180° - 62.11° \approx 117.89°$

$A\hat{D}P \approx 180° - (45° + 117.89°) \approx 17.11°$

$\theta \approx 180° - (17.11° + 55.78°) \approx 107.11°$

Therefore, the bearing of *B* from *D* is approximately 107° to an accuracy of 3 s.f.

Compass bearings are measured **clockwise** from north.

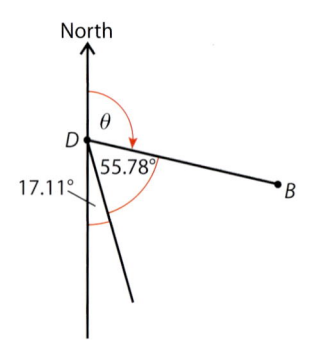

Three-dimensional trigonometry problems

Of course, not all applications of triangle trigonometry are restricted to just two dimensions. In many problems, it is necessary to calculate lengths and angles in three-dimensional structures. As in the preceding section, it is very important to carefully analyze the three-dimensional diagram and to extract any relevant triangles in order to solve for the necessary angle or length.

Example 30

The diagram shows a vertical pole GH that is supported by two wires fixed to the horizontal ground at C and D. The following measurements are indicated in the diagram: $CD = 50$ m, $G\hat{D}H = 32°$, $H\hat{D}C = 26°$ and $H\hat{C}D = 80°$.

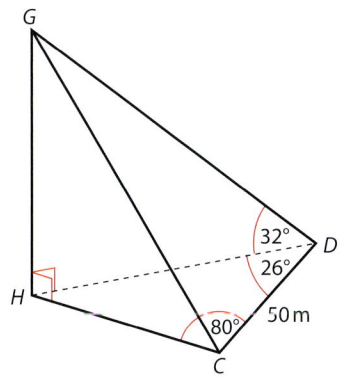

Find a) the distance between H and D, and b) the height of the pole GH.

Solution

a) In triangle HDC: $D\hat{H}C = 180° - (80° + 26°) = 74°$.

Now apply the law of sines:

$$\frac{\sin 80°}{HD} = \frac{\sin 74°}{50} \Rightarrow HD = \frac{50 \sin 80°}{\sin 74°} \approx 51.225 \text{ m}$$

Therefore, the distance from H to D is 51.2 m accurate to 3 s.f.

b) Using the right triangle GHD:

$$\tan 32° = \frac{GH}{51.225} \Rightarrow GH = 51.225 \tan 32° \approx 32.009 \text{ m}$$

Therefore, the height of the pole is 32.0 m accurate to 3 s.f.

Example 31

The figure shown is a pyramid with a square base. It is a *right* pyramid, so the line segment (i.e. the height) drawn from the top vertex A perpendicular to the base will intersect the square base at its centre C. If each side of the square base has a length of 2 cm and the height of the pyramid is also 2 cm, find:

a) the measure of $A\hat{G}F$

b) the total surface area of the pyramid.

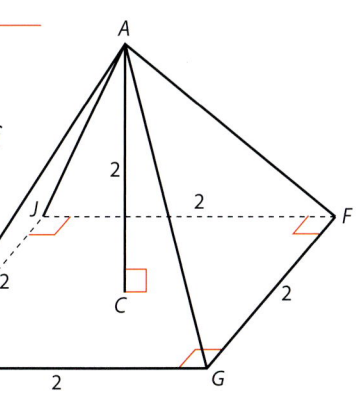

241

Solution

a) Label the midpoint of $[GF]$ as point M and draw two line segments, $[CM]$ and $[AM]$. Since C is the centre of the square base then $CM = 1$ cm. Extract right triangle ACM and use Pythagoras' theorem to find the length of $[AM]$.

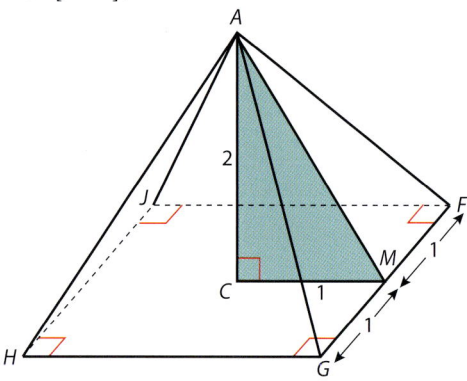

$AM = \sqrt{1^2 + 2^2} = \sqrt{5}$ $[AM]$ is perpendicular to $[GF]$

Extract right triangle AMG and use the tangent ratio to find $A\hat{G}M$ (same as $A\hat{G}F$):

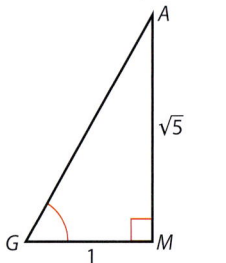

$$\tan(A\hat{G}M) = \frac{\sqrt{5}}{1}$$

$$A\hat{G}M = \tan^{-1}(\sqrt{5}) \approx 65.905°$$

Therefore, $A\hat{G}M = A\hat{G}F \approx 65.9°$.

b) The total surface area comprises the square base plus four identical lateral faces that are all isosceles triangles. Triangle AGM is one-half the area of one of these triangular faces.

Area of triangle $AGM = \frac{1}{2}(1)(\sqrt{5}) = \frac{\sqrt{5}}{2}$

$\Rightarrow$ Area of triangle $AGF = 2\left(\frac{\sqrt{5}}{2}\right) = \sqrt{5}$

Surface area = area of square base + area of four lateral faces
= $2^2 + 4\sqrt{5} = 4 + 4\sqrt{5} \approx 12.94$ cm²

Example 32

For the rectangular box shown, find a) the measure of $A\hat{B}C$, and b) the area of triangle ABC.

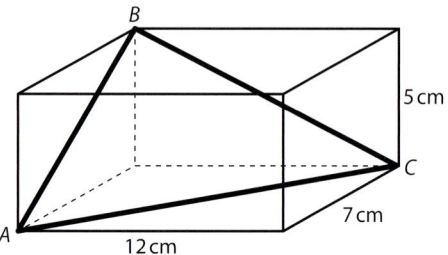

Solution

a) Each of the three sides of triangle *ABC* is the hypotenuse of a right triangle. Using Pythagoras' theorem:

$AC = \sqrt{7^2 + 12^2} = \sqrt{49 + 144} = \sqrt{193} = 13.892$

$AB = \sqrt{5^2 + 7^2} = \sqrt{25 + 49} = \sqrt{74} \approx 8.602$

$BC = \sqrt{5^2 + 12^2} = \sqrt{25 + 144} = \sqrt{169} = 13$

Apply the law of cosines to find $A\hat{B}C$, using exact lengths of the sides of the triangle.

$\cos A\hat{B}C = \dfrac{(\sqrt{74})^2 + 13^2 - (\sqrt{193})^2}{2(\sqrt{74})(13)} \Rightarrow A\hat{B}C = \cos^{-1}\left[\dfrac{74 + 169 - 193}{2(\sqrt{74})(13)}\right] \approx 77.082°$

Therefore, the measure of $A\hat{B}C$ is approximately 77.1° to 3 s.f.

b) Area of triangle $= \frac{1}{2}(AB)(BC)\sin A\hat{B}C = \frac{1}{2}(\sqrt{74})(13)\sin(77.082°)$
$\approx 54.499\,96\text{ cm}^2$

Therefore, the area of triangle *ABC* is approximately 54.5 cm².

Exercise 7.5

In questions 1–4, determine:
  a) the slope (gradient) of the line (approximate to 3 s.f. if not exact)
  b) the equation of the line.

1 (line through origin at 70° from positive x-axis)

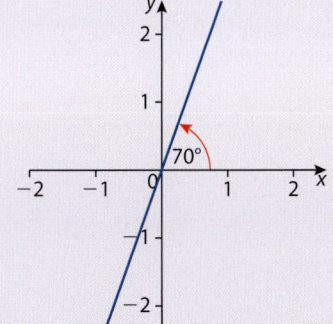

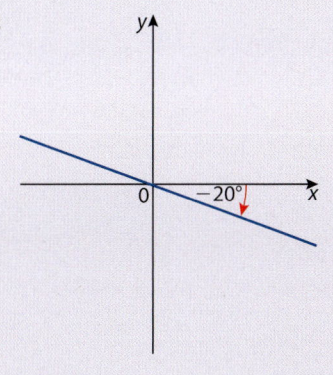

2 (line through origin at −20°)

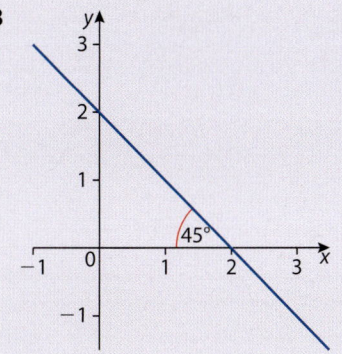

3 (line at 45° descending through x-axis near x = 2)

4 (line through $(0, -\tfrac{3}{2})$ at 68°)

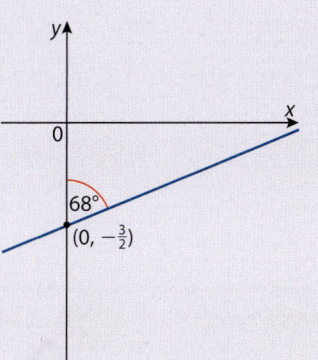

7 Triangle Trigonometry

In questions 5–7, find the acute angle that the line through the given pair of points makes with the *x*-axis.

5 (1, 4) and (−1, 2)

6 (−3, 1) and (6, −5)

7 $\left(2, \frac{1}{2}\right)$ and (−4, −10)

In questions 8 and 9, find the acute angle between the two given lines.

8 $y = -2x$ and $y = x$

9 $y = -3x + 5$ and $y = 2x$

10 a) Find the exact equation of line L_1 that passes through the origin and makes an angle of 30° with the positive direction of the *x*-axis.
b) The equation of line L_2 is $x + 2y = 6$. Find the acute angle between L_1 and L_2.

11 Calculate *AB* given *CD* = 30 cm, and the angle measures given in the diagram.

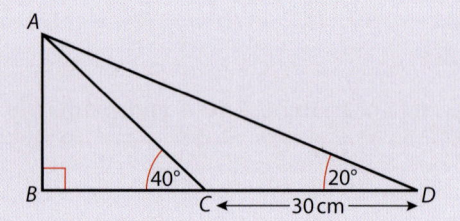

12 The circle with centre *O* and radius of 8 cm has two chords *PR* and *RS*, such that *PR* = 5 cm and *RS* = 10 cm. Find each of the angles $P\hat{R}O$ and $S\hat{R}O$, and then calculate the area of the triangle *PRS*.

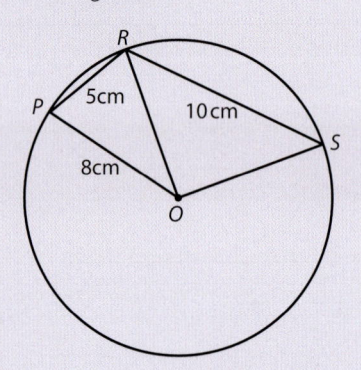

13 A pilot measures the angles of depression to two ships to be 40° and 52° (see diagram). If the pilot is flying at an elevation of 10 000 m, find the distance between the two ships.

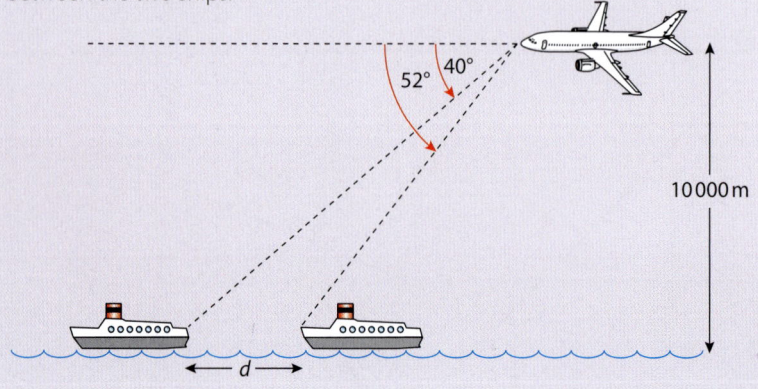

244

14 A forester was conducting a survey of a tropical jungle that was mostly inaccessible on foot. The points F and G indicate the location of two rare trees. To find the distance between points F and G, a line AB of length 250 m is measured out so that F and G are on opposite sides of AB. The angles between the line segment AB and the line of sight from each endpoint of AB to each tree are measured, and are shown in the diagram. Calculate the distance between F and G.

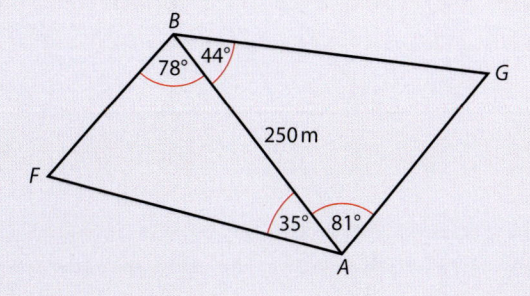

15 Calculate the distance between the tips of the hands of a large clock on a building at 10 o'clock if the minute hand is 3 m long and the hour hand is 2.25 m long.

16 An airplane takes off from point A. It flies 850 km on a bearing of 030°. It then changes direction to a bearing of 065° and flies a further 500 km and lands at point B.
   a) What is the straight line distance from A to B?
   b) What is the bearing **from** A **to** B?

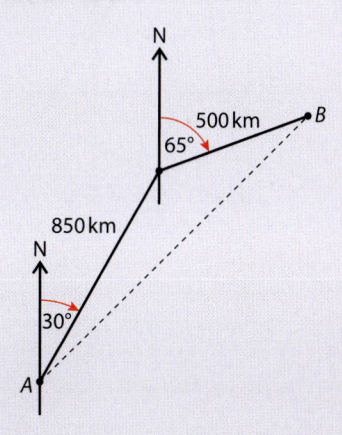

17 The traditional bicycle frame consists of tubes connected together in the shape of a triangle and a quadrilateral (four-sided polygon). In the diagram, AB, BC, CD and AD represent the four tubes of the quadrilateral section of the frame. A frame maker has prepared three tubes such that $AD = 53$ cm, $AB = 55$ cm and $BC = 11$ cm. If $D\hat{A}B = 76°$ and $A\hat{B}C = 97°$, what must be the length of tube CD? Give your answer to the nearest tenth of a centimetre.

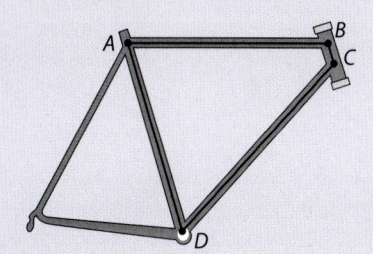

18 The tetrahedron shown in the diagram has the following measurements.
$AB = 12$ cm, $DC = 10$ cm, $A\hat{C}B = 45°$ and $A\hat{D}B = 60°$

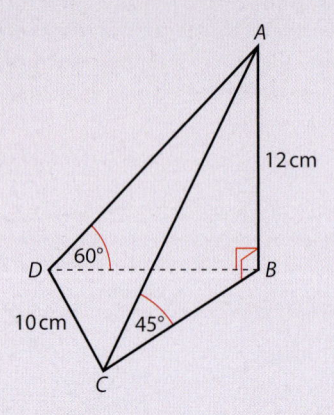

AB is perpendicular to the triangle BCD. Find the area of each of the four triangular faces: ABC, ABD, BCD and ACD.

19 Find the measure of angle DEF in the rectangular box.

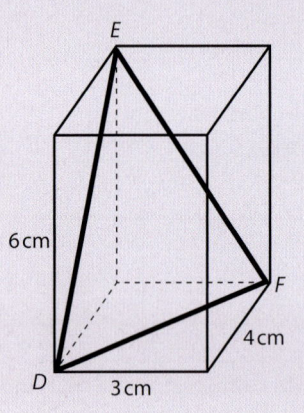

20 At a point A, due south of a building, the angle of elevation from the ground to the top of a building is 58°. At a point B (on level ground with A), 80 m due west of A, the angle of elevation to the top of the building is 27°. Find the height of the building.

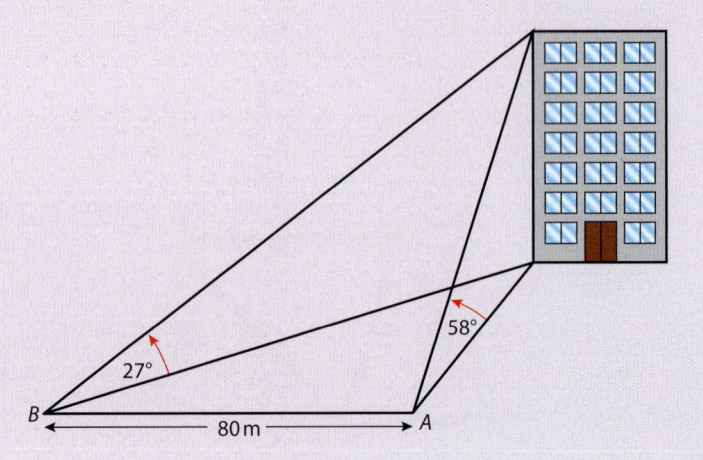

Practice questions

1 The shortest distance from a chord [AB] to the centre O of a circle is 3 units. The radius of the circle is 5 units. Find the exact value of sin $A\hat{O}B$.

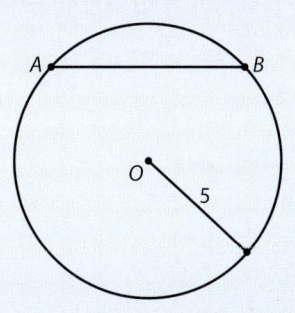

2 In a right triangle, $\tan\theta = \frac{3}{7}$. Find the exact value of $\sin 2\theta$ and $\cos 2\theta$.

3 A triangle has sides of length 4, 5 and 7 units. Find, to the nearest tenth of a degree, the size of the largest angle.

4 If A is an obtuse angle in a triangle and $\sin A = \frac{5}{13}$, calculate the exact value of $\sin 2A$.

5 The diagram shows a vertical pole PQ, which is supported by two wires fixed to the horizontal ground at A and B.

$BQ = 40$ m
$P\hat{B}Q = 36°$
$B\hat{A}Q = 70°$
$A\hat{B}Q = 30°$

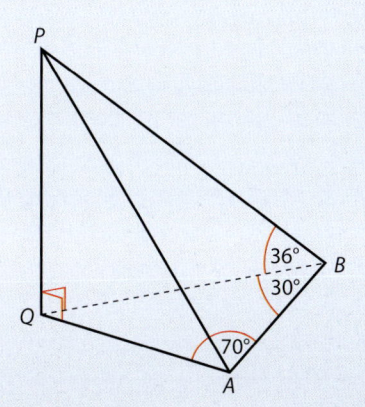

Find: **a)** the height of the pole PQ
b) the distance between A and B.

© International Baccalaureate Organization, 2000

6 Town A is 48 km from town B and 32 km from town C, as shown in the diagram.

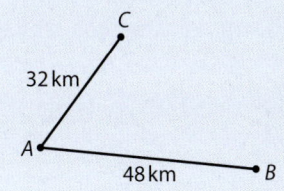

Given that town B is 56 km from town C, find the size of the angle $C\hat{A}B$ to the nearest tenth of a degree.

© International Baccalaureate Organization, 2003

7 The following diagram shows a triangle with sides 5 cm, 7 cm and 8 cm.

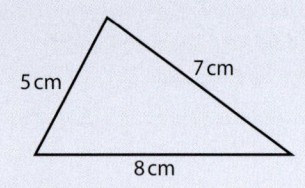

Find: **a)** the size of the smallest angle, in degrees

b) the area of the triangle.

© International Baccalaureate Organization, 2001

8 The diagrams below show two different triangles, both satisfying the conditions: $AB = 20$ cm, $AC = 17$ cm, $A\hat{B}C = 50°$.

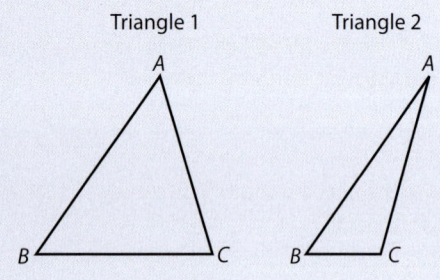

Triangle 1   Triangle 2

a) Calculate the size of $A\hat{C}B$ in Triangle 2.

b) Calculate the area of Triangle 1.

© International Baccalaureate Organization, 2001

9 Two boats A and B start moving from the same point P. Boat A moves in a straight line at 20 km/h and boat B moves in a straight line at 32 km/h. The angle between their paths is 70°. Find the distance between the two boats after 2.5 hours.

© International Baccalaureate Organization, 2002

10 In triangle JKL, $JL = 25$, $KL = 38$ and $K\hat{J}L = 51°$, as shown in the diagram.

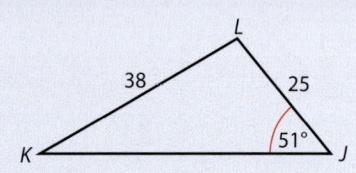

Find $J\hat{K}L$, giving your answer correct to the nearest degree.

11 The following diagram shows a triangle ABC, where $BC = 5$ cm, $A\hat{B}C = 60°$ and $A\hat{C}B = 40°$.

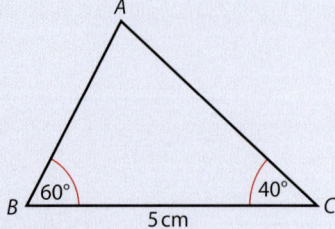

a) Calculate AB.

b) Find the area of the triangle.

© International Baccalaureate Organization, 2001

12 Find the measure of the acute angle between a pair of diagonals of a cube.

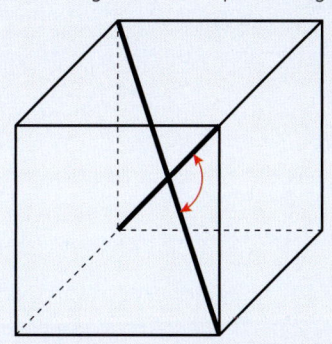

13 A farmer owns a triangular field *ABC*. One side of the triangle, [*AC*], is 104 m, a second side, [*AB*], is 65 m and the angle between these two sides is 60°.
  a) Use the cosine rule to calculate the length of the third side, [*BC*], of the field.
  b) Given that $\sin 60° = \frac{\sqrt{3}}{2}$, find the area of the field in the form $p\sqrt{3}$, where *p* is an integer.

Let *D* be a point on [*BC*] such that [*AD*] bisects the 60° angle. The farmer divides the field into two parts, A_1 and A_2, by constructing a straight fence [*AD*] of length *x* m, as shown in the diagram.

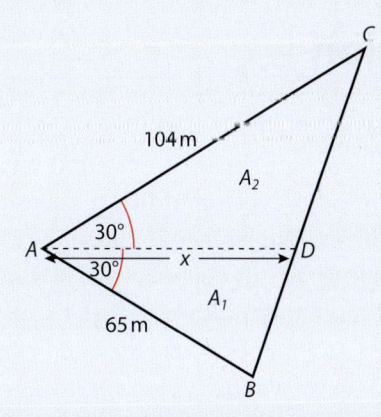

  c) **(i)** Show that the area of A_1 is given by $\frac{65x}{4}$.
   (ii) Find a similar expression for the area of A_2.
   (iii) Hence, find the value of *x* in the form $q\sqrt{3}$, where *q* is an integer.
  d) **(i)** Explain why $\sin A\hat{D}C = \sin A\hat{D}B$.
   (ii) Use the result of part **(i)** and the sine rule to show that $\frac{BD}{DC} = \frac{5}{8}$.

© International Baccalaureate Organization, 2002

8 Vectors I

Assessment statements

5.1 Vectors as displacements in the plane.
Components of a vector; column representation.
$$\mathbf{v} = \begin{pmatrix} v_1 \\ v_2 \\ v_3 \end{pmatrix} = v_1\mathbf{i} + v_2\mathbf{j} + v_3\mathbf{k}$$
Algebraic and geometric approaches to the following topics:
the sum and difference of two vectors; the zero vector; the vector $-\mathbf{v}$;
multiplication by a scalar, $k\mathbf{v}$;
magnitude of a vector, $|\mathbf{v}|$;
unit vectors; base vectors, $\mathbf{i}$, $\mathbf{j}$ and $\mathbf{k}$;
position vectors $\overrightarrow{OA} = \mathbf{a}$.

5.2 The scalar product of two vectors.
Perpendicular vectors; parallel vectors.
The angle between two vectors.

Introduction

Vectors are an essential tool in physics and a very significant part of mathematics. Historically, their primary application was to represent forces, and the operation called '**vector addition**' corresponds to the combining of various forces. Many other applications in physics and other fields have been found since. In this chapter, we will discuss what vectors are and how to add, subtract and multiply them by scalars; we will also examine why vectors are useful in everyday life and how they are used in real-life applications. Then we will discuss scalar products.

Control panel of a passenger jet cockpit.

8.1 Vectors as displacements in the plane

We can represent physical quantities like temperature, distance, area, speed, density, pressure and volume by a single number indicating magnitude or size. These are called **scalar quantities**. Other physical quantities possess the properties of magnitude and direction. We define the force needed to pull a truck up a 10° slope by its **magnitude** and **direction**. Force, displacement, velocity, acceleration, lift, drag, thrust and weight are quantities that cannot be described by a single number. These are called **vector quantities**. Distance and displacement, for example, have distinctly different meanings; so do speed and velocity. Speed is a scalar quantity that refers to 'how fast an object is moving'.

Velocity is a vector quantity that refers to 'the rate at which an object *changes its position*'. When evaluating the velocity of an object, we must keep track of direction. It would not be enough to say that an object has a velocity of 55 km/h; we must include direction information in order to fully describe the velocity of the object. For instance, you must describe the object's velocity as being 55 km/h east. This is one of the essential differences between speed and velocity. Speed is a **scalar** quantity and does not keep track of direction; velocity is a **vector** quantity and is direction-conscious.

Thus, an airplane moving westward with a speed of 600 km/h has a velocity of 600 km/h west. Note that speed has no direction (it is scalar) and velocity, at any instant, is simply the speed with a direction.

We represent vector quantities with **directed line segments** (Figure 8.1). The directed line segment $\overrightarrow{AB}$ has **initial point** A and **terminal point** B. We use the notation $\overrightarrow{AB}$ to indicate that the line segment represents a vector quantity. We use $|\overrightarrow{AB}|$ to represent the **magnitude** of the directed line segment. The terms **size**, **length** or **norm** are also used. The direction of $\overrightarrow{AB}$ is from A to B. $\overrightarrow{BA}$ has the same length but the opposite direction to $\overrightarrow{AB}$ and hence cannot be equal to it.

Two directed line segments that have the same magnitude and direction are equivalent. For example, the directed line segments in Figure 8.2 are all equivalent.

We call the set of all directed line segments equivalent to a given directed line segment $\overrightarrow{AB}$ a **vector v**, and write $\mathbf{v} = \overrightarrow{AB}$. We denote vectors by lower-case, boldface letters such as **a**, **u**, and **v**.

We say that two vectors **a** and **b** are equal if their corresponding directed line segments are equivalent.

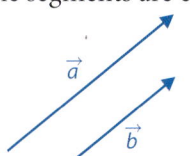

Vectors $\vec{a}$ and $\vec{b}$ have the same direction but different magnitudes $\Rightarrow \vec{a} \neq \vec{b}$.

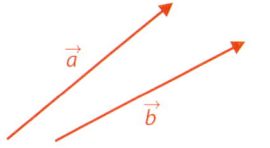

Vectors $\vec{a}$ and $\vec{b}$ have equal magnitudes but different directions $\Rightarrow \vec{a} \neq \vec{b}$.

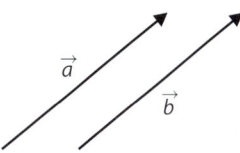

Vectors $\vec{a}$ and $\vec{b}$ have equal magnitudes and the same direction $\Rightarrow \vec{a} = \vec{b}$.

The notion of vector, as presented here, is due to the mathematician-physicist J. Williard Gibbs (1839–1903) of Yale University. His book *Vector Analysis* (1881) made these ideas accessible to a wide audience.

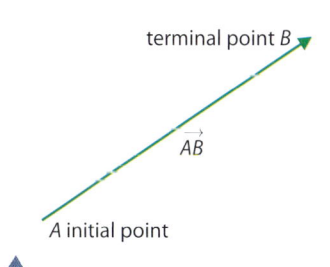

Figure 8.1

terminal point B
$\overrightarrow{AB}$
A initial point

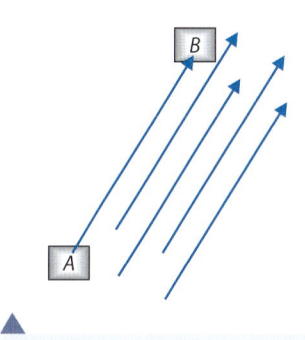

Figure 8.2

Figure 8.3

• **Hint:** Note: When we handwrite vectors, we cannot use boldface, so the convention is to use the arrow notation.

251

8 Vectors I

Definition 1: Two vectors **u** and **v** are equal if they have the same magnitude and the same direction.

Definition 2: The negative of a vector **u**, denoted by −**u**, is a vector with the same magnitude but opposite direction.

Example 1

Marco walked around the park as shown in the diagram. What is Marco's displacement at the end of his walk?

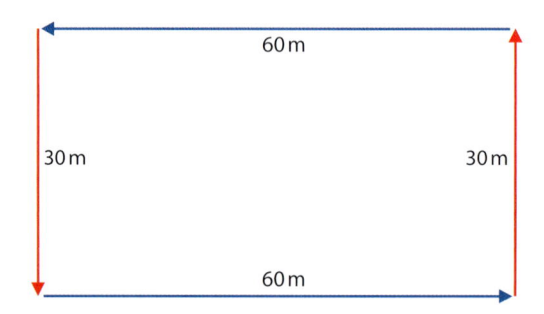

Solution

Even though he walked a total distance of 180 m, his displacement is zero since he returned to his original position. So, his displacement is **0**.

This is a displacement and hence direction is also important, not only magnitude. The 30 m south 'cancelled' the 30 m north, and the 60 m east is cancelled by the 60 m west.

Vectors can also be looked at as displacement/translation in the plane. Take, for example, the directed segments PQ and RS as representing the vectors **u** and **v**, respectively. The points $P(0, 0)$, $Q(2, 5)$, $R(3, 1)$ and $S(5, 6)$ are shown in Figure 8.4.

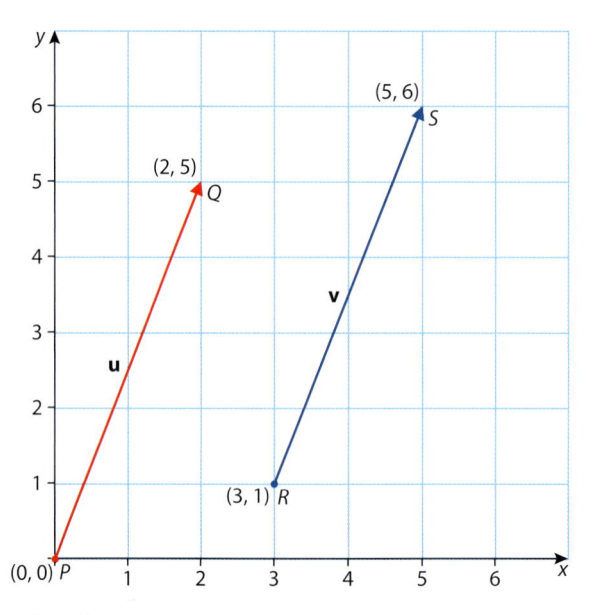

Figure 8.4

We can prove that these two vectors are equal.

The directed line segments representing the vectors have the same direction, since they both have a slope of $\frac{5}{2}$.

They also have the same magnitude, as:
$$|\overrightarrow{PQ}| = \sqrt{5^2 + 2^2} = \sqrt{29} \text{ and}$$
$$|\overrightarrow{RS}| = \sqrt{(5-3)^2 + (6-1)^2} = \sqrt{29}$$

Component form

The directed line segment with the origin as its initial point is the most convenient way of representing a vector. This representation of the vector is said to be in **standard position**. In Figure 8.4, **u** is in standard position. A vector in standard position can be uniquely represented by the coordinates of its terminal point (u_1, u_2). This is called the **component form of a vector u**, written as $\mathbf{u} = (u_1, u_2)$.

The coordinates u_1 and u_2 are the **components** of the vector **u**. In Figure 8.4, the components of the vector **u** are 2 and 5.

If the initial and terminal points of the vector are the same, the vector is a **zero vector** and is denoted by $\mathbf{0} = (0, 0)$.

> If **u** is a vector in the plane with initial point $(0, 0)$ and terminal point (u_1, u_2), the **component form** of **u** is $\mathbf{u} = (u_1, u_2)$.
> Note: The component form is also written as $\begin{pmatrix} u_1 \\ u_2 \end{pmatrix}$.

So, a vector in the plane is also an ordered pair (u_1, u_2) of real numbers. The numbers u_1 and u_2 are the components of **u**. The vector $\mathbf{u} = (u_1, u_2)$ is also called the **position vector** of the point (u_1, u_2).

If the vector **u** is not in standard position and is represented by a directed segment AB, then it can be written in its component form, observing the following fact:

$\mathbf{u} = (u_1, u_2) = (x_2 - x_1, y_2 - y_1)$, where $A(x_1, y_1)$ and $B(x_2, y_2)$ (Figure 8.5).

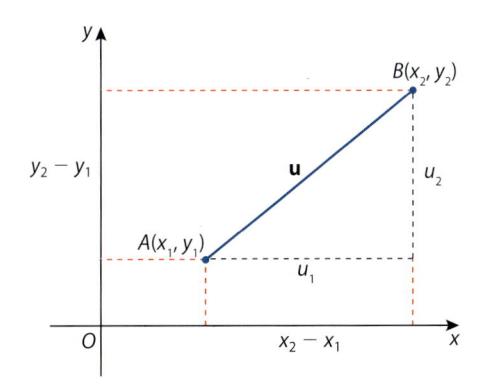

◀ Figure 8.5

The length of vector **u** can be given using Pythagoras' theorem and/or the distance formula:
$$|\mathbf{u}| = \sqrt{u_1^2 + u_2^2} = \sqrt{(x_2 - x_1)^2 + (y_2 - y_1)^2}$$

8 Vectors I

Example 2

Find the components and the length of the vector between the points $P(-2, 3)$ and $Q(4, 7)$.

Solution

$$\vec{PQ} = (4 - (-2), 7 - 3) = (6, 4)$$
$$|\vec{PQ}| = \sqrt{36 + 16} = \sqrt{52} = 2\sqrt{13}$$

Example 3

The directed segment from $(-1, 2)$ to $(3, 5)$ represents a vector **v**. Find the length of vector **v**, draw the vector in standard position and find the opposite of the vector in component form.

Solution

The length of vector **v** can be found using the distance formula:

$$|\mathbf{v}| = \sqrt{(3 + 1)^2 + (5 - 2)^2} = 5$$

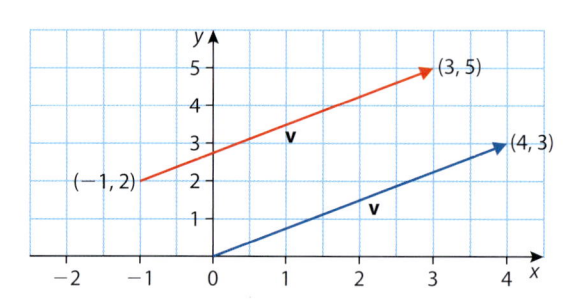

The opposite of this vector can be represented by $-\mathbf{v} = (-4, -3)$.

8.2 Vector operations

Two of the most basic and important operations are scalar multiplication and vector addition.

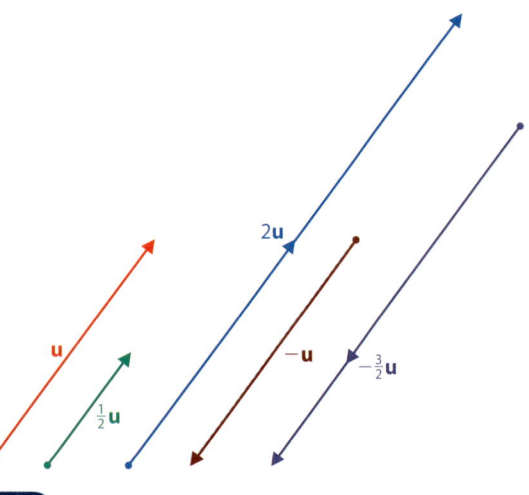

Scalar multiplication

In working with vectors, numbers are considered scalars. In this discussion, scalars will be limited to real numbers only. Geometrically, the product of a vector **u** and a scalar k, $\mathbf{v} = k\mathbf{u}$, is a vector that is $|k|$ times as long as **u**. If k is positive, **v** has the same direction as **u**, and when k is negative, **v** has the opposite direction to **u** (Figure 8.6).

◀ **Figure 8.6**

254

Consequence: It becomes clear from this discussion that for two vectors to be parallel, it is necessary and sufficient that one of them is a scalar multiple of the other. That is, if **v** and **u** are parallel, then **v** = k**u**; and vice versa, if **v** = k**u**, then **v** and **u** are parallel.

In terms of their components, the operation of scalar multiplication is straightforward.

If $\mathbf{u} = (u_1, u_2)$ then $\mathbf{v} = k\mathbf{u} = k(u_1, u_2) = (ku_1, ku_2)$.

Vector addition

There are two equivalent ways of looking at the addition of vectors geometrically. One is the triangular method and the other is the parallelogram method.

Let **u** and **v** denote two vectors. Draw the vectors such that the terminal point of **u** and initial point of **v** coincide. The vector joining the initial point of **u** to the terminal point of **v** is the sum (resultant) of vectors **u** and **v** and is denoted by **u** + **v** (Figure 8.7).

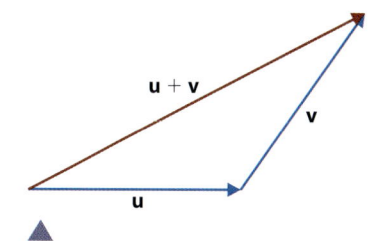

Figure 8.7

Another equivalent way of looking at the sum also gives us the grounds to say that vector addition is commutative.

Let **u** and **v** denote two vectors. Draw the vectors such that the initial point of **u** and initial point of **v** coincide. The vector joining the common initial point of **u** and **v** to the opposite corner of the parallelogram, formed by the vectors as its adjacent sides, is the sum (resultant) of vectors **u** and **v** and is denoted by **u** + **v** (Figure 8.8).

The difference of two vectors is an extremely important rule that will be used later in the chapter.

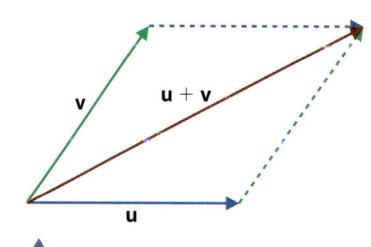

Figure 8.8

As Figure 8.9 shows, it is an extension of the addition rule. An easy way of looking at it is through a combination of the parallelogram rule and the triangle rule. We draw the vectors **u** and **v** in the usual way, then we draw −**v** starting at the terminal point of **u** and we add **u** + (−**v**) to get the difference **u** − **v**. As it turns out, the difference of the two vectors **u** and **v** is the diagonal of the parallelogram with its initial point the terminal of **v** and its terminal point the terminal point of **u**.

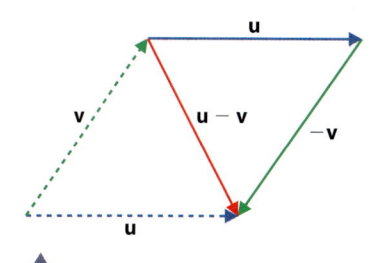

Figure 8.9

Example 4

Consider the vectors $\mathbf{u} = (2, -3)$ and $\mathbf{w} = (1, 3)$.
a) Write down the components of $\mathbf{v} = 2\mathbf{u}$.
b) Find $|\mathbf{u}|$ and $|\mathbf{v}|$ and compare them.
c) Draw the vectors **u**, **v**, **w**, 2**w**, **u** + **w**, **v** + 2**w**, **u** − **w**, **v** − 2**w**.
d) Comment on the results of c) above.

Solution

a) $\mathbf{v} = 2(2, -3) = (4, -6)$
b) $|\mathbf{u}| = \sqrt{4 + 9} = \sqrt{13}$, $|\mathbf{v}| = \sqrt{16 + 36} = \sqrt{52} = 2\sqrt{13}$. Clearly, $|\mathbf{v}| = 2|\mathbf{u}|$.

8 Vectors I

c)

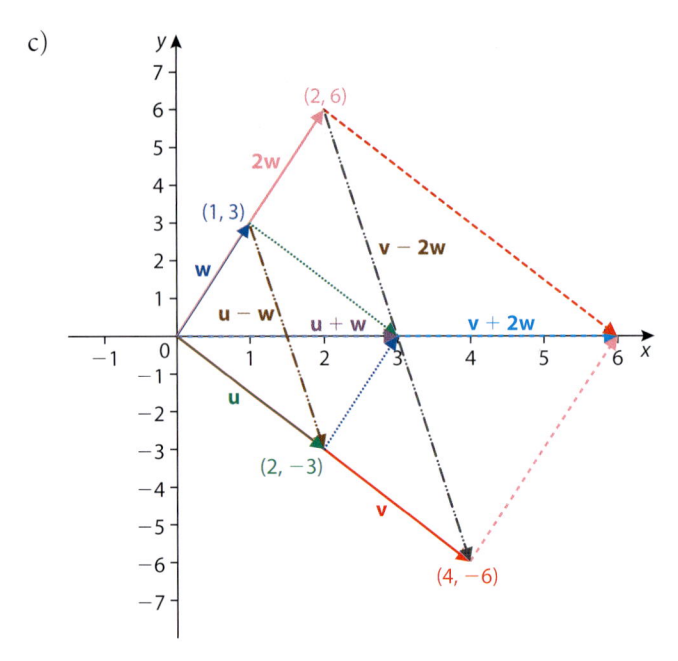

[Figure: coordinate plane showing vectors u, w, v, 2w, u+w, u−w, v−2w, v+2w with points (1,3), (2,6), (2,−3), (4,−6)]

d) We observe that **u** + **w** = (3, 0) which turns out to be (1 + 2, 3 − 3), the sum of the corresponding components. We observe the same for **v** + **2w** = (6, 0), which in turn is (2 + 4, 6 − 6).

We also observe that **v** + **2w** = 2**u** + 2**w** = 2(**u** + **w**), and

v − **2w** is parallel to **u** − **w** and is twice its length!

Can you draw more observations?

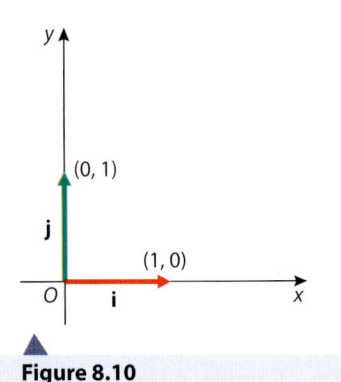

Base vectors in the coordinate plane

As you have seen before, vectors can also be represented in a coordinate system using their component form. This is a very useful tool that helps make many applications of vectors simple and easy. At the heart of the component approach to vectors we find the 'base' vectors **i** and **j**.

i is a vector of magnitude 1 with the direction of the positive *x*-axis and **j** is a vector of magnitude 1 with the direction of the positive *y*-axis. These vectors and any vector that has a magnitude of 1 are called **unit vectors**. Since vectors of same direction and length are equal, each vector **i** and **j** may be drawn at any point in the plane, but it is usually more convenient to draw them at the origin, as shown in Figure 8.10.

Now, the vector *k***i** has magnitude *k* and is parallel to the vector **i**. Similarly, the vector *m***j** has magnitude *m* and is parallel to **j**.

Consider the vector **u** = (u_1, u_2). This vector, in standard position, has an *x*-component u_1 and *y*-component u_2 (Figure 8.11).

Since the vector **u** is the diagonal of the parallelogram with adjacent sides u_1**i** and u_2**j**, then it is the sum of the two vectors, i.e. **u** = u_1**i** + u_2**j**. It is customary to say that u_1**i** is the **horizontal component** and u_2**j** is the **vertical component** of **u**.

Figure 8.10

Figure 8.11

The previous discussion shows that it is always possible to express any vector in the plane as a linear combination of the unit vectors **i** and **j**.

This form of representation of vectors opens the door to a rich world of vector applications.

Vector addition and subtraction in component form

Consider the two vectors $\mathbf{u} = u_1\mathbf{i} + u_2\mathbf{j}$ and $\mathbf{v} = v_1\mathbf{i} + v_2\mathbf{j}$.

(i) Vector sum $\mathbf{u} + \mathbf{v}$

$$\mathbf{u} + \mathbf{v} = (u_1\mathbf{i} + u_2\mathbf{j}) + (v_1\mathbf{i} + v_2\mathbf{j}) = (u_1\mathbf{i} + v_1\mathbf{i}) + (u_2\mathbf{j} + v_2\mathbf{j})$$
$$= (u_1 + v_1)\mathbf{i} + (u_2 + v_2)\mathbf{j}$$

For example, to add the two vectors $\mathbf{u} = 2\mathbf{i} + 4\mathbf{j}$ and $\mathbf{v} = 5\mathbf{i} - 3\mathbf{j}$, it is enough to add the corresponding components:

$$\mathbf{u} + \mathbf{v} = (2 + 5)\mathbf{i} + (4 - 3)\mathbf{j} = 7\mathbf{i} + \mathbf{j}$$

(ii) Vector difference $\mathbf{u} - \mathbf{v}$

$$\mathbf{u} - \mathbf{v} = (u_1\mathbf{i} + u_2\mathbf{j}) - (v_1\mathbf{i} + v_2\mathbf{j}) = (u_1\mathbf{i} - v_1\mathbf{i}) + (u_2\mathbf{j} - v_2\mathbf{j})$$
$$= (u_1 - v_1)\mathbf{i} + (u_2 - v_2)\mathbf{j}$$

For example, to subtract the two vectors $\mathbf{u} = 2\mathbf{i} + 4\mathbf{j}$ and $\mathbf{v} = 5\mathbf{i} - 3\mathbf{j}$, it is enough to subtract the corresponding components:

$$\mathbf{u} - \mathbf{v} = (2 - 5)\mathbf{i} + (4 + 3)\mathbf{j} = -3\mathbf{i} + 7\mathbf{j}$$

This interpretation of the difference gives us another way of finding the components of any vector in the plane, even if it is not in standard position (Figure 8.12).

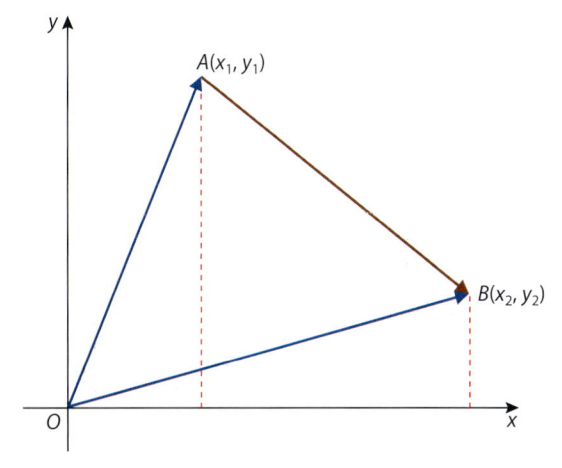

Figure 8.12

Consider the vector $\overrightarrow{AB}$ where the position vectors of its endpoints are given by the vectors $\overrightarrow{OA} = x_1\mathbf{i} + y_1\mathbf{j}$ and $\overrightarrow{OB} = x_2\mathbf{i} + y_2\mathbf{j}$.

As we have seen in section 8.1, $\overrightarrow{AB} = \overrightarrow{OB} - \overrightarrow{OA} = (x_2 - x_1)\mathbf{i} + (y_2 - y_1)\mathbf{j}$. This result was given in Section 8.1 as a definition.

- Many of the laws of ordinary algebra are also valid for vector algebra. These laws are:
  - Commutative law for addition: $\mathbf{a} + \mathbf{b} = \mathbf{b} + \mathbf{a}$
  - Associative law for addition: $(\mathbf{a} + \mathbf{b}) + \mathbf{c} = \mathbf{a} + (\mathbf{b} + \mathbf{c})$

Vectors I

The verification of the associative law is shown in Figure 8.13.

Figure 8.13

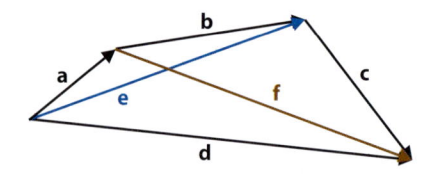

If we add **a** and **b** we get a vector **e**. And similarly, if **b** is added to **c**, we get **f**.

Now **d** = **e** + **c** = **a** + **f**. Replacing **e** with (**a** + **b**) and **f** with (**b** + **c**), we get (**a** + **b**) + **c** = **a** + (**b** + **c**) and we see that the law is verified.

- Commutative law for multiplication: $m\mathbf{a} = \mathbf{a}m$
- Distributive law (1): $(m + n)\mathbf{a} = m\mathbf{a} + n\mathbf{a}$, where m and n are two different scalars.
- Distributive law (2): $m(\mathbf{a} + \mathbf{b}) = m\mathbf{a} + m\mathbf{b}$

These laws allow the manipulation of vector quantities in much the same way as ordinary algebraic equations.

Exercise 8.1 and 8.2

1 Consider the vectors **u** and **v** given. Sketch each indicated vector.
   a) 2**u**
   b) −**v**
   c) **u** + **v**
   d) 2**u** − **v**
   e) **v** − 2**u**

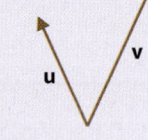

For questions 2 and 3, consider the points A and B given and answer the following questions:

   a) Find $|\vec{AB}|$.
   b) Find the components of the vector $\mathbf{u} = \vec{AB}$ and sketch it in standard position.
   c) Write the vector $\mathbf{v} = \dfrac{1}{|\vec{AB}|} \cdot \mathbf{u}$ in component form.
   d) Find $|\mathbf{v}|$.
   e) Sketch the vector **v** and compare it to **u**.

2 $A(3, 4)$ and $B(7, -1)$

3 $A(-2, 3)$ and $B(5, 1)$

4 Consider the vector shown.
   a) Write down the component representation of the vector.
   b) Find the length of the vector.
   c) Sketch the vector in standard position.
   d) Find a vector equal to this one with initial point $(-1, 1)$.

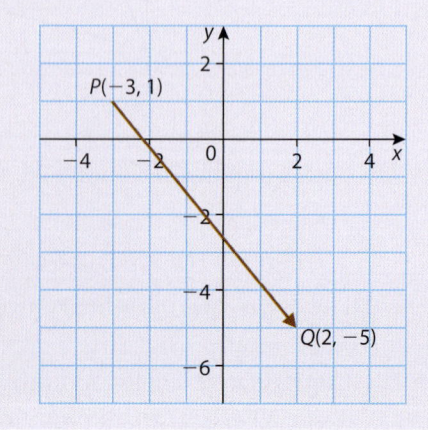

For questions 5–7, the initial point P and terminal point Q are given. Answer the same questions as in question 4.

5. $P(3, 2), Q(7, 8)$

6. $P(2, 2), Q(7, 7)$

7. $P(-6, -8), Q(-2, -2)$

8. Find the terminal point of $\mathbf{v} = 3\mathbf{i} - 2\mathbf{j}$ if the initial point is $(-2, 1)$.

9. Find the initial point of $\mathbf{v} = (-3, 1)$ if the terminal point is $(5, 0)$.

10. Find the terminal point of $\mathbf{v} = (6, 7)$ if the initial point is $(-2, 1)$.

11. Find the initial point of $\mathbf{v} = 2\mathbf{i} + 7\mathbf{j}$ if the terminal point is $(-3, 2)$.

12. Consider the vectors $\mathbf{u} = 3\mathbf{i} - \mathbf{j}$ and $\mathbf{v} = -\mathbf{i} + 3\mathbf{j}$.
    a) Find $\mathbf{u} + \mathbf{v}$, $\mathbf{u} - \mathbf{v}$, $2\mathbf{u} + 3\mathbf{v}$ and $2\mathbf{u} - 3\mathbf{v}$.
    b) Find $|\mathbf{u} + \mathbf{v}|$, $|\mathbf{u} - \mathbf{v}|$, $|\mathbf{u}| + |\mathbf{v}|$ and $|\mathbf{u}| - |\mathbf{v}|$.
    c) Find $|2\mathbf{u} + 3\mathbf{v}|$, $|2\mathbf{u} - 3\mathbf{v}|$, $2|\mathbf{u}| + 3|\mathbf{v}|$ and $2|\mathbf{u}| - 3|\mathbf{v}|$.

13. Let $\mathbf{u} = (1, 5)$ and $\mathbf{v} = (3, -4)$. Find the vector $\mathbf{x}$ such that
    $2\mathbf{u} - 3\mathbf{x} + \mathbf{v} = 5\mathbf{x} - 2\mathbf{v}$.

14. Find $\mathbf{u}$ and $\mathbf{v}$ if $\mathbf{u} - 2\mathbf{v} = 2\mathbf{j} - 3\mathbf{j}$ and $\mathbf{u} + 3\mathbf{v} = \mathbf{i} + \mathbf{j}$.

15. Find the lengths of the diagonals of the parallelogram whose sides are the vectors $2\mathbf{i} - 3\mathbf{i}$ and $\mathbf{i} + \mathbf{j}$.

16. Vectors $\mathbf{u}$ and $\mathbf{v}$ form two sides of parallelogram $PQRS$, as shown. Express each of the following vectors in terms of $\mathbf{u}$ and $\mathbf{v}$.
    a) $\overrightarrow{PR}$
    b) $\overrightarrow{PM}$, where M is the midpoint of $[RS]$
    c) $\overrightarrow{QS}$
    d) $\overrightarrow{QN}$

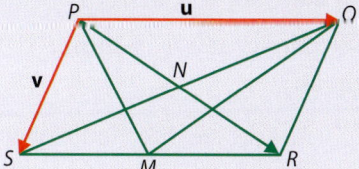

8.3 Unit vectors and direction angles

Consider the vector $\mathbf{u} = 3\mathbf{i} + 4\mathbf{j}$. To find the magnitude of this vector, $|\mathbf{u}|$, we use the distance formula:

$$|\mathbf{u}| = \sqrt{3^2 + 4^2} = 5$$

If we divide the vector $\mathbf{u}$ by $|\mathbf{u}| = 5$, i.e. we multiply the vector $\mathbf{u}$ by the reciprocal of its magnitude, we get another vector that is parallel to $\mathbf{u}$, since they are scalar multiples of each other. The new vector is

$$\frac{\mathbf{u}}{5} = \frac{3}{5}\mathbf{i} + \frac{4}{5}\mathbf{j}$$

This vector is a unit vector in the same direction as $\mathbf{u}$, because

$$\left|\frac{\mathbf{u}}{5}\right| = \sqrt{\left(\frac{3}{5}\right)^2 + \left(\frac{4}{5}\right)^2} = 1$$

Therefore, to find a unit vector in the same direction as a given vector, we divide that vector by its own magnitude.

8 Vectors I

This is tightly connected to the concept of the **direction angle** of a given vector. The **direction angle** of a vector (in standard position) is the angle it makes with the positive *x*-axis (Figure 8.14).

Figure 8.14

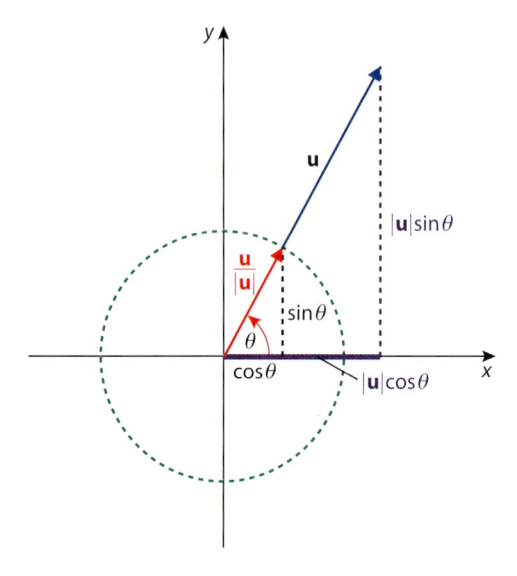

So, the vector **u** can be expressed in terms of the unit vector parallel to it in the following manner:

$\mathbf{u} = u_1\mathbf{i} + u_2\mathbf{j} = (|\mathbf{u}|\cos\theta)\mathbf{i} + (|\mathbf{u}|\sin\theta)\mathbf{j} = |\mathbf{u}|(\cos\theta\mathbf{i} + \sin\theta\mathbf{j})$, where $u_1 = |\mathbf{u}|\cos\theta$ and $u_2 = |\mathbf{u}|\sin\theta$. This fact implies two important tools that help us:

1. find the direction of a given vector
2. find vectors of any magnitude parallel to a given vector.

Given a vector $\mathbf{u} = u_1\mathbf{i} + u_2\mathbf{j}$, find the direction angle of this vector and another vector, whose magnitude is m, that is parallel to the vector **u**.

1. To help determine the direction angle, we observe the following:

   $u_1 = |\mathbf{u}|\cos\theta$ and $u_2 = |\mathbf{u}|\sin\theta$

   This implies that $\dfrac{u_2}{u_1} = \dfrac{|\mathbf{u}|\sin\theta}{|\mathbf{u}|\cos\theta} = \tan\theta$.

   So, $\tan^{-1}\theta$ is the reference angle for the direction angle in question. To know what the direction angle is, it is best to look at the numbers u_1 and u_2 in order to determine which quadrant the vector is in. The following example (Example 5) will clarify this point.

2. To find a vector of magnitude m parallel to **u**, we must first find the unit vector in the direction of **u** and then we multiply it by the scalar m.

   The unit vector in the direction of **u** is $\dfrac{\mathbf{u}}{|\mathbf{u}|} = \dfrac{1}{|\mathbf{u}|}(u_1\mathbf{i} + u_2\mathbf{j})$, and the vector of magnitude m in this direction will be

   $m\dfrac{\mathbf{u}}{|\mathbf{u}|} = \dfrac{m}{\sqrt{u_1^2 + u_2^2}}(u_1\mathbf{i} + u_2\mathbf{j})$.

Example 5

Find the direction angle (to the nearest degree) of each vector, and find a vector of magnitude 7 that is parallel to each.

a) $\mathbf{u} = 2\mathbf{i} + 2\mathbf{j}$
b) $\mathbf{v} = -3\mathbf{i} + 3\mathbf{j}$
c) $\mathbf{w} = 3\mathbf{i} - 4\mathbf{j}$

Solution

a) The direction angle for $\mathbf{u}$ is θ, as shown in Figure 8.15.

$\tan \theta = \frac{2}{2} = 1 \Rightarrow \theta = 45°$

A vector of magnitude 7 that is parallel to $\mathbf{u}$ is

$$7\frac{\mathbf{u}}{|\mathbf{u}|} = \frac{7}{\sqrt{2^2 + 2^2}}(2\mathbf{i} + 2\mathbf{j}) = \frac{7}{2\sqrt{2}}(2\mathbf{i} + 2\mathbf{j}) = \frac{7}{\sqrt{2}}(\mathbf{i} + \mathbf{j})$$

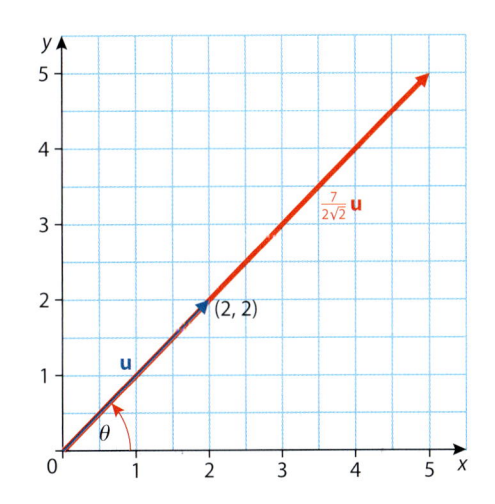

◀ Figure 8.15

b) The direction angle for $\mathbf{v}$ is $180° - \theta$, as shown in Figure 8.16.

$\tan \theta = \frac{-3}{3} = -1 \Rightarrow \theta = 180° - 45° = 135°$

A vector of magnitude 7 that is parallel to $\mathbf{v}$ is

$$7\frac{\mathbf{v}}{|\mathbf{v}|} = \frac{7}{\sqrt{3^2 + 3^2}}(-3\mathbf{i} + 3\mathbf{j}) = \frac{7}{3\sqrt{2}}(-3\mathbf{i} + 3\mathbf{j}) = \frac{7}{\sqrt{2}}(-\mathbf{i} + \mathbf{j})$$

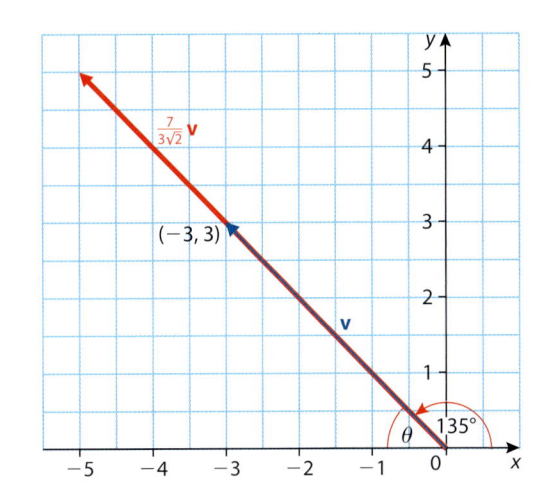

◀ Figure 8.16

8 Vectors I

c) The direction angle for **w** is θ, as shown in Figure 8.17.

$$\tan \theta = \frac{-4}{3} \Rightarrow \theta \approx -53°$$

A vector of magnitude 7 that is parallel to **w** is

$$7\frac{\mathbf{u}}{|\mathbf{u}|} = \frac{7}{\sqrt{3^2 + (-4)^2}}(3\mathbf{i} - 4\mathbf{j}) = \frac{7}{5}(3\mathbf{i} - 4\mathbf{j})$$

Figure 8.17

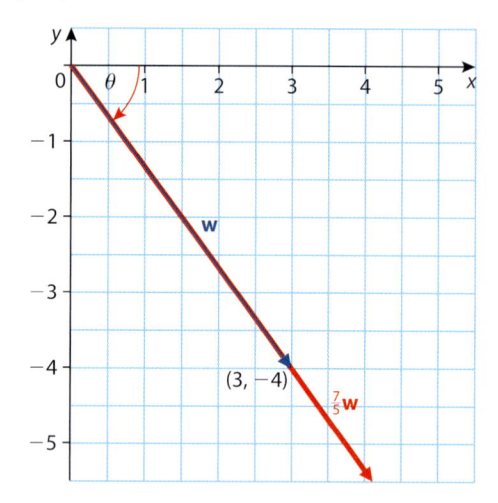

Example 6

What force is required to pull a boat of 800 N up a ramp inclined at 15° from the horizontal? Friction is ignored in this case.

Solution

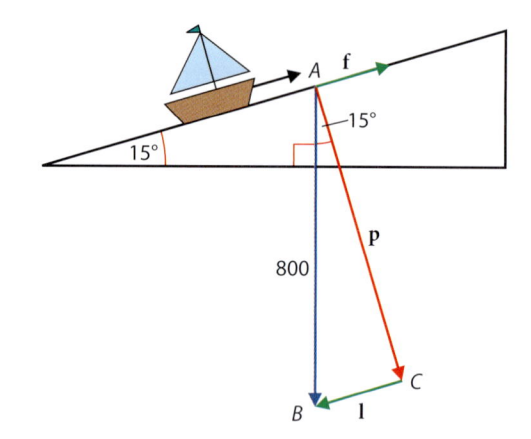

The situation can be shown on a diagram. The weight is represented by the vector $\overrightarrow{AB}$. The weight of the boat has two components – one perpendicular to the ramp, which is the force responsible for keeping the boat on the ramp and preventing it from tumbling down (**p**). The other force is parallel to the ramp, and is the force responsible for pulling the boat down the ramp (**l**). Therefore, the force we need, **f**, must counter **l**.

In triangle ABC:

$$\sin \angle A = |\mathbf{l}|/800 \Rightarrow |\mathbf{l}| = 800 \sin \angle A = 800 \sin 15° = 207.06.$$

We need an upward force of 207.06 N along the ramp to move the boat.

> The process of 'breaking-up' the vector into its components, as we did in the example, is called **resolving** the vector into its components. Notice that the process of resolving a vector is not unique. That is, you can resolve a vector into several pairs of directions.

Example 7

In many countries, it is a requirement that disabled people have access to all places without needing the help of others. Consider an office building whose entrance is 40 cm above ground level. Assuming, on average, that the weight of a person including the equipment used is 1200 N, answer the following questions:
a) At what angle should the ramp designed for disabled persons be set if, on average, the force that a person can apply using their hands is 300 N?
b) How long should the ramp be?

Solution
a)

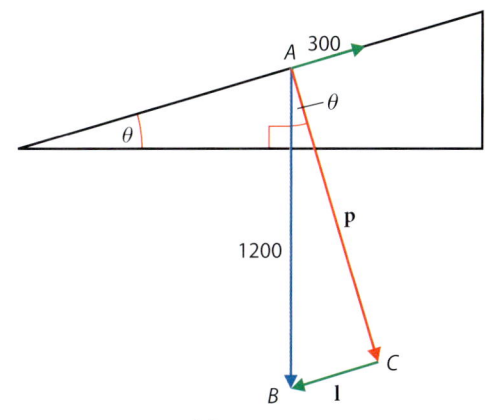

As the diagram above shows, $|\mathbf{l}| = 300$, and
$$\sin \angle A = \frac{|\mathbf{l}|}{1200} = \frac{300}{1200} \Rightarrow \angle A = \sin^{-1} 0.25 \approx 14.47°.$$

b) The length d of the ramp can be found using right triangle trigonometry:

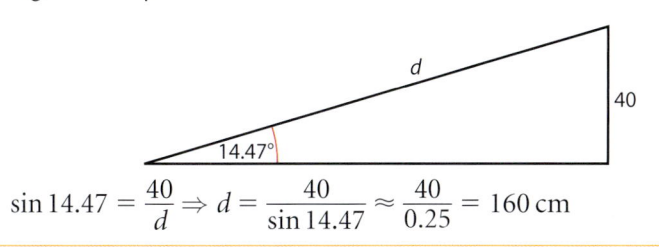

$$\sin 14.47 = \frac{40}{d} \Rightarrow d = \frac{40}{\sin 14.47} \approx \frac{40}{0.25} = 160 \text{ cm}$$

> Vectors can be used to help tackle displacement situations. For example, an object at a position defined by the position vector (**a**, **b**) and a velocity vector (**c**, **d**) has a position vector (**a**, **b**) + *t*(**c**, **d**) after time *t*.

8 Vectors I

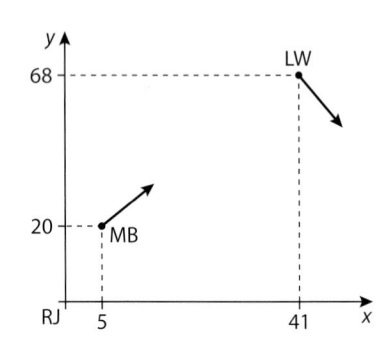

Example 8

The position vector of a ship (MB) from its starting position at a port RJ is given by $\begin{pmatrix} x \\ y \end{pmatrix} = \begin{pmatrix} 5 \\ 20 \end{pmatrix} + t\begin{pmatrix} 12 \\ 16 \end{pmatrix}$. Distances are in kilometres and speeds are in km/h. t is time after 00 hour.

a) Find the position of the MB after 2 hours.

b) What is the speed of the MB?

c) Another ship (LW) is at sea in a location $\begin{pmatrix} 41 \\ 68 \end{pmatrix}$ relative to the same port. LW has stopped for some reason. Show that if LW does not start to move, the two ships will collide. Find the time of the potential collision.

d) To avoid collision, LW is ordered to leave its position and start moving at a velocity of $\begin{pmatrix} 15 \\ -36 \end{pmatrix}$ one hour after MB started. Find the position vector of LW.

e) How far apart are the two ships after two hours since the start of MB?

Solution

a) MB is at a position with vector $\begin{pmatrix} x \\ y \end{pmatrix} = \begin{pmatrix} 5 \\ 20 \end{pmatrix} + 2\begin{pmatrix} 12 \\ 16 \end{pmatrix} = \begin{pmatrix} 29 \\ 52 \end{pmatrix}$.

b) Since the velocity of the ship is $\begin{pmatrix} 12 \\ 16 \end{pmatrix}$, the speed is $\left|\begin{pmatrix} 12 \\ 16 \end{pmatrix}\right| = \sqrt{12^2 + 16^2} = 20 \text{ km/h}$.

c) The collision can happen if the position vectors of the two ships are equal:
$\begin{pmatrix} 5 \\ 20 \end{pmatrix} + t\begin{pmatrix} 12 \\ 16 \end{pmatrix} = \begin{pmatrix} 41 \\ 68 \end{pmatrix} \Rightarrow 5 + 12t = 41$ and $20 + 16t = 68 \Rightarrow 12t = 36$
and $16t = 48 \Rightarrow t = 3$. After 3 hours, at 03:00, a collision could happen.

d) Since LW started one hour later, its position vector is
$\begin{pmatrix} x \\ y \end{pmatrix} = \begin{pmatrix} 41 \\ 68 \end{pmatrix} + (t-1)\begin{pmatrix} 15 \\ -36 \end{pmatrix}, t \geqslant 1$.

e) MB is at $\begin{pmatrix} 29 \\ 52 \end{pmatrix}$ and LW is at $\begin{pmatrix} 41 \\ 68 \end{pmatrix} + (2-1)\begin{pmatrix} 15 \\ -36 \end{pmatrix} = \begin{pmatrix} 56 \\ 32 \end{pmatrix}$. The distance between them is $\sqrt{(56-29)^2 + (32-52)^2} = \sqrt{1129} = 33.6 \text{ km}$.

Exercise 8.3

1 Find the direction angle for each vector
 a) **u** = (2, 0)
 b) **v** = (0, 3)
 c) **w** = (−3, 0)
 d) **u** + **v**
 e) **v** + **w**

2 Find the magnitude and direction angle for each vector.
   a) **u** = (3, 2)
   b) **v** = (−3, −2)
   c) 2**u**
   d) 3**v**
   e) 2**u** + 3**v**
   f) 2**u** − 3**v**

3 Write each of the following vectors in component form. θ is the angle that the vector makes with the positive horizontal axis.
   a) |**u**| = 310, θ = 62°
   b) |**u**| = 43.2, θ = 19.6°

4 Find the unit vector in the same direction as **u** in each of the following cases:
   a) **u** = (3, 4)
   b) **u** = 2**i** − 5**j**

5 Find a vector of magnitude 7 that is parallel to **u** = 3**i** − 4**j**.

6 A plane is flying on a bearing of 170° at a speed of 840 km/h. The wind is blowing in the direction N 120° E with a strength of 60 km/h.
   a) Find the vector components of the plane's still-air velocity and the wind's velocity.
   b) Determine the true velocity (ground) of the plane in component form.
   c) Write down the true speed and direction of the plane.

7 A plane is flying on a compass heading of 340° at 520 km/h. The wind is blowing with the bearing 320° at 64 km/h.
   a) Find the component form of the velocities of the plane and the wind.
   b) Find the actual ground speed and direction of the plane.

8

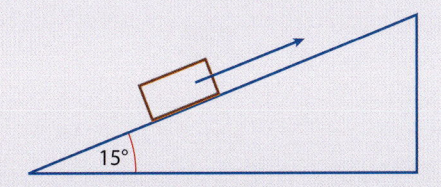

A box is being pulled up a 15° inclined plane. The force needed is 25 N. Find the horizontal and vertical components of the force vector and interpret each of them.

9 A motor boat with the power to steer across a river at 30 km/h is moving such that the bow is pointed in a northerly direction. The stream is moving eastward at 6 km/h. The river is 1 km wide. Where on the opposite side will the boat meet the land?

> Note: In navigation, the convention is that the **course** or **bearing** of a moving object is the angle that its direction makes with the north direction measured clockwise. So, for example, a ship going east has a bearing of 90°.

10 A force of 2500 N is applied at an angle of 38° to pull a 10 000 N ship in the direction given. What force **F** is needed to achieve this?

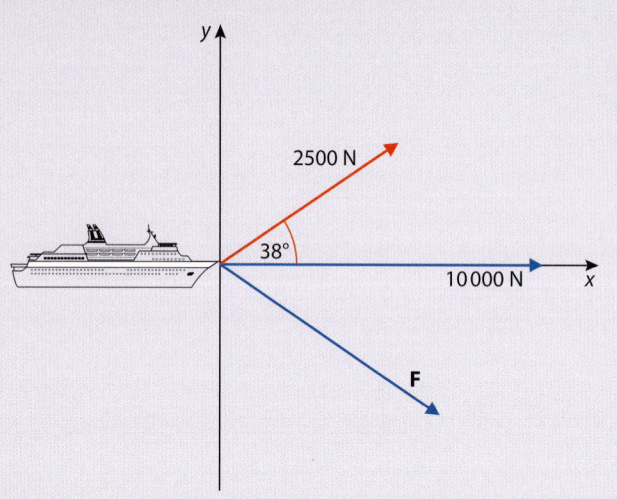

11 A boat is observed to have a bearing of 072°. The speed of the boat relative to still water is 40 km/h. Water is flowing directly south. The boat appears to be heading directly east.
  a) Express the velocity of the boat with respect to the water in component form.
  b) Find the speed of the water stream and the true speed of the boat.

12 A 50 N weight is suspended by two strings as shown. Find the tensions **T** and **S** in the strings.

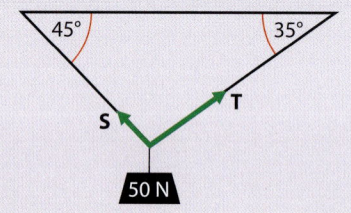

13 A runner runs in a westerly direction on the deck of a cruise ship at 8 km/h. The cruise ship is moving north at a speed of 35 km/h. Find the velocity of the runner relative to the water.

14 The boat in question 9 wants to reach a point exactly north of the starting point. In which direction should the boat be steered in order to achieve this objective?

8.4 Scalar product of two vectors

The multiplication of two vectors is not uniquely defined: in other words, it is unclear whether the product will be a vector or not. For this reason there are two types of vector multiplication:

The **scalar** or **dot product** of two vectors, which results in a scalar; and the **vector** or **cross product** of two vectors, which results in a vector.

In this book, we shall discuss only the scalar or dot product.

The **scalar product of two vectors**, **a** and **b** denoted by **a** · **b**, is defined as the product of the magnitudes of the vectors times the cosine of the angle between them:
a · **b** = |**a**||**b**| cos θ

This is illustrated in Figure 8.18.

Note that the result of a dot product is a scalar, not a vector. The rules for scalar products are given in the following list:

$$\mathbf{a} \cdot \mathbf{b} = \mathbf{b} \cdot \mathbf{a}$$
$$0 \cdot \mathbf{a} = \mathbf{a} \cdot 0 = 0$$
$$\mathbf{a} \cdot (\mathbf{b} + \mathbf{c}) = \mathbf{a} \cdot \mathbf{b} + \mathbf{a} \cdot \mathbf{c}$$
$$\mathbf{a} \cdot \mathbf{a} = |\mathbf{a}|^2$$
$$k(\mathbf{a} \cdot \mathbf{b}) = k\mathbf{a} \cdot \mathbf{b} = \mathbf{a} \cdot k\mathbf{b}, \text{ with k any scalar.}$$

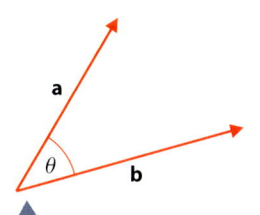

Figure 8.18

The first properties follow directly from the definition:

a · **b** = |**a**||**b**| cos θ, and **b** · **a** = |**b**||**a**| cos θ, and, since multiplication of real numbers is commutative, it follows that **a** · **b** = **b** · **a** The third property will be proved later in this section. Proofs of the rest of the properties are left as exercises.

Using the definition, it is immediately clear that for two non-zero vectors **u** and **v**, if **u** and **v** are perpendicular, the dot product is zero. This is so, because
u · **v** = |**u**||**v**| cos θ = |**u**||**v**| cos 90° = |**u**||**v**| × 0 = 0.
The converse is also true: if **u** · **v** = 0, the vectors are perpendicular,
u · **v** = 0 ⇒ |**u**||**v**| cos θ = 0 ⇒ cos θ = 0 ⇒ θ = 90°.

Using the definition, it is also clear that for two non-zero vectors **u** and **v**, if **u** and **v** are parallel then the dot product is equal to ± |**u**||**v**|. This is so, because
u · **v** = |**u**||**v**| cos θ = |**u**||**v**| cos 0° = |**u**||**v**| × 1 = |**u**||**v**|, or
u · **v** = |**u**||**v**| cos θ = |**u**||**v**| cos 180° = |**u**||**v**| × (−1) = −|**u**||**v**|.
The converse is also true: if **u** · **v** = ± |**u**||**v**|, the vectors are parallel, since
u · **v** = |**u**||**v**| cos θ ⇒ |**u**||**v**| cos θ = ± |**u**||**v**| ⇒ cos θ = ±1 ⇒ θ = 0° or θ = 180°.

Another interpretation of the dot product
Projection

(This subsection is optional – it is beyond the scope of IB/SL syllabus, but very helpful in clarifying the concept of dot products.)

The quantity |**a**| cos θ is called the projection of the vector **a** on vector **b** (Figure 8.19). So, the dot product **b** · **a** = |**b**||**a**| cos θ = |**b**|(|**a**| cos θ) = |**b**| × (the projection of **a** on **b**).

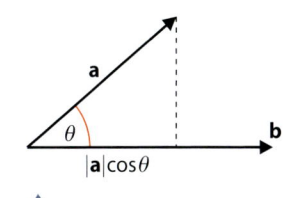

Figure 8.19

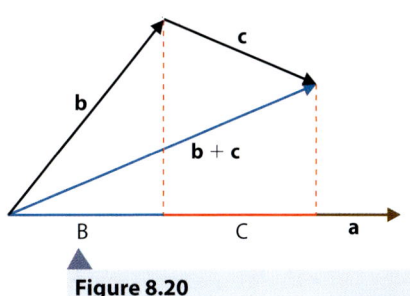

This fact is used in proving the third property on the list above.

If we let B and C stand for the projections of **b** and **c** on **a**, we have

a(**b** + **c**) = |**a**|(B + C) = |**a**|B + |**a**|C = **a** · **b** + **a** · **c**

See Figure 8.20 right.

With this result, we can develop another definition for the dot product that is more useful in the calculation of this product.

Figure 8.20

267

Theorem

If vectors are expressed in component form, $\mathbf{u} = u_1\mathbf{i} + u_2\mathbf{j}$ and $\mathbf{v} = v_1\mathbf{i} + v_2\mathbf{j}$, then $\mathbf{u} \cdot \mathbf{v} = (u_1\mathbf{i} + u_2\mathbf{j}) \cdot (v_1\mathbf{i} + v_2\mathbf{j}) = u_1v_1 + u_2v_2$.

Proof

$\mathbf{u} \cdot \mathbf{v} = (u_1\mathbf{i} + u_2\mathbf{j}) \cdot (v_1\mathbf{i} + v_2\mathbf{j}) = u_1v_1\mathbf{i}^2 + u_1v_2\mathbf{ij} + u_2v_1\mathbf{ji} + u_2v_2\mathbf{j}^2$

However, $\mathbf{i}^2 = \mathbf{j}^2 = 1$ and $\mathbf{ij} = \mathbf{ji} = 0$. (Proof is left as an exercise for you.)

Therefore, $\mathbf{u} \cdot \mathbf{v} = (u_1\mathbf{i} + u_2\mathbf{j}) \cdot (v_1\mathbf{i} + v_2\mathbf{j}) = u_1v_1 + u_2v_2$.

For example, to find the scalar product of the two vectors $\mathbf{u} = 2\mathbf{i} + 4\mathbf{j}$ and $\mathbf{v} = 5\mathbf{i} - 3\mathbf{j}$, it is enough to add the products' corresponding components:

$$\mathbf{u} \cdot \mathbf{v} = 2 \times 5 + 4 \times (-3) = -2$$

Example 9

Find the dot product of $\mathbf{u} = 2\mathbf{i} - 3\mathbf{j}$ and $\mathbf{v} = 3\mathbf{i} + 2\mathbf{j}$.

Solution

$$\mathbf{u} \cdot \mathbf{v} = 2 \times 3 - 3 \times 2 = 0$$

What does this tell us about the two vectors?

The angle between two vectors

The basic definition of the scalar product offers us a method for finding the angle between two vectors.

Since $\mathbf{a} \cdot \mathbf{b} = |\mathbf{a}||\mathbf{b}|\cos\theta$, then $\cos\theta = \dfrac{\mathbf{a} \cdot \mathbf{b}}{|\mathbf{a}||\mathbf{b}|}$.

Example 10

Find the angle between the following two vectors:

$$\mathbf{v} = -3\mathbf{i} + 3\mathbf{j} \text{ and } \mathbf{w} = 2\mathbf{i} - 4\mathbf{j}$$

Solution

$$\cos\theta = \frac{\mathbf{v} \cdot \mathbf{w}}{|\mathbf{v}||\mathbf{w}|} = \frac{-3 \times 2 + 3 \times -4}{\sqrt{(-3)^2 + 3^2} \times \sqrt{2^2 + 4^2}} = \frac{-18}{\sqrt{18}\sqrt{20}} \Rightarrow \theta = 161.57°$$

Example 11

The instrument panel in a plane indicates that its airspeed (the speed of the plane relative to the surrounding air) is 200 km/h and that its compass heading (the direction in which the plane's nose is pointing) is due at N 45° E. There is a steady wind blowing from the west at 50 km/h. Because of the wind, the plane's *true* velocity is different from the panel reading. Find the true velocity of the plane. Also, find its true speed and direction.

Solution

A diagram can help clarify the situation.

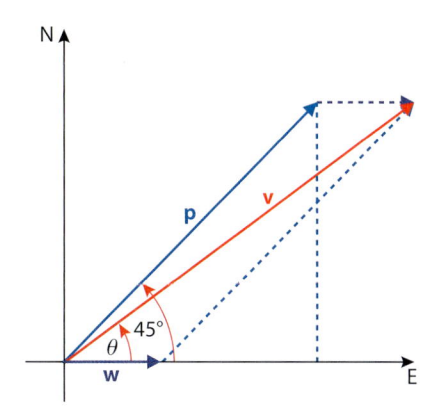

The plane velocity **p** can be expressed in its component form:

$x = |\mathbf{p}|\cos 45° = 200 \cos 45° = 100\sqrt{2},$

$y = |\mathbf{p}|\sin 45° = 200 \sin 45° = 100\sqrt{2},$

so **p** can be written as $\mathbf{p} = (100\sqrt{2}, 100\sqrt{2})$.

The wind velocity **w** can also be expressed in component form:

$\mathbf{w} = (50, 0)$

So, the true velocity, $\mathbf{v} = (100\sqrt{2} + 50, 100\sqrt{2})$.

To find the true speed, we find the magnitude of the resultant found above:

$|\mathbf{v}| = \sqrt{(100\sqrt{2} + 50)^2 + (100\sqrt{2})^2} \approx 238 \text{ km/h}$

To find the true direction, we find θ and calculate the *heading* of the plane:

$\tan \theta = \dfrac{100\sqrt{2}}{100\sqrt{2} + 50} \approx 0.739 \Rightarrow \theta \approx 36.5°,$

so the true direction is N 53.5° E.

Example 12

Consider the segment $[AB]$ with $A(-2, -3)$ and $B(3, 1)$. Use dot products to find the equation of the circle whose diameter is AB.

Solution

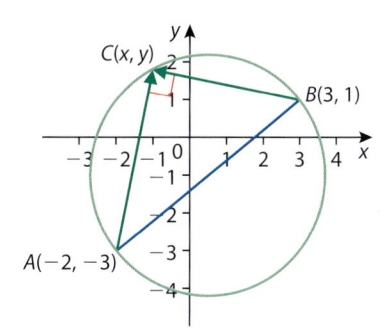

Consider any point $C(x, y)$ on the graph. Find the vectors $\overrightarrow{AC}$ and $\overrightarrow{BC}$.

269

For the point C to be on the circle, the angle at C must be a right angle. Hence, the vectors $\overrightarrow{AC}$ and $\overrightarrow{BC}$ are perpendicular.

For perpendicular vectors, the dot product must be zero.

$$\overrightarrow{AC} = (x+2, y+3), \overrightarrow{BC} = (x-3, y-1)$$
$$\overrightarrow{AC} \cdot \overrightarrow{BC} = 0 \Rightarrow (x+2)(x-3) + (y+3)(y-1) = 0$$
$$\Rightarrow x^2 - x + y^2 + 2y = 9$$

Exercise 8.4

1. Find (i) $\mathbf{u} \cdot \mathbf{v}$ and (ii) the angle between $\mathbf{u}$ and $\mathbf{v}$ to the nearest degree.
   a) $\mathbf{u} = \mathbf{i} + \sqrt{3}\mathbf{j}, \mathbf{v} = \sqrt{3}\mathbf{i} - \mathbf{j}$
   b) $\mathbf{u} = (2, 5), \mathbf{v} = (4, 1)$
   c) $\mathbf{u} = 2\mathbf{i} - 3\mathbf{j}, \mathbf{v} = 4\mathbf{i} - \mathbf{j}$
   d) $\mathbf{u} = 2\mathbf{j}, \mathbf{v} = -\mathbf{i} + \sqrt{3}\mathbf{j}$

2. Using the vectors $\mathbf{u} = 3\mathbf{i} - 2\mathbf{j}, \mathbf{v} = \mathbf{i} + 3\mathbf{j}$ and $\mathbf{w} = 4\mathbf{i} + 5\mathbf{j}$, find each of the indicated results.
   a) $\mathbf{u} \cdot (\mathbf{v} + \mathbf{w})$
   b) $\mathbf{u} \cdot \mathbf{v} + \mathbf{u} \cdot \mathbf{w}$
   c) $\mathbf{u}(\mathbf{v} \cdot \mathbf{w})$
   d) $(\mathbf{u} \cdot \mathbf{v})\mathbf{w}$
   e) $(\mathbf{u} \cdot \mathbf{v})(\mathbf{u} \cdot \mathbf{w})$
   f) $(\mathbf{u} + \mathbf{v}) \cdot (\mathbf{u} - \mathbf{v})$
   g) Looking at a)–d) write one paragraph to summarize what you learned!

3. Find the work done by the force $\mathbf{F}$ in moving an object between points M and N.
   a) $\mathbf{F} = 400\mathbf{i} - 50\mathbf{j}, M(2, 3), N(12, 43)$
   b) $\mathbf{F} = 30\mathbf{i} + 150\mathbf{j}, M(0, 30), N(15, 70)$

4. Find the interior angles of the triangle ABC.
   a) $A(1, 2), B(3, 4), C(2, 5)$
   b) $A(3, 4), B(-1, -7), C(-8, -2)$
   c) $A(3, -5), B(1, -9), C(-7, -9)$

5. Find a vector perpendicular to $\mathbf{u}$ in each case below. (Answers are not unique!)
   a) $\mathbf{u} = (3, 5)$
   b) $\mathbf{u} = \frac{1}{2}\mathbf{i} - \frac{3}{4}\mathbf{j}$

6. Use the dot product to find the equation of a circle whose diameter is $[AB]$.
   a) $A(1, 2), B(3, 4)$.
   b) $A(3, 4), B(-1, -7)$.

7. Decide whether the triangle ABC is right angled using vector algebra: $A(1, -3), B(2, 0), C(6, -2)$

8. Find t such that $\mathbf{a} = t\mathbf{i} - 3\mathbf{j}$ is perpendicular to $\mathbf{b} = 5\mathbf{i} + 7\mathbf{j}$.

9. For what value(s) are the vectors $(-6, b)$ and (b, b^2) perpendicular?

10. Find a unit vector that makes an angle of $60°$ with $\mathbf{u} = (3, 4)$.

11. Find t such that $\mathbf{a} = t\mathbf{i} - \mathbf{j}$ and $\mathbf{b} = \mathbf{i} + \mathbf{j}$ make an angle of $\frac{3}{4}\pi$ radians.

12. Use the dot product to prove that the diagonals of a rhombus are perpendicular to each other.

• **Hint:** The work done by any force is defined as the product of the force multiplied by the distance it moves a certain object. In other words, it is the product of the force multiplied by the displacement of the object. As such, work is the dot product between the force and displacement $\mathbf{W} = \mathbf{F} \cdot \mathbf{D}$

Practice questions

1. *ABCD* is a rectangle with *M* the midpoint of [*AB*]. **u** and **v** represent the vectors joining *M* to *D* and *C* respectively. Express each of the following vectors in terms of **u** and **v**.
   a) $\overrightarrow{DC}$
   b) $\overrightarrow{AM}$
   c) $\overrightarrow{BC}$
   d) $\overrightarrow{AC}$

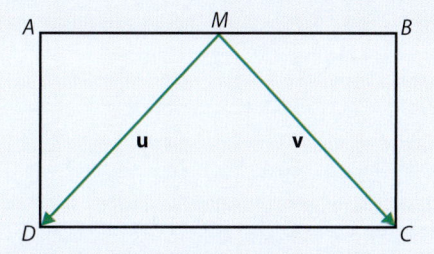

2. Consider the vectors **u** = **i** − 2**j** and **v** = 4**i** + 3**j**.
   a) Find the component form of the vector **w** = 2**u** + **v**.
   b) Find the vector **z** which has a magnitude of 6 units and same direction as **w**.

3. *M* and *A* are the ends of the diameter of a circle with centre at the origin. The radius of the circle is 15 cm and $\overrightarrow{OR} = \begin{pmatrix} 10 \\ 5\sqrt{5} \end{pmatrix}$.
   a) Verify that *C* lies on the circle.
   b) Find the vector $\overrightarrow{AR}$.
   c) Find the cosine of ∠*OAR*.
   d) Find the area of △*MAR*.

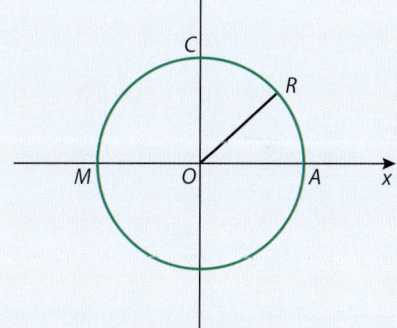

4. Quadrilateral *MARC* has vertices with coordinates *M*(0, 0), *A*(6, 2), *R*(11, 4) and *C*(3, 8).
   a) Find the vectors $\overrightarrow{MR}$ and $\overrightarrow{AC}$.
   b) Find the angle between the diagonals of quadrilateral *MARC*.
   c) Let the vector **u** be the vector joining the midpoints of [*MA*] and [*AR*], and **v** be the vector joining the midpoints of [*RC*] and [*CM*]. Compare **u** and **v** to $\overrightarrow{MR}$, and hence show that the quadrilateral connecting the midpoints of the sides of *MARC* form a parallelogram.

5. Vectors **u** = 5**i** + 3**j** and **v** = **i** − 4**j** are given. Find the scalars *m* and *n* such that *m*(**u** + **v**) − 5**i** + 7**j** = *n*(**u** − **v**).

6. Vector $\begin{pmatrix} 1 \\ 0 \end{pmatrix}$ represents a displacement in the eastern direction while vector $\begin{pmatrix} 0 \\ 1 \end{pmatrix}$ represents a displacement north. Distances are in kilometres.
   Two crews of workers are laying gas pipes in a north-south direction across the North Sea. Consider the base port where the crews leave to start work as the origin (0, 0).
   At 07:00 the crews left the base port with their motor boats to two different locations. The crew called 'Marco' travel at a velocity of $\begin{pmatrix} 9 \\ 12 \end{pmatrix}$ and the crew called 'Tony' travel at a velocity of $\begin{pmatrix} 18 \\ -8 \end{pmatrix}$. Speeds are in km/h.
   a) Find the speed of each boat.
   b) Find the position vectors of each crew at 07:30.

c) Hence, or otherwise, find the distance between the vehicles at 07:30.
d) At 07:30 'Tony' stops and the crew begins laying pipes towards the north. 'Marco' continues travelling in the same direction at the same speed until it is exactly north of 'Tony'. At this point, 'Marco' stops and the crew then begins laying pipes towards the south. At what time does 'Marco' start work?
e) Each crew lays an average of 400 m of pipe in an hour. If they work non-stop until their lunch break at 12:30, what is the distance between them at this time?
f) How long would 'Marco' take to return to base port from its lunchtime position, assuming it travelled in a straight line and with the same average speed as on the morning journey? (Give your answer to the nearest minute.)

7 Triangle *TRI* is defined as follows:
$\overrightarrow{OT} = \begin{pmatrix} 3 \\ -1 \end{pmatrix}$, $\overrightarrow{TR} = \begin{pmatrix} 5 \\ 6 \end{pmatrix}$, $\overrightarrow{TR} \cdot \overrightarrow{IR} = 0$, and $\overrightarrow{TI} = k\mathbf{j}$ where k is a scalar and $\mathbf{j}$ is the unit vector in the *y*-direction.
a) Draw an accurate diagram of $\triangle TRI$.
b) Write the vector $\overrightarrow{IR}$.

8 The position vector of a plane for AUA airlines from its starting position in Vienna is given by $\begin{pmatrix} x \\ y \end{pmatrix} = \begin{pmatrix} 25 \\ 40 \end{pmatrix} + t\begin{pmatrix} 360 \\ 480 \end{pmatrix}$. Distances are in kilometres and speeds are in km/h. *t* is time after 00 hour.
a) Find the position of the AUA plane after 2 hours.
b) What is the speed of the plane?
c) A plane for LH airline started at the same time from a location $\begin{pmatrix} -155 \\ 1300 \end{pmatrix}$ relative to Vienna and moving with a velocity vector $\begin{pmatrix} 480 \\ -360 \end{pmatrix}$, flying at the same height as the AUA plane. Show that if the LH plane does not change route, the two planes will collide. Find the time of the potential collision.
d) To avoid collision, the LH plane is ordered to leave its position and start moving at a velocity of $\begin{pmatrix} 450 \\ -390 \end{pmatrix}$ one hour after it started. Find the position vector of the LH plane at that time.
e) How far apart are the two planes after two hours?

9 For what value(s) of *n* are the vectors $\begin{pmatrix} 3n \\ 2n+3 \end{pmatrix}$ and $\begin{pmatrix} 2n-1 \\ 4-2n \end{pmatrix}$ perpendicular. Otherwise, show that it is not possible.

9 Statistics

Assessment statements

6.1 Concepts of population, sample, random sample and frequency distribution of discrete and continuous data.

6.2 Presentation of data: frequency tables and diagrams, box-and-whisker plots.
Grouped data: mid-interval values, interval width, upper and lower interval boundaries, frequency histograms.

6.3 Mean, median, mode; quartiles, percentiles.
Range; interquartile range; variance; standard deviation.

6.4 Cumulative frequency; cumulative frequency graphs; use to find median, quartiles, percentiles.

Introduction

You will almost inevitably encounter statistics in one form or another on a daily basis. Here is an example:

The World Health Organization (WHO) collects and reports data pertaining to worldwide population health on all 192 UN member countries. Among the indicators reported is the **health-adjusted life expectancy** (HALE), which is based on life expectancy at birth, but includes an adjustment for time spent in poor health. It is most easily understood as the equivalent number of years in full health that a newborn can expect to live, based on current rates of ill-health and mortality. According to WHO rankings, lost years due to disability are substantially higher in poorer countries. Several factors contribute to this trend including injury, blindness, paralysis, and the debilitating effects of tropical disease.

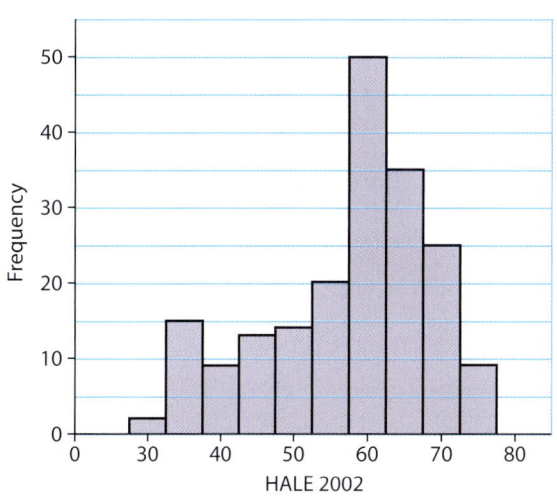

More information on HALE can be found by visiting www.heinemann.co.uk/hotlinks, entering the express code 4235P, then clicking on weblink 2.

9 Statistics

Of the 192 countries ranked by WHO, Japan has the highest life expectancy (75 years) and the lowest ranking country is Sierra Leone (29 years).

Reports similar to this one are commonplace in publications of several organizations, newspapers and magazines, and on the internet.

Questions that come to mind as we read such a report include: How did the researchers collect the data? How can we be sure that these results are reliable? What conclusions should be drawn from this report? The increased frequency with which statistical techniques are used in all fields, from business to agriculture to social and natural sciences, leads to the need for statistical literacy – familiarity with the goals and methods of these techniques – to be a part of any well-rounded educational programme.

Since statistical methods for summary and analysis provide us with powerful tools for making sense out of the data we collect, in this chapter we will first start by introducing two basic components of most statistical problems – population and sample – and then delve into the methods of presenting and making sense of data.

In the language of statistics, one of the most basic concepts is sampling. In most statistical problems, we draw a specified number of measurements or data – a sample – from a much larger body of measurements, called the population. On the basis of our observation of the data in the well-chosen sample, we try to describe or predict the behaviour of the population.

A **population** is any entire collection of people, animals, plants or things from which we may collect data. It is the entire group we are interested in, which we wish to describe or draw conclusions about. In order to make any generalizations about a population, a **sample**, that is meant to be representative of the population, is often studied. For each population there are many possible samples.

For example, a report on the effect the economic status (ES) has on healthy children's postures stated that:

> '…ES, independent of overt malnutrition, affects height, weight, … with some gender differences in healthy children. Influence of income on height and weight show sexual dimorphism, a slight but significant effect is observed only in boys. MPH (mid-parental height) is the most prominent variable effecting height in healthy children. Higher height … observed in higher income groups suggest that secular trend in growth still exists, at least in boys, in a country of favorable economic development.'
>
> **Source:** *European Journal of Clinical Nutrition* (2007) **61**, 752–758

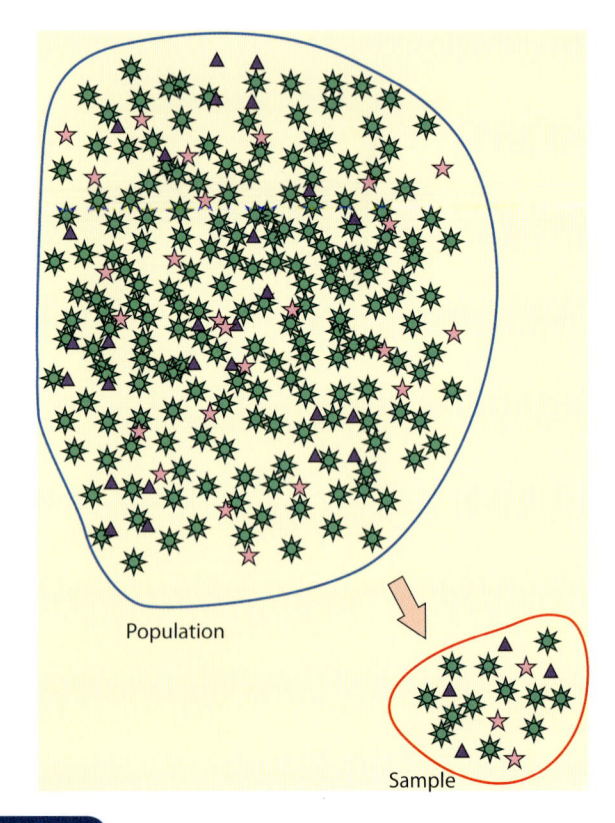

The population is the 3-tuple measurement (economic status, height, weight) of all children of age 3–18 in Turkey. The sample is the set of measurements of the 428 boys and 386 girls that took part in the study. Notice that the population and sample are the measurements and not the people! The boys and girls are 'experimental units' or subjects in this study.

In this chapter we will present some basic techniques in **descriptive statistics** – the branch of statistics concerned with describing sets of measurements, both samples and populations.

9.1 Graphical tools

Once you have collected a set of measurements, how can you display this set in a clear, understandable and readable form? First, you must be able to define what is meant by measurement or 'data' and to categorize the types of data you are likely to encounter. We begin by introducing some definitions of the new terms in the statistical language that you need to know.

> A **variable** is a characteristic that changes or varies over time and/or for different objects under consideration.

For example, if you are measuring the height of adults in a certain area, the height is a variable that changes with time for an individual and from person to person. When a variable is actually measured, a set of measurements or **data** will result. So, if you gather the heights of the students at your school, the set of measurements you get is a **data set**.

As the process of data collection begins, it becomes clear that often the number of data collected is so large that it is difficult for the statistician to see the findings of the data. The statistician's objective is to summarize succinctly, bringing out the important characteristics of the numbers and values in such a way that a clear and accurate picture emerges.

There are several ways of summarizing and describing data. Among them are tables and graphs and numerical measures.

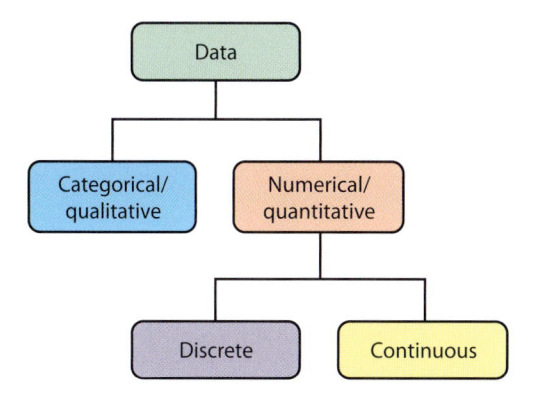

```
                    Data
                   /    \
        Categorical/    Numerical/
        qualitative    quantitative
                        /      \
                    Discrete  Continuous
```

9 Statistics

Classification of variables
Numerical or categorical

When classifying data, there are two major classifications: numerical or categorical data.

NUMERICAL (QUANTITATIVE) DATA – Quantitative variables measure a numerical quantity or amount on each experimental unit. Quantitative data yields a numerical response.

Examples: Yearly income of company presidents, the heights of students at school, the length of time it takes students to finish their lunch at school, and the total score you receive on exams are all numerical.

Moreover, there are two types of numerical data:

> DISCRETE – responses which arise from counting.
>> Example: Number of courses students take in a day.
>
> CONTINUOUS – responses which arise from measuring.
>> Example: Time it takes a student to travel from home to school.

CATEGORICAL (QUALITATIVE) DATA – Qualitative variables measure a quality or characteristic of the experimental unit. Categorical data yields a qualitative response, i.e. data is kind or type rather than quantity.

Examples: Categorizing students into first year IB or second year IB; into Maths Studies SL, Maths SL, Further Maths SL, or Maths HL; or political affiliation will result in qualitative variables and data.

When data is first collected, there are some simple ways of beginning to organize the data. These include an ordered array and the stem-and-leaf display – not required.
- Data in raw form (as collected):
  24, 26, 24, 21, 27, 27, 30, 41, 32, 38
- Data in ordered array from smallest to largest (an ordered array is an arrangement of data in either ascending or descending order):
  21, 24, 24, 26, 27, 27, 30, 32, 38, 41

Suppose a consumer organization was interested in studying weekly food and living expenses of college students. A survey of 80 students yielded the following expenses to the nearest euro:

Table 9.1

38	50	55	60	46	51	58	64	50	49	48	65	58	61	65	53
39	51	56	61	48	53	59	65	54	54	54	59	65	66	47	49
40	51	56	62	47	55	60	63	60	59	59	50	46	45	54	47
41	52	57	64	50	53	58	67	67	66	65	58	54	52	55	52
44	52	57	64	51	55	61	68	67	54	55	48	57	57	66	66

The first step in the analysis is a summary of the data, which should show the following information:
- What values of the variable have been measured?
- How often has each value occurred?

Such summaries can be done in many ways. The most useful are the frequency distribution and the histogram. There are other methods of presenting data, some of which we will discuss later. The rest are not within the scope of this book.

Frequency distribution (table)

A **frequency distribution** is a table used to organize data. The left column (called classes or groups) includes numerical intervals on a variable being studied. The right column is a list of the frequencies, or number of observations, for each class. Intervals normally are of equal size, must cover the range of the sample observations, and are non-overlapping (Table 2).

There are some general rules for preparing frequency distributions that make it easier to summarize data and to communicate results.

Construction of a frequency distribution (table)

Rule 1: Intervals (classes) must be inclusive and non-overlapping; each observation must belong to one and only one class interval. Consider a frequency distribution for the living expenses of the 80 college students. If the frequency distribution contains the intervals '35–40' and '40–45', to which of these two classes would a person spending €40 belong?

The boundaries, or endpoints, of each class must be clearly defined. For our example, appropriate intervals would be '35 but less than 40' and '40 but less than 45'.

Rule 2: Determine k, the number of classes. Practice and experience are the best guidelines for deciding on the number of classes. In general, the number of classes could be between 5 and 10. But this is not an absolute rule. Practitioners use their judgement in these issues. If the number of classes is too few, some characteristics of the distribution will be hidden, and if too many, some characteristics will be lost with the detail.

Rule 3: Intervals should be the same width, w. The width is determined by the following:

$$\text{interval width} = \frac{\text{largest number} - \text{smallest number}}{\text{number of intervals}}$$

Both the number of intervals and the interval width should be rounded upward, possibly to the next largest integer. The above formula can be used when there are no natural ways of grouping the data. If this formula is used, the interval width is generally rounded to a convenient whole number to provide for easy interpretation.

In the example of the weekly living expenses of students, a reasonable grouping with nice round numbers was that of '35 but less than 40' and '40 but less than 45', etc.

Table 9.2 Frequency and percentage frequency distributions of the weekly expenses of 80 students.

Living expenses (€)	Number of students	Percentage of students
35 but < 40	2	2.50
40 but < 45	3	3.75
45 but < 50	11	13.75
50 but < 55	21	26.25
55 but < 60	19	23.75
60 but < 65	11	13.75
65 but < 70	13	16.25
Total	**80**	**100.00**

Grouping the data in a table like this one enables us to see some of its characteristics. For example, we can observe that there are few students who spend as little as 35 to 45 euros, while the majority of the students spend more than €45. Grouping the data will also cause some loss of detail, as we do not see from the table what the real values in each class are.

In the table above, the impression we get is that the class midpoint, also known as the mid-interval value, will represent the data in that interval. For example, 37.5 will represent the data in the first class, while 62.5 will represent the data in the 60 to 65 class. 35 and 40 are known as the **interval boundaries**.

Graphically, we have a tool that helps visualize the distribution. This tool is the **histogram**.

Histogram

A histogram is a graph that consists of vertical bars constructed on a horizontal line that is marked off with intervals for the variable being displayed. The intervals correspond to those in a frequency distribution table. The height of each bar is proportional to the number of observations in that interval. The number of observations can also be displayed above the bars.

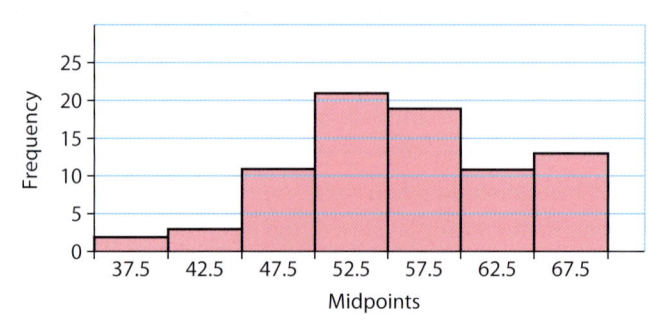

By looking at the histogram, it becomes visually clear that our observations above are true. From the histogram we can also see that the distribution is not symmetric.

To get a histogram on your GDC:
- Enter your data into a list
- Go to **StatPlot** and change it as shown below
- Graph

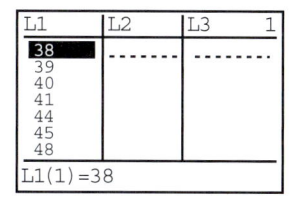

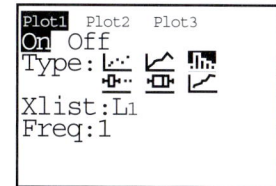

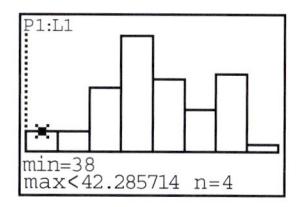

Cumulative and relative cumulative frequency distributions

A **cumulative frequency distribution** contains the total number of observations whose values are less than the upper limit for each interval. It is constructed by adding the frequencies of all frequency distribution intervals up to and including the present interval. A **relative cumulative frequency distribution** converts all cumulative frequencies to cumulative percentages.

In our example above, the following is a cumulative distribution and a relative (percentage) cumulative distribution.

Living expenses (€)	Number of students	Cumulative number of students	Percentage of students	Cumulative percentage of students
35 but < 40	2	2	2.50	2.50
40 but < 45	3	5	3.75	6.25
45 but < 50	11	16	13.75	20.00
50 but < 55	21	37	26.25	46.25
55 but < 60	19	56	23.75	70.00
60 but < 65	11	67	13.75	83.75
65 but < 70	13	80	16.25	100.00
	80		100.00	

◀ **Table 9.3** Cumulative frequency and cumulative relative frequency distributions of the weekly expenses of 80 students.

Notice how every cumulative frequency is added to the frequency in the next interval to give you the next cumulative frequency. The same is true for the relative frequencies.

As we will see later, cumulative frequencies and their graphs help in analyzing data that are given in group form.

Cumulative line graph/cumulative frequency graph

Sometimes called an **ogive**, this is a line that connects points that are the cumulative percentage of observations below the upper limit of each class in a cumulative frequency distribution.

9 Statistics

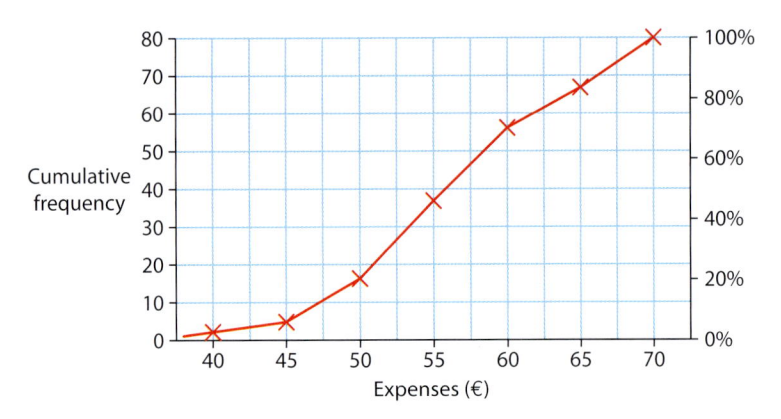

Notice how the height of each line at the upper boundary represents the cumulative frequency for that interval. For example, at 50 the height is 16 and at 60 it is 56.

Example 1

Here is the WHO data in raw form.

29	36	40	44	48	52	54	56	59	60	61	61	62	63	64	66	68	71	72	73	63	64	66	68
31	36	41	44	49	52	54	57	59	60	61	62	62	64	64	66	68	71	72	75	63	64	66	68
33	36	41	44	49	52	55	57	59	60	61	62	62	64	65	66	69	71	72	35	38	43	47	71
34	37	41	45	49	53	55	58	59	60	61	62	63	64	65	66	69	71	73	36	40	44	48	71
34	37	42	45	50	53	55	58	59	60	61	62	63	64	65	67	70	71	73	50	54	56	59	72
35	37	42	45	50	53	55	58	59	60	61	62	63	64	65	67	70	71	73	51	54	56	59	72
35	37	43	46	50	54	55	58	59	60	61	62	63	64	65	67	70	71	73	60	60	61	62	73
35	38	43	46	50	54	55	58	59	60	61	62	63	64	65	67	70	72	73	60	61	61	62	73

Prepare a frequency table starting with 25 and with a class interval of 5. Then draw a histogram of the data and a cumulative frequency graph.

Solution

We first sort the data and then make sure we count every number in one class only.

Life expectancy (years)[1]	Number of countries	Life expectancy (years)	Number of countries
25–30	1	55–60	26
30–35	4	60–65	54
35–40	14	65–70	22
40–45	14	70–75	27
45–50	11	75–80	1
50–55	18		

[1] 25–30 contains all observations larger than or equal to 25 but less than 30.

The histogram created by Excel is shown on the next page. Since we have classes of equal width, the height and the area give the same impression about the frequency of the class interval. For example, the class of 60–65 contains almost twice as much as the class of 55–60, and the height of the histogram is also twice as high. So is the area. Similarly, the height of the 65–70 class is double that of the 45–50 class.

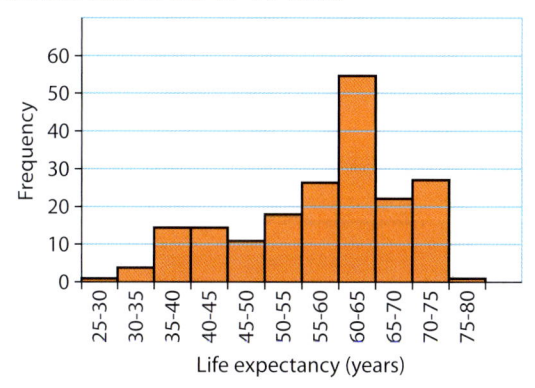

Life expectancy (years)	Number of countries	Cumulative number of countries	Life expectancy (years)	Number of countries	Cumulative number of countries
25–30	1	1	55–60	26	88
30–35	4	5	60–65	54	142
35–40	14	19	65–70	22	164
40–45	14	33	70–75	27	191
45–50	11	44	75–80	1	192
50–55	18	62			

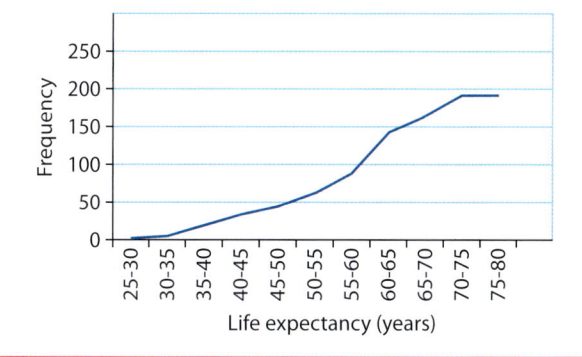

Exercise 9.1

1 Identify the experimental units, sensible population and sample on which each of the following variables is measured. Then indicate whether the variable is quantitative or qualitative.
   a) Gender of a student
   b) Number of errors on a final exam for 10th-grade students
   c) Height of a newborn child
   d) Eye colour for children aged less than 14
   e) Amount of time it takes to travel to work
   f) Rating of a country's leader: excellent, good, fair, poor
   g) Country of origin of students at international schools

9 Statistics

2 State what you expect the shapes of the distributions of the following variables to be: uniform, unimodal, bimodal, symmetric, etc. Explain why.
   a) Number of goals shot by football players during last season.
   b) Weights of newborn babies in a major hospital during the course of 10 years.
   c) Number of countries visited by a student at an international school.
   d) Number of emails received by a high school student at your school per week.

3 Identify each variable as quantitative or qualitative:
   a) Amount of time to finish your extended essay.
   b) Number of students in each section of IB Maths SL.
   c) Rating of your textbook as excellent, good, satisfactory, terrible.
   d) Country of origin of each student on Maths SL courses.

4 Identify each variable as discrete or continuous:
   a) Population of each country represented by SL students in your session of the exam.
   b) Weight of IB Maths SL exams printed every May since 1976.
   c) Time it takes to mark an exam paper by an examiner.
   d) Number of customers served at a bank counter.
   e) Time it takes to finish a transaction at a bank counter.
   f) Amount of sugar used in preparing your favourite cake.

5 Grade point averages (GPA) in several colleges are on a scale of 0–4. Here are the GPAs of 45 students at a certain college.

1.8	1.9	1.9	2.0	2.1	2.1	2.1	2.2	2.2	2.3	2.3	2.4	2.4	2.4	2.5
2.5	2.5	2.5	2.5	2.5	2.6	2.6	2.6	2.6	2.6	2.7	2.7	2.7	2.7	2.7
2.8	2.8	2.8	2.9	2.9	2.9	3.0	3.0	3.0	3.1	3.1	3.1	3.2	3.2	3.4

Prepare a frequency histogram, a relative frequency histogram and a cumulative frequency graph. Describe the data in two to three sentences.

6 The following are the grades of an IB course with 40 students (two sections) on a 100-point test. Use the graphical methods you have learned so far to describe the grades.

61	62	93	94	91	92	86	87	55	56
63	64	86	87	82	83	76	77	57	58
94	95	89	90	67	68	62	63	72	73
87	88	68	69	65	66	75	76	84	85

7 The length of time (months) between repeated speeding violations of 50 young drivers are given in the table below:

2.1	1.3	9.9	0.3	32.3	8.3	2.7	0.2	4.4	7.4
9	18	1.6	2.4	3.9	2.4	6.6	1	2	14.1
14.7	5.8	8.2	8.2	7.4	1.4	16.7	24	9.6	8.7
19.2	26.7	1.2	18	3.3	11.4	4.3	3.5	6.9	1.6
4.1	0.4	13.5	5.6	6.1	23.1	0.2	12.6	18.4	3.7

   a) Construct a histogram for the data.
   b) Would you describe the shape as symmetric?
   c) The law in this country requires that the driving licence be taken away if the driver repeats the violation within a period of 10 months. Use a cumulative frequency graph to estimate the fraction of drivers who may lose their licence.

8 To decide on the number of counters needed to be open during rush hours in a supermarket, the management collected data from 60 customers for the time they spent waiting to be served. The times, in minutes, are given in the following table.

3.6	0.7	5.2	0.6	1.3	0.3	1.8	2.2	1.1	0.4
1	1.2	0.7	1.3	0.7	1.6	2.5	0.3	1.7	0.8
0.3	1.2	0.2	0.9	1.9	1.2	0.8	2.1	2.3	1.1
0.8	1.7	1.8	0.4	0.6	0.2	0.9	1.8	2.8	1.8
0.4	0.5	1.1	1.1	0.8	4.5	1.6	0.5	1.3	1.9
0.6	0.6	3.1	3.1	1.1	1.1	1.1	1.4	1	1.4

a) Construct a relative frequency histogram for the times.
b) Construct a cumulative frequency graph and estimate the number of customers who have to wait 2 minutes or more.

9 The histogram below shows the number of days spent by heart patients in Austrian hospitals in the 2003–2005 period.

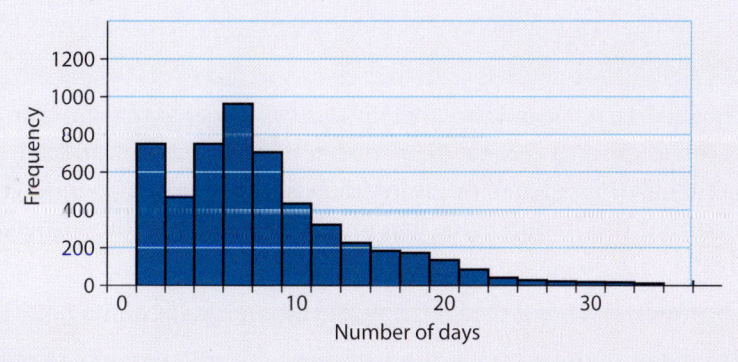

a) Describe the data in a few sentences.
b) Draw a cumulative frequency graph for the data.
c) What percentage of the patients stayed less than 6 days?

10 One of the authors exercises on almost a daily basis. He records the length of time he exercises on most of the days. Here is what he recorded for 2006.

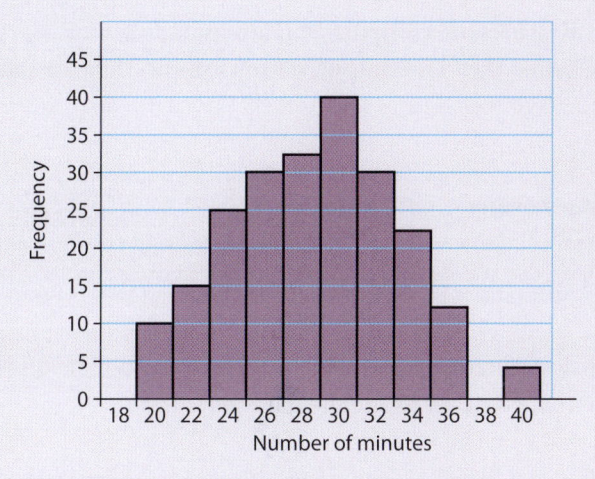

a) What is the longest time he has spent doing his exercises?
b) What percentage of the time did he exercise more than 30 minutes?
c) Draw a cumulative frequency graph for his exercise time.

9 Statistics

9.2 Measures of central tendency

Summarizing data can help us understand them, especially when the number of data is large. This section presents several ways to summarize quantitative data by a typical value (a measure of location, such as the **mean**, **median** or **mode**) and a measure of how well the typical value represents the list (a measure of spread, such as the **range**, **interquartile range** or **standard deviation**). When looking at raw data, rather than looking at tables and graphs, it may be of interest to use summary measures to describe the data. The farthest we can reduce a set of data, and still retain any information at all, is to summarize the data with a single value. Measures of location do just that – they try to capture with a single number what is typical of the data. What single number is most representative of an entire list of numbers? We cannot say without defining 'representative' more precisely. We will study three common measures of location: the mean, the median and the mode. The mean, median and mode are all 'most representative', but for different, related notions of representativeness.

- The **median** is the number that divides the (ordered) data in half. At least half the data is equal to or smaller than the median, and at least half the data is equal to or greater than the median. (In a histogram, the median is that middle value that divides the histogram into two equal areas.)
- The **mode** of a set of data is the most common value among the data.
- The **mean** (more precisely, the arithmetic mean) is commonly called the average. It is the sum of the data, divided by the number of data:

$$\text{mean} = \frac{\text{sum of data}}{\text{number of data}} = \frac{\text{total}}{\text{number of data}}$$

• **Hint:** It is rare that several data coincide exactly, unless the variable is discrete, or the measurements are reported with low precision.

When these measures are computed for a population, they are called **parameters**. When they are computed for a sample, they are called **statistics**.

> **Statistic and parameter**
> A statistic is a descriptive measure computed from a sample of data. A parameter is a descriptive measure computed from an entire population of data.

Measures of central tendency provide information about a 'typical' observation in the data, or locate the data set.

> **The mean and the median**
> The most common measure of central tendency is the arithmetic mean, usually referred to simply as the 'mean' or the 'average'.

Example 2

The following are the five closing prices of the NASDAQ Index for the first business week in November 2007. This is a sample of size $n = 5$ for the closing prices from the entire 2007 population: 2794.83, 2810.38, 2795.18, 2825.18, 2748.76.

What is the average closing price?

Solution

$$\text{Average} = \frac{2794.83 + 2810.38 + 2795.18 + 2825.18 + 2748.76}{5} = 2794.87.$$

This is called the sample mean. A second measure of central tendency is the median, which is the value in the middle position when the measurements are ordered from smallest to largest. The median of this data can only be calculated if we first sort them in ascending order:

2748.76 2794.83 **2795.18** 2810.38 2825.18

The **arithmetic mean** or **average** of a set of n measurements (data set) is equal to the sum of the measurements divided by n.

Notation

The sample mean: $\bar{x} = \dfrac{\sum_{i=1}^{n} x_i}{n} = \dfrac{x_1 + x_2 + x_3 + \ldots + x_n}{n}$, where n is the sample size.

This is a **statistic**.

The population mean: $\mu = \dfrac{\sum_{i=1}^{N} x_i}{N} = \dfrac{x_1 + x_2 + x_3 + \ldots + x_N}{N}$, where N is the population size. This is a **parameter**.

It is important to observe that you normally do not know the mean of the population μ and that you usually estimate it with the sample mean $\bar{x}$.

The **median** of a set of n measurements is the value of x that falls in the middle position when the data is sorted in ascending order.

In the previous example, we calculated the sample median by finding the third measurement to be in the middle position. However, in a different situation, where the number of measurements is even, the process is slightly different.

Let us assume that you took six tests last term and your marks were, in ascending order, 52, 63, 74, 78, 80, 89.

52   63 | 74   78 | 80   89

There are two 'middle' observations here. To find the median, choose a value halfway between the two middle observations:

$$m = \frac{74 + 78}{2} = 76$$

9 Statistics

Note: The position of the median can be given by $\frac{n+1}{2}$. If this number ends with a decimal, you need to average the adjacent values.

In the NASDAQ Index case, we have five observations, the position of the median is then at $\frac{5+1}{2} = 3$, which we found. In the grades example, the position of the median score is at $\frac{6+1}{2} = 3.5$, and hence we average the numbers at positions three and four.

Although both the mean and median are good measures for the centre of a distribution, the median is less sensitive to extreme values or *outliers*. For example, the value 52 in the previous example is lower than all your test scores and is the only failing score you have. The median, 76, is not affected by this outlier even if it were much lower than 52. Assume, for example, that your lowest score is 12 rather than 52. The median calculation

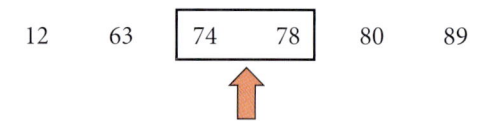

12   63 | 74   78 | 80   89

still gives the same median of 76. If we were to calculate the mean of the original set, we would get

$$\bar{x} = \frac{\sum x}{6} = \frac{436}{6} = 72.\overline{6}$$

While the new mean, with 12 as the lowest score, is

$$\bar{x} = \frac{\sum x}{6} = \frac{396}{6} = 66$$

Clearly, the low outlier 'pulled' the mean towards it while leaving the median untouched. However, because the mean depends on every observation and uses all the information in the data, it is generally, wherever possible, the preferred measure of central tendency.

A third way to locate the centre of a distribution is to look for the value of x that occurs with the highest frequency. This measure of the centre is called the **mode**.

Example 3

Here is a table listing the frequency distribution of 25 families in Lower Austria that were polled in a marketing survey to list the number of litres of milk consumed during a particular week.

Number of litres	Frequency	Relative frequency
0	2	0.08
1	5	0.20
2	9	0.36
3	5	0.20
4	3	0.12
5	1	0.04

Find the frequency histogram.

Solution

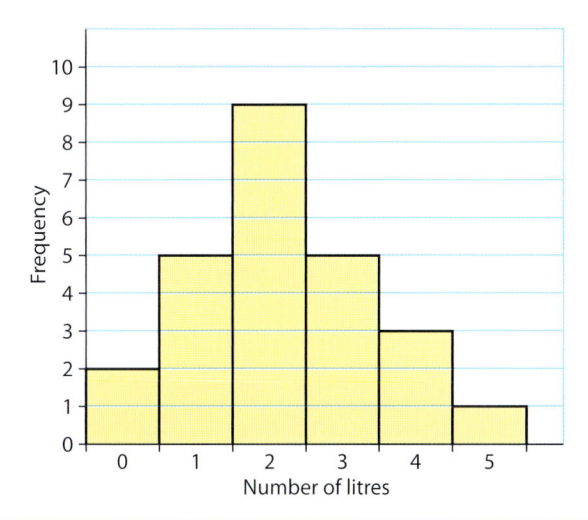

The histogram (Example 3) shows a relatively symmetric shape with a modal class at $x = 2$. Apparently, the mean and median are not far from each other. The median is the 13th observation, which is 2, and the mean is calculated to be 2.2.

> For lists, the **mode** is the most common (frequent) value. A list can have more than one mode. For histograms, the mode is a relative maximum.

Shape of the distribution

An examination of the shape of a distribution will illustrate how the distribution is centred around the mean. Distributions are either symmetric or they are not symmetric, in which case the shape of the distribution is described as asymmetric or skewed.

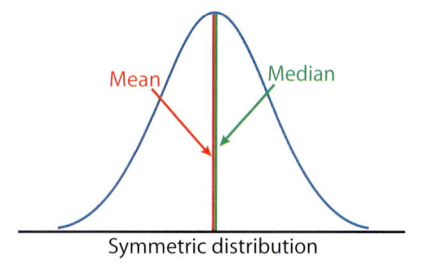

Symmetric distribution

Symmetry

The shape of a distribution is said to be **symmetric** if the observations are balanced, or evenly distributed, about the mean. In a symmetric distribution, *the mean and the median are equal.*

Skewness

A distribution is **skewed** if the observations are not symmetrically distributed above and below the mean.

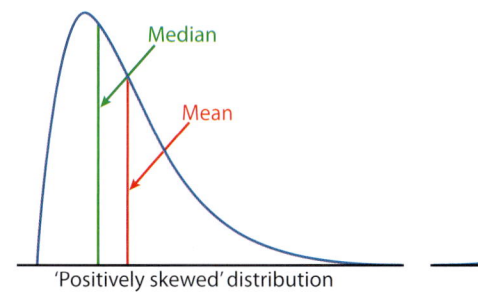

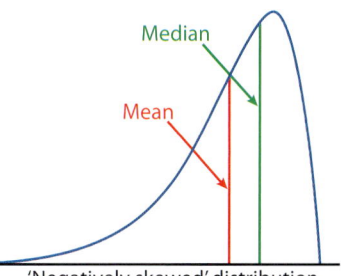

'Positively skewed' distribution   'Negatively skewed' distribution

A **positively skewed** (or skewed to the right) distribution has a tail that extends to the right in the direction of positive values. A **negatively skewed** (or skewed to the left) distribution has a tail that extends to the left in the direction of negative values.

Looking back at the WHO data, we can clearly see that the data is skewed to the left. Few countries have low life expectancies. The bulk of the countries have life expectancies between approximately 50 and 65 years.

The average HALE is $\mu = \frac{\sum x}{n} = \frac{11028}{192} = 57.44$. Looking at the raw data, it does not appear sensible to search for the mode, as there are very few of them (61, 59, 60 or 62). However, after grouping the data into classes, we can see that the modal class is 60–65.

As there are 192 observations, which means that the median is at $\frac{n+1}{2} = \frac{192+1}{2} = 96.5$, we take the average of the 96th and 97th observations, which are Palau and Moldova with 60 each. So, the median is 60!

Knowing the median, we could say that a typical life expectancy is 60 years. How much does this really tell us? How well does this median describe the real situation? After all, not all countries have the same 60 years HALE. Whenever we find the centre of data, the next step is always to ask how well it actually summarizes the data.

When we describe a distribution numerically, we always report a measure of its spread along with its centre.

9.3 Measures of variability

Measures of location summarize what is typical of elements of a list, but not every element is typical. Are all the elements close to each other? Are most of the elements close to each other? What is the biggest difference between elements? On average, how far are the elements from each other? The answer lies in the measures of spread or variability.

It is possible that two data sets have the same mean, but the individual observations in one set could vary more from the mean than do the observations in the second set. It takes more than the mean alone to describe data. Measures of variability (also called measures of dispersion or spread), which include the range, the variance, the standard deviation, interquartile range and the coefficient of variation, will help to summarize the data.

Range

The range in a set of data is the difference between the largest and smallest observations.

Consider the expense data given at the beginning of this chapter. Also consider the same data when the largest value of 68 is replaced by 120. What is the range for these two sets of data?

Table 9.4

	Expense data	Expense data with outlier
Minimum	38	38
Maximum	68	120
Range	30	82

Notice that the range is a *single number*, not an interval of values as you might think from its use in common speech. The maximum of the HALE data is 79 and the minimum is 29, so the range is 50.

Range doesn't take into account how the data is distributed and is, of course, affected by extreme values (outliers) as you see above.

Variance and standard deviation

The most comprehensive measures of dispersion are those in terms of the average deviation from some location parameter.

Variance

The sample variance, s^2, is the sum of the squared differences between each observation and the sample mean divided by the sample size minus 1.

$$s^2 = \frac{\sum_{i=1}^{n}(x_i - \bar{x})^2}{n - 1}$$

> Discussing the reason we define the sample variance in this manner is beyond the scope of this book. The use of $n - 1$ in the denominator has to deal with the use of the sample variance as an estimate of the population variance. Such an estimate has to be unbiased, and this sample variance is the most unbiased estimate of the population variance. However, since the IB syllabus uses a different definition of this variance, we will use the IB's definition in our calculations. You should also be careful with use of your calculator, as the listed s_x in TI GDC's is this one and not the IB's definition. So, when you use your GDC, make sure you use what is called σ_x.
> The IB variance is listed as s_n^2 and is evaluated as follows:
>
> $$s_n^2 = \frac{\sum_{i=1}^{n}(x_i - \bar{x})^2}{n}$$
>
> From this point on, we will use this statistic to denote the sample variance!

The population variance, σ^2, is the sum of the squared differences between each observation and the population mean divided by the population size, N.

$$\sigma^2 = \frac{\sum_{i=1}^{N}(x_i - \mu)^2}{N}$$

The variance is a measure of the variation about the mean squared. In order to bring the measure down to the data measurements, the square root is taken and the measure looked at is the standard deviation.

The standard deviation measures the **standard amount of deviation** or **spread** around the mean.

Standard deviation

The sample standard deviation, s_n, is the (positive) square root of the variance, and is defined as:

$$s_n = \sqrt{s_n^2} = \sqrt{\frac{\sum_{i=1}^{n}(x_i - \bar{x})^2}{n}}$$

The population standard deviation is:

$$\sigma = \sqrt{\sigma^2} = \sqrt{\frac{\sum_{i=1}^{N}(x_i - \mu)^2}{N}}$$

> When is $s = 0$? Answer: When all the data takes on the same value and there is no variability about the mean.
>
> When is s large? Answer: When there is a large amount of variability about the mean.

These are measures of variation about the mean.

Consider the following example:

In business, investors invest their money in stocks whose prices fluctuate with market conditions. Stocks are considered risky if they have high fluctuations. Here are the closing prices of two stocks traded on Vienna's stock market for the first seven business days in September 2007:

Stock A	Stock B
4	1
4.25	3
5	2.5
4.75	5
5.75	7
5.25	6.5
6	10
$\bar{x}_A = 5$	$\bar{x}_B = 5$
Median (A) = 5	Median (B) = 5

Even though the two stocks have similar central values, they do behave differently. It is obvious that stock B is more variable and it becomes more obvious when we calculate the standard deviations.

We will calculate the standard deviation manually in this example to demonstrate the process. You do not have to do this manually all the time!

$$s_A^2 = \frac{\sum_{i=1}^{7}(x_i - 5)^2}{7} = \frac{(4-5)^2 + (4.25-5)^2 + (5-5)^2 + (4.75-5)^2 + (5.57-5)^2 + (5.25-5)^2 + (6-5)^2}{7} = 0.464$$

$$s_B^2 = \frac{\sum_{i=1}^{7}(x_i - 5)^2}{7} = \frac{(1-5)^2 + (3-5)^2 + (2.5-5)^2 + (5-5)^2 + (7-5)^2 + (6.5-5)^2 + (10-5)^2}{7} = 8.21$$

This means that the standard deviations are $s_A = 0.681$ and $s_B = 2.866$. Stock B is four times as variable as stock A.

Note: When computing the sample variance manually, you may find the following shortcut of some use:

$$s_n^2 = \frac{\sum_{i=1}^{n}(x_i - \bar{x})^2}{n} = \frac{\sum_{i=1}^{n}(x_i^2 - 2x_i\bar{x} + \bar{x}^2)}{n} = \frac{\sum_{i=1}^{n}x_i^2 - 2\sum_{i=1}^{n}x_i\bar{x} + \sum_{i=1}^{n}\bar{x}^2}{n}$$

$$= \frac{\sum_{i=1}^{n}x_i^2}{n} - \frac{2\bar{x}\sum_{i=1}^{n}x_i}{n} + \frac{\sum_{i=1}^{n}\bar{x}^2}{n} = \frac{\sum_{i=1}^{n}x_i^2}{n} - 2\bar{x}\sum_{i=1}^{n}\frac{x_i}{n} + \frac{n\bar{x}^2}{n} = \frac{\sum_{i=1}^{n}x_i^2}{n} - \bar{x}^2$$

However, remember that once you have a good understanding of the standard deviation, you will rely on a GDC or software to do most of the calculation for you.

Here is how you can use your TI GDC:

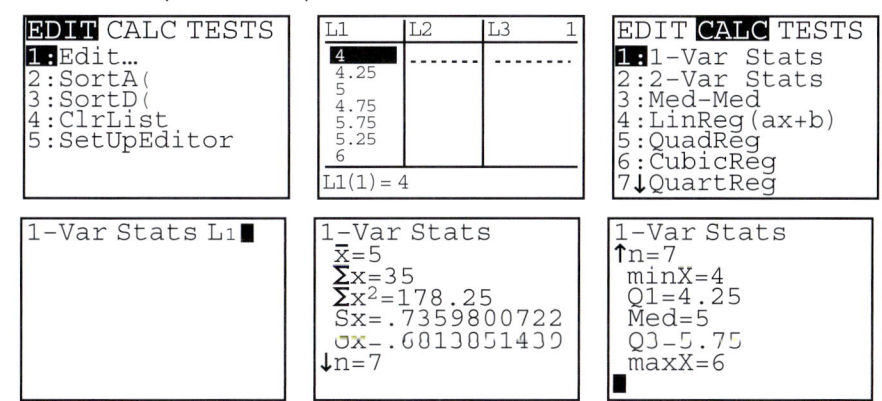

```
EDIT CALC TESTS
1:Edit…
2:SortA(
3:SortD(
4:ClrList
5:SetUpEditor
```

```
L1      L2      L3    1
4
4.25
5
4.75
5.75
5.25
6
L1(1)=4
```

```
EDIT CALC TESTS
1:1-Var Stats
2:2-Var Stats
3:Med-Med
4:LinReg(ax+b)
5:QuadReg
6:CubicReg
7↓QuartReg
```

```
1-Var Stats L1
```

```
1-Var Stats
x̄=5
Σx=35
Σx²=178.25
Sx=.7359800722
σx=.6013051439
↓n=7
```

```
1-Var Stats
↑n=7
minX=4
Q1=4.25
Med=5
Q3=5.75
maxX=6
```

Notice that the standard deviation you read from this output is called σ_x rather than s_x.

The S_x used by your GDC gives $\sqrt{\dfrac{\sum_{i=1}^{n}(x_i - \bar{x})^2}{n-1}}$ instead of $\sqrt{\dfrac{\sum_{i=1}^{n}(x_i - \bar{x})^2}{n}}$, which is officially used on exam papers.

The screenshots also show you that the GDC gives you $\sum x^2$, which can be used if you want to find the variance by hand.

$$s_n^2 = \frac{\sum_{i=1}^{n}x_i^2}{n} - \bar{x}^2 = \frac{178.25}{7} - 5^2 = 0.464 \Rightarrow s_n = 0.681$$

The interquartile range and measures of non-central tendency

To understand another measure of spread known as the **interquartile range**, it is first necessary to define percentiles and quartiles.

Percentiles and quartiles

Data must first be in ascending order.

Percentiles separate large ordered data sets into hundredths. The pth percentile is a number such that p per cent of the observations are at or below that number.

9 Statistics

Quartiles are descriptive measures that separate large ordered data sets into four quarters.

To score in the 90th percentile indicates 90% of the tests scores were less than or equal to your score. An excellent performance! You scored in the upper 10% of all persons taking the test.

- **First quartile, Q_1**

  The first quartile, Q_1, is another name for the 25th percentile. The first quartile divides the ordered data such that 25% of the observations are at or below this value. Q_1 is located in the $0.25(n + 1)$st position when the data is in ascending order. That is,

  $$Q_1 = \frac{n+1}{4} \text{ ordered observation}$$

- **Third quartile, Q_3**

  The third quartile, Q_3, is another name for the 75th percentile. The third quartile divides the ordered data such that 75% of the observations are at or below this value. Q_3 is located in the $0.75(n + 1)$st position when the data is in ascending order. That is,

  $$Q_3 = \frac{3(n+1)}{4} \text{ ordered observation}$$

- **The median**

  The median is the 50th percentile, or the second quartile, Q_2.

A measure which helps to measure variability and is not affected by extreme values is the interquartile range. It avoids the problem of extreme values by just looking at the range of the middle 50% of the data.

Interquartile range

The interquartile range (IQR) measures the spread in the middle 50% of the data; it is the difference between the observations at the 25th and the 75th percentiles:

$$\text{IQR} = Q_3 - Q_1$$

If we consider the student expense data in Table 1 and once again look at that same data with the outlier 120 replacing the largest value 68, we have the following results:

	Expense data	Expense data with outlier
Minimum	38	38
Q_1	50	50
Median	55	55
Q_3	61	61
Maximum	68	120
Range	30	82
IQR	11	11

- **Hint:** The first quartile is also called the lower quartile. The third quartile is also called the upper quartile.

A practical method to calculate the quartiles is to split the data into two halves at the median. (When n is odd, include the median in both halves!) The first quartile is the median of the first half and the third quartile is the median of the second half. For example, with the stocks data, {4, 4.25, 4.75, 5, 5.25, 5.75, 6}, $n = 7$, the median is the fourth observation, 5. The first quartile is then the median of {4, 4.25, 4.75, 5}, which is 4.5, and the third quartile is the median of {5, 5.25, 5.75, 6}, which is 5.5.

Range doesn't take into account how the data is distributed and is, of course, affected by extreme values. We clearly saw that in Table 4. However, the IQR evidently does not have that problem.

Five-number summary

Five-number summary refers to the five descriptive measures: minimum, first quartile, median, third quartile, maximum.
Clearly, $X_{minimum} < Q_1 < \text{Median} < Q_3 < X_{maximum}$.

Box-and-whisker plot

Whenever we have a five-number summary, we can put the information together in one graphical display called a **box plot**, also known as a **box-and-whisker** plot. In the student expenditure data, the IQR is €11. This is evident in the box plot below, where the IQR is the difference between 50 and 61.

Let us make a box plot with the student expense data.
- Draw an axis spanning the range of the data. Mark the numbers corresponding to the median, minimum, maximum, and the lower and upper quartiles.
- Draw a rectangle with lower end at Q1 and upper end at Q3, as shown below.
- To help us consider outliers, mark the points corresponding to lower and upper fences. Mark them with a dotted line since they are not part of the box. The fences are constructed at the following positions:
  - Lower fence: $Q_1 - 1.5 \times IQR$ (Here it is $50 - 1.5(11) = 33.5$.)
  - Upper fence: $Q_3 + 1.5 \times IQR$ (Here it is $61 + 1.5(11) = 77.5$.)

Any point beyond the lower or upper fence is considered an **outlier**.
- Mark any outlier with an asterisk (*) on the graph. (Shown below).
- Extend horizontal lines called 'whiskers' from the ends of the box to the smallest and largest observations that are not outliers. In the first case these are 38 and 68, while in the second they are 38 and 67.

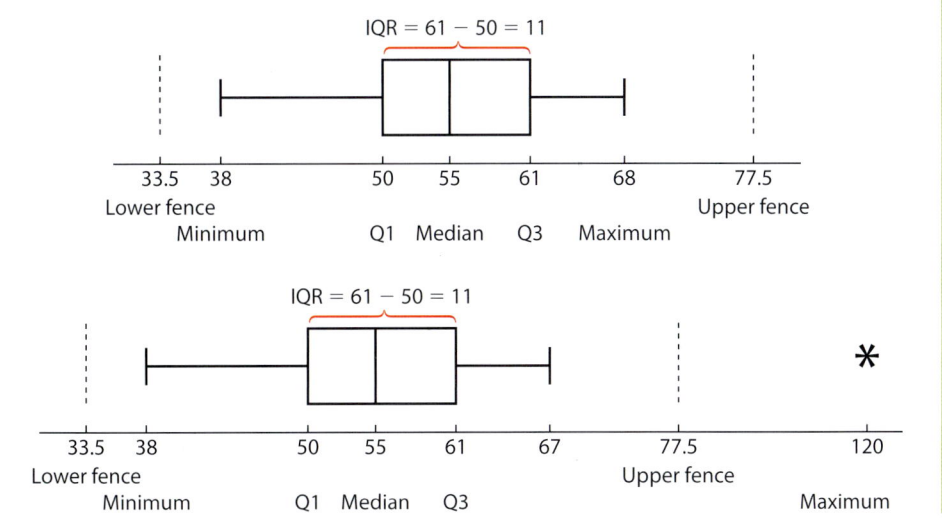

> An outlier is an *unusual* observation. It lies at an abnormal distance from the rest of the data. There is no unique way of describing what an outlier is. A common practice is to consider any observation that is further than 1.5 IQR from the first or third quartile as an outlier. Outliers are important in statistical analysis. Outliers may contain important information not shared with the rest of the data. Statisticians look very carefully at outliers because of their influence on the shapes of distributions and their effect on the values of the other statistics, such as the mean and standard deviation.

9 Statistics

Here is a box plot of the data done by a software package.

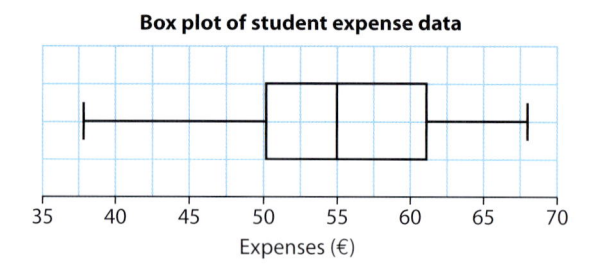

Box plot of student expense data

As you can see, the box contains the middle 50% of the data. The width of the box is nothing but the IQR! Now we know that the middle 50% of the students' expenditure is €11. This seems, at times, as a reasonable summary of the spread of the distribution, as you can see in the histogram below.

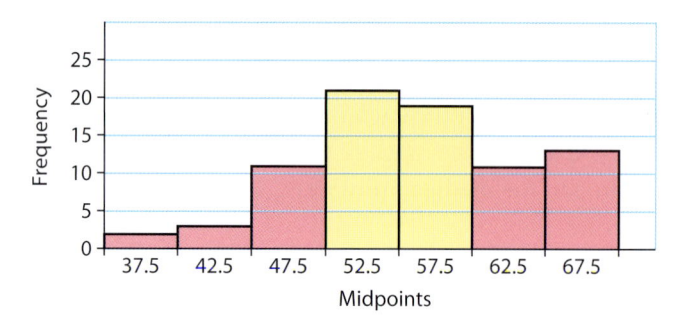

If you locate the IQR on the histogram, you can also get another visual indication of the spread of the data.

How to use your GDC for histograms and box plots:

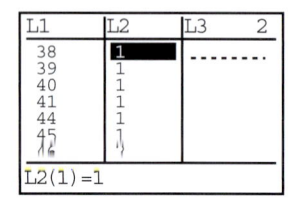

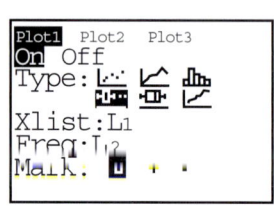

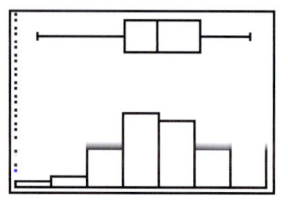

For grouped data:

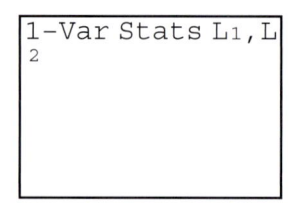

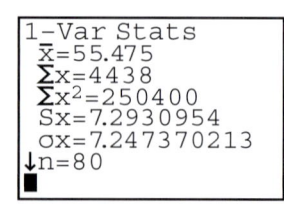

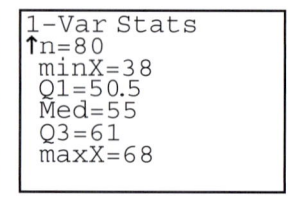

```
1-Var Stats L1,L2
```

```
1-Var Stats
x̄=55.475
Σx=4438
Σx²=250400
Sx=7.2930954
σx=7.247370213
↓n=80
```

```
1-Var Stats
↑n=80
minX=38
Q1=50.5
Med=55
Q3=61
maxX=68
```

An ogive can also be produced:

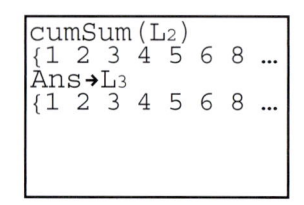

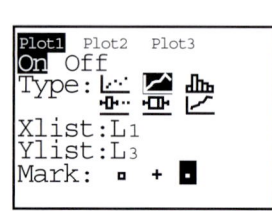

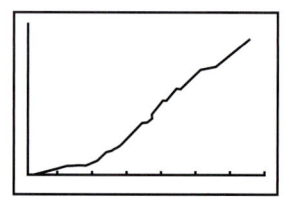

```
cumSum (L2)
{1 2 3 4 5 6 8 …
Ans→L3
{1 2 3 4 5 6 8 …
```

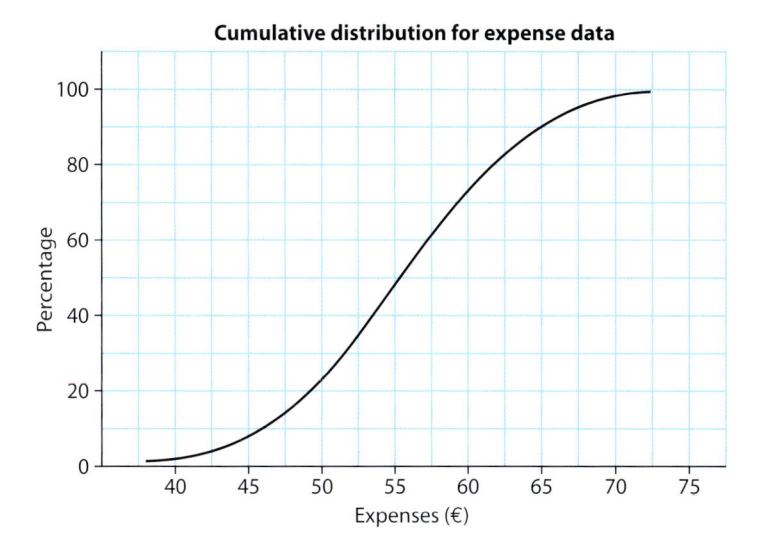

This is a realistic ogive.

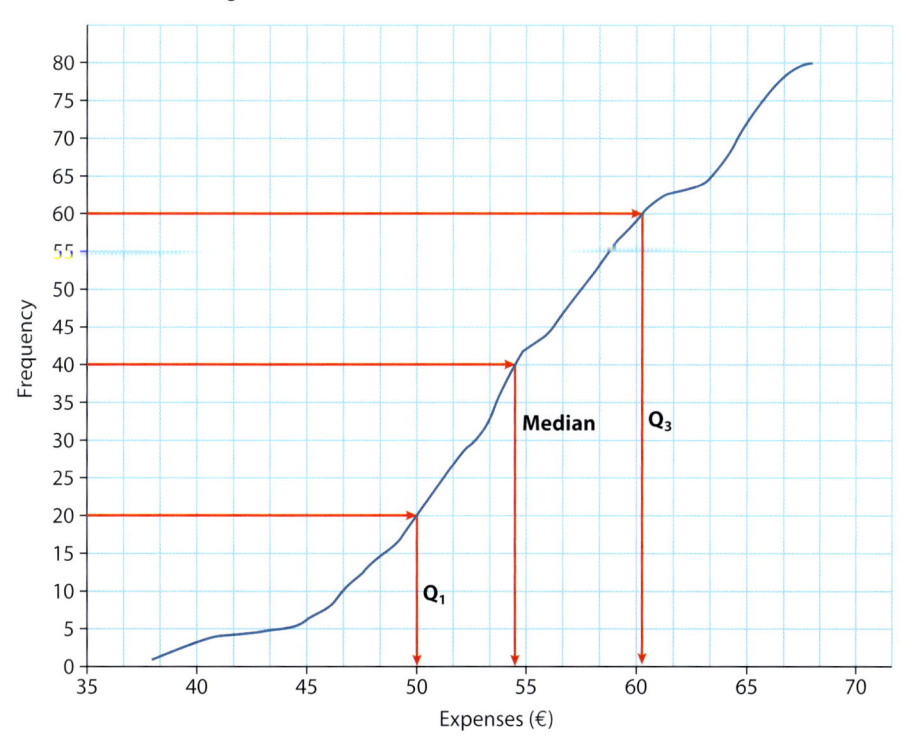

Notice how we locate the first quartile. Since there are 80 observations, the first quartile is approximately at the $\frac{n+1}{4} = \frac{81}{4} \approx$ 20th position, which appears to be around 50.

The median is at the $\frac{n+1}{2} = \frac{81}{2} \approx$ 40th–41st position, i.e. approximately at 55.

Similarly, the third quartile is at $\frac{3(n+1)}{4} = \frac{243}{4} \approx$ 61st, which happens here at approximately 61!

The calculation of the mean and variance for grouped data is essentially the same as for raw data. The difference lies in the use of frequencies to save typing (writing) all numbers. Here is a comparison:

Statistic	Raw data	Grouped data	Grouped data with intervals
$\bar{x}$	$\bar{x} = \dfrac{\sum_{\text{all } x} x}{n}$	$\bar{x} = \dfrac{\sum_{\text{all } x} x_i \cdot f(x_i)}{n} = \dfrac{\sum_{\text{all } x} x_i \cdot f(x_i)}{\sum f(x_i)}$	$\bar{x} = \dfrac{\sum_{\text{all } x} m_i \cdot f(m_i)}{n} = \dfrac{\sum_{\text{all } x} m_i \cdot f(m_i)}{\sum f(m_i)}$
s_n^2	$s_n^2 = \dfrac{\sum_{\text{all } x}(x_i - \bar{x})^2}{n}$	$s_n^2 = \dfrac{\sum_{\text{all } x}(x_i - \bar{x})^2 \cdot f(x_i)}{n}$ $= \dfrac{\sum_{\text{all } x}(x_i - \bar{x})^2 \cdot f(x_i)}{\sum f(x_i)}$	$s_n^2 = \dfrac{\sum_{\text{all } x}(m_i - \bar{x})^2 \cdot f(m_i)}{n}$ $= \dfrac{\sum_{\text{all } x}(m_i - \bar{x})^2 \cdot f(m_i)}{\sum f(m_i)}$

where x_i = data point
$f(x_i)$ = frequency of x_i
m_i = interval midpoint (mid mark or mid value)
$f(m_i)$ = frequency of interval i
$\sum f(x_i), \sum f(m_i)$ = total number of data points

For the grouped data reproduced here, this is how we estimate the mean and variance:

Living expenses	Midpoint m	Number of students $f(m)$	$m_i \times f(m_i)$	$(m_i - \bar{x})^2$	$(m_i - \bar{x})^2 \times f(m_i)$
35 but <40	37.5	2	75	344.5	688.9
40 but <45	42.5	3	127.5	183.9	551.6
45 but <50	47.5	11	522.5	73.3	806.0
50 but <55	52.5	21	1102.5	12.7	266.1
55 but <60	57.5	19	1092.5	2.1	39.4
60 but <65	62.5	11	687.5	41.5	456.2
65 but <70	67.5	13	877.5	130.9	1701.4
Totals		$\sum f(m_i) = 80$	$\sum_{\text{all } x} m_i \cdot f(m_i) = 4485$	$\sum_{\text{all } x} (m_i - \bar{x})^2 \cdot f(m_i) = 4509.6$	
		Mean	$\dfrac{4485}{80} = 56.06$	Variance	$\dfrac{4509.6}{80} = 56.37$
				Standard deviation	7.51

The numbers here are estimates of the mean and the variance and eventually the standard deviation. As you will notice, they are not equal to the values we calculated earlier, but are close. The reason for this is that, with grouping, we lost the detail in each interval. For example, the interval between 45 and 50 is represented by the midpoint 47.5. In essence, we are assuming that every number in the interval is equal to 47.5.

Shape, centre and spread

Statistics is about variation, so spread is an important fundamental concept. Measures of spread help us to precisely analyze what we do not know! If the values we are looking at are scattered very far from the centre, the IQR and the standard deviation will be large. If these are large, our central values will not represent the data well. That is why we always report spread with any central value.

A practical way of seeing the significance of the standard deviation can be demonstrated with the following (optional) observations:

Empirical rule:

If the data is close to being symmetrical, as in the figure right, the following is true:

- The interval $\mu \pm \sigma$ contains approximately 68% of the measurements.
- The interval $\mu \pm 2\sigma$ contains approximately 95% of the measurements.
- The interval $\mu \pm 3\sigma$ contains approximately 99.7% of the measurements.

The empirical rule usually indicates if an observation is very far from the expected or not. Take the following example:

I have recorded my car's fuel efficiency over the last 98 times that I have filled the tank with gasoline. Here is the data expressing how many kilometeres per litre the car travelled:

Symmetric distribution

km/litre	Frequency	km/litre	Frequency
6.0	1	10.0	14
7.0	1	10.5	7
7.5	4	11.0	9
8.0	8	11.5	5
8.5	14	12.0	1
9.0	21	12.5	2
9.5	11		

The summary measures are:

Mean	9.454
σ	1.223
Median	9.25
Q_1	8.5
Q_3	10.125
IQR	1.625

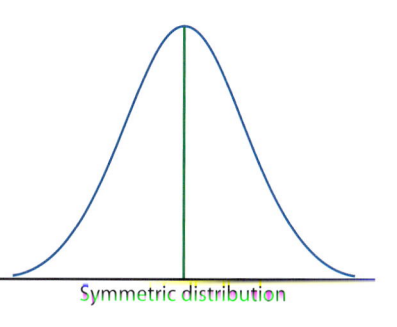

The histogram shows that the distribution is almost symmetric. The possible outlier has little effect on the mean and standard deviation. That is why the mean and median are almost the same. Looking at the box plot, you can see that there is one outlier. The confirmation is below:

$9.25 - 1.5 \times 1.625 = 6.8$, which is why 6 is considered as an outlier.

$10.125 + 1.5 \times 1.625 = 12.6$, and hence no outliers on this side.

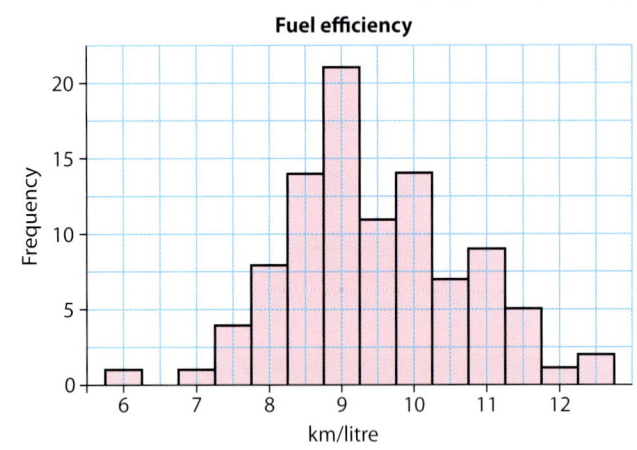

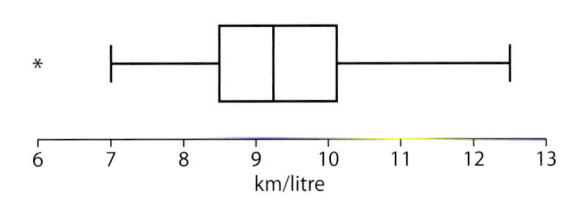

If we use the empirical rule, we can expect about 99.7% of the data to lie within three standard deviations of the mean, i.e. $9.454 - 3 \times 1.223 = 5.8$ and $9.454 + 3 \times 1.223 = 13.1$. In fact, you see all the data is within the specified interval, including the potential outlier!

Question: What should you be able to tell about a quantitative variable?

Answer: Report the shape of its distribution, and include a centre and a spread.

Question: Which central measure and which measure of spread?

Answer: The rules are:
- If the shape is skewed, report the median and IQR. You *may* want to include the mean and standard deviation, but you should point out that the mean and median differ as this difference is a sign that the data is skewed. A histogram can help.
- If the shape is symmetrical, report the mean and standard deviation. You may report the median and IQR as well.
  - If there are clear outliers, report the data with and without the outliers. The differences may be revealing.

Example 4

The records of a large high school show the heights of their students for the year 2006.

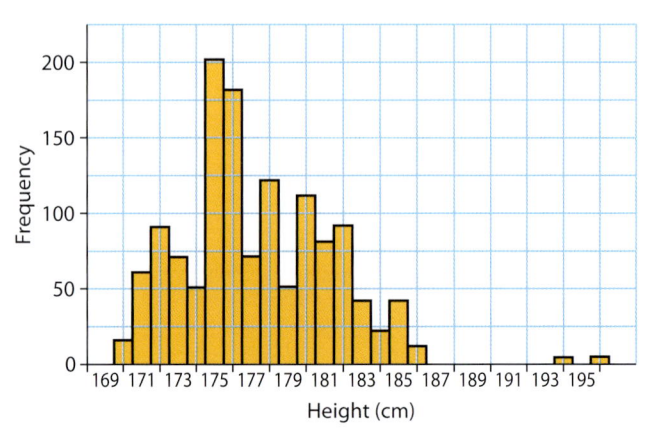

a) Which statistics would best represent the data here? Why?
b) Calculate the mean and standard deviation.
c) Develop a cumulative frequency graph of the data.
d) Use your result of c) above to estimate the median, Q_1, Q_3 and IQR.
e) Are there any outliers in the data? Why?
f) Write a few sentences describing the distribution.

Solution

a) The data appears to have outliers and is slightly skewed to the right. The most appropriate measure is the median, since the mean is influenced by the extreme values.

b) To calculate the mean and standard deviation, we will set up a table that will facilitate the calculation.

Height (cm) x_i	Number of students $f(x)$	$x_i \times f(x_i)$	$(x_i - \bar{x})^2$	$(x_i - \bar{x})^2 \times f(x_i)$
170	15	2550	51.84	777.6
171	60	10 260	38.44	2306.4
172	90	15 480	27.04	2433.6
173	70	12 110	17.64	1234.8
174	50	8700	10.24	512
175	200	35 000	4.84	968
176	180	31 680	1.44	259.2
177	70	12 390	0.04	2.8
178	120	21 360	0.64	76.8
179	50	8950	3.24	162
180	110	19 800	7.84	862.4
181	80	14 480	14.44	1155.2
182	90	16 380	23.04	2073.6
183	40	7320	33.64	1345.6
184	20	3680	46.24	924.8
185	40	7400	60.84	2433.6
186	10	1860	77.44	774.4
194	2	388	282.24	564.5
196	3	588	353.44	1060.3
Totals	$\sum f(x_i)$ = 1300	$\sum_{\text{all } x} x_i \cdot f(x_i)$ = 230 376		$\sum_{\text{all } x} (x_i - \bar{x})^2 \cdot f(x_i)$ = 19 927.6
	Mean	$\dfrac{230\,376}{1300}$ = 177.2	Variance	$\dfrac{19\,927.4}{1300}$ = 15.33
			Standard deviation	3.92

Note: Using the alternative formula for the variance will also give the same result. (Due to rounding, answers will differ slightly.)

$$s_n^2 = \frac{\sum_{i=1}^{n} x_i^2 \times f(x_i)}{n} - \bar{x}^2 = \frac{40\,845\,390}{1300} - 177.2123^2 = 15.3315 \Rightarrow s_n = 3.92$$

c) To develop the cumulative frequency graph, we first need to develop the cumulative frequency table. This is done by accumulating the frequencies as shown below.

x	f(x)	Cum f(x)	x	f(x)	Cum f(x)
170	15	15	184	20	1245
171	60	75	185	40	1285
172	90	165	186	10	1295
173	70	235	187	0	1295
174	50	285	188	0	1295
175	200	485	189	0	1295
176	180	665	190	0	1295
177	70	735	191	0	1295
178	120	855	192	0	1295
179	50	905	193	0	1295
180	110	1015	194	2	1297
181	80	1095	195	0	1297
182	90	1185	196	3	1300
183	40	1225			

The cumulative frequency table is constructed such that the cumulative frequency corresponding to any measurement is the number of observations that are less than or equal to its value. So, for example, the cumulative frequency corresponding to a height of 174 cm is 285, which consists of the 50 observations with height 174 cm and the 235 observations for heights less than 174 cm.

The cumulative frequency graph plots the observations on the horizontal axis against their cumulative frequencies on the vertical axis, as shown below.

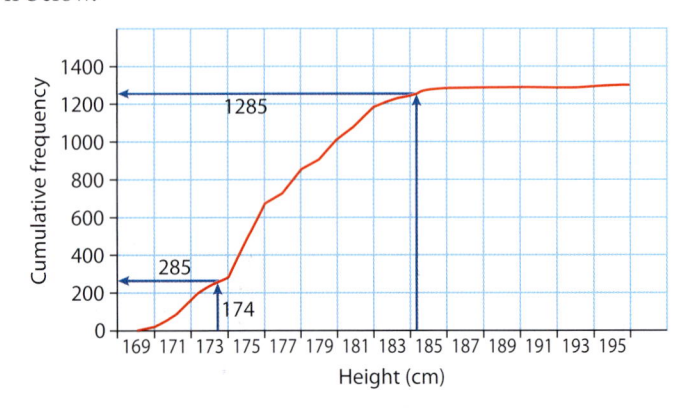

d) The median is the observation between $\frac{1300}{2} = 650$th and 651st observations, since the number is even. From the cumulative table, we can see that the median is in the 176 interval. So the median is 176.

Q_1 is at $\frac{1301}{4} \approx 325$th observation. From the table, as 174 has a cumulative frequency of 285, and 175 has a cumulative frequency of 485, then Q1 has to be **175**.

Also, Q_3 is at $\frac{3 \times 1301}{4} \approx 976$th observation. So, similarly, it is **180**.

IQR $= 180 - 175 = $ **5**.

e) To check for outliers, we can calculate the lengths of the whiskers.

Lower fence: $175 - 1.5 \times 5 = 167.5$, which is lower than the minimum value, so there are no outliers on the left.

Upper fence: $180 + 1.5 \times 5 = 187.5$. So we have five outliers, two at 194 cm and three at 196 cm.

f) The distribution appears to be bimodal with two modes at 175 and 176. It is slightly skewed to the right with a few extreme values at 194 and 196. This is further confirmed by the fact that the mean of 177.2 is higher than the median of 176.

Note: Here are the calculations using a GDC:

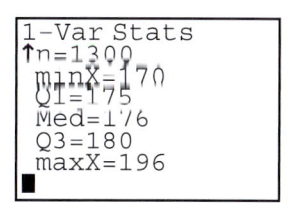

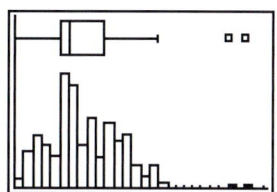

Exercise 9.2 and 9.3

1. You are given eight measurements: 5, 4, 7, 8, 6, 6, 5, 7.
   a) Find $\bar{x}$.
   b) Find the median.
   c) Based on the previous results, is the data symmetric or skewed? Explain and support your conclusion with an appropriate graph.

2. You are given ten measurements: 5, 7, 8, 6, 12, 7, 8, 11, 4, 10.
   a) Find $\bar{x}$.
   b) Find the median.
   c) Find the mode.

3. The following table gives the number of DVD players owned by a sample of 50 typical families in a large city in Germany.

Number of DVD players	0	1	2	3
Number of households	12	24	8	6

   Find the average and the median number of DVD players. Which measure is more appropriate here? Explain.

9 Statistics

4 Ten of the Fortune 500 large businesses that lost money in 2006 are listed below:

Company	Loss ($ million)	Company	Loss ($ million)
Vodafone	39 093	General Motors	10 567
Kodak	1362	Japan Airlines	417
UAL	21 167	Japan Post	3
Mitsubishi Motors	814	AMR	861
Visteon	270	Karstadt Quelle	393

Calculate the mean and median of the losses. Which measure is more appropriate in this case? Explain.

5 Even on a crucial examination, students tend to lose focus while writing their tests. In a psychology experiment, 20 students were given a 10-minute quiz and the number of seconds they spent 'on task' were recorded. Here are the results:

350	380	500	460	480	400	370	380	450	530
520	460	390	360	410	470	470	490	390	340

Find the mean and median of the time spent on task. If you were writing a report to describe these times, which measure of central tendency would you use and why?

6 At 5:30 p.m. during the holiday season, a toy shop counted the number of items sold and the revenue collected for that day. The result was $n = 90$ toys with a total revenue of $\sum x = €4460$.
a) Find the average amount spent on each toy that day.

Shortly before the shop closed at 6 p.m., two new purchases of €74 and €60 were made.
b) Calculate the new mean of the sales per toy that day.

7 Cats is a famous musical. In a large theatre in Vienna (1744 capacity), during a period of 10 years, it played 1000 performances. The manager of the group kept a record of the empty seats on the days it played. Here is the table.

Number of empty seats	1–10	11–20	21–30	31–40	41–50	51–60	61–70	71–80	81–90	91–100
Days	15	50	100	170	260	220	90	45	30	20

a) Copy and complete the following cumulative frequency table for the above information.

Number of empty seats	$x \leq 10$	$x \leq 20$	$x \leq 30$	$x \leq 40$	$x \leq 50$	$x \leq 60$	$x \leq 70$	$x \leq 80$	$x \leq 90$	$x \leq 100$
Days	15		165		815					1000

b) Draw a cumulative frequency graph of this distribution. Use 1 unit on the vertical axis to represent the number of 100 days and 1 unit on the horizontal axis to represent every 10 seats.
c) Use the graph from b) to answer the following questions:

(i) Find an estimate of the median number of empty seats.
(ii) Find an estimate for the first quartile, third quartile and the IQR.
(iii) The days the number of empty seats was less than 35 seats were considered bumper days (lots of profit). How many days were considered bumper days?
(iv) The highest 15% of the days with empty seats were categorized as loss days. What is the number of empty seats above which a day is claimed as a loss?

8 A farmer has 144 bags of new potatoes weighing 2.15 kg each. He also has 56 bags of potatoes from last year with an average weight of 1.80 kg. Find the mean weight of a bag of potatoes available from this farmer.

9 The heights of football players at a given school are given in the table below:

Height	Frequency	Height	Frequency	Height	Frequency	Height	Frequency	Height	Frequency	Height	Frequency
152	2	160	7	168	18	175	5	183	9	191	4
155	6	163	5	170	7	178	11	185	4	193	1
157	9	165	20	173	12	180	8	188	2		

a) Find the five-number summary for this data.
b) Display the data with a box plot and a histogram
c) Find the mean and standard deviation of the data.
d) Describe the data with a few sentences.
e) Draw a cumulative frequency graph and estimate the height of the player that is in the 90th percentile.
f) 10 players' data was missing when we collected the data. The average height of the 10 players is 182. Find the average height of all the players, including the last 10.

Practice questions

1 Given that μ is the mean of a data set $y_1, y_2, \ldots, y_{30}$, and you know that
$$\sum_{i=1}^{30} y_i = 360 \text{ and } \sum_{i=1}^{30} (y_i - \mu)^2 = 925, \text{ find}$$
a) the value of μ
b) the standard deviation of the set.

2 Laura made a survey of some students at school asking them about the time it takes each of them to come to school every morning. She scribbled the numbers on a piece of paper and, unfortunately, could not read the number of students who spend 40 minutes on their trip to school. The average number of minutes she had originally found was 34 minutes. Find out how many students spend 40 minutes on their trip.

Time in minutes	10	20	30	40	50
Number of students with this time	1	2	5	?	3

3 The following table gives 50 measurements of the time it took a certain reaction to be done in a laboratory experiment.

3.1	5.1	4.9	1.8	2.8	5.6	3.6	2.2	2.5	3.4
4.5	2.5	3.5	3.6	3.7	5.1	4.1	4.8	4.9	1.6
2.9	3.6	2.1	6.1	3.5	4.7	4	3.9	3.7	3.9
2.7	4.3	4	5.7	4.4	3.7	3.7	4.6	4.2	4
3.8	5.6	6.2	4.9	2.5	4.2	2.9	3.1	2.8	3.9

a) Construct a frequency table and histogram starting at 1.6 and with interval length of 0.5.
b) What fraction of the measurements is less than 5.1?
c) Estimate, from your histogram, the median of this data set.
d) Estimate the mean and standard deviation using your frequency table.
e) Construct a cumulative frequency graph.
f) From your cumulative frequency graph, estimate each of the five numbers in the five-number summary.

4 In large cities around the world, governments offer parking facilities for public use. The histogram below gives a picture of the number of parking sites available with the capacity of each, in a number of cities chosen at random.

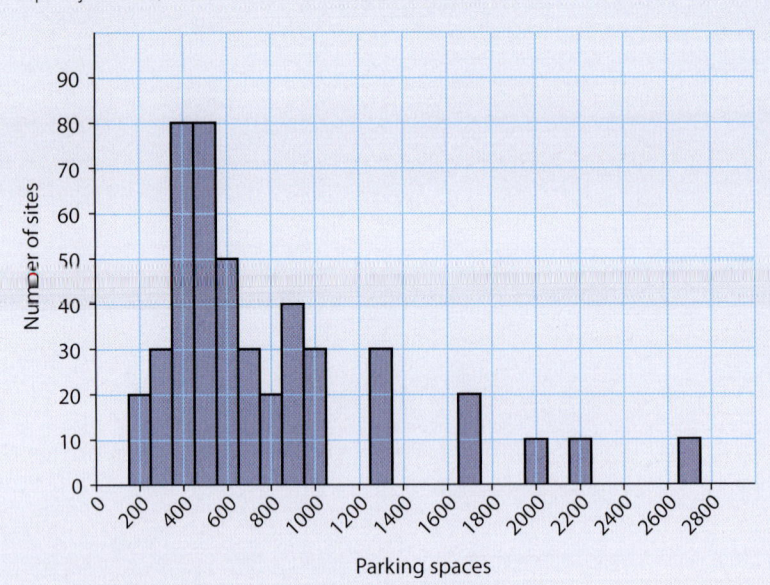

a) Which statistics would best represent the data here? Why?
b) Calculate the mean and standard deviation.
c) Develop a cumulative frequency graph of the data.
d) Use your result from c) above to estimate the median, Q_1, Q_3 and IQR.
e) Are there any outliers in the data? Why?
f) Write a few sentences describing the distribution.

5 The box plots display the case prices (in €) of red wines produced in France, Italy and Spain.

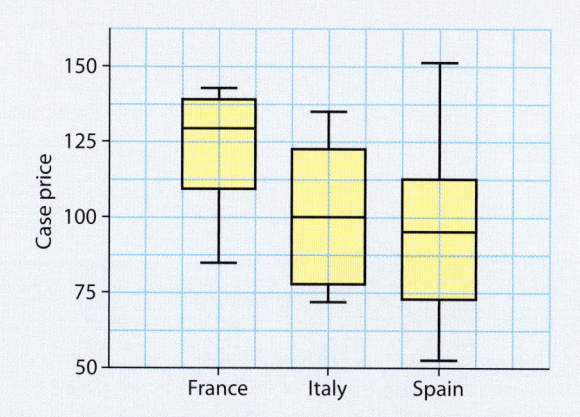

a) Which country appears to produce the most expensive red wine? The cheapest?
b) In which country are the red wines generally more expensive?
c) Write a few sentences comparing the pricing of red wines in the three countries.

6

112.72	53.55	54.12	54.33	58.79	59.26	60.39	62.45	52.22	52.52	52.58	52.85
54.06	51.34	51.93	52.09	52.14	52.24	52.24	52.53	53.5	51.82	51.93	52
52.78	52.82	50.28	50.49	51.28	51.28	51.52	51.62	52.4	52.43	49.83	50.46
50.95	51.07	51.11	49.45	49.45	49.73	49.76	49.93	50.19	50.32	50.63	48.64
49.79	50.19	50.62	50.96	49.09	49.16	49.29	49.74	49.74	49.75	49.84	49.76
52.9	52.91	53.4	52.18	52.57	52.72	50.56	50.87	50.9	49.32	49.7	

The table shows the record for the times (seconds) of the 71 male swimmers in the 100 m swim on the first day during the Summer Olympics 2000 in Sydney.

a) Calculate the mean time and the standard deviation.
b) Calculate the median and IQR.
c) Explain the differences between these two sets of measures.

7 In a survey of universities in major cities in the world, the percentage of first-year students who graduate on time (some require 4 years and some 5 years) was reported. The summary statistics are given below.

Number of universities surveyed	120	Mean percentage	69
Median percentage	70	Standard deviation	9.8
Minimum	42	Maximum	86
Range	44	Q_1	60.25
Q_3	75.75		

a) Is this distribution symmetric? Explain.
b) Check for outliers.
c) Create a box plot of the data.
d) Describe the data in a short paragraph.

Statistics

8 The International Heart Association studies, among other factors, the influence of cholesterol level (in mg/dl) on the conditions of heart patients. In a study of 2000 subjects, the following cumulative relative frequency graph was recorded.

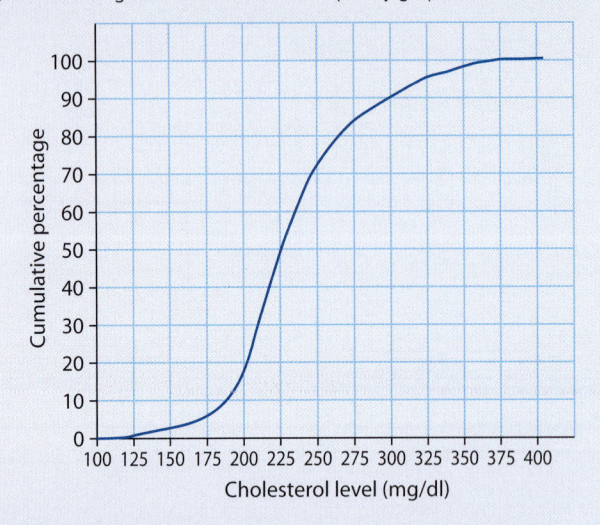

a) Estimate the median cholesterol level of heart patients in the study.
b) Estimate the first and third quartiles, and the 90th and 10th percentiles.
c) Estimate the IQR. Also estimate the number of patients in the middle 50% of this distribution.
d) Create a box plot of the data.
e) Give a short description of the distribution.

9 Many of the streets in Vienna, Austria have a speed limit of 30 km/h. On one Sunday evening the police registered the speed of cars passing an important intersection, in order to give speeding tickets when drivers exceeded the limit. Here is a random sample of 100 cars recorded that evening.

26	46	39	41	44	37	38	35	34	31
27	47	39	41	44	37	38	35	34	32
27	47	39	41	44	37	38	35	34	32
27	48	39	41	44	27	38	35	34	32
29	48	40	41	45	37	38	36	34	33
30	48	40	41	45	37	38	36	35	33
30	48	40	42	45	38	39	36	35	33
30	49	40	42	46	38	39	36	35	33
30	50	41	42	46	38	39	36	35	33
31	54	41	43	46	38	39	36	35	33

a) Prepare a frequency table for the data.
b) Draw a histogram of the data and describe the shape.
c) Calculate, showing all work, the mean and standard deviation of the data.
d) Prepare a cumulative frequency table of the data.
e) Find the median, Q_1, Q_3 and IQR.
f) Are there any outliers in the data? Explain using an appropriate diagram.

10 The following is the data collected from 50 industrial countries chosen at random in 2001. The data represents the per capita gasoline consumption in these countries. The Netherlands' consumption was at 1123 litres per capita while Italy stood at 2220 litres per capita.

2062	2076	1795	1732	2101	2211	1748	1239	1936	1658
1639	1924	2086	1970	2220	1919	1632	1894	1934	1903
1714	1689	1123	1671	1950	1705	1822	1539	1976	1999
2017	2055	1943	1553	1888	1749	2053	1963	2053	2117
1600	1795	2176	1445	1727	1751	1714	2024	1714	2133

a) Calculate the mean, median, standard deviation, Q1, Q3 and IQR.
b) Are there any outliers?
c) Draw a box plot.
d) What consumption levels are within 1 standard deviation from the mean?
e) Germany, with a consumption level of 2758 litres per capita, was not included in the sample. What effect on the different statistics calculated would adding Germany have? Do not recalculate the statistics.

11 90 students on a statistics course were given an experiment where each reported, to the nearest minute, the time, x, it took them to commute to school on a specific day. The teachers then reported back that the total travelling time for the course participants was $\sum x = 4460$ minutes.

a) Find the mean number of minutes the students spent travelling to school that day.

Four students who were absent when the data was first collected reported that they spent 35, 39, 28 and 32 minutes, respectively.

b) Calculate the new mean including these four students.

12 Two thousand students at a large university take the final statistics examination, which is marked on a 100-scale, and the distribution of marks received is given in the table below.

Marks	1–10	11–20	21–30	31–40	41–50	51–60	61–70	71–80	81–90	91–100
Number of candidates	30	100	200	340	520	440	180	90	60	40

a) Complete the table below so that it represents the cumulative frequency for each interval.

Marks	⩽10	⩽20	⩽30	⩽40	⩽50	⩽60	⩽70	⩽80	⩽90	⩽100
Number of candidates	30	130				1630				

b) Draw a cumulative frequency graph of the distribution, using a scale of 1 cm for 100 students on the vertical axis and 1 cm for 10 marks on the horizontal axis.
c) Use your graph from b) to answer parts (i)–(iii) below.
  (i) Find an estimate for the median score.
  (ii) Candidates who scored less than 35 were required to retake the examination. How many candidates had to retake the exam?
  (iii) The highest-scoring 15% of candidates were awarded a distinction. Find the mark above for which a distinction was awarded.

9 Statistics

13 At a conference of 100 mathematicians there are 72 men and 28 women. The men have a mean height of 1.79 m and the women have a mean height of 1.62 m. Find the mean height of the 100 mathematicians.

14 The mean of the population $x_1, x_2, \ldots, x_{25}$ is m. Given that $\sum_{i=1}^{25} x_i = 300$ and $\sum_{i=1}^{25}(x_i - m)^2 = 625$, find

   a) the value of m
   b) the standard deviation of the population.

15 A survey is carried out to find the waiting times for 100 customers at a supermarket.

Waiting time (seconds)	Number of customers
0–30	5
30–60	15
60–90	33
90–120	21
120–150	11
150–180	7
180–210	5
210–240	3

   a) Calculate an estimate for the mean of the waiting times, by using an appropriate approximation to represent each interval.
   b) Construct a cumulative frequency table for this data.
   c) Use the cumulative frequency table to draw, on graph paper, a cumulative frequency graph, using a scale of 1 cm per 20 seconds waiting time for the horizontal axis and 1 cm per 10 customers for the vertical axis.
   d) Use the cumulative frequency graph to find estimates for the median and the lower and upper quartiles (i.e. first and third quartiles).

16 The following diagram represents the lengths, in cm, of 80 plants grown in a laboratory.

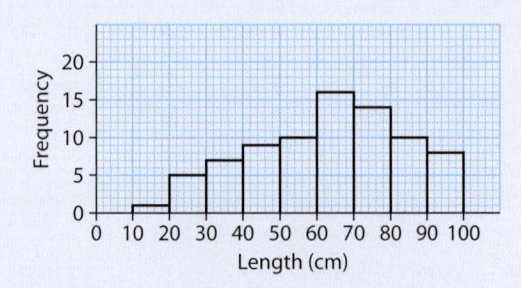

   a) How many plants have lengths in cm between
      (i) 50 and 60?
      (ii) 70 and 90?

b) Calculate estimates for the mean and the standard deviation of the lengths of the plants.

c) Explain what feature of the diagram suggests that the median is different from the mean.

d) The following is an extract from the cumulative frequency table.

Length in cm less than	Cumulative frequency
.	.
50	22
60	32
70	48
80	62
.	.

Use the information in the table to estimate the median. Give your answer to 2 significant figures.

17 The table below represents the weights, W, in grams, of 80 packets of roasted peanuts.

Weight (W)	$80 < W \leqslant 85$	$85 < W \leqslant 90$	$90 < W \leqslant 95$	$95 < W \leqslant 100$	$100 < W \leqslant 105$	$105 < W \leqslant 110$	$110 < W \leqslant 115$
Number of packets	5	10	15	26	13	7	4

a) Use the midpoint of each interval to find an estimate for the standard deviation of the weights.

b) Copy and complete the following cumulative frequency table for the above data.

Weight (W)	$W \leqslant 85$	$W \leqslant 90$	$W \leqslant 95$	$W \leqslant 100$	$W \leqslant 105$	$W \leqslant 110$	$W \leqslant 115$
Number of packets	5	15					80

c) A cumulative frequency graph of the distribution is shown below, with a scale of 2 cm for 10 packets on the vertical axis and 2 cm for 5 grams on the horizontal axis.

Use the graph to estimate
  (i) the median
  (ii) the upper quartile (that is, the third quartile).
Give your answers to the nearest gram.

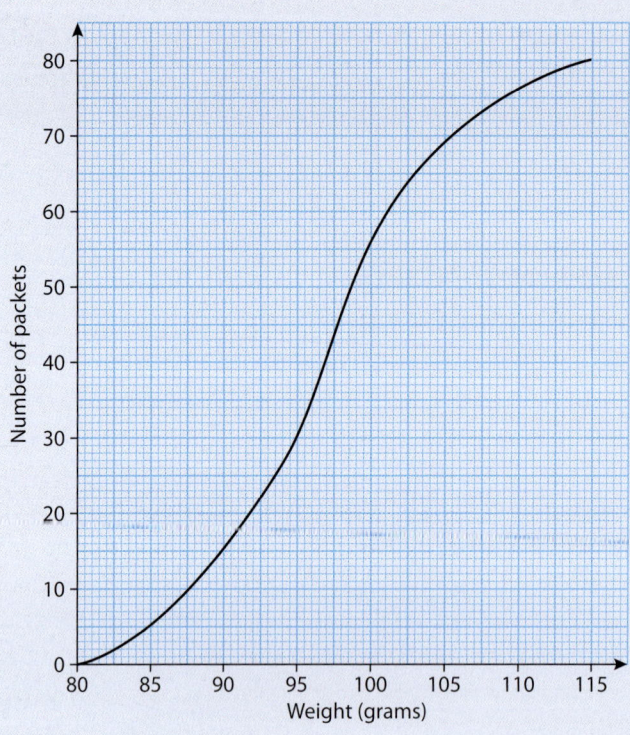

d) Let $W_1, W_2, \ldots, W_{80}$ be the individual weights of the packets, and let $\overline{W}$ be their mean. What is the value of the sum
$$(W_1 - \overline{W}) + (W_2 - \overline{W}) + (W_3 - \overline{W}) + \ldots + (W_{79} - \overline{W}) + (W_{80} - \overline{W})?$$

e) One of the 80 packets is selected at random. Given that its weight satisfies $85 < W \leq 110$, find the probability that its weight is greater than 100 grams.

18 The speeds, in km h^{-1}, of cars passing a point on a highway are recorded in the following table.

Speed v	Number of cars
$v \leq 60$	0
$60 < v \leq 70$	7
$70 < v \leq 80$	25
$80 < v \leq 90$	63
$90 < v \leq 100$	70
$100 < v \leq 110$	71
$110 < v \leq 120$	39
$120 < v \leq 130$	20
$130 < v \leq 140$	5
$v > 140$	0

a) Calculate an estimate of the mean speed of the cars.

b) The following table gives some of the cumulative frequencies for the information above.

Speed v	Cumulative frequency
$v \leqslant 60$	0
$v \leqslant 70$	7
$v \leqslant 80$	32
$v \leqslant 90$	95
$v \leqslant 100$	a
$v \leqslant 110$	236
$v \leqslant 120$	b
$v \leqslant 130$	295
$v \leqslant 140$	300

(i) Write down the values of *a* and *b*.
(ii) On graph paper, construct a cumulative frequency **curve** to represent this information. Use a scale of 1 cm for 10 km h^{-1} on the horizontal axis and a scale of 1 cm for 20 cars on the vertical axis.

c) Use your graph to determine
(i) the percentage of cars travelling at a speed in excess of 105 km h^{-1}
(ii) the speed which is exceeded by 15% of the cars.

19 A taxi company has 200 taxi cabs. The cumulative frequency curve below shows the fares in dollars ($) taken by the cabs on a particular morning.

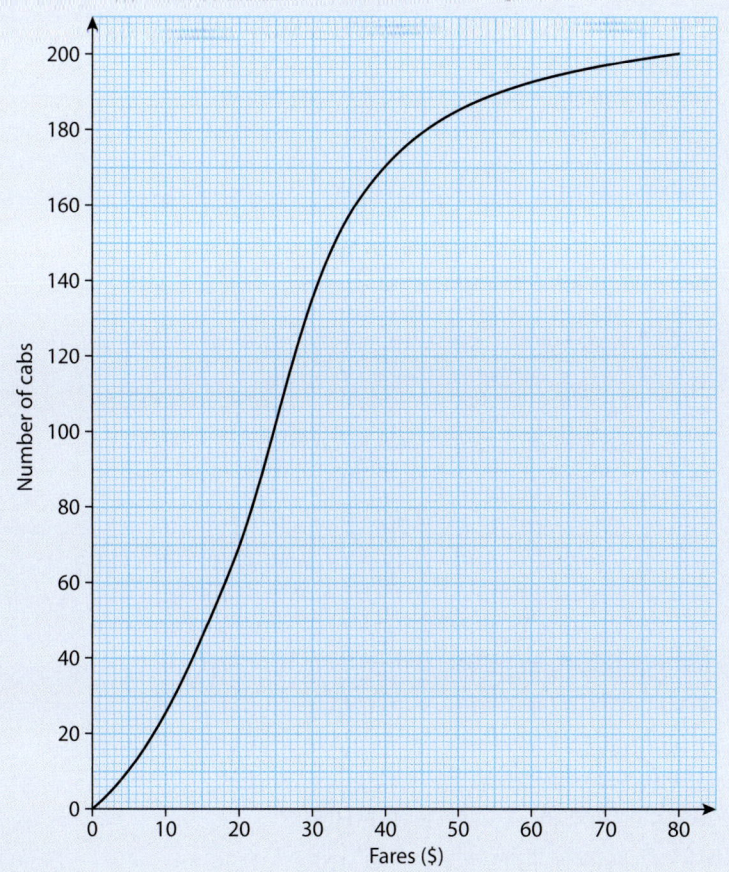

311

a) Use the curve to estimate
   (i) the median fare
   (ii) the number of cabs in which the fare taken is $35 or less.
The company charges 55 cents per kilometre for distance travelled. There are no other charges. Use the curve to answer the following.
b) On that morning, 40% of the cabs travel less than a km. Find the value of a.
c) What percentage of the cabs travel more than 90 km on that morning?

20 Three positive integers a, b and c, where $a < b < c$, are such that their median is 11, their mean is 9 and their range is 10. Find the value of a.

21 In a suburb of a large city, 100 houses were sold in a three-month period. The following **cumulative frequency table** shows the distribution of selling prices (in thousands of dollars).

Selling price P ($ thousand)	$P \leqslant 100$	$P \leqslant 200$	$P \leqslant 300$	$P \leqslant 400$	$P \leqslant 500$
Total number of houses	12	58	87	94	100

a) Represent this information on a cumulative frequency **curve**, using a scale of 1 cm to represent $50 000 on the horizontal axis and 1 cm to represent 5 houses on the vertical axis.
b) Use your curve to find the interquartile range.

The information above is represented in the following frequency distribution.

Selling price P ($ thousand)	$0 < P \leqslant 100$	$100 < P \leqslant 200$	$200 < P \leqslant 300$	$300 < P \leqslant 400$	$400 < P \leqslant 500$
Total number of houses	12	46	29	a	b

c) Find the values of a and b.
d) Use mid-interval values to calculate an estimate for the mean selling price.
e) Houses which sell for more than $350 000 are described as *De Luxe*.
   (i) Use your graph to estimate the number of *De Luxe* houses sold. Give your answer to the nearest integer.
   (ii) Two *De Luxe* houses are selected at random. Find the probability that **both** have a selling price of more than $400 000.

22 A student measured the diameters of 80 snail shells. His results are shown in the following cumulative frequency graph. The lower quartile (LQ) is 14 mm and is marked clearly on the graph.
a) On the graph, mark clearly and write down the value of
   (i) the median   (ii) the upper quartile.
b) Write down the interquartile range.

23 The cumulative frequency curve right shows the marks obtained in an examination by a group of 200 students.

a) Use the cumulative frequency curve to complete the frequency table below.

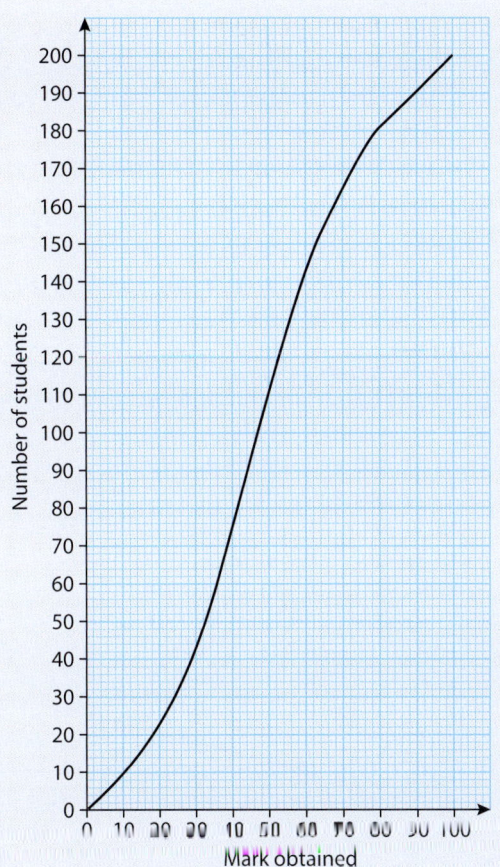

Mark (x)	$0 \leq x < 20$	$20 \leq x < 40$	$40 \leq x < 60$	$60 \leq x < 80$	$80 \leq x < 100$
Number of students	22				20

b) Forty per cent of the students fail. Find the pass mark.

24 The cumulative frequency curve right shows the heights (in centimetres) of 120 basketball players.
Use the curve to estimate
a) the median height
b) the interquartile range.

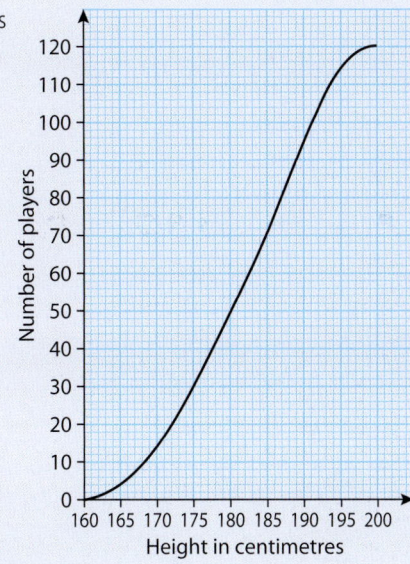

25 Let *a*, *b*, *c* and *d* be integers such that $a < b$, $b < c$ and $c = d$.
The mode of these four numbers is 11.
The range of these four numbers is 8.
The mean of these four numbers is 8.
Calculate the value of each of the integers *a*, *b*, *c*, *d*.

26 A test, to be marked out of 100, is completed by 800 students. The cumulative frequency graph for the marks is given below.

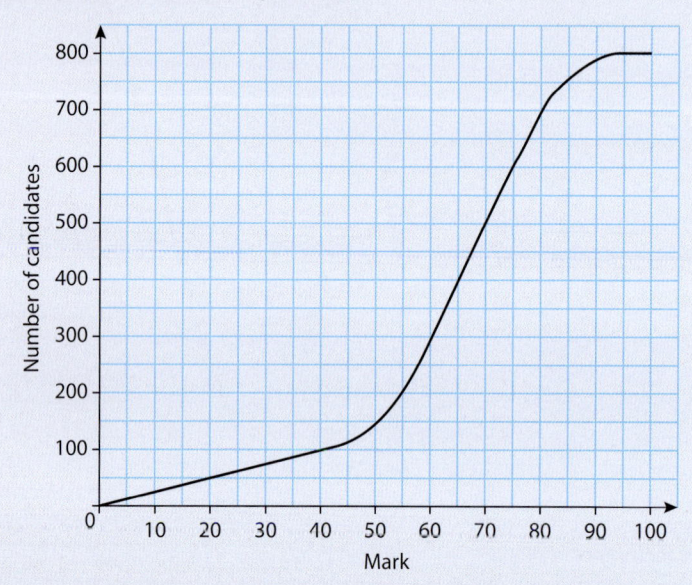

a) Write down the number of students who scored 40 marks or less on the test.
b) The middle 50% of test results lie between marks *a* and *b*, where $a < b$. Find *a* and *b*.

Questions 13–26: © International Baccalaureate Organization

10 Probability

Assessment statements

6.5 Concepts of trial, outcome, equally likely outcomes, sample space (*U*) and event.
The probability of an event *A* as P(*A*) = n(*A*)/n(*U*).
The complementary events as *A* and *A'* (not *A*);
P(*A*) + P(*A'*) = 1.

6.6 Combined events, the formula: P(*A* ∪ *B*) = P(*A*) + P(*B*) − P(*A* ∩ *B*).
P(*A* ∩ *B*) = 0 for mutually exclusive events.

6.7 Conditional probability; the definition: P(*A*|*B*) = P(*A* ∩ *B*)/P(*B*).
Independent events; the definition: P(*A*|*B*) = P(*A*) = P(*A*|*B'*).

6.8 Use of Venn diagrams, tree diagrams and tables of outcomes to solve problems.

6.9 Concept of discrete random variables and their probability distributions.
Expected value (mean), E(*X*) for discrete data.

Introduction

Now that you have learned to describe a data set in Chapter 9, how can you use sample data to draw conclusions about the populations from which you drew your samples? The techniques we use in drawing conclusions are part of what we call **probability**. To use this tool properly, you must first understand how it works. This chapter will introduce you to the language and basic tools of probability.

The variables we discussed in Chapter 9 can now be redefined as *random variables*, whose values depend on the chance selection of the elements in the sample. Using probability as a tool, you will be able to create **probability distributions** that serve as models for random variables. You can then describe these using a mean and a standard deviation as you did in Chapter 9.

10.1 Randomness

Probability is the study of randomness.

The reasoning in statistics rests on asking, 'How often would this method give a correct answer if I used it very many times?' When we produce data by random sampling or by experiments, the laws of probability enable us to answer the question, 'What would happen if we did this many times?'

10 Probability

What does 'random' mean? In ordinary speech, we use 'random' to denote things that are unpredictable. Events that are **random** are not perfectly predictable, but *they have long-term regularities* that we can describe and quantify using probability. In contrast, **haphazard** events *do not necessarily have long-term regularities*. Take, for example, the tossing of an unbiased coin and observing the number of heads that appear.

When you throw the coin, there are only two outcomes, heads or tails. Figure 10.1 shows the results of the first 50 tosses of an experiment that tossed the coin 5000 times. Two sets of trials are shown. The red graph shows the result of the first trial: the first toss was a head followed by a tail, making the proportion of heads to be 0.5. The third toss was also a tail, so the proportion of heads is 0.33, then 0.25. On the other hand, the other set of trials, shown in green, starts with a series of tails, then a head, which raises the proportion to 0.2, etc.

The proportion of heads is quite variable at first. However, in the long run, and as the number of tosses increases, the proportion of heads stabilizes around 0.5. We say that 0.5 is the **probability** of a head.

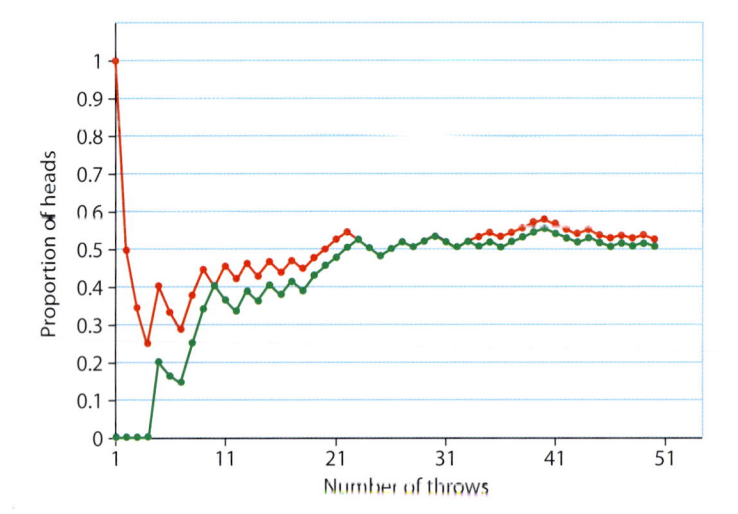

> Please distinguish between random and haphazard (chaos). At first glance they might seem to be the same because neither of their outcomes can be anticipated with certainty.

Figure 10.1

It is important that you know that the proportion of heads in a small number of tosses can be far from the probability. Probability describes only what happens in the long run. How a fair coin lands when it is tossed is an example of a random event. One cannot predict perfectly whether the coin will land heads or tails. However, in repeated tosses, the fraction of times the coin lands heads will tend to settle down to a limit of 50%. The outcome of an individual toss is not perfectly predictable, but the long-term average behaviour is predictable. Thus, it is reasonable to consider the outcome of tossing a fair coin to be random.

Imagine the following scenario:

I drive every day to school. Shortly before school, there is a traffic light. It appears that it is always red when I get there. I collected data over the course of one year (180 school days) and considered the green light to be a 'success'. Here is a partial table of the collected data.

Day	1	2	3	4	5	6	7	...
Light	red	green	red	green	red	red	red	...
Percentage green	0	50	33.3	50	40	33.3	28.6	...

The first day it was red, so the proportion of success is 0% (0 out of 1); the second day it was green, so the frequency is now 50% (1 out of 2); the third day it was red again, so 33.3% (1 out of 3), and so on. As we collect more data, the new measurement becomes a smaller and smaller fraction of the accumulated frequency, so, in the long run, the graph settles to the real chance of finding it green, which in this case is about 30%. The graph is shown below.

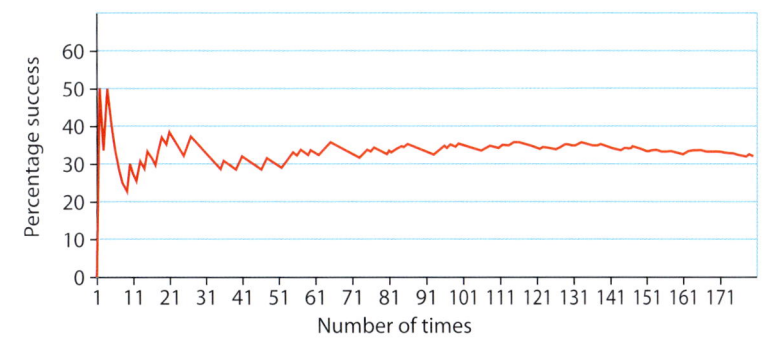

Actually, if you run a simulation for a longer period, you can see that it really stabilizes around 30%. See graph below.

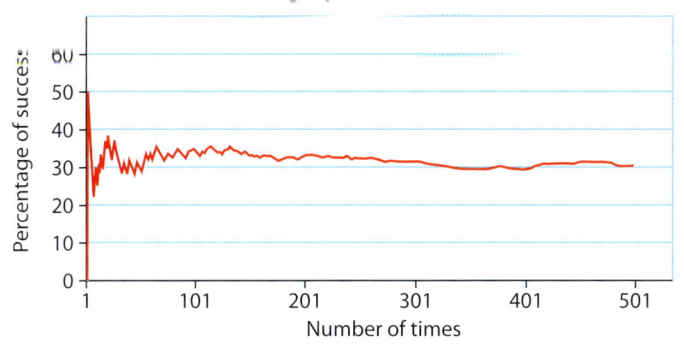

You have to observe here that the randomness in the experiment is not in the traffic light itself, as it is controlled by a timer. In fact, if the system works well, it may turn green at the same time every day. The randomness of the event is the time I arrive at the traffic light.

> The French Count Buffon (1707–1788) tossed a coin 4040 times and received 2048 heads, i.e. a proportion of 50.69%. Also, the English statistician Karl Pearson (1857–1936) tossed a coin 24 000 times and received 12 012 heads, a 50.05% proportion for heads.

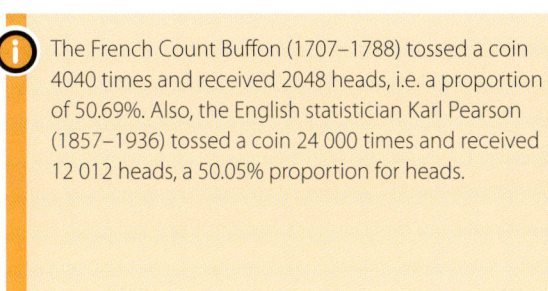

The French Count Buffon

If we ask for the probability of finding the traffic light green in the above example, our answer will be about 30%. We base our answer on knowing that, in the long run, the fraction of time that the traffic light was green is 30%. We could also say that the **long-run relative frequency** of the green light settles down to about 30%.

10.2 Basic definitions

Data is obtained by observing either uncontrolled events in nature or controlled situations in a laboratory. We use the term **experiment** to describe either method of data collection.

> An **experiment** is the process by which an observation (or measurement) is obtained. A **random** (chance) **experiment** is an experiment where there is uncertainty concerning which of two or more possible outcomes will result.

Tossing a coin, rolling a die and observing the number on the top surface, counting cars at a traffic light when it turns green, measuring daily rainfall in a certain area, etc. are a few experiments in this sense of the word.

A description of a random phenomenon in the language of mathematics is called a **probability model**. For example, when we toss a coin, we cannot know the outcome in advance. What *do* we know? We are willing to say that the outcome will be either heads or tails. Because the coin appears to be balanced, we believe that each of these outcomes has probability 0.50. This description of coin tossing has two parts:
- A list of possible outcomes.
- A probability for each outcome.

This two-part description is the starting point for a probability model. We will begin by describing the outcomes of a random phenomenon and learn how to assign probabilities to the outcomes by using one of the definitions of probability.

> The **sample space S** of a random experiment (or phenomenon) is the set of all possible outcomes.

For example, for one toss of a coin, the sample space is

$$S = \{\text{heads, tails}\}, \text{ or simply } \{h, t\}$$

Example 1

Toss a coin twice (or two coins once) and record the results. What is the sample space?

Solution

$$S = \{hh, ht, th, tt\}$$

Example 2

Toss a coin twice (or two coins once) and count the number of heads showing. What is the sample space?

Solution

$$S = \{0, 1, 2\}$$

A **simple event** is the outcome we observe in a single repetition (trial) of the experiment.

For example, an experiment is throwing a die and observing the number that appears on the top face. The simple events in this experiment are $\{1\}$, $\{2\}$, $\{3\}$, $\{4\}$, $\{5\}$ and $\{6\}$. Of course, the set of all these simple events is the sample space of the experiment.

We are now ready to define an **event**. There are several ways of looking at it, which in essence are all the same.

An **event** is an **outcome** or a **set of outcomes** of a random experiment.

With this understanding, we can also look at the event as a subset of the sample space or as a collection of simple events.

Example 3

When rolling a standard six-sided die, what are the sets of event A 'observe an odd number', and event B 'observe a number less than 5'.

Solution
Event A is the set $\{1, 3, 5\}$. Event B is the set $\{1, 2, 3, 4\}$.

Sometimes it helps to visualize an experiment using some tools of set theory. Basically, there are several similarities between the ideas of set theory and probability, and it is very helpful when we see the connection. A simple but powerful diagram is the **Venn diagram**. The diagram shows the outcomes of the die rolling experiment.

> Set theory provides a foundation for all of mathematics. The language of probability is much the same as the language of set theory. Logical statements can be interpreted as statements about sets. This will enable us later to introduce a much better understanding how to set up probability problems that we need to tackle.

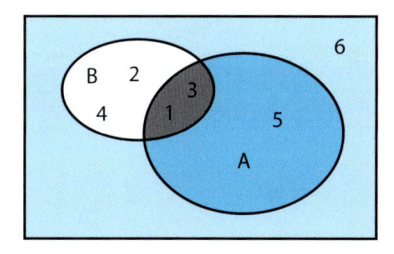

In general, in this book, we will use a rectangle to represent the sample space and closed curves to represent events, as shown in Example 3.

To understand the definitions more clearly, let's look at the following additional example.

Example 4

Suppose we choose one card at random from a deck of 52 playing cards, what is the sample space S?

Solution

S = {A♣, 2♣, … K♣, A♦, 2♦, … K♦, A♥, 2♥, … K♥, A♠, 2♠,… K♠}

Some events of interest:

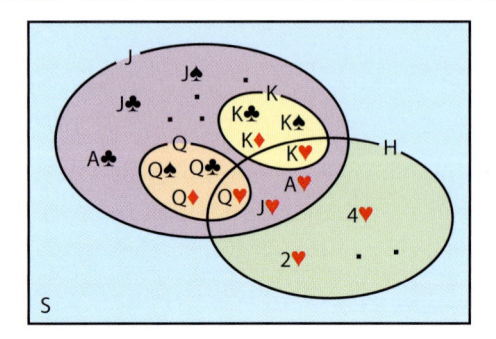

K = event of king = {K♣, K♦, K♥, K♠}

H = event of heart = {A♥, 2♥, … K♥}

J = event of jack or better
  = {J♣, J♦, J♥, J♠, Q♣, Q♦, Q♥, Q♠, K♣, K♦, K♥, K♠, A♣, A♦, A♥, A♠}

Q = event of queen = {Q♣, Q♦, Q♥, Q♠}

Example 5

Toss a coin three times and record the results. Show the event 'observing two heads' as a Venn diagram.

Solution

The sample space is made up of 8 possible outcomes such as hhh, hht, tht, etc.

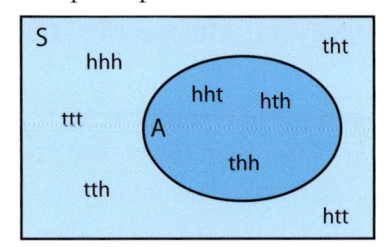

Observing exactly two heads is an event with three elements: {hht, hth, thh}.

Exercise 10.1 and 10.2

1. In a large school, a student is selected at random. Give a reasonable sample space for answers to each of the following questions:
   a) Are you left-handed or right-handed?
   b) What is your height in centimetres?
   c) How many minutes did you study last night?

2. We throw a coin and a standard six-sided die and we record the number and the face that appear in that order. For example, (5, h) represents a 5 on the die and a head on the coin. Find the sample space.

3 We draw cards from a deck of 52 playing cards.
  a) List the sample space if we draw one card at a time.
  b) List the sample space if we draw two cards at a time.
  c) How many outcomes do you have in each of the experiments above?

4 Tim carried out an experiment where he tossed 20 coins together and observed the number of heads showing. He repeated this experiment 10 times and got the following results:
   11, 9, 10, 8, 13, 9, 6, 7, 10, 11
  a) Use Tim's data to get the probability of obtaining a head.
  b) He tossed the 20 coins for the 11th time. How many heads should he expect to get?
  c) He tossed the coins 1000 times. How many heads should he expect to see?

5 In the game 'Dungeons and Dragons', a four-sided die with sides marked with 1, 2, 3 and 4 spots is used. The intelligence of the player is determined by rolling the die twice and adding 1 to the sum of the spots.
  a) What is the sample space for rolling the die twice? (Record the spots on the 1st and 2nd throws.)
  b) What is the sample space for the intelligence of the player?

6 A box contains three balls, blue, green and yellow. You run an experiment where you draw a ball, look at its colour and then replace it and draw a second ball.
  a) What is the sample space of this experiment?
  b) What is the event of drawing yellow first?
  c) What is the event of drawing the same colour twice?

7 Repeat the same exercise as in question 6 above, without replacing the first ball.

8 Nick flips a coin three times and each time he notes whether it is heads or tails.
  a) What is the sample space of this experiment?
  b) What is the event that heads occur more often than tails?

9 Franz lives in Vienna. He and his family decided that their next vacation will be to either Italy or Hungary. If they go to Italy, they can fly, drive or take the train. If they go to Hungary, they will drive or take a boat. Letting the outcome of the experiment be the location of their vacation and their mode of travel, list all the points in the sample space. Also list the sample space of the event 'fly to destination.'

10 A hospital codes patients according to whether they have health insurance or no insurance, and according to their condition. The condition of the patient is rated as good (g), fair (f), serious (s), or critical (c). The clerk at the front desk marks 0, for non-insured patients, and 1 for insured, and uses one of the letters for the condition. So, (1, c) means an insured patient with critical condition.
  a) List the sample space of this experiment.
  b) What is the event 'not insured, in serious or critical condition'?
  c) What is the event 'patient in good or fair condition'?
  d) What is the event 'patient has insurance'?

10 Probability

10.3 Probability assignments

There are a few theories of probability that assign meaning to statements like 'the probability that A occurs is $p\%$'. In this book, we will primarily examine only the **relative frequency theory**. In essence, we will follow the idea that probability is 'the long-run proportion of repetitions on which an event occurs'. This allows us to 'merge' two concepts into one.

- Equally likely outcomes
  In the theory of equally likely outcomes, probability has to do with symmetries and the indistinguishability of outcomes. If a given experiment or trial has n possible outcomes among which there is no preference, they are equally likely. The probability of each outcome is then $\frac{100\%}{n}$ or $\frac{1}{n}$. For example, if a coin is balanced well, there is no reason for it to land heads in preference to tails when it is tossed, so, accordingly, the probability that the coin lands heads is equal to the probability that it lands tails, and both are $\frac{100\%}{2} = 50\%$. Similarly, if a die is fair, the chance that when it is rolled it lands with the side with 1 on top is the same as the chance that it shows 2, 3, 4, 5 or 6: $\frac{100\%}{6}$ or $\frac{1}{6}$.

  In the theory of equally likely outcomes, probabilities are between 0% and 100%. If an event consists of more than one possible outcome, the chance of the event is the number of ways it can occur divided by the total number of things that could occur. For example, the chance that a die lands showing an even number on top is the number of ways it could land showing an even number (2, 4 or 6) divided by the total number of things that could occur (6, namely showing 1, 2, 3, 4, 5 or 6).

- Frequency theory
  In the frequency theory, probability is the limit of the relative frequency with which an event occurs in repeated trials. Relative frequencies are always between 0% and 100%. According to the frequency theory of probability, 'the probability that A occurs is $p\%$' means that if you repeat the experiment over and over again, independently and under essentially identical conditions, the percentage of the time that A occurs will converge to p. For example, to say that the chance a coin lands heads is 50% means that if you toss the coin over and over again, independently, the ratio of the number of times the coin lands heads to the total number of tosses approaches a limiting value of 50%, as the number of tosses grows. Because the ratio of heads to tosses is always between 0% and 100%, when the probability exists it must be between 0% and 100%.

In all theories, probability is on a scale of 0% to 100%. 'Probability' and 'chance' are synonymous.

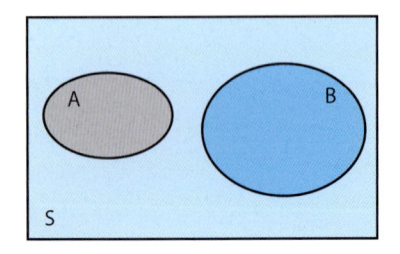

Using Venn diagrams and the 'equally likely' concept, we can say that the probability of any event is the number of elements in an event A divided by the total number of elements in the sample space S. This is equivalent to saying: $P(A) = \frac{n(A)}{n(S)}$, where $n(A)$ represents the number of outcomes in A and $n(S)$ represents the total number of outcomes. So, in Example 5, the probability of observing exactly two heads is: $P(\text{2 heads}) = \frac{3}{8}$.

Probability rules

Regardless of which theory we subscribe to, the probability rules apply.

Rule 1

Any probability is a number between 0 and 1, i.e. the probability $P(A)$ of any event A satisfies $0 \leq P(A) \leq 1$. If the probability of any event is 0, the event *never* occurs. Likewise, if the probability is 1, it *always* occurs. In rolling a standard die, it is impossible to get the number 9, so $P(9) = 0$. Also, the probability of observing any integer between 1 and 6, inclusive, is 1.

Rule 2

All possible outcomes together must have a probability of 1, i.e. the probability of the sample space **S** is 1: $P(S) = 1$. Informally, this is sometimes called the 'something has to happen rule'.

Rule 3

If two events have no outcomes in common, the probability that one or the other occurs is the sum of their individual probabilities. Two events that have no outcomes in common, and hence can never occur together, are called **disjoint** events or **mutually exclusive** events.

$$P(A \text{ or } B) = P(A) + P(B)$$

This is the **addition rule for mutually exclusive events**.

For example, in tossing three coins, the events of getting exactly two heads or exactly two tails are disjoint, and hence the probability of getting exactly two heads or two tails is $\frac{3}{8} + \frac{3}{8} = \frac{6}{8} = \frac{3}{4}$.

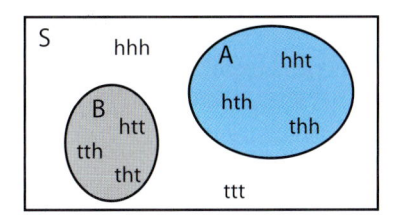

> No matter how little a chance you think an event has, there is **no** such thing as **negative** probability.

> No matter how large a chance you think an event has, there is no such thing as a probability larger than 1!

Additionally, we can always add the probabilities of **outcomes** because they are always disjoint. A trial cannot come out in two different ways at the same time. This will give you a way to check whether the probabilities you assigned are *legitimate*.

Rule 4

Suppose that the probability that you receive a 7 on your IB exam is 0.2, then the probability of *not* receiving a 7 on the exam is 0.8. The event that contains the outcomes **not in A** is called the **complement** of A, and is denoted by A'.

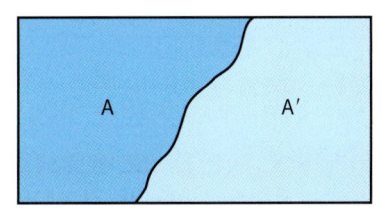

> You have to be careful with these rules. By the 'something has to happen' rule, the total of the probabilities of all possible outcomes **must be 1**. This is so because they are disjoint, and their sum covers all the elements of the sample space. Suppose someone reports the following probabilities for students in your high school (4 years). If the probability that a grade 1, 2, 3 or 4 student is chosen at random from the high school is 0.24, 0.24, 0.25 and 0.19 respectively, with no other possibilities, you should know immediately that there is something wrong. These probabilities add up to 0.92. Similarly, if someone claims that these probabilities are 0.24, 0.28, 0.25, 0.26 respectively, there is also something wrong. These probabilities add up to 1.03, which is more than 1.

$P(A') = 1 - P(A)$, or $P(A) = 1 - P(A')$.

Example 6

Data collected for traffic violations was collected in a certain country and a summary is given below:

Age group	18–20 years	21–29 years	30–39 years	Over 40 years
Probability	0.06	0.47	0.29	0.18

What is the probability that the offender is a) in the youngest age group, b) between 21 and 40, and c) younger than 40?

Solution

Each probability is between 0 and 1, and the probabilities add up to 1. Therefore, this is a legitimate assignment of probabilities.

a) The probability that the offender is in the youngest group is 6%.
b) The probability that the driver is in the group 21 to 39 years is $0.47 + 0.29 = 0.76$.
c) The probability that a driver is younger than 40 years is $1 - 0.18 = 0.82$.

Example 7

It is a striking fact that when people create codes for their cellphones, the first digits follow distributions very similar to the following one:

First digit	0	1	2	3	4	5	6	7	8	9
Probability	0.009	0.300	0.174	0.122	0.096	0.078	0.067	0.058	0.051	0.045

a) Find the probabilities of the following three events:

$A = \{\text{first digit is } 1\}$
$B = \{\text{first digit is more than } 5\}$
$C = \{\text{first digit is an odd number}\}$

b) Find the probability that the first digit is (i) 1 or greater than 5, (ii) not 1, and (iii) an odd number or a number larger than 5.

Solution

a) From the table:
$P(A) = 0.300$
$P(B) = P(6) + P(7) + P(8) + P(9)$
$= 0.067 + 0.058 + 0.051 + 0.045$
$= 0.221$
$P(C) = P(1) + P(3) + P(5) + P(7) + P(9)$
$= 0.300 + 0.122 + 0.078 + 0.058 + 0.045$
$= 0.603$

b) (i) Since A and B are mutually exclusive, by the addition rule, the probability that the first digit is 1 or greater than 5 is

$$P(A \text{ or } B) = 0.300 + 0.221 = 0.521$$

(ii) Using the complement rule, the probability that the first digit is not 1 is

$$P(A') = 1 - P(A) = 1 - 0.300 = 0.700$$

(iii) The probability that the first digit is an odd number or a number larger than 5:

$$P(B \text{ or } C) = P(1) + P(3) + P(5) + P(6) + P(7) + P(8) + P(9)$$
$$= 0.300 + 0.122 + 0.078 + 0.067 + 0.058 + 0.051$$
$$+ 0.045$$
$$= 0.721$$

• **Hint:** Notice here that $P(B$ or $C)$ is *not* the sum of $P(B)$ and $P(C)$ because B and C are not disjoint.

Equally likely outcomes

In some cases we are able to assume that individual outcomes are equally likely because of some balance in the experiment. Tossing a balanced coin renders heads or tails equally likely, with each having a probability of 50%, and rolling a standard balanced die gives the numbers from 1 to 6 as equally likely, with each having a probability of $\frac{1}{6}$.

Suppose in Example 7 we consider all the digits to be equally likely to happen, then our table would be

First digit	0	1	2	3	4	5	6	7	8	9
Probability	0.1	0.1	0.1	0.1	0.1	0.1	0.1	0.1	0.1	0.1

$P(A) = 0.1$

$P(B) = P(6) + P(7) + P(8) + P(9) = 4 \times 0.1 = 0.4$

$P(C) = P(1) + P(3) + P(5) + P(7) + P(9) = 5 \times 0.1 = 0.5$

Also, by the complement rule, the probability that the first digit is not 1 is

$$P(A') = 1 - P(A) = 1 - 0.1 = 0.9$$

Tree diagrams

In an experiment to check the blood types of patients, the experiment can be thought of as a two-stage experiment: first we identify the type of the blood and then we classify the Rh factor $+$ or $-$.

The simple events in this experiment can be counted using another tool, the **tree diagram**, which is extremely powerful and helpful in solving probability problems.

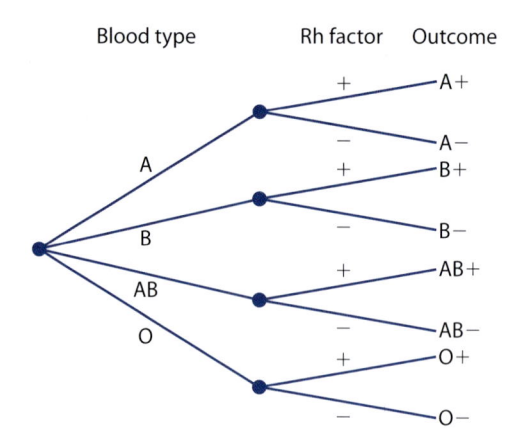

Our sample space in this experiment is the set {A+, A−, B+, B−, AB+, AB−, O+, O−} as we can read from the last column.

This data can also be arranged in a **probability table**:

	Blood type			
Rh factor	A	B	AB	O
Positive	A+	B+	AB+	O+
Negative	A−	B−	AB−	O−

Exercise 10.3

1 In a simple experiment, chips with integers 1–20 inclusive were placed in a box and one chip was picked at random.
   a) What is the probability that the number drawn is a multiple of 3?
   b) What is the probability that the number drawn is not a multiple of 4?

2 The probability an event *A* happens is 0.37.
   a) What is the probability that it does not happen?
   b) What is the probability that it may or may not happen?

3 You are playing with an ordinary deck of 52 cards by drawing cards at random and looking at them.
   a) Find the probability that the card you draw is
      (i) the ace of hearts
      (ii) the ace of hearts or any spade
      (iii) an ace or any heart
      (iv) not a face card.
   b) Now you draw the ten of diamonds, put it on the table and draw a second card. What is the probability that the second card is
      (i) the ace of hearts?
      (ii) not a face card?
   c) Now you draw the ten of diamonds, return it to the deck and draw a second card. What is the probability that the second card is
      (i) the ace of hearts?
      (ii) not a face card?

4 On Monday morning, my class wanted to know how many hours students spent studying on Sunday night. They stopped schoolmates at random as they arrived and asked each, 'How many hours did you study last night?' Here are the answers of the sample they chose on Monday, 14 January, 2008.

Number of hours	0	1	2	3	4	5
Number of students	4	12	8	3	2	1

a) Find the probability that a student spent less than three hours studying Sunday night.
b) Find the probability that a student studied for two or three hours.
c) Find the probability that a student studied less than six hours.

5 We throw a coin and a standard six-sided die and we record the number and the face that appear. Find
a) the probability of having a number larger than 3
b) the probability that we receive a head and a 6.

6 A die is constructed in a way that a 1 has the chance to occur twice as often as any other number.
a) Find the probability that a 5 appears.
b) Find the probability an odd number will occur.

7 You are given two fair dice to roll in an experiment.
a) Your first task is to report the numbers you observe.
   (i) What is the sample space of your experiment?
   (ii) What is the probability that the two numbers are the same?
   (iii) What is the probability that the two numbers differ by 2?
   (iv) What is the probability that the two numbers are not the same?
b) In a second stage, your task is to report the sum of the numbers that appear.
   (i) What is the probability that the sum is 1?
   (ii) What is the probability that the sum is 9?
   (iii) What is the probability that the sum is 8?
   (iv) What is the probability that the sum is 13?

8 The blood types of people can be one of four types: O, A, B or AB. The distribution of people with these types differs from one group of people to another. Here are the distributions of blood types for randomly chosen people in the US, China and Russia.

Country \ Blood type	O	A	B	AB
US	0.43	0.41	0.12	?
China	0.36	0.27	0.26	0.11
Russia	0.39	0.34	?	0.09

a) What is the probability of type AB in the US?
b) Dirk lives in the US and has type B blood. What is the probability that a randomly chosen US citizen can donate blood to Dirk? (Type B can only receive from O and B.)
c) What is the probability of randomly choosing an American and a Chinese (independently) with type O blood?
d) What is the probability of randomly choosing an American, a Chinese and a Russian (independently) with type O blood?
e) What is the probability of randomly choosing an American, a Chinese and a Russian (independently) with the same blood type?

9 In each of the following situations, state whether or not the given assignment of probabilities to individual outcomes is legitimate. Give reasons for your answer.
   a) A die is loaded such that the probability of each face is according to the following assignment (x is the number of spots on the upper face and P(x) is its probability.)

x	1	2	3	4	5	6
P(x)	0	$\frac{1}{6}$	$\frac{1}{3}$	$\frac{1}{3}$	$\frac{1}{6}$	0

   b) A student at your school categorized in terms of gender and whether they are diploma candidates or not.
   P(female, diploma candidate) = 0.57, P(female, not a diploma candidate) = 0.23, P(male, diploma candidate) = 0.43, P(male, not a diploma candidate) = 0.18.
   c) Draw a card from a deck of 52 cards (x is the suit of the card and P(x) is its probability).

x	Hearts	Spades	Diamonds	Clubs
P(x)	$\frac{12}{52}$	$\frac{15}{52}$	$\frac{12}{52}$	$\frac{13}{52}$

10 In Switzerland, there are three 'official' mother tongues, German, French and Italian. You choose a Swiss at random and ask, 'What is your mother tongue?' Here is the distribution of responses:

Language	German	French	Italian	Other
Probability	0.58	0.24	0.12	?

   a) What is the probability that a Swiss person's mother tongue is not one of the official ones?
   b) What is the probability that a Swiss person's mother tongue is not German?
   c) What is the probability that you choose two Swiss independent of each other and they both have German mother tongue?
   d) What is the probability that you choose two Swiss independent of each other and they both have the same mother tongue?

11 The majority of email messages are now 'spam'. Choose a spam email message at random. Here is the distribution of topics:

Topic	Adult	Financial	Health	Leisure	Products	Scams
Probability	0.165	0.142	0.075	0.081	0.209	0.145

   a) What is the probability of choosing a spam message that does not concern these topics?
   Parents are usually concerned with spam messages with 'adult' content and scams.
   b) What is the probability that a randomly chosen spam email falls into one of the other categories?

10.4 Operations with events

In Example 7, we talked about the following events:
   $B = \{$first digit is more than 5$\}$
   $C = \{$first digit is an odd number$\}$

We also claimed that these two events are not disjoint. This brings us to another concept for looking at combined events.

The **intersection** of two events B and C, denoted by the symbol B∩C or simply BC, is the event containing all outcomes common to B and C.

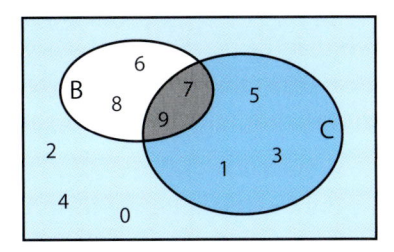

Here $B \cap C = \{7, 9\}$ because these outcomes are in both B and C. Since the intersection has outcomes common to the two events B and C, they are not mutually exclusive.

The probability of $B \cap C$ is $0.058 + 0.045 = 0.103$. Recall from Example 7 that we said that the probability of B or C is not simply the sum of the probabilities. That brings us to the next concept. How can we find the probability of B or C when they are not mutually exclusive? To answer this question, we need to define another operation.

The **union** of two events B and C, denoted by the symbol $B \cup C$, is the event containing all the outcomes that belong to B or to C or to both.

Here $B \cup C = \{1, 3, 5, 6, 7, 8, 9\}$. In calculating the probability of $B \cup C$, we observe that the outcomes 7 and 9 are counted twice. To remedy the situation, if we decide to add the probabilities of B and C, we subtract one of the incidents of double counting. So, $P(B \cup C) = 0.221 + 0.603 - 0.103 = 0.721$, which is the result we received with direct calculation. In general, we can state the following probability rule:

Rule 5

For any two events A and B, $P(A \cup B) = P(A) + P(B) - P(A \cap B)$.

As you see from the diagram below, $P(A \cap B)$ has been added twice, so the 'extra' one is subtracted to give the probability of $(A \cup B)$.

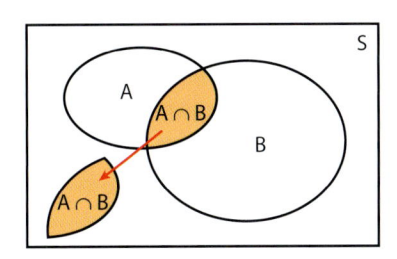

This general probability addition rule applies to the case of mutually exclusive events too. Consider any two events A and B. The probability of A or B is given by

$$P(A \cup B) = P(A) + P(B) - P(A \cap B)$$
$$= P(A) + P(B), \text{ since } P(A \cap B) = 0.$$

Rule 6

The simple multiplication rule.

Consider the following situation: In a large school, 55% of the students are male. It is also known that the percentage of smokers among males and females in this school is the same, 22%. What is the probability of selecting a student at random from this population and the student is a male smoker?

Applying common sense only, we can think of the problem in the following manner. Since the proportion of smokers is the same in both groups, smoking and gender are independent of each other in the sense that knowing that the student is a male does not influence the probability that he smokes!

The chance we pick a male student is 55%. From those 55% of the population, we know that 22% are smokers, so by simple arithmetic the chance that we select a male smoker is $0.22 \times 0.55 = 12.1\%$.

This is an example of the multiplication rule for independent events.

> Two events A and B are **independent** if knowing that one of them occurs does not change the probability that the other occurs.

> The **multiplication rule for independent events:** If two events A and B are independent, then $P(A \cap B) = P(A) \times P(B)$.

Example 8

Reconsider the situation with the traffic light at the beginning of this chapter. The probability that I find the light green is 30%. What is the probability that I find it green on two consecutive days?

Solution

We will assume that my arrival and finding the light green is a random event, and that if it turns green on one day it does not influence how it turns the next day. In that case our calculation is very simple:

P(green the first and second day) = P(green first day) × P(green second day)
= 0.30 × 0.30 = 0.09.

This rule can also be extended to more than two independent events. For example, on the assumption of independence, what is the chance that I find the light green five days of the week?

$$P(\text{green on five days}) = 0.3 \times 0.3 \times 0.3 \times 0.3 \times 0.3 = 0.00243$$

Do not confuse independent with disjoint. 'Disjoint' means that if one of the events occurs then the other does not occur; while 'independent' means that knowing one of the events occurs does not influence the probability of whether the other occurs or not!

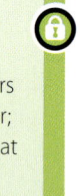

Example 9

Computers bought from a well-known producer require repairs quite frequently. It is estimated that 17% of computers bought from the company require one repair job during the first month of purchase, 7% will need repairs twice during the first month, and 4% require three or more repairs.

a) What is the probability that a computer chosen at random from this producer will need
   (i) no repairs?
   (ii) no more than one repair?
   (iii) some repair?
b) If you buy two such computers, what is the probability that
   (i) neither will require repair?
   (ii) both will need repair?

Solution

a) Since all of the events listed are disjoint, the addition rule can be used.
   (i) P(no repairs) = 1 − P(some repairs) = 1 − (0.17 + 0.07 + 0.04)
       = 1 − (0.28) = 0.72
   (ii) P(no more than one repair) = P(no repairs or one repair)
       = 0.72 + 0.17 = 0.89
   (iii) P(some repairs) = P(one or two or three or more repairs)
       = 0.17 + 0.07 + 0.04 = 0.28

b) Since repairs on the two computers are independent from one another, the multiplication rule can be used. Use the probabilities of events from part a) in the calculations.
   (i) P(neither will need repair) = (0.72)(0.72) = 0.5184
   (ii) P(both will need repair) = (0.28)(0.28) = 0.0784

Conditional probability

In probability, conditioning means incorporating new restrictions on the outcome of an experiment: updating probabilities to take into account new information. This section describes conditioning, and how conditional probability can be used to solve complicated problems. Let us start with an example.

Example 10

A public health department wanted to study the smoking behaviour of high school students. They interviewed 768 students from grades 10–12 and asked them about their smoking habits. They categorized the students into three categories: smokers (more than 1 pack of 20 cigarettes per week), occasional smokers (less than 1 pack per week), and non-smokers. The results are summarized below:

	Smoker	Occasional	Non-smoker	Total
Male	127	73	214	414
Female	99	66	189	354
Total	226	139	403	768

If we select a student at random from this study, what is the probability that we select a) a girl, b) a male smoker, and c) a non-smoker?

Solution

a) P(female) = $\frac{354}{768}$ = 0.461

So, 46.1% of our sample are females.

b) Since we have 127 boys categorized as smokers, the chance of a male smoker will be

P(male smoker) = $\frac{127}{768}$ = 0.165

c) P(non-smoker) = $\frac{403}{768}$ = 0.525

In the above example, what if we know that the selected student is a girl? Does that influence the probability that the selected student is a non-smoker? Yes, it does!

Knowing that the selected student is a female changes our choices. The 'revised' sample space is not made up of all students anymore. It is only the female students. The chance of finding a non-smoker among the females is $\frac{189}{354}$ = 0.534, i.e. 53.4% of the females are non-smokers as compared to the 52.5% of non-smokers in the whole population.

This probability is called a conditional probability, and we write this as

P(non-smoker|female) = $\frac{189}{354}$.

We read this as, '*Probability of selecting a non-smoker **given that** we have selected a female*'.

The conditional probability of A given B, P($A|B$), is the probability of the event A, updated on the basis of the knowledge that the event B occurred. Suppose that A is an event with probability P(A) = $p \neq 0$, and that $A \cap B = \emptyset$ (A and B are disjoint). Then if we learn that B occurred we know A did not occur, so we should revise the probability of A to be zero, P($A|B$) = 0 (the conditional probability of A given B is zero).

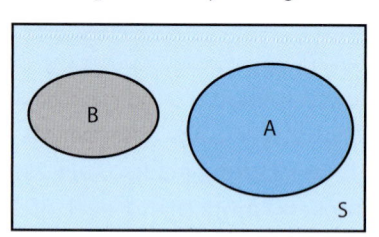

On the other hand, suppose that $A \cap B = B$ (B is a subset of A, so B implies A). Then if we learn that B occurred we know A must have occurred as well, so we should revise the probability of A to be 100%, P($A|B$) = 1 (the conditional probability of A given B is 100%).

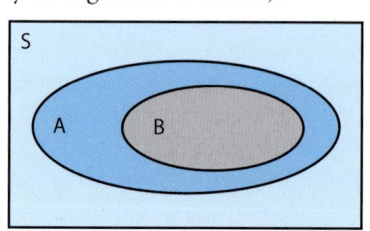

Remember that the probability we assign to an event can change if we know that some other event has occurred. This idea is the key to understanding conditional probability.

Imagine the following scenario:
You are playing cards and your opponent is about to give you a card. What is the probability that the card you receive is a queen?

As you know, there are 52 cards in the deck, 4 of these cards are queens. So, assuming that the deck was thoroughly shuffled, the probability of receiving a queen is

$$P(\text{queen}) = \frac{4}{52} = \frac{1}{13}$$

This calculation assumes that you know nothing about any cards already dealt from the deck.

Suppose now that you are looking at the five cards you have in your hand, and one of them is a queen. You know nothing about the other 47 cards except that exactly three queens are among them. The probability of being given a queen as the next card, given what you know, is

$$P(\text{queen} \mid 1 \text{ queen in hand}) = \frac{3}{47} \neq \frac{1}{13}$$

So, knowing that there is one queen among your five cards changes the probability of the next card being a queen.

Consider Example 10 again. We want to express the table frequencies as relative frequencies or probabilities. Our table will look like this:

	Smoker	Occasional	Non-smoker
Male	0.165	0.095	0.279
Female	0.129	0.086	0.246

To find the probability of selecting a student at random and finding that student is a female non-smoker, we look at the intersection of the female row with the non-smoking column and find that this probability is 0.246.

Looking at this calculation from a different perspective, we can think about it in the following manner:

We know that the percentage of females in our sample is 46.1, and among those females, in Example 10, we found that 53.4% of those are non-smokers. So, the percentage of female non-smokers in the population is the 53.4% of those 46.1% females, i.e. $0.534 \times 0.461 = 0.246$.

In terms of events, this can be read as:

$$P(\text{non-smoker} \mid \text{female}) \times P(\text{female}) = P(\text{female and non-smoker})$$
$$= P(\text{female} \cap \text{non-smoker}).$$

The previous discussion is an example of the **multiplication rule** of any two events A and B.

> **Multiplication rule**
> Given any events A and B, the probability that both events happen is given by
> $$P(A \cap B) = P(A \mid B) \times P(B)$$

Example 11

In a psychology lab, researchers are studying the colour preferences of young children. Six green toys and four red toys (identical apart from colour) are placed in a container. The child is asked to select two toys at random. What is the probability that the child chooses two red toys?

Solution

To solve this problem, we will use a tree diagram.

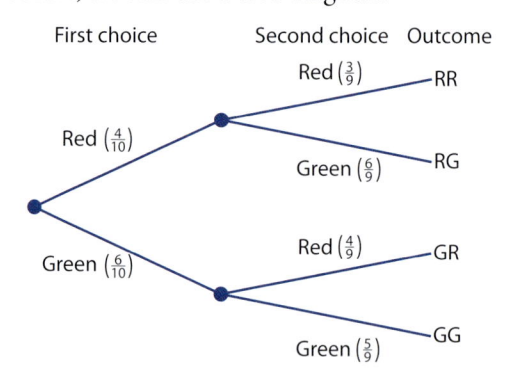

```
                First choice        Second choice    Outcome
                                    Red (3/9) ─────── RR
                Red (4/10) ●
                                    Green (6/9) ───── RG
          ●
                                    Red (4/9) ─────── GR
                Green (6/10) ●
                                    Green (5/9) ───── GG
```

As you notice, every entry on the 'branches' has a conditional probability. So, Red on the second choice is actually either Red|Red or Red|Green. We are interested in RR, so the probability is

$$P(RR) = P(R) \times P(R|R) = \frac{4}{10} \times \frac{3}{9} = 13.3\%$$

If $P(A \cap B) = P(A|B) \times P(B)$, as discussed above, and if $P(B) \neq 0$, we can rearrange the multiplication rule to produce a definition of the conditional probability $P(A|B)$ in terms of the 'unconditional' probabilities $P(A \cap B)$ and $P(B)$.

> When $P(B) \neq 0$, the **conditional probability** of A given B is $P(A|B) = \dfrac{P(A \cap B)}{P(B)}$

Why does this formula make sense?

First of all, note that it does agree with the intuitive answers we found above. If $A \cap B = \emptyset$, $P(A \cap B) = 0$, so $P(A|B) = 0/P(B) = 0$;

and if $A \cap B = B$, $P(A|B) = P(B)/P(B) = 100\%$.

Now, if we learn that B occurred, we can restrict attention to just those outcomes that are in B, and disregard the rest of S, so we have a new sample space that is just B (see diagram below).

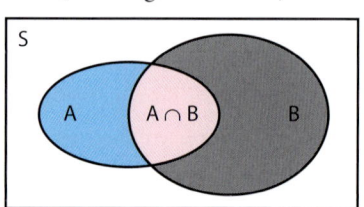

For A to have occurred in addition to B, requires that $A \cap B$ occurred, so the conditional probability of A given B is $P(A \cap B)/P(B)$, just as we defined it above.

Example 12

In an experiment to study the phenomenon of colour blindness, researchers collected information concerning 1000 people in a small town and categorized them according to colour blindness and gender. Here is a summary of the findings:

	Male	Female	Total
Colour-blind	40	2	42
Not colour-blind	470	488	958
Total	510	490	1000

What is the probability that a person is colour-blind given that the person is a woman?

Solution

To answer this question, we notice that we do not have to search the whole population for this event. We limit our search to the women. We have 490 women. As we only need to consider women, then when we search for colour blindness, we only look for the women who are colour-blind, i.e. the intersection. Here we only have two women. Therefore, the chance we get a colour-blind person given the person is a woman is

$$P(C|W) = \frac{P(C \cap W)}{P(W)} = \frac{n(C \cap W)}{n(W)} = 0.004,$$ where C is for colour-blind and W for woman.

Notice here that we used the frequency rather than the probability. However, these are equivalent since dividing by $n(S)$ will transform the frequency into a probability.

$$\frac{n(C \cap W)}{n(W)} = \frac{\frac{n(C \cap W)}{n(S)}}{\frac{n(W)}{n(S)}} = \frac{P(C \cap W)}{P(W)} = P(C|W).$$

Example 13

AUA, a national airline, are known for their punctuality. The probability that a regularly scheduled flight departs on time is $P(D) = 0.83$, the probability that it arrives on time is $P(A) = 0.92$, and the probability that it arrives and departs on time, $P(A \cap D) = 0.78$. Find the probability that a flight
a) arrives on time given that it departed on time
b) departs on time given that it arrived on time.

Solution
a) The probability that a flight arrives on time given that it departed on time is
$$P(A|D) = \frac{P(A \cap D)}{P(D)} = \frac{0.78}{0.83} = 0.94$$

b) The probability that a flight departs on time given that it arrived on time
$$P(D|A) = \frac{P(D \cap A)}{P(A)} = \frac{0.78}{0.92} = 0.85$$

10 Probability

Independence

Two events are **independent** if learning that one occurred does not affect the chance that the other occurred. That is, if $P(A|B) = P(A)$, and vice versa.

This means that if we apply our definition to the general multiplication rule, then

$$P(A \cap B) = P(A|B) \times P(B) = P(A) \times P(B)$$

which is the multiplication rule for independent events we studied earlier.

These results give us some helpful tools in checking the independence of events.

> Two events are **independent** if and only if either $P(A \cap B) = P(A) \times P(B)$, or $P(A|B) = P(A)$. Otherwise, the events are **dependent**.

Example 14

Take another look at the AUA situation in Example 13. Are the events of arriving on time (A) and departing on time (D) independent?

Solution

We can answer this question in two different ways:
a) $P(A) = 0.92$ and we found that $P(A|D) = 0.94$. Since the two values are not the same, we can say that the two events are not independent.
b) Alternately, $P(A \cap D) = 0.78$ and
$P(A) \times P(D) = 0.92 \times 0.83 = 0.76 \neq P(A \cap D)$.

Example 15

In many countries, the police stop drivers on suspicion of drunk driving. The stopped drivers are given a breath test, a blood test or both. In a country where this problem is vigorously dealt with, the police records show the following:

81% of the drivers stopped are given a breath test, 40% a blood test, and 25% both tests.
a) What is the probability that a suspected driver is given
  (i) a test?
  (ii) exactly one test?
  (iii) no test?
b) Are giving the two tests independent?

Solution

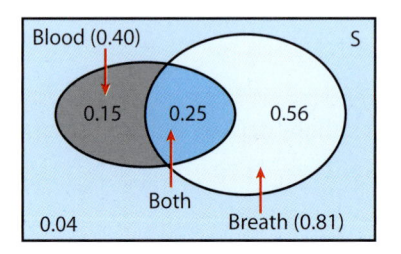

A Venn diagram can help explain the solution.

a) (i) The probability that a driver receives a test means that he/she receives either a blood test, a breath test or both tests. The probability as such can be calculated directly from the diagram, or by applying the addition rule. The diagram shows that if 81% receive the breath test and 25% are also given the blood test, then 56% do not receive a blood test. Similarly, 15% of the blood test receivers do not get a breath test. So, the probability of receiving a test is $0.56 + 0.25 + 0.15 = 0.96$.

Also, if we apply the addition rule,

$$P(\text{breath or blood}) = P(\text{breath}) + P(\text{blood}) - P(\text{both})$$
$$= 0.81 + 0.40 - 0.25 = 0.96.$$

(ii) To receive exactly one test is to receive a blood test or a breath test, but not both! So, from the Venn diagram it is clear that this probability is $0.15 + 0.56 = 0.71$. To approach it differently, since we know that the union of the two events still contains the intersection, we can subtract the probability of the intersection from that of the union: $0.96 - 0.25 = 0.71$.

(iii) To receive no test is equivalent to the complement of the union of the events. Hence, $P(\text{no test}) = 1 - P(1 \text{ test}) = 1 - 0.96 = 0.04$.

b) To check for independence, we can use any of the two methods we tried before. Since all the necessary probabilities are given, we can use the product rule. If they were independent, then

$$P(\text{both tests}) = P(\text{breath}) \times P(\text{blood}) = 0.81 \times 0.40 = 0.324,$$

but $P(\text{both tests}) = 0.25$. Therefore, the events of receiving a breath test and a blood test are not independent.

Exercise 10.4

1 Events A and B are given such that $P(A) = \frac{3}{4}$, $P(A \cup B) = \frac{4}{5}$ and $P(A \cap B) = \frac{3}{10}$. Find $P(B)$.

2 Events A and B are given such that $P(A) = \frac{7}{10}$, $P(A \cup B) = \frac{9}{10}$ and $P(A \cap B) = \frac{3}{10}$. Find
   a) $P(B)$
   b) $P(B' \cap A)$
   c) $P(B \cap A')$
   d) $P(B' \cap A')$
   e) $P(B|A')$

3 People with O-negative blood type are universal donors, i.e. they can donate blood to individuals with any blood type. Only 8% of people have O-negative.
   a) One person randomly appears to give blood. What is the probability that he/she does not have O-negative?
   b) Two people appear independently to give blood. What is the probability that
      (i) both have O-negative?
      (ii) at least one of them has O-negative?
      (iii) only one of them has O-negative?
   c) Eight people appear randomly to give blood. What is the probability that at least one of them has O-negative?

4 PIN numbers for cellular phones usually consist of four digits that are not necessarily different.
  a) How many possible PINs are there?
  b) You don't want to consider the pins that start with 0. What is the probability that a PIN chosen at random does not start with a zero?
  c) What is the probability that a PIN contains at least one zero?
  d) Given a PIN with at least one zero, what is the probability that it starts with a zero?

5 An urn contains six red balls and two blue ones. We make two draws and each time we put the ball back after marking its colour.
  a) What is the probability that at least one of the balls is red?
  b) Given that at least one is red, what is the probability that the second one is red?
  c) Given that at least one is red, what is the probability that the second one is blue?

6 Two dice are rolled and the numbers on the top face are observed.
  a) List the elements of the sample space.
  b) Let x represent the sum of the numbers observed. Copy and complete the following table.

x	2	3	4	5	6	7	8	9	10	11	12
$P(x)$		$\frac{1}{18}$									

  c) What is the probability that at least one die shows a 6?
  d) What is the probability that the sum is at most 10?
  e) What is the probability that a die shows 4 or the sum is 10?
  f) Given that the sum is 10, what is the probability that one of the dice is a 4?

7 A large school has the following numbers categorized by class and gender:

Gender \ Grade	Grade 9	Grade 10	Grade 11	Grade 12	Total
Male	180	170	230	220	800
Female	200	130	190	180	700

  a) What is the probability that a student chosen at random will be a female?
  b) What is the probability that a student chosen at random is a male grade 12 student?
  c) What is the probability that a female student chosen at random is a grade 12 student?
  d) What is the probability that a student chosen at random is a grade 12 or female student?
  e) What is the probability that a grade 12 student chosen at random is a male?
  f) Are gender and grade independent of each other? Explain.

8 Some young people do not like to wear glasses. A survey considered a large number of teenage students as to whether they needed glasses to correct their vision and whether they used the glasses when they needed to. Here are the results.

		Used glasses when needed	
		Yes	No
Need glasses for correct vision	Yes	0.41	0.15
	No	0.04	0.40

a) Find the probability that a randomly chosen young person from this group
   (i) is judged to need glasses
   (ii) needs to use glasses but does not use them.
b) From those who are judged to need glasses, what is the probability that he/she does not use them?
c) Are the events of using and needing glasses independent?

9 Fill in the missing entries in the following table.

P(A)	P(B)	Conditions for events A and B	P(A∩B)	P(A∪B)	P(A|B)
0.3	0.4	Mutually exclusive			
0.3	0.4	Independent			
0.1	0.5			0.6	
0.2	0.5		0.1		

10 In a large graduating class, there are 100 students taking the IB examination. 40 students are doing Maths/SL, 30 students are doing Physics/SL and 12 are doing both.
   a) A student is chosen at random. Find the probability that this student is doing Physics/SL given that he/she is doing Maths/SL.
   b) Are doing Physics/SL and Maths/SL independent?

11 A market chain in Germany accepts only Mastercard and Visa. It estimates that 21% of its customers use Mastercard, 57% use Visa and 13% use both cards.
   a) What is the probability that a customer will have an acceptable credit card?
   b) What proportion of their customers has neither card?
   c) What proportion of their customers has exactly one acceptable card?

Practice questions

1 Two independent events A and B are given such that $P(A) = k$, $P(B) = k + 0.3$ and $P(A \cap B) = 0.18$.
   a) Find k.
   b) Find $P(A \cup B)$.
   c) Find $P(A' | B')$.

2 Many airport authorities test prospective employees for drug use, with the intent of improving efficiency and reducing accidents. This procedure has plenty of opponents who claim that it creates difficulties for some classes of people and that it prevents others from getting these jobs even if they were not drug users. The claim depends on the fact that these tests are not 100% accurate. To test this claim, let us assume that a

test is 98% accurate in the sense that it identifies a person as a user or non-user 98% of the time. Each job applicant takes this test twice. The tests are done at separate times and are designed to be independent of each other. What is the probability that
   a) a non-user fails both tests?
   b) a drug user is detected (i.e. he/she fails at least one test)?
   c) a drug user passes both tests?

3 Communications satellites are difficult to repair when something goes wrong. One satellite works on solar energy and has two systems that provide electricity: the main system with a probability of failure of 0.002, and a back-up system that works independently of the main one. It has a failure rate of 0.01. What is the probability that the systems do not fail at the same time?

4 In a group of 200 students taking the IB examination, 120 take Spanish, 60 take French and 10 take both.
   a) If a student is selected at random, what is the probability that he/she
      (i) takes either French or Spanish?
      (ii) takes either French or Spanish but not both?
      (iii) does not take any French or Spanish?
   b) Given that a student takes the Spanish exam, what is the chance that he/she takes French?

5 In a factory producing disk drives for computers, there are three machines that work independently to produce one of the components. In any production process, machines are not 100% fault free. The production after one 'run' from these machines is listed below.

	Defective	Non-defective
Machine I	6	120
Machine II	4	80
Machine III	10	150

   a) A component is chosen at random from the produced lot. Find the probability that the chosen component is
      (i) from machine I
      (ii) a defective component from machine II
      (iii) non-defective or from machine I
      (iv) from machine I given that it is defective.
   b) Is the quality of the component dependent on the machine used?

6 At a school, the students are organizing a lottery to raise money for the needy in their community. The lottery tickets they have consist of small coloured envelopes inside which there is a small note. The note says: 'You won a prize!' or 'Sorry, try another ticket.' The envelopes have several colours. They have 70 red envelopes that contain two prizes, and the rest (130 tickets) contain four other prizes.
   a) You want to help this class and you buy a ticket hoping that it does not have a prize. Additionally, you don't like the red colour. You pick your ticket at random by closing your eyes. What is the probability that your wish comes true?
   b) You are surprised – you picked a red envelope. What is the probability that you did not win a prize?

7 You are given two events A and B with the following conditions

$P(A|B) = 0.30$, $P(B|A) = 0.60$, $P(A \cap B) = 0.18$

a) Find P(B).

b) Are A and B independent? Why?

c) Find $P(B \cap A')$.

8 In several ski resorts in Austria and Switzerland, the local sports authorities use high school students as 'ski instructors' to help deal with the surge in demand during vacations. However, to become an instructor, you have to pass a test and be a senior at your school. Here are the results of a survey of 120 students in a Swiss school who are training to become instructors. In this group, there are 70 boys and 50 girls. 74 students took the test, 32 boys and 16 girls passed the test, and the rest, including 12 girls, failed the test. 10 of the students, including 6 girls, were too young to take the ski test.

a) Copy and complete the table.

	Boys	Girls
Passed the ski test	32	16
Failed the ski test		12
Training, but did not take the test yet		
Too young to take the test		

b) Find the probability that
  (i) a student chosen at random has taken the test
  (ii) a girl chosen at random has taken the test
  (iii) a randomly chosen boy and randomly chosen girl have both passed the ski test.

9 Two events A and B are such that $P(A) = \frac{9}{16}$, $P(B) = \frac{3}{8}$, and $P(A|B) = \frac{1}{4}$. Find the probability that

a) both events will happen

b) only one of the events will happen

c) neither event will happen.

10 Martina plays tennis. When she serves, she has a 60% chance of succeeding with her first serve and continuing the game. She has a 95% chance on the second serve. Of course if both serves are not successful, she loses the point.

a) Find the probability that she misses both serves.

If Martina succeeds with the first serve, her chances of gaining the point against Steffy is 75%. If she is only successful with the second serve, her chances against Steffy for that point go down to 50%.

b) Find the probability that Martina wins a point against Steffy.

11 For the events A and B, $P(A) = 0.6$, $P(B) = 0.8$ and $P(A \cup B) = 1$. Find

a) $P(A \cap B)$

b) $P(A' \cup B')$

12 In a survey, 100 students were asked, 'Do you prefer to watch television or play sport?' Of the 46 boys in the survey, 33 said they would choose sport, while 29 girls made this choice.

	Boys	Girls	Total
Television			
Sport	33	29	
Total	46		100

By completing this table or otherwise, find the probability that
a) a student selected at random prefers to watch television
b) a student prefers to watch television given that the student is a boy.

13 Two ordinary, six-sided dice are rolled and the total score is noted.
a) Complete the tree diagram by entering probabilities and listing outcomes.

b) Find the probability of getting one or more sixes.

14 The following Venn diagram shows a sample space U and events A and B.

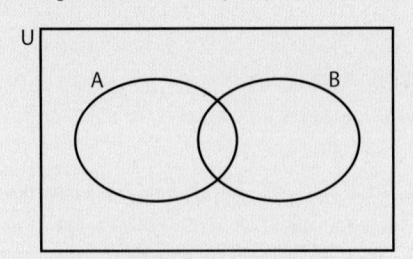

$n(U) = 36$, $n(A) = 11$, $n(B) = 6$ and $n(A \cup B)' = 21$.

a) On the diagram, shade the region $(A \cup B)'$.
b) Find
   (i) $n(A \cap B)$
   (ii) $P(A \cap B)$.
c) Explain why events A and B are not mutually exclusive.

15 In a survey of 200 people, 90 of whom were female, it was found that 60 people were unemployed, including 20 males.

   a) Using this information, complete the table below.

	Males	Females	Totals
Unemployed			
Employed			
Totals			200

   b) If a person is selected at random from this group of 200, find the probability that this person is
      (i) an unemployed female
      (ii) a male given that the person is employed.

16 A bag contains 10 red balls, 10 green balls and 6 white balls. Two balls are drawn at random from the bag without replacement. What is the probability that they are of different colours?

17 The following Venn diagram shows the universal set U and the sets A and B.

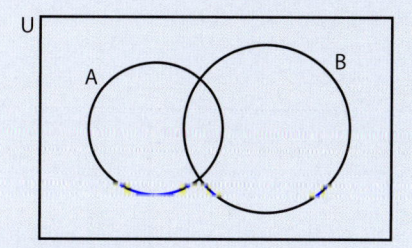

   a) Shade the area in the diagram which represents the set $B \cap A'$.

$$n(U) = 100, \ n(A) = 30, \ n(B) = 50, \ n(A \cup B) = 65.$$

   b) Find $n(B \cap A')$.
   c) An element is selected at random from U. What is the probability that this element is in $B \cap A'$?

18 The events B and C are dependent, where C is the event 'a student takes chemistry', and B is the event 'a student takes biology'. It is known that
$$P(C) = 0.4, \ P(B|C) = 0.6, \ P(B|C') = 0.5.$$

   a) Complete the following tree diagram.
   b) Calculate the probability that a student takes biology.
   c) Given that a student takes biology, what is the probability that the student takes chemistry?

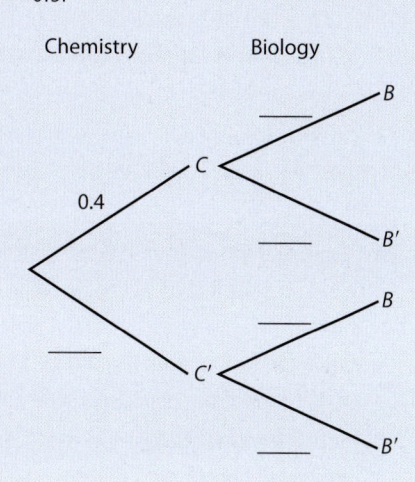

19 Two fair dice are thrown and the number showing on each is noted. The sum of these two numbers is S. Find the probability that
   a) S is less than 8
   b) at least one die shows a 3
   c) at least one die shows a 3 given that S is less than 8.

20 For events A and B, the probabilities are $P(A) = \frac{3}{11}$ and $P(B) = \frac{4}{11}$.
   Calculate the value of $P(A \cap B)$ if
   a) $P(A \cup B) = \frac{6}{11}$
   b) events A and B are independent.

21 Consider events A and B such that $P(A) \neq 0$, $P(A) \neq 1$, $P(B) \neq 0$ and $P(B) \neq 1$.
   In each of the situations **a)**, **b)**, **c)** below, state whether A and B are mutually exclusive (M), independent (I), or neither (N).
   a) $P(A|B) = P(A)$
   b) $P(A \cap B) = 0$
   c) $P(A \cap B) = P(A)$

22 In a school of 88 boys, 32 study economics (E), 28 study history (H) and 39 do not study either subject. This information is represented in the following Venn diagram.

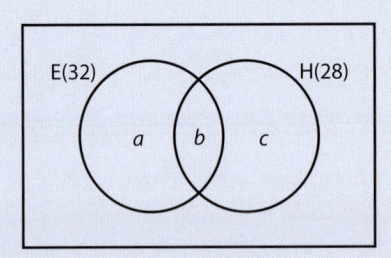

   a) Calculate the values a, b, c.
   b) A student is selected at random.
      (i) Calculate the probability that he studies both economics and history.
      (ii) Given that he studies economics, calculate the probability that he does not study history.
   c) A group of three students is selected at random from the school.
      (i) Calculate the probability that none of these students studies economics.
      (ii) Calculate the probability that at least one of these students studies economics.

23 A painter has 12 tins of paint. Seven tins are red and five tins are yellow. Two tins are chosen at random. Calculate the probability that both tins are the same colour.

24 Dumisani is a student at IB World College.
   The probability that he will be woken by his alarm clock is $\frac{7}{8}$.
   If he is woken by his alarm clock, the probability he will be late for school is $\frac{1}{4}$.
   If he is not woken by his alarm clock, the probability he will be late for school is $\frac{3}{5}$.
   Let W be the event 'Dumisani is woken by his alarm clock'.
   Let L be the event 'Dumisani is late for school'.

a) Copy and complete the tree diagram below.

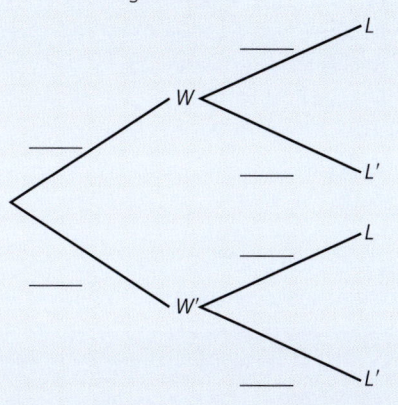

b) Calculate the probability that Dumisani will be late for school.
c) Given that Dumisani is late for school, what is the probability that he was woken by his alarm clock?

25 The diagram shows a circle divided into three sectors A, B and C. The angles at the centre of the circle are 90°, 120° and 150°. Sectors A and B are shaded as shown.

The arrow is spun. It cannot land on the lines between the sectors. Let A, B, C and S be the events defined by

   A : Arrow lands in sector A
   B : Arrow lands in sector B
   C : Arrow lands in sector C
   S : Arrow lands in a shaded region.

Find
a) $P(B)$ b) $P(S)$ c) $P(A|S)$.

26 A packet of seeds contains 40% red seeds and 60% yellow seeds. The probability that a red seed grows is 0.9, and that a yellow seed grows is 0.8. A seed is chosen at random from the packet.
a) Complete the probability tree diagram below.

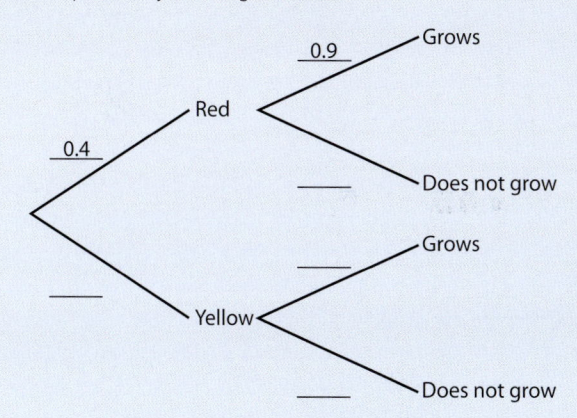

b) (i) Calculate the probability that the chosen seed is red and grows.
   (ii) Calculate the probability that the chosen seed grows.
   (iii) Given that the seed grows, calculate the probability that it is red.

27 Two unbiased six-sided dice are rolled, a red one and a black one. Let E and F be the events

   E: the same number appears on both dice
   F: the sum of the numbers is 10.

Find
a) $P(E)$
b) $P(F)$
c) $P(E \cup F)$.

28 The table below shows the subjects studied by 210 students at a college.

	Year 1	Year 2	Totals
History	50	35	85
Science	15	30	45
Art	45	35	80
Totals	110	100	210

a) A student from the college is selected at random.

   Let A be the event the student studies art.
   Let B be the event the student is in year 2.

   (i) Find $P(A)$.
   (ii) Find the probability that the student is a year 2 art student.
   (iii) Are the events A and B independent? Justify your answer.

b) Given that a history student is selected at random, calculate the probability that the student is in year 1.

c) Two students are selected at random from the college. Calculate the probability that one student is in year 1 and the other in year 2.

Questions 11–28: © International Baccalaureate Organization

11 Differential Calculus I: Fundamentals

Assessment statements
7.1 Informal ideas of limits and convergence.
Definition of derivative as $f'(x) = \lim_{h \to 0} \dfrac{f(x + h) - f(x)}{h}$.
Derivative interpreted as gradient function and as rate of change.
7.3 Local maximum and minimum points.
7.6 Kinematic problems involving displacement, s, velocity, v, and acceleration, a.
7.7 Graphical behaviour of functions: tangents and normals, behaviour for large $|x|$, horizontal and vertical asymptotes.
The significance of the second derivative; distinction between maximum and minimum points.
Points of inflexion with zero and non-zero gradients.
(See also Chapters 13 and 14.)

Introduction

Calculus is the branch of mathematics that was developed to analyze and model change – such as velocity and acceleration. We can also apply it to study change in the context of slope, area, volume and a wide range of other real-life phenomena. Although mathematical techniques that you have studied previously deal with many of these concepts, the ability to model change was restricted. For example, consider the curve in Figure 11.1. This shows the motion of an object by indicating the distance (y metres) travelled after a certain amount of time (t seconds). Pre-calculus mathematics will only allow us to compute the **average velocity** between two different times (Figure 11.2). With calculus – specifically, techniques of differential calculus – we will be able to find the velocity of the object at a particular instant, known as its **instantaneous velocity** (Figure 11.3). The starting point for our study of calculus is the idea of a limit.

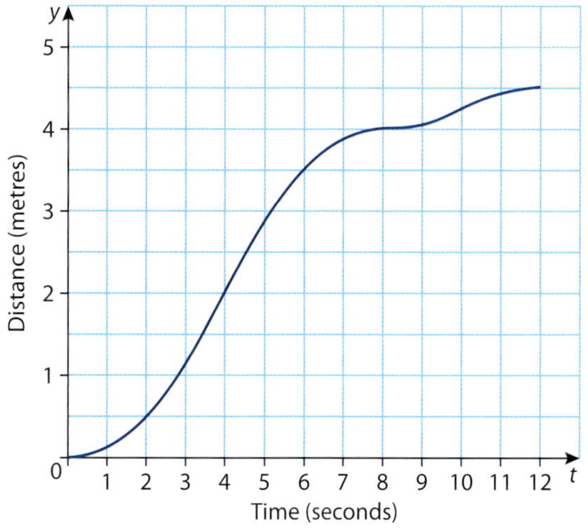

Figure 11.1

11 Differential Calculus I: Fundamentals

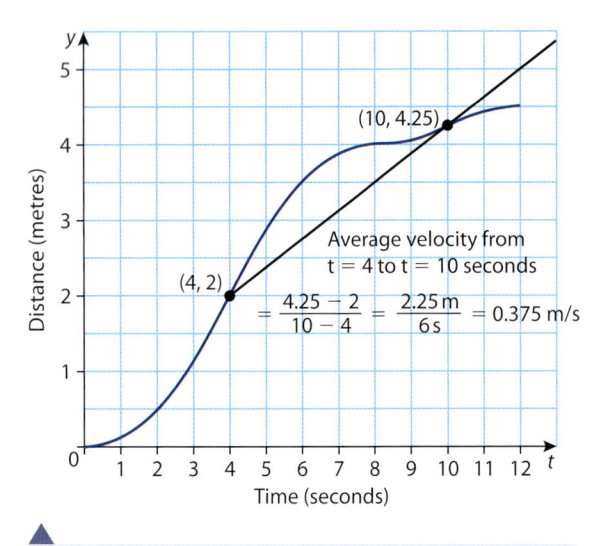

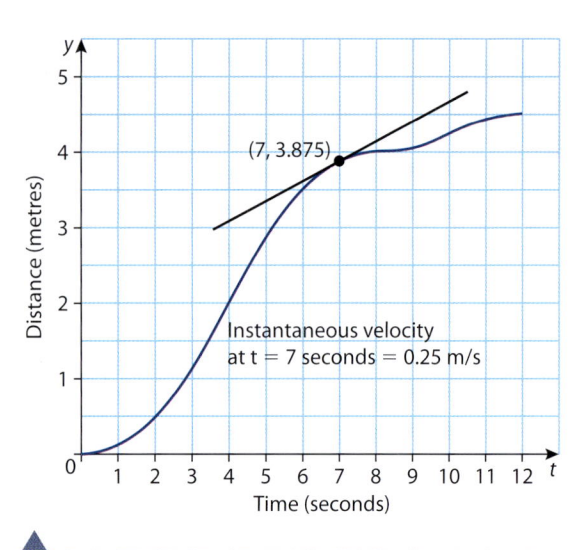

Figure 11.2

Average velocity from t = 4 to t = 10 seconds
$$= \frac{4.25 - 2}{10 - 4} = \frac{2.25\,m}{6\,s} = 0.375 \text{ m/s}$$

Figure 11.3

Instantaneous velocity at t = 7 seconds = 0.25 m/s

11.1 Limits of functions

A **limit** is one of the ideas that distinguish calculus from algebra, geometry and trigonometry. The notion of a limit is a fundamental concept of calculus. Limits are not new to us. We often use the idea of a 'limit' in many non-mathematical situations. Mathematically speaking, we have encountered limits on at least two occasions previously in this book – finding the sum of an infinite geometric series (Section 3.4) and computing the irrational number e (Section 4.3).

Recall from Section 3.4 that we established that if the sequence of partial sums for an infinite series **converges** to a finite number L we say that the infinite series has a 'sum' of L. Further on in that section, we used limits to algebraically confirm that the infinite series $2 + 1 + \frac{1}{2} + \frac{1}{4} + \frac{1}{8} + \ldots$ has a sum of 4. As part of the algebra for this, we reasoned that as the value of n increases in the positive direction without bound (i.e. $n \to +\infty$) the expression $\left(\frac{1}{2}\right)^n$ converges to zero – in other words, the **limit** of $\left(\frac{1}{2}\right)^n$ as n goes to positive infinity is zero. We express this result more efficiently using limit notation, as we did in Chapter 3, by writing $\lim_{n \to \infty} \left(\frac{1}{2}\right)^n = 0$.

It is beyond the requirements of this course to establish a precise formal definition of a limit, but a closer look at justifying this limit and a couple of others can lead us to a useful informal definition.

Example 1

Evaluate $\lim_{n \to \infty} \left(\frac{1}{2}\right)^n$ by using your GDC to analyze the behaviour of the function $f(x) = \left(\frac{1}{2}\right)^x$ for large positive values of x.

Solution

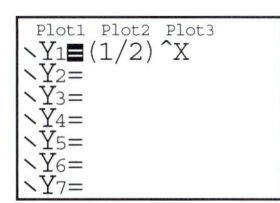

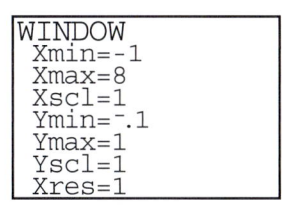

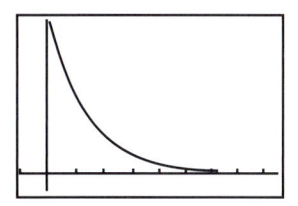

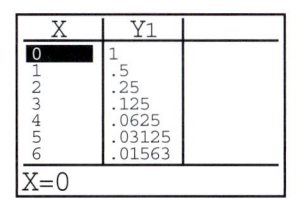

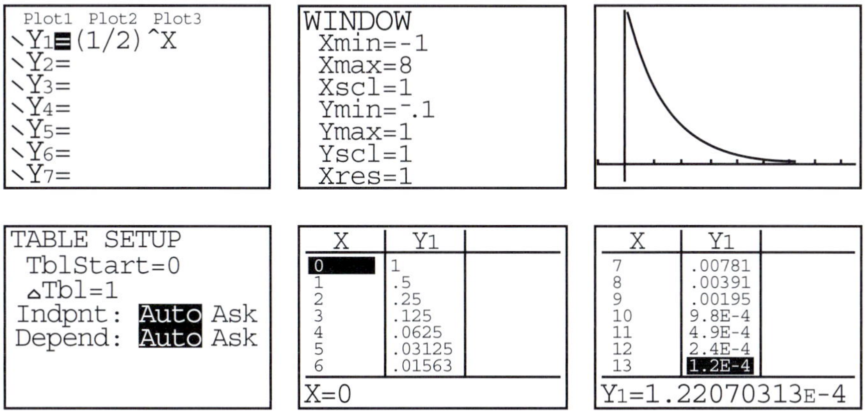

The GDC screen images show the graph and table of values for $y = \left(\frac{1}{2}\right)^x$. Clearly, the larger the value of x, the closer that y gets to zero. Although there is no value of x that will produce a value of y equal to zero, we can get as close to zero as we wish. For example, if we wish to produce a value of y within 0.001 of zero, then we could choose $x = 10$ and $y = \left(\frac{1}{2}\right)^{10} = \frac{1}{1024} \approx 0.000\,976\,56$; and if we want a result within 0.000 0001 of zero, then we could choose $x = 24$ and $y = \left(\frac{1}{2}\right)^{24} = \frac{1}{16\,777\,216} \approx 0.000\,000\,059\,605$; and so on. Therefore, we can conclude that $\lim_{n \to \infty} \left(\frac{1}{2}\right)^n = 0$.

In calculus we are interested in limits of functions of real numbers. Although many of the limits of functions that we will encounter can only be approached and not actually reached (as in Example 1), this is not always the case. For example, if asked to evaluate the limit of the function $f(x) = \frac{x}{2} - 1$ as x approaches 6, we simply need to evaluate the function for $x = 6$. Since $f(6) = 2$, then $\lim_{n \to 6} \left(\frac{x}{2} - 1\right) = 2$.

However, it is more common that we are unable to evaluate the limit of $f(x)$ as x approaches some number c because $f(c)$ does not exist.

> The line $y = c$ is a **horizontal asymptote** of the graph of a function $y = f(x)$ if either $\lim_{x \to \infty} f(x) = c$ or $\lim_{x \to -\infty} f(x) = c$. For example, the line $y = 0$ (x-axis) is a horizontal asymptote of the graph of $y = \left(\frac{1}{2}\right)^x$ because $\lim_{n \to \infty} \left(\frac{1}{2}\right)^n = 0$.

Example 2

Evaluate $\lim_{x \to 0} \frac{\sin x}{x}$.

- **Hint:** x must be in radians because in calculus we are interested in functions of real numbers.

Solution

We are not able to evaluate this limit by direct substitution because when $x = 0$, $\frac{\sin x}{x} = \frac{0}{0}$ and is therefore undefined. Let's use our GDC again to analyze the behaviour of the function $f(x) = \frac{\sin x}{x}$ as x approaches zero from the right side and the left side.

Differential Calculus I: Fundamentals

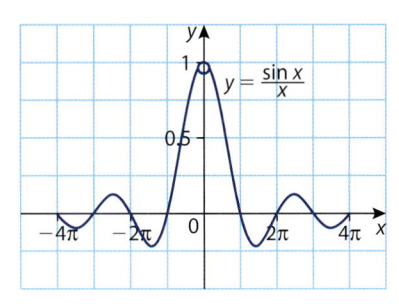

Although there is no point on the graph of $y = \frac{\sin x}{x}$ corresponding to $x = 0$, it is clear from the graph that as x approaches zero (from either direction) the value of $\frac{\sin x}{x}$ converges to one. We can get the value of $\frac{\sin x}{x}$ arbitrarily close to 1 depending on our choice of x. If we want $\frac{\sin x}{x}$ to be within 0.001 of 1, we choose $x = 0.05$ giving $\frac{\sin 0.05}{0.05} \approx 0.999\,583$ and $1 - 0.999\,583 = 0.000\,417 < 0.001$; and if we want $\frac{\sin x}{x}$ to be within 0.000\,001 of 1, then we choose $x = 0.002$ giving $\frac{\sin 0.02}{0.02} \approx 0.999\,999\,3333$ and $1 - 0.999\,999\,3333 = 0.000\,000\,6667 < 0.000\,001$; and so on. Therefore, $\lim_{x \to 0} \frac{\sin x}{x} = 1$.

Functions do not necessarily converge to a finite value at every point – it's possible for a limit not to exist.

Example 3

Find $\lim_{x \to 0} \frac{1}{x^2}$, if it exists.

Solution

As x approaches zero, the value of $\frac{1}{x^2}$ becomes increasingly large in the positive direction. The graph of the function (left) seems to indicate that we can make the values of $y = \frac{1}{x^2}$ arbitrarily large by choosing x close enough to zero. Therefore, the values of $y = \frac{1}{x^2}$ do not approach a finite number, so $\lim_{x \to 0} \frac{1}{x^2}$ does not exist.

Although we can describe the behaviour of the function $y = \frac{1}{x^2}$ by writing $\lim_{x \to 0} \frac{1}{x^2} = \infty$, this does not mean that we consider ∞ to represent a number – it does not. This notation is simply a convenient way to indicate in what manner the limit does not exist.

> The line $x = c$ is a **vertical asymptote** of the graph of a function $y = f(x)$ if either $\lim_{x \to c} f(x) = \infty$ or $\lim_{x \to c} f(x) = -\infty$.
> For example, the line $x = 0$ (y-axis) is a vertical asymptote of the graph of $y = \frac{1}{x^2}$ because $\lim_{x \to 0} \frac{1}{x^2} = \infty$.

> **Limit of a function**
> If $f(x)$ becomes arbitrarily close to a unique finite number L as x approaches c from either side, then the **limit** of $f(x)$ as x approaches c is L. The notation for indicating this is $\lim_{x \to c} f(x) = L$.
> When a function $f(x)$ becomes *arbitrarily close* to a finite number L, we say that $f(x)$ **converges** to L.

For our purposes in this course, it is also important to be able to apply some basic algebraic manipulation in order to evaluate the limits of some functions algebraically, rather than by conjecturing from a graph or table.

Example 4

Evaluate each limit algebraically.

a) $\lim_{x \to \infty} \dfrac{5x - 3}{x}$

b) $\lim_{p \to 0} (3x^2 - 4px + p^2)$

c) $\lim_{h \to 0} \dfrac{[(x + h)^2 - 6] - (x^2 - 6)}{h}$

Solution

a) $\lim_{x \to \infty} \dfrac{5x - 3}{x} = \lim_{x \to \infty} \left(\dfrac{5x}{x} - \dfrac{3}{x} \right)$ 　　Split the fraction into two terms and …

$\qquad = \lim_{x \to \infty} 5 - \lim_{x \to \infty} \dfrac{3}{x}$ 　　… evaluate the limit of each term separately.

$\qquad = 5 - 0 = 5$ 　　Therefore, $\lim_{x \to \infty} \dfrac{5x - 3}{x} = 5$.

b) $\lim_{p \to 0} (3x^2 - 4px + p^2) = \lim_{p \to 0} 3x^2 - \lim_{p \to 0} 4px + \lim_{p \to 0} p^2$ 　　Evaluate the limit of each term separately.

$\qquad = 3x^2 - 0 + 0 = 3x^2$ 　　Therefore, $\lim_{p \to 0} (3x^2 - 4px + p^2) = 3x^2$.

c) $\lim_{h \to 0} \dfrac{[(x + h)^2 - 6] - (x^2 - 6)}{h} = \lim_{h \to 0} \dfrac{x^2 + 2xh + h^2 - 6 - x^2 + 6}{h}$

$\qquad = \lim_{h \to 0} \dfrac{2xh + h^2}{h}$

$\qquad = \lim_{h \to 0} \dfrac{h(2x + h)}{h}$

$\qquad = \lim_{h \to 0} 2x + \lim_{h \to 0} h$

$\qquad = 2x + 0 = 2x$

Therefore, $\lim_{h \to 0} \dfrac{[(x + h)^2 - 6] - (x^2 - 6)}{h} = 2x$.

The limits in parts b) and c) of Example 4 show that in some cases the limit of a function is itself a function.

Exercise 11.1

In questions 1–4, evaluate each limit algebraically and then confirm your result by means of a table or graph on your GDC.

1 $\lim_{n \to \infty} \dfrac{1 + 4n}{n}$

2 $\lim_{h \to 0} (3x^2 + 2hx + h^2)$

3 $\lim_{d \to 0} \dfrac{(x + d)^2 - x^2}{d}$

4 $\lim_{x \to 3} \dfrac{x^2 - 9}{x - 3}$

In questions 5–7, investigate the limit of the expression (if it exists) as $x \to \infty$ by evaluating the expression for the following values of x: 10, 50, 100, 1000, 10000 and 1 000 000. Hence, make a conjecture for the value of each limit.

5 $\lim\limits_{x \to \infty} \dfrac{3x + 2}{x^2 - 3}$

6 $\lim\limits_{x \to \infty} \dfrac{5x - 6}{2x + 5}$

7 $\lim\limits_{x \to \infty} \dfrac{3x^2 + 2}{x - 3}$

8 Use the graphing or table capabilities of your GDC to investigate the values of the expression $\left(1 + \dfrac{1}{c}\right)^c$ as c increases without bound (i.e. $c \to \infty$). Explain the significance of the result.

9 If it is known that the line $y = 3$ is a horizontal asymptote for the function $f(x)$, state the value of each of the following two limits: $\lim\limits_{x \to \infty} f(x)$ and $\lim\limits_{x \to -\infty} f(x)$.

10 If it is known that the line $x = a$ is a vertical asymptote for the function $g(x)$ and $g(x) > 0$, what conclusion can be made about $\lim\limits_{x \to a} g(x)$?

11 State the equations of all horizontal and vertical asymptotes for the following functions.

a) $f(x) = \dfrac{3x - 1}{1 + x}$

b) $g(x) = \dfrac{1}{(x - 2)^2}$

c) $g(x) = \dfrac{1}{x - a} + b$

11.2 The derivative of a function: definition and basic rules

Tangent lines and the slope (gradient) of a curve

In Section 1.6, we reviewed linear equations in two variables. And, later in Section 2.1, we established that any non-vertical line represents a function for which we typically assign the variables x and y for values in the domain and range of the function, respectively. Any linear function can be written in the form $y = mx + c$. This is the slope-intercept form for a linear equation, where m is the slope (or gradient) of the graph and c is the y-coordinate of the point at which the graph intersects the y-axis (i.e. the y-intercept). The value of the slope m, defined as $m = \dfrac{y_2 - y_1}{x_2 - x_1} = \dfrac{\text{vertical change}}{\text{horizontal change}}$, will be the same for any pair of points, (x_1, y_1) and (x_2, y_2), on the line. An essential characteristic of the graph of a linear function is that it has a constant slope. This is not true for the graphs of non-linear functions.

Consider a person walking up the side of a pitched roof as shown in Figure 11.4. At *any* point along the line segment PQ the person is experiencing a slope of $\frac{3}{4}$. Now consider someone walking up the curve shown in Figure 11.5, which passes through the three points A, B and C. As the person walks along the curve from *A* to *C*, he/she will experience a steadily increasing slope. The slope is continually changing from one point to the next along the curve. Therefore, it is incorrect to say that a non-linear function, whose graph is a curve, has *a* slope – it has *infinitely many* slopes. We need a means to determine the slope of a non-linear function *at a specific point* on its graph.

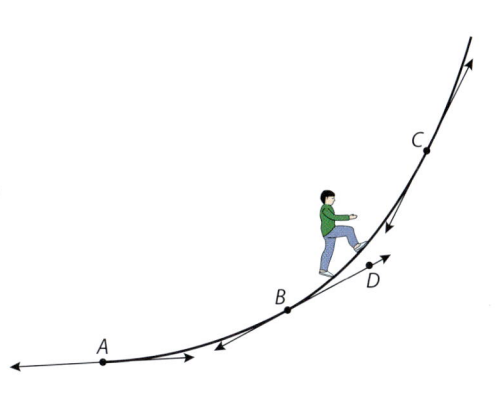

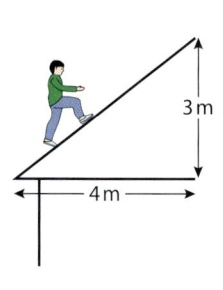

Figure 11.4

Imagine if the slope of the curve in Figure 11.5 stopped increasing (remained constant) after point *B*. From that point on, a person walking up the curve would move along a line with a slope equal to the slope of the curve at point *B*. This line – containing point *D* in the diagram – only 'touches' the curve once at *B*. Line (*BD*) is **tangent** to the curve at point *B*. Therefore, finding the slope of the line that is tangent to a curve at a certain point will give us the slope of the curve at that point.

Figure 11.5

Finding the slope of a curve at a point – or better – finding a rule (function) that gives us the slope at any point on the curve is very useful information in many applications. The slope of a line, or of a curve at a point, is a measure of how fast variable *y* is changing as variable *x* changes. **The slope represents the rate of change of *y* with respect to *x*.** To find the slope of a tangent line, we first need to clarify what it means to say that a line is tangent to a curve at a point. Then we can establish a method to find the tangent line at a point.

> The slope (gradient) of a curve at a point is the slope of the line that is tangent to the curve at that point.

• **Hint:** The word 'curve' can often mean the same as 'function', even if the function is linear.

The three graphs in Figure 11.6 show different configurations of tangent lines. A tangent line may cross or intersect the graph at one or more points.

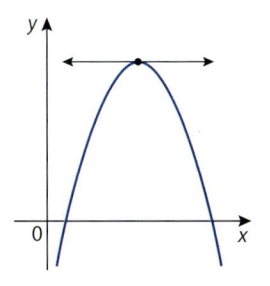

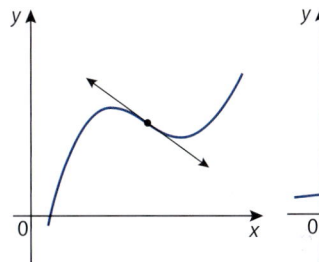

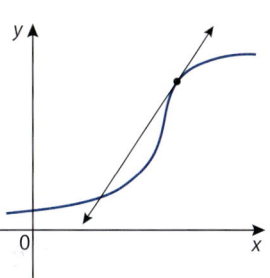

Figure 11.6

Figure 11.7

For many functions, the graph has a tangent at *every* point. Informally, a function is said to be *smooth* if it has this property. Any linear function is certainly smooth, since the tangent at each point coincides with the original graph. However, some graphs are not smooth at every point. Consider the point $(0, 0)$ on the graph of the function $y = |x|$ (Figure 11.7). Zooming in on $(0, 0)$ will always produce a V-shape rather than smoothing out to appear more and more linear. Therefore, there is no tangent to the graph at this point.

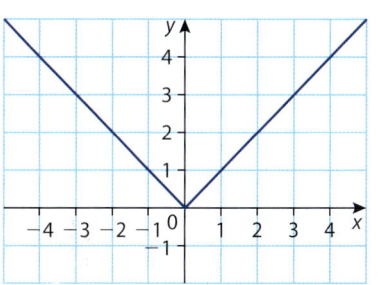

353

11 Differential Calculus I: Fundamentals

Figure 11.8

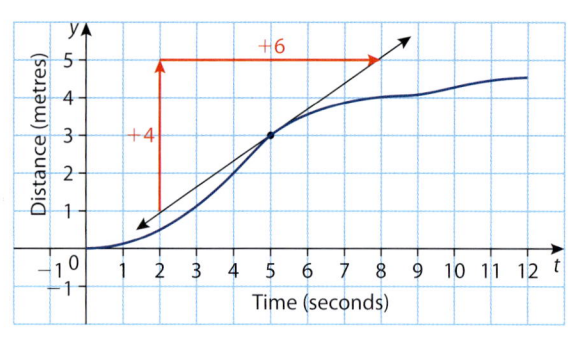

One way to find the tangent line of a graph at a particular point is to make a visual estimate. Figure 11.8 reproduces the time-distance graph for an object's motion from the previous section (Figure 11.1). The slope at any point (t, y) on the curve will give us the rate of change of the distance y with respect to time t, in other words the object's **instantaneous velocity** at time t. In the figure, an estimate of the line tangent to the curve at $(5, 3)$ has been drawn. Reading from the graph, the slope appears to be $\frac{4}{6} = \frac{2}{3}$. Or, in other words, the object has a velocity of approximately 0.667 m/s at the instant when $t = 5$ seconds.

A more precise method of finding tangent lines makes use of a secant line and a limit process. Suppose that f is any smooth function, so the tangent to its graph exists at all points. A **secant line** (or chord) is drawn through the point for which we are trying to find a tangent to f and a second point on the graph of f, as shown in Figure 11.9a. If P is the point of tangency with coordinates $(x, f(x))$, choose a point Q to be horizontally some h units away. Hence, the coordinates of point Q are $(x + h, f(x + h))$. Then the slope of the secant line (PQ) is $m_{\text{sec}} = \dfrac{f(x + h) - f(x)}{(x + h) - x} = \dfrac{f(x + h) - f(x)}{h}$.

The right side of this equation is often referred to as a **difference quotient**. The numerator is the change in y, and the denominator h is the change in x. The limit process of achieving better and better approximations for the slope of the tangent at P consists of finding the slope of the secant (PQ) as Q moves ever closer to P, as shown in the graphs in Figure 11.9b and Figure 11.9c. In doing so, the value of h will approach zero.

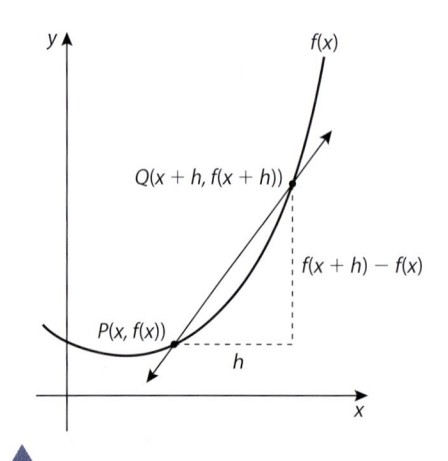

Figure 11.9a

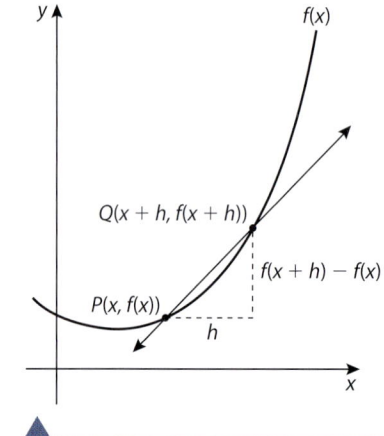

Figure 11.9b

Figure 11.9c

Figure 11.9d

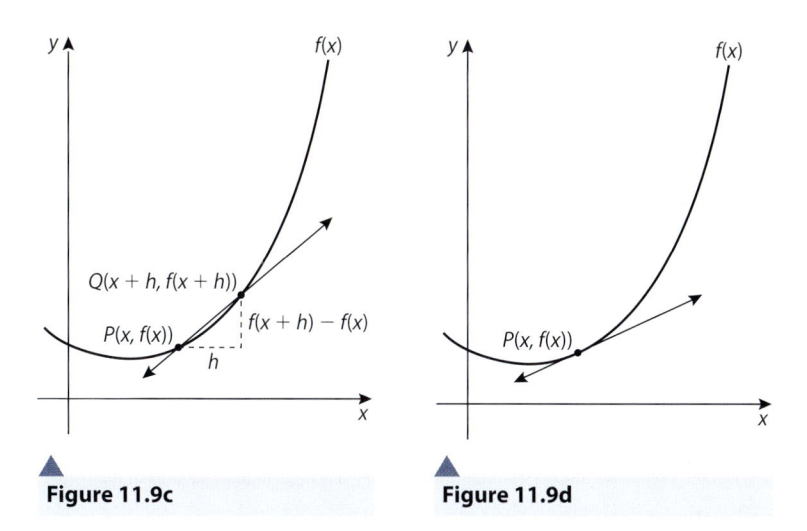

By evaluating a limit of the slope of the secant lines as h approaches zero, we can find the exact slope of the tangent line at $P(x, f(x))$.

> **The slope (gradient) of a curve at a point**
> The slope of the curve $y = f(x)$ at the point $(x, f(x))$ is equal to the slope of its tangent line at $(x, f(x))$, and is given by
> $$m_{\tan} = \lim_{h \to 0} m_{\sec} = \lim_{h \to 0} \frac{f(x + h) - f(x)}{h}$$
> provided that this limit exists.

The word 'secant', as applied to a line, comes from the Latin word secare, *meaning to cut. The word 'tangent' comes from the Latin verb* tangere, *meaning to touch.*

Let's apply the definition of the slope of a curve at a point to find a rule, or function, for the slope of all of the tangent lines to a curve.

Example 5

Find a rule for the slopes of the tangent lines to the graph of $f(x) = x^2 + 1$. Use this rule to find the exact slope of the curve at the point where $x = 0$ and at the point where $x = 1$.

Solution

Let $(x, f(x))$ represent any point on the graph of f. By definition, the slope of the tangent line at $(x, f(x))$ is:

$$m = \lim_{h \to 0} \frac{f(x + h) - f(x)}{h} = \lim_{h \to 0} \frac{[(x + h)^2 + 1] - [x^2 + 1]}{h}$$

$$= \lim_{h \to 0} \frac{[x^2 + 2xh + h^2 + 1] - [x^2 + 1]}{h}$$

$$= \lim_{h \to 0} \frac{\cancel{x^2} - \cancel{x^2} + 2xh + h^2 + \cancel{1} - \cancel{1}}{h}$$

$$= \lim_{h \to 0} \frac{\cancel{h}(2x + h)}{\cancel{h}}$$

$$= \lim_{h \to 0} (2x + h)$$

$$= 2x$$

Therefore, the slope at any point $(x, f(x))$ on the graph of f is $2x$.

355

11 Differential Calculus I: Fundamentals

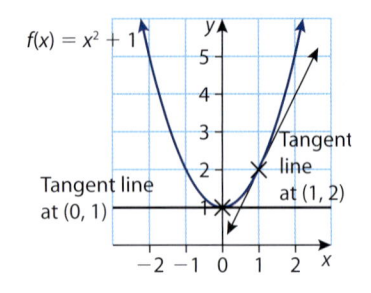

At the point where $x = 0$, the slope is $2(0) = 0$. This makes visual sense because the point $(0, 1)$ is the vertex of the parabola $y = x^2 + 1$, and we expect that the tangent at this point is a horizontal line with a slope of zero. At the point where $x = 1$, the slope is $2(1) = 2$. This also makes visual sense because moving along the curve from $(0, 1)$ to $(1, 2)$ the slope is steadily increasing.

In Example 5, from the function $f(x) = x^2 + 1$ we used the limit process to derive another function with the rule $2x$. With this derived function we can compute the slope (gradient) of the graph of $f(x)$ at a point from simply inputting the x-coordinate of the point. This *derived* function is called the **derivative** of f at x. It is given the notation $f'(x)$, which is commonly read as 'f prime of x', or simply, 'the derivative of f of x.'

> **The derivative and differentiation**
> - The **derivative**, $f'(x)$, at a point x in the domain of f is the slope (gradient) of the graph of f at $(x, f(x))$, and is given by
> $$f'(x) = \lim_{h \to 0} \frac{f(x+h) - f(x)}{h}$$
> provided that this limit exists.
> - If the derivative exists at each point of the domain of f, we say that f is **smooth**.
> - The process of finding the derivative, $f'(x)$, is called **differentiation**.
> - If $y = f(x)$, then $f'(x)$ is a formula for the instantaneous **rate of change** of y with respect to x.

> If finding the derivative of a function indicated with the function notation $f(x)$, then – as shown already – the derivative is usually denoted as $f'(x)$. However, there are two other notations with which you should be familiar. Commonly, if a function is given as y in terms of x, then the derivative is denoted as y', read as 'y prime.' The notation $\frac{dy}{dx}$ is also often used to indicate a derivative, and is read as 'the derivative of y with respect to x.' Note: $\frac{dy}{dx}$ is not a fraction. If, for example, $y = x^2 + 1$, the derivative can be denoted by writing $\frac{d}{dx}(x^2 + 1) = 2x$. This is read as 'the derivative of $x^2 + 1$ with respect to x is $2x$.'

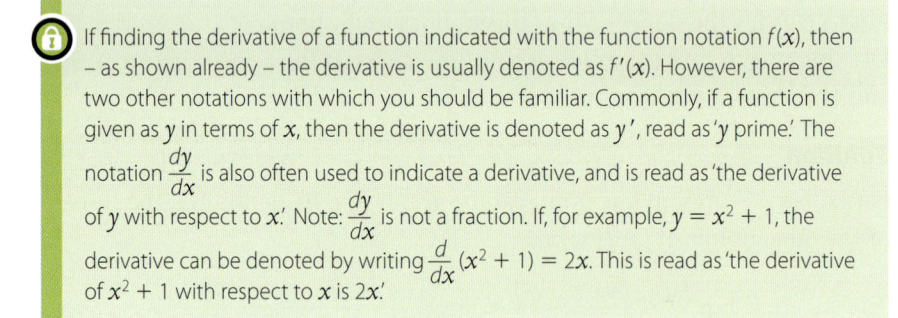

Differentiating from first principles

Depending on the particular purpose that you have in differentiating a function, you can consider the derivative as giving the slope of the graph of the function *or* the rate of change of the dependent variable (commonly y) with respect to the independent variable (commonly x). Both interpretations are useful and widely applied.

Using the limit definition directly to find the derivative of a function (as we did in Example 5) is often called 'differentiating from first principles'.

Example 6

Differentiating from first principles, find the derivative of $f(x) = x^3$.

Solution

$$f'(x) = \lim_{h \to 0} \frac{f(x+h) - f(x)}{h} = \lim_{h \to 0} \frac{(x+h)^3 - x^3}{h}$$

$$= \lim_{h \to 0} \frac{(x+h)(x+h)^2 - x^3}{h}$$

$$= \lim_{h \to 0} \frac{(x+h)(x^2 + 2hx + h^2) - x^3}{h}$$

$$= \lim_{h \to 0} \frac{x^3 + 3hx^2 + 3h^2 x + h^3 - x^3}{h}$$

$$= \lim_{h \to 0} \frac{h(3x^2 + 3hx + h^2)}{h}$$

$$= \lim_{h \to 0} (3x^2 + 3hx + h^2)$$

$$= 3x^2$$

Therefore, the derivative of $f(x) = x^3$ is $f'(x) = 3x^2$.

As in Example 5, the result for Example 6 is a function that gives us the slope at any point on the graph of $y = x^3$. For example, the points $(1, 1)$ and $(-1, -1)$ both lie on $y = x^3$, and the slopes at these points are respectively $f'(1) = 3(1)^2 = 3$ and $f'(-1) = 3(-1)^2 = 3$. Hence, the tangents at these points will be parallel, as shown in Figure 11.10.

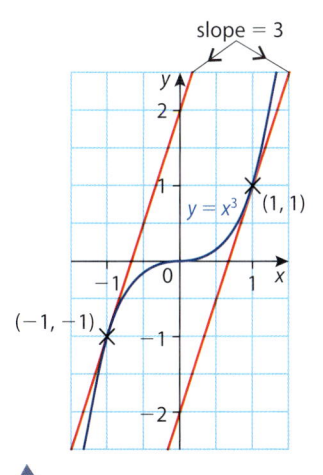

Figure 11.10

Let's examine the relationship between the slopes of tangents to the curve $f(x) = x^2 + 1$ (Example 5) and slopes of tangents to $g(x) = x^2$. Recall that we found the derivative of $f(x)$ to be $f'(x) = 2x$. It appears from the graphs of f and g, in Figure 11.11, that the slopes of tangents at points with the same x-coordinate will be equal. For example, the tangent to g at the point $(1, 2)$ looks parallel to the tangent to f at $(1, 1)$, as shown in Figure 11.11. This implies that the derivatives of the two functions are equal. Rather than confirming this conjecture by finding the derivative of $g(x) = x^2$ by first principles (i.e. using the limit definition), let's use the graphical and computing power of our GDC. Any GDC model is capable of computing the slope of a curve at a point – either on the GDC's 'home' screen, or its graphing screen. The screen images on page 358 show computing derivative values for $y = x^2$ on the 'home' screen.

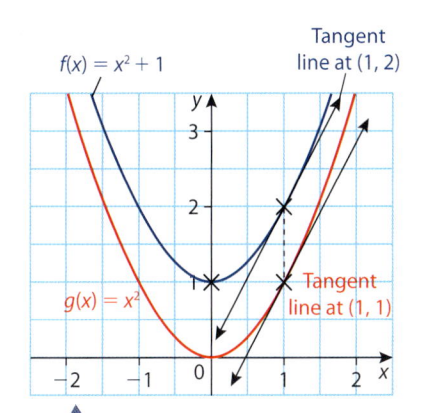

Figure 11.11

Differential Calculus I: Fundamentals

The exact command name and syntax for computing the value of a derivative at a point may vary from one GDC model to another.

This command finds the value of the derivative of $y = x^2$ in terms of x, at the point $x = 1$.

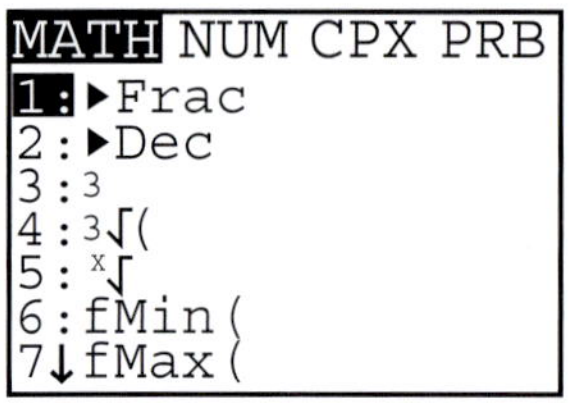

```
MATH NUM CPX PRB
1:▶Frac
2:▶Dec
3:3
4:3√(
5:ˣ√
6:fMin(
7↓fMax(
```

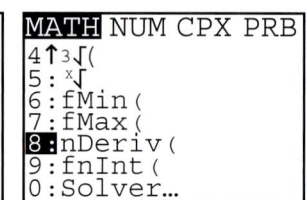

```
MATH NUM CPX PRB
4↑3√(
5:ˣ√
6:fMin(
7:fMax(
8:nDeriv(
9:fnInt(
0:Solver...
```

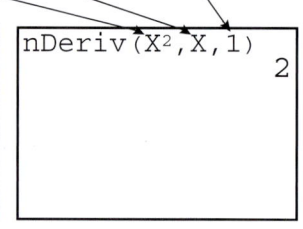

```
nDeriv(X²,X,1)
                2
```

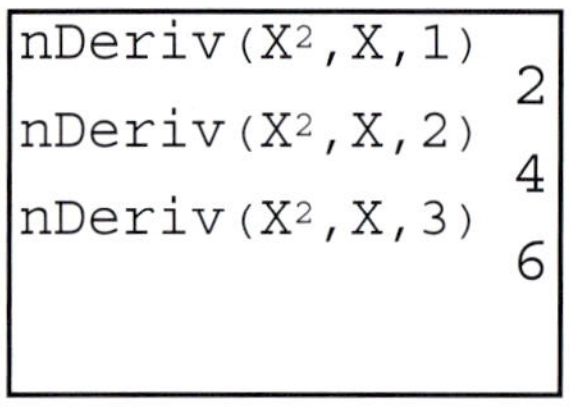

```
nDeriv(X²,X,1)
                2
nDeriv(X²,X,2)
                4
nDeriv(X²,X,3)
                6
```

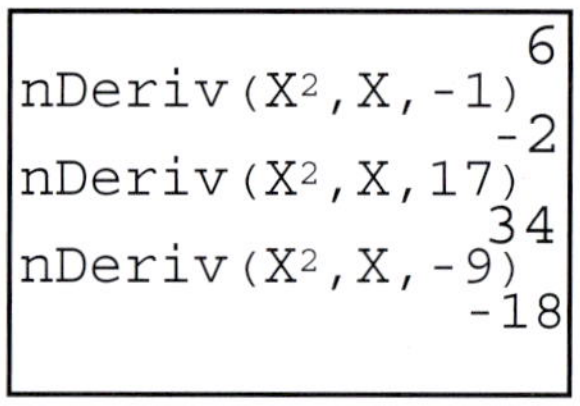

```
nDeriv(X²,X,-1)
                6
                -2
nDeriv(X²,X,17)
                34
nDeriv(X²,X,-9)
                -18
```

Our GDC results confirm our conjecture that the derivative of $g(x) = x^2$ is $g'(x) = 2x$.

Example 7

From first principles, find:

a) y' given $y = 3x^2 + 2x$

b) $\dfrac{dy}{dx}$ given $y = \dfrac{1}{x}$

Solution

We will apply the definition of the derivative, $f'(x) = \lim\limits_{h \to 0} \dfrac{f(x+h) - f(x)}{h}$, in both a) and b).

a) $y' = \lim\limits_{h \to 0} \dfrac{[3(x+h)^2 + 2(x+h)] - (3x^2 + 2x)}{h}$

$= \lim\limits_{h \to 0} \dfrac{(3x^2 + 6hx + 3h^2 + 2x + 2h) - (3x^2 + 2x)}{h}$

$= \lim\limits_{h \to 0} \dfrac{6hx + 3h^2 + 2h}{h}$

$= \lim\limits_{h \to 0} (6x + 3h + 2) \quad \Rightarrow \quad y' = 6x + 2$

b) $\dfrac{dy}{dx} = \dfrac{d}{dx}\left(\dfrac{1}{x}\right) = \lim\limits_{h \to 0} \dfrac{\dfrac{1}{x+h} - \dfrac{1}{x}}{h}$

$= \lim\limits_{h \to 0} \dfrac{\dfrac{x}{x(x+h)} - \dfrac{x+h}{x(x+h)}}{h}$

$= \lim\limits_{h \to 0} \left(\dfrac{\dfrac{-h}{x(x+h)}}{\dfrac{h}{1}}\right)$

$= \lim\limits_{h \to 0} \left(\dfrac{-h}{x(x+h)} \cdot \dfrac{1}{h}\right)$

$= \lim\limits_{h \to 0} \left(\dfrac{-1}{x^2 + hx}\right) \Rightarrow \dfrac{d}{dx}\left(\dfrac{1}{x}\right) = -\dfrac{1}{x^2}$ or $\dfrac{d}{dx}(x^{-1}) = -x^{-2}$

Basic differentiation rules

We have now established the following results:
- If $f(x) = x^2$, then $f'(x) = 2x$.
- If $f(x) = x^2 + 1$, then $f'(x) = 2x$.
- If $f(x) = 3x^2 + 2x$, then $f'(x) = 6x + 2$.
- If $f(x) = x^3$, then $f'(x) = 3x^2$.
- If $f(x) = x^{-1}$, then $f'(x) = -x^{-2}$.

In addition, we know that if $f(x) = x$, then $f'(x) = 1$, since the line $y = x$ has a constant slope equal to 1; and that if $f(x) = 1$, then $f'(x) = 0$ because the line $y = 1$ is horizontal and thus has a constant slope equal to 0. Furthermore, the graph of any function $f(x) = c$, where c is a constant, is a horizontal line, confirming that if $f(x) = c$, $c \in \mathbb{R}$, then $f'(x) = 0$. In other words, the derivative of a constant is zero. This leads to our first basic rule of differentiation.

> **The constant rule**
> The derivative of a constant function is zero. That is, given c is a real number, and if $f(x) = c$, then $f'(x) = 0$.

These following results:
$$f(x) = x^{-1} \Rightarrow f'(x) = -x^{-2}$$
$$f(x) = x^0 = 1 \Rightarrow f'(x) = 0$$
$$f(x) = x^1 = x \Rightarrow f'(x) = 1$$
$$f(x) = x^2 \Rightarrow f'(x) = 2x$$
$$f(x) = x^3 \Rightarrow f'(x) = 3x^2$$

can be summarized in the single statement:

if $f(x) = x^n$ then $f'(x) = nx^{n-1}$ for $n = -1, 0, 1, 2, 3$

In fact, this statement is true not just for these values but for *any* value of n that is a rational number ($n \in \mathbb{Q}$). This leads to our second basic rule of differentiation.

> **The derivative of x^n**
> Given n is a rational number, and if $f(x) = x^n$, then $f'(x) = nx^{n-1}$.

Another basic rule of differentiation is suggested by our result that the derivative of $f(x) = x^2 + 1$ is $f'(x) = 2x$. The derivative of a sum of a number of terms is obtained by differentiating each term separately – i.e. differentiating 'term-by-term'. That is,

$$\frac{d}{dx}(x^2 + 1) = \frac{d}{dx}(x^2) + \frac{d}{dx}(1) = 2x + 0 = 2x.$$

> **The sum and difference rule**
> If $f(x) = g(x) \pm h(x)$ then $f'(x) = g'(x) \pm h'(x)$.

A fourth basic rule of differentiation is illustrated by our result that the derivative of $f(x) = 3x^2 + 2x$ is $f'(x) = 6x + 2$. Using the sum rule, $f'(x) = \frac{d}{dx}(3x^2 + 2x) = \frac{d}{dx}(3x^2) + \frac{d}{dx}(2x) = 6x + 2$. The fact that $\frac{d}{dx}(3x^2) = 6x$ suggests that $3 \cdot \frac{d}{dx}(x^2) = 3 \cdot 2x = 6x$. In other words, the derivative of a function being multiplied by a constant is equal to the constant multiplying the derivative of the function.

> Functions of the form $f(x) = x^n$ are called **power functions**, so the differentiation rule $\frac{d}{dx}(x^n) = nx^{n-1}$ gives the rule for differentiating power functions – and is often referred to as the **power rule**.

359

11 Differential Calculus I: Fundamentals

> **The constant multiple rule**
> If $f(x) = c \cdot g(x)$ then $f'(x) = c \cdot g'(x)$.

As mentioned before, and as you have seen, there are different notations used for indicating a derivative or differentiation. These can be traced back to the fact that calculus was first developed by Isaac Newton (1642–1727) and Gottfried Leibniz (1646–1716) independently of each other – and hence, introduced different symbols for methods of calculus. The 'prime' notations y' and $f'(x)$ come from notations that Newton used for derivatives. The $\frac{dy}{dx}$ notation is similar to that used by Leibniz for indicating differentiation. Each has its advantages and disadvantages. For example, it is often easier to write our four basic rules of differentiation using Leibniz notation as shown below.

Constant rule: $\quad \frac{d}{dx}(c) = 0, \; c \in \mathbb{R}$

Power rule: $\quad \frac{d}{dx}(x^n) = nx^{n-1}, \; n \in \mathbb{Q}$

Sum and difference rule: $\quad \frac{d}{dx}[g(x) + h(x)] = \frac{d}{dx}[g(x)] + \frac{d}{dx}[h(x)]$

Constant multiple rule: $\quad \frac{d}{dx}[c \cdot f(x)] = c \cdot \frac{d}{dx}[f(x)], \; c \in \mathbb{R}$

Example 8

For each function: (i) find the derivative using the basic differentiation rules; (ii) find the slope of the graph of the function at the indicated points; and (iii) use your GDC to confirm your answer for (ii).

Function	Points
a) $f(x) = x^3 + 2x^2 - 15x - 13$	$(-3, 23), (3, -13)$
b) $f(x) = (2x - 7)^2$	$(2, 9), (\frac{7}{2}, 0)$
c) $f(x) = 3\sqrt{x} - 6$	$(4, 0), (9, 3)$
d) $f(x) = \frac{x^4}{4} - \frac{3x^3}{2} - 2x^2 + \frac{15x}{2} + \frac{3}{4}$	$(5, -43), (0, 0)$

Solution

a) (i) $\frac{d}{dx}(x^3 + 2x^2 - 15x - 13) = \frac{d}{dx}(x^3) + 2 \cdot \frac{d}{dx}(x^2) - 15 \cdot \frac{d}{dx}(x) - \frac{d}{dx}(13)$

$\qquad = 3x^2 + 2(2x) - 15(1) - 0$

$\qquad = 3x^2 + 4x - 15$

Therefore, the derivative of $f(x) = x^3 + 2x^2 - 15x - 13$ is $f'(x) = 3x^2 + 4x - 15$.

(ii) Slope of curve at $(-3, 23)$ is $f'(-3) = 3(-3)^2 + 4(-3) - 15$
$= 27 - 12 - 15 = 0$.
We should observe a horizontal tangent (slope $= 0$) to the curve at $(-3, 23)$.
Slope of curve at $(3, -13)$ is $f'(3) = 3(3)^2 + 4(3) - 15$
$= 27 + 12 - 15 = 24$.
We should observe a very steep tangent (slope $= 24$) to the curve at $(3, -13)$.

(iii) Not only can we use the GDC to compute the value of the derivative at a particular value of x on the 'home' screen, but we can also do it on the graph screen.

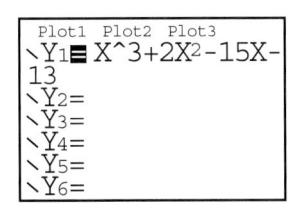

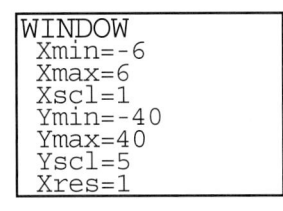

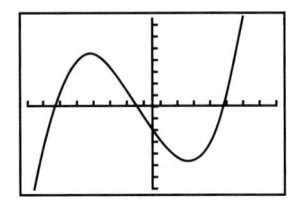

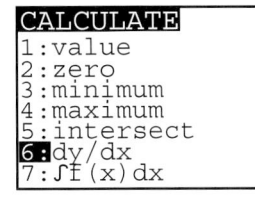

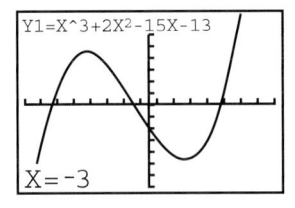

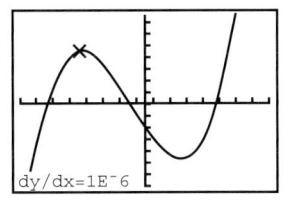

The GDC computes a slope of 1E⁻6 at the point $(-3, 23)$.
$(1\text{E}^{-6} = 1 \times 10^{-6} = 0.000\,001)$

Although the method the GDC uses is very accurate, sometimes there is a small amount of error in its calculation. This most commonly occurs when performing calculus computations (e.g. the value of the derivative at a point). $1\text{E}^{-6} = 0.000\,001$ is very close to zero which is the exact value of the derivative. Observe that the graph of $y = x^3 + 2x^2 - 15x - 13$ appears to have a 'turning point' at $(-3, 23)$, confirming that a line tangent to the curve at that point would be horizontal.

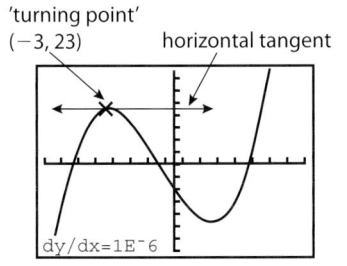

Let's check on our GDC that the slope of the curve is 24 at $(3, -13)$. Again, the GDC exhibits a small amount of error in its result.

Most GDCs are also capable of drawing a tangent at a point and displaying its equation as shown in the final screen image below.

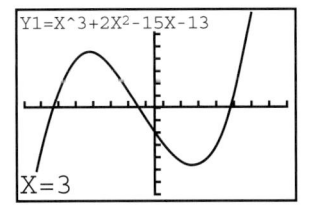

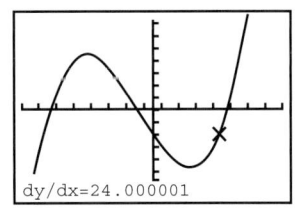

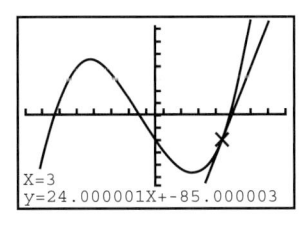

The equation of the tangent line at $(3, -13)$ is $y = 24x - 85$. We will look at finding the equations of tangent lines analytically in the last section of the chapter.

b) (i) $\dfrac{d}{dx}[(2x-7)^2] = \dfrac{d}{dx}[(2x-7)(2x-7)]$ differentiate term-by-term after expanding

$$= \dfrac{d}{dx}(4x^2 - 28x + 49)$$

$$= 4\dfrac{d}{dx}(x^2) - 28\dfrac{d}{dx}(x) + \dfrac{d}{dx}(49)$$

$$= 8x - 28 + 0$$

Therefore, the derivative of $f(x) = (2x - 7)^2$ is $f'(x) = 8x - 28$.

(ii) Slope of curve at $(2, 9)$ is $f'(2) = 8(2) - 28 = -12$.

Slope of curve at $\left(\dfrac{7}{2}, 0\right)$ is $f'\left(\dfrac{7}{2}\right) = 8\left(\dfrac{7}{2}\right) - 28 = 0$.

Thus, we should observe a horizontal tangent to the curve at $\left(\dfrac{7}{2}, 0\right)$.

(iii)

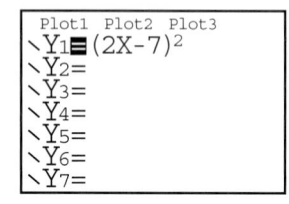

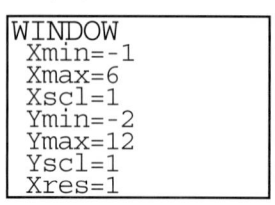

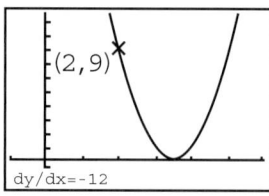

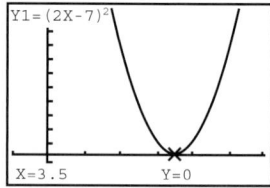

There's no error this time in the GDC's computation of the slope at $(2, 9)$. The vertex of the parabola is at $\left(\dfrac{7}{2}, 0\right)$, confirming that it has a horizontal tangent at that point.

c) (i) $\dfrac{d}{dx}(3\sqrt{x} - 6) = 3\dfrac{d}{dx}(x^{\frac{1}{2}}) - \dfrac{d}{dx}(6)$

$= 3\left(\dfrac{1}{2}x^{-\frac{1}{2}}\right) - 0$

$= 3\left(\dfrac{1}{2}x^{-\frac{1}{2}}\right) - 0 = \dfrac{3}{2x^{\frac{1}{2}}}$

Therefore, the derivative of $f(x) = 3\sqrt{x} - 6$ is $f'(x) = \dfrac{3}{2x^{\frac{1}{2}}}$ or $f'(x) = \dfrac{3}{2\sqrt{x}}$.

(ii) Slope of curve at $(4, 0)$ is $f'(4) = \dfrac{3}{2\sqrt{4}} = \dfrac{3}{4}$.

Slope of curve at $(9, 3)$ is $f'(9) = \dfrac{3}{2\sqrt{9}} = \dfrac{1}{2}$.

Thus, because the slope at $x = 9$ is less than that at $x = 4$, we should observe the graph of the equation becoming less steep as we move along the curve from $x = 4$ to $x = 9$.

(iii)

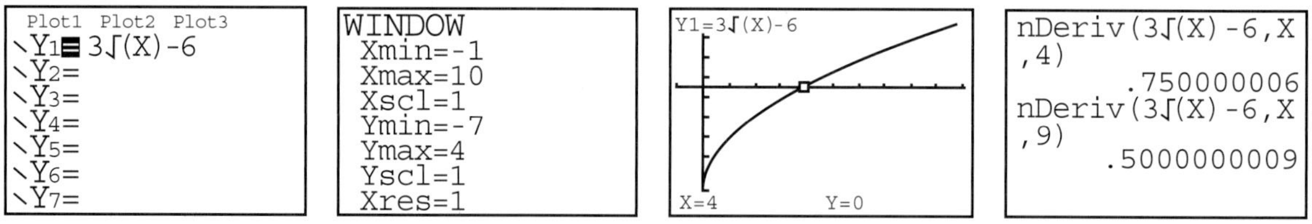

The slope of the graph of $y = 3\sqrt{x} - 6$ appears to steadily decrease as x increases. Let's check the results for (ii) by evaluating the derivative at a point on the 'home' screen. The GDC confirms the slopes for the curve when $x = 4$ and $x = 9$, but again the GDC computations have incorporated a small amount of error.

d) (i) $\dfrac{d}{dx}\left(\dfrac{x^4}{4} - \dfrac{3x^3}{2} - 2x^2 + \dfrac{15x}{2} + \dfrac{3}{4}\right)$

$= \dfrac{1}{4}\dfrac{d}{dx}(x^4) - \dfrac{3}{2}\dfrac{d}{dx}(x^3) - 2\dfrac{d}{dx}(x^2) + \dfrac{15}{2}\dfrac{d}{dx}(x) + \dfrac{d}{dx}\left(\dfrac{3}{4}\right)$

$= \dfrac{1}{4}(4x^3) - \dfrac{3}{2}(3x^2) - 2\dfrac{d}{dx}(2x) + \dfrac{15}{2}(1) + 0$

$= x^3 - \dfrac{9x^2}{2} - 4x + \dfrac{15}{2}$

Therefore, the derivative of $f(x) = \dfrac{x^4}{4} - \dfrac{3x^3}{2} - 2x^2 + \dfrac{15x}{2} + \dfrac{3}{4}$
is $f'(x) = x^3 - \dfrac{9x^2}{2} - 4x + \dfrac{15}{2}$.

(ii) Slope of curve at $(5, -43)$ is $f'(5) = 5^3 - \dfrac{9(5)^2}{2} - 4(5) + \dfrac{15}{2} = 0$.
Thus, there should be a horizontal tangent to the curve at $(5, -43)$.

Slope of curve at $(0, 0)$ is $f'(0) = \dfrac{15}{2}$.

(iii) Your GDC is not capable of computing the derivative function – only the specific value of the derivative for a given value of x. However, we can have the GDC graph the values of the derivative over a given *interval* of x. We can then graph the derivative function found from differentiation rules (result from (i)) and see if the two graphs match.

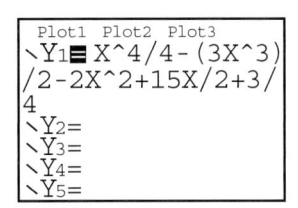

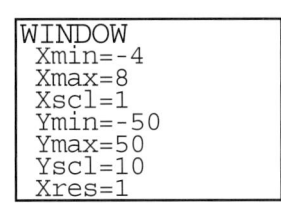

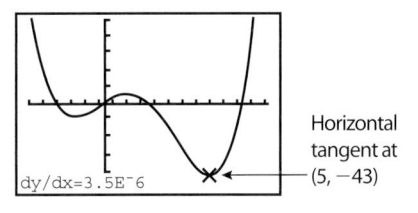

The command nDeriv(Y_1, X, X) computes the value of the derivative of function Y_1 in terms of x for all x.

Values of the derivative of $f(x)$ will be graphed as Y_2, and the derivative function, $f'(x) = x^3 - \dfrac{9x^2}{2} - 4x + \dfrac{15}{2}$, determined by manual application of differentiation rules (part (i)), will be graphed as Y_3. Note that the graph of Y_3 will be in bold style to distinguish it from Y_2, and that the equation Y_1 has been turned 'off.'

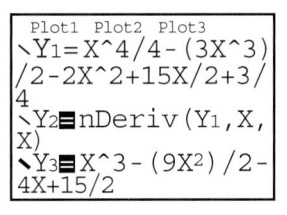

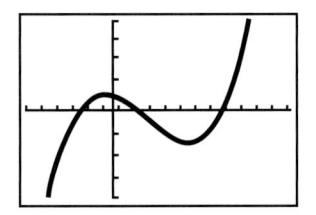

$Y_1 = \dfrac{x^4}{4} - \dfrac{3x^3}{2} - 2x^2 + \dfrac{15x}{2} + \dfrac{3}{4}$ $Y_2 = \text{nDeriv}(Y_1, X, X)$ $Y_3 = x^3 - \dfrac{9x^2}{2} - 4x + \dfrac{15}{2}$

Since the two graphs match, this confirms that the derivative found in part (i) using differentiation rules is correct.

Example 9

The curve $y = ax^3 + 7x^2 - 8x - 5$ has a turning point at the point where $x = -2$. Determine the value of a.

Solution

There must be a horizontal tangent, and a slope of zero, at the point where the graph has a turning point.

$$\frac{dy}{dx} = \frac{d}{dx}(ax^3 + 7x^2 - 8x - 5)$$
$$= a\frac{d}{dx}(x^3) + 7\frac{d}{dx}(x^2) - 8\frac{d}{dx}(x) + \frac{d}{dx}(-5) = 3ax^2 + 14x - 8$$
$$\frac{dy}{dx} = 0 \text{ when } x = -2: \ 3a(-2)^2 + 14(-2) - 8 = 0$$
$$\Rightarrow 12a - 28 - 8 = 0 \Rightarrow 12a = 36 \Rightarrow a = 3$$

Recall that the derivative of a function is a formula for the **instantaneous rate of change** of the dependent variable (commonly y) with respect to the dependent variable (x). In other words, as illustrated earlier in this section, the slope of the tangent at a point gives the slope, or rate of change, of the curve at that point. The slope of a **secant line** (that crosses the curve at two points) gives the **average rate of change** between the two points.

Example 10

Boiling water is poured into a cup. The temperature of the water in degrees Celsius, C, after t minutes is given by $C = 19 + \dfrac{182}{t^{\frac{3}{2}}}$, for times $t \geq 1$ minute.

a) Find the average rate of change of the temperature from $t = 2$ to $t = 6$.
b) Find the rate of change of the temperature at the instant that $t = 4$.

Solution

a)

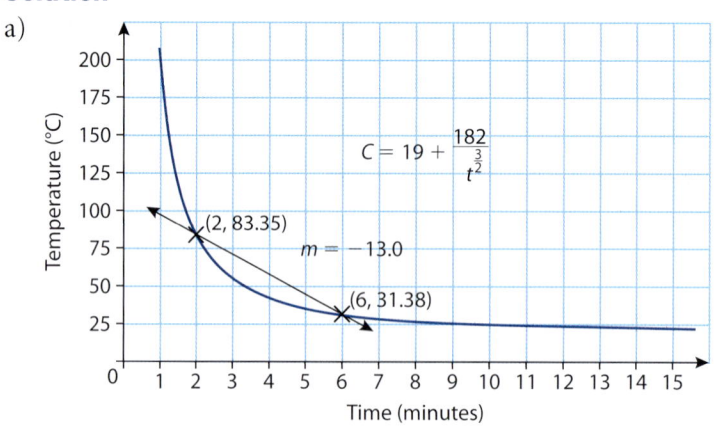

When $t = 2$, $C \approx 83.35°$ and when $t = 6$, $C \approx 31.38°$. The average rate of change from $t = 2$ to $t = 6$ is the slope of the line through the points $(2, 83.35)$ and $(6, 31.38)$.

$$\text{Average rate of change} = \frac{83.35 - 31.38}{2 - 6} = \frac{51.97}{-4} = -12.9925.$$

To an accuracy of 3 significant figures, the average rate of change from

$t = 2$ to $t = 6$ is $-13.0\,°C$ per minute. During that period of time the water is, on average, becoming 13 degrees cooler every minute.

b) Let's compute the derivative $\frac{dC}{dt}$, i.e. the rate of change of degrees C with respect to time t, from which we can compute the rate the temperature is changing at the moment when $t = 4$.

$$\frac{dC}{dt} = \frac{d}{dt}\left(19 + \frac{182}{t^{\frac{3}{2}}}\right) = \frac{d}{dt}(19 + 182t^{-\frac{3}{2}}) = \frac{d}{dt}(19) + 182\frac{d}{dt}(t^{-\frac{3}{2}})$$

$$= 0 + 182 \cdot -\frac{3}{2}t^{-\frac{3}{2}-1} = -273t^{-\frac{5}{2}}$$

$$\frac{dC}{dt} = -\frac{273}{t^{\frac{5}{2}}} = -\frac{273}{\sqrt{t^5}}$$

At $t = 4$:

$$\frac{dC}{dt} = -\frac{273}{\sqrt{4^5}} = -\frac{273}{32} \approx -8.53$$

Therefore, the temperature's instantaneous rate of change at $t = 4$ minutes is $-8.53\,°C$ per minute.

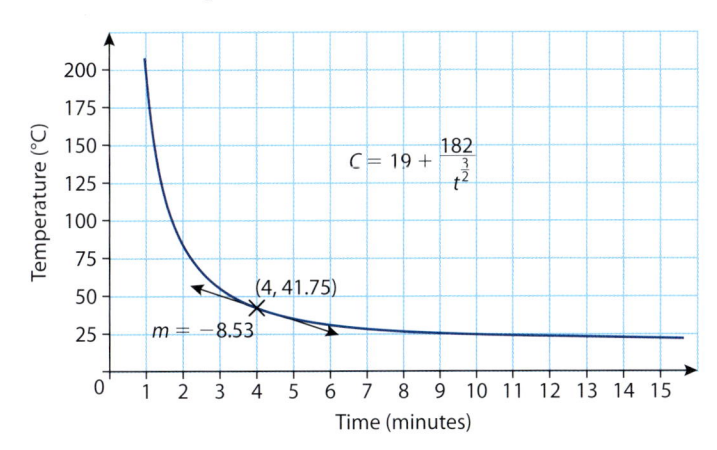

Exercise 11.2

In questions 1–4, find the derivative of the function by applying the limit definition
$f'(x) = \lim_{h \to 0} \frac{f(x+h) - f(x)}{h}$.

1. $f(x) = 1 - x^2$
2. $g(x) = x^3 + 2$
3. $h(x) = \sqrt{x}$
4. $r(x) = \frac{1}{x^2}$

5. Using your results from questions 1–4, find the slope of the graph of each function in 1–4 at the point where $x = 1$. Sketch each function and draw a line tangent to the graph at $x = 1$.

In questions 6–12, a) find the derivative of the function, and b) compute the slope of the graph of the function at the indicated point. Use a GDC to confirm your results.

6 $y = 3x^2 - 4x$ point $(0, 0)$

7 $y = 1 - 6x - x^2$ point $(-3, 10)$

8 $y = \dfrac{2}{x^3}$ point $(-1, 2)$

9 $y = x^5 - x^3 - x$ point $(1, -1)$

10 $y = (x + 2)(x - 6)$ point $(2, -16)$

11 $y = 2x + \dfrac{1}{x} - \dfrac{3}{x^3}$ point $(1, 0)$

12 $y = \dfrac{x^3 + 1}{x^2}$ point $(-1, 0)$

13 The slope of the curve $y = x^2 + ax + b$ at the point $(2, -4)$ is -1. Find the value of a and the value of b.

In questions 14–17, find the coordinates of any points on the graph of the function where the slope is equal to the given value.

14 $y = x^2 + 3x$ slope $= 3$

15 $y = x^3$ slope $= 12$

16 $y = x^2 - 5x + 1$ slope $= 0$

17 $y = x^2 - 3x$ slope $= -1$

18 Use the graph of f to answer each question.

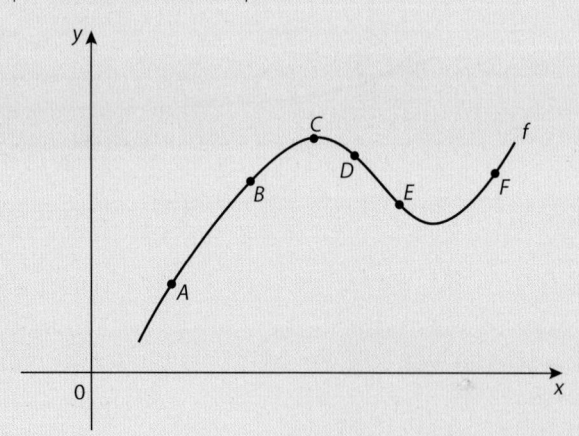

a) Between which two consecutive points is the average rate of change of the function greatest?

b) At what points is the instantaneous rate of change of f positive, negative and zero?

c) For which two pairs of points is the average rate of change approximately equal?

19 The slope of the curve $y = x^2 - 4x + 6$ at the point $(3, 3)$ is equal to the slope of the curve $y = 8x - 3x^2$ at (a, b). Find the value of a and the value of b.

20 The graph of the equation $y = ax^3 - 2x^2 - x + 7$ has a slope of 3 at the point where $x = 2$. Find the value of a.

21 Find the coordinates of the point on the graph of $y = x^2 - x$ at which the tangent is parallel to the line $y = 5x$.

22 Let $f(x) = x^3 + 1$.
   a) Evaluate $\dfrac{f(2 + h) - f(2)}{h}$ for $h = 0.1$.
   b) What number does $\dfrac{f(2 + h) - f(2)}{h}$ approach as h approaches zero?

23 From first principles, find the derivative for the general quadratic function, $f(x) = ax^2 + bx + c$. Confirm your result by checking that it produces:
   (i) the derivative of x^2 when $a = 1, b = 0, c = 0$
   (ii) the derivative of $3x^2 - 4x + 2$ when $a = 3, b = -4, c = 2$.

24 A car is parked with the windows and doors closed for five hours. The temperature inside the car in degrees Celsius, C, is given by $C = 2\sqrt{t^3} + 17$ with t representing the number of hours since the car was first parked.
   a) Find the average rate of change of the temperature from $t = 1$ to $t = 4$.
   b) Find the function that gives the instantaneous rate of change of the temperature for any time $t, 0 < t < 5$.
   c) Find the time t at which the instantaneous rate of change of the temperature is equal to the average rate of change from $t = 1$ to $t = 4$.

11.3 Maxima and minima – first and second derivatives

The relationship between a function and its derivative

The derivative, written in Newton notation as $f'(x)$ or in Leibniz notation as $\dfrac{dy}{dx}$, is a function derived from a function f that gives the slope of the graph of f at any x in the function's domain (given that the curve is 'smooth' at the value of x). The derivative is a slope, or rate of change, function. Knowing the slope of a function at different values in its domain tells us about properties of the function and the shape of its graph.

In the previous section, we observed that if a graph 'turns' at a particular point (for example, at the vertex of a parabola), then it has a horizontal tangent (slope = 0) at the point. Hence, the derivative will equal zero at a 'turning point'. In Section 2.5, we found the vertex of the graph of a quadratic function by using the technique of completing the square to write its equation in vertex form. We can also find the vertex by means of differentiation. As we look at the graph of a parabola moving from left to right (i.e. domain values increasing), it either turns from going down to going up (decreasing to increasing), or from going up to going down (increasing to decreasing) (Figure 11.12).

> If the graph of a function is 'smooth' at a particular point, the function is considered to be *differentiable* at this point. In other words, a tangent line exists at this point. All functions that will be differentiated in this course will be differentiable at all values in the function's domain.

Figure 11.12

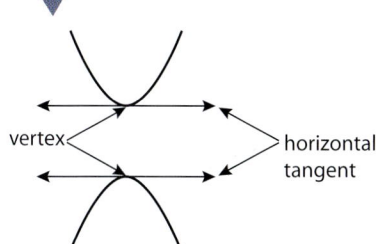

11 Differential Calculus I: Fundamentals

Example 11

Using differentiation, find the vertex of the parabola with the equation $y = x^2 - 8x + 14$.

Solution

Find the value of x for which the derivative, $\dfrac{dy}{dx}$, is zero.

$$\frac{dy}{dx} = \frac{d}{dx}(x^2 - 8x + 14) = 2x - 8 = 0 \Rightarrow x = 4$$

Thus, the x-coordinate of the vertex is 4.

To find the y-coordinate of the vertex, we substitute $x = 4$ into the equation, giving $y = 4^2 - 8(4) + 14 = -2$. Therefore, the vertex has coordinates $(4, -2)$.

Figure 11.13 x increases from left to right along the x-axis.

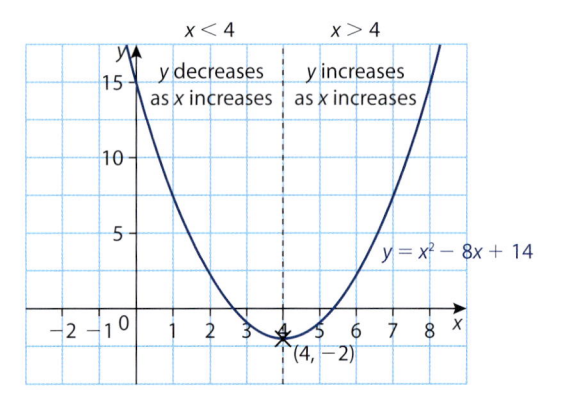

We know that the parabola in Example 11 will 'open up' because the coefficient of the quadratic term, x^2, is positive. The parabola has a negative slope (decreasing) to the left of the vertex and a positive slope (increasing) to the right of the vertex (Figure 11.13). As the values of x increase, the derivative of $y = x^2 - 8x + 14$ will change from negative to zero to positive, accordingly.

$$\frac{dy}{dx} = 2x - 8 \Rightarrow \frac{dy}{dx} < 0 \text{ for } x < 4 \text{ and } \frac{dy}{dx} = 0 \text{ for } x = 4 \text{ and } \frac{dy}{dx} > 0 \text{ for } x > 4$$

In other words, the function $f(x) = x^2 - 8x + 14$ is decreasing for all $x < 4$; it is neither decreasing nor increasing at $x = 4$; and it is increasing for all $x > 4$. A point at which a function is neither increasing nor decreasing (i.e. there is a horizontal tangent) is called a **stationary point**. A convenient way to demonstrate where a function is increasing or decreasing and the location of any stationary points is with a **sign chart** for the function and its derivative, as shown in Figure 11.14 for $f(x) = x^2 - 8x + 14$. The derivative $f'(x) = 2x - 8$ is zero only at $x = 4$, thereby dividing the domain of f (i.e. $\mathbb{R}$) into two intervals: $x < 4$ and $x > 4$. $f'(x) = 2x - 8$ is a **continuous** function (i.e. no 'gaps' in the domain) so it is only necessary to test one point in each interval in order to determine the sign of all the values of the derivative in that interval. $f'(x)$ can only change sign at $x = 4$. For example, the fact that $f'(3) = 2(3) - 8 = -2 < 0$ means that $f'(x) < 0$ for all x when $x < 4$. Therefore, f is decreasing for all x in the open interval $(-\infty, 4)$.

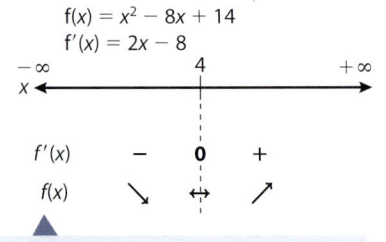

Figure 11.14 Sign chart for $f'(x)$ and $f(x)$.

368

Increasing and decreasing functions and stationary points
If $f'(x) > 0$ for $a < x < b$, then $f(x)$ is **increasing** on the interval $a < x < b$.
If $f'(x) < 0$ for $a < x < b$, then $f(x)$ is **decreasing** on the interval $a < x < b$.
If $f'(x) = 0$ for $a < x < b$, then $f(x)$ is **constant** on the interval $a < x < b$.
If $f'(x) = 0$ for a single value $x = c$ on some interval $a < c < b$, then $f(x)$ has a **stationary point** at $x = c$. The corresponding point $(c, f(c))$ on the graph of f is called a stationary point.

> Geometrically speaking, a function is **continuous** if there is no break in its graph; and a function is **differentiable** (i.e. a derivative exists) at any points where it is 'smooth'.

It is at stationary points, or endpoints of the domain if the domain is not all real numbers, where a function may have a maximum or minimum value. These points at which extreme values of a function *may* occur are often referred to as **critical points**. Whether a function is increasing or decreasing on either side of a stationary point will indicate whether the stationary point is a maximum, minimum or neither.

Example 12

Consider the function $f(x) = 2x^3 + 3x^2 - 12x - 4, \; x \in \mathbb{R}$.
a) Find any stationary points of f.
b) Using the derivative of f, classify any stationary points as a maximum or minimum.

Solution

a) $f'(x) = 6x^2 + 6x - 12 = 0 \Rightarrow 6(x^2 + x - 2) = 0$
$\Rightarrow 6(x + 2)(x - 1) = 0 \Rightarrow x = -2 \text{ or } x = 1$

With a domain of all real numbers there are no domain endpoints that may be an extreme value. Thus, f has two critical points: one at $x = -2$ and the other at $x = 1$.

When $x = -2$: $y = 2(-2)^3 + 3(-2)^2 - 12(-2) - 4 = 16 \Rightarrow f$ has a stationary point at $(-2, 16)$.

When $x = 1$: $y = 2(1)^3 + 3(1)^2 - 12(1) - 4 = -11 \Rightarrow f$ has a stationary point at $(1, -11)$.

b) Construct a sign chart for $f'(x)$ and $f(x)$ (right) to show where f is increasing or decreasing. The derivative $f'(x)$ has two zeros, at $x = -2$ and $x = 1$, thereby dividing the domain of f into three intervals that need to be tested. Since $f'(-3) = 6(-1)(-4) = 24 > 0$, then $f'(x) > 0$ for all $x < -2$. Likewise, since $f'(2) = 6(4)(1) = 24 > 0$, then $f'(x) > 0$ for all $x > 1$. Thus, f is increasing on the open intervals $(-\infty, -2)$ and $(1, \infty)$. Since $f'(0) = -12 < 0$, then $f'(x) < 0$ for all x such that $-2 < x < 1$. Thus, f is decreasing on the open interval $(-2, 1)$, i.e. $-2 < x < 1$. From this information, we can visualize for increasing values of x that the graph of f is going up for all $x < -2$, then turning down at $x = -2$, then going down for values of x from -2 to 1, then turning up at $x = 1$, and then going up for all $x > 1$. The basic shape of the graph of f will look something like the rough sketch shown right. Clearly, the stationary point $(-2, 16)$ is a maximum and the stationary point $(1, -11)$ is a minimum.

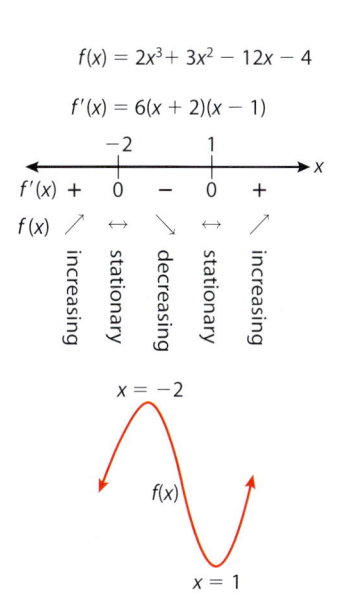

The graph of $f(x) = 2x^3 + 3x^2 - 12x - 4$ from Example 12 (Figure 11.15) visually confirms the results acquired from analyzing the derivative of f.

Figure 11.15

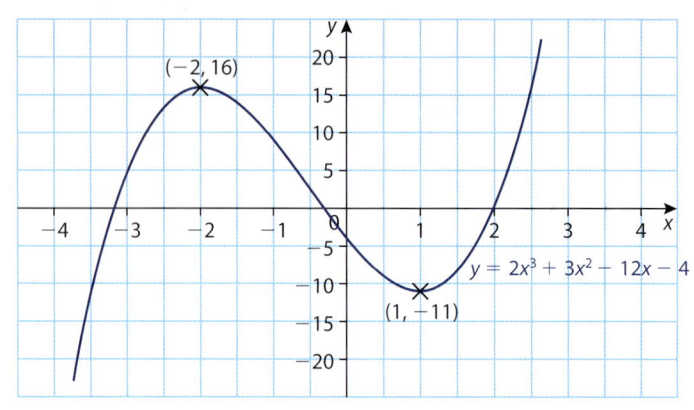

For Example 12, we can express the result for part b) most clearly by saying that $f(x)$ has a **relative maximum** value of 16 at $x = -2$, and $f(x)$ has a **relative minimum** value of -11 at $x = 1$. The reason that these *extreme* values are described as 'relative' (sometimes described as 'local') is because they are a maximum or minimum for the function in the immediate vicinity of the point, but not for the entire domain of the function. A point that is a maximum/minimum for the entire domain is called an **absolute**, or **global**, **maximum/minimum**.

> The plural of 'maximum' is 'maxima', and the plural of 'minimum' is 'minima'. Maxima and minima are collectively referred to as 'extrema' – the plural of 'extremum' (extreme value). Extrema of a function that do not occur at domain endpoints will be 'turning points' of the graph of the function.

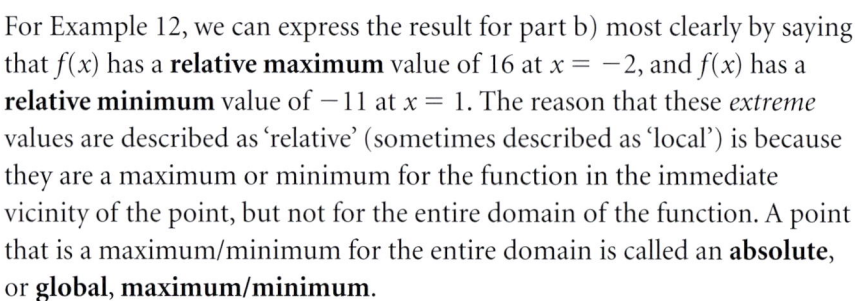

The first derivative test

From Example 12, we can see that a function f has a maximum at some $x = c$ if $f'(c) = 0$ and f is *increasing* immediately to the left of $x = c$ and *decreasing* immediately to the right of $x = c$. Similarly, f has a minimum at some $x = c$ if $f'(c) = 0$ and f is *decreasing* immediately to the left of $x = c$ and *increasing* immediately to the right of $x = c$. It is important to understand, however, that not all stationary points are either a maximum or minimum.

Example 13

For the function $f(x) = x^4 - 2x^3$, find all stationary points and describe them completely.

Solution

$$f'(x) = \frac{d}{dx}(x^4 - 2x^3) = 4x^3 - 6x^2 = 0 \Rightarrow 2x^2(2x - 3) = 0$$
$$\Rightarrow x = 0 \text{ or } x = \frac{3}{2}$$

The implied domain is all real numbers, so $x = 0$ and $x = \frac{3}{2}$ are the critical points of f.

When $x = 0$, $y = f(0) = 0$.

When $x = \frac{3}{2}$, $y = f\left(\frac{3}{2}\right) = \left(\frac{3}{2}\right)^4 - 2\left(\frac{3}{2}\right)^3 = \frac{81}{16} - \frac{54}{8} = -\frac{27}{16}$.

Therefore, f has stationary points at $(0, 0)$ and $\left(\frac{3}{2}, -\frac{27}{16}\right)$.

Because f has two stationary points, there are three intervals for which to test the sign of the derivative. We could use some form of a sign chart as shown previously, or we can use a more detailed table that summarizes the testing of the three intervals and the two critical points as shown below.

Interval/point	$x<0$	$x=0$	$0<x<\frac{3}{2}$	$x=\frac{3}{2}$	$x>\frac{3}{2}$
Test value	$x=-1$		$x=1$		$x=2$
Sign of $f'(x)$	$f'(-1)=-10<0$	0	$f'(1)=-2<0$	0	$f'(2)=8>0$
Conclusion	f decreasing ↘	none	f decreasing ↘	abs. min.	f increasing ↗

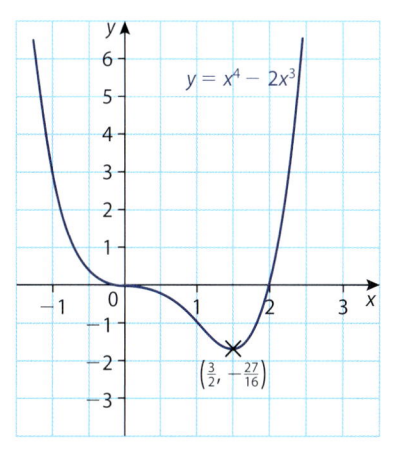

On either side of $x=0$, f does not change from either decreasing to increasing or from increasing to decreasing. Although there is a horizontal tangent at $(0, 0)$, it is *not* an extreme value (turning point). The function steadily decreases as x approaches zero, then at $x = 0$ the function has a rate of change (slope) of zero for an instant and then continues on decreasing. As x approaches $\frac{3}{2}$, f is decreasing and then switches to increasing at $x = \frac{3}{2}$.

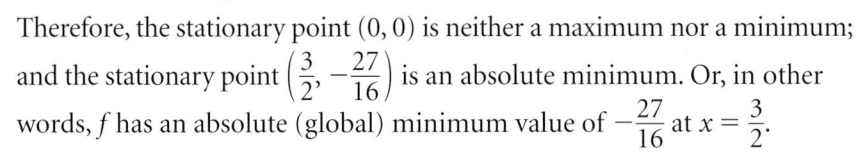

Therefore, the stationary point $(0, 0)$ is neither a maximum nor a minimum; and the stationary point $\left(\frac{3}{2}, -\frac{27}{16}\right)$ is an absolute minimum. Or, in other words, f has an absolute (global) minimum value of $-\frac{27}{16}$ at $x = \frac{3}{2}$.

The reason that an *absolute*, rather than a *relative*, minimum value occurs at $x = \frac{3}{2}$ is because for all $x < \frac{3}{2}$ the function f is either decreasing or constant (at $x = 0$) and for all $x < \frac{3}{2}$ f is increasing.

First derivative test for maxima and minima of a function

Suppose that $x = c$ is a critical point of a continuous and smooth function f. That is, $f(c) = 0$ and $x = c$ is a stationary point or $x = c$ is an endpoint of the domain.

I. At a stationary point $x = c$:
   1. If $f'(x)$ changes sign from positive to negative as x increases through $x = c$, then f has a relative maximum at $x = c$.

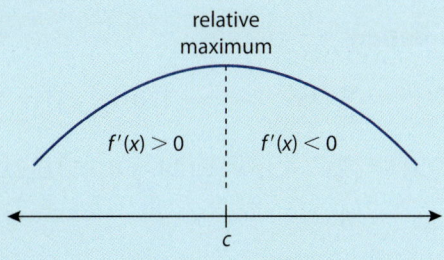

   relative maximum
   $f'(x) > 0$ $f'(x) < 0$
   c

   2. If $f'(x)$ changes sign from negative to positive as x increases through $x = c$, then f has a relative minimum at $x = c$.

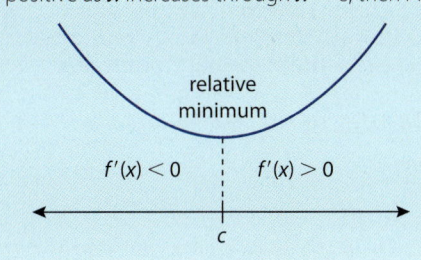

   relative minimum
   $f'(x) < 0$ $f'(x) > 0$
   c

3. If $f'(x)$ does not change sign as x increases through $x = c$, then f has neither a relative maximum nor a relative minimum at $x = c$.

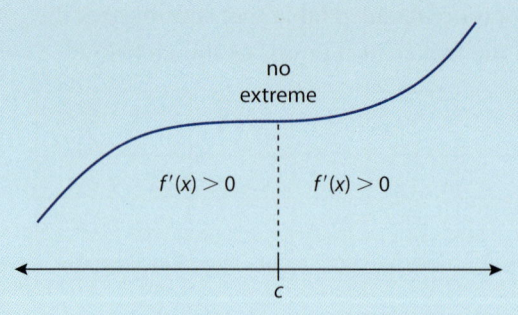

II. At a domain endpoint $x = c$:
If $x = c$ is an endpoint of the domain, then $x = c$ will be a relative maximum or minimum of f if the sign of $f'(x)$ is always positive or always negative for $x > c$ (at a left endpoint), or for $x < c$ (at a right endpoint), as illustrated below.

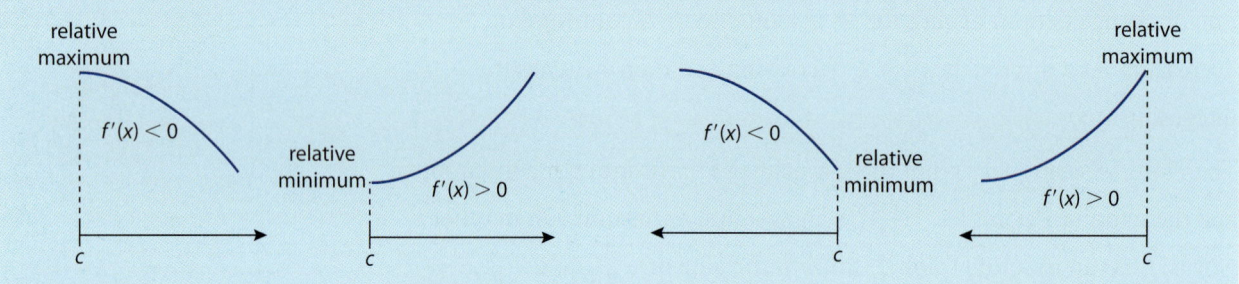

If it is possible to show that a relative maximum/minimum at $x = c$ is the greatest/least value for the entire domain of f, then it is classified as an absolute maximum/minimum.

Example 14

Apply the first derivative test to find any local extreme values for $f(x)$. Identify any absolute extrema.

$$f(x) = 4x^3 - 9x^2 - 120x + 25$$

Solution

$$f'(x) = \frac{d}{dx}(4x^3 - 9x^2 - 120x + 25) = 12x^2 - 18x - 120$$

$$f'(x) = 12x^2 - 18x - 120 = 0 \Rightarrow 6(2x^2 - 3x - 20) = 0$$
$$\Rightarrow 6(2x + 5)(x - 4) = 0$$

Thus, f has stationary points at $x = -\frac{5}{2}$ and $x = 4$.

To classify the stationary point at $x = -\frac{5}{2}$, we need to choose test points on either side of $-\frac{5}{2}$, for example, $x = -3$ (left) and $x = 0$ (right). Then we have

$$f'(-3) = 6(-1)(-7) = 42 > 0$$
$$f'(0) = 6(5)(-4) = -120 < 0$$

So f has a relative maximum at $x = -\frac{5}{2}$.

$f\left(-\frac{5}{2}\right) = 4\left(-\frac{5}{2}\right)^3 - 9\left(-\frac{5}{2}\right)^2 - 120\left(-\frac{5}{2}\right) + 25 = 206.25$

Therefore, f has a relative maximum value of 206.25 at $x = -\frac{5}{2}$.

To classify the stationary point at $x = 4$, we need to choose test points on either side of 4, for example, $x = 0$ (left) and $x = 5$ (right). Then we have

$$f'(0) = -120 < 0$$
$$f'(5) = 6(15)(1) = 90 > 0$$

So f has a relative minimum at $x = 4$.

$f(4) = 4(4)^3 - 9(4)^2 - 120(4) + 25 = -343$

Therefore, f has a relative minimum value of -343 at $x = 4$.

Change in displacement and velocity

Consider the motion of an object such that we know its position s relative to a reference point or line as a function of time t given by $s(t)$. The **displacement** of the object over the time interval from t_1 to t_2 is:

$$\text{change in } s = \text{displacement} = s(t_2) - s(t_1)$$

The **average velocity** of the object over the time interval is:

$$v_{\text{avg}} = \frac{\text{displacement}}{\text{change in time}} = \frac{s(t_2) - s(t_1)}{t_2 - t_1}$$

The object's **instantaneous velocity** at a particular time, t, is the value of the derivative of the position function, s, with respect to time at t.

$$\text{velocity} = \frac{ds}{dt} = s'(t)$$

Example 15

A rocket is launched upwards into the air. Its vertical position, s metres, above the ground at t seconds is given by $s(t) = -5t^2 + 18t + 1$.

a) Find the average velocity over the time interval from $t = 1$ second to $t = 2$ seconds.

b) Find the instantaneous velocity at $t = 1$ second.

c) Find the maximum height reached by the rocket and the time at which this occurs.

Solution

a) $v_{avg} = \dfrac{s(2) - s(1)}{2 - 1} = \dfrac{[-5(2)^2 + 18(2) + 1] - [-5 + 18 + 1]}{1}$

$= 3$ metres per second (or m s^{-1})

b) $s'(t) = -10t + 18 \Rightarrow s'(1) = -10 + 18 = 8$ m s^{-1}

c) $s'(t) = -10t + 18 = 0 \Rightarrow t = 1.8$

Thus, s has a stationary point at $t = 1.8$. t must be positive and ranges from time of launch ($t = 0$) to when the rocket hits the ground, i.e. $h = 0$.

$s(t) = -5t^2 + 18t + 1 = 0 \Rightarrow t = \dfrac{-18 \pm \sqrt{18^2 - 4(-5)(1)}}{2(-5)}$

$\Rightarrow t \approx -0.5472$ or $t \approx 3.655$

So, the rocket hits the ground about 3.66 seconds after the time of launch. Hence, the domain for the position (s) and velocity (v) functions is $0 \leq t \leq 3.66$. Therefore, the function s has three critical points: $t = 0$, $t = 1.8$ and $t \approx 3.66$.

The maximum of the function, i.e. the maximum height, most likely occurs at the critical point $t = 1.8$. Let's confirm this.

Applying the first derivative test, we determine the sign of the derivative, $s'(t)$, for values on either side of $t = 1.8$, for example, $t = 0$ and $t = 2$. $s'(0) = 18 > 0$ and $s'(2) = -2 < 0$. Neither of the domain endpoints, $t = 0$ and $t \approx 3.66$, are at a maximum or minimum because the function is not constantly increasing or constantly decreasing before or after the endpoint. Since the function changes from increasing to decreasing at $t = 1.8$ and $s(1.8) = -5(1.8)^2 + 18(1.8) + 1 = 17.2$, then the rocket reaches a maximum height of 17.2 metres 1.8 seconds after it was launched.

The relationship between a function and its second derivative

You may have wondered why the strategy we are applying to locate and classify extrema for a function focuses on using the *first* derivative of the function. This implies that we are interested in using some other type of derivative, namely the *second* derivative. There is another useful test for the purpose of analyzing the stationary point of a function that makes use of the derivative of the derivative, i.e. the second derivative, of the function.

When we differentiate a function $y = f(x)$, we obtain the first derivative $f'(x)$ (also denoted as $\dfrac{dy}{dx}$). Often this is a function that can also be differentiated. The result of doing so is the derivative of $f'(x)$, which is denoted in Newton notation as $f''(x)$ or in Leibniz notation as $\dfrac{d^2y}{dx^2}$ and called the second derivative of f with respect to x. For example, if $f(x) = x^3$, then $f'(x) = 3x^2$ and $f''(x) = 6x$ (or $\dfrac{d^2y}{dx^2} = 6x$).

Second derivatives, like first derivatives, occur often in methods of applying calculus. In Example 15, the function $s(t)$ gave the position, in metres above the ground, of a projectile (toy rocket) where t, in seconds, is the time since the projectile was launched. The function $s'(t)$, the first derivative of the position function, then gives the rate of change of the object's position, i.e. its velocity, in metres per second (m s^{-1}). Differentiation of this function gives the rate of change of the object's velocity, i.e. its *acceleration*, measured in metres per second per second (m s^{-2}).

The graphs of the position, velocity and acceleration functions for Example 15 aligned vertically (Figure 11.16) nicely illustrate the relationships between a function, its first derivative and its second derivative. The slope of the graph of $s(t)$ is initially a large positive value (graph is steep), but steadily decreases until it is zero (horizontal tangent) at $t = 1.8$ and then continues to decrease, becoming a large negative value (again, steep, but in the other direction). This corresponds to the real-life situation in which the rocket is launched with a high initial velocity ($v(0) = 18$ m s^{-1}) and then its velocity decreases steadily due to gravity. The rocket's velocity is zero for just an instant when it reaches its maximum height at $t = 1.8$ and then its velocity becomes more and more negative because it has changed direction and is moving back (negative direction) to the ground. The rate of change of the velocity, $v'(t)$, is constant and it is negative because the velocity is decreasing from positive values to zero to negative values. This is clear from the fact that the graph of the velocity function, $v(t)$, is a straight line with a negative slope. It follows then that the acceleration function – the rate of change of velocity – is a negative constant, $a = -10$ in this case, and its graph is a horizontal line.

In Example 15, it is not possible to have a negative function value for $s(t)$ because the rocket's position is always above, or at, ground level. In many motion problems in calculus, we consider a simplified version by limiting an object's motion to a line with its position given as its **displacement** from a fixed point (usually the origin). At a position left of the fixed point, the object's displacement is negative, and at a position right of the fixed point, the displacement is positive. Velocity can also be positive or negative depending on the direction of travel (i.e. the sign of the rate of change of the object's displacement). Likewise, acceleration is positive if velocity is increasing (i.e. rate of change of velocity is positive) and negative if velocity is decreasing.

Position function:
$s(t) = -5t^2 + 18t + 1$

Velocity function:
$v(t) = s'(t) = -10t + 18$

Acceleration function:
$a(t) = v'(t) = s''(t) = -10$

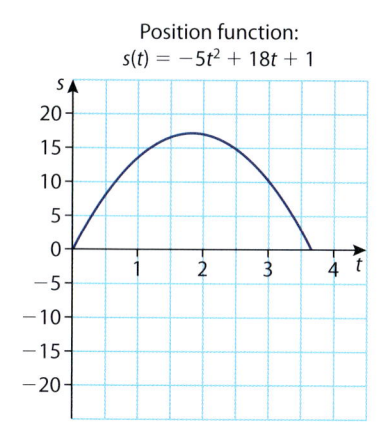

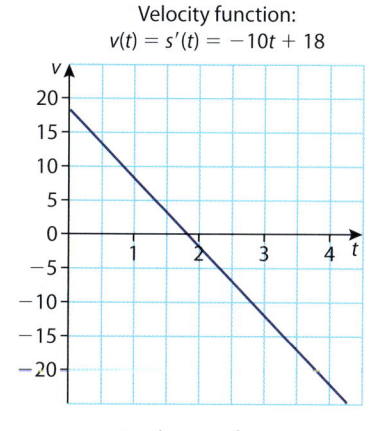

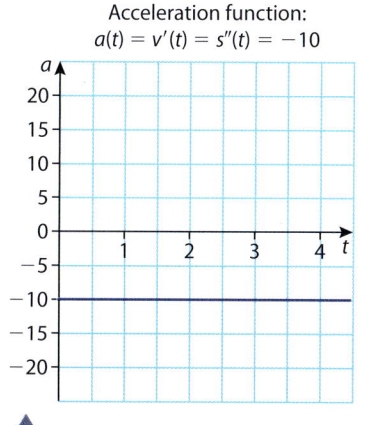

Figure 11.16

It would be incorrect to graph a function and its first and/or second derivative on the same axes. For example, the position $s(t)$, velocity $v(t)$ and acceleration $a(t)$ functions graphed on separate axes in Figure 11.16 will have different units on each vertical axis: metres for $s(t)$, metres per second for $v(t)$ and metres per second per second for $a(t)$.

A common misconception is that acceleration is positive for motion in the positive direction (usually 'right' or 'up') and negative for motion in the negative direction (usually 'left' or 'down'). Acceleration indicates how velocity is changing. Even though an object may be moving in a positive direction (e.g. to the right) if it is slowing down, then its acceleration is acting in the opposite direction and would be negative. In Example 15, the rocket was always accelerating in the negative direction, -10 m s^{-2}, due to the force of gravity. Note: A more accurate value for the acceleration of a free-falling object due to gravity is -9.8 m s^{-2}.

375

Differential Calculus I: Fundamentals

> **Displacement** can be negative, positive or zero.
> **Distance** is the absolute value of displacement. **Velocity** can be negative, positive or zero.
> **Speed** is the absolute value of velocity.

Motion along a line

If an object moves in a straight line such that at time t its displacement (position) from a fixed point is $s(t)$, then the first derivative $s'(t)$, also written as $\dfrac{ds}{dt}$, gives the velocity $v(t)$ at time t.

The second derivative $s''(t)$, also written as $\dfrac{d^2s}{dt^2}$, is the first derivative of $v(t)$. Hence, the second derivative of the displacement, or position, function is a measure of the rate at which the velocity is changing, i.e. it represents the acceleration of the object, which we express as

$$a(t) = v'(t) = s''(t) \quad \text{or} \quad a(t) = \frac{dv}{dt} = \frac{d^2s}{dt^2}$$

Example 16

An object moves along a straight line so that after t seconds its displacement from the origin is s metres. Given that $s(t) = -2t^3 + 6t^2$, answer the following:

a) Find expressions for the (i) velocity and (ii) acceleration at time t seconds.

b) Find the (i) initial velocity and (ii) initial acceleration of the object (i.e. at time when $t = 0$).

c) Find the (i) maximum displacement and (ii) maximum velocity for the interval $0 \leq t \leq 3$.

Solution

a) (i) $v(t) = \dfrac{ds}{dt} = \dfrac{d}{dt}(-2t^3 + 6t^2) = -6t^2 + 12t$

   (ii) $a(t) = \dfrac{d^2s}{dt^2} = \dfrac{dv}{dt} = \dfrac{d}{dt}(-6t^2 + 12t) = -12t + 12$

b) (i) $v(0) = -6(0)^2 + 12(0) = 0 \Rightarrow$ The object's initial velocity is 0 m s^{-1}.

   (ii) $a(0) = -12(0) + 12 = 12 \Rightarrow$ The object's initial acceleration is 12 m s^{-2}.

c) (i) To find the maximum displacement, we can apply the first derivative test to $s(t)$. Since the first derivative of displacement, $s(t)$, is velocity, $v(t)$, then the critical points of $s(t)$ are where the velocity is zero (stationary points) and domain endpoints.

$$s'(t) = v(t) = -6t^2 + 12t = 0 \Rightarrow 6t(-t + 2) = 0$$
$$\Rightarrow v(t) = 0 \text{ when } t = 0 \text{ or } t = 2$$

For the interval $0 \leq t \leq 3$, the critical points to be tested for finding the maximum displacement are at $t = 0$, $t = 2$ and $t = 3$. Check whether the velocity is increasing or decreasing on either side of the stationary point at $t = 2$ by finding the sign of $v(t)$ for $t = 1$ and $t = 2.5$.

$v(1) = -6(1)^2 + 12(1) = 6$ and $v(2.5) = -6(2.5)^2 + 12(2.5) = -7.5$

Hence, the displacement s is increasing for $0 < t < 2$ and decreasing for $2 < t < 3$. This indicates that the stationary point at $t = 2$ must be an absolute maximum for s in the interval $0 \leq t \leq 3$.

$$s(2) = -2(2)^3 + 6(2)^2 = 8$$

Therefore, the object has a maximum displacement of 8 metres at $t = 2$ seconds.

(ii) To find the maximum velocity, we can apply the first derivative test to $v(t)$. The first derivative of $v(t)$ is acceleration $a(t)$, which is the *second* derivative of $s(t)$. Hence, where $s''(t) = 0$ (acceleration is zero) indicates critical points for $v(t)$, i.e. where velocity may change from increasing to decreasing, or vice versa.

$$s''(t) = a(t) = \frac{d}{dt}(-6t^2 + 12t) = -12t + 12$$

$$\Rightarrow 12(-t + 1) = 0 \Rightarrow a(t) = 0 \text{ when } t = 1$$

For the interval $0 \leq t \leq 3$, the critical points to be tested for finding the maximum velocity are at $t = 0$, $t = 1$ and $t = 3$. Check whether the velocity is increasing or decreasing on either side of $t = 1$ by finding the sign of $a(t)$ for $t = 0.5$ and $t = 2$.
$a(0.5) = -12(0.5) + 12 = 6$ and $a(2) = -12(2) + 12 = -12$
Hence, the velocity v is increasing for $0 < t < 1$ and decreasing for $1 < t < 3$. This indicates that the point at $t = 1$ must be an absolute maximum for v in the interval $0 \leq t \leq 3$.

$$v(1) = -6(1)^2 + 12(1) = 6$$

Therefore, the object has a maximum velocity of 6 metres per second at $t = 1$ second.

The second derivative of a function tells us how the first derivative of the function changes. From this we can use the second derivative, as we did the first derivative, to reveal information about the shape of the graph of a function. Note in Example 16 that the object's velocity changed from increasing to decreasing when the object's acceleration was zero at $t = 1$. Let's examine graphically the significance of the point where acceleration is zero (i.e. velocity changing from increasing to decreasing) in connection to the displacement graph for Example 16. In other words, what can the second derivative of a function tell us about the shape of the function's graph?

Figure 11.17 shows the graphs of the displacement, velocity and acceleration functions for the motion of the object in Example 16. A dashed vertical line highlights the nature of the three graphs where $t = 1$. At this point, velocity has a maximum value and acceleration is zero. It is also where velocity changes from increasing to decreasing, which has a corresponding effect on the shape of the displacement function $s(t)$.

Figure 11.17

Displacement function:
$s(t) = -2t^3 + 6t^2$

Velocity function:
$v(t) = s'(t) = -6t^2 + 12t$

Acceleration function:
$a(t) = v'(t) = s''(t) = -12t + 12$

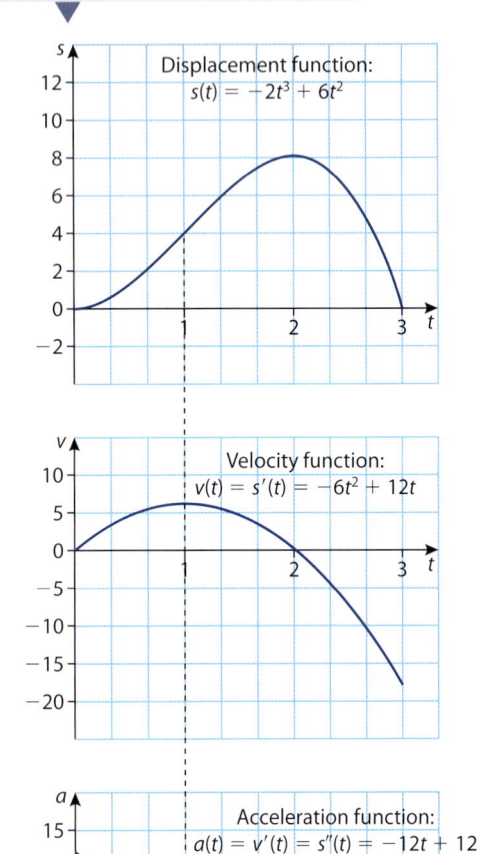

At the point where $t = 1$, the graph of $s(t)$ changes from curving 'upwards' (*concave up*) to curving 'downwards' (*concave down*) because its slope (corresponding to velocity) changes from increasing to decreasing. This can only occur when velocity (first derivative) has a maximum and, hence, where acceleration (second derivative) is zero. We can see from this illustration that for a general function $f(x)$, finding intervals where the first derivative $f'(x)$ is increasing (positive acceleration) or decreasing (negative acceleration) can be used to determine where the graph of $f(x)$ is curving upward or curving downward. A point at which a function's curvature (concavity) changes – as at $t = 1$ for the graph of $s(t)$ above – is called a **point of inflexion**.

Concavity and the second derivative

The graph of $f(x)$ is **concave up** where $f'(x)$ is increasing and **concave down** where $f'(x)$ is decreasing. It follows that:

(i) if $f''(x) > 0$ for all x in some interval of the domain of f, the graph of f is concave up in the interval

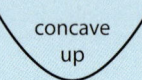

concave up

(ii) if $f''(x) < 0$ for all x in some interval of the domain of f, the graph of f is concave down in the interval.

concave down

If $f(x)$ is a continuous function, its graph can only change concavity (up to down, or down to up) where $f''(x) = 0$. Hence, for a continuous function, an **inflexion point** may only occur where $f''(x) = 0$.

Note: Concavity is not defined for a line – it is neither concave up nor concave down.

Example 17

Determine the intervals on which the graph of $y = x^4 - 4x^3$ is concave up or concave down and identify any inflexion points.

Solution

We first note that the function is continuous for its domain of all real numbers. To locate points of inflexion, we then find for what value(s) the second derivative is zero.

$$\frac{dy}{dx} = \frac{d}{dx}(x^4 - 4x^3) = 4x^3 - 12x^2$$

$$\Rightarrow \frac{d^2y}{dx^2} = \frac{d}{dx}(4x^3 - 12x^2) = 12x^2 - 24x = 12x(x - 2)$$

Setting $\frac{d^2y}{dx^2} = 0$, it follows that inflexion points may occur at $t = 0$ and $t = 2$. These two values divide the domain of the function into three intervals that we need to test. Let's choose $t = -1$, $t = 1$ and $t = 3$ as our test values. At $t = -1$, $\frac{d^2y}{dx^2} = 36 > 0$; at $t = 1$, $\frac{d^2y}{dx^2} = -12 < 0$; and at $t = 3$, $\frac{d^2y}{dx^2} = 36 > 0$. These results can be organized in a sign chart, illustrating that the graph of $y = x^4 - 4x^3$ is concave up for the open intervals $(-\infty, 0)$ and $(2, \infty)$, and concave down on the open interval $(0, 2)$. At $t = 0$, $y = 0$ and at $t = 2$, $y = 2^4 - 4(2)^3 = -16$. Therefore, $(0, 0)$ and $(2, -16)$ are inflexion points because it is at these points the concavity of the graph changes.

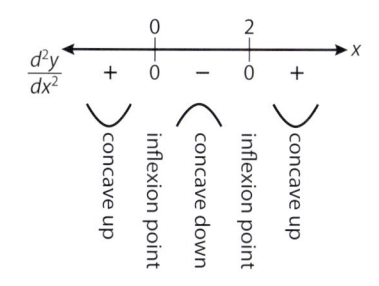

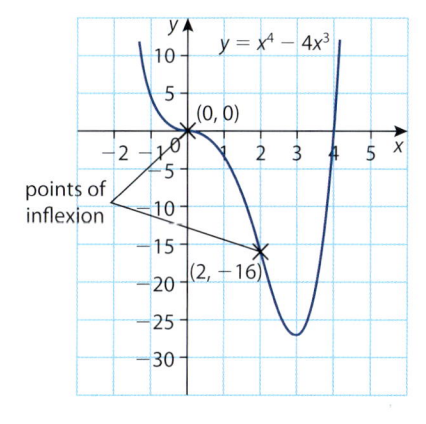

Figure 11.18

The graph of the function (Figure 11.18) from Example 17 reveals two different types of inflexion points. The slope of the curve at $(0, 0)$ is zero – i.e. it is a stationary point. The slope of the curve at the other inflexion point, $(2, -16)$, is negative.

For either type of inflexion point, the graph crosses its tangent line at the point of inflexion, as shown in Figure 11.19.

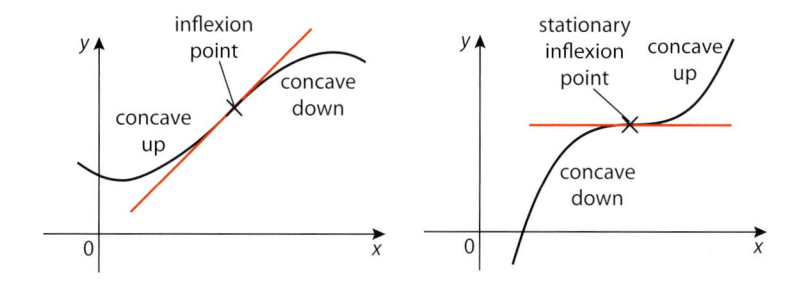

Figure 11.19 The concavity of a graph changes at a point of inflexion.

The fact that the second derivative of a function is zero at a certain point does not guarantee that an inflexion point exists at the point.

The functions $y = x^3$ and $y = x^4$ will serve to illustrate that $\frac{d^2y}{dx^2} = 0$ is a necessary but not sufficient condition for the existence of an inflexion point.

379

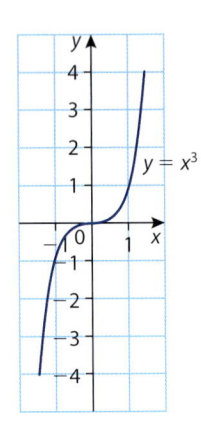

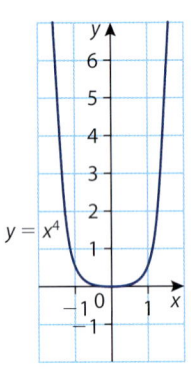

- For $y = x^3$: $\dfrac{dy}{dx} = \dfrac{d}{dx}(x^3) = 3x^2 \Rightarrow \dfrac{d^2y}{dx^2} = \dfrac{d}{dx}(3x^2) = 6x \Rightarrow \dfrac{d^2y}{dx^2} = 0$
  at $x = 0$. We can conclude from this that there may be an inflexion point at $x = 0$. We need to investigate further by checking to see if $\dfrac{d^2y}{dx^2}$ changes sign at $x = 0$. At $x = -1$, $\dfrac{d^2y}{dx^2} = -6$ and at $x = 1$, $\dfrac{d^2y}{dx^2} = 6$.
  Thus, there is an inflexion point at $x = 0$ (confirmed by graph) because the second derivative changes sign at $x = 0$.

- For $y = x^4$: $\dfrac{dy}{dx} = \dfrac{d}{dx}(x^4) = 4x^3 \Rightarrow \dfrac{d^2y}{dx^2} = \dfrac{d}{dx}(4x^3) = 12x^2 \Rightarrow \dfrac{d^2y}{dx^2} = 0$
  at $x = 0$. Again, we need to see if $\dfrac{d^2y}{dx^2}$ changes sign at $x = 0$.
  At $x = -1$, $\dfrac{d^2y}{dx^2} = 12$ and at $x = 1$, $\dfrac{d^2y}{dx^2} = 12$. Thus, there is *no* inflexion point at $x = 0$ (confirmed by graph) because the second derivative does *not* change sign at $x = 0$.

The second derivative test

Earlier in this section, we developed the first derivative test for locating maxima and minima of a function. Instead of using the first derivative to check whether a function changes from increasing to decreasing (maximum) or decreasing to increasing (minimum) at a stationary point, we can simply evaluate the second derivative at the stationary point. If the graph is concave up at the stationary point then it will be a minimum, and if it is concave down then it will be a maximum. If the second derivative is zero at a stationary point (as for $y = x^3$ and $y = x^4$), no conclusion can be made and we need to go back to the first derivative test. Using the second derivative in this way is a very efficient method for telling us whether a stationary point is a relative maximum or minimum.

The second derivative test

1. If $f'(c) = 0$ and $f''(c) < 0$, then f has a relative maximum at $x = c$.

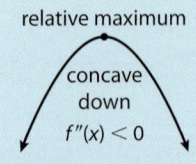

   relative maximum
   concave down
   $f''(x) < 0$

2. If $f'(c) = 0$ and $f''(c) > 0$, then f has a relative minimum at $x = c$.

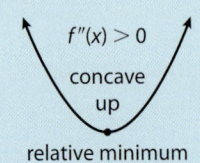

   $f''(x) > 0$
   concave up
   relative minimum

If $f''(c) = 0$, the test fails and the first derivative test should be applied.

Example 18

Find any relative extrema for $f(x) = 3x^5 - 25x^3 + 60x + 20$.

Solution

The implied domain of f is all real numbers. Solve $f'(x) = 0$ to obtain possible extrema.

$$f'(x) = 15x^4 - 75x^2 + 60 = 0$$
$$15(x^4 - 5x^2 + 4) = 0$$
$$15(x^2 - 4)(x^2 - 1) = 0$$
$$15(x + 2)(x - 2)(x + 1)(x - 1) = 0$$

Therefore, f has four stationary points: $x = -2$, $x = -1$, $x = 1$ and $x = 2$.

Applying the second derivative test:

$$f''(x) = 60x^3 - 150x = 30x(2x^2 - 5)$$
$$f''(-2) = -180 < 0 \Rightarrow f \text{ has a relative maximum at } x = -2$$
$$f''(-1) = 90 > 0 \Rightarrow f \text{ has a relative minimum at } x = -1$$
$$f''(1) = -90 < 0 \Rightarrow f \text{ has a relative maximum at } x = 1$$
$$f''(2) = 180 > 0 \Rightarrow f \text{ has a relative minimum at } x = 2$$

Exercise 11.3

In questions 1–3, find the vertex of the parabola using differentiation.

1 $y = x^2 - 2x - 6$ **2** $y = 4x^2 + 12x + 17$ **3** $y = -x^2 + 6x - 7$

For questions 4–7, a) find the derivative, $f'(x)$, b) indicate the interval(s) for which $f(x)$ is increasing, and c) the interval(s) for which $f(x)$ is decreasing.

4 $y = x^2 - 5x + 6$ **5** $y = 7 - 4x - 3x^2$
6 $y = \frac{1}{3}x^3 - x$ **7** $y = x^4 - 4x^3$

For questions 8–13:
  a) find the coordinates of any stationary points for the graph of the equation
  b) state, with reasoning, whether each stationary point is a minimum, maximum or neither
  c) sketch a graph of the equation and indicate the coordinates of each stationary point on the graph.

8 $y = 2x^3 + 3x^2 - 72x + 5$ **9** $y = \frac{1}{6}x^3 - 5$
10 $y = x(x - 3)^2$ **11** $y = x^4 - 2x^3 - 5x^2 + 6$
12 $y = x^3 - 2x^2 - 7x + 10$ **13** $y = x - \sqrt{x}$

14 An object moves along a line such that its displacement, s metres, from the origin O is given by $s(t) = t^3 - 4t^2 + t$.
  a) Find expressions for the object's velocity and acceleration in terms of t.
  b) For the interval $-1 \leq t \leq 3$, sketch the displacement-time, velocity-time, and acceleration-time graphs on separate sets of axes, vertically aligned as in Figure 11.17.
  c) For the interval $-1 \leq t \leq 3$, find the time at which the displacement is a maximum and find its value.
  d) For the interval $-1 \leq t \leq 3$, find the time at which the velocity is a minimum and find its value.
  e) In words, accurately describe the motion of the object during the interval $-1 \leq t \leq 3$.

For each function $f(x)$ in questions 15–20, find any relative extrema and points of inflexion. State the coordinates of any such points. Use your GDC to assist you in sketching the function.

15 $f(x) = x^3 - 12x$

16 $f(x) = \frac{1}{4}x^4 - 2x^2$

17 $f(x) = x + \frac{4}{x}$

18 $y = x^2 - \frac{1}{x}$

19 $f(x) = -3x^5 + 5x^3$

20 $f(x) = 3x^4 - 4x^3 - 12x^2 + 5$

21 An object moves along a line such that its displacement, s metres, from a fixed point P is given by $s(t) = t(t-3)(8t-9)$.
  a) Find the initial velocity and initial acceleration of the object.
  b) Find the velocity and acceleration of the object at $t = 3$ seconds.
  c) Find for what values of t the object changes direction. What significance do these times have in connection to the displacement of the object?
  d) Find for what value of t the object's velocity is a minimum. What significance does this time have in connection to the acceleration of the object?

22 The delivery cost per tonne of bananas, D (in thousands of dollars), when x tonnes of bananas are shipped is given by $D = 3x + \frac{100}{x}, x > 0$. Find the value of x for which the delivery cost per tonne of bananas is a minimum, and find the value of the minimum delivery cost. Explain why this cost is a minimum rather than a maximum.

23 The curve $y = x^4 + ax^2 + bx + c$ passes through the point $(-1, -8)$ and at that point $\frac{d^2y}{dx^2} = \frac{dy}{dx} = 6$. Find the values of a, b and c and sketch the curve.

24 Find any maxima, minima or stationary points of inflexion of the function $f(x) = \frac{x^3 + 3x - 1}{x^2}$, stating, with explanation, the nature of each point.
Sketch the curve, indicating clearly what happens as $x \to \pm\infty$.

11.4 Tangents and normals

In many areas of mathematics and physics, it is useful to have an accurate description of a line that is tangent or normal (perpendicular) to a curve. The most complete mathematical description we can obtain is to find the algebraic equation of such lines. In this chapter, much of our work has been in connection to the slopes of tangent lines, so this will be our starting point.

Finding equations of tangents

We now make use of the basic differentiation rules that we established earlier to determine the equation of lines that are tangent to a curve at a point. The first example shows how we can approximate the square root of a number quite accurately without a calculator by making use of a tangent line.

Example 19

a) Find the equation of the line tangent to $y = \sqrt{x}$ at $x = 9$.
b) Use this tangent line to approximate $\sqrt{10}$.

Solution

a) We can find the equation of any line if we know its slope and a point it passes through. Since $y = 3$ when $x = 9$, the point of tangency is $(9, 3)$. We differentiate to find the slope of the curve at $x = 9$, thus giving us the slope of the tangent line.

$$\frac{dy}{dx} = \frac{d}{dx}(\sqrt{x}) = \frac{d}{dx}(x^{\frac{1}{2}}) = \frac{1}{2}x^{-\frac{1}{2}} = \frac{1}{2\sqrt{x}}$$

At $x = 9$: $\frac{dy}{dx} = \frac{1}{2\sqrt{9}} = \frac{1}{6}$ ⇒ The slope of the curve and tangent line at $x = 9$ is $\frac{1}{6}$.

Now that we have a point and a slope for the line we can substitute in the point-slope form for the equation of a line.

$$y - 3 = \tfrac{1}{6}(x - 9) \quad \Rightarrow \quad y = \tfrac{1}{6}x + \tfrac{3}{2}$$

The equation of the line tangent to $y = \sqrt{x}$ at $x = 9$ is $y = \frac{x}{6} + \frac{3}{2}$.

b) For values of x near 9, $y = \sqrt{x} \approx \frac{x}{6} + \frac{3}{2}$.

$$\sqrt{10} \approx \frac{10}{6} + \frac{3}{2} = \frac{19}{6} \qquad \begin{array}{r} 3.1\overline{6} \\ 6\overline{)19.00} \end{array}$$

The actual value of $\sqrt{10}$ to 4 significant figures is 3.162. Our approximation expressed to 3 significant figures is 3.167. The percentage error is less than 0.2%.

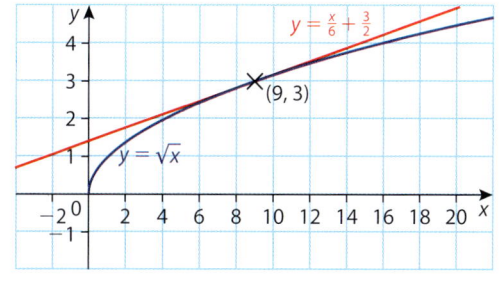

◀ Figure 11.20

The graphs of $y = \sqrt{x}$ and its tangent at $x = 9$, $y = \frac{x}{6} + \frac{3}{2}$, in Figure 11.20 illustrate that the tangent is a very good approximation to the curve in the interval $5 < x < 13$ centred on the point of tangency $(9, 3)$.

> Finding the tangent to a curve was a challenge that motivated many of the initial developments of calculus in the 17th century. In one of his books on mathematics, Descartes wrote the following about the problem of how to find a tangent to a curve:
>
> *And I dare say that this is not only the most useful and most general problem in geometry that I know, but even that I have ever desired to know.*

Differential Calculus I: Fundamentals

Example 20

Find the equation of the tangent to $f(x) = x + \frac{1}{x}$ at the point $\left(\frac{1}{2}, \frac{5}{2}\right)$.

Solution

$$f(x) = x + \frac{1}{x} = x + x^{-1}$$

$$f'(x) = 1 - x^{-2} = 1 - \frac{1}{x^2}$$

When $x = \frac{1}{2}$, $f'\left(\frac{1}{2}\right) = 1 - \frac{1}{\left(\frac{1}{2}\right)^2} = -3$. Hence, the slope of the tangent is -3.

$$y - \frac{5}{2} = -3\left(x - \frac{1}{2}\right) \Rightarrow y = -3x + \frac{3}{2} + \frac{5}{2} \Rightarrow y = -3x + 4$$

The equation of the line tangent to $f(x) = x + \frac{1}{x}$ at $x = \frac{1}{2}$ is $y = -3x + 4$.

Example 21

Consider the function $g(x) = x^2(x - 1)$.
a) Find the two points on the graph of g at which the slope of the curve is 8.
b) Find the equations of the tangents at both of these points.

Solution

a) In order to differentiate by applying the power rule term-by-term, we first need to write the equation for g in expanded form:
$$g(x) = x^2(x - 1) = x^3 - x^2$$

$$g'(x) = \frac{d}{dx}(x^3 - x^2) = 3x^2 - 2x$$

$$g'(x) = 3x^2 - 2x = 8 \Rightarrow 3x^2 - 2x - 8 = 0$$

$$(3x + 4)(x - 2) = 0 \Rightarrow x = -\frac{4}{3} \text{ or } x = 2$$

$$g\left(-\frac{4}{3}\right) = \left(-\frac{4}{3}\right)^3 - \left(-\frac{4}{3}\right)^2 = -\frac{112}{27} \text{ and } g(2) = 2^3 - 2^2 = 4$$

Thus, the slope of the curve is equal to 8 at the points $\left(-\frac{4}{3}, -\frac{112}{27}\right)$ and $(2, 4)$.

b) Tangent at $\left(-\frac{4}{3}, -\frac{112}{27}\right)$:

$$y - \left(-\frac{112}{27}\right) = 8\left[x - \left(-\frac{4}{3}\right)\right] \Rightarrow y = 8x + \frac{32}{3} - \frac{112}{27}$$

$$\Rightarrow y = 8x + \frac{176}{27}$$

Therefore, the equation of the tangent at $\left(-\frac{4}{3}, -\frac{112}{27}\right)$ is $y = 8x + \frac{176}{27}$.

Tangent at $(2, 4)$:

$$y - 4 = 8(x - 2) \Rightarrow y = 8x - 16 + 4 \Rightarrow y = 8x - 12$$

Therefore, the equation of the tangent at $(2, 4)$ is $y = 8x - 12$.

Figure 11.21 shows the results for Example 21 – the graph of the function g and the two tangent lines to the graph of the function that have a slope of 8. Note that the scales on the x- and y-axes are not equal which causes the slope of the tangent lines to appear less than 8 for this particular graph.

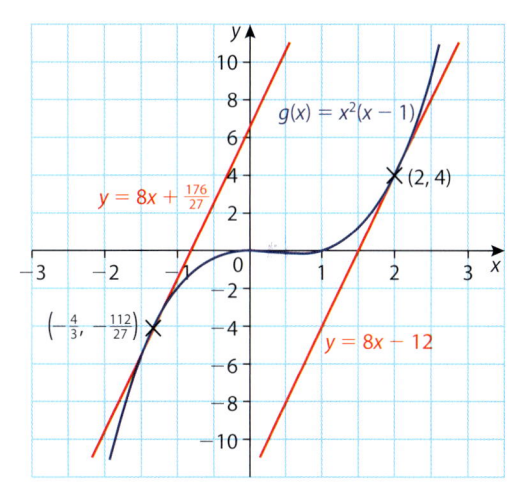

The normal to a curve at a point

Another line we often need to find is the line that is 'perpendicular' to a curve at a certain point, which we define to be the line that is perpendicular to the tangent at that point. In this particular context, we apply the adjective 'normal' rather than 'perpendicular' to denote that two lines are at right angles to one another.

A **normal** to a graph of a function at a point is the line through the point that is at a right angle to the tangent at the point. In other words, the tangent and normal to a curve at a certain point are perpendicular.

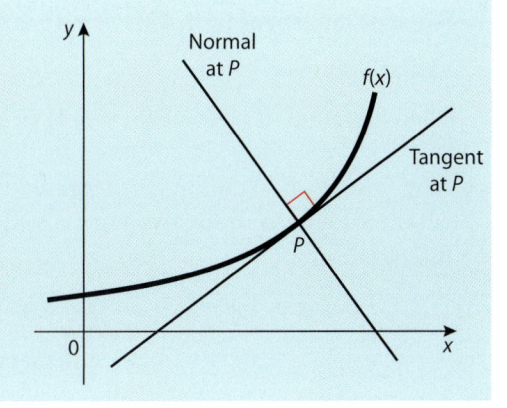

Figure 11.21

Recall that two perpendicular lines have slopes that are opposite reciprocals. If the slopes of two perpendicular lines are m_1 and m_2, then $m_1 = -\dfrac{1}{m_2}$ or $m_1 m_2 = -1$. The exception is if one of the lines is horizontal (slope is zero) and the other is vertical (slope is undefined).

Example 22

Find the equation of the normal to the graph of $y = 2x^2 - 6x + 3$ at the point $(1, -1)$.

Solution

$$\frac{dy}{dx} = \frac{d}{dx}(2x^2 - 6x + 3) = 4x - 6$$

Slope of tangent at $(1, -1)$ is $4(1) - 6 = -2$. Hence, slope of normal is $+\dfrac{1}{2}$.

Equation of normal: $y - (-1) = \dfrac{1}{2}(x - 1) \Rightarrow y = \dfrac{1}{2}x - \dfrac{3}{2}$

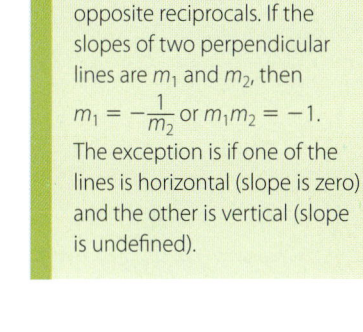

Figure 11.22 shows the results for Example 22 with the curve at both its tangent and normal at the point $(1, -1)$. Please be aware that if you graph a function with its tangent and normal at a certain point, the normal will only appear perpendicular if the scales on both the x- and y-axes are equal. Regardless of whether the scales are equal or not, the tangent will always appear tangent to the curve.

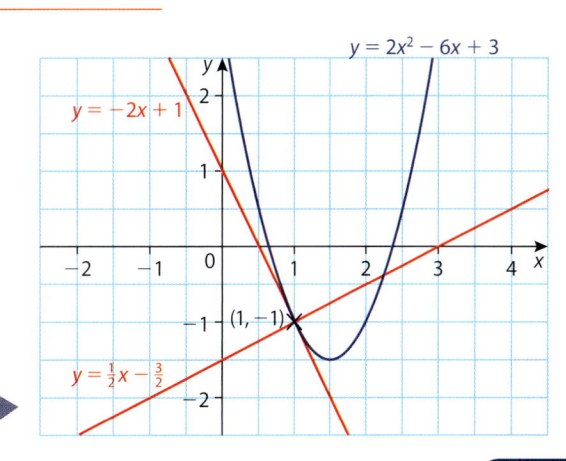

Figure 11.22

Differential Calculus I: Fundamentals

Example 23

Consider the parabola with equation $y = \frac{1}{4}x^2$.

a) Find the equation of the normals at the points $(-2, 1)$ and $(-4, 4)$.
b) Show that the point of intersection of these two normals lies on the parabola.

Solution

a) $\dfrac{dy}{dx} = \dfrac{1}{2}x$

Slope of tangent at $(-2, 1)$ is $\frac{1}{2}(-2) = -1$, so the slope of the normal at that point is $+1$.

Then equation of normal at $(-2, 1)$ is: $y - 1 = x - (-2) \Rightarrow y = x + 3$

Slope of tangent at $(-4, 4)$ is $\frac{1}{2}(-4) = -2$, so the slope of the normal at that point is $\frac{1}{2}$.

Then equation of normal at $(-4, 4)$ is: $y - 4 = \frac{1}{2}[x - (-4)]$
$$\Rightarrow y = \frac{1}{2}x + 6$$

b) Set the equations of the two normals equal to each other to find their intersection.

$$x + 3 = \tfrac{1}{2}x + 6 \Rightarrow \tfrac{1}{2}x = 3 \Rightarrow x = 6 \text{ then } y = 9$$
$$\Rightarrow \text{intersection point is } (6, 9)$$

Substitute the coordinates of the points into the equation for the parabola.

$$y = \tfrac{1}{4}x^2 \Rightarrow 9 = \tfrac{1}{4}(6)^2 \Rightarrow 9 = \tfrac{1}{4} \cdot 36 \Rightarrow 9 = 9$$

This confirms that the intersection point, $(6, 9)$, of the normals is also a point on the parabola.

Exercise 11.4

1. Find an equation of the tangent line to the graph of the equation at the indicated value of x.
   a) $y = x^2 + 2x + 1$ $x = -3$
   b) $y = x^3 + x^2$ $x = -\frac{2}{3}$
   c) $y = 3x^2 - x + 1$ $x = 0$
   d) $y = 2x + \frac{1}{x}$ $x = \frac{1}{2}$

2. Find the equations of the normal to the functions in question 1 at the indicated value of x.

3. Find the equations of the lines tangent to the curve $y = x^3 - 3x^2 + 2x$ at any point where the curve intersects the x-axis.

4. Find the equation of the tangent to the curve $y = x^2 - 2x$ that is perpendicular to the line $x - 2y = 1$.

5 Using your GDC for assistance, make accurate sketches of the curves $y = x^2 - 6x + 20$ and $y = x^3 - 3x^2 - x$ on the same set of axes. The two curves have the same slope at an integer value for x somewhere in the interval $0 \leq x \leq \frac{3}{2}$.
   a) Find this value of x.
   b) Find the equation for the line tangent to each curve at this value of x.

6 Find the equation of the normal to the curve $y = x^2 + 4x - 2$ at the point where $x = -3$. Find the coordinates of the other point where this normal intersects the curve again.

7 Consider the function $g(x) = \frac{1-x^3}{x^4}$. Find the equation of both the tangent and the normal to the graph of g at the point $(1, 0)$.

8 The normal to the curve $y = ax^{\frac{1}{2}} + bx$ at the point where $x = 1$ has a slope of 1 and intersects the y-axis at $(0, -4)$. Find the value of a and the value of b.

9 a) Find the equation of the tangent to the function $f(x) = x^3 + \frac{1}{2}x^2 + 1$ at the point $\left(-1, \frac{1}{2}\right)$.
   b) Find the coordinates of another point on the graph of f where the tangent is parallel to the tangent found in a).

10 Find the equation of both the tangent and the normal to the curve $y = \sqrt{x}(1 - \sqrt{x})$ at the point where $x = 4$.

Practice questions

1 The function f is defined as $f(x) = x^2$.
   a) Find the gradient (slope) of f at the point P, where $x = 1.5$.
   b) Find an equation for the tangent to f at the point P.
   c) Draw a diagram to show clearly the graph of f and the tangent at P.
   d) The tangent of part b) intersects the x-axis at the point Q and the y-axis at the point R. Find the coordinates of Q and R.
   e) Verify that Q is the midpoint of $[PR]$.
   f) Find an equation, in terms of a, for the tangent to f at the point $S(a, a^2)$, $a \neq 0$.
   g) The tangent of part f) intersects the x-axis at the point T and the y-axis at the point U. Find the coordinates of T and U.
   h) Prove that, whatever the value of a, T is the midpoint of SU.

2 The curve with equation $y = Ax + B + \frac{C}{x}$, $x \in \mathbb{R}$, $x \neq 0$, has a minimum at $P(1, 4)$ and a maximum at $Q(-1, 0)$. Find the value of each of the constants A, B and C.

3 Differentiate:
   a) $x^2(2 - 3x^3)$
   b) $\frac{1}{x}$

4 Consider the function $f(x) = \frac{8}{x} + 2x$, $x > 0$.
   a) Solve the equation $f'(x) = 0$. Show that the graph of f has a turning point at $(2, 8)$.
   b) Find the equations of the asymptotes to the graph of f, and hence sketch the graph.

5 Find the coordinates of the stationary point on the curve with equation $y = 4x^2 + \frac{1}{x}$.

6 The curve $y = ax^3 - 2x^2 - x + 7$ has a gradient (slope) of 3 at the point where $x = 2$. Determine the value of a.

387

11 Differential Calculus I: Fundamentals

7 If $f(2) = 3$ and $f'(2) = 5$, find an equation of **a)** the line tangent to the graph of f at $x = 2$, and **b)** the line normal to the graph of f at $x = 2$.

8 The function $g(x)$ is defined for $-3 \leq x \leq 3$. The behaviour of $g'(x)$ and $g''(x)$ is given in the tables below.

x	$-3 < x < -2$	-2	$-2 < x < 1$	1	$1 < x < 3$
$g'(x)$	negative	0	positive	0	negative

x	$-3 < x < -\frac{1}{2}$	$-\frac{1}{2}$	$-\frac{1}{2} < x < 3$
$g''(x)$	positive	0	negative

Use the information above to answer the following. In each case, justify your answer.
a) Write down the value of x for which g has a maximum.
b) On which intervals is the value of g decreasing?
c) Write down the value of x for which the graph of g has a point of inflexion.
d) Given that $g(-3) = 0$, sketch the graph of g. On the sketch, clearly indicate the position of the maximum point, the minimum point and the point of inflexion.

© International Baccalaureate Organization, 2005

9 Given the function $f(x) = x^2 - 3bx + (c + 2)$, determine the values of b and c such that $f(1) = 0$ and $f'(3) = 0$.

10 Figure 1 shows the graphs of the functions f_1, f_2, f_3, f_4. **Figure 2** includes the graphs of the derivatives of the functions shown in **Figure 1**.

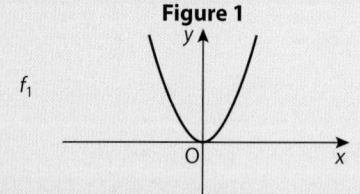

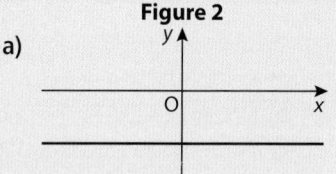

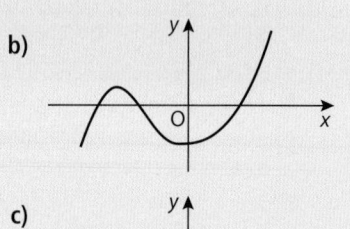

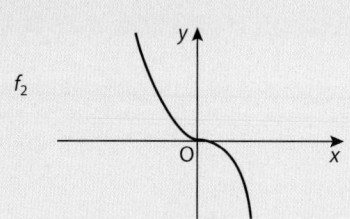

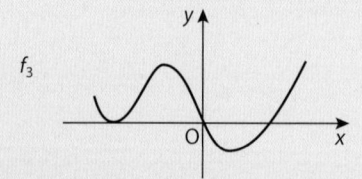

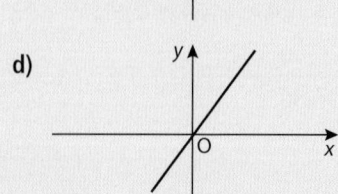

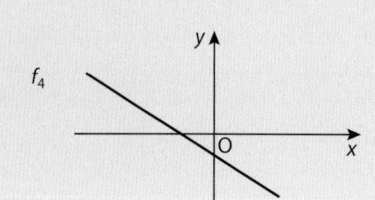

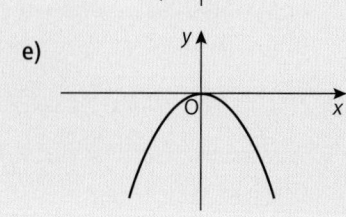

Complete the table below by matching each function with its derivative.

Function	Derivative diagram
f_1	
f_2	
f_3	
f_4	

© International Baccalaureate Organization, 2002

11 Consider the function $y = \dfrac{3x - 2}{x}$. The graph of this function has a vertical and a horizontal asymptote.
  a) Write down the equation of
    (i) the vertical asymptote
    (ii) the horizontal asymptote.
  b) Find $\dfrac{dy}{dx}$.
  c) Indicate the intervals for which the curve is increasing or decreasing.
  d) How many stationary points does the curve have? Explain using your result to **b)**.

12 Show that there are two points at which the function $h(x) = 2x^2 - x^4$ has a maximum value, and one point at which h has a minimum value. Find the coordinates of these three points, indicating whether it is a maximum or minimum.

13 The normal to the curve $y = x^{\frac{1}{2}} + x^{\frac{1}{3}}$ at the point (1, 2) meets the axes at $(a, 0)$ and $(0, b)$.
Find a and b.

14 The displacement, s metres, of a car, t seconds after leaving a fixed point A, is given by $s(t) = 10t - \dfrac{1}{2}t^2$.
  a) Calculate the velocity when $t = 0$.
  b) Calculate the value of t when the velocity is zero.
  c) Calculate the displacement of the car from A when the velocity is zero.

15 A ball is thrown vertically upwards from ground level such that its height h metres at t seconds is given by $h = 14t - 4.9t^2$.
  a) Write expressions for the ball's velocity and acceleration.
  b) Find the maximum height the ball reaches and the time it takes to reach the maximum.
  c) At the moment the ball reaches its maximum height, what is the ball's velocity and acceleration?

12 Vectors II

Assessment statements

5.1 Vectors as displacements in the plane and in three dimensions. Components of a vector; column representation.
$$\mathbf{v} = \begin{pmatrix} v_1 \\ v_2 \\ v_3 \end{pmatrix} = v_1\mathbf{i} + v_2\mathbf{j} + v_3\mathbf{k}$$
Algebraic and geometric approaches to the following topics:
the sum and difference of two vectors; the zero vector; the vector $-\mathbf{v}$;
multiplication by a scalar, $k\mathbf{v}$;
magnitude of a vector, $|\mathbf{v}|$;
unit vectors; base vectors, $\mathbf{i}$, $\mathbf{j}$ and $\mathbf{k}$;
position vectors $\overrightarrow{OA} = \mathbf{a}$.

5.2 The scalar product of two vectors.
Perpendicular vectors; parallel vectors.
The angle between two vectors.

5.3 Representation of a line as $\mathbf{r} = \mathbf{a} + t\mathbf{b}$.
The angle between two lines.

5.4 Distinguishing between coincident and parallel lines.
Finding points where lines intersect.

Introduction

You have seen vectors in the plane in Chapter 8. We will limit our discussion to mainly three-dimensional space in this chapter. If you need to refresh your knowledge of the plane case, refer to Chapter 8.

Because we live in a three-dimensional world, it is essential that we study objects in three dimensions. To that end, we consider in this section a three-dimensional coordinate system in which points are determined by ordered triples. We construct the coordinate system in the following manner: Choose three mutually perpendicular axes, as shown in Figure 12.1, to serve as our reference. The orientation of the system is *right-handed* in the sense that if you stand with your back to the *z*-axis and stretch your arms out with a right angle between them, then the right hand will point towards the *x*-axis and the left hand towards the *y*-axis. That is, if you are looking straight at the system, the *yz*-plane is the plane facing you, and the *xz*-plane is perpendicular to it and extending out of the page towards you, and the *xy*-plane is the bottom part of that picture. The *xy*-, *xz*- and *yz*-planes are called the **coordinate planes**. Points in space are assigned coordinates in the same manner as in the plane. So, the point *P* (next page) is assigned the ordered triple (*x*, *y*, *z*) to indicate that it is *x*, *y* and *z* units from the *yz*-, *xz*- and *xy*-planes (Figure 12.2).

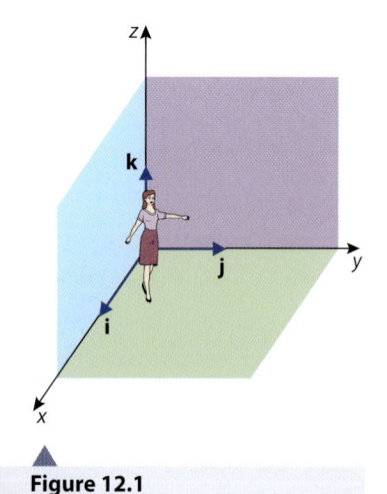

Figure 12.1

◀ **Figure 12.2**

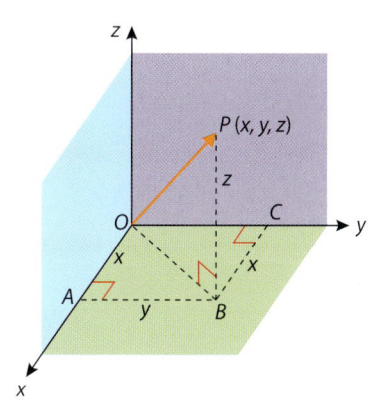

In this chapter, we will extend our study of vectors to space. The good news is that many of the rules you know from the plane also apply to vectors in space. So, we will only have to introduce few new concepts. Some of the material will either be a repeat of what you have learned for two-dimensional space or an extension.

12.1 Vectors from a geometric viewpoint

Vectors can be represented geometrically by arrows in two- or three-dimensional space; the direction of the arrow specifies the direction of the vector, and the length of the arrow describes its magnitude. The first point on the arrow is called the **initial point** of the vector and the tip is called the **terminal point**. We shall denote vectors in lower-case boldface type, such as **v**, when using one letter to name the vector, and we will use $\overrightarrow{AB}$ to denote the vector from A to B. The handwritten notation will be the latter too.

If the initial point of a vector is at the origin, the vector is said to be in standard position. It is also called the **position vector** of point P. The terminal point will have coordinates of the form (x, y, z). We call these coordinates the **components** of **v** and we write $\mathbf{v} = (x, y, z)$.

The length (magnitude) of a vector **v** is also known as its **modulus** or its **norm** and it is written as $|\mathbf{v}|$.

Using Pythagoras' theorem, we can show that the magnitude of a vector **v**,

$|\mathbf{v}| = \sqrt{x^2 + y^2 + z^2}$

Let $\overrightarrow{OP} = \mathbf{v}$, then

$|\mathbf{v}| = |\overrightarrow{OP}| = \sqrt{OB^2 + BP^2}$, since the triangle OBP is right-angled at B. Now, consider triangle OAB, which is right-angled at A:

$OB^2 = OA^2 + AB^2 = x^2 + y^2$, and, therefore,

$|\mathbf{v}| = \sqrt{OB^2 + BP^2} = \sqrt{(x^2 + y^2) + z^2} = \sqrt{x^2 + y^2 + z^2}$

Two vectors like **v** and $\overrightarrow{AB}$ are equal (equivalent) if they have the same length (magnitude) and the same direction; we write $\mathbf{v} = \overrightarrow{AB}$. Geometrically, two vectors are equal if they are translations of one another as you see in Figures 12.3 and 12.4. Notice in Figure 12.4 that the four vectors are equal, even though they are in different positions.

12 Vectors II

Figure 12.3

Figure 12.4

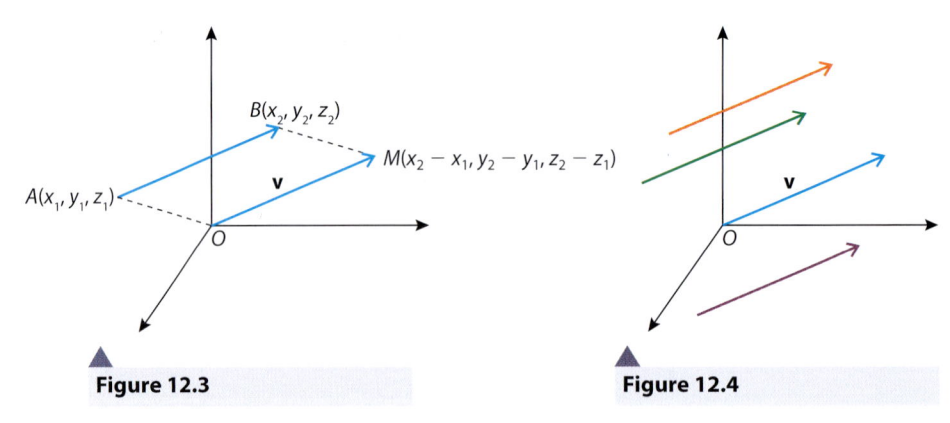

Because vectors are not affected by translation, the initial point of a vector **v** can be moved to any convenient position by making an appropriate translation.

Two vectors are said to be opposite if they have equal modulus but opposite direction (Figure 12.5).

If the initial and terminal points of a vector coincide, the vector has length zero; we call this the **zero vector** and denote it by **0**.

The zero vector does not have a specific direction, so we will agree that it can be assigned any convenient direction in a specific problem.

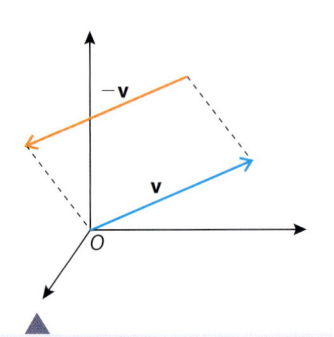

Figure 12.5

Addition and subtraction of vectors

As you recall from Chapter 8, according to the **triangular rule**, if **u** and **v** are vectors, the sum **u** + **v** is the vector from the initial point of **u** to the terminal point of **v**, when the vectors are positioned so that the initial point of **v** is the terminal point of **u**, as shown in Figure 12.6.

Equivalently, **u** + **v** is also the diagonal of the parallelogram whose sides are **u** and **v**, as shown in Figure 12.7.

The difference of the two vectors **u** and **v** can be dealt with in the same manner. So, the vector **w** = **u** − **v** is a vector such that **u** = **v** + **w**.

In Figure 12.8, we can clearly see that the difference is along the diagonal joining the two terminal points of the vectors and in the direction from **v** to **u**.

• **Hint:** When we discuss vectors, we will refer to real numbers as scalars.

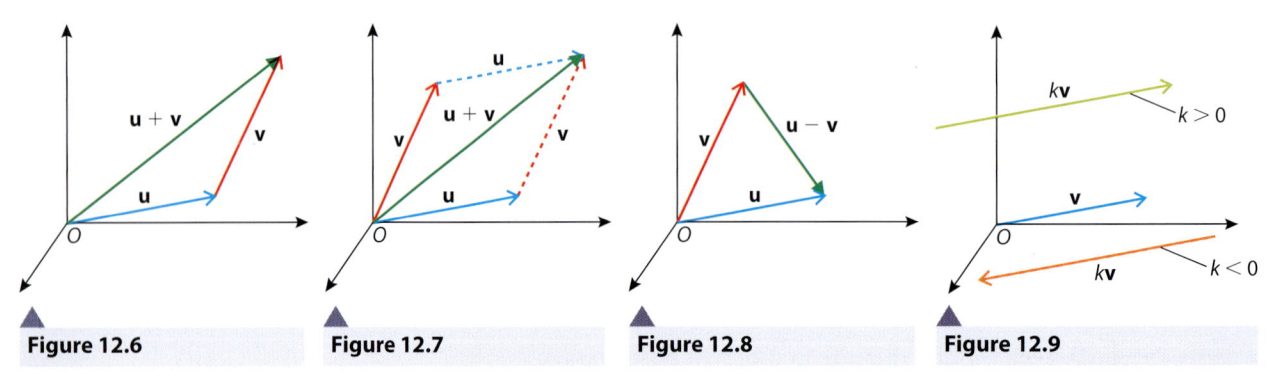

If k is a real positive number, $k\mathbf{v}$ is a vector of magnitude $k|\mathbf{v}|$ and in the same direction as **v**. It follows that when k is negative, $k\mathbf{v}$ has magnitude $|k| \times |\mathbf{v}|$ and is in the opposite direction to **v** (Figure 12.9).

Figure 12.6

Figure 12.7

Figure 12.8

Figure 12.9

A result of the previous situation is the necessary and sufficient condition for two vectors to be parallel:

> Two vectors are parallel if one of them is a scalar multiple of the other.
> For example, the vector $(-3, 4, -2)$ is parallel to the vector $(4.5, -6, 3)$ since $(-3, 4, -2) = -\frac{2}{3}(4.5, -6, 3)$.

Components provide a simple way to algebraically perform several operations on vectors. First, by definition, we know that two vectors are equal if they have the same length and the same magnitude. So, if we choose to draw the two equal vectors $\mathbf{u} = (u_1, u_2, u_3)$ and $\mathbf{v} = (v_1, v_2, v_3)$ from the origin, their terminal points must coincide, and hence $u_1 = v_1, u_2 = v_2$ and $u_3 = v_3$. So, we showed that equal vectors have the same components. The converse is obviously true, i.e. if $u_1 = v_1, u_2 = v_2$ and $u_3 = v_3$, the two vectors are equal. The following results are also obvious from the simple geometry of similar figures:

If $\mathbf{u} = (u_1, u_2, u_3)$ and $\mathbf{v} = (v_1, v_2, v_3)$ and k is any real number, then

$$\mathbf{u} + \mathbf{v} = (u_1 + v_1, u_2 + v_2, u_3 + v_3) \text{ and } k\mathbf{u} = (ku_1, ku_2, ku_3)$$

If the initial point of the vector is not at the origin, the following theorem generalizes the previous notation to any position:

If $\overrightarrow{AB}$ is a vector with initial point $A(x_1, y_1, z_1)$ and terminal point $B(x_2, y_2, z_2)$, then $\overrightarrow{AB} = \overrightarrow{OB} - \overrightarrow{OA} = (x_2 - x_1, y_2 - y_1, z_2 - z_1)$, as you see in Figure 12.10.

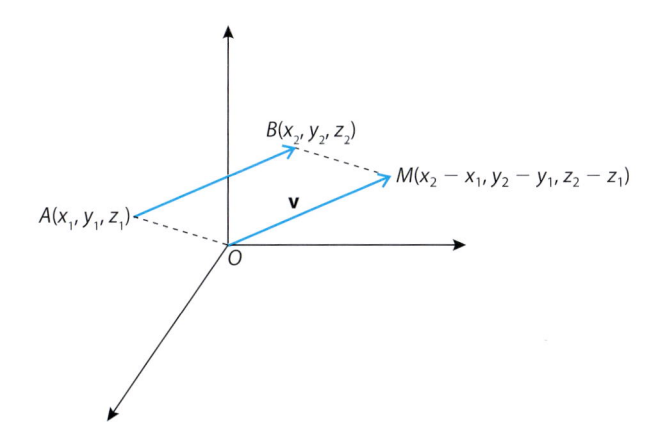

Figure 12.10

As illustrated in Figure 12.10, either by applying the distance formula or by using the equality of vectors $\mathbf{v}$ and $\overrightarrow{AB}$,

$$|\overrightarrow{AB}| = \sqrt{(x_2 - x_1)^2 + (y_2 - y_1)^2 + (z_2 - z_1)^2}$$

Additionally, the following results can follow easily from properties of real numbers: $\mathbf{u} + \mathbf{v} = \mathbf{v} + \mathbf{u}$; $(\mathbf{u} + \mathbf{v}) + \mathbf{w} = \mathbf{u} + (\mathbf{v} + \mathbf{w})$; $k(\mathbf{u} + \mathbf{v}) = k\mathbf{u} + k\mathbf{v}$; and the other obvious relationships.

12 Vectors II

Example 1

Given the points $A(-2, 3, 5)$ and $B(1, 0, -4)$,
a) find the components of vector $\overrightarrow{AB}$
b) find the components of vector $\overrightarrow{BA}$
c) find the components of vector $3\overrightarrow{AB}$
d) find the components of vector $\overrightarrow{OA} + \overrightarrow{OB}$
e) calculate $|\overrightarrow{AB}|$ and $|\overrightarrow{BA}|$
f) calculate $|3\overrightarrow{AB}|$ and $|\overrightarrow{OA} + \overrightarrow{OB}|$.

Solution

a) $\overrightarrow{AB} = \overrightarrow{OB} - \overrightarrow{OA} = (x_2 - x_1, y_2 - y_1, z_2 - z_1)$
$= (1 - (-2), 0 - 3, -4 - 5) = (3, -3, -9)$

b) Since $\overrightarrow{BA}$ is the opposite of $\overrightarrow{AB}$, then $\overrightarrow{BA} = (-3, 3, 9)$.

c) $3\overrightarrow{AB} = 3(3, -3, -9) = (9, -9, -27)$

d) $\overrightarrow{OA} + \overrightarrow{OB} = (-2 + 1, 3 + 0, 5 - 4) = (-1, 3, 1)$

e) $|\overrightarrow{AB}| = \sqrt{(x_2 - x_1)^2 + (y_2 - y_1)^2 + (z_2 - z_1)^2} = \sqrt{9 + 9 + 81} = 3\sqrt{11}$
$|\overrightarrow{BA}| = \sqrt{(x_2 - x_1)^2 + (y_2 - y_1)^2 + (z_2 - z_1)^2} = \sqrt{9 + 9 + 81} = 3\sqrt{11}$

f) $|3\overrightarrow{AB}| = \sqrt{(x_2 - x_1)^2 + (y_2 - y_1)^2 + (z_2 - z_1)^2} = \sqrt{81 + 81 + 729}$
$= \sqrt{891} = 9\sqrt{11}$

Obviously, $|3\overrightarrow{AB}| = 3|\overrightarrow{AB}|$!

$|\overrightarrow{OA} + \overrightarrow{OB}| = |(-1, 3, 1)| = \sqrt{1 + 9 + 1} = \sqrt{11}$

Notice that $|\overrightarrow{OA} + \overrightarrow{OB}| = \sqrt{11} \neq |\overrightarrow{OA}| + |\overrightarrow{OB}|$
$= \sqrt{4 + 9 + 25} + \sqrt{1 + 0 + 16} = \sqrt{38} + \sqrt{17}$

Example 2

Determine the relationship between the coordinates of point $M(x, y, z)$ so that the points M, $A(0, -1, 5)$ and $B(1, 2, 3)$ are collinear.

Solution

For the points to be collinear, it is enough to make $\overrightarrow{AM}$ parallel to $\overrightarrow{AB}$.
If the two vectors are parallel, then one of them is a scalar multiple of the other. Say $\overrightarrow{AM} = t\overrightarrow{AB}$.

$\overrightarrow{AM} = (x, y + 1, z - 5) = t(1, 2, 3) = (t, 2t, 3t)$

So, $x = t, y + 1 = 2t$, and $z - 5 = 3t$.

Unit vectors

A vector of length 1 is called a **unit vector**. So, in two-dimensional space, the vectors $\mathbf{i} = (1, 0)$ and $\mathbf{j} = (0, 1)$ are unit vectors along the x- and y-axes, and in three-dimensional space, the unit vectors along the axes are $\mathbf{i} = (1, 0, 0)$, $\mathbf{j} = (0, 1, 0)$ and $\mathbf{k} = (0, 0, 1)$. The vectors $\mathbf{i}, \mathbf{j}$ and $\mathbf{k}$ are called the **base vectors** of the 3-space.

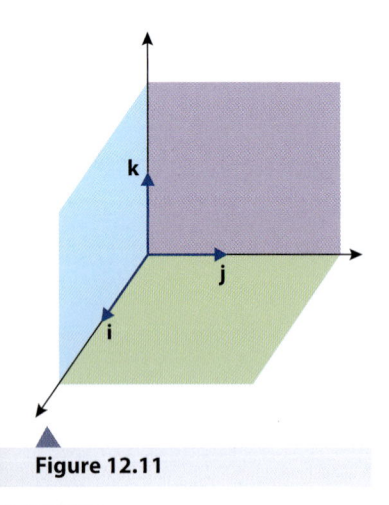

Figure 12.11

It follows immediately that each vector in 3-space can be expressed uniquely in terms of **i**, **j** and **k** as follows:

$$\mathbf{u} = (x, y, z) = (x, 0, 0) + (0, y, 0) + (0, 0, z)$$
$$= x(1, 0, 0) + y(0, 1, 0) + z(0, 0, 1) = x\mathbf{i} + y\mathbf{j} + z\mathbf{k}$$

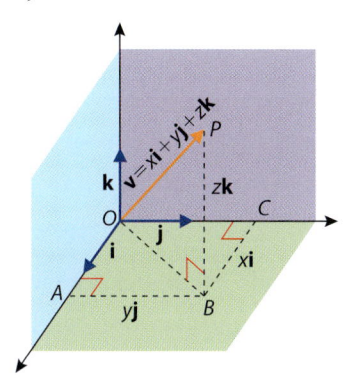

• **Hint:** The terms '2-space' and '3-space' are short forms for two-dimensional space and three-dimensional space respectively.

◀ Figure 12.12

So, in the previous example, $\overrightarrow{AB} = (3, -3, -9) = 3\mathbf{i} - 3\mathbf{j} - 9\mathbf{k}$.

Unit vectors can be found in any direction, not only in the direction of the axes. For example, if we want to find the unit vector in the same direction as **u**, we need to find a vector parallel to **u**, which has a magnitude of 1. Since **u** has a magnitude of |**u**|, it is enough to multiply this vector by 1/|**u**| to 'normalize' it. So, the unit vector **v** in the same direction as **u** is $\mathbf{v} = \frac{1}{|\mathbf{u}|}\mathbf{u} = \frac{\mathbf{u}}{|\mathbf{u}|}$. This is a unit vector since its length is 1. This is why:

Recall that |**u**| is a real number (scalar), and so is 1/|**u**|.

Let $1/|\mathbf{u}| = k \Rightarrow \mathbf{v} = \frac{1}{|\mathbf{u}|}\mathbf{u} = k\mathbf{u} \Rightarrow |\mathbf{v}| = |k\mathbf{u}| = k|\mathbf{u}| = \frac{1}{|\mathbf{u}|} \cdot |\mathbf{u}| = 1$.

Example 3

Find a unit vector in the direction of $\mathbf{v} = \mathbf{i} - 2\mathbf{j} + 3\mathbf{k}$.

Solution

The length of the vector **v** is $\sqrt{1^2 + 2^2 + 3^2} = \sqrt{14}$, so the unit vector is

$$\frac{1}{\sqrt{14}}(\mathbf{i} - 2\mathbf{j} + 3\mathbf{k}) = \frac{\mathbf{i}}{\sqrt{14}} - \frac{2\mathbf{j}}{\sqrt{14}} + \frac{3\mathbf{k}}{\sqrt{14}}$$

To verify that this is a unit vector, we find its length:

$$\sqrt{\left(\frac{1}{\sqrt{14}}\right)^2 + \left(\frac{2}{\sqrt{14}}\right)^2 + \left(\frac{3}{\sqrt{14}}\right)^2} = \sqrt{\frac{1}{14} + \frac{4}{14} + \frac{9}{14}} = 1$$

The unit vector plays another important role: it determines the direction of the given vector.

Recall from Chapter 8 that, in 2-space, we can write the vector in a form that gives us its direction (in terms of the angle it makes with the horizontal axis, called the direction angle) and its magnitude.

12 Vectors II

In the diagram below, θ is the angle with the horizontal axis.
The unit vector $\mathbf{v}$, in the same direction as $\mathbf{u}$, is:
$$\mathbf{v} = 1\cos\theta\,\mathbf{i} + 1\sin\theta\,\mathbf{j}$$
and from the results above,
$$\mathbf{v} = \frac{1}{|\mathbf{u}|}\mathbf{u} \Rightarrow$$
$$\mathbf{u} = |\mathbf{u}|\,(\mathbf{v})$$
$$= |\mathbf{u}|\cos\theta\,\mathbf{i} + |\mathbf{u}|\sin\theta\,\mathbf{j}$$
$$= |\mathbf{u}|(\cos\theta\,\mathbf{i} + \sin\theta\,\mathbf{j}).$$

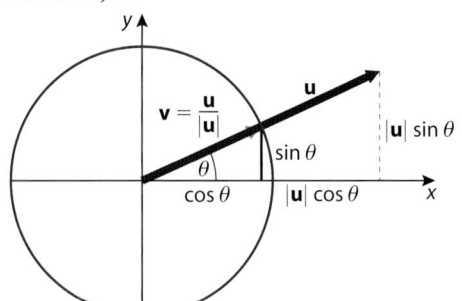

Example 4
Find the vector with magnitude 2 that makes an angle of 60° with the positive x-axis.

Solution
$$\mathbf{v} = |\mathbf{v}|(\cos 60°\,\mathbf{i} + \sin 60°\,\mathbf{j}) = 2\left(\frac{1}{2}\mathbf{i} + \frac{\sqrt{3}}{2}\mathbf{j}\right) = \mathbf{i} + \sqrt{3}\mathbf{j}$$

Example 5
Find the direction and magnitude of the vector $\mathbf{v} = 2\sqrt{3}\mathbf{i} - 2\mathbf{j}$.

Solution
$$|\mathbf{v}| = \sqrt{(2\sqrt{3})^2 + 4} = 4$$
$$\cos\theta = \frac{2\sqrt{3}}{4} = \frac{\sqrt{3}}{2},\ \sin\theta = \frac{-2}{4} = -\frac{1}{2} \Rightarrow \theta = -\frac{\pi}{6}$$

Example 6
a) Find the unit vector that has the same direction as $\mathbf{v} = \mathbf{i} + 2\mathbf{j} - 2\mathbf{k}$.
b) Find a vector of length 6 that is parallel to $\mathbf{v} = \mathbf{i} - 2\mathbf{j} + 3\mathbf{k}$.

Solution
a) The vector $\mathbf{v}$ has magnitude $|\mathbf{v}| = \sqrt{1 + 2^2 + (-2)^2} = 3$,
so the unit vector ν in the same direction as $\mathbf{v}$ is
$$\nu = \frac{1}{3}\mathbf{v} = \frac{1}{3}\mathbf{i} + \frac{2}{3}\mathbf{j} - \frac{2}{3}\mathbf{k}$$
b) Let $\mathbf{u}$ be the vector in question and ν be the unit vector in the direction of $\mathbf{v}$.
$$\mathbf{u} = 6\cdot\nu = 6 \times \frac{1}{\sqrt{14}}(\mathbf{i} - 2\mathbf{j} + 3\mathbf{k}) = \frac{6\mathbf{i}}{\sqrt{14}} - \frac{12\mathbf{j}}{\sqrt{14}} + \frac{18\mathbf{k}}{\sqrt{14}}$$

Exercise 12.1

1. Write the vector $\overrightarrow{AB}$ in component form in each of the following cases.
   a) $A\left(-\frac{3}{2}, -\frac{1}{2}, 1\right); B\left(1, -\frac{5}{2}, 1\right)$
   b) $A\left(-2, -\sqrt{3}, -\frac{1}{2}\right); B\left(1, \sqrt{3}, -\frac{1}{2}\right)$
   c) $A(2, -3, 5); B(1, -1, 3)$
   d) $A(a, -a, 2a); B(-a, -2a, a)$

2. Given the coordinates of point P or Q and the components of $\overrightarrow{PQ}$, find the missing items.
   a) $P\left(-\frac{3}{2}, -\frac{1}{2}, 1\right); \overrightarrow{PQ}\left(1, -\frac{5}{2}, 1\right)$
   b) $\overrightarrow{PQ}\left(-\frac{3}{2}, -\frac{1}{2}, 1\right); Q\left(1, -\frac{5}{2}, 1\right)$
   c) $P(a, -2a, 2a); \overrightarrow{PQ}(-a, -2a, a)$

3 Determine the relationship between the coordinates of point $M(x, y, z)$ so that the points M, A and B are collinear.
   a) $A(0, 0, 5)$; $B(1, 1, 0)$
   b) $A(-1, 0, 1)$; $B(3, 5, -2)$
   c) $A(2, 3, 4)$; $B(-2, -3, 5)$

4 Given the coordinates of the points A and B, find the symmetric image C of B with respect to A.
   a) $A(3, -4, 0)$; $B(-1, 0, 1)$
   b) $A(-1, 3, 5)$; $B\left(-1, \frac{1}{2}, \frac{1}{3}\right)$
   c) $A(1, 2, -1)$; $B(a, 2a, b)$

5 Given a triangle ABC and a point G such that $\overrightarrow{GA} + \overrightarrow{GB} + \overrightarrow{GC} = 0$, find the coordinates of G in each of the following cases.
   a) $A(-1, -1, -1)$; $B(-1, 2, -1)$; $C(1, 2, 3)$
   b) $A(2, -3, 1)$; $B(1, -2, -5)$; $C(0, 0, 1)$
   c) $A(a, 2a, 3a)$; $B(b, 2b, 3b)$; $C(c, 2c, 3c)$

6 Determine the fourth vertex D of the parallelogram $ABCD$ having AB and BC as adjacent sides.
   a) $A(\sqrt{3}, 2, -1)$; $B(1, 3, 0)$; $C(-\sqrt{3}, 2, -5)$
   b) $A(\sqrt{2}, \sqrt{3}, \sqrt{5})$; $B(3\sqrt{2}, -\sqrt{3}, 5\sqrt{5})$; $C(-2\sqrt{2}, \sqrt{3}, -3\sqrt{5})$
   c) $A\left(-\frac{1}{2}, \frac{1}{3}, 0\right)$; $B\left(\frac{1}{2}, \frac{2}{3}, 5\right)$; $C\left(\frac{7}{2}, -\frac{1}{3}, 1\right)$

7 Determine the values of m and n such that the vectors $\mathbf{v}(m - 2, m + n, -2m + n)$ and $\mathbf{w}(2, 4, -6)$ have the same direction.

8 Find a unit vector in the same direction as each vector.
   a) $\mathbf{v} = 2\mathbf{i} + 2\mathbf{j} - \mathbf{k}$
   b) $\mathbf{v} = 6\mathbf{i} - 4\mathbf{j} + 2\mathbf{k}$
   c) $\mathbf{v} = 2\mathbf{i} - \mathbf{j} - 2\mathbf{k}$

9 Let $\mathbf{u} = \mathbf{i} + 3\mathbf{j} - 2\mathbf{k}$ and $\mathbf{v} = 2\mathbf{i} + \mathbf{j}$. Find
   a) $|\mathbf{u} + \mathbf{v}|$
   b) $|\mathbf{u}| + |\mathbf{v}|$
   c) $|-3\mathbf{u}| + |3\mathbf{v}|$
   d) $\frac{1}{|\mathbf{u}|}\mathbf{u}$
   e) $\left|\frac{1}{\|\mathbf{u}\|}\mathbf{u}\right|$

10 Find the terminal points for each vector.
   a) $\mathbf{w} = 4\mathbf{i} + 2\mathbf{j} - 2\mathbf{k}$, given the initial point $(-1, 2, -3)$
   b) $\mathbf{v} = 2\mathbf{i} - 3\mathbf{j} + \mathbf{k}$, given the initial point $(-2, 1, 4)$

11 Find vectors that satisfy the stated conditions:
   a) opposite direction of $\mathbf{u} = (-3, 4)$ and third the magnitude of $\mathbf{u}$
   b) length of 12 and same direction as $\mathbf{w} = 4\mathbf{i} + 2\mathbf{j} - 2\mathbf{k}$
   c) of the form $x\mathbf{i} + y\mathbf{j} - 2\mathbf{k}$ and parallel to $\mathbf{w} = \mathbf{i} - 4\mathbf{j} + 3\mathbf{k}$

12 Let $\mathbf{u}$, $\mathbf{v}$ and $\mathbf{w}$ be the vectors from each vertex of a triangle to the midpoint of the opposite side. Find the value of $\mathbf{u} + \mathbf{v} + \mathbf{w}$.

12 Vectors II

12.2 Scalar (dot) product

If $\mathbf{u} = (u_1, u_2, u_3)$ and $\mathbf{v} = (v_1, v_2, v_3)$ are two vectors, the dot product (scalar) is written as $\mathbf{u} \cdot \mathbf{v}$ and is defined as

$$\mathbf{u} \cdot \mathbf{v} = u_1 v_1 + u_2 v_2 + u_3 v_3$$

Result 1: $\mathbf{u}^2 = \mathbf{u} \cdot \mathbf{u} = u_1 \cdot u_1 + u_2 \cdot u_2 + u_3 \cdot u_3 = u_1^2 + u_2^2 + u_3^2 = |\mathbf{u}|^2$

From this definition, we can deduce another geometric 'definition' of the dot product:

$\mathbf{u} \cdot \mathbf{v} = |\mathbf{u}||\mathbf{v}| \cos \theta$, where θ is the angle between the two vectors

Proof:

Let $\mathbf{u}$ and $\mathbf{v}$ be drawn from the same point, as shown in Figure 12.13. Then

$$|\mathbf{u} - \mathbf{v}|^2 = (\mathbf{u} - \mathbf{v}) \cdot (\mathbf{u} - \mathbf{v}) = \mathbf{u}^2 + \mathbf{v}^2 - 2\mathbf{u} \cdot \mathbf{v}$$
$$= |\mathbf{u}|^2 + |\mathbf{v}|^2 - 2\mathbf{u} \cdot \mathbf{v}$$

Also, using the law of cosines,

$$|\mathbf{u} - \mathbf{v}|^2 = |\mathbf{u}|^2 + |\mathbf{v}|^2 - 2|\mathbf{u}| \cdot |\mathbf{v}| \cdot \cos \theta$$

By comparing the two results above, we can conclude that $\mathbf{u} \cdot \mathbf{v} = |\mathbf{u}||\mathbf{v}| \cos \theta$.

The scalar product can be used, among other things, to find angles between vectors.

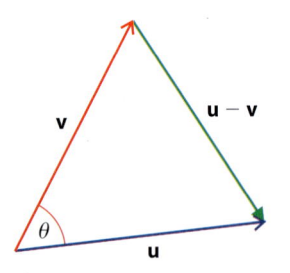

Figure 12.13

Example 7

Find the angle between the vectors $\mathbf{u} = \mathbf{i} - 2\mathbf{j} + 2\mathbf{k}$ and $\mathbf{v} = -3\mathbf{i} + 6\mathbf{j} + 2\mathbf{k}$.

Solution

From the previous results, we have

$$\cos \theta = \frac{\mathbf{u} \cdot \mathbf{v}}{|\mathbf{u}| \cdot |\mathbf{v}|} = \frac{-3 - 12 + 4}{\sqrt{1 + 4 + 4}\sqrt{9 + 36 + 4}} = \frac{-11}{21}$$

$$\Rightarrow \theta = \cos^{-1}\left(\frac{-11}{21}\right) \approx 2.12 \text{ radians}$$

Result 2: A direct conclusion of the previous definitions is that if two vectors are *perpendicular*, the dot product is *zero*.

This is so because when the two vectors are perpendicular the angle between them is $\pm 90°$ and, therefore,

$$\mathbf{u} \cdot \mathbf{v} = |\mathbf{u}||\mathbf{v}| \cos \theta = |\mathbf{u}||\mathbf{v}| \cos 90° = |\mathbf{u}||\mathbf{v}| \cdot 0 = 0$$

Result 3: If two vectors $\mathbf{u}$ and $\mathbf{v}$ are parallel, then $\mathbf{u} \cdot \mathbf{v} = \pm |\mathbf{u}||\mathbf{v}|$.

Again, this is so because when the vectors are parallel the angle between them is either 0° or 180° and, therefore,

$$\mathbf{u} \cdot \mathbf{v} = |\mathbf{u}||\mathbf{v}| \cos \theta = |\mathbf{u}||\mathbf{v}| \cos 0° = |\mathbf{u}||\mathbf{v}| \cdot 1 = |\mathbf{u}||\mathbf{v}|, \text{ or}$$
$$\mathbf{u} \cdot \mathbf{v} = |\mathbf{u}||\mathbf{v}| \cos \theta = |\mathbf{u}||\mathbf{v}| \cos 180° = |\mathbf{u}||\mathbf{v}| \cdot (-1) = -|\mathbf{u}||\mathbf{v}|$$

Example 8
Determine which, if any, of the following vectors are orthogonal.

$$\mathbf{u} = 7\mathbf{i} + 3\mathbf{j} + 2\mathbf{k}, \mathbf{v} = -3\mathbf{i} + 5\mathbf{j} + 3\mathbf{k}, \mathbf{w} = \mathbf{i} + \mathbf{k}$$

Solution
$\mathbf{u} \cdot \mathbf{v} = 7(-3) + 3 \times 5 + 2 \times 3 = 0$; orthogonal vectors
$\mathbf{u} \cdot \mathbf{w} = 7 \times 1 + 3 \times 0 + 2 \times 1 = 9$; not orthogonal
$\mathbf{v} \cdot \mathbf{w} = -3 \times 1 + 5 \times 0 + 3 \times 1 = 0$; orthogonal vectors

Example 9
$A(1, 2, 3)$, $B(-3, 2, 4)$ and $C(1, -4, 3)$ are the vertices of a triangle. Show that the triangle is right-angled and find its area.

Solution
$\overrightarrow{AB} = (-3 - 1)\mathbf{i} + (2 - 2)\mathbf{j} + (4 - 3)\mathbf{k} = -4\mathbf{i} + \mathbf{k}$
$\overrightarrow{AC} = (1 - 1)\mathbf{i} + (-4 - 2)\mathbf{j} + (3 - 3)\mathbf{k} = -6\mathbf{j}$
$\overrightarrow{BC} = (1 - (-3))\mathbf{i} + (-4 - 2)\mathbf{j} + (3 - 4)\mathbf{k} = 4\mathbf{i} - 6\mathbf{j} - \mathbf{k}$

Since $\overrightarrow{AB} \cdot \overrightarrow{AC} = -4 \times 0 + 0 \times -6 + 1 \times 0 = 0$, the vectors are perpendicular. So the triangle is right-angled at A.

The area of this right triangle is half the product of the legs.

$$\text{Area} = \frac{1}{2}|\overrightarrow{AB}||\overrightarrow{AC}| = \frac{1}{2} \cdot \sqrt{(-4)^2 + 1} \cdot 6 - 3\sqrt{17}$$

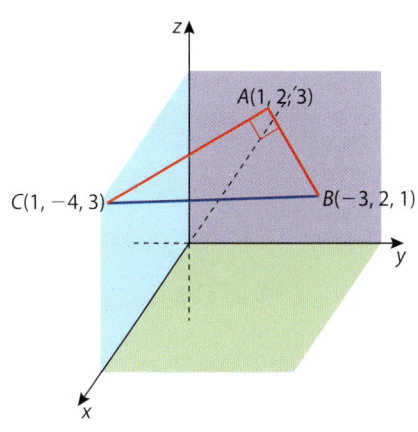

Direction angles, direction cosines

Figure 12.14 shows a non-zero vector $\mathbf{v}$. The angles α, β and γ that the vector makes with the unit coordinate vectors are called the **direction angles** of $\mathbf{v}$, and $\cos\alpha$, $\cos\beta$ and $\cos\gamma$ are called the **direction cosines**.

Let $\mathbf{v} = x\mathbf{i} + y\mathbf{j} + z\mathbf{k}$. Considering the right triangles OAP, OCP and ODP, the hypotenuse in each of these triangles is OP, i.e. $|\mathbf{v}|$. From your trigonometry chapters, you know that the side adjacent to an angle θ in a right triangle is related to it by

$$\cos\theta = \frac{\text{adjacent}}{\text{hypotenuse}} \Leftrightarrow \text{adjacent} = \text{hypotenuse} \cdot \cos\theta, \text{ so in this case}$$

$x = |\mathbf{v}|\cos\alpha$, $y = |\mathbf{v}|\cos\beta$, $z = |\mathbf{v}|\cos\gamma$, and so
$\mathbf{v} = (|\mathbf{v}|\cos\alpha)\mathbf{i} + (|\mathbf{v}|\cos\beta)\mathbf{j} + (|\mathbf{v}|\cos\gamma)\mathbf{k} = |\mathbf{v}|(\cos\alpha\,\mathbf{i} + \cos\beta\,\mathbf{j} + \cos\gamma\,\mathbf{k})$

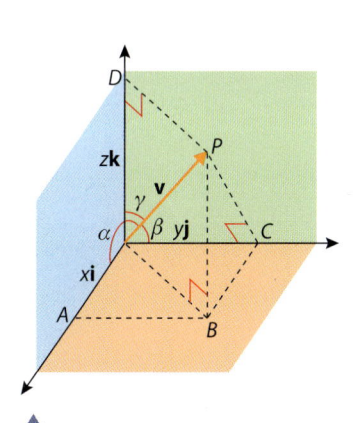

Figure 12.14

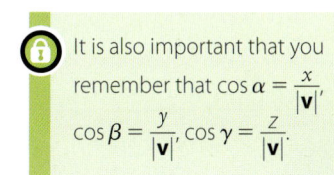

Taking the magnitude of both sides,

$$|\mathbf{v}| = |\mathbf{v}|\sqrt{\cos^2\alpha + \cos^2\beta + \cos^2\gamma}$$

Therefore,
$\cos^2\alpha + \cos^2\beta + \cos^2\gamma = 1$, i.e. the sum of the squares of the direction cosines is always 1. For a unit vector, the expression will be of the form

$$\mathbf{u} = |\mathbf{u}|(\cos\alpha\,\mathbf{i} + \cos\beta\,\mathbf{j} + \cos\gamma\,\mathbf{k}) = \cos\alpha\,\mathbf{i} + \cos\beta\,\mathbf{j} + \cos\gamma\,\mathbf{k} \ (|\mathbf{u}| = 1)$$

This means that for a unit vector its x-, y- and z-coordinates are its direction cosines.

It is also important that you remember that $\cos\alpha = \frac{x}{|\mathbf{v}|}$, $\cos\beta = \frac{y}{|\mathbf{v}|}$, $\cos\gamma = \frac{z}{|\mathbf{v}|}$.

Example 10
Find the direction cosines of the vector $\mathbf{v} = 4\mathbf{i} - 2\mathbf{j} + 4\mathbf{k}$, and then approximate the direction angles to the nearest degree.

Solution

$$|\mathbf{v}| = \sqrt{4^2 + (-2)^2 + 4^2} = 6 \Rightarrow \mathbf{v} = \frac{\mathbf{v}}{|\mathbf{v}|} = \frac{2}{3}\mathbf{i} - \frac{1}{3}\mathbf{j} + \frac{2}{3}\mathbf{k}, \text{ thus}$$

$$\cos\alpha = \frac{2}{3}, \cos\beta = -\frac{1}{3}, \cos\gamma = \frac{2}{3}$$

From your GDC you will obtain

$$\alpha = \cos^{-1}\left(\frac{2}{3}\right) \approx 48°, \beta = \cos^{-1}\left(-\frac{1}{3}\right) \approx 109°, \gamma = \cos^{-1}\left(\frac{2}{3}\right) \approx 48°$$

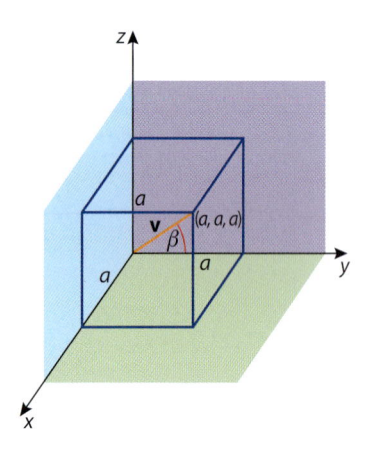

Example 11
Find the angle that a main diagonal of a cube with side a makes with the adjacent edges.

Solution

We can place the cube in a coordinate system such that three of its adjacent edges lie on the coordinate axes as shown (left). The diagonal, represented by the vector $\mathbf{v}$ has a terminal point (a, a, a). Hence,

$|\mathbf{v}| = \sqrt{a^2 + a^2 + a^2} = a\sqrt{3}$. Take angle β, for example:

$$\beta = \cos^{-1}\left(\frac{a}{a\sqrt{3}}\right) = \cos^{-1}\left(\frac{1}{\sqrt{3}}\right) \approx 54.7°$$

Exercise 12.2

1. Find the dot product and the angle between the vectors.
    a) $\mathbf{u} = (3, -2, 4), \mathbf{v} = 2\mathbf{i} - \mathbf{j} - 6\mathbf{k}$
    b) $\mathbf{u} = \begin{pmatrix} 2 \\ -6 \\ 0 \end{pmatrix}, \mathbf{v} = \begin{pmatrix} -1 \\ 3 \\ 5 \end{pmatrix}$
    c) $\mathbf{u} = 3\mathbf{i} - \mathbf{j}, \mathbf{v} = 5\mathbf{i} + 2\mathbf{j}$
    d) $\mathbf{u} = \mathbf{i} - 3\mathbf{j}, \mathbf{v} = 5\mathbf{j} + 2\mathbf{k}$
    e) $|\mathbf{u}| = 3, |\mathbf{v}| = 4$, the angle between $\mathbf{u}$ and $\mathbf{v}$ is $\frac{\pi}{3}$
    f) $|\mathbf{u}| = 3, |\mathbf{v}| = 4$, the angle between $\mathbf{u}$ and $\mathbf{v}$ is $\frac{2\pi}{3}$

2. State whether the following vectors are orthogonal. If not orthogonal, is the angle acute?
    a) $\mathbf{u} = \begin{pmatrix} 2 \\ -6 \\ 4 \end{pmatrix}, \mathbf{v} = \begin{pmatrix} -1 \\ 3 \\ 5 \end{pmatrix}$
    b) $\mathbf{u} = 3\mathbf{i} - 7\mathbf{j}, \mathbf{v} = 5\mathbf{i} + 2\mathbf{j}$
    c) $\mathbf{u} = \mathbf{i} - 3\mathbf{j} + 6\mathbf{k}, \mathbf{v} = 6\mathbf{j} + 3\mathbf{k}$

3. a) Show that the vectors $\mathbf{v} = -y\mathbf{i} + x\mathbf{j}$ and $\mathbf{w} = y\mathbf{i} - x\mathbf{j}$ are both perpendicular to $\mathbf{u} = x\mathbf{i} + y\mathbf{j}$.
    b) Find two unit vectors that are perpendicular to $\mathbf{u} = 2\mathbf{i} - 3\mathbf{j}$. Plot the three vectors in the same coordinate system.

• **Hint:** Orthogonal means 'at right angles to each other'.

4 (i) Find the direction cosines of **v**.
   (ii) Show that they satisfy $\cos^2 \alpha + \cos^2 \beta + \cos^2 \gamma = 1$.
   (iii) Approximate the direction angles to the nearest degree.
   a) $\mathbf{v} = 2\mathbf{i} - 3\mathbf{j} + \mathbf{k}$
   b) $\mathbf{v} = \mathbf{i} - 2\mathbf{j} + \mathbf{k}$
   c) $\mathbf{v} = 3\mathbf{i} - 2\mathbf{j} + \mathbf{k}$
   d) $\mathbf{v} = 3\mathbf{i} - 4\mathbf{k}$

5 Determine m so that **u** and **v** are perpendicular.
   a) $\mathbf{u} = (3, 5, 0); \mathbf{v} = (m - 2, m + 3, 0)$
   b) $\mathbf{u} = (2m, m - 1, m + 1); \mathbf{v} = (m - 1, m, m - 1)$

6 Given the vectors $\mathbf{u} = (-3, 1, 2)$, $\mathbf{v} = (1, 2, 1)$, and $\mathbf{w} = \mathbf{u} + m\mathbf{v}$, determine the value of m so that the vectors **u** and **w** are orthogonal.

7 Given the vectors $\mathbf{u} = (-2, 5, 4)$ and $\mathbf{v} = (6, -3, 0)$, find, to the nearest degree, the measures of the angles between the following vectors.
   a) **u** and **v**
   b) **u** and **u** + **v**
   c) **v** and **u** + **v**

8 Consider the following three points: $A(1, 2, -3)$, $B(3, 5, -2)$ and $C(m, 1, -10m)$. Determine m so that
   a) A, B and C are collinear
   b) $\overrightarrow{AB}$ and $\overrightarrow{AC}$ are perpendicular.

9 Consider the triangle with vertices $A(4, -2, -1)$, $B(3, -5, -1)$ and $C(3, 1, 2)$. Find the vector equations of each of its medians and then find the coordinates of its centroid (i.e. where the medians meet).

10 Consider the tetrahedron $ABCD$ with vertices as shown in the diagram. Find, to the nearest degree, all the angles in the tetrahedron.

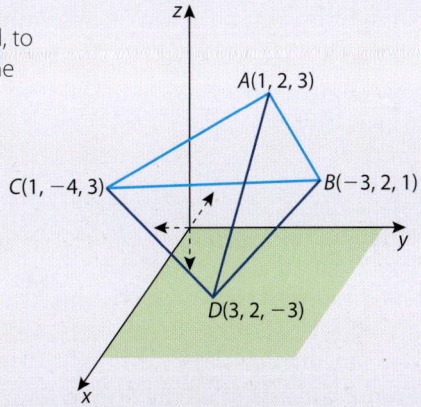

$A(1, 2, 3)$
$B(-3, 2, 1)$
$C(1, -4, 3)$
$D(3, 2, -3)$

11 In question 10 above, use the angles you found to calculate the total surface area of the tetrahedron.

12 In question 10, what angles does $\overrightarrow{DC}$ make with each of the coordinate axes?

13 In question 10, find $(\overrightarrow{DA} - \overrightarrow{DB}) \cdot \overrightarrow{AC}$.

14 Find k such that the angle between the vectors $\begin{pmatrix} 3 \\ -k \\ -1 \end{pmatrix}$ and $\begin{pmatrix} 1 \\ -3 \\ k \end{pmatrix}$ is $\frac{\pi}{3}$.

15 Find x and y such that $\begin{pmatrix} 2 \\ x \\ y \end{pmatrix}$ is perpendicular to both $\begin{pmatrix} 3 \\ 1 \\ -1 \end{pmatrix}$ and $\begin{pmatrix} 4 \\ -1 \\ 2 \end{pmatrix}$.

16 Consider the vectors $\begin{pmatrix} 1 - x \\ 2x - 2 \\ 3 + x \end{pmatrix}$ and $\begin{pmatrix} 2 - x \\ 1 + x \\ 1 + x \end{pmatrix}$. Find the value(s) of x such that the two vectors are parallel.

17 In triangle ABC, $\overrightarrow{OA} = \begin{pmatrix} 2 \\ 3 \\ 1 \end{pmatrix}$, $\overrightarrow{OB} = \begin{pmatrix} 3 \\ 5 \\ 4 \end{pmatrix}$ and $\overrightarrow{BC} = \begin{pmatrix} -1 \\ 4 \\ 0 \end{pmatrix}$.

Find the measure of $A\hat{B}C$.
Find $\overrightarrow{AC}$ and use it to find the measure of $B\hat{A}C$.

18 Find the value(s) of b such that the vectors are orthogonal.
   a) $(b, 3, 2)$ and $(1, b, 1)$
   b) $(4, -2, 7)$ and $(b^2, b, 0)$

19 If two vectors **p** and **q** are such that $|\mathbf{p}| = |\mathbf{q}|$, show that $\mathbf{p} + \mathbf{q}$ and $\mathbf{p} - \mathbf{q}$ are perpendicular. (This proves that the diagonals of a rhombus are perpendicular to each other!)

20 Shortly after take-off, a plane is rising at a rate of 300 m/min. It is heading at an angle of 45° north-west with an airspeed of 200 km/h. Find the components of its velocity vector. The x-axis is in the east direction, the y-axis north and the z-axis is the elevation.

12.3 Equations of lines

A straight line in space can be determined uniquely by specifying a point on it and a direction given by a non-zero vector parallel to it. The following theorem gives parametric equations of the line through a point A and parallel to the vector **v**.

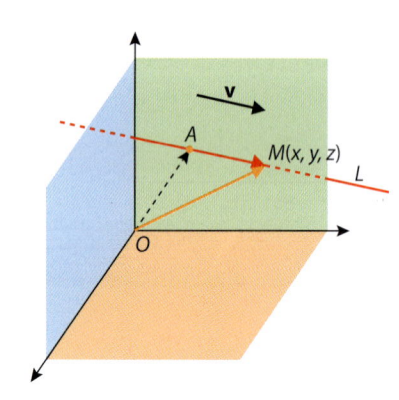

> The line that passes through the point $A(x_0, y_0, z_0)$ and parallel to the vector $\mathbf{v} = (a, b, c)$ has parametric equations:
> $$x = x_0 + at, \; y = y_0 + bt, \; z = z_0 + ct$$

If L is the line that passes through A and is parallel to the non-zero vector **v**, then L consists of all the points $M(x, y, z)$ for which the vector $\overrightarrow{AM}$ is parallel to **v**. Also, a result we established in Chapter 8 and in this chapter enables us to state:

Two vectors are parallel if one of them is a scalar multiple of the other.

This means that for the point M to be on L, $\overrightarrow{AM}$ must be a scalar multiple of **v**, i.e. $\overrightarrow{AM} = t\mathbf{v}$, where t is a scalar.

The previous equation can be written in coordinate form as
$$(x - x_0, y - y_0, z - z_0) = t(a, b, c) = (ta, tb, tc)$$

For two vectors to be equal, their components must be the same, then
$$x - x_0 = ta, \; y - y_0 = tb, \; z - z_0 = tc$$

This leads to the previous result:
$$x = x_0 + at, \; y = y_0 + bt, \; z = z_0 + ct$$

Example 12

a) Find parametric equations of the line through $A(1, -2, 3)$ and parallel to $\mathbf{v} = 5\mathbf{i} + 4\mathbf{j} - 6\mathbf{k}$.

b) Find parametric equations of the line through the points $A(1, -2, 3)$ and $B(2, 4, -2)$.

Solution

a) From the previous theorem, $x = 1 + 5t, y = -2 + 4t, z = 3 - 6t$.

b) We need to find a vector parallel to the given line. The vector $\overrightarrow{AB}$ provides a good choice: $\overrightarrow{AB} = (1, 6, -5)$. So the equations are

$$x = 1 + t, y = -2 + 6t, z = 3 - 5t$$

Another set of equations could be

$$x = 2 + t, y = 4 + 6t, z = -2 - 5t$$

Other sets are possible by considering any vector parallel to $\overrightarrow{AB}$.

Symmetric (Cartesian) equations of lines (optional)

If $a \neq 0$, $b \neq 0$ and $c \neq 0$, the set of parametric equations can be arranged differently:

$$\left. \begin{array}{l} x - x_0 = ta \Leftrightarrow \dfrac{x - x_0}{a} = t \\ y - y_0 = tb \Leftrightarrow \dfrac{y - y_0}{b} = t \\ z - z_0 = tc \Leftrightarrow \dfrac{z - z_0}{c} = t \end{array} \right\} \Leftrightarrow \dfrac{x - x_0}{a} = \dfrac{y - y_0}{b} = \dfrac{z - z_0}{c}$$

If any of the components a, b or c is zero, the equations are written in a mixed form. For example, if $c = 0$, then we write

$$\dfrac{x - x_0}{a} = \dfrac{y - y_0}{b}, z = z_0$$

Intersecting, parallel and skew straight lines

In the plane, lines can coincide, intersect or be parallel. This is not necessarily so in space. In addition to the three cases above, there is the case of skew straight lines. Although these lines are not parallel, they do not intersect either. They lie in different planes.

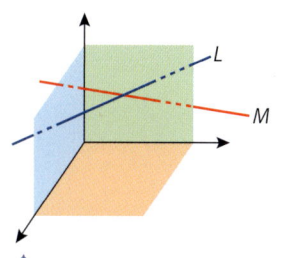

Figure 12.15a Two intersecting straight lines.

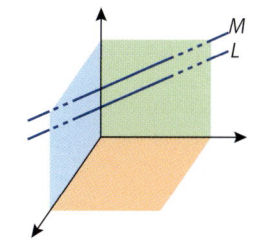

Figure 12.15b Two parallel lines.

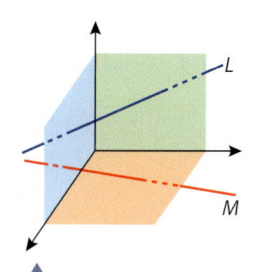

Figure 12.15c Two skew straight lines.

403

Vectors II

How do we know whether two lines are parallel?

If the 'direction' vectors are parallel, then the lines are. Check to see if one of the vectors is a scalar multiple of the other. Alternatively, you can find the angle between them, and if it is 0° or 180°, the lines are either parallel or coincident. The case for coincidence is always there, and you need to check it by examining a point on one of the lines to see whether it is also on the other line.

Example 13

Show that the following two lines are parallel.

$$L_1: x = 2 - 3t, y = t, z = -1 + 2t$$
$$L_2: x = 1 + 6s, y = 2 - 2s, z = 2 - 4s$$

Solution

Let $\mathbf{l}_1$ be the vector parallel to L_1 and $\mathbf{l}_2$ be the vector parallel to L_2.

$$\mathbf{l}_1 = -3\mathbf{i} + \mathbf{j} + 2\mathbf{k} \text{ and } \mathbf{l}_2 = 6\mathbf{i} - 2\mathbf{j} - 4\mathbf{k}.$$

Now you can easily see that $\mathbf{l}_2 = -2\mathbf{l}_1$, and hence the vectors are parallel.

To check whether the lines coincide, we examine the point $(2, 0, -1)$, which is on the first line, and see whether it lies on the second line too.

If we choose $y = 0$, then $0 = 2 - 2s$, so $s = 1$; and when we substitute $s = 1$ into $x = 1 + 6s$ we find out that x must be 7 in order for the point $(2, 0, -1)$ to be on L_2. Therefore, the lines cannot intersect, and their 'direction' vectors are parallel, so they must be parallel.

● **Hint:** There are more elegant methods, but they are beyond the scope of this course.

Are the lines intersecting or skew?

If the direction vectors are not parallel, the lines either intersect or are skew. For the purposes of this course, the method starts by examining whether the lines intersect. If they do, we can find the coordinates of the point of intersection; if they do not intersect, we cannot find the coordinates of the point of intersection. Finding the coordinates of the point of intersection is a straightforward method that you already know: solving systems of equations. This can best be explained with an example.

Example 14

The lines L_1 and L_2 have the following equations:

$$L_1: x = 1 + 4t, y = 5 - 4t, z = -1 + 5t$$
$$L_2: x = 2 + 8s, y = 4 - 3s, z = 5 + s$$

Show that the lines are skew.

Solution

We first examine whether the lines are parallel. Since the vector parallel to L_1 is $\mathbf{l}_1 = (4, -4, 5)$ and the vector parallel to L_2 is $\mathbf{l}_2 = (8, -3, 1)$, they are not scalar multiples of each other and the vectors and consequently the lines are not parallel.

For the lines to intersect, there should be some point $M(x_0, y_0, z_0)$ which satisfies the equations of both lines for some values of t and s. That is

$$x_0 = 1 + 4t = 2 + 8s;\ y_0 = 5 - 4t = 4 - 3s;\ z_0 = -1 + 5t = 5 + s$$

This leads to a set of three simultaneous equations in two unknowns: s and t.

By solving the first two equations:

$$\left.\begin{array}{l} 1 + 4t = 2 + 8s \\ 5 - 4t = 4 - 3s \end{array}\right\} \Rightarrow 6 = 6 + 5s \Rightarrow s = 0,\ t = \tfrac{1}{4}$$

For the system to be consistent, these values must satisfy the third equation, i.e. $-1 + \dfrac{5}{4} = 5 + 0$, which is false. Hence, the system is inconsistent and the lines are skew.

Example 15

The lines L_1 and L_2 have the following equations:

$$L_1 : x = 1 + 2t,\ y = 3 - 4t,\ z = -2 + 4t$$
$$L_2 : x = 4 + 3s,\ y = 4 + s,\ z = -4 - 2s$$

Show that the lines intersect.

Solution

We first examine whether the lines are parallel. Since the vector parallel to L_1 is $\mathbf{l}_1 = (2, -4, 4)$ and the vector parallel to L_2 is $\mathbf{l}_2 = (3, 1, -2)$, they are not scalar multiples of each other and the vectors and consequently the lines are not parallel.

For the lines to intersect, there should be some point $M(x_0, y_0, z_0)$ which satisfies the equations of both lines for some values of t and s. That is,

$$x_0 = 1 + 2t = 4 + 3s;\ y_0 = 3 - 4t = 4 + s;\ z_0 = -2 + 4t = -4 - 2s$$

This leads to a set of three simultaneous equations in two unknowns: s and t.

By solving the first two equations:

$$\left.\begin{array}{l} 1 + 2t = 4 + 3s \\ 3 - 4t = 4 + s \end{array}\right\} \Rightarrow 5 = 12 + 7s \Rightarrow s = -1,\ t = 0$$

For the system to be consistent, these values must satisfy the third equation, i.e. $-2 + 4(0) = -4 - 2(-1) \Rightarrow -2 = -2$, which is a correct statement. Hence, the two lines intersect.

The point of intersection can be found through substitution of the value of the parameter into the corresponding line equation:

$$L_1\text{: } (1, 3, -2) \text{ and } L_2\text{: } (4 - 3, 4 - 1, -4 - 2(-1)) = (1, 3, -2)$$

12 Vectors II

Vector equation of a line

A concept that is closely related to the parametric equation approach is the use of vector notation in expressing the equation of a line. In the previous part, we studied the parametric form of the equation of a line and found

$$(x - x_0, y - y_0, z - z_0) = t(a, b, c) = (ta, tb, tc)$$

where the left-hand side represented the vector from a fixed point on the line to any point on the line. The right-hand side gives a scalar multiple of the vector **v** parallel to the given line. This equation can be transformed into another form:

$$(x - x_0, y - y_0, z - z_0) = t(a, b, c)$$
$$\Leftrightarrow (x, y, z) - (x_0, y_0, z_0) = t(a, b, c)$$
$$\Leftrightarrow (x, y, z) = (x_0, y_0, z_0) + t(a, b, c)$$
$$\Leftrightarrow \mathbf{r} = \mathbf{r}_0 + t\mathbf{v}$$

The last equation is the vector equation of the line: **r** is the position vector of any point on the line, while $\mathbf{r}_0$ is the position vector of a fixed point (A in this case) on the line and **v** is the vector parallel to the given one. See Figure 12.16.

You can interpret this equation in several ways. One of these has to do with displacement. That is, to reach point M from point O, you first arrive at A, and then go towards M along the line a multiple of **v**, t**v**.

By observing Figure 12.17, you will notice, for example, that for each value of t you describe a point on the line. When $t > 0$, the points are in the same direction as **v**. When $t < 0$, the points are in the opposite direction.

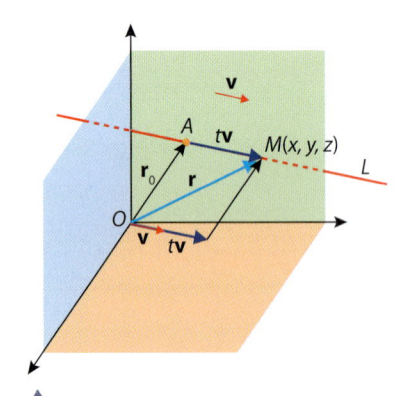

Figure 12.16

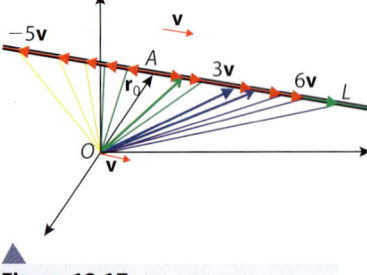

Figure 12.17

Example 16

Find a vector equation of the line that contains $(-1, 3, 0)$ and is parallel to $\mathbf{v} = 3\mathbf{i} - 2\mathbf{j} + \mathbf{k}$.

Solution

From the discussion above,

$$\mathbf{r} = (-\mathbf{i} + 3\mathbf{j}) + t(3\mathbf{i} - 2\mathbf{j} + \mathbf{k})$$

When $t = 0$, the equation gives the point $(-1, 3, 0)$. When $t = 1$, the equation yields $\mathbf{r} = (-\mathbf{i} + 3\mathbf{j}) + (3\mathbf{i} - 2\mathbf{j} + \mathbf{k}) = 2\mathbf{i} + \mathbf{j} + \mathbf{k}$, a point shifted by 1**v** down the line. Similarly, when $t = 3$,

$$\mathbf{r} = (-\mathbf{i} + 3\mathbf{j}) + 3(3\mathbf{i} - 2\mathbf{j} + \mathbf{k}) = 8\mathbf{i} - 3\mathbf{j} + 3\mathbf{k},$$ a point 3**v** down the line, etc.

Alternatively, the equation can be written as

$$\mathbf{r} = (-1 + 3t)\mathbf{i} + (3 - 2t)\mathbf{j} + t\mathbf{k}$$

This last form allows us to recognize the parametric equations of the line by simply reading the components of the vector on the right-hand side of the equation.

Example 17

Find a vector equation of the line passing through $A(2, 7)$ and $B(6, 2)$.

Solution

We let the vector $\overrightarrow{AB} = (6 - 2, 2 - 7) = (4, -5)$ be the vector giving the direction of the line, so

$\mathbf{r} = (2, 7) + t(4, -5)$ or, equivalently,

$\mathbf{r} = 2\mathbf{i} + 7\mathbf{j} + t(4\mathbf{i} - 5\mathbf{j})$

Application of lines to motion

The vector equation, as discussed before, gives rise to an interpretation of the equation that describes motions of objects placed in an appropriate coordinate system. See Figure 12.18. Generally speaking, you find an object at an initial location A, represented by $\mathbf{r}_0$. The object moves on its path with a velocity vector $\mathbf{v} = (a, b, c)$. The object's position at any point in time after the start can then be described by $\mathbf{r} = \mathbf{r}_0 + t\mathbf{v}$.

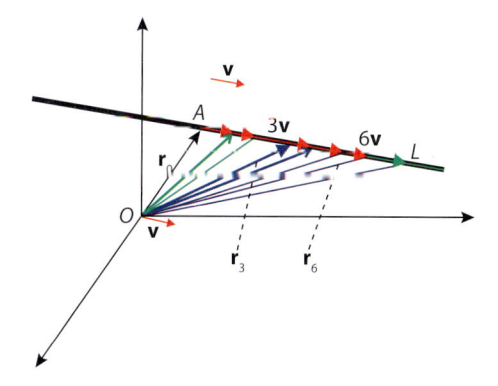

◀ Figure 12.18

Assuming the unit of time is seconds, the equation tells us that for every second, the object moves a units in the x direction, b in the y direction and c in the z direction. So, for example, after 2 seconds you find the object at $\mathbf{r} = \mathbf{r}_0 + 2\mathbf{v}$.

The speed of the object is then $|\mathbf{v}|$ in the $\mathbf{v}$ direction. Let us clarify this with an example.

Example 18

An object is moving in the plane of an appropriately fitted coordinate system such that its position is given by

$\mathbf{r} = (3, 1) + t(-2, 3)$,

where t stands for time in hours after start and distances are measured in km.

a) Find the initial position of the object.
b) Show the position of the object on a graph at start, 1 hour and 3 hours after start.
c) Find the velocity and speed of the object.

Solution

a) Initial position is when $t = 0$. This is the point $(3, 1)$.
b) See graph.

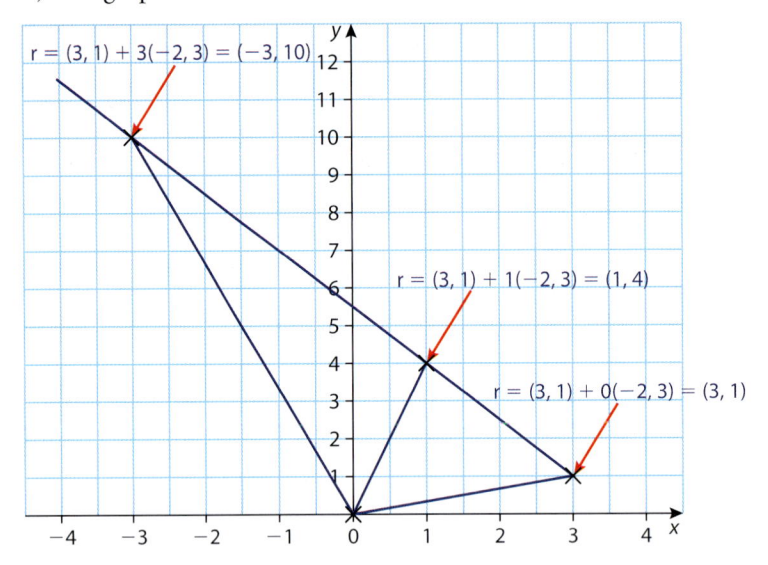

$r = (3, 1) + 3(-2, 3) = (-3, 10)$
$r = (3, 1) + 1(-2, 3) = (1, 4)$
$r = (3, 1) + 0(-2, 3) = (3, 1)$

c) The velocity vector is $\mathbf{v} = (-2, 3)$, which means that every hour the object moves 2 units west and 3 units north.

The speed is $|\mathbf{v}| = \sqrt{(-2)^2 + 3^2} = \sqrt{13}$ km/h.

We can also express the velocity as $\sqrt{13}$ km/h in the direction of $(-2, 3)$.

Note: We can also express the direction in terms of the unit vector in the direction of $\mathbf{v}$ instead. That is, we can say that the speed is $\sqrt{13}$ km/h in the direction of $\left(\frac{-2}{\sqrt{13}}, \frac{3}{\sqrt{13}}\right)$, or, equivalently, at an angle of $\cos^{-1}\left(\frac{-2}{\sqrt{13}}\right) \approx 124°$ to the positive x-direction.

Example 19

At 12:00 midday a plane A is passing in the vicinity of an airport at a height of 12 km and a speed of 800 km/h. The direction of the plane is $(4, 3, 0)$. [Consider that $(1, 0, 0)$ is a displacement of 1 km due east, $(0, 1, 0)$ due north, and $(0, 0, 1)$ is an altitude of 1 km.]

a) Using the airport as the origin, find the position vector $\mathbf{r}$ of the plane t hours after midday.
b) Find the position of the plane 1 hour after midday.
c) Another plane B is heading towards the airport with velocity vector $(-300, -400, 0)$ from a location $(600, 480, 12)$. Is there a danger of collision?

Solution

a) The position vector at midday is $(0, 0, 12)$. The direction of the velocity vector is given by the unit vector $\frac{1}{5}(4, 3, 0)$. So, the velocity vector of this plane is $800 \cdot \frac{1}{5}(4, 3, 0) = (640, 480, 0)$.

The position vector of the plane is $\mathbf{r} = (0, 0, 12) + t(640, 480, 0)$.

b) $\mathbf{r} = (0, 0, 12) + (640, 480, 0) = (640, 480, 12)$

c) A collision can happen if the two planes pass the same point at the same time.

The position vector for the second plane is $\mathbf{r} = (600, 480, 12) + t(-300, -400, 0)$.

If the two paths intersect, they may intersect at instances corresponding to t_1 and t_2 and they should have the same position, i.e.

$$(0, 0, 12) + t_1(640, 480, 0) = (600, 480, 12) + t_2(-300, -400, 0)$$

This gives rise to a set of three equations in two variables:

$$\left.\begin{array}{r} 640t_1 = 600 - 300t_2 \\ 480t_1 = 480 - 400t_2 \\ 12 = 12 \end{array}\right\}$$

Solving the system of equations simultaneously will give $t_1 = \frac{6}{7}$ and $t_2 = \frac{6}{35}$.

This means that the planes' paths will cross at $(548.57, 411.43, 12)$. There is no collision though because plane A will pass that point at 12:51 while plane B will pass this point at 12.10!

Exercise 12.3

1 Find the vector equation as well as the parametric equations of the line containing the point A and parallel to the vector $\mathbf{u}$.
  a) $A(-1, 0, 2)$, $\mathbf{u} = (1, 5, -4)$
  b) $A(3, -1, 2)$, $\mathbf{u} = (2, 5, -1)$
  c) $A(1, -2, 6)$, $\mathbf{u} = (3, 5, -11)$

2 Find the equation of the line that passes through the points A and B.
  a) $A(-1, 4, 2)$, $B(7, 5, 0)$
  b) $A(4, 2, -3)$, $B(0, -2, 1)$
  c) $A(1, 3, -3)$, $B(5, 1, 2)$

3 a) Write the equation of the line through the points $(3, -2)$ and $(5, 1)$ in the form $\mathbf{r} = \mathbf{a} + t\mathbf{b}$.
  b) Write the equation of the line through the points $(0, -2)$ and $(5, 0)$ in the form $\mathbf{r} = \mathbf{a} + t\mathbf{b}$.

4 The equation of a line in 2-space is given by $\mathbf{r} = (2, 1) + t(3, -2)$. Write the equation in the form $ax + by = c$.

5 Find the equation of a line through $(2, -3)$ that is parallel to the line with equation $\mathbf{r} = 3\mathbf{i} - 7\mathbf{j} + \lambda(4\mathbf{i} - 3\mathbf{j})$.

6 Find the equation of a line through $(-2, 1, 4)$ and parallel to the vector $3\mathbf{i} - 4\mathbf{j} + 7\mathbf{k}$.

Vectors II

7 In each of the following, find the point of intersection of the two given lines, and if they do not intersect, explain why.
  a) $L_1: \mathbf{r} = (2, 2, 3) + t(1, 3, 1)$
     $L_2: \mathbf{r} = (2, 3, 4) + t(1, 4, 2)$
  b) $L_1: \mathbf{r} = (-1, 3, 1) + t(4, 1, 0)$
     $L_2: \mathbf{r} = (-13, 1, 2) + t(12, 6, 3)$
  c) $L_1: \mathbf{r} = (1, 3, 5) + t(7, 1, -3)$
     $L_2: \mathbf{r} = (4, 6, 7) + t(-1, 0, 2)$
  d) $L_1: \begin{pmatrix} x \\ y \\ z \end{pmatrix} = \begin{pmatrix} 3 \\ 4 \\ 6 \end{pmatrix} + t\begin{pmatrix} -2 \\ 1 \\ -1 \end{pmatrix}$
     $L_2: \begin{pmatrix} x \\ y \\ z \end{pmatrix} = \begin{pmatrix} 5 \\ -2 \\ 7 \end{pmatrix} + s\begin{pmatrix} -4 \\ 2 \\ -2 \end{pmatrix}$

8 Find the vector and parametric equations of each line:
  a) through the points $(2, -1)$ and $(3, 2)$
  b) through the point $(2, -1)$ and parallel to the vector $\begin{pmatrix} -3 \\ 7 \end{pmatrix}$
  c) through the point $(2, -1)$ and perpendicular to the vector $\begin{pmatrix} -3 \\ 7 \end{pmatrix}$
  d) with y-intercept $(0, 2)$ and in the direction of $2\mathbf{i} - 4\mathbf{j}$

9 Consider the line with equation
$$\begin{pmatrix} x \\ y \\ z \end{pmatrix} = \begin{pmatrix} 3 \\ 4 \\ 6 \end{pmatrix} + t\begin{pmatrix} -2 \\ 1 \\ -1 \end{pmatrix}$$
  a) For what value of t does this line pass through the point $\left(0, \frac{11}{2}, \frac{9}{2}\right)$?
  b) Does the point $(-1, 4, 6)$ lie on this line?
  c) For what value of m does the point $\left(\frac{1-2m}{2}, 2m, 3\right)$ lie on the given line?

10 Consider the following equations representing the paths of cars after starting time $t \geq 0$, where distances are measured in km and time in hours. For each car, determine
  (i) starting position
  (ii) the velocity vector
  (iii) the speed.
  a) $\mathbf{r} = (3, -4) + t\begin{pmatrix} 7 \\ 24 \end{pmatrix}$
  b) $\begin{pmatrix} x \\ y \end{pmatrix} = \begin{pmatrix} -3 \\ 1 \end{pmatrix} + t\begin{pmatrix} 5 \\ -12 \end{pmatrix}$
  c) $(x, y) = (5, -2) + t(24, -7)$

11 Find the velocity vector of each of the following racing cars taking part in the Paris–Dakar rally:
  a) direction $\begin{pmatrix} -3 \\ 4 \end{pmatrix}$ with a speed of 160 km/h
  b) direction $\begin{pmatrix} 12 \\ -5 \end{pmatrix}$ with a speed of 170 km/h

12 After leaving an intersection of roads located at 3 km east and 2 km north of a city, a car is moving towards a traffic light 7 km east and 5 km north of the city at a speed of 30 km/h. (Consider the city as the origin for an appropriate coordinate system.)
  a) What is the velocity vector of the car?
  b) Write down the equation of the position of the car after t hours.
  c) When will the car reach the traffic light?

13 Consider the vectors $\mathbf{u} = (1, a, b)$, $\mathbf{v} = \mathbf{i} - 3\mathbf{j} + 2\mathbf{k}$ and $\mathbf{w} = -2\mathbf{i} + \mathbf{j} - \mathbf{k}$.
   a) Find a and b so that $\mathbf{u}$ is perpendicular to both $\mathbf{v}$ and $\mathbf{w}$.
   b) If O is the origin, P a point whose position vector is $\mathbf{v}$ and Q is with position vector $\mathbf{w}$, find the cosine of the angle between $\mathbf{v}$ and $\mathbf{w}$.
   c) Hence, find the sine of the angle and use it to find the area of the triangle OPQ.

14 The triangle ABC has vertices at the points $A(-1, 2, 3)$, $B(-1, 3, 5)$ and $C(0, -1, 1)$.
   a) Find the size of the angle θ between the vectors $\overrightarrow{AB}$ and $\overrightarrow{AC}$.
   b) Hence, or otherwise, find the area of triangle ABC.
   Let L_1 be the line parallel to $\overrightarrow{AB}$ which passes through $D(2, -1, 0)$, and L_2 be the line parallel to $\overrightarrow{AC}$ which passes through $E(-1, 1, 1)$.
   c) (i) Find the equations of the lines L_1 and L_2.
      (ii) Hence, show that L_1 and L_2 do not intersect.
   © International Baccalaureate Organization, 2001

15 Consider the points $A(1, 3, -17)$ and $B(6, -7, 8)$ which lie on the line l.
   a) Find an equation of line l, giving the answer in parametric form.
   b) The point P is on l such that $\overrightarrow{OP}$ is perpendicular to l. Find the coordinates of P.

Practice questions

1 $ABCD$ is a rectangle and O is the midpoint of $[AB]$. Express each of the following vectors in terms of $\overrightarrow{OC}$, $\overrightarrow{OD}$ and
   a) $\overrightarrow{CD}$
   b) $\overrightarrow{OA}$
   c) $\overrightarrow{AD}$

2 The vectors $\mathbf{i}$ and $\mathbf{j}$ are unit vectors along the x-axis and y-axis respectively. The vectors $\mathbf{u} = -\mathbf{i} + \mathbf{j}$ and $\mathbf{v} = 3\mathbf{i} + 5\mathbf{j}$ are given.
   a) Find $\mathbf{u} + 2\mathbf{v}$ in terms of $\mathbf{i}$ and $\mathbf{j}$.
   A vector $\mathbf{w}$ has the same direction as $\mathbf{u} + 2\mathbf{v}$, and has a magnitude of 26.
   b) Find $\mathbf{w}$ in terms of $\mathbf{i}$ and $\mathbf{j}$.

3 The circle shown has centre O and radius 6. $\overrightarrow{OA}$ is the vector $\begin{pmatrix} 6 \\ 0 \end{pmatrix}$, $\overrightarrow{OB}$ is the vector $\begin{pmatrix} -6 \\ 0 \end{pmatrix}$ and $\overrightarrow{OC}$ is the vector $\begin{pmatrix} 5 \\ \sqrt{11} \end{pmatrix}$.
   a) Verify that A, B and C lie on the circle.
   b) Find the vector $\overrightarrow{AC}$.
   c) Using an appropriate scalar product, or otherwise, find the cosine of angle OAC.
   d) Find the area of triangle ABC, giving your answer in the form $a\sqrt{11}$, where $a \in \mathbb{N}$.

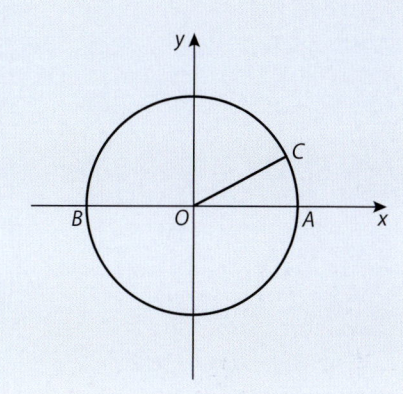

4 The quadrilateral *OABC* has vertices with coordinates *O*(0, 0), *A*(5, 1), *B*(10, 5) and *C*(2, 7).
  a) Find the vectors $\overrightarrow{OB}$ and $\overrightarrow{AC}$.
  b) Find the angle between the diagonals of the quadrilateral *OABC*.

5 The vectors **u** and **v** are given by **u** = 3**i** + 5**j** and **v** = **i** − 2**j**.
  Find scalars *a* and *b* such that *a*(**u** + **v**) = 8**i** + (*b* − 2)**j**.

6 Find a vector equation of the line passing through (−1, 4) and (3, −1). Give your answer in the form **r** = **p** + *t***d**, where $t \in \mathbb{R}$.

7 In this question, the vector $\begin{pmatrix} 1 \\ 0 \end{pmatrix}$ represents a displacement due east and the vector $\begin{pmatrix} 0 \\ 1 \end{pmatrix}$ a displacement due north. Distances are in kilometres and time in hours.
  Two crews of workers are laying an underground cable in a north-south direction across a desert. At 06:00 each crew sets out from their base camp, which is situated at the origin (0, 0). One crew is in a Toyundai vehicle and the other in a Chryssault vehicle.
  The Toyundai has velocity vector $\begin{pmatrix} 18 \\ 24 \end{pmatrix}$ and the Chryssault has velocity vector $\begin{pmatrix} 36 \\ -16 \end{pmatrix}$.
  a) Find the speed of each vehicle.
  b) **(i)** Find the position vectors of each vehicle at 06:30.
      (ii) Hence, or otherwise, find the distance between the vehicles at 06:30.
  c) At this time (06:30) the Chryssault stops and its crew begin their day's work, laying cable in a northerly direction. The Toyundai continues travelling in the same direction, at the same speed, until it is exactly north of the Chryssault. The Toyundai crew then begin their day's work, laying cable in a southerly direction. At what time does the Toyundai crew begin laying cable?
  d) Each crew lays an average of 800 m of cable in an hour. If they work non-stop until their lunch break at 11:30, what is the distance between them at this time?
  e) How long would the Toyundai take to return to base camp from its lunchtime position, assuming it travelled in a straight line and with the same average speed as on the morning journey? (Give your answer to the nearest minute.)

8 The line *L* passes through the origin and is parallel to the vector 2**i** + 3**j**.
  Write down a vector equation for *L*.

9 The triangle *ABC* is defined by the following information:
  $\overrightarrow{OA} = \begin{pmatrix} 2 \\ -3 \end{pmatrix}$, $\overrightarrow{AB} = \begin{pmatrix} 3 \\ 4 \end{pmatrix}$, $\overrightarrow{AB} \cdot \overrightarrow{BC} = 0$, $\overrightarrow{AC}$ is parallel to $\begin{pmatrix} 0 \\ 1 \end{pmatrix}$.
  a) On the grid below, draw an accurate diagram of triangle *ABC*.

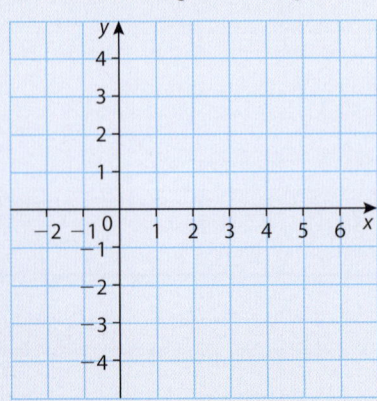

  b) Write down the vector $\overrightarrow{OC}$.

10 In this question, the vector $\begin{pmatrix} 1 \\ 0 \end{pmatrix}$ represents a displacement due east and the vector $\begin{pmatrix} 0 \\ 1 \end{pmatrix}$ represents a displacement due north.

The point (0, 0) is the position of Shipple Airport. The position vector $\mathbf{r}_1$ of an aircraft, *Air One*, is given by

$$\mathbf{r}_1 = \begin{pmatrix} 16 \\ 12 \end{pmatrix} + t\begin{pmatrix} 12 \\ -5 \end{pmatrix},$$

where *t* is the time in minutes since 12:00.

a) Show that *Air One*
   (i) is 20 km from Shipple Airport at 12:00
   (ii) has a speed of 13 km/min.
b) Show that a Cartesian equation of the path of *Air One* is:
   $$5x + 12y = 224.$$

The position vector $\mathbf{r}_2$ of an aircraft, *Air Two*, is given by

$$\mathbf{r}_2 = \begin{pmatrix} 23 \\ -5 \end{pmatrix} + t\begin{pmatrix} 2.5 \\ 6 \end{pmatrix},$$

where *t* is the time in minutes since 12:00.

c) Find the angle between the paths of the two aircraft.
d) (i) Find a Cartesian equation for the path of *Air Two*.
   (ii) Hence, find the coordinates of the point where the two paths cross.
e) Given that the two aircraft are flying at the same height, show that they do not collide.

11 Find the size of the angle between the two vectors $\begin{pmatrix} 1 \\ 2 \end{pmatrix}$ and $\begin{pmatrix} 6 \\ -8 \end{pmatrix}$. Give your answer to the nearest degree.

12 A line passes through the point (4, −1) and its direction is perpendicular to the vector $\begin{pmatrix} 2 \\ 3 \end{pmatrix}$. Find the equation of the line in the form $ax + by = p$, where *a*, *b* and *p* are integers to be determined.

13 In this question, the vector $\begin{pmatrix} 1 \\ 0 \end{pmatrix}$ represents a displacement due east and the vector $\begin{pmatrix} 0 \\ 1 \end{pmatrix}$ represents a displacement due north. Distances are in kilometres.

The diagram shows the path of the oil tanker *Aristides* relative to the port of Orto, which is situated at the point (0, 0).

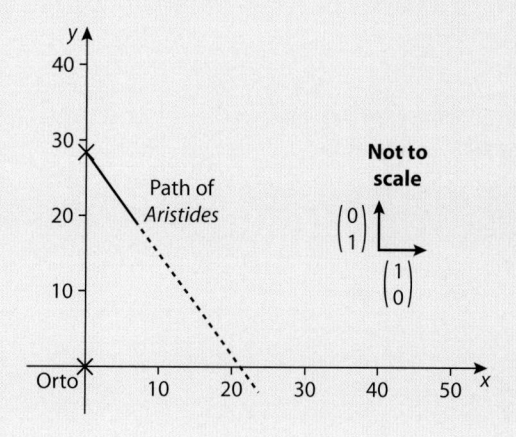

The position of the *Aristides* is given by the vector equation

$$\begin{pmatrix} x \\ y \end{pmatrix} = \begin{pmatrix} 0 \\ 28 \end{pmatrix} + t \begin{pmatrix} 6 \\ -8 \end{pmatrix}$$

at a time t hours after 12:00.

a) Find the position of the *Aristides* at 13:00.

b) Find
   (i) the velocity vector
   (ii) the speed of the *Aristides*.

c) Find a Cartesian equation for the path of the *Aristides* in the form $ax + by = g$.

Another ship, the cargo vessel *Boadicea*, is stationary, with position vector $\begin{pmatrix} 18 \\ 4 \end{pmatrix}$.

d) Show that the two ships will collide, and find the time of collision.

To avoid collision, the *Boadicea* starts to move at 13:00 with velocity vector $\begin{pmatrix} 5 \\ 12 \end{pmatrix}$.

e) Show that the position of the *Boadicea* for $t \geq 1$ is given by

$$\begin{pmatrix} x \\ y \end{pmatrix} = \begin{pmatrix} 13 \\ -8 \end{pmatrix} + t \begin{pmatrix} 5 \\ 12 \end{pmatrix}$$

f) Find how far apart the two ships are at 15:00.

14 Find the angle between the following vectors **a** and **b**, giving your answer to the nearest degree.

$$\mathbf{a} = -4\mathbf{i} - 2\mathbf{j}$$
$$\mathbf{b} = \mathbf{i} - 7\mathbf{j}$$

15 In this question, a unit vector represents a displacement of 1 metre.

A miniature car moves in a straight line, starting at the point (2, 0). After t seconds, its position, (x, y), is given by the vector equation

$$\begin{pmatrix} x \\ y \end{pmatrix} = \begin{pmatrix} 2 \\ 0 \end{pmatrix} + t \begin{pmatrix} 0.7 \\ 1 \end{pmatrix}$$

a) How far from the point (0, 0) is the car after 2 seconds?

b) Find the speed of the car.

c) Obtain the equation of the car's path in the form $ax + by = c$.

Another miniature vehicle, a motorcycle, starts at the point (0, 2) and travels in a straight line with constant speed. The equation of its path is

$$y = 0.6x + 2, \quad x \geq 0$$

Eventually, the two miniature vehicles collide.

d) Find the coordinates of the collision point.

e) If the motorcycle left point (0, 2) at the same moment the car left point (2, 0), find the speed of the motorcycle.

16 The diagram right shows a line passing through the points (1, 3) and (6, 5).

Find a vector equation for the line, giving your answer in the form

$$\begin{pmatrix} x \\ y \end{pmatrix} = \begin{pmatrix} a \\ b \end{pmatrix} + t \begin{pmatrix} c \\ d \end{pmatrix},$$

where t is any real number.

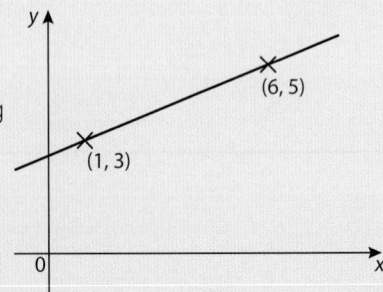

17 The vectors $\begin{pmatrix} 2x \\ x-5 \end{pmatrix}$ and $\begin{pmatrix} x+1 \\ 5 \end{pmatrix}$ are perpendicular for two values of x.

   a) Write down the quadratic equation which the two values of x must satisfy.
   b) Find the two values of x.

18 The diagram below shows the positions of towns O, A, B and X.

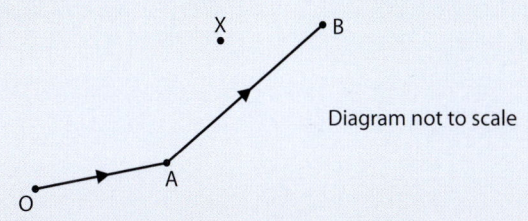

Diagram not to scale

Town A is 240 km east and 70 km north of O.
Town B is 480 km east and 250 km north of O.
Town X is 339 km east and 238 km north of O.

A plane flies at a constant speed of 300 km h^{-1} from O towards A.

   a) (i) Show that a unit vector in the direction of $\overrightarrow{OA}$ is $\begin{pmatrix} 0.96 \\ 0.28 \end{pmatrix}$.

   (ii) Write down the velocity vector for the plane in the form $\begin{pmatrix} v_1 \\ v_2 \end{pmatrix}$.

   (iii) How long does it take for the plane to reach A?

At A the plane changes direction so it now flies towards B. The angle between the original direction and the new direction is θ, as shown in the following diagram. This diagram also shows the point Y, between A and B, where the plane comes closest to X.

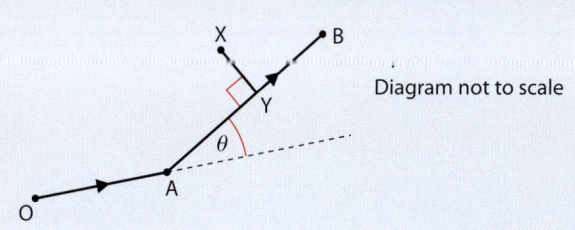

Diagram not to scale

   b) Use the scalar product of two vectors to find the value of θ in degrees.
   c) (i) Write down the vector $\overrightarrow{AX}$.

   (ii) Show that the vector $\mathbf{n} = \begin{pmatrix} -3 \\ 4 \end{pmatrix}$ is perpendicular to $\overrightarrow{AB}$.

   (iii) By finding the projection of $\overrightarrow{AX}$ in the direction of $\mathbf{n}$, calculate the distance XY.

   d) How far is the plane from A when it reaches Y?

19 A vector equation of a line is $\begin{pmatrix} x \\ y \end{pmatrix} = \begin{pmatrix} 1 \\ 2 \end{pmatrix} + t\begin{pmatrix} -2 \\ 3 \end{pmatrix}$, $t \in \mathbb{R}$.

Find the equation of this line in the form $ax + by = c$, where a, b and $c \in \mathbb{Z}$.

20 Three of the coordinates of the parallelogram STUV are S(−2, −2), T(7, 7) and U(5, 15).

   a) Find the vector $\overrightarrow{ST}$ and hence the coordinates of V.
   b) Find a vector equation of the line (UV) in the form $\mathbf{r} = \mathbf{p} + \lambda\mathbf{d}$, where $\lambda \in \mathbb{R}$.
   c) Show that the point E with position vector $\begin{pmatrix} 1 \\ 11 \end{pmatrix}$ is on the line (UV), and find the value of λ for this point.

The point W has position vector $\begin{pmatrix} a \\ 17 \end{pmatrix}$, $a \in \mathbb{R}$.

   d) (i) If $\overrightarrow{EW} = 2\sqrt{13}$, show that one value of a is −3 and find the other possible value of a.

   (ii) For $a = -3$, calculate the angle between $\overrightarrow{EW}$ and $\overrightarrow{ET}$.

12 Vectors II

21 Calculate the acute angle between the lines with equations

$$\mathbf{r} = \begin{pmatrix} 4 \\ -1 \end{pmatrix} + s\begin{pmatrix} 4 \\ 3 \end{pmatrix} \text{ and } \mathbf{r} = \begin{pmatrix} 2 \\ 4 \end{pmatrix} + t\begin{pmatrix} 1 \\ -1 \end{pmatrix}.$$

22 The following diagram shows the point O with coordinates $(0, 0)$, the point A with position vector $\mathbf{a} = 12\mathbf{i} + 5\mathbf{j}$, and the point B with position vector $\mathbf{b} = 6\mathbf{i} + 8\mathbf{j}$. The angle between (OA) and (OB) is θ.

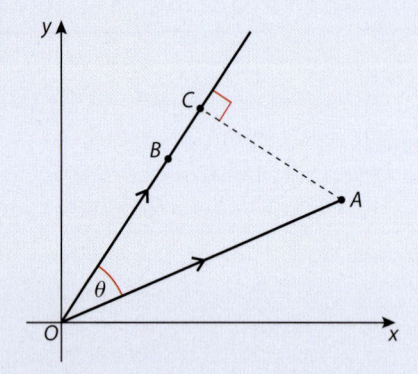

Find

a) $|\mathbf{a}|$

b) a unit vector in the direction of $\mathbf{b}$

c) the exact value of $\cos \theta$ in the form $\dfrac{p}{q}$, where $p, q \in \mathbb{Z}$.

23 The vector equations of two lines are given below.

$$\mathbf{r}_1 = \begin{pmatrix} 5 \\ 1 \end{pmatrix} + \lambda \begin{pmatrix} 3 \\ -2 \end{pmatrix}, \mathbf{r}_2 = \begin{pmatrix} -2 \\ 2 \end{pmatrix} + t\begin{pmatrix} 4 \\ 1 \end{pmatrix}$$

The lines intersect at the point P. Find the position vector of P.

24 The diagram shows a parallelogram $OPQR$ in which $\overrightarrow{OP} = \begin{pmatrix} 7 \\ 3 \end{pmatrix}$ and $\overrightarrow{OQ} = \begin{pmatrix} 10 \\ 1 \end{pmatrix}$.

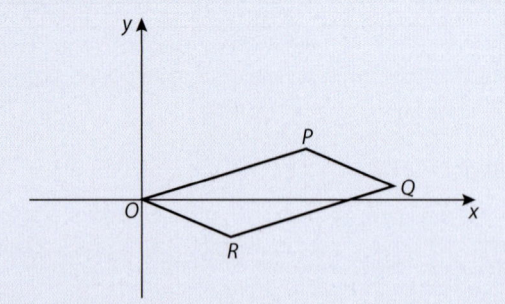

a) Find the vector $\overrightarrow{OR}$.

b) Use the scalar product of two vectors to show that $\cos O\hat{P}Q = -\dfrac{15}{\sqrt{754}}$.

c) (i) Explain why $\cos P\hat{Q}R = -\cos O\hat{P}Q$.

(ii) Hence, show that $\sin P\hat{Q}R = \dfrac{23}{\sqrt{754}}$.

(iii) Calculate the area of the parallelogram $OPQR$, giving your answer as an integer.

25 The diagram shows points A, B and C, which are three vertices of a parallelogram ABCD. The point A has position vector $\begin{pmatrix} 2 \\ 2 \end{pmatrix}$.

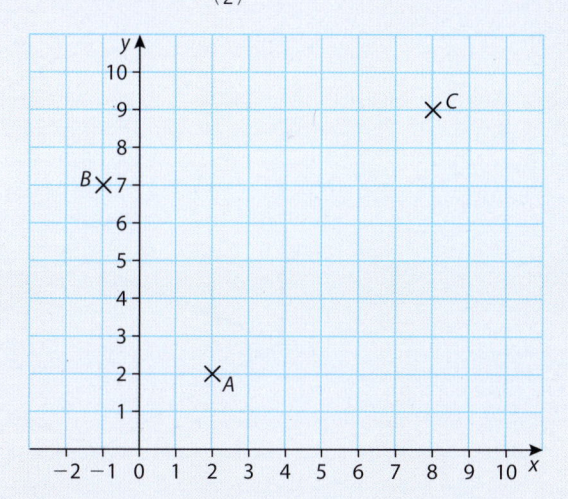

a) Write down the position vector of B and C.
b) The position vector of point D is $\begin{pmatrix} d \\ 4 \end{pmatrix}$. Find d.
c) Find $\overrightarrow{BD}$.

The line L passes through B and D.

d) (i) Write down a vector equation of L in the form $\begin{pmatrix} x \\ y \end{pmatrix} = \begin{pmatrix} -1 \\ 7 \end{pmatrix} + t\begin{pmatrix} m \\ n \end{pmatrix}$.
   (ii) Find the value of t at point B.
e) Let P be the point (7, 5). By finding the value of t at P, show that P lies on the line L.
f) Show that $\overrightarrow{CP}$ is perpendicular to $\overrightarrow{BD}$.

26 The points A and B have the position vectors $\begin{pmatrix} 2 \\ -2 \end{pmatrix}$ and $\begin{pmatrix} -3 \\ -1 \end{pmatrix}$ respectively.

a) (i) Find the vector $\overrightarrow{AB}$.
   (ii) Find $|\overrightarrow{AB}|$.

The point D has position vector $\begin{pmatrix} d \\ 23 \end{pmatrix}$.

b) Find the vector $\overrightarrow{AD}$ in terms of d.

The angle $B\hat{A}D$ is 90°.

c) (i) Show that d = 7.
   (ii) Write down the position vector of the point D.

The quadrilateral ABCD is a rectangle.

d) Find the position vector of the point C.
e) Find the area of the rectangle ABCD.

27 Points A, B and C have position vectors $4\mathbf{i} + 2\mathbf{j}$, $\mathbf{i} - 3\mathbf{j}$ and $-5\mathbf{i} - 5\mathbf{j}$, respectively. Let D be a point on the x-axis such that ABCD forms a parallelogram.

a) (i) Find $\overrightarrow{BC}$.
   (ii) Find the position vector of D.
b) Find the angle between $\overrightarrow{BD}$ and $\overrightarrow{AC}$.

The line L_1 passes through A and is parallel to $\mathbf{i} + 4\mathbf{j}$. The line L_2 passes through B and is parallel to $2\mathbf{i} + 7\mathbf{j}$. A vector equation of L_1 is $\mathbf{r} = (4\mathbf{i} + 2\mathbf{j}) + s(\mathbf{i} + 4\mathbf{j})$.

c) Write down a vector equation of L_2 in the form $\mathbf{r} = \mathbf{b} + t\mathbf{q}$.
d) The lines L_1 and L_2 intersect at the point P. Find the position vector of P.

28 The diagram shows a cube, *OABCDEFG*, where the length of each edge is 5 cm. Express the following vectors in terms of **i**, **j** and **k**.

   a) $\overrightarrow{OG}$

   b) $\overrightarrow{BD}$

   c) $\overrightarrow{EB}$

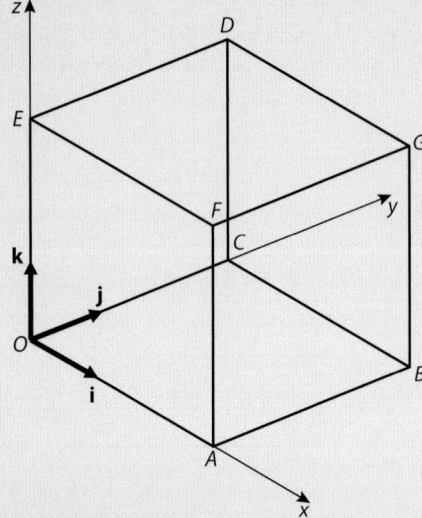

29 In this question, distance is in kilometres and time is in hours.

A balloon is moving at a constant height with a speed of 18 km h^{-1}, in the direction of the vector $\begin{pmatrix} 3 \\ 4 \\ 0 \end{pmatrix}$.

At time $t = 0$, the balloon is at point *B* with coordinates (0, 0, 5).

   a) Show that the position vector **b** of the balloon at time *t* is given by
   $$\mathbf{b} = \begin{pmatrix} x \\ y \\ z \end{pmatrix} = \begin{pmatrix} 0 \\ 0 \\ 5 \end{pmatrix} + \frac{18t}{5}\begin{pmatrix} 3 \\ 4 \\ 0 \end{pmatrix}$$

At time $t = 0$, a helicopter goes to deliver a message to the balloon. The position vector **h** of the helicopter at time *t* is given by
$$\mathbf{h} = \begin{pmatrix} x \\ y \\ z \end{pmatrix} = \begin{pmatrix} 49 \\ 32 \\ 0 \end{pmatrix} + t\begin{pmatrix} -48 \\ -24 \\ 6 \end{pmatrix}$$

   b) (i) Write down the coordinates of the starting position of the helicopter.

   (ii) Find the speed of the helicopter.

   c) The helicopter reaches the balloon at point *R*.

   (i) Find the time the helicopter takes to reach the balloon.

   (ii) Find the coordinates of *R*.

30 In this question, the vector $\begin{pmatrix} 1 \\ 0 \end{pmatrix}$ represents a displacement due east and the vector $\begin{pmatrix} 0 \\ 1 \end{pmatrix}$ represents a displacement of 1 km north.

The diagram right shows the positions of towns A, B and C in relation to an airport O, which is at the point (0, 0). An aircraft flies over the three towns at a constant speed of 250 km h^{-1}.

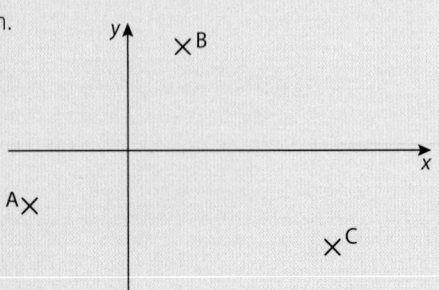

Town A is 600 km west and 200 km south of the airport.
Town B is 200 km east and 400 km north of the airport.
Town C is 1200 km east and 350 km south of the airport.

a) (i) Find $\vec{AB}$.

   (ii) Show that the vector of length one unit in the direction of $\vec{AB}$ is $\begin{pmatrix} 0.8 \\ 0.6 \end{pmatrix}$.

An aircraft flies over town A at 12:00, heading towards town B at 250 km h^{-1}.

Let $\begin{pmatrix} p \\ q \end{pmatrix}$ be the velocity vector of the aircraft. Let t be the number of hours in flight after 12:00.

The position of the aircraft can be given by the vector equation

$$\begin{pmatrix} x \\ y \end{pmatrix} = \begin{pmatrix} -600 \\ -200 \end{pmatrix} + t \begin{pmatrix} p \\ q \end{pmatrix}$$

b) (i) Show that the velocity vector is $\begin{pmatrix} 200 \\ 150 \end{pmatrix}$.

   (ii) Find the position of the aircraft at 13:00.

   (iii) At what time is the aircraft flying over town B?

Over town B the aircraft changes direction so it now flies towards town C. It takes five hours to travel the 1250 km between B and C. Over town A the pilot noted that she had 17 000 litres of fuel left. The aircraft uses 1800 litres of fuel per hour when travelling at 250 km h^{-1}. When the fuel gets below 1000 litres a warning light comes on.

c) How far from town C will the aircraft be when the warning light comes on?

Questions 1–30 © International Baccalaureate Organization

13 Differential Calculus II: Further Techniques and Applications

Assessment statements
7.1 Derivative of x^n, $\sin x$, $\cos x$, $\tan x$, e^x and $\ln x$.
7.2 Differentiation of a sum and a real multiple of the functions in 7.1.
The chain rule for composite functions.
The product and quotient rules.
The second derivative.
7.3 Use of the first and second derivative in optimization problems.

Introduction

The primary purpose of the earlier chapter on calculus, Chapter 11, was to establish some fundamental concepts and techniques of differential calculus. Chapter 11 also introduced some applications involving the differentiation of functions: finding maxima and minima of a function; kinematic problems involving displacement, velocity and acceleration; and finding equations of tangents and normals. The focus of this chapter is to expand our set of differentiation rules and techniques and to deepen and extend the applications introduced in Chapter 11 – particularly using methods of finding extrema in the context of finding an 'optimum' solution to a problem. We start by investigating the derivatives of some important functions.

13.1 Derivatives of trigonometric, exponential and logarithmic functions

Throughout Sections 11.2 and 11.3, we used information from the values, or the sign, of the derivative of a function to deduce the behaviour of the graph of the function. For example, if the derivative of a function is positive for all x in some interval, we know that the graph of the function is sloping upwards (increasing) in that interval. And if the derivative is equal to zero at some point, we know that the graph will have a horizontal tangent (momentarily 'flat' or 'stationary') at that point. To conjecture a rule for the derivative of a function, we will use an informal approach, which is essentially the reverse process of what we did in Sections 11.2 and 11.3, by studying the behaviour of the graph of a function to obtain information to sketch the graph of its derivative. Effective use of technology can then help confirm our conjecture. We will follow the same 'conjecture → confirm' approach to assist in determining the derivatives of the trigonometric functions sine and cosine, and then the functions e^x and $\ln x$.

The derivative of the sine function

We start with the graph of $f(x) = \sin x$ (Figure 13.1). The graph of $y = \sin x$ is periodic, with period 2π, so the same will be true of its derivative that gives the slope at each point on the graph. Therefore, it's only necessary for us to consider the portion of the graph in the interval $0 \leqslant x \leqslant 2\pi$.

Figure 13.2 shows two pairs of axes having equal scales on the x- and y-axes and corresponding x-coordinates aligned vertically. On the top pair of axes, $y = \sin x$ is graphed with tangent lines drawn at nine selected points. The points were chosen such that the slopes of the tangents at those points, in order, appear to be equal to $1, \frac{1}{2}, 0, -\frac{1}{2}, -1, -\frac{1}{2}, 0, \frac{1}{2}, 1$. The values of these slopes were then plotted in the bottom graph with the y-coordinate of each point indicating the slope of the curve for that particular value of x. Hence, the points in the bottom pair of axes should be on the graph of the derivative of $y = \sin x$.

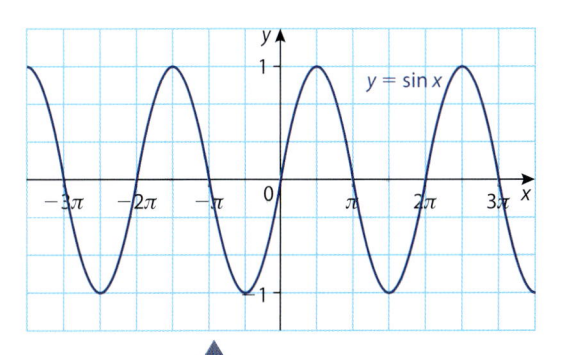

Figure 13.1

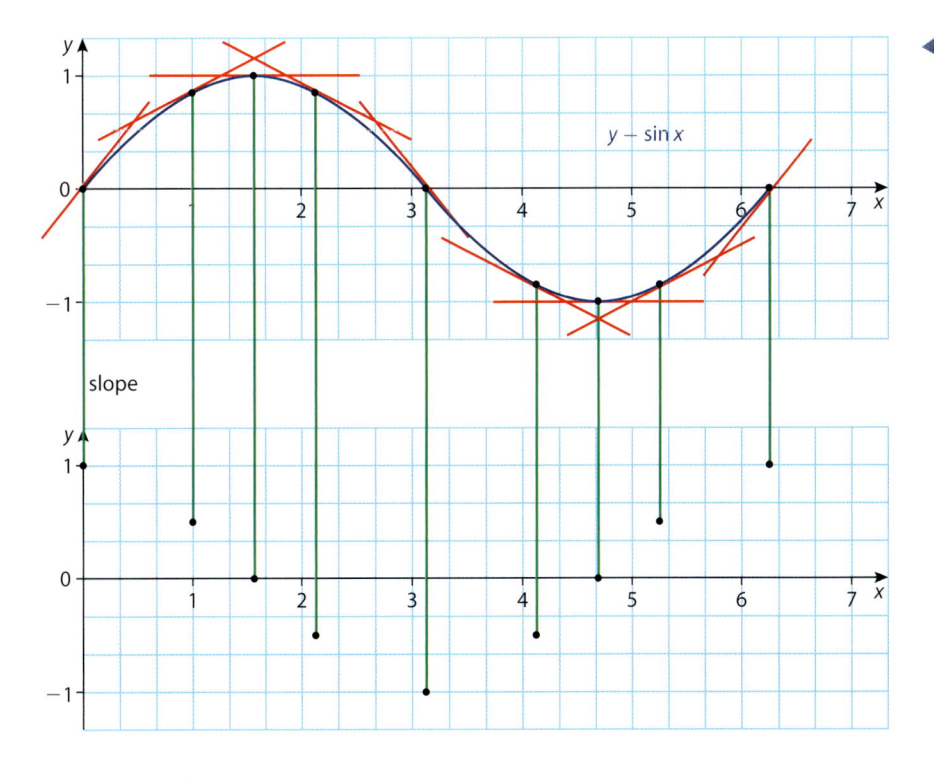

Figure 13.2

13   Differential Calculus II: Further Techniques and Applications

Figure 13.3 is the same as Figure 13.2 except with the graph of $y = \sin x$, the grid lines and the lines connecting points between the two graphs removed.

Figure 13.3a Lines tangent to $y = \sin x$.

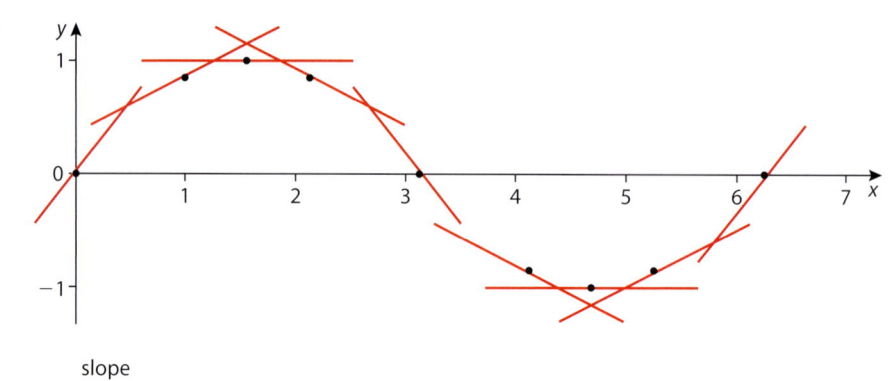

Figure 13.3b Points on the graph of the derivative of $y = \sin x$.

Note that the graphs in Figures 13.1, 13.2 and 13.3 have x in radians. As mentioned previously, we must use only radian measure when trigonometric functions are involved in calculus.

This leads to an obvious choice for our conjecture of the derivative function for sine. For $f(x) = \sin x$, it appears that $f'(x) = \cos x$. Let's use our GDC to provide confirmation of this conjecture.

The GDC screen images below show the derivative of $\sin x$ being evaluated for various values of x, and then showing that this is equivalent (to 6 significant figures) to $\cos x$ for the same value of x.

```
NORMAL   SCI   ENG
FLOAT   0 1 2 3 4 5 6 7 8 9
RADIAN  DEGREE
FUNC   PAR   POL   SEQ
CONNECTED   DOT
SEQUENTIAL   SIMUL
REAL   a+bi   re^θi
FULL   HORIZ   G-T
SET CLOCK  13/09/07  13:13
```

```
nDeriv(sin(X),X,
π/4)
          .7071066633
cos(π/4)
          .7071067812
```

```
nDeriv(sin(X),X,
5π/6)
         -.8660252595
cos(5π/6)
         -.8660254038
```

```
nDeriv(sin(X),X,
5.25)
          .5120853919
cos(5.25)
          .5120854772
```

The discrepancies beyond 6 significant figures is due to the small amount of error incorporated in the algorithm used by the GDC to compute the derivative of a function at a point.

Note that the calculator must be in radian mode.

> **The derivative of the sine function**
> If $f(x) = \sin x$, then $f'(x) = \cos x$. Or, in Leibniz notation, $\frac{d}{dx}(\sin x) = \cos x$.
> This result is only true when x is in radian measure.

The derivative of the cosine function

We could take the same approach as we did for investigating the derivative of $\sin x$ to make a conjecture for the derivative of $\cos x$, but using our result $\frac{d}{dx}(\sin x) = \cos x$ there is a constructive approach that uses our knowledge about the transformations of graphs (Section 2.4).

Figure 13.4

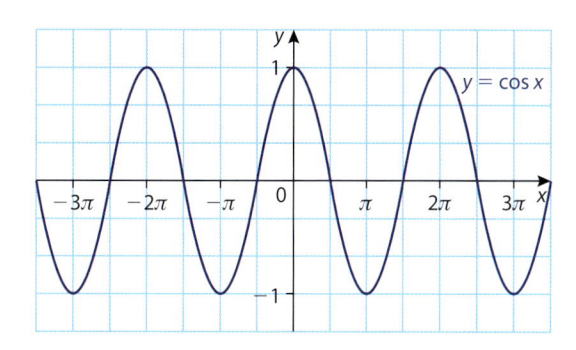

The graph of $y = \cos x$ (the derivative of $\sin x$) is shown in Figure 13.4. This graph is the same *shape* as that of $y = \sin x$, but translated $\frac{\pi}{2}$ units to the left. Recall that in Example 13 of Chapter 6 (Section 6.3) we came to this same conclusion and established the identity $\cos x = \sin\left(x + \frac{\pi}{2}\right)$. Therefore, given that for $f(x) = \sin x \Rightarrow f'(x) = \sin\left(x + \frac{\pi}{2}\right) = \cos x$, we anticipate that for $f(x) = \cos x \Rightarrow f'(x) = \cos\left(x + \frac{\pi}{2}\right)$.

Figure 13.5 shows the graph of $y = \cos x$ translated $\frac{\pi}{2}$ units to the left, i.e. the graph of $y = \cos\left(x + \frac{\pi}{2}\right)$, which we expect to be the graph of the derivative of the function $\cos x$.

Figure 13.5

The graph of $y = \cos\left(x + \frac{\pi}{2}\right)$ is the graph of $y = \sin x$ but reflected in the x-axis. Knowing that the graph of $y = -f(x)$ is obtained by reflecting the graph of $y = f(x)$ in the x-axis (Section 2.4), then the graph of $y = \cos\left(x + \frac{\pi}{2}\right)$ is the graph of $y = -\sin x$. Therefore, our conjecture is that the derivative of $\cos x$ is $-\sin x$.

Again, let's utilize our GDC to help confirm our conjecture.

```
nDeriv(cos(X),X,
π/6)
           -.4999999167
-sin(π/6)
                   -.5
```

```
nDeriv(cos(X),X,
3π/2)
            .9999998333
-sin(3π/2)
                     1
```

```
nDeriv(cos(X),X,
9)
           -.4121184166
-sin(9)
           -.4121184852
```

423

13 Differential Calculus II: Further Techniques and Applications

> **The derivative of the cosine function**
> If $f(x) = \cos x$, then $f'(x) = -\sin x$. Or, in Leibniz notation, $\frac{d}{dx}(\cos x) = -\sin x$.
> This result is only true when x is in radian measure.

To demonstrate that our 'conjecture → confirm' approach to finding derivatives does not always work so smoothly, consider the other trigonometric function, $y = \tan x$. We can use the GDC command that evaluates the derivative of a function at a specified point to graph the value of the derivative at all points on a graph. We used this technique in Chapter 11 for Example 8 (Section 11.2). The GDC screen images below show the graph of $y = \tan x$ and then the GDC graphing its derivative (in bold) on the same set of axes. Although, as pointed out in Section 11.3, in general it is incorrect to graph a function and its derivative on the same pair of axes (units on the vertical axis will not be the same), it is helpful in seeing the connection between the graph of a function and that of its derivative.

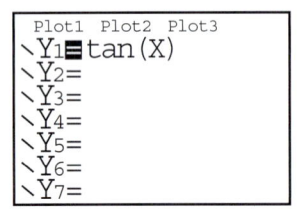

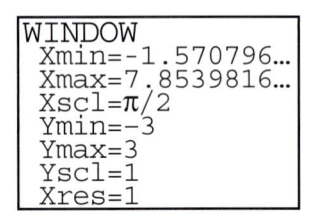

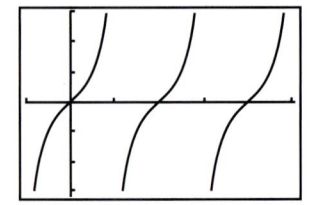

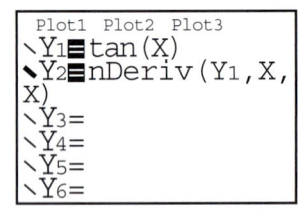

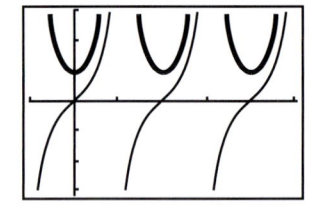

The graph of the derivative of $\tan x$ is always above the x-axis meaning that the derivative is always positive. This clearly agrees with the fact that the tangent function, except for where it is undefined, is always increasing (moving upwards) as the values of x increase. However, the shape of the graph does not bring to mind an easy conjecture for a rule for the derivative of $\tan x$.

Later in this chapter we will learn how to differentiate quotients which we can use to find the derivative of $\tan x$ given that $\tan x = \frac{\sin x}{\cos x}$.

The derivative of the exponential function e^x

Let's review some important facts about exponential functions in general. An exponential function with base b is defined as $f(x) = b^x$, $b > 0$ and $b \neq 1$. The graph of f passes through $(0, 1)$, has the x-axis as a horizontal asymptote and, depending on the value of the base b of the exponential function, will be either a continually increasing exponential growth curve or a continually decreasing exponential decay curve, as shown in Figure 13.6.

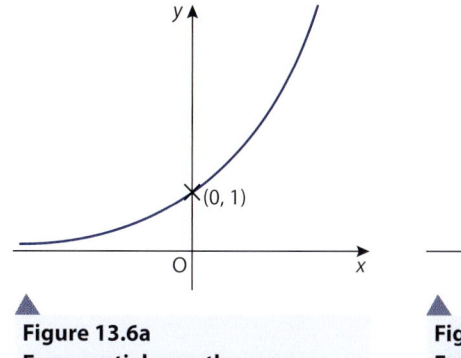

Figure 13.6a
Exponential growth curve:
$f(x) = b^x$ for $b > 1$
as $x \to \infty$, $f(x) \to \infty$
f is an increasing function

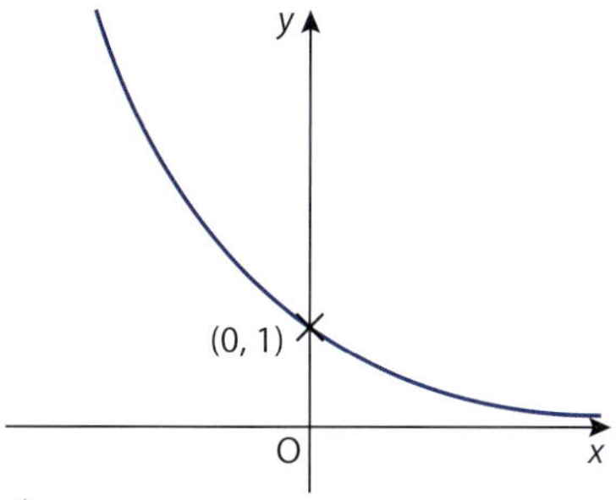

Figure 13.6b
Exponential decay curve:
$f(x) = b^x$ for $0 < b < 1$
as $x \to \infty$, $f(x) \to 0$
f is a decreasing function

In Chapter 4 we learned that *the* exponential function e^x, sometimes written as 'exp x', is a particularly important function for modelling exponential growth and decay. The number e was defined in Section 4.3 as the limit of $\left(1 + \dfrac{1}{x}\right)^x$ as $x \to \infty$. Although the method was not successful in coming up with a conjecture for the derivative of the tangent function, let's try to guess the derivative of e^x by having our GDC graph its derivative.

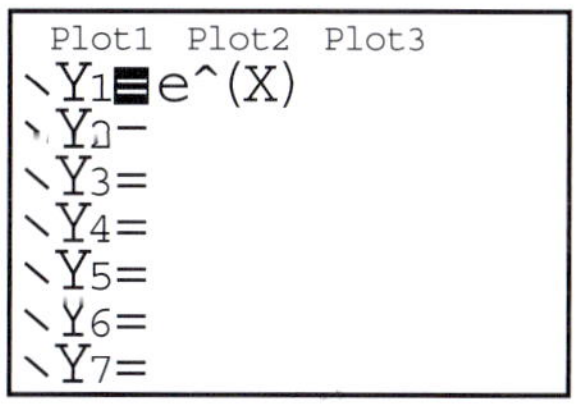

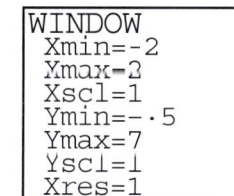

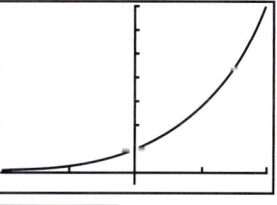

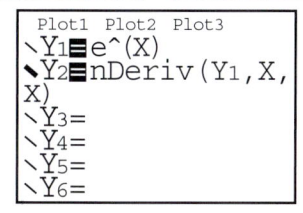

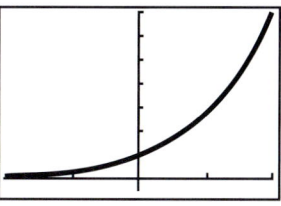

The graph of the derivative of e^x appears to be identical to e^x itself! This is a very interesting result, but one which we will see fits in exactly with the nature of exponential growth/decay. Given that the result is so intriguing, let's try to apply the limit definition of the derivative to provide a more formal justification.

Recall the definition of the derivative of a function f is

$f'(x) = \lim\limits_{h \to 0} \dfrac{f(x+h) - f(x)}{h}$ and that the slope of the graph of f at a specific point where $x = c$ is defined as $\lim\limits_{h \to 0} \dfrac{f(c+h) - f(c)}{h}$.

$\dfrac{d}{dx}(e^x) = \lim\limits_{h \to 0} \dfrac{e^{x+h} - e^x}{h}$ definition of derivative, i.e. 'first principles'

$= \lim\limits_{h \to 0} \dfrac{e^x \cdot e^h - e^x}{h}$ reverse of law of exponents: $a^m \cdot a^n = a^{m+n}$

$= \lim\limits_{h \to 0} \dfrac{e^x(e^h - 1)}{h}$ factorizing

$= e^x \cdot \lim\limits_{h \to 0} \dfrac{e^h - 1}{h}$ e^x is not affected by the value of h

A closer look at the limit that is multiplying e^x reveals that it is equivalent to the slope of the graph of $y = e^x$ at $x = 0$: $\lim\limits_{h \to 0} \dfrac{e^{0+h} - e^0}{h} = \lim\limits_{h \to 0} \dfrac{e^h - 1}{h}$.

To finish our differentiation of e^x by first principles, we need to evaluate this limit. It is beyond the scope of this course to give a formal algebraic proof for the limit. Nevertheless, we can provide a convincing informal justification by evaluating the expression $\dfrac{e^h - 1}{h}$ for values of h approaching zero, as shown in the table.

h	$\dfrac{e^h - 1}{h}$
0.1	1.051 709 181
0.01	1.005 016 708
0.0001	1.000 050 002
0.000 001	1.000 000 005

Thus, $\lim\limits_{h \to 0} \dfrac{e^h - 1}{h} = 1$, and we can complete our algebraic work for the derivative of e^x.

$$\dfrac{d}{dx}(e^x) = e^x \cdot \lim\limits_{h \to 0} \dfrac{e^h - 1}{h} = e^x \cdot 1 = e^x$$

The derivative of the exponential function *is* the exponential function. More precisely, the slope of the graph of $f(x) = e^x$ at any point (x, e^x) is equal to the y-coordinate of the point.

The derivative of the exponential function

If $f(x) = e^x$, then $f'(x) = e^x$. Or, in Leibniz notation, $\dfrac{d}{dx}(e^x) = e^x$.

What about exponential functions with bases other than e? We now differentiate the general exponential function $f(x) = b^x$, $b > 1$, $b \neq 0$, repeating the same steps we did with $f(x) = e^x$.

$$\dfrac{d}{dx}(b^x) = \lim\limits_{h \to 0} \dfrac{b^{x+h} - b^x}{h} \qquad \text{definition of derivative}$$

$$= \lim\limits_{h \to 0} \dfrac{b^x \cdot b^h - b^x}{h} \qquad \text{reverse of } a^m \cdot a^n = a^{m+n}$$

$$= \lim\limits_{h \to 0} \dfrac{b^x(b^h - 1)}{h} \qquad \text{factorizing}$$

$$= b^x \cdot \lim\limits_{h \to 0} \dfrac{b^h - 1}{h} \qquad b^x \text{ is not affected by the value of } h$$

```
nDeriv(2^X,X,0)
      .6931472361
nDeriv(3^X,X,0)
      1.09861251
nDeriv((1/2)^X,X
,0)
     -.6931472361
```

As with e^x, $\lim\limits_{h \to 0} \dfrac{b^h - 1}{h}$ is equivalent to the slope of the graph of $f(x) = b^x$ at $x = 0$, i.e. $f'(0)$. Therefore, the derivative of the general exponential function $f(x) = b^x$ is $b^x \cdot f'(0)$. Although the value of $f'(0)$ will be a constant, it will depend on the value of the base b. The GDC screen image left shows the value of $f'(0)$ for $b = 2, 3$ and $\tfrac{1}{2}$.

The differentiation rule that we will learn in the next section will give us the means to determine the exact value of $f'(0)$ given b for the function $f(x) = b^x$. Then we will be able to state a general rule for the derivative of exponential functions $f(x) = b^x$.

> If $f(x) = b^x$, then $f'(x) = b^x \cdot f'(0)$. The value of $f'(0)$ is the slope of the graph of $f(x) = b^x$ at the point $(0, 1)$. Hence, this will be a particular constant for each value of b ($b > 1, b \neq 0$). Therefore, if $f(x) = b^x$, then $f'(x) = kb^x$, where k is a constant dependent on the value of b. If the amount of a quantity y at a time t is given by $y = b^t$ then $\frac{dy}{dt} = kb^t = ky$. In other words, the rate of change of the quantity y at time t is proportional to the amount of y at time t. This is the essential behaviour of exponential growth/decay. It is because of this property that exponential functions have so many applications to real-life phenomena. Here are some good examples:
>
> (1) The rate of population growth for many living organisms is proportional to the size of the population p: $\frac{dp}{dt} = kp$.
>
> (2) The rate at which a radioactive substance decays is proportional to the amount A of the substance present: $\frac{dA}{dt} = kA$.
>
> (3) Newton's law of cooling states that if a substance is placed in cooler surroundings then its temperature decreases at a rate proportional to the temperature difference T between the temperature of the substance and the temperature of its surroundings: $\frac{dT}{dt} = -k(T - T_s)$, where T_s is the temperature of the surroundings and k is an experimental constant.

The derivative of the natural logarithm function ln x

Now that we have found the derivative of $y = e^x$, let's find the derivative of its inverse, $y = \ln x, x > 0$. We start by using our GDC to view a graph of the derivative of $f(x) = \ln x$ and also to construct a table of ordered pairs $(x, f'(x))$.

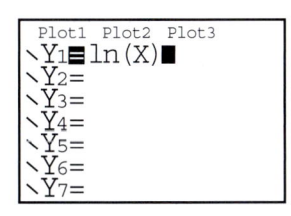

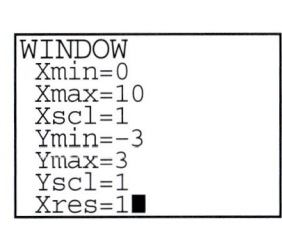

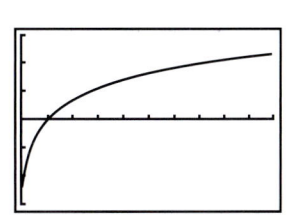

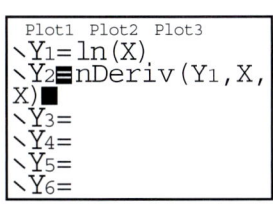

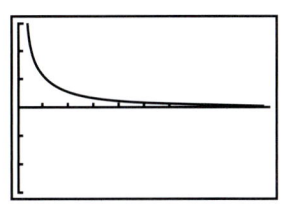

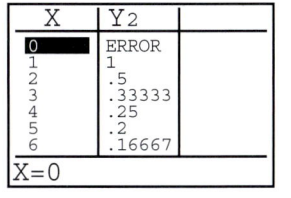

In the table, each value in the Y₂ column is the slope of the curve (derivative) at the particular value of x for $y = \ln x$. From the graph of the derivative and especially from the table, we conjecture that the derivative of $\ln x$ is $\frac{1}{x}$. As we did with the function $y = e^x$, let's try to justify our conjecture with a more formal approach.

$$\frac{d}{dx}(\ln x) = \lim_{h \to 0} \frac{\ln(x + h) - \ln x}{h} \quad \text{definition of derivative}$$

13 Differential Calculus II: Further Techniques and Applications

$$= \lim_{h \to 0} \frac{\ln\left(\frac{x+h}{x}\right)}{h}$$

law of logarithms:
$$\log_b M - \log_b N = \log_b\left(\frac{M}{N}\right)$$

$$= \lim_{h \to 0} \left(\frac{1}{x} \cdot x \cdot \frac{\ln\left(\frac{x+h}{x}\right)}{h}\right)$$

'trick': multiply argument of limit by $1 = \frac{x}{x} = \frac{1}{x} \cdot x$

$$= \frac{1}{x} \cdot \lim_{h \to 0} \left(\frac{x}{h} \cdot \ln\left(1 + \frac{h}{x}\right)\right)$$

x is independent of h; and
$$\frac{x+h}{x} = \frac{x}{x} + \frac{h}{x} = 1 + \frac{h}{x}$$

$$= \frac{1}{x} \cdot \lim_{h \to 0} \left(\ln\left(1 + \frac{h}{x}\right)^{\frac{x}{h}}\right)$$

law of logarithms: $k \log_b M = \log_b(M^k)$

The argument of **ln** in the limit, $\left(1 + \frac{h}{x}\right)^{\frac{x}{h}}$, looks very much like the expression $\left(1 + \frac{1}{n}\right)^n$ whose limit as $n \to \infty$ is the number e (Section 4.3). For $\left(1 + \frac{h}{x}\right)^{\frac{x}{h}}$, let's apply the substitution $n = \frac{x}{h}$. It follows that as $h \to 0$ then $\frac{x}{h} = n \to \infty$, and also that $\frac{1}{n} = \frac{h}{x}$.

Hence, $\frac{1}{x} \cdot \lim_{h \to 0}\left(\ln\left(1 + \frac{h}{x}\right)^{\frac{x}{h}}\right) = \frac{1}{x} \cdot \lim_{n \to \infty}\left(\ln\left(1 + \frac{1}{n}\right)^n\right)$

with substitution, $\lim_{h \to 0}$ changes to $\lim_{n \to \infty}$

$$= \frac{1}{x} \cdot \lim_{n \to \infty}(\ln e)$$

$\lim_{n \to \infty}\left(1 + \frac{1}{n}\right)^n = e$

$$= \frac{1}{x} \cdot \ln e$$

$\ln e$ is not affected by the value of n

$$= \frac{1}{x} \cdot 1$$

$\ln e = 1$

$$= \frac{1}{x}$$

Therefore, $\frac{d}{dx}(\ln x) = \frac{1}{x}$.

The derivative of the natural logarithm function

If $f(x) = \ln x$, then $f'(x) = \frac{1}{x}$. Or, in Leibniz notation, $\frac{d}{dx}(\ln x) = \frac{1}{x}$.

Summary of differentiation rules

Derivative of x^n: $f(x) = x^n$ ⇒ $f'(x) = nx^{n-1}$
Derivative of $\sin x$: $f(x) = \sin x$ ⇒ $f'(x) = \cos x$
Derivative of $\cos x$: $f(x) = \cos x$ ⇒ $f'(x) = -\sin x$
Derivative of e^x: $f(x) = e^x$ ⇒ $f'(x) = e^x$
Derivative of $\ln x$: $f(x) = \ln x$ ⇒ $f'(x) = \frac{1}{x}$

Example 1

Differentiate each of the following functions.
a) $f(x) = 3 \sin x - 4 \cos x$
b) $g(x) = 5 \ln x - 2e^x$
c) $y = 6 - \ln\left(\dfrac{e^3}{x}\right)$

Solution

a) $f'(x) = 3\dfrac{d}{dx}(\sin x) - 4\dfrac{d}{dx}(\cos x)$
$= 3 \cos x - 4(-\sin x)$
$= 3 \cos x + 4 \sin x$

b) $g'(x) = 5\dfrac{d}{dx}(\ln x) - 2\dfrac{d}{dx}(e^x)$
$= 5 \cdot \dfrac{1}{x} - 2e^x$
$= \dfrac{5}{x} - 2e^x$

c) $\dfrac{dy}{dx} = \dfrac{d}{dx}(6) - \dfrac{d}{dx}\left(\ln\left(\dfrac{e^3}{x}\right)\right)$
$= 0 - \dfrac{d}{dx}(\ln e^3 - \ln x)$
$= -\dfrac{d}{dx}(\ln e^3) + \dfrac{d}{dx}(\ln x)$
$= -\dfrac{d}{dx}(3) + \dfrac{1}{x}$
$= \dfrac{1}{x}$

Example 2

Find the equation of the line tangent to the given function at the specified value of x. Express the equation exactly.
a) $y = \cos x \quad x = \dfrac{5\pi}{6}$
b) $y = e^x + 1 \quad x = 1$
c) $y = \ln x \quad x = 4$

Solution

a) When $x = \dfrac{5\pi}{6}$, $y = \cos\left(\dfrac{5\pi}{6}\right) = -\dfrac{\sqrt{3}}{2}$. Thus, point of tangency is $\left(\dfrac{5\pi}{6}, -\dfrac{\sqrt{3}}{2}\right)$.

$\dfrac{dy}{dx} = -\sin x \Rightarrow$ slope of tangent line $= -\sin\left(\dfrac{5\pi}{6}\right) = -\dfrac{1}{2}$

Substitute into point-slope form of a line: $y - y_1 = m(x - x_1)$

$y - \left(-\dfrac{\sqrt{3}}{2}\right) = -\dfrac{1}{2}\left(x - \dfrac{5\pi}{6}\right) \Rightarrow$ equation of tangent line is

$y = -\dfrac{1}{2}x + \dfrac{5\pi}{12} - \dfrac{\sqrt{3}}{2}$

A graphical check on our GDC confirms the equation of our tangent line.

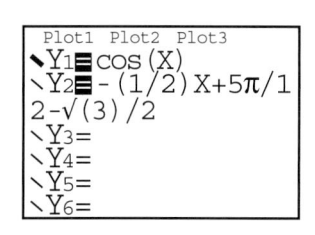

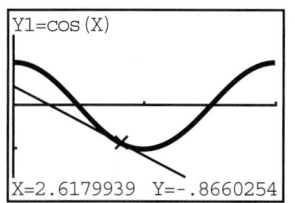

b) When $x = 1$, $y = e^1 + 1 = e + 1$. The point of tangency is $(1, e + 1)$.
$\dfrac{dy}{dx} = e^x \Rightarrow$ slope of tangent line $= e^1 = e$

Substitute into point-slope form of a line: $y - y_1 = m(x - x_1)$
$y - (e + 1) = e(x - 1) \Rightarrow$ equation of tangent line is $y = ex + 1$

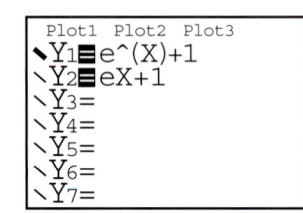

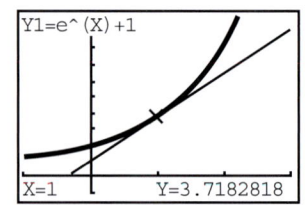

Again, a graph of the curve and our result for the tangent lines gives confirmation that our result appears correct.

c) When $x = 4$, $y = \ln 4$. The point of tangency is $(4, \ln 4)$.
$\dfrac{dy}{dx} = \dfrac{1}{x} \Rightarrow$ slope of tangent line $= \dfrac{1}{4}$

Substitute into point-slope form of a line: $y - y_1 = m(x - x_1)$
$y - \ln 4 = \dfrac{1}{4}(x - 4) \Rightarrow$ equation of tangent line is
$y = \dfrac{1}{4}x - 1 + \ln 4$

Equivalent answer:
$y = \dfrac{1}{4}x - 1 + \ln 4 = \dfrac{1}{4}x + \ln 4 - 1 = \dfrac{1}{4}x + \ln 4 - \ln e \Rightarrow y = \dfrac{1}{4}x + \ln\left(\dfrac{4}{e}\right)$

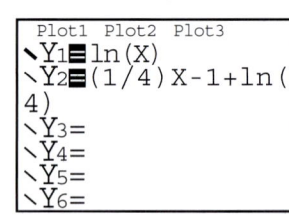

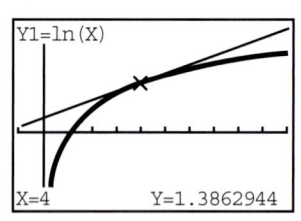

Exercise 13.1

1 Write down the derivative of each function.
 a) $y = \sin x - \cos x$
 b) $y = 5 - e^x$
 c) $y = x + \ln x$
 d) $y = \dfrac{2e^x}{5}$
 e) $y = x^3 + 2\cos x$
 f) $y = 2e \ln x$

2 Find the equation of the line tangent to the given curve at the specified value of x. Express the equation exactly in the form $y = mx + c$.
 a) $y = \sin x$ $\quad x = \dfrac{\pi}{3}$
 b) $y = x + e^x$ $\quad x = 0$
 c) $y = \dfrac{1}{2}\ln x$ $\quad x = e$

3 Consider the function $g(x) = x + 2\cos x$. For the interval $0 \leqslant x \leqslant 2\pi$,
 a) find the exact x-coordinates of any stationary points
 b) determine whether each stationary point is a maximum, minimum or neither and give a brief explanation.

4 Find the coordinates of any stationary points on the curve $y = x - e^x$. Classify any such points as a maximum, minimum or neither and explain.

5 Show that the curve $y = x - \ln x$ has no inflexion points.

6 Find the equation of the normal line to the curve $y = 3 + \sin x$ at the point where $x = \dfrac{\pi}{2}$.

7 Consider the function $f(x) = e^x - x^3$.
   a) Find $f'(x)$ and $f''(x)$.
   b) Find the x-coordinates (accurate to 3 significant figures) for any points where $f'(x) = 0$.
   c) Indicate the intervals for which $f(x)$ is increasing, and indicate the intervals for which $f(x)$ is decreasing.
   d) For the values of x found in part b), state whether that point on the graph of f is a maximum, minimum or neither.
   e) Find the x-coordinate of any inflexion point(s) for the graph of f.
   f) Indicate the intervals for which $f(x)$ is concave up, and indicate the intervals for which $f(x)$ is concave down.

8 A line with slope m passes through the origin and is tangent to the graph of $y = \ln x$. What is the value of m?

9 Use the **change of base formula** for logarithms (Section 4.4) to derive a general rule for the derivative of a logarithmic function of any base b ($b > 0$, $b \neq 1$). That is, find the derivative of the general logarithm function $y = \log_b x$ expressed in terms of b.   • **Hint:** Rewrite $\log_b x$ in terms of the natural logarithm **ln**.

13.2 The chain rule

We know how to differentiate functions such as $f(x) = x^3 + 2x - 3$ and $g(x) = \sqrt{x}$, but how do we differentiate the composite function $f(g(x)) = \sqrt{x^3 + 2x - 3}$? The rule for computing the derivative of the composite of two functions, i.e. the 'function of a function', is called the **chain rule**. Because most functions that we encounter in applications are composites of other functions, it can be argued that the chain rule is the most important, and most widely used, rule of differentiation.

Below are some examples of functions that we can differentiate with the rules that we have learned thus far in Sections 11.2 and 13.1, and further examples of functions which are best differentiated with the chain rule.

Differentiate *without* the chain rule	Differentiate *with* the chain rule
$y = \cos x$	$y = \cos 2x$
$y = 3x^2 + 5x$	$x = \sqrt{3x^2 + 5x}$
$y = \ln x$	$y = \ln(1 - 3x)$
$y = \dfrac{1}{3x^2}$	$y = \dfrac{1}{3x^2 + x}$

The derivative of composite functions

The chain rule says, in a very basic sense, that given two functions, the derivative of their composite is the product of their derivatives –

remembering that a derivative is a rate of change of one quantity (variable) with respect to another quantity (variable). For example, the function $y = 8x + 6 = 2(4x + 3)$ is the composite of the functions $y = 2u$ and $u = 4x + 3$. Note that the function y is in terms of u, and the function u is in terms of x. How are the derivatives of these three functions related?

Clearly, $\dfrac{dy}{dx} = 8$, $\dfrac{dy}{du} = 2$ and $\dfrac{du}{dx} = 4$. Since $8 = 2 \cdot 4$, the derivatives relate such that $\dfrac{dy}{dx} = \dfrac{dy}{du} \cdot \dfrac{du}{dx}$. In other words, rates of change multiply.

Again, if we think of derivatives as rates of change, the relationship $\dfrac{dy}{dx} = \dfrac{dy}{du} \cdot \dfrac{du}{dx}$ can be illustrated by a practical example. Consider the pair of levers in Figure 13.7 with lever endpoints U and U′ connected by a segment that can shrink and stretch but always remains horizontal. Hence, points U and U′ are always the same distance u from the ground.

Figure 13.7 Two levers with horizontal connection between U′ and U.

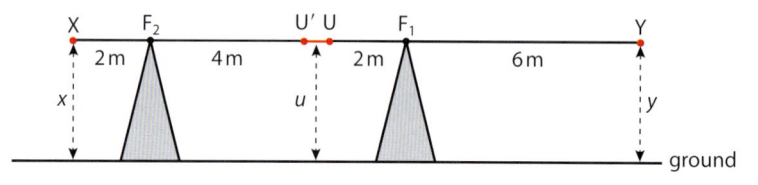

As point Y moves down, points U and U′ move up, and point X moves down but at different rates. Let dy, du and dx represent the change in distance from the ground for the points Y, U and X, respectively. Because $YF_1 = 6$ and $UF_1 = 2$, if point Y moves such that $dy = 3$, then $du = 1$. Since $U'F_2 = 4$ and $XF_2 = 2$, if point U′ moves so that $du = 2$, then $dx = 1$.

Hence, $\dfrac{dy}{du} = 3$ and $\dfrac{du}{dx} = 2$.

Figure 13.8 dx, du and dy represent the change in distance from the ground for X, U and Y.

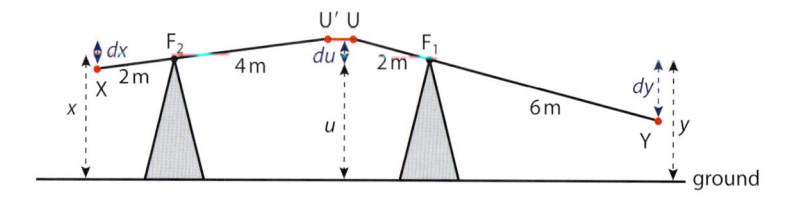

Combining these two results, we can see that for every 6 units that Y's distance changes, X's distance will change 1 unit. That is, $\dfrac{dy}{dx} = 6$. Therefore, we can write $\dfrac{dy}{dx} = \dfrac{dy}{du} \cdot \dfrac{du}{dx} = 3 \cdot 2 = 6$. In other words, the rate of change of y with respect to x is the product of the rate of change of y with respect to u and the rate of change of u with respect to x.

Example 3

The polynomial function $y = 16x^4 - 8x^2 + 1 = (4x^2 - 1)^2$ is the composite of $y = u^2$ and $u = 4x^2 - 1$. Use the chain rule to find $\dfrac{dy}{dx}$, the derivative of y with respect to x.

Solution

$y = u^2 \Rightarrow \dfrac{dy}{du} = 2u$

$u = 4x^2 - 1 \Rightarrow \dfrac{du}{dx} = 8x$

Applying the chain rule: $\dfrac{dy}{dx} = \dfrac{dy}{du} \cdot \dfrac{du}{dx} = 2u \cdot 8x$

$\qquad\qquad\qquad\qquad\qquad\qquad\quad = 2(4x^2 - 1) \cdot 8x$

$\qquad\qquad\qquad\qquad\qquad\qquad\quad = 64x^3 - 16x$

In this particular case, we could have differentiated the function in expanded form by differentiating term-by-term rather than differentiating the factored form by the chain rule. $\dfrac{dy}{dx} = \dfrac{d}{dx}(16x^4 - 8x^2 + 1) = 64x^3 - 16x$; confirming the result above. It is not always easier to differentiate powers of polynomials by expanding and then differentiating term-by-term. For example, it is far better to find the derivative of $y = (3x + 5)^8$ by the chain rule.

In Section 2.2, we often wrote composite functions using nested function notation. For example, the notation $f(g(x))$ denotes a function composed of functions f and g such that g is the 'inside' function and f is the 'outside' function. For the composite function $y = (4x^2 - 1)^2$ in Example 3, the 'inside' function is $g(x) = 4x^2 - 1$ and the 'outside' function is $f(u) = u^2$. Looking again at the solution for Example 3, we see that we can choose to express and work out the chain rule in function notation rather than Leibniz notation.

For $y = f(g(x)) = (4x^2 - 1)^2$ and $y = f(u) = u^2$, $u = g(x) = 4x^2 - 1$,

Leibniz notation	Function notation
$\dfrac{dy}{dx} = \dfrac{dy}{du} \cdot \dfrac{du}{dx} = 2u \cdot 8x$	$\dfrac{d}{dx}[f(g(x))] = f'(u) \cdot g'(x) = 2u \cdot 8x$
$\quad = 2(4x^2 - 1) \cdot 8x$	$\quad = f'(g(x)) \cdot g'(x) = 2(4x^2 - 1) \cdot 8x$
$\quad = 64x^3 - 16x$	$\quad = 64x^3 - 16x$

This leads us to formally state the chain rule in two different notations.

The chain rule

If $y = f(u)$ is a function in terms of u and $u = g(x)$ is a function in terms of x, the function $y = f(g(x))$ is differentiated as follows:

$\dfrac{dy}{dx} = \dfrac{dy}{du} \cdot \dfrac{du}{dx}$ \qquad (Leibniz form)

or, equivalently,

$\dfrac{dy}{dx} = \dfrac{d}{dx}[f(g(x))] = f'(g(x)) \cdot g'(x)$ \qquad (function notation form)

The chain rule needs to be applied carefully. Consider the function notation form for the chain rule $\dfrac{d}{dx}[f(g(x))] = f'(g(x)) \cdot g'(x)$. Although it is the product of two derivatives, it is important to point out that the first derivative involves the function f differentiated at $g(x)$ and the second is

13 Differential Calculus II: Further Techniques and Applications

function g differentiated at x. The chain rule written in Leibniz form, $\frac{dy}{dx} = \frac{dy}{du} \cdot \frac{du}{dx}$, is easily remembered because it appears to be an obvious statement about fractions – but, they are *not* fractions. The expressions $\frac{dy}{dx}, \frac{dy}{du}$ and $\frac{du}{dx}$ are derivatives or, more precisely, limits and although du and dx essentially represent very small changes in the variables u and x, we cannot guarantee that they are non-zero.

The function notation form of the chain rule offers a very useful way of saying the rule 'in words', and, thus, a very useful structure for applying it.

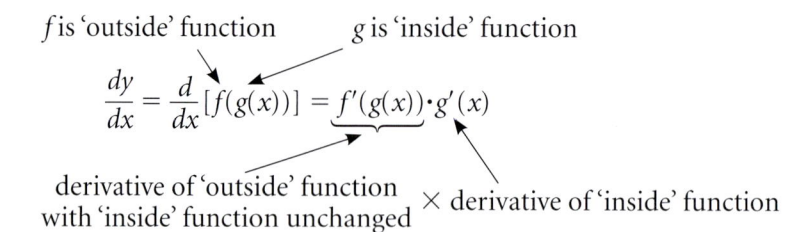

f is 'outside' function $\qquad g$ is 'inside' function

$$\frac{dy}{dx} = \frac{d}{dx}[f(g(x))] = \underbrace{f'(g(x))}_{\text{derivative of 'outside' function with 'inside' function unchanged}} \cdot \underbrace{g'(x)}_{\text{derivative of 'inside' function}}$$

The chain rule in words:

$$\begin{pmatrix} \text{derivative of} \\ \text{composite} \end{pmatrix} = \begin{pmatrix} \text{derivative of 'outside' function} \\ \text{with 'inside' function unchanged} \end{pmatrix} \times \begin{pmatrix} \text{derivative of} \\ \text{'inside' function} \end{pmatrix}$$

Although this is taking some liberties with mathematical language, the mathematical interpretation of the phrase "with 'inside' function unchanged" is that the derivative of the 'outside' function f is evaluated at $g(x)$, the 'inside' function.

- **Hint:** The chain rule is our most important rule of differentiation. It is an indispensable tool in differential calculus. Forgetting to apply the chain rule when it needs to be applied, or by applying it improperly, is a common source of errors in calculus computations. It is important to understand it, practise it and master it.

> The chain rule acquired its name because we use it to take derivatives of composites of functions by 'chaining' together their derivatives. A function could be the composite of more than two functions. If a function were the composite of three functions, we would take the product of three derivatives 'chained' together. For example, if $y = f(u)$, $u = g(v)$ and $v = h(x)$, the derivative of the function $y = f(g(h(x)))$ is $\frac{dy}{dx} = \frac{dy}{du} \cdot \frac{du}{dv} \cdot \frac{dv}{dx}$.

Example 4

Differentiate each function by applying the chain rule. Start by 'decomposing' the composite function into the 'outside' function and the 'inside' function.

a) $y = \cos 2x$

b) $y = \sqrt{3x^2 + 5x}$

c) $y = \ln(1 - 3x)$

d) $y = \dfrac{1}{3x^2 + x}$

Solution

a) $y = f(g(x)) = \cos 2x \Rightarrow$ 'outside' function is $f(u) = \cos u$
$\Rightarrow$ 'inside' function is $g(x) = 2x$

In Leibniz form: $\dfrac{dy}{dx} = \dfrac{dy}{du} \cdot \dfrac{du}{dx} = (-\sin u) \cdot 2 = -2\sin(2x)$

Or, alternatively, in function notation form:

$\dfrac{dy}{dx} = f'(g(x)) \cdot g'(x) = [\underbrace{-\sin(2x)}] \cdot 2 = -2\sin(2x)$

derivative of 'outside' function with 'inside' function unchanged $\times$ derivative of 'inside' function

b) $y = f(g(x)) = \sqrt{3x^2 + 5x} \Rightarrow$ 'outside' function is $f(u) = \sqrt{u} = u^{\frac{1}{2}}$
$f'(u) = \dfrac{1}{2}u^{-\frac{1}{2}}$
$\Rightarrow$ 'inside' function is $g(x) = 3x^2 + 5x$

$\dfrac{dy}{dx} = f'(g(x)) \cdot g'(x) = \dfrac{1}{2}(3x^2 + 5x)^{-\frac{1}{2}} \cdot (6x + 5)$

$\dfrac{dy}{dx} = \dfrac{6x + 5}{2(3x^2 + 5x)^{\frac{1}{2}}}$ or $\dfrac{6x + 5}{2\sqrt{3x^2 + 5x}}$

c) $y = f(g(x)) = \ln(1 - 3x) \Rightarrow$ 'outside' function is $f(u) = \ln u$ $f'(u) = \dfrac{1}{u}$
$\Rightarrow$ 'inside' function is $g(x) = 1 - 3x$

$\dfrac{dy}{dx} = f'(g(x)) \cdot g'(x) = \dfrac{1}{1 - 3x} \cdot (-3)$

$\dfrac{dy}{dx} = -\dfrac{3}{1 - 3x}$ or $\dfrac{3}{3x - 1}$

d) $y = f(g(x)) = \dfrac{1}{3x^2 + x} \Rightarrow$ 'outside' function is $f(u) = \dfrac{1}{u} = u^{-1}$
$f'(u) = -u^{-2}$
$\Rightarrow$ 'inside' function is $g(x) = 3x^2 + x$

$\dfrac{dy}{dx} = f'(g(x)) \cdot g'(x) = -(3x^2 + x)^{-2} \cdot (6x + 1)$

$\dfrac{dy}{dx} = -\dfrac{6x + 1}{(3x^2 + x)^2}$

Example 5

Find the derivative of the function $y = (2x + 3)^3$ by:
a) expanding the binomial and differentiating term-by-term
b) the chain rule.

Solution

a) $y = (2x + 3)^3 = (2x + 3)(2x + 3)^2$
$= (2x + 3)(4x^2 + 12x + 9)$
$= 8x^3 + 24x^2 + 18x + 12x^2 + 36x + 27$
$= 8x^3 + 36x^2 + 54x + 27$

$\dfrac{dy}{dx} = 24x^2 + 72x + 54$

b) $y = f(g(x)) = (2x + 3)^3 \Rightarrow y = f(u) = u^3; u = g(x) = 2x + 3$
$\Rightarrow f'(u) = 3u^2; g'(x) = 2$

$\dfrac{dy}{dx} = \dfrac{dy}{du} \cdot \dfrac{du}{dx} = 3u^2 \cdot 2 = 6u^2$

$= 6(2x + 3)^2$

$= 6(4x^2 + 12x + 9)$

$= 24x^2 + 72x + 54$

Example 6

For each function $f(x)$, find $f'(x)$.

a) $f(x) = \sin^2 x$
b) $f(x) = \sin x^2$
c) $f(x) = e^{\sin x}$
d) $f(x) = \sqrt[3]{(7 - 5x)^2}$

Solution

a) The expression $\sin^2 x$ is an abbreviated way of writing $(\sin x)^2$.
Hence, if $f(x) = g(h(x)) = (\sin x)^2$, the 'outside' function is $g(u) = u^2$, and the 'inside' function is $h(x) = \sin x$.
By the chain rule, $f(x) = g'(h(x)) \cdot h'(x)$
$= 2(\sin x)^1 \cdot \cos x$.
Therefore, $f'(x) = 2 \sin x \cos x$.

b) The expression $\sin x^2$ is equivalent to $\sin(x^2)$, and is *not* $(\sin x)^2$.
Hence, if $f(x) = g(h(x)) = \sin(x^2)$, the 'outside' function is $g(u) = \sin u$, and the 'inside' function is $h(x) = x^2$.
By the chain rule, $f(x) = g'(h(x)) \cdot h'(x)$
$= \cos(x^2) \cdot 2x$.
Therefore, $f'(x) = 2x \cos(x^2)$.

c) $f(x) = g(h(x)) = e^{\sin x} \Rightarrow$ 'outside' function is $g(u) = e^u$;
'inside' function is $h(x) = \sin x$
By the chain rule, $f'(x) = g'(h(x)) \cdot h'(x)$.
Therefore, $f'(x) = e^{\sin x} \cdot \cos x$.

d) First change from radical (surd) form to rational exponent form:
$f(x) = \sqrt[3]{(7 - 5x)^2} = (7 - 5x)^{\frac{2}{3}}$
$f(x) = g(h(x)) = (7 - 5x)^{\frac{2}{3}}$
$\Rightarrow$ 'outside' function $g(u) = u^{\frac{2}{3}}$; 'inside' function $h(x) = 7 - 5x$
By the chain rule, $f'(x) = g'(h(x)) \cdot h'(x)$
$= \dfrac{2}{3}(7 - 5x)^{-\frac{1}{3}} \cdot (-5)$.

Therefore, $f'(x) = -\dfrac{10}{3(7 - 5x)^{\frac{1}{3}}}$ or $-\dfrac{10}{3(\sqrt[3]{7 - 5x})}$.

• **Hint:** Aim to write a function in a way that eliminates any confusion regarding the argument of the function. For example, write $\sin(x^2)$ rather than $\sin x^2$; $1 + \ln x$ rather than $\ln x + 1$; $5 + \sqrt{x}$ rather than $\sqrt{x + 5}$.

The derivative of exponential functions b^x ($b > 0$, $b \neq 1$)

In the previous section, we established the derivative rule for the exponential function when the base b is equal to the quite special irrational number e: $\frac{d}{dx}(e^x) = e^x$. What about an exponential function with a base b other than e? Remember that b must be positive and not equal to 1. We can use the laws of logarithms to write b^x in terms of e^x. Recall from Section 4.5 that $b^{\log_b x} = x$, and if $b = e$ then $e^{\ln x} = x$. Hence, $b^x = e^{x \ln b}$ because $e^{x \ln b} = e^{\ln(b^x)} = b^x$. We can now find the derivative of b^x by applying the chain rule to its equivalent expression $e^{x \ln b}$.

$y = f(g(x)) = e^{x \ln b} \Rightarrow$ 'outside' function is $f(u) = e^u \quad f'(u) = e^u$

$\Rightarrow$ 'inside' function is $g(x) = x \ln b \quad g'(x) = \ln b$
[$\ln b$ is a constant]

$$\frac{dy}{dx} = f'(g(x)) \cdot g'(x) = e^{x \ln b} \cdot \ln b$$

$$\frac{dy}{dx} = b^x \ln b$$

Therefore, $\frac{d}{dx}(b^x) = b^x \ln b$.

This result agrees with the fact that $\frac{d}{dx}(e^x) = e^x$. Using this 'new' general rule, $\frac{d}{dx}(b^x) = b^x \ln b$, then $\frac{d}{dx}(e^x) = e^x \ln e$. Since $\ln e = 1$, then $\frac{d}{dx}(e^x) = e^x$.

> **The derivative of b^x**
>
> For $b > 0$ and $b \neq 1$, if $f(x) = b^x$, then $f'(x) = b^x \ln b$. Or, in Leibniz notation, $\frac{d}{dx}(b^x) = b^x \ln b$.

This result now answers the question we posed near the end of Section 13.1. In that section, we used the definition of the derivative to determine that the derivative of the general exponential function $f(x) = b^x$ is $b^x \cdot f'(0)$, where $f'(0)$ is the slope of the graph at $x = 0$. From our result above, we can see that for a specific base b, the slope of the curve $y = b^x$ when $x = 0$ is $\ln b$. The first screen image below is from Section 13.1 and shows the value of $f'(0)$ for $b = 2, 3$ and $\frac{1}{2}$. Evaluating $\ln 2$, $\ln 3$ and $\ln\left(\frac{1}{2}\right)$ on a GDC confirms that $f'(0)$ is equal to $\ln b$.

```
nDeriv(2^X,X,0)
       .6931472361
nDeriv(3^X,X,0)
       1.09861251
nDeriv((1/2)^X,X
,0)
      -.6931472361
```

```
ln(2)
       .6931471806
ln(3)
       1.098612289
ln(1/2)
      -.6931471806
```

13 Differential Calculus II: Further Techniques and Applications

Exercise 13.2

1 Find the derivative of each function.
   a) $y = (3x - 8)^4$
   b) $y = \sqrt{1 - x}$
   c) $y = \ln(x^2)$
   d) $y = 2\sin\left(\dfrac{x}{2}\right)$
   e) $y = (x^2 + 4)^{-2}$
   f) $y = e^{-3x}$
   g) $y = \dfrac{1}{\sqrt{x + 2}}$
   h) $y = \cos^2 x$
   i) $y = e^{x^2} - 2x$
   j) $y = \dfrac{1}{3x^2 - 5x + 7}$
   k) $y = \sqrt[3]{2x + 5}$
   l) $y = \ln(x^2 - 9)$

2 Find the equation of the line tangent to the given curve at the specified value of x. Express the equation exactly in the form $y = mx + c$.
   a) $y = (2x^2 - 1)^3$ $x = -1$
   b) $y = \sqrt{3x^2 - 2}$ $x = 3$
   c) $y = \sin 2x$ $x = \pi$

3 An object moves along a line so that its position s relative to a starting point at any time $t \geq 0$ is given by $s(t) = \cos(t^2 - 1)$.
   a) Find the velocity of the object as a function of t.
   b) What is the object's velocity at $t = 0$?
   c) In the interval $0 < t < 2.5$, find any times (values of t) for which the object is stationary.
   d) Describe the object's motion during the interval $0 < t < 2.5$.

4 In a) – f), find $\dfrac{dy}{dx}$. Use your GDC to check your answer.
   a) $y = \sqrt{x^2 + 2x + 1}$
   b) $y = \dfrac{1}{\sin x}$
   c) $y = (x + \sqrt{x})^3$
   d) $y = e^{\cos x}$
   e) $y = (\ln x)^2$
   f) $y = \dfrac{3}{\sqrt{2x + 1}}$

For questions 5–7, find the equation of a) the tangent, and b) the normal to the curve at the given point.

5 $y = \dfrac{2}{x^2 - 8}$ at $(3, 2)$

6 $y = \sqrt{1 + 4x}$ at $(2, 3)$

7 $y = \ln(4x - 3)$ at $(1, 0)$

8 Consider the exponential function $f(x) = 2^x$.
   a) Find $f'(x)$.
   b) Find the equation of the tangent to the graph of f at the point $(0, 1)$.
   c) Explain why the graph of f has no stationary points.

9 Consider the trigonometric curve $y = \sin\left(2x - \dfrac{\pi}{2}\right)$.
   a) Find $\dfrac{dy}{dx}$ and $\dfrac{d^2y}{dx^2}$.
   b) Find the exact coordinates of any inflexion points for the curve in the interval $0 < x < \pi$.

13.3 The product and quotient rules

The product rule

With the differentiation rules that we have learned thus far we can differentiate some functions that are products. For example, we can differentiate the function $f(x) = (x^2 + 3x)(2x - 1)$ by expanding and then differentiating the polynomial term-by-term. In doing so, we are applying the sum and difference, constant multiple and power rules from Section 11.2.

$$f(x) = (x^2 + 3x)(2x - 1) = 2x^3 + 5x^2 - 3x$$

$$f'(x) = 2\frac{d}{dx}(x^3) + 5\frac{d}{dx}(x^2) - 3\frac{d}{dx}(x)$$

$$f'(x) = 6x^2 + 10x - 3$$

The sum and difference rule states that the derivative of a sum/difference of two functions is the sum/difference of their derivatives. Perhaps the derivative of the product of two functions is the product of their derivatives. Let's try this with the above example.

$$f(x) = (x^2 + 3x)(2x - 1)$$

$$f'(x) = \frac{d}{dx}(x^2 + 3x) \cdot \frac{d}{dx}(2x - 1)?$$

$$f'(x) = (2x + 3) \cdot 2?$$

$$f'(x) = 4x + 6? \quad \text{However, } 4x + 6 \neq 6x^2 + 10x - 3.$$

Thus, one important fact we have learned from this example is that the derivative of a product of two functions is *not* the product of their derivatives. However, there are many products, such as $y = (4x - 3)^3(x - 1)^4$ and $f(x) = x^2 \sin x$, for which it is either difficult or impossible to write the function as a polynomial. In order to differentiate functions like this, we need a '**product**' rule.

> **Gottfried Wilhelm Leibniz (1646–1716)**
>
> Leibniz was a German philosopher, mathematician, scientist and professional diplomat – and, although self-taught in mathematics, was a major contributor to the development of mathematics in the 17th century. He developed the elementary concepts of calculus independent of, but slightly after, Newton. Nevertheless, the notation that Leibniz created for differential and integral calculus is still in use today. Leibniz' approach to the development of calculus was more purely mathematical, whereas Newton's was more directly connected to solving problems in physics. Leibniz created the idea of differentials (infinitely small differences in length), which he used to define the slope of a tangent, before the modern concept of limits was fully developed. Thus, Leibniz considered the derivative $\frac{dy}{dx}$ as the quotient of two differentials, dy and dx. Though it caused some confusion and consternation in his time (and to some extent still), Leibniz manipulated differentials algebraically to establish many of the important differentiation rules – including the product rule.

13 Differential Calculus II: Further Techniques and Applications

> **The product rule**
> If y is a function in terms of x that can be expressed as the product of two functions u and v that are also in terms of x, the product $y = uv$ can be differentiated as follows:
> $$\frac{dy}{dx} = \frac{d}{dx}(uv) = u\frac{dv}{dx} + v\frac{du}{dx}$$
> or, equivalently, if $y = f(x) \cdot g(x)$, then
> $$\frac{dy}{dx} = \frac{d}{dx}[f(x) \cdot g(x)] = f(x) \cdot g'(x) + g(x) \cdot f'(x)$$

Although a formal proof of the product rule is beyond the scope of this book, we can provide sufficient support for the result by considering the relationship between the functions u, v and y when there is a small change in the variable x. Recall that the definition of the derivative (Section 11.2) is essentially the limit of $\dfrac{\text{change in } y}{\text{change in } x}$ as the 'change in x' goes to zero.

Let δx (read 'delta x') and δy represent small changes in x and y, respectively. As $\delta x \to 0$, then $\dfrac{\delta y}{\delta x} \to \dfrac{dy}{dx}$, i.e. the derivative of y with respect to x. Any small change in x, i.e. δx, will cause small changes, δu and δv, in the values of functions u and v respectively. Since $y = uv$, these changes will also cause a small change, δy, in the value of function y.

Now consider the rectangles in Figure 13.9.

The area of the first smaller rectangle is $y = uv$.

The values of u and v then increase by δu and δv respectively.

The area of the larger rectangle is $y + \delta y = uv + u\delta v + v\delta u + \delta u \delta v$.

The product uv changes by the amount $\delta y = u\delta v + v\delta u + \delta u \delta v$.

Dividing through by δx: $\dfrac{\delta y}{\delta x} = u\dfrac{\delta v}{\delta x} + v\dfrac{\delta u}{\delta x} + \delta u \dfrac{\delta v}{\delta x}$.

Let $\delta x \to 0$ and $\delta u \to 0$, then:

$$\frac{\delta y}{\delta x} = u\frac{\delta v}{\delta x} + v\frac{\delta u}{\delta x} + \delta u \frac{\delta v}{\delta x} \implies \frac{dy}{dx} = u\frac{dv}{dx} + v\frac{du}{dx} + 0 \cdot \frac{dv}{dx}$$

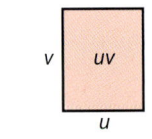

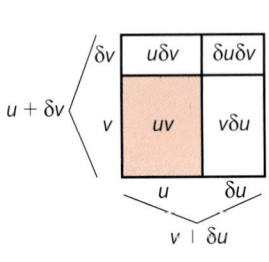

Giving, $\dfrac{dy}{dx} = u\dfrac{dv}{dx} + v\dfrac{du}{dx}$, the product rule.

Figure 13.9

Example 7

Use the product rule to compute the derivative of the function $y = (x^2 + 3x)(2x - 1)$, from the start of this section.

Solution

Recall $y = (x^2 + 3x)(2x - 1) = 2x^3 + 5x^2 - 3x \implies \dfrac{dy}{dx} = 6x^2 + 10x - 3$.

Let $u(x) = x^2 + 3x$ and $v(x) = 2x - 1$, then $y = u(x) \cdot v(x)$ or simply $y = uv$.

By the product rule (in Leibniz form),

$$\frac{dy}{dx} = \frac{d}{dx}(uv) = u\frac{dv}{dx} + v\frac{du}{dx} = (x^2 + 3x) \cdot 2 + (2x - 1) \cdot (2x + 3)$$
$$= (2x^2 + 6x) + (4x^2 + 4x - 3)$$
$$= 6x^2 + 10x - 3$$

This result agrees with the derivative we obtained earlier from differentiating the expanded polynomial.

Example 8

Given $y = x^2 \sin x$, find $\dfrac{dy}{dx}$.

Solution

Let $y = f(x) \cdot g(x) = x^2 \sin x \;\Rightarrow\; f(x) = x^2$ and $g(x) = \sin x$

By the product rule (function notation form),

$$\dfrac{dy}{dx} = \dfrac{d}{dx}[f(x) \cdot g(x)] = f(x) \cdot g'(x) + g(x) \cdot f'(x)$$
$$= x^2 \cdot \cos x + (\sin x) \cdot 2x$$
$$\dfrac{dy}{dx} = x^2 \cos x + 2x \sin x$$

As with the chain rule, it is very helpful to remember the structure of the product rule in words.

$$\dfrac{dy}{dx} = \dfrac{d}{dx}[\underbrace{f(x)}_{\text{first factor}} \cdot \underbrace{g(x)}_{\text{second factor}}] = f(x) \cdot g'(x) + g(x) \cdot f'(x)$$

product of two functions, i.e. factors = first factor × derivative of second factor + second factor × derivative of first factor

Example 9

Find the exact coordinates of any stationary points, and any inflexion points, for the curve $y = xe^x$. Classify any stationary points as a maximum, minimum or neither.

Solution

Recall from Chapter 11 that stationary points occur where the first derivative is zero and that inflexion points (where concavity changes) *may* occur where the second derivative is zero.

$$\dfrac{dy}{dx} = \dfrac{d}{dx}[f(x) \cdot g(x)] = f(x) \cdot g'(x) + g(x) \cdot f'(x)$$
$$= \dfrac{d}{dx}(xe^x) = x\dfrac{d}{dx}(e^x) + e^x \dfrac{d}{dx}(x)$$
$$\dfrac{dy}{dx} = xe^x + e^x = 0 \;\Rightarrow\; e^x(x+1) = 0 \;\Rightarrow\; \dfrac{dy}{dx} = 0 \text{ when } x = -1$$

When $x = -1$, $y = -e^{-1} = -\dfrac{1}{e}$ $\;\Rightarrow\;$ Therefore, the curve has a stationary point at $\left(-1, -\dfrac{1}{e}\right)$.

$$\dfrac{d^2y}{dx^2} = \dfrac{d}{dx}(xe^x + e^x) = \dfrac{d}{dx}(xe^x) + \dfrac{d}{dx}e^x$$
$$= (xe^x + e^x) + e^x$$
$$\dfrac{d^2y}{dx^2} = xe^x + 2e^x = 0 \;\Rightarrow\; e^x(x+2) = 0 \;\Rightarrow\; \dfrac{d^2y}{dx^2} = 0 \text{ when } x = -2$$

An inflexion point will occur at $x = -2$ if the sign of the second derivative changes (i.e. concavity changes) at that point. Find the sign of $\dfrac{d^2y}{dx^2}$ at test points $x = -3$ and $x = 0$.

At $x = -3$, $\dfrac{d^2y}{dx^2} = e^{-3}(-3 + 2) = -\dfrac{1}{e^3} < 0$;

and at $x = 0$, $\dfrac{d^2y}{dx^2} = e^0(0 + 2) = 2 > 0$.

The second derivative undergoes a sign change at $x = -2$; hence, there is an inflexion point on the curve at that point. When $x = -2$, $y = -2e^{-2} = -\dfrac{2}{e^2}$.

Therefore, the curve has an inflexion point at $\left(-2, -\dfrac{2}{e^2}\right)$.

We can use the second derivative test to classify the stationary point $\left(-1, -\dfrac{1}{e}\right)$.

At $x = -1$, $\dfrac{d^2y}{dx^2} = e^{-1}(-1 + 2) = \dfrac{1}{e} > 0 \Rightarrow$ curve is concave up at $x = -1$.

Therefore, the stationary point $\left(-1, -\dfrac{1}{e}\right)$ is a minimum point for the curve.

It's good practice to perform a graphical check of our results on a GDC.

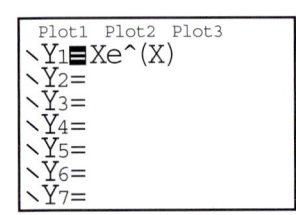

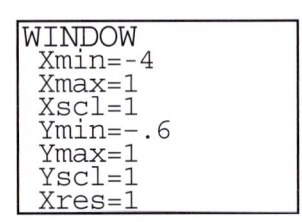

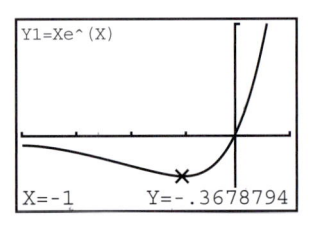

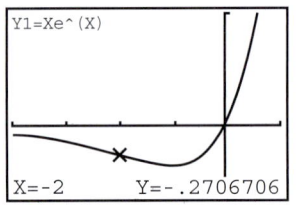

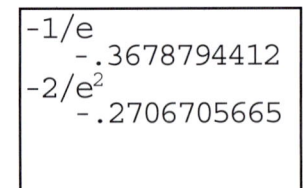

The graph on the GDC not only visually confirms our results but also informs us that $\left(-1, -\dfrac{1}{e}\right)$ is an *absolute* minimum.

The quotient rule

Recall that at the end of Section 13.1 we attempted to conjecture the derivative of the tangent function. We graphed its derivative on a GDC (shown again here) but could not think of a function that matched the graph. We do know the derivatives of $\sin x$ and $\cos x$, so perhaps we can use these results and the fact that $\tan x = \dfrac{\sin x}{\cos x}$ to differentiate $\tan x$. We will need a 'quotient' rule.

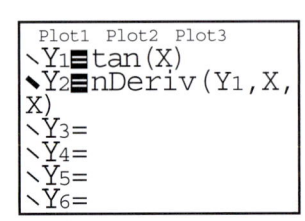

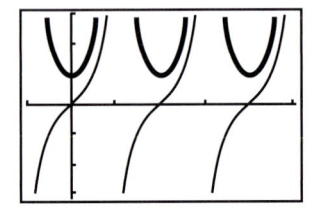

The quotient rule

If y is a function in terms of x that can be expressed as the quotient of two functions u and v that are also in terms of x, the quotient $y = \frac{u}{v}$ can be differentiated as follows:

$$\frac{dy}{dx} = \frac{d}{dx}\left(\frac{u}{v}\right) = \frac{v\frac{du}{dx} - u\frac{dv}{dx}}{v^2}$$

or, equivalently, if $y = \frac{f(x)}{g(x)}$, then

$$\frac{dy}{dx} = \frac{d}{dx}\left[\frac{f(x)}{g(x)}\right] = \frac{g(x) \cdot f'(x) - f(x) \cdot g'(x)}{[g(x)]^2}$$

For the sake of remaining within the scope of this course, we will forego a detailed justification of the quotient rule. As with the chain rule and the product rule, it is helpful to recognize the structure of the quotient rule by remembering it in words:

$$\begin{pmatrix}\text{derivative} \\ \text{of quotient}\end{pmatrix} = \frac{(\text{denominator}) \times \begin{pmatrix}\text{derivative of} \\ \text{numerator}\end{pmatrix} - (\text{numerator})\begin{pmatrix}\text{derivative of} \\ \text{denominator}\end{pmatrix}}{(\text{denominator})^2}$$

Example 10

Determine the derivative of $f(x) = \tan x$ by using the quotient rule to differentiate $\frac{\sin x}{\cos x}$.

Solution

$y = \frac{u}{v} = \frac{\sin x}{\cos x} \Rightarrow u = \sin x$ and $v = \cos x$

By the quotient rule (Leibniz form),

$$\frac{dy}{dx} = \frac{d}{dx}\left(\frac{u}{v}\right) = \frac{v\frac{du}{dx} - u\frac{dv}{dx}}{v^2} = \frac{\cos x \cdot \cos x - \sin x(-\sin x)}{\cos^2 x}$$

$$= \frac{\cos^2 x + \sin^2 x}{\cos^2 x}$$

$\frac{dy}{dx} = \frac{1}{\cos^2 x}$ applying Pythagorean identity $\sin^2 x + \cos^2 x = 1$

Therefore, the derivative of $\tan x$ is $\frac{1}{\cos^2 x}$.

Let's see if the graph of $y = \frac{1}{\cos^2 x}$ matches with the graph of the derivative that we produced on our GDC. Remember to use radian mode.

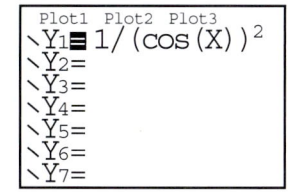

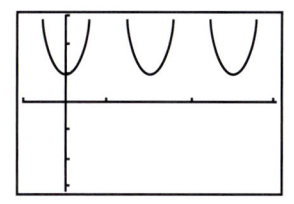

```
Plot1 Plot2 Plot3
\Y1■1/(cos(X))2
\Y2=
\Y3=
\Y4=
\Y5=
\Y6=
\Y7=
```

```
WINDOW
 Xmin=-1.570796…
 Xmax=7.8539816…
 Xscl=π/2
 Ymin=-3
 Ymax=3
 Yscl=1
 Xres=1
```

It matches … and a nice application of the quotient rule that completes our derivative rules for trigonometric functions.

13 Differential Calculus II: Further Techniques and Applications

> **The derivative of the tangent function**
> If $f(x) = \tan x$, then $f'(x) = \dfrac{1}{\cos^2 x}$. Or, in Leibniz notation, $\dfrac{d}{dx}(\tan x) = \dfrac{1}{\cos^2 x}$.
> This result is only true when x is in radian measure.

Example 11

Given $y = \dfrac{1}{2x - 3}$, find $\dfrac{dy}{dx}$ by:

a) the quotient rule
b) the chain rule.

Solution

a) $y = \dfrac{f(x)}{g(x)} = \dfrac{1}{2x - 3} \;\Rightarrow\; f(x) = 1$ and $g(x) = 2x - 3$

By the quotient rule (function notation form),

$$\dfrac{dy}{dx} = \dfrac{d}{dx}\left[\dfrac{f(x)}{g(x)}\right] = \dfrac{g(x) \cdot f'(x) - f(x) \cdot g'(x)}{[g(x)]^2}$$

$$= \dfrac{(2x - 3) \cdot 0 - 1 \cdot (2)}{(2x - 3)^2}$$

$$\dfrac{dy}{dx} = \dfrac{-2}{(2x - 3)^2}$$

b) $y = f(g(x)) = \dfrac{1}{2x - 3} = (2x - 3)^{-1} \;\Rightarrow\;$ 'outside' function is $f(u) = u^{-1}$
$\;\Rightarrow\; f'(u) = -u^{-2}$
$\;\Rightarrow\;$ 'inside' function is $g(x) = 2x - 3$

By the chain rule (function notation form),

$$\dfrac{dy}{dx} = f'(g(x)) \cdot g'(x) = -(2x - 3)^{-2} \cdot 2$$

$$\dfrac{dy}{dx} = -\dfrac{2}{(2x - 3)^2}$$

As Example 11 illustrates, when required to differentiate a quotient, you can choose to rewrite the quotient $y = \dfrac{u}{v}$ as $y = uv^{-1}$, thereby applying the product rule instead of the quotient rule.

Example 12

For each function, find its derivative (i) by the quotient rule, and (ii) by another method.

a) $f(x) = \dfrac{3x - 2}{2x - 5}$
b) $g(x) = \dfrac{5x - 1}{3x^2}$

Solution

a) (i) $f(x) = y = \dfrac{u}{v} = \dfrac{3x - 2}{2x - 5}$

$$f'(x) = \dfrac{dy}{dx} = \dfrac{v\dfrac{du}{dx} - u\dfrac{dv}{dx}}{v^2} = \dfrac{(2x - 5) \cdot 3 - (3x - 2) \cdot 2}{(2x - 5)^2}$$

$$= \dfrac{6x - 15 - 6x + 4}{(2x - 5)^2}$$

$$f'(x) = \dfrac{-11}{(2x - 5)^2}$$

(ii) Rewrite $f(x)$ as a product and apply the product rule (with chain rule embedded):
$$f(x) = y = \frac{3x-2}{2x-5} = (3x-2)(2x-5)^{-1}$$
$\Rightarrow$ for $y = uv$, $u = 3x - 2$ and $v = (2x-5)^{-1}$

Note: $v = (2x-5)^{-1}$ is a composite function, so we'll need the chain rule to find $\frac{dv}{dx}$.

$$f'(x) = \frac{d}{dx}(uv) = u\frac{dv}{dx} + v\frac{du}{dx}$$
$$= (3x-2) \cdot \frac{d}{dx}[(2x-5)^{-1}] + (2x-5)^{-1} \cdot 3$$
$$= (3x-2)[-(2x-5)^{-2} \cdot 2] + 3(2x-5)^{-1}$$

chain rule applied for $\frac{d}{dx}[(2x-5)^{-1}]$

$$= (-6x+4)(2x-5)^{-2} + 3(2x-5)^{-1}$$
$$= (2x-5)^{-2}[(-6x+4) + 3(2x-5)]$$
factorizing out GCF of $(2x-5)^{-2}$
$$= (2x-5)^{-2}[-6x+4+6x-15]$$
$$f'(x) = \frac{-11}{(2x-5)^2}$$

b) (i) $g(x) = y = \frac{u}{v} = \frac{5x-1}{3x^2}$

$$g'(x) = \frac{dy}{dx} = \frac{v\frac{du}{dx} - u\frac{dv}{dx}}{v^2} = \frac{3x^2 \cdot 5 - (5x-1) \cdot 6x}{(3x^2)^2}$$
$$= \frac{15x^2 - 30x^2 + 6x}{9x^4}$$
$$= \frac{3x(-5x+2)}{9x^4}$$
$$g'(x) = \frac{-5x+2}{3x^3}$$

(ii) Using algebra, 'split' the numerator:
$$g(x) = \frac{5x-1}{3x^2} = \frac{5x}{3x^2} - \frac{1}{3x^2} = \frac{5}{3x} - \frac{1}{3x^2} = \frac{5}{3}x^{-1} - \frac{1}{3}x^{-2}$$

Now, differentiate term-by-term using the power rule:
$$g'(x) = \frac{5}{3}\frac{d}{dx}(x^{-1}) - \frac{1}{3}\frac{d}{dx}(x^{-2})$$
$$= \frac{5}{3}(-x^{-2}) - \frac{1}{3}(-2x^{-3})$$
$$g'(x) = -\frac{5}{3x^2} + \frac{2}{3x^3}$$

$\left[\text{Results for (i) and (ii) are equivalent:} \right.$
$\left. -\frac{5}{3x^2} + \frac{2}{3x^3} = -\frac{5}{3x^2} \cdot \frac{x}{x} + \frac{2}{3x^3} = \frac{-5x}{3x^3} + \frac{2}{3x^3} = \frac{-5x+2}{3x^3} \right]$

As Example 12 demonstrates, before differentiating a quotient, it is worthwhile to consider if performing some algebra may allow other more efficient differentiation techniques to be used.

- **Hint:** The function $h(x) = \dfrac{3x^2}{5x-1}$ initially looks similar to the function g in Example 12, part b) (they're reciprocals). However, it is *not* possible to 'split' the denominator and express as two fractions. Recognize that $\dfrac{3x^2}{5x-1}$ is *not* equivalent to $\dfrac{3x^2}{5x} - \dfrac{3x^2}{1}$. Hence, in order to differentiate $h(x) = \dfrac{3x^2}{5x-1}$ we would apply either the quotient rule, or the product rule with the function rewritten as $h(x) = 3x^2(5x-1)^{-1}$ and using the chain rule to differentiate the factor $(5x-1)^{-1}$.

Exercise 13.3

1 Find the derivative of each function.
  a) $y = x^2 e^x$
  b) $y = x\sqrt{1-x}$
  c) $y = x \ln x$
  d) $y = \sin x \cos x$
  e) $y = \dfrac{e^x}{x}$
  f) $y = \dfrac{x+1}{x-1}$
  g) $y = (2x-1)^3(x^4+1)$
  h) $y = \dfrac{\sin x}{x}$
  i) $y = \dfrac{x}{e^x - 1}$
  j) $y = \dfrac{6x-7}{3x+2}$
  k) $y = (x^2 - 1)\ln(3x)$
  l) $y = \dfrac{1}{\sin^2 x + \cos^2 x}$

2 Find the equation of the line tangent to the given curve at the specified value of x. Express the equation exactly in the form $y = mx + c$.
  a) $y = \dfrac{8}{4+x^2}$ $x = 2$
  b) $y = \dfrac{x^3+1}{2x}$ $x = 1$
  c) $y = x\sqrt{x^2-3}$ $x = 2$

3 Consider the function $h(x) = \dfrac{x^2 - 3}{e^x}$.
  a) Find the exact coordinates of any stationary points.
  b) Determine whether each stationary point is a maximum, minimum or neither.
  c) What do the function values approach as (i) $x \to \infty$, and (ii) $x \to -\infty$.
  d) Write down the equation of any asymptotes for the graph of $h(x)$.
  e) Make an accurate sketch of the curve indicating any extrema and points where the graph intersects the x- and y-axis.

4 Use the product rule to prove the constant multiple rule for differentiation. That is, show that $\dfrac{d}{dx}(c \cdot f(x)) = c \cdot \dfrac{d}{dx}(f(x))$ for any constant c.

5 A curve has equation $y = x(x-4)^2$.
  a) For this curve, find
   (i) the x-intercepts
   (ii) the coordinates of the maximum point
   (iii) the coordinates of the point of inflexion.
  b) Use your answers to part a) to sketch a graph of the curve for $0 \leq x \leq 4$, clearly indicating the features you have found in part a).

6 Consider the function $f(x) = \dfrac{x^2 - 3x + 4}{(x+1)^2}$.
  a) Show that $f'(x) = \dfrac{5x - 11}{(x+1)^3}$.
  b) Show that $f''(x) = \dfrac{-10x + 38}{(x+1)^4}$.
  c) Does the graph of f have an inflexion point at $x = 3.8$? Explain.

13.4 Optimization

Many problems in science and mathematics involve finding the maximum or minimum value (**optimum** value) of a function over a specified or implied domain. The development of the calculus in the seventeenth century was motivated to a large extent by maxima and minima (**optimization**) problems. One such problem lead Pierre de Fermat (1601–1665) to develop his Principle of Least Time: a ray of light will follow the path that takes the least (or minimum) time. The solution to Fermat's principle lead to Snell's law, or law of refraction (see the investigation at end of this section). The solution is found by applying techniques of differential calculus – which can also be used to solve other optimization problems involving ideas such as least cost, maximum profit, minimum surface area and greatest volume.

Previously, we learned the theory of how to use the derivative of a function to locate points where the function has a maximum or minimum (i.e. extreme) value. It is important to remember that if the derivative of a function is zero at a certain point it does not *necessarily* follow that the function has an extreme value (relative or absolute) at that point – it only ensures that the function has a horizontal tangent (stationary point) at that point. An extreme value *may* occur where the derivative is zero or at the endpoints of the function's domain.

The graph of $f(x) = x^4 - 8x^3 + 18x^2 - 16x - 2$ is shown right. The derivative of $f(x)$ is $f'(x) = 4x^3 - 24x^2 + 36x - 16 = 4(x - 4)(x - 1)^2$. The function has horizontal tangents at both $x = 1$ and $x = 4$, since the derivative is zero at these points. However, an extreme value (absolute minimum) occurs only at $x = 4$. It is important to confirm – graphically or algebraically – the precise nature of a point on a function where the derivative is zero. Some different algebraic methods for confirming that a value is a maximum or minimum will be illustrated in the examples that follow.

It is also useful to remember that one can often find extreme values (extrema) without calculus (e.g. using a 'minimum' command on a graphics calculator, as shown). Calculator or computer technology can be very helpful in modelling, solving or confirming solutions to optimization problems. However, it is important to learn how to apply algebraic methods of differentiation to optimization problems because it may be the only efficient way to obtain an accurate solution.

Let's start with a relatively straightforward example. We can use the steps in the solution to develop a general strategy that can be applied to more sophisticated problems.

Example 13

(Developing a general strategy.)

Find the maximum area of a rectangle inscribed in an isosceles right triangle whose hypotenuse is 20 cm long.

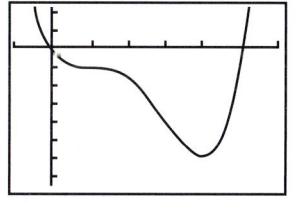

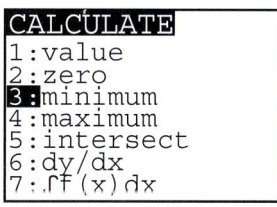

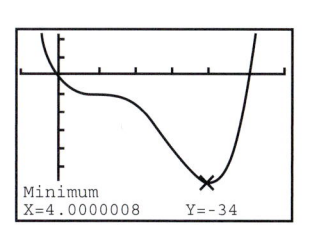

Differential Calculus II: Further Techniques and Applications

Solution

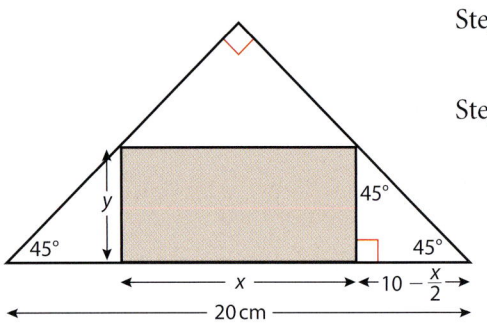

Step 1: Draw an accurate diagram. Let the base of the rectangle be x cm and the height y cm. Then the area of the rectangle is $A = xy$ cm^2.

Step 2: Express area as a function in terms of only one variable.

It can be deduced from the diagram that $y = 10 - \frac{x}{2}$.

Therefore, $A(x) = x\left(10 - \frac{x}{2}\right) = 10x - \frac{x^2}{2}$.

x must be positive and from the diagram it is clear that x must be less than 20 (domain of A: $0 < x < 20$).

Step 3: Find the derivative of the area function and find for what value(s) of x it is zero.

$A'(x) = 10 - x$ $A'(x) = 0$ when $x = 10$

Step 4: Analyze $A(x)$ at $x = 10$ and also at the endpoints of the domain, $x = 0$ and $x = 20$.

The second derivative test (Section 11.3) provides information about the concavity of a function. The second derivative is $A''(x) = -1$ and since $A''(x)$ is always negative then $A(x)$ is always concave down, indicating $A(x)$ has a maximum at $x = 10$.

$A(0) = 0$ and $A(20) = 0$, indicating $A(x)$ has an absolute maximum at $x = 10$.

Therefore, the rectangle has a maximum area equal to

$A(10) = 10\left(10 - \frac{10}{2}\right) = 50$ cm^2.

General strategy for solving optimization problems

Step 1: Draw a diagram that accurately illustrates the problem. Label all known parts of the diagram. Using variables, label the important unknown quantity (or quantities) (for example, x for base and y for height in Example 13).

Step 2: For the quantity that is to be optimized (area in Example 13), express this quantity as a function in terms of a single variable. From the diagram and/or information provided, determine the domain of this function.

Step 3: Find the derivative of the function from Step 2, and determine where the derivative is zero. This value (or values) of the derivative, along with any domain endpoints, are the **critical values** ($x = 0$, $x = 10$ and $x = 20$ in Example 13) to be tested.

Step 4: Using algebraic (e.g. second derivative test) or graphical (e.g. GDC) methods, analyze the nature (maximum, minimum, neither) of the points at the critical values for the optimized function. Be sure to answer the precise question that was asked in the problem.

Example 14

(Finding a minimum length – two posts problem.)

Two vertical posts, with heights of 7 m and 13 m, are secured by a rope going from the top of one post to a point on the ground between the posts and then to the top of the other post. The distance between the two posts is 25 m. Where should the point at which the rope touches the ground be located so that the least amount of rope is used?

Solution

Step 1: An accurate diagram is drawn. The posts are drawn as line segments PQ and TS and the point where the rope touches the ground is labelled R. The optimum location of point R can be given as a distance from the base of the shorter post, QR, or from the taller post, SR. It is decided to give the answer as the distance from the shorter post – and this is labelled x. There are two other important unknown quantities: the lengths of the two portions of the rope, PR and TR. These are labelled a and b, respectively.

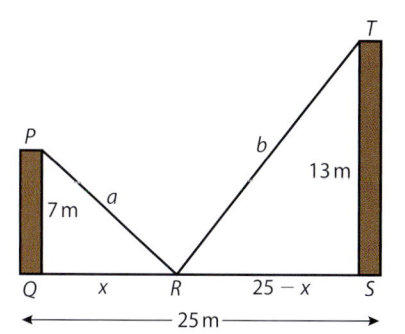

Step 2: The quantity to be minimized is the length L of the rope, which is the sum of a and b. From Pythagoras' theorem, $a = \sqrt{x^2 + 49}$ and $b = \sqrt{(25-x)^2 + 169}$. Therefore, the function for length (L) can be expressed in terms of the single variable x as

$$L(x) = \sqrt{x^2 + 49} + \sqrt{(25-x)^2 + 169}$$

$$= \sqrt{x^2 + 49} + \sqrt{x^2 - 50x + 625 + 169}$$

$$\Rightarrow L(x) = \sqrt{x^2 + 49} + \sqrt{x^2 - 50x + 794}$$

From the given information and diagram, the domain of $L(x)$ is $0 \leq x \leq 25$.

Step 3: To facilitate differentiation, express $L(x)$ using fractional exponents:

$$L(x) = (x^2 + 49)^{\frac{1}{2}} + (x^2 - 50x + 794)^{\frac{1}{2}}$$

Then apply the chain rule for differentiation:

$$\frac{dL}{dx} = \tfrac{1}{2}(x^2 + 49)^{-\frac{1}{2}}(2x) + \tfrac{1}{2}(x^2 - 50x + 794)^{-\frac{1}{2}}(2x - 50) \Rightarrow$$

$$\frac{dL}{dx} = \frac{x}{\sqrt{x^2 + 49}} + \frac{x - 25}{\sqrt{x^2 - 50x + 794}}$$

By setting $\dfrac{dL}{dx} = 0$, we obtain

$$x\sqrt{x^2 - 50x + 794} = -(x - 25)\sqrt{x^2 + 49}$$

$$x^2(x^2 - 50x + 794) = (25 - x)^2(x^2 + 49)$$

$$x^4 - 50x^3 + 794x^2 = x^4 - 50x^3 + 674x^2 - 2450x + 30\,625$$

$$120x^2 + 2450x - 30\,625 = 0$$

$$5(4x - 35)(6x + 175) = 0$$

$$x = \frac{35}{4} \quad \text{or} \quad x = -\frac{175}{6}$$

Step 4: Since $x = -\dfrac{175}{6}$ is not in the domain for $L(x)$, then the critical values are $x = 0$, $x = \dfrac{35}{4}$ and $x = 25$. Simply evaluate $L(x)$ for these critical values.

$$L(0) = 7 + \sqrt{794} \approx 35.18, \quad L(25) = \sqrt{674} + 13 \approx 38.96,$$

$$L\left(\dfrac{35}{4}\right) = 5\sqrt{41} \approx 32.02$$

Therefore, the rope should touch the ground at a distance of $\dfrac{35}{4} = 8.75$ m from the base of the shorter post, to give a minimum rope length of approximately 32.02 m.

The minimum value could also be confirmed from the graph of $L(x)$, but it would be difficult to confirm using the second derivative test because of the algebra required. From this example, we can see that applied optimization problems can involve a high level of algebra. If you have access to suitable graphing technology, you could perform Steps 3 and 4 graphically rather than algebraically.

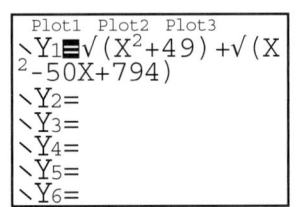

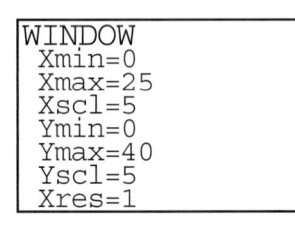

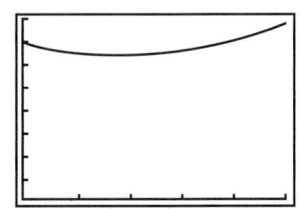

```
Plot1 Plot2 Plot3
\Y1=√(X²+49)+√(X
²-50X+794)
\Y2=
\Y3=
\Y4=
\Y5=
\Y6=
```

```
WINDOW
Xmin=0
Xmax=25
Xscl=5
Ymin=0
Ymax=40
Yscl=5
Xres=1
```

```
CALCULATE
1:value
2:zero
3:minimum
4:maximum
5:intersect
6:dy/dx
7:∫f(x)dx
```

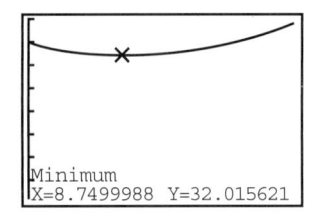

```
Minimum
X=8.7499988  Y=32.015621
```

It is interesting to observe that the result for x produced by the calculator does not appear to be exact. Why is that? Algebraic techniques using differentiation give us the certainty of an exact solution while also allowing us to deal with the abstract nature of optimization problems involving parameters rather than fixed measurements (e.g. the heights of the posts).

In both Example 13 and 14, the extreme value occurred at a point where the derivative was zero. Although this often happens, an extreme value may occur at the endpoint of the domain.

Example 15

(An endpoint maximum.)

A supply of four metres of wire is to be used to form a square and a circle. How much of the wire should be used to make the square and how much should be used to make the circle in order to enclose the greatest amount of area? Guess the answer before looking at the following solution.

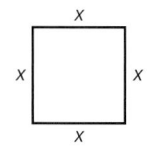

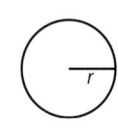

Solution

Step 1: Let x = length of each edge of the square and r = radius of the circle.

Step 2: The total area is given by $A = x^2 + \pi r^2$. The task is to write the area A as a function of a single variable. Therefore, it is necessary to express r in terms of x, or vice versa, and perform a substitution.

The perimeter of the square is $4x$ and the circumference of the circle is $2\pi r$. The total amount of wire is 4 m which gives

$$4 = 4x + 2\pi r \implies 2\pi r = 4 - 4x \implies r = \frac{2(1-x)}{\pi}$$

Substituting gives $A(x) = x^2 + \pi \left[\frac{2(1-x)}{\pi}\right]^2 = x^2 + \frac{4(1-x)^2}{\pi}$

$$= \frac{1}{\pi}[(\pi + 4)x^2 - 8x + 4]$$

Because the square's perimeter is $4x$, then the domain for $A(x)$ is $0 \leq x \leq 1$.

Step 3: Differentiate the function $A(x)$, set equal to zero, and solve.

$$\frac{d}{dx}\left(\frac{1}{\pi}[(\pi+4)x^2 - 8x + 4]\right) = \frac{1}{\pi}[2(\pi+4)x - 8] = 0$$

$$2(\pi+4)x - 8 = 0 \implies (\pi+4)x = 4 \implies x = \frac{4}{\pi+4} \approx 0.5601$$

The critical values are $x = 0$, $x \approx 0.5601$ and $x = 1$.

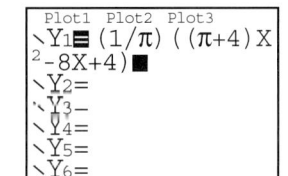

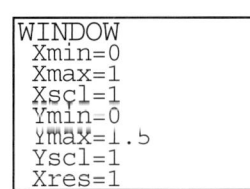

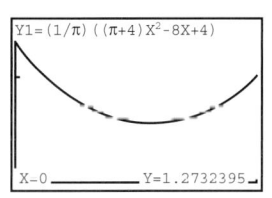

Step 4: Evaluating $A(x)$: $A(0) \approx 1.273$, $A(0.5601) \approx 0.5601$ and $A(1) = 1$. Therefore, the maximum area occurs when $x = 0$ which means <u>all</u> the wire is used for the circle.

What would the answer be if Example 15 asked for the dimensions of the square and circle to enclose the *least* total area?

Example 16

(Minimizing time.)

A pipeline needs to be constructed to link an offshore drilling rig to an onshore refinery depot. The oil rig is located at a distance (perpendicular to the coast) of 140 km from the coast. The depot is located inland at a distance (perpendicular) of 60 km from the coast. For modelling purposes, the coastline is assumed to follow a straight line. The point on the coastline nearest to the oil rig is 160 km from the point on the coastline nearest to the depot. The rate at which crude oil is pumped through the pipeline varies according to several variables, including pipe dimensions, materials, temperature, etc. On average, oil flows through the offshore section of the pipeline at a rate of 9 km per hour and 5 km per hour through the onshore section. Assume that both sections of pipeline can travel straight from one point to another. At what point should the pipeline intersect with the coastline in order for the oil to take a minimum amount of time to flow from the rig to the depot?

Solution

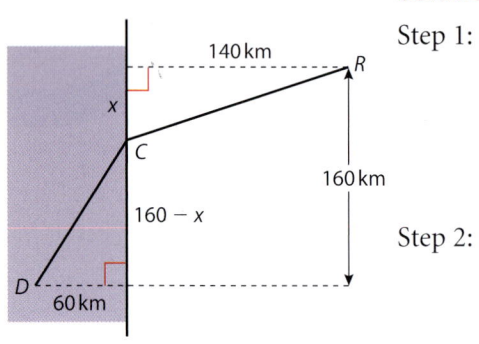

Step 1: The optimum location of the point, C, where the pipeline comes ashore will be designated by the distance, x, it is from the point on the coast that is a minimum distance (perpendicular) from the rig, R (140 km). The distance from R to C is $\sqrt{x^2 + 140^2}$ and the distance from D (depot) to C is $\sqrt{(160-x)^2 + 60^2}$.

Step 2: The quantity to be minimized is time, so it is necessary to express the total time it takes the oil to flow from R to D in terms of a single variable.

$$\text{time} = \frac{\text{distance}}{\text{rate}} \Rightarrow \text{time (offshore)} = \frac{\sqrt{x^2 + 19\,600}}{9} \text{ km/hr};$$

$$\text{time (onshore)} = \frac{\sqrt{x^2 - 320x + 29\,200} \text{ km}}{5 \text{ km/hr}}$$

The function for time T in terms of x is:

$$T(x) = \frac{\sqrt{x^2 + 19\,600}}{9} + \frac{\sqrt{x^2 - 320x + 29\,200}}{5}$$

and the domain for $T(x)$ is $0 \leq x \leq 160$.

Steps 3/4: The algebra for finding the derivative of $T(x)$ is similar to that of Step 3 in Example 14. Let's use graphing technology to find the value of x that produces a minimum for $T(x)$.

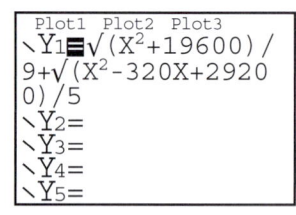

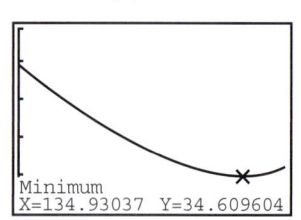

Therefore, the optimum point for the pipeline to intersect with the coast is approximately 134.9 km from the point on the coast nearest to the drilling rig.

The result could also be obtained by having a calculator or computer graph the derivative of $T(x)$ and compute any zeros for $T'(x)$ in the domain.

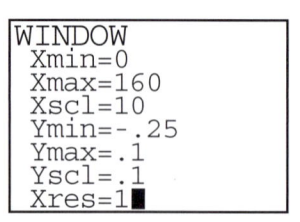

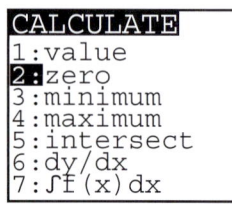

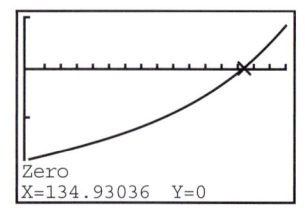

See the **Investigation** and how solving a problem similar to Example 16 derives Snell's law (or law of refraction).

Investigation – Snell's law

The speed of light depends on the medium through which light travels and is generally slower in denser media. The speed of light in a vacuum is an important physical constant and is exactly 299 792 458 m/s. A metre is defined to be the distance that light travels in a vacuum in $\frac{1}{299\,792\,458}$ of a second. Typically, the speed of light in a vacuum (denoted by the letter c) is given the approximate value of 3×10^8 m/s, but in the Earth's atmosphere light travels more slowly than that and even more slowly through glass and water.

Fermat's principle in optics states that light travels from one point to another along a path for which time is a minimum. Investigate the path that a ray of light will follow in going from a point A in a transparent medium, where the speed of light is c_1, to a point B in a different transparent medium, where its speed is c_2, as illustrated in the diagram right. Using algebra and differentiation, prove that for time to be a minimum the following relationship must hold: $\frac{\sin \theta_1}{c_1} = \frac{\sin \theta_2}{c_2}$. This equation is known as Snell's law or the law of refraction. Why is a graphics calculator not helpful?

Assume that the two points, A and B, lie in the xy-plane and the x-axis (interface) separates the two media. A light ray is refracted (deflected) when it passes from one medium to another. θ_1 is the **angle of incidence** and θ_2 is the **angle of refraction** (both angles measured between ray and normal to the interface).

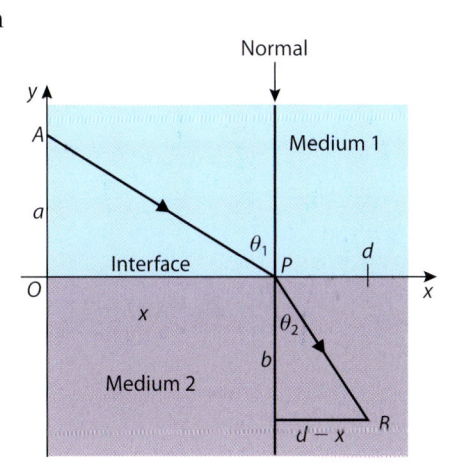

Exercise 13.4

1 Find the dimensions of the rectangle with maximum area that is inscribed in a semicircle with radius 1 cm. Two vertices of the rectangle are on the semicircle and the other two vertices are on the x-axis, as shown in the diagram.

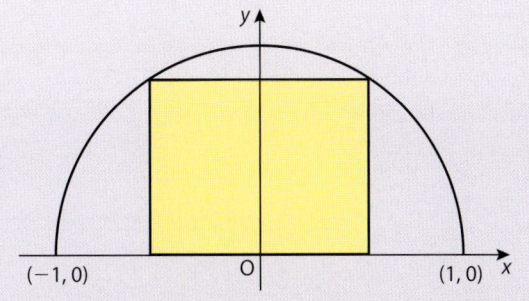

2 A rectangular piece of aluminium is to be rolled to make a cylinder with open ends (a tube). Regardless of the dimensions of the rectangle, the perimeter of the rectangle must be 40 cm. Find the dimensions (length and width) of the rectangle that gives a maximum volume for the cylinder.

3 A rectangular box has height h cm, width x cm and length $2x$ cm. It is designed to have a volume equal to 1 litre (1000 cm³).
   a) Show that $h = \frac{500}{x^2}$ cm.
   b) Find an expression for the total surface area, S cm², of the box in terms of x.
   c) Find the dimensions of the box that produces a minimum surface area.

4 The figure right consists of a rectangle *ABCD* and two semicircles on either end. The rectangle has an area of 100 cm². If x represents the length of the rectangle *AB*, find the value of x that makes the perimeter of the entire figure a minimum.

- **Hint:** Write an equation for θ in terms of x and find the value of x which makes θ a maximum by using your GDC.

5 Two vertical posts, with heights 12 metres and 8 metres, are 10 metres apart on horizontal ground. A rope that stretches is attached to the top of both posts and is stretched down so that it touches the ground at point *A* between the two posts. The distance from the base of the taller post to point *A* is represented by x and the angle between the two sections of rope is θ. What value of x makes θ a maximum?

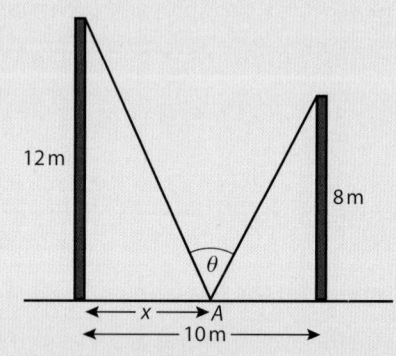

6 A ladder is to be carried horizontally down an L-shaped hallway. The first section of the hallway is 2 metres wide and then there is a right-angled turn into a 3 metre wide section. What is the longest ladder that can be carried around the corner?

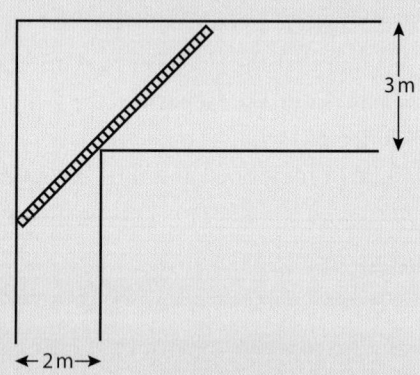

7 Charlie is walking from the wildlife observation tower (point *T*) to the Big Desert Park office (point *O*). The tower is 7 km due west and 10 km due south from the office. There is a road that goes to the office that Charlie can get to if she walks 10 km due north from the tower. Charlie can walk at a rate of 2 kilometres per hour (kph) through the sandy terrain of the park, but she can walk a faster rate of 5 kph on the road. To what point, *A*, on the road should Charlie walk to in order to take the least time to walk from the tower to the office? Find the value of d such that point *A* is d km from the office.

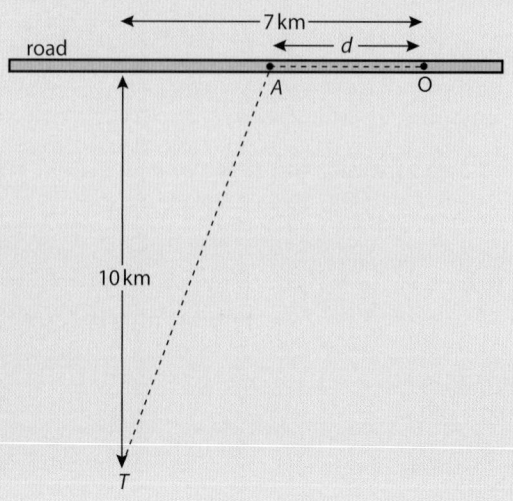

13.5 Summary of differentiation rules and applications

Summary of differentiation rules

Derivative of $f(x)$: $\quad y = f(x) \quad \Rightarrow \quad f'(x) = \lim_{h \to 0} \dfrac{f(x+h) - f(x)}{h}$ (Limit definition – 'first principles')

Derivative of x^n: $\quad f(x) = x^n \quad \Rightarrow \quad f'(x) = nx^{n-1}$

Derivative of $\sin x$: $\quad f(x) = \sin x \quad \Rightarrow \quad f'(x) = \cos x$

Derivative of $\cos x$: $\quad f(x) = \cos x \quad \Rightarrow \quad f'(x) = -\sin x$

Derivative of $\tan x$: $\quad f(x) = \tan x \quad \Rightarrow \quad f'(x) = \dfrac{1}{\cos^2 x}$

Note: Derivative rules for trigonometric functions only apply if x is in radian measure.

Derivative of e^x: $\quad f(x) = e^x \quad \Rightarrow \quad f'(x) = e^x$

Derivative of b^x: $\quad f(x) = b^x \quad \Rightarrow \quad f'(x) = b^x \ln b$

Derivative of $\ln x$: $\quad f(x) = \ln x \quad \Rightarrow \quad f'(x) = \dfrac{1}{x}$

Chain rule for composite functions: $\quad \dfrac{dy}{dx} = \dfrac{d}{dx}[f(g(x))] = f'(g(x)) \cdot g'(x)$

Product rule: $\quad \dfrac{dy}{dx} = \dfrac{d}{dx}[f(x) \cdot g(x)] = f(x) \cdot g'(x) + g(x) \cdot f'(x)$

Quotient rule: $\quad \dfrac{dy}{dx} = \dfrac{d}{dx}\left[\dfrac{f(x)}{g(x)}\right] = \dfrac{g(x) \cdot f'(x) - f(x) \cdot g'(x)}{[g(x)]^2}$

Summary of derivative tests

First derivative test

$f'(c) = 0$ or $x = c$ is an endpoint of the domain.

I. For $f'(c) = 0$:

   1. If $f'(x)$ changes sign from positive to negative as x increases through $x = c$, then f has a relative maximum at $x = c$.

   2. If $f'(x)$ changes sign from negative to positive as x increases through $x = c$, then f has a relative minimum at $x = c$.

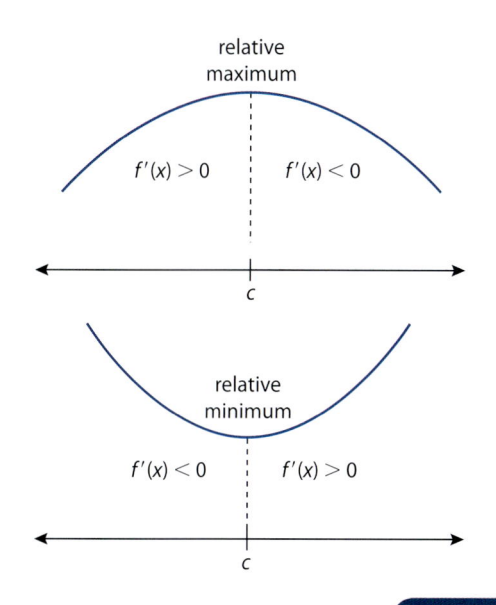

3. If $f'(x)$ does not change sign as x increases through $x = c$, then f has neither a relative maximum nor a relative minimum at $x = c$.

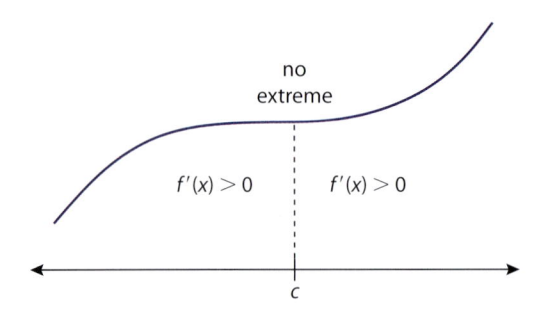

II. For $x = c$ and endpoint of the domain of f:

If $x = c$ is an endpoint of the domain, then $x = c$ will be a relative maximum or minimum of f if the sign of $f'(x)$ is always positive or always negative for $x > c$ (at a left endpoint), or for $x < c$ (at a right endpoint), as illustrated below.

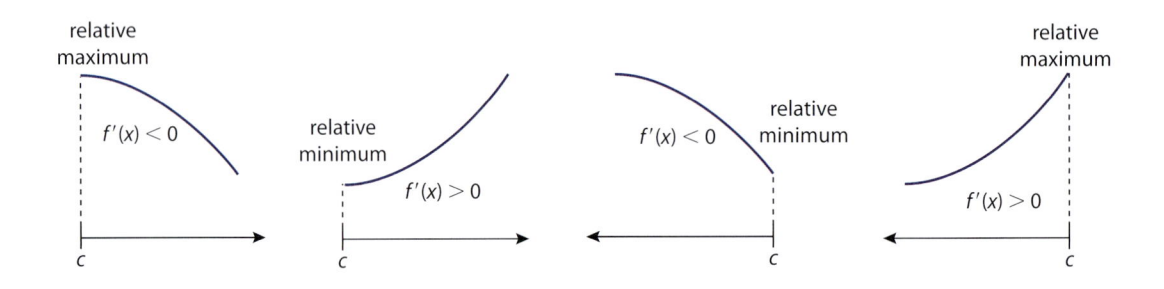

Second derivative test

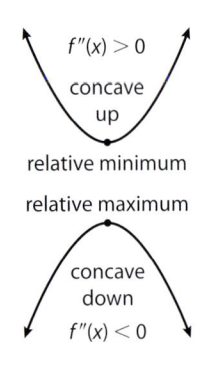

1. If $f'(c) = 0$ and $f''(c) < 0$, then f has a relative maximum at $x = c$.
2. If $f'(c) = 0$ and $f''(c) > 0$, then f has a relative minimum at $x = c$.

If $f''(c) = 0$, the test fails and the first derivative test should be applied.

The rules, methods and applications covered in Chapters 11 and 13 provide a firm foundation for differential calculus. It is important to gain sufficient practice with applying the algebraic rules and techniques *and* their application to problems. With that in mind, the exercise set here provides additional preparation in these areas while incorporating the material covered in Chapters 11 and 13. The rules that are covered were listed in the summary above, and the applications covered include:

- slopes (gradients) and rates of change
- properties of graphs of functions
- equations of tangents and normals
- displacement, velocity and acceleration
- maxima and minima – optimization.

Exercise 13.5

In questions 1–20, find the derivative of each function.

1. $y = (x - 1)^7$
2. $y = \dfrac{1}{5x}$
3. $y = 2x(3x + 4)^4$
4. $y = \sin x \tan x$
5. $y = e^{6x}$
6. $y = x^3 \ln x$
7. $y = \sin\left(4x - \dfrac{\pi}{4}\right)$
8. $y = \dfrac{1}{e^{x^2}}$
9. $y = (\ln x)^3$
10. $y = x\sqrt{1 - x^2}$
11. $y = 3^x$
12. $y = \dfrac{x^2 - 1}{2x + 3}$
13. $y = x^2 \tan x$
14. $y = \dfrac{e^x}{\cos x}$
15. $y = \sqrt[3]{x^2 + 2x - 4}$
16. $y = \sin(\cos 2x)$
17. $y = e^{1 - 2x}$
18. $y = e^x \ln(x^2)$
19. $y = \ln(e^x - 2x)$
20. $y = 2^{3x - 1}$

21. The temperature outside a house during a 24-hour period is given by
$$C(t) = 25 - 6\cos\left(\dfrac{\pi}{12}t - \dfrac{\pi}{6}\right), 0 \leqslant t \leqslant 24$$
where $C(t)$ is measured in degrees Celsius and t is measured in hours.
a) Make a sketch of the graph of the function C for the interval $0 \leqslant t \leqslant 24$.
b) Find the average rate of change of the temperature from $t = 2$ to $t = 8$.
c) Find $C'(t)$.
d) Find the instantaneous rate of change of the temperature at $t = 5$.
e) At what time is the temperature a maximum? What is the maximum temperature?

22. Find the points on the graph of the equation $y = 4 - x^2$ that are nearest to the point (0, 2).

23. The normal to the curve $y = x^2 - 4x$ at the point $(3, -3)$ intersects the x-axis at point P and the y-axis at point Q. Find the equation of the normal and the coordinates of P and Q.

24. The rate at which cars on a road pass a certain point is known as the flow rate and is in units of cars per minute. The flow rate, F, of a certain road is given by
$F(x) = \dfrac{2x}{18 + 0.015x^2}$, where x is the speed of the traffic in kilometres per hour.
What speed will maximize the flow rate on the road?

25. Determine the constant a such that the function $f(x) = x^2 + \dfrac{a}{x}$ has a) a local minimum at $x = 2$, b) a local minimum at $x = -3$, and c) show that the function cannot have a local maximum for any value of a.

26. A line passes through the point (3, 2) and intersects both the x-axis and the y-axis forming a triangular region in the first quadrant, bounded by the x-axis, the y-axis and the line. Find the equation of such a line that creates a triangle of minimum area.

27. A very important function in statistics is the equation for the *standard normal curve*
(mean = 0, standard deviation = 1) given by $f(x) = \dfrac{e^{-\frac{x^2}{2}}}{\sqrt{2\pi}}$.
a) Find the coordinates of any stationary points and of any inflexion points.
b) What happens when $x \to \infty$ and when $x \to -\infty$. Give the equation for any asymptotes.
c) Sketch a graph of $f(x)$ and indicate the location of any of the points found in part a).

28 Find the equation of both the tangent and normal to the curve $y = x \tan x$ at the point where $x = \dfrac{\pi}{4}$.

29 A window is in the shape of a rectangle with a semicircle on top. Find the dimensions of the rectangular section of the window when the perimeter of the entire window is 4 m and the area of the entire window is a maximum.

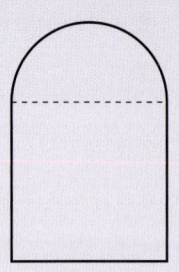

30 An object is moving along a line such that that its displacement, s metres, from a fixed point at any time t, in seconds, is given by $s(t) = 3\sin t + 4\cos t$ for the interval $0 \leq t \leq 6$.
   a) What is the object's initial displacement?
   b) At what time is the object's displacement a maximum, and what is the displacement?
   c) At what time is the object's displacement a minimum, and what is the displacement?
   d) Find an expression for the object's velocity and an expression for its acceleration?
   e) At what time is the object's velocity a maximum, and what is the velocity?
   f) In the interval $0 \leq t \leq 3$, at what time is the object's acceleration zero? Comment on the object's displacement and velocity at that moment when the acceleration is zero.

Practice questions

1 The diagram shows the graph of $y = f(x)$.

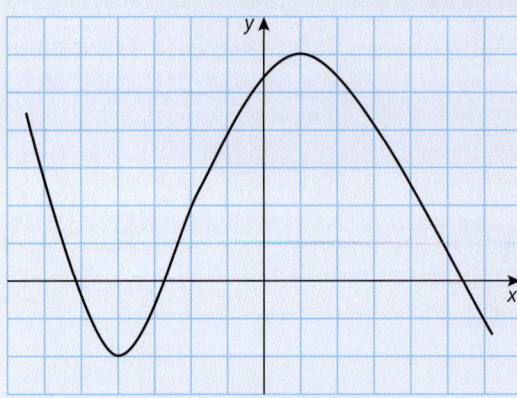

On the grid right sketch the graph of $y = f'(x)$.

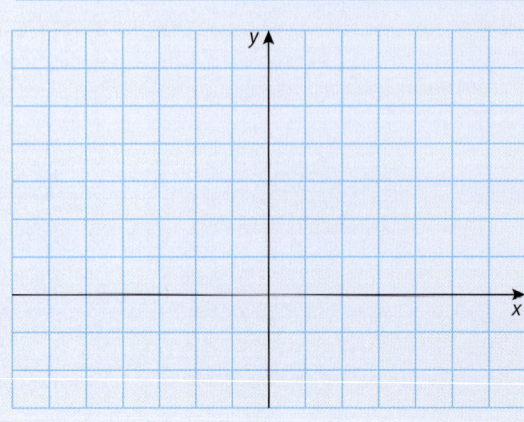

2 A curve has equation $y = -x(x+5)^2$.
   a) For this curve find:
   - **(i)** the x-intercepts
   - **(ii)** the exact coordinates of the maximum point
   - **(iii)** the exact coordinates of the point of inflexion.

   b) Use your answers to part **a)** to sketch a graph of the curve for $-5 \leq x \leq 0$, clearly indicating the features you have found in part **a)**.

3 Find the coordinates of the point on the graph of $y = 3x^2 + 2x$ at which the tangent is parallel to the line $y = 4x$.

4 Find the equation of the line tangent to the curve of $y = \sin(3x + 1)$ at the point $\left(-\frac{1}{3}, 0\right)$.

5 The diagram right shows part of the graph of the function $f: x \mapsto -x^3 - 2x^2 + 8x$.

The graph intersects the x-axis at $(-4, 0)$, $(0, 0)$ and $(2, 0)$. There is a minimum point at C and a maximum point at D.

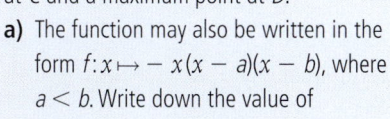

a) The function may also be written in the form $f: x \mapsto -x(x-a)(x-b)$, where $a < b$. Write down the value of
   - **(i)** a
   - **(ii)** b.

b) Find
   - **(i)** $f'(x)$
   - **(ii)** the exact values of x at which $f'(x) = 0$
   - **(iii)** the value of the function at D.

c) **(i)** Find the equation of the tangent to the graph of f at $(0, 0)$.
   (ii) This tangent cuts the graph of f at another point. Give the x-coordinate of this point.

6 In a controlled experiment, a tennis ball is dropped from the uppermost observation deck (447 metres high) of the CN Tower in Toronto. The tennis ball's velocity is given by
$$v(t) = 66 - 66e^{-0.15t}$$
where v is in metres per second and t is in seconds.

a) Find the value of v when
   - **(i)** $t = 0$
   - **(ii)** $t = 10$.

b) **(i)** Find an expression for the acceleration, a, as a function of t.
   (ii) What is the value of a when $t = 0$?

c) **(i)** As t becomes large, what value does v approach?
   (ii) As t becomes large, what value does a approach?
   (iii) Explain the relationship between the answers to parts **c)(i)** and **(ii)**.

7 Given the function $f(x) = x^3 + 7x^2 + 8x - 3$,
  a) identify any points as a relative maximum or minimum and find their exact coordinates
  b) find the exact coordinates of any inflexion point(s).

8 Consider the function $g(x) = 2 + \dfrac{1}{e^{3x}}$.
  a) **(i)** Find $g'(x)$.
      (ii) Explain briefly how this shows that $g(x)$ is a decreasing function for all values of x (i.e. that $g(x)$ always decreases in value as x increases).
  Let P be the point on the graph of g where $x = -\dfrac{1}{3}$.
  b) Find an expression in terms of e for
      (i) the y-coordinate of P
      (ii) the gradient of the tangent to the curve at P.
  c) Find the equation of the tangent to the curve at P, giving your answer in the form $y = mx + c$.

9 Consider the function f given by $f(x) = \dfrac{2x^2 - 13x + 20}{(x-1)^2}$, $x \neq 1$.
  a) Show that $f'(x) = \dfrac{9x - 27}{(x-1)^3}$, $x \neq 1$.
  The second derivative is given by $f''(x) = \dfrac{72 - 18x}{(x-1)^4}$, $x \neq 1$.
  b) Using values of $f'(x)$ and $f''(x)$, explain why a minimum must occur at $x = 3$.
  c) There is a point of inflexion on the graph of f. Write down the coordinates of this point.

10 Differentiate with respect to x:
  a) $\dfrac{1}{(2x+3)^2}$
  b) $e^{\sin 5x}$

11 The curve with equation $y = Ax + B + \dfrac{C}{x}$, $x \in \mathbb{R}$, $x \neq 0$, has a minimum at $P(1, 4)$ and a maximum at $Q(-1, 0)$. Find the value of each of the constants A, B and C.

12 a) Differentiate:
      (i) $\ln x$
      (ii) $\dfrac{1}{x}$
  b) The curve C has equation $y = \dfrac{\ln x}{x}$, $0 < x < \infty$.
      (i) Show that $\dfrac{dy}{dx} = \dfrac{1}{x^2}(1 - \ln x)$.
      (ii) Show that y has a maximum value of $\dfrac{1}{e}$ and justify that this is a maximum value.
  c) Assuming $y \to 0$ as $x \to \infty$, draw a sketch of the graph of the curve C. Find the two values of x for which $\dfrac{\ln x}{x} = \dfrac{1}{2}\ln 2$.

13 Differentiate with respect to x:
  a) $\dfrac{x^3}{x^2 + 1}$
  b) $e^x \sin 2x$

14 The curve $y = ax^3 - 2x^2 - x + 7$ has a gradient (slope) of 3 at the point where $x = 2$. Determine the value of a.

15 Let $y = h(x)$ be a function of x for $0 \leqslant x \leqslant 6$. The graph of h has an inflexion point at P and a maximum point at M.
Partial sketches of the curves of $h'(x)$ and $h''(x)$ are shown below.

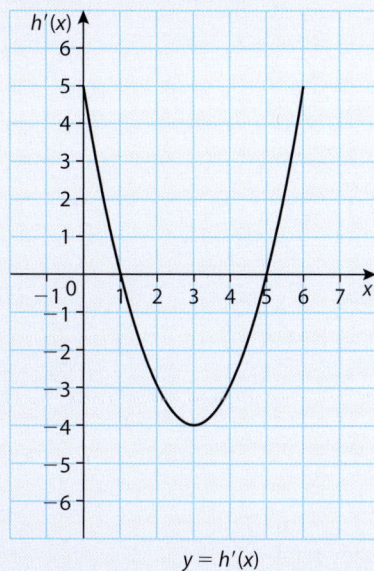

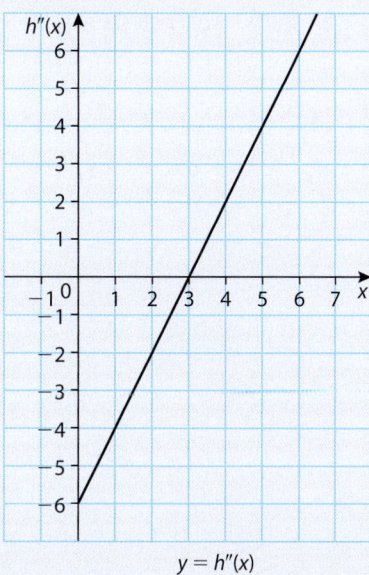

$y = h'(x)$

$y = h''(x)$

Use the above information to answer the following.
a) Write down the x-coordinate of P and justify your answer.
b) Write down the x-coordinate of M and justify your answer.
c) Given that $h(3) = 0$, sketch the graph of h. On the sketch, mark the points P and M.

16 Find the equation of the tangent to the curve $y = xe^x$ at the point on the curve where $x = 1$.

17 A cylinder is to be made with an exact volume of 128π cm³. What should be the height h and the radius r of the cylinder's base so that the cylinder's surface area is a minimum?

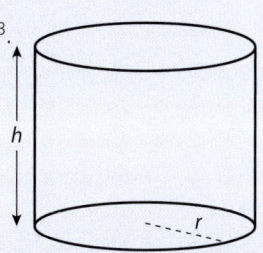

18 A rectangle has its base on the x-axis and its upper two vertices on the parabola $y = 12 - x^2$, as shown in the diagram. What is the largest area that the rectangle can have, and what are its dimensions (i.e. length and width)?

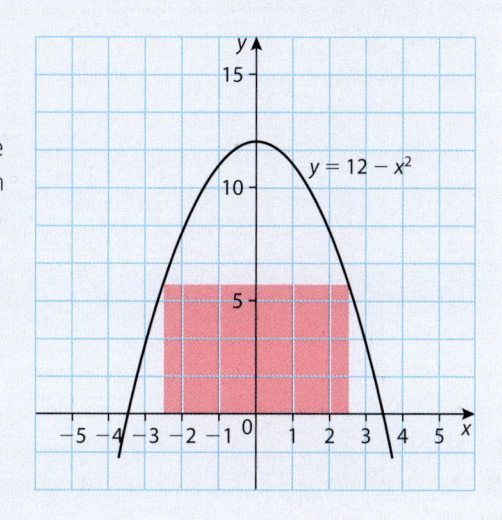

461

19 The figure below shows the graph of a function $y = f(x)$. At which one of the five points on the graph:
  a) are $f'(x)$ and $f''(x)$ both negative?
  b) is $f'(x)$ negative and $f''(x)$ positive?
  c) is $f'(x)$ positive and $f''(x)$ negative?

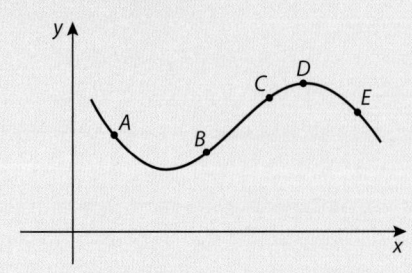

20 Find the equation of the normal to the curve with equation $y = \dfrac{2x - 1}{x + 2}$ at the point $(-3, 7)$.

14 Integral Calculus

Assessment statements
7.4 Indefinite integration as anti-differentiation.
   Indefinite integral of x^n, $\sin x$, $\cos x$, $1/x$ and e^x.
   The composites of any of these with the linear function $ax + b$.
7.5 Anti-differentiation with a boundary condition to determine the constant term.
   Definite integrals.
   Areas under curves (between the curve and the x-axis), areas between curves.
   Volumes of revolution.

Introduction

In Chapters 11 and 13 you learned about the process of differentiation. That is, given a function, how you can find its derivative. In this chapter, we will look at the reverse process. That is, given a function $f(x)$, how can we find a function $F(x)$ whose derivative is $f(x)$. This process is the opposite of differentiation and is therefore called **anti-differentiation**.

14.1 Anti-derivative

An **anti-derivative** of the function $f(x)$ is a function $F(x)$ such that
$$\frac{d}{dx}F(x) = F'(x) = f(x) \text{ wherever } f(x) \text{ is defined.}$$

For instance, let $f(x) = x^2$. It is not difficult to discover an anti-derivative of $f(x)$. Keep in mind that this is a power function. Since the power rule reduces the power of the function by 1, we examine the derivative of x^3:
$$\frac{d}{dx}(x^3) = 3x^2.$$

This derivative, however, is 3 times $f(x)$. To 'compensate' for the 'extra' 3, we have to multiply by $\frac{1}{3}$, so that the anti-derivative is now $\frac{1}{3}x^3$. Now,
$$\frac{d}{dx}\left(\frac{1}{3}x^3\right) = x^2$$

And, therefore, $\frac{1}{3}x^3$ is an anti-derivative of x^2.

Table 14.1 shows some examples of functions, each paired with one of its anti-derivatives.

Table 14.1

Function $f(x)$	Anti-derivative $F(x)$
1	x
x	$\frac{x^2}{2}$
$3x^2$	x^3
x^4	$\frac{x^5}{5}$
$\cos x$	$\sin x$
$\cos 2x$	$\frac{1}{2}\sin 2x$
e^x	e^x
$\sin x$	$-\cos x$
$2x$	x^2

Integral Calculus

The diagrams below show the relationship between the derivative and the integral as opposite operations.

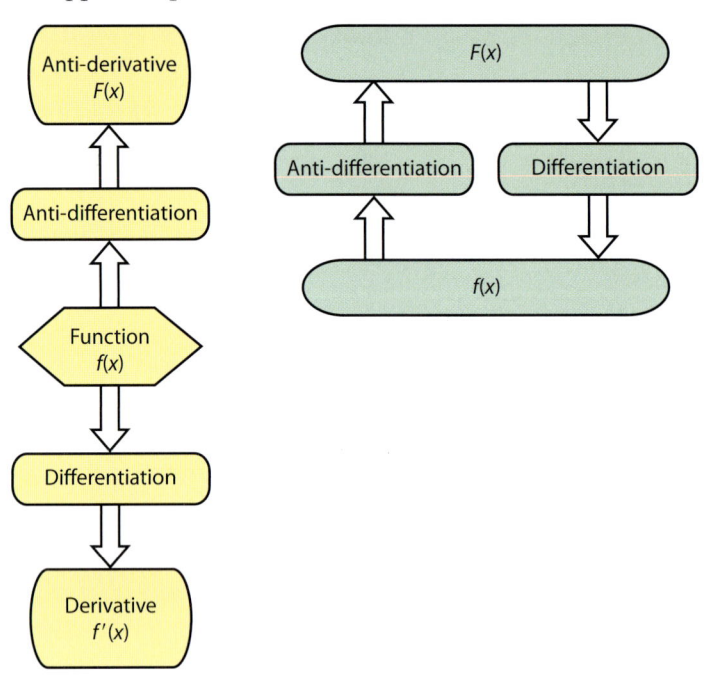

Example 1

Given the function $f(x) = 3x^2$, find an anti-derivative of $f(x)$.

Solution

$F_1(x) = x^3$ is such an anti-derivative because $\frac{d}{dx}(F_1(x)) = 3x^2$.

The following functions are also anti-derivatives because the derivative of each one of them is also $3x^2$.

$$H_1(x) = x^3 + 27, \; H_2(x) = x^3 - \pi, \text{ or } H_3(x) = x^3 + \sqrt{5}$$

Indeed, $F(x) = x^3 + c$ is an anti-derivative of $f(x) = 3x^2$ for any choice of the constant c.

This is so simply because

$$(F(x) + c)' = F'(x) + c' = F'(x) + 0 = f(x)!$$

Thus, we can say that any single function $f(x)$ has many anti-derivatives, whereas a function can have only one derivative.

> If $F(x)$ is an anti-derivative of $f(x)$, then so is $F(x) + c$ for any choice of the constant c.

Stated slightly differently, this observation says:

> If $F(x)$ is an anti-derivative of $f(x)$ over a certain interval I, then every anti-derivative of $f(x)$ on I is of the form $F(x) + c$.

This statement is an indirect conclusion of one of the results of the mean value theorem.

> Two functions with the same derivative on an interval differ only by a constant on that interval.

We will state the mean value theorem here in order to establish the general rule for anti-derivatives.

> **Mean value theorem**
> A function $H(x)$, continuous over an interval $[a, b]$ and differentiable over $]a, b[$, satisfies
> $$H(b) - H(a) = (b - a)H'(c) \text{ for some } c \in]a, b[$$

Let $F(x)$ and $G(x)$ be any anti-derivatives of $f(x)$, i.e. $F'(x) = G'(x)$.
Take $H(x) = F(x) - G(x)$ and any two numbers x_1 and x_2 in the interval $[a, b]$ such that $x_1 < x_2$, then

$$H(x_2) - H(x_1) = (x_2 - x_1)H'(c) = (x_2 - x_1) \cdot (F'(c) - G'(c))$$
$$= (x_2 - x_1) \cdot 0 = 0 \Rightarrow H(x_1) = H(x_2)$$

which means that $H(x)$ is a constant function.
Hence, $H(x) = F(x) - G(x) =$ constant. That is, any two anti-derivatives of a function differ by a constant.

Notation:
The notation
$$\int f(x)\,dx = F(x) + c \qquad (1)$$
where c is an arbitrary constant, means that $F(x) + c$ is an anti-derivative of $f(x)$.

Equivalently, $F(x)$ satisfies the condition that
$$\frac{d}{dx}(F(x)) = F'(x) = f(x) \qquad (2)$$
for all x in the domain of $f(x)$.

It is important to note that (1) and (2) are just different notations to express the same fact. For example,

$$\int x^2 dx = \tfrac{1}{3}x^3 + c \text{ is equivalent to } \frac{d}{dx}\left(\tfrac{1}{3}x^3\right) = x^2$$

> ⓘ Note that if we differentiate an anti-derivative of $f(x)$, we obtain $f(x)$ back again.
> Thus, $\frac{d}{dx}(\int f(x)dx) = f(x)$.
>
> The expression $\int f(x)dx$ is called an **indefinite integral** of $f(x)$. The function $f(x)$ is called the **integrand** and the constant c is called the **constant of integration**.
>
> The integral symbol $\int$ is made like an elongated capital S. It is, in fact, a medieval S, used by Leibniz as an abbreviation for the Latin word *summa*.
>
> We think of the combination $\int [\ ]dx$ as a single symbol; we fill in the 'blank' with the formula of the function whose anti-derivative we seek. We may regard the differential dx as specifying the independent variable x both in the function $f(x)$ and in its anti-derivatives.
>
> If an independent variable other than x is used, say t, the notation must be adjusted appropriately.
> Thus, $\frac{d}{dt}(\int f(t)dt) = f(t)$ and $\int f(t)dt = F(t) + c$ are equivalent statements.

14 Integral Calculus

Derivative formula	Equivalent integration formula
$\frac{d}{dx}(x^3) = 3x^2$	$\int 3x^2 dx = x^3 + c$
$\frac{d}{dx}(\sqrt{x}) = \frac{1}{2\sqrt{x}}$	$\int \frac{1}{2\sqrt{x}} dx = \sqrt{x} + c$
$\frac{d}{dt}(\tan t) = \sec^2 t$	$\int \sec^2 t \, dt = \tan t + c$
$\frac{d}{dv}\left(v^{\frac{3}{2}}\right) = \frac{3}{2} v^{\frac{1}{2}}$	$\int \frac{3}{2} v^{\frac{1}{2}} dv = v^{\frac{3}{2}} + c$

Note: The integral sign and differential serve as delimiters, adjoining the integrand on the left and right, respectively. In particular, we do not write $\int dx f(x)$ when we mean $\int f(x) \, dx$.

Basic integration formulae

Integration is essentially educated guesswork – given the derivative $f(x)$ of a function $F(x)$, we try to guess what the function $F(x)$ is. However, many basic integration formulae can be obtained directly from their companion differentiation formulae. Some of the most important are given in Table 14.2.

Table 14.2

	Differentiation formula	Integration formula
1	$\frac{d}{dx}(x) = 1$	$\int dx = x + c$
2	$\frac{d}{dx}(x^{n+1}) = (n+1)x^n, n \neq -1$	$\int x^n dx = \frac{x^{n+1}}{n+1} + c, n \neq -1$
3	$\frac{d}{dx}(\sin x) = \cos x$	$\int \cos x \, dx = \sin x + c$
4	$\frac{d}{dx}(\cos x) = -\sin x$	$\int \sin v \, dv = -\cos v + c$
5	$\frac{d}{dt}(\tan t) = \sec^2 t$	$\int \sec^2 t \, dt = \tan t + c$
6	$\frac{d}{dv}(e^v) = e^v$	$\int e^v dv = e^v + c$
7	$\frac{d}{dx}(\ln|x|) = \frac{1}{x}$	$\int \frac{1}{x} dx = \ln|x| + c$

The last formula (7) is a special case of the 'power' rule formula (2), but needs some modification.

If we are given the task to integrate $\frac{1}{x}$, we may attempt to do it using the power rule:

$$\int \frac{1}{x} dx = \int x^{-1} dx = \frac{1}{(-1)+1} x^{(-1)+1} + c = \frac{1}{0} x^0 + c, \text{ which is undefined.}$$

However, the solution is clearly found by observing what you learned in Chapter 13.

In Section 13.1 you learned that

$$\frac{d}{dx}(\ln x) = \frac{1}{x}, x > 0$$

This implies
$$\int \frac{1}{x} dx = \ln x + c, x > 0$$
However, the function $\frac{1}{x}$ is differentiable for $x < 0$ too. So, we must be able to find its integral.

The solution lies in the chain rule!

If $x < 0$, we can write $x = -u$ where $u > 0$. Then $dx = -du$, and
$$\int \frac{1}{x} dx = \int \frac{1}{-u}(-du) = \int \frac{1}{u} du = \ln u + c, u > 0.$$
But $u = -x$, therefore when $x < 0$
$$\int \frac{1}{x} dx = \ln u + c = \ln(-x) + c,$$ and, combining the two results, we have
$$\int \frac{1}{x} dx = \ln|x| + c, x \neq 0.$$

Suppose that $f(x)$ and $g(x)$ are differentiable functions and k is a constant, then:
1. A constant factor can be moved through an integral sign, i.e.
$$\int kf(x)dx = k\int f(x)dx$$
2. An anti-derivative of a sum (difference) is the sum (difference) of the anti-derivatives, i.e.
$$\int (f(x) + g(x))dx = \int f(x)dx + \int g(x)dx, \text{ or}$$
$$\int (f(x) - g(x))dx = \int f(x)dx - \int g(x)dx$$

Example 2

Evaluate

a) $\int 3\cos x\, dx$
b) $\int (x^3 + x^2)dx$

Solution

a) $\int 3\cos x\, dx = 3\int \cos x\, dx = 3\sin x + c$

b) $\int (x^3 + x^2)dx = \int x^3\, dx + \int x^2\, dx = \frac{x^4}{4} + \frac{x^3}{3} + c$

Sometimes it is useful to rewrite the integrand in a different form before performing the integration.

Example 3

Evaluate

a) $\int \frac{t^3 - 3t^5}{t^5} dt$
b) $\int \frac{x + 5x^4}{x^2} dx$

Solution

a) $\int \frac{t^3 - 3t^5}{t^5} dt = \int \frac{t^3}{t^5} dt - \int \frac{3t^5}{t^5} dt = \int t^{-2}\, dt - \int 3\, dt = \frac{t^{-1}}{-1} - 3t + c$

$\qquad = \frac{-1}{t} - 3t + c$

b) $\int \frac{x + 5x^4}{x^2} dx = \int \frac{x}{x^2} dx + \int \frac{5x^4}{x^2} dx = \int \frac{1}{x} dx + \int 5x^2\, dx = \ln|x| + 5\cdot\frac{x^3}{3} + c$

Integration by simple substitution

In this section, we will study a technique called substitution that can often be used to transform complicated integration problems into simpler ones.

The method of substitution depends on our understanding of the chain rule as well as the use of variables in integration. Two facts to recall:

1. When we find an anti-derivative, we established earlier that the use of x is arbitrary. We can use any other variable as you have seen in several exercises and examples so far.

   So, $\int f(u)\,du = F(x) + c$, where u is a 'dummy' variable in the sense that it can be replaced by any other variable.

2. The chain rule enables us to say
$$\frac{d}{dx}(F(u(x))) = F'(u(x)) \cdot u'(x)$$
   This can be written in integral form as
$$\int F'(u(x)) \cdot u'(x)\,dx = F(u(x)) + c$$
   or, equivalently, since $F(x)$ is an anti-derivative of $f(x)$,
$$\int f(u(x)) \cdot u'(x)\,dx = F(u(x)) + c$$

For our purposes, it will be useful and simpler to let $u(x) = u$ and to write $\frac{du}{dx} = u'(x)$ in its 'differential' form $du = u'(x)\,dx$, or, simply, $du = u'dx$.

With this notation, the integral can now be written as
$$\int f(u(x)) \cdot u'(x)\,dx = \int f(u)\,du = F(x) + c$$

The following example explains how the method works.

Example 4

Evaluate
a) $\int (x^3 + 2)^{10} \cdot 3x^2\,dx$
b) $\int \tan x\,dx$
c) $\int \cos 5x\,dx$
d) $\int \cos x^2 \cdot x\,dx$
e) $\int e^{3x+1}\,dx$

Solution

a) To integrate this function, it is simplest to make the following substitution.
   Let $u = x^3 + 2$, and so $du = 3x^2 dx$. Now the integral can be written as
$$\int (x^3 + 2)^{10} \cdot 3x^2\,dx = \int u^{10}\,du = \frac{u^{11}}{11} + c = \frac{(x^3+2)^{11}}{11} + c$$

b) This integrand has to be rewritten first and then we make the substitution.
$$\int \tan x\,dx = \int \frac{\sin x}{\cos x}\,dx = \int \frac{1}{\cos x} \cdot \sin x\,dx$$
   We now let $u = \cos x \Rightarrow du = -\sin x\,dx$, and
$$\int \tan x\,dx = \int \frac{1}{\cos x} \cdot \sin x\,dx = \int \frac{1}{u} \cdot (-du) = -\int \frac{1}{u}\,du = -\ln|u| + c$$

This last result can be then expressed in one of two ways:
$$\int \tan x \, dx = -\ln|\cos x| + c, \text{ or}$$
$$\int \tan x \, dx = -\ln|\cos x| + c = \ln|(\cos x)^{-1}| + c$$
$$= \ln\left|\frac{1}{(\cos x)}\right| + c = \ln|\sec x| + c$$

c) We let $u = 5x$, then $du = 5dx \Rightarrow dx = \frac{1}{5} du$, and so
$$\int \cos 5x \, dx = \int \cos u \cdot \tfrac{1}{5} du = \tfrac{1}{5} \int \cos u \, du = \tfrac{1}{5} \sin u + c$$
$$= \tfrac{1}{5} \sin 5x + c$$

Another method can be applied here:

The substitution $u = 5x$ requires $du = 5dx$. As there is no factor of 5 in the integrand, and since 5 is a constant, we can multiply and divide by 5 so that we group the 5 and dx to form the du required by the substitution:
$$\int \cos 5x \, dx = \tfrac{1}{5}\int \cos x \cdot \mathbf{5dx} = \tfrac{1}{5}\int \cos u \, \mathbf{du} = \tfrac{1}{5} \sin u + c$$
$$= \tfrac{1}{5} \sin 5x + c$$

d) By letting $u = x^2$, $du = 2x \, dx$ and so
$$\int \cos x^2 \cdot x \, dx = \tfrac{1}{2}\int \cos x^2 \cdot 2x \, dx = \tfrac{1}{2}\int \cos u \, du = \tfrac{1}{2}\sin u + c$$
$$= \tfrac{1}{2}\sin x^2 + c$$

e) $\int e^{3x+1} \, dx = \tfrac{1}{3}\int e^{3x+1} \, \mathbf{3dx} = \tfrac{1}{3}\int e^u \, \mathbf{du} = \tfrac{1}{3}e^u + c = \tfrac{1}{3}e^{3x+1} + c$

> In integration, multiplying by a constant 'inside' the integral and 'compensating' for that with the reciprocal 'outside' the integral depends on theorem 1 (page 467). That is,
> $$\int kf(x)dx = k\int f(x)dx$$
> However, you *cannot* multiply with a variable. So, you cannot, say, for example,
> $$\int \cos x^2 \, dx = \frac{1}{2x} \int \cos x^2 \cdot 2x \, dx$$

Example 5

Some of the exercises in this example go beyond what is required in the present syllabus – these will be marked by an asterisk.

Evaluate each integral

a) $\int e^{-3x} \, dx$
b)* $\int \sin^2 x \cos x \, dx$
c) $\int 2 \sin(3x - 5) \, dx$
d) $\int e^{mx+n} \, dx$
e) $\int x\sqrt{x} \, dx$, and $F(1) = 2$

Solution

a) Let $u = -3x$, then $du = -3dx$, and
$$\int e^{-3x} \, dx = -\tfrac{1}{3}\int e^{-3x}(-3dx) = -\tfrac{1}{3}\int e^u \, du = -\tfrac{1}{3}e^u + c$$
$$= -\tfrac{1}{3}e^{-3x} + c$$

b) Let $u = \sin x$, then $du = \cos x \, dx$, and
$$\int \sin^2 x \cos x \, dx = \int u^2 \, du = \tfrac{1}{3}u^3 + c = \tfrac{1}{3}\sin^3 x + c$$

c) Let $u = 3x - 5$, then $du = 3dx$, and
$$\int 2\sin(3x-5) \, dx = 2 \cdot \tfrac{1}{3}\int \sin(3x-5) \mathbf{3dx} = \tfrac{2}{3}\int \sin u \, du$$
$$= -\tfrac{2}{3}\cos u + c = -\tfrac{2}{3}\cos(3x - 5) + c$$

14 Integral Calculus

d) Let $u = mx + n$, then $du = m\, dx$, and

$$\int e^{mx+n}\, dx = \frac{1}{m}\int e^{mx+n}\, m\, dx = \frac{1}{m}\int e^u\, du$$

$$= \frac{1}{m}e^u + c = \frac{1}{m}e^{mx+n} + c$$

e) $F(x) = \int x\sqrt{x}\, dx = \int x^{\frac{3}{2}}\, dx = \dfrac{x^{\frac{5}{2}}}{\left(\frac{5}{2}\right)} + c = \frac{2}{5}x^{\frac{5}{2}} + c$, but $F(1) = 2$

$F(1) = \frac{2}{5}1^{\frac{5}{2}} + c = \frac{2}{5} + c = 2 \Rightarrow c = \frac{8}{5}$

Therefore, $F(x) = \frac{2}{5}x^{\frac{5}{2}} + \frac{8}{5}$.

Exercise 14.1

In questions 1–15, find the most general anti-derivative of the function.

1. $f(x) = x + 2$
2. $f(t) = 3t^2 + 1$
3. $g(x) = \frac{1}{3} - \frac{2}{7}x^3$
4. $f(t) = (t-1)(2t+3)$
5. $g(u) = u^{\frac{2}{3}} - 4u^3$
6. $f(x) = 2\sqrt{x} - \frac{3}{2\sqrt{x}}$
7. $h(\theta) = 3\sin\theta + 4\cos\theta$
8. $f(t) = 3t^2 - 2\sin t$
9. $f(x) = \sqrt{x}(2x - 5)$
10. $g(\theta) = 3\cos\theta - 2\sec^2\theta$
11. $h(t) = e^{3t-1}$
12. $f(t) = \frac{2}{t}$
13. $h(u) = \dfrac{t}{3t^2 + 5}$
14. $h(\theta) = e^{\sin\theta}\cos\theta$
15. $f(x) = (3 + 2x)^2$

In questions 16–20, find f.

16. $f''(x) = 4x - 15x^2$
17. $f''(x) = 1 + 3x^2 - 4x^3;\ f'(0) = 2,\ f(1) = 2$
18. $f''(t) = 8t - \sin t$
19. $f'(x) = 12x^3 - 8x + 7,\ f(0) = 3$
20. $f'(\theta) = 2\cos\theta - \sin(2\theta)$

14.2 Area and definite integral

The main goal of this section is to introduce you to the following major problem of calculus.

The area problem: Given a function $f(x)$ that is continuous and non-negative on an interval $[a, b]$, find the area between the graph of $f(x)$ and the interval $[a, b]$ on the x-axis.

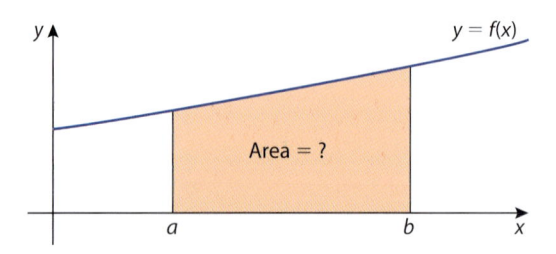

• **Hint:** This is only an expository treatment that explains to you how the definite integral is developed. You will not be required to reproduce this calculation yourself.

Figure 14.1

We divide the base interval $[a, b]$ into n equal sub-intervals, and over each sub-interval construct a rectangle that extends from the x-axis to any point on the curve $y = f(x)$ that is above the sub-interval. The particular point does not matter – it can be above the centre, above one endpoint, or any other point in the sub-interval. In Figure 14.1 it is above the centre.

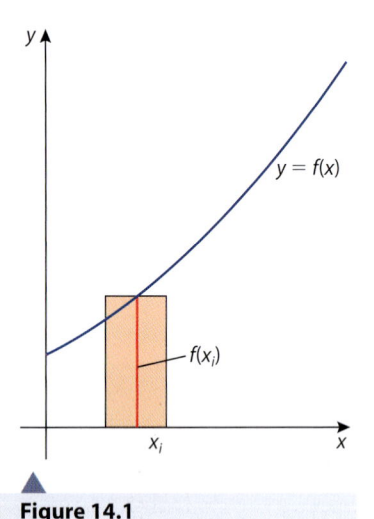

For each n, the total area of the rectangles can be viewed as an approximation to the exact area in question. Moreover, it is evident intuitively that as n increases, these approximations will get better and better and will eventually approach the exact area as a limit. See Figure 14.2.

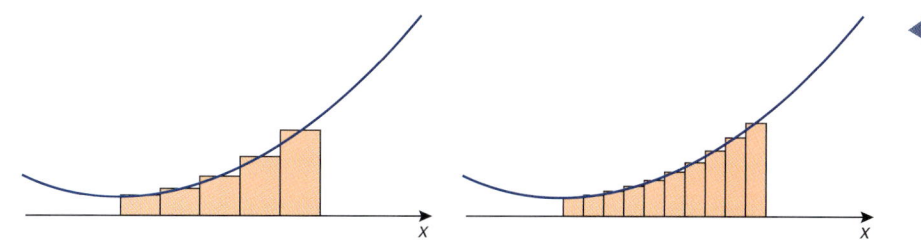

Figure 14.2

A traditional approach to this would be to study how the choice of where to erect the rectangular strip does not affect the approximation as the number of intervals increases. You can construct 'inscribed' rectangles, which, at the start, give you an underestimate of the area. On the other hand, you can construct 'circumscribed' rectangles that, at the start, overestimate the area. See Figure 14.3.

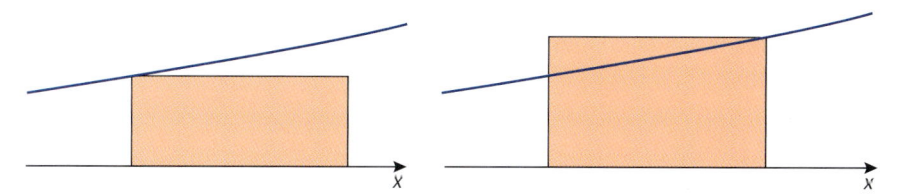

Figure 14.3

As the number of intervals increases, the difference between the overestimates and the underestimates will approach 0.

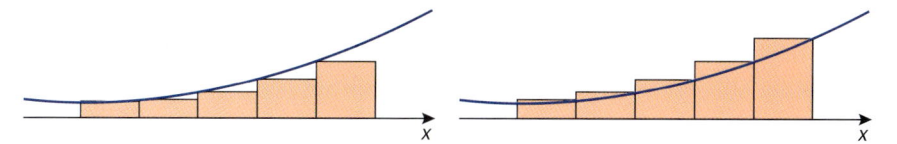

Figure 14.4

Figure 14.4 above shows n inscribed and subscribed rectangles and Figure 14.5 shows us the difference between the overestimates and the underestimates.

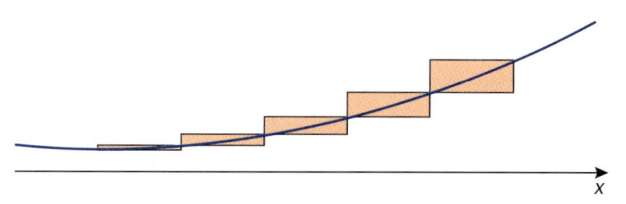

Figure 14.5

Figure 14.6 demonstrates that as the number n increases, the difference between the estimates will approach 0. Since we set up our rectangles by choosing a point inside the interval, the areas of the rectangles will lie between the overestimates and the underestimates, and hence, as the difference between the extremes approaches zero, the rectangles we constructed will give the area of the region required.

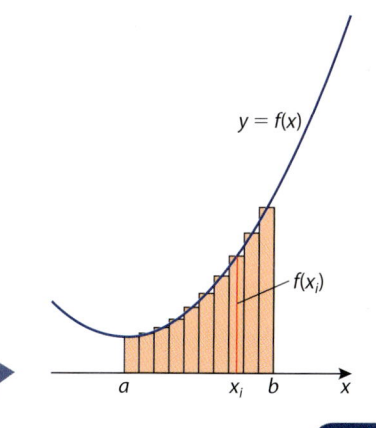

Figure 14.6

14 Integral Calculus

If we consider the width of each interval to be Δx, the area of any rectangle is given as

$$A_i = f(x_i^*)\Delta x$$

The total area of the rectangles so constructed is

$$A_n = \sum_{i=0}^{n} f(x_i^*)\Delta x$$

x_i^* is an arbitrary point within any sub-interval $[x_{i-1}, x_i]$, $x_0 = a$ and $x_n = b$.

In the case of a function $f(x)$ that has both positive and negative values on $[a, b]$, it is necessary to consider the *signs* of the areas in the following sense.

Figure 14.7

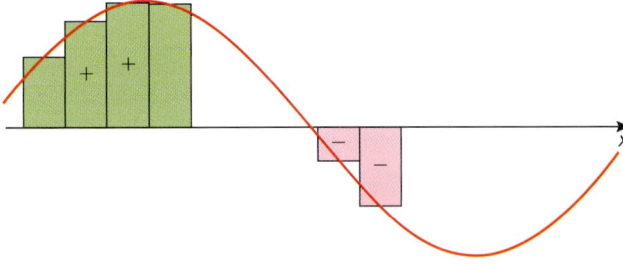

On each sub-interval, we have a rectangle with width Δx and height $f(x^*)$. If $f(x^*) > 0$, this rectangle is above the x-axis; if $f(x^*) < 0$, this rectangle is below the x-axis. We will consider the sum defined above as the sum of the signed areas of these rectangles. That means the total area on the interval is the sum of the areas above the x-axis minus the sum of the areas of the rectangles below the x-axis.

We are now ready to look at a 'loose' definition of the definite integral[1]:

> For a list of recommended resources about definite integrals, visit www.heinemann.co.uk/hotlinks and enter express code 4235P, then click on weblink 3.

If $f(x)$ is a continuous function defined for $a \leq x \leq b$, we divide the interval $[a, b]$ into n sub-intervals of equal width $\Delta x = (b - a)/n$. We let $x_0 = a$ and $x_n = b$ and we choose $x_1^*, x_2^*, \ldots, x_n^*$ in these sub-intervals, so that x_i^* lies in the ith sub-interval $[x_{i-1}, x_i]$. Then the definite integral of $f(x)$ from a to b is

$$\int_a^b f(x)\,dx = \lim_{n \to \infty} \sum_{i=1}^{n} f(x_i^*)\Delta x$$

In the notation $\int_a^b f(x)\,dx$, in addition to the known integrand and differential, a and b are called the limits of integration: a is the lower limit and b is the upper limit.

Note: Because we have assumed that $f(x)$ is continuous, it can be proved that the limit definition above always exists and gives the same value no matter how we choose the points x_i^*. If we take these points at the centre, at two-thirds the distance from the lower endpoint or at the upper endpoint, the value is the same. This is why we will state the definition of the integral from now on as

$$\int_a^b f(x)\,dx = \lim_{n \to \infty} \sum_{i=1}^{n} f(x_i)\Delta x$$

---

[1]For a rigorous treatment of the definition of definite integrals using Riemann sums, refer to university calculus books. Such a treatment is beyond the scope of the SL syllabus and this book.

Calling the area under the function an integral is no coincidence. To make the point, let us take the following example.

Example 6(I)

Find the area, $A(x)$, between the graph of the function $f(x) = 3$ and the interval $[-1, x]$, and find the derivative $A'(x)$ of this area function.

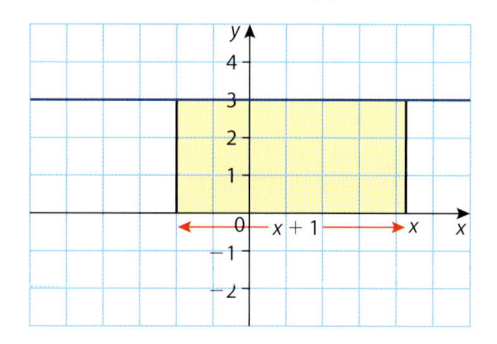

Solution

The area in question is

$$A(x) = 3(x - (-1)) = 3x + 3, \text{ and}$$
$$A'(x) = 3 = f(x)$$

Example 6(II)

Find the area, $A(x)$, between the graph of the function $f(x) = 3x + 2$ and the interval $[-2/3, x]$, and find the derivative $A'(x)$ of this area function.

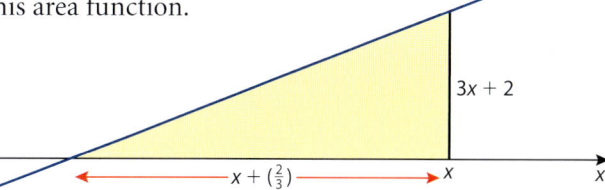

Solution

The area in question is

$$A(x) = \tfrac{1}{2}\left(x + \tfrac{2}{3}\right)(3x + 2) = \tfrac{1}{6}(3x + 2)^2, \text{ since this is the area of a triangle. Hence,}$$
$$A'(x) = \tfrac{1}{6} \times 2(3x + 2) \times 3 = 3x + 2 = f(x)$$

Example 6(III)

Find the area, $A(x)$, between the graph of the function $f(x) = x + 2$ and the interval $[-1, x]$, and find the derivative $A'(x)$ of this area function.

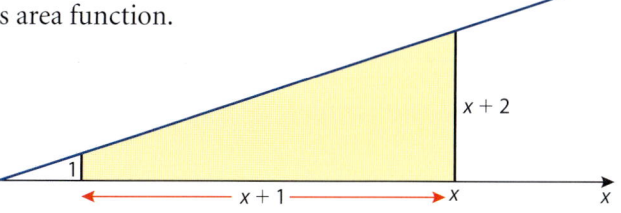

Solution

This is a trapezium, so the area is

$$A(x) = \tfrac{1}{2}(1 + (x + 2))(x + 1) = \tfrac{1}{2}(x^2 + 4x + 3), \text{ and}$$
$$A'(x) = \tfrac{1}{2} \times (2x + 4) = x + 2 = f(x)$$

Note that, in every case, $A'(x) = f(x)$.

14 Integral Calculus

The derivative of the area function $A(x)$ is the function whose graph forms the upper boundary of the region. It can be shown that this relation is true, not only for linear functions but for all continuous functions. Thus, to find the area function $A(x)$, we can look instead for a particular function whose derivative is $f(x)$. This is, of course, the anti-derivative of $f(x)$.

Properties of the definite integral

1. $\int_a^b f(x)\,dx = -\int_b^a f(x)\,dx$

   When we defined the definite integral $\int_a^b f(x)\,dx$, we implicitly assumed that $a < b$. When we reverse a and b, then Δx changes from $(b - a)/n$ to $(a - b)/n$. Therefore, the result above follows.

2. $\int_a^a f(x)\,dx = 0$

   When $a = b$, then $\Delta x = 0$ and so the result above follows.

The following are a few straightforward properties:

3. $\int_a^b c\,dx = c(b - a)$

4. $\int_a^b [f(x) \pm g(x)]\,dx = \int_a^b f(x)\,dx \pm \int_a^b g(x)\,dx$

5. $\int_a^b cf(x)\,dx = c\int_a^b f(x)\,dx$, where c is any constant

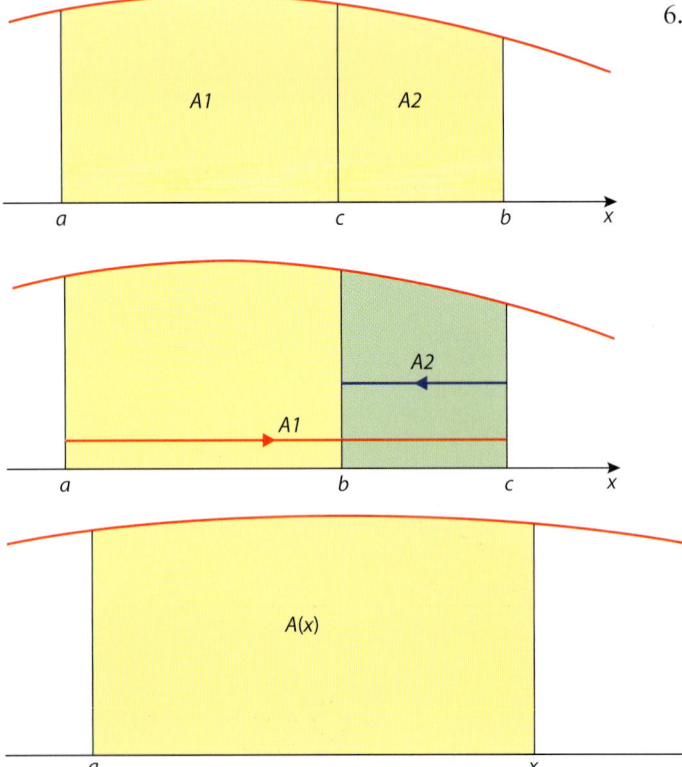

6. $\int_a^b f(x)\,dx = \int_a^c f(x)\,dx + \int_c^b f(x)\,dx$

Property 6 can be demonstrated with a diagram (Figure 14.8) where the area from a to b is the sum of the two areas, i.e. $A(x) = A1 + A2$. Additionally, even if $c > b$ the relationship holds because the area from c to b in this case will be negative.

Finally, the **fundamental theorem of calculus**:

Intuitively, as we have seen above, we define the area function as

$$A(x) = \int_a^x f(t)\,dt, \text{ that is, } A'(x) = f(x)$$

This definition says that $A(x)$ is an anti-derivative of $f(x)$. Now, suppose that $F(x)$ is another anti-derivative of $f(x)$. In Section 14.1, we proved that

$$A(x) = F(x) + c$$

because two anti-derivatives of the same function differ by a constant.

Figure 14.8

474

Now, $A(x) = \int_a^x f(t)dt$ implies that

$$A(a) = \int_a^a f(t)dt = 0, \text{ and}$$

$$A(b) = \int_a^b f(t)dt, \text{ so it follows that}$$

$$\int_a^b f(t)dt = A(b) - 0 = A(b) - A(a), \text{ but}$$

$$A(b) = F(b) + c, \text{ and } A(a) = F(a) + c, \text{ and therefore:}$$

Fundamental theorem of calculus

$$\int_a^b f(t)dt = A(b) - A(a) = (F(b) + c) - (F(a) + c) = F(b) - F(a)$$

The fundamental theorem is also referred to as the **evaluation theorem**. Also, since we know that $F'(x)$ is the rate of change in $F(x)$ with respect to x and that $F(b) - F(a)$ is the change in y when x changes from a to b, we can reformulate the theorem in words.

The integral of a rate of change is the **total change**

$$\int_a^b F'(x)dx = F(b) - F(a)$$

Here are a few instances where this applies:

1. If $V'(t)$ is the rate at which a liquid flows into or out of a container at time t, then

$$\int_{t_1}^{t_2} V'(t)dt = V(t_2) - V(t_1)$$

   is the change in the amount of liquid in the container between time t_1 and t_2.

2. If the rate of growth of a population is $n'(t)$, then

$$\int_{t_1}^{t_2} n'(t)dt = n(t_2) - n(t_1)$$

   is the increase (decrease!) in population during the time period from t_1 to t_2.

3. Displacement situations are described separately later in the chapter.

This theorem has many other applications in calculus and several other fields. It is a very powerful tool to deal with problems of area, volume and work among other applications. In this book, we will apply it to finding areas between functions and volumes of revolution as well displacement problems.

Notation:

We will use the following notation:

$$\int_a^b f(t)\,dt = F(x)\Big|_a^b = F(b) - F(a)$$

Integral Calculus

Example 7

a) Evaluate the integral $\int_{-1}^{3} x^5 \, dx$

b) Evaluate the integral $\int_{0}^{4} \sqrt{x} \, dx$

c) Evaluate the integral $\int_{\pi}^{2\pi} \cos \theta \, d\theta$

d) Evaluate the integral $\int_{1}^{2} \frac{4 + u^2}{u^3} \, du$

Solution

a) $\int_{-1}^{3} x^5 \, dx = \left[\frac{x^6}{6}\right]_{-1}^{3} = \frac{3^6}{6} - \frac{1}{6} = \frac{364}{3}$

b) $\int_{0}^{4} \sqrt{x} \, dx = \left[\frac{2}{3} x^{\frac{3}{2}}\right]_{0}^{4} = \frac{2}{3} 4^{\frac{3}{2}} - 0 = \frac{16}{3}$

c) $\int_{\pi}^{2\pi} \cos \theta \, d\theta = \left[\sin \theta\right]_{\pi}^{2\pi} = 0 - 0 = 0$

d) $\int_{1}^{2} \frac{4 + u^2}{u^3} \, du = \int_{1}^{2} \left(\frac{4}{u^3} + \frac{1}{u}\right) du = \left[4 \cdot \frac{u^{-2}}{-2} + \ln|u|\right]_{1}^{2}$

$= \left[-2u^{-2} + \ln u\right]_{1}^{2}$

$= (-2 \cdot 2^{-2} + \ln 2) - (-2 \cdot 1 + \ln 1) = -\frac{1}{2} + \ln 2 + 2$

$= \frac{3}{2} + \ln 2$

Exercise 14.2

In questions 1–15, evaluate the integral.

1 $\int_{-2}^{1} (3x^2 - 4x^3) \, dx$

2 $\int_{2}^{7} 8 \, dx$

3 $\int_{1}^{5} \frac{2}{t^3} \, dt$

4 $\int_{2}^{2} (\cos t - \tan t) \, dt$

5 $\int_{1}^{7} \frac{2x^2 - 3x + 5}{\sqrt{x}} \, dx$

6 $\int_{0}^{\pi} \cos \theta \, d\theta$

7 $\int_{0}^{\pi} \sin \theta \, d\theta$

8 $\int_{3}^{1} (5x^4 + 3x^2) \, dx$

9 $\int_{1}^{3} \frac{u^5 + 2}{u^2} \, du$

10 $\int_{1}^{e} \frac{2 \, dx}{x}$

11 $\int_{1}^{3} \frac{2x}{x^2 + 2} \, dx$

12 $\int_{1}^{3} (2 - \sqrt{x})^2 \, dx$

13 $\int_{0}^{\frac{\pi}{4}} 3 \sec^2 \theta \, d\theta$

14 $\int_{0}^{1} (8x^7 + \sqrt{\pi}) \, dx$

15 a) $\int_{0}^{2} |3x| \, dx$ b) $\int_{-2}^{0} |3x| \, dx$ c) $\int_{-2}^{2} |3x| \, dx$

14.3 Areas

We have seen how the area between a curve, defined by $y = f(x)$, and the x-axis can be computed by the integral $\int_a^b f(x)\,dx$ on an interval $[a, b]$, where $f(x) \geq 0$. In this section, we shall find that integration can be used to find the area of more general regions between curves.

Areas between curves of functions of the form $y = f(x)$ and the x-axis

If the function $y = f(x)$ is always above the x-axis, finding the area is a straightforward computation of the integral $\int_a^b f(x)\,dx$.

Example 8

Find the area under the curve $f(x) = x^3 - x + 1$ and the x-axis over the interval $[-1, 2]$.

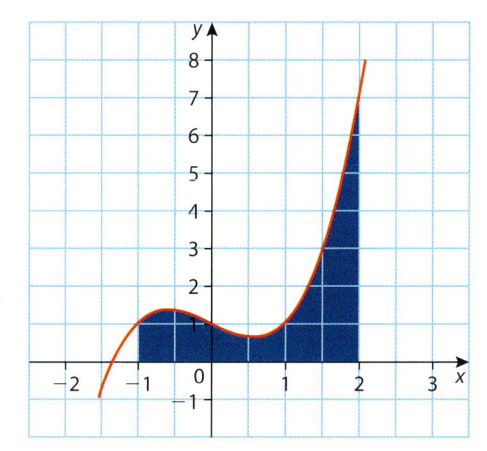

Solution

This area is simply

$$\int_{-1}^{2} (x^3 - x + 1)\,dx = \left[\frac{x^4}{4} - \frac{x^2}{2} + x\right]_{-1}^{2}$$

$$= (4 - 2 + 2) - \left(\frac{1}{4} - \frac{1}{2} - 1\right) = 5\frac{1}{4}$$

Using your GDC, this is done by simply choosing the 'MATH' menu, then the 'fnInt' menu item.

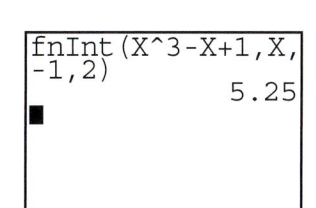

Or, you can type in your function and then go to the 'CALC' menu, where you choose '$\int f(x)\,dx$' and type in your integration limits. Here is what you see.

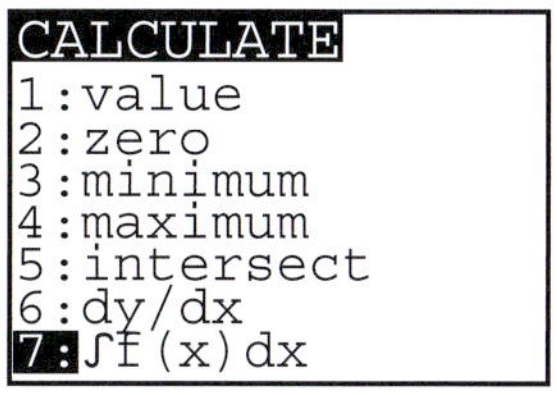

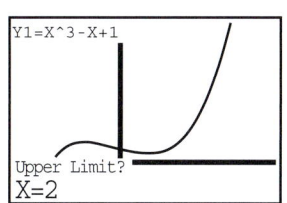

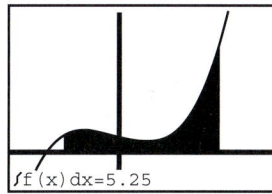

In some cases, you will have to adjust how you work. This is the case when the graph intersects the *x*-axis. Since you are interested in the area bounded by the curve and the interval [*a*, *b*] on the *x*-axis, you do not want the 'signed' areas to cancel each other. This is why you have to split the process into different sub-intervals where you take the absolute values of the areas found and add them.

Example 9

Find the area under the curve $f(x) = x^3 - x - 1$ and the *x*-axis over the interval $[-1, 2]$.

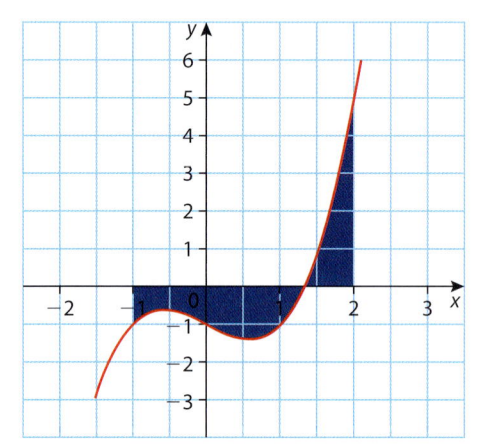

Solution

As you see from the diagram, a part of the graph is below the *x*-axis, and its area will be negative. If you try to integrate this function without paying attention to the intersection with the *x*-axis, this is what you get:

$$\int_{-1}^{2} (x^3 - x - 1)\, dx = \left[\frac{x^4}{4} - \frac{x^2}{2} - x\right]_{-1}^{2}$$

$$= (4 - 2 - 2) - \left(\frac{1}{4} - \frac{1}{2} + 1\right) = -\frac{3}{4}$$

This integration has to be split before we start. However, this is a function where you cannot find the intersection point. So, we either use our GDC to find the intersection, or we just take the absolute values of the different parts of the region. This is done by integrating the absolute value of the function:

$$\text{Area} = \int_{a}^{b} |f(x)|\, dx$$

Hence, area $= \int_{-1}^{2} |(x^3 - x - 1)|\, dx$

As we said earlier, this is not easy to find given the difficulty with the *x*-intercept. It is best if we make use of a GDC.

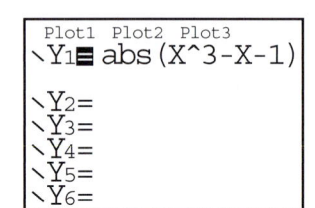

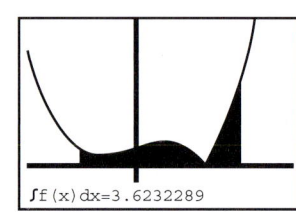

Or, using 'fnInt' directly:

```
fnInt(Y₁,X,-1,2)
        3.614515798
```

The difference between them is that the latter is more of a rough approximation than the first.

Example 10

Find the area enclosed by the graph of the function $f(x) = x^3 - 4x^2 + x + 6$ and the x-axis.

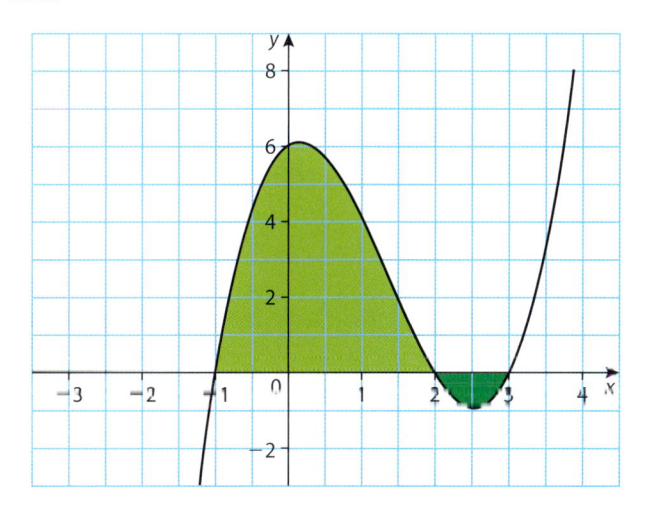

Solution

This function intersects the x-axis at three points where $x = -1, 2$ and 3. To find the area, we split it into two and then add the absolute values:

$$\text{Area} = \int_{-1}^{3} |f(x)|\, dx = \int_{-1}^{2} f(x)\, dx + \int_{2}^{3} (-f(x))\, dx$$

$$= \int_{-1}^{2} (x^3 - 4x^2 + x + 6)\, dx + \int_{2}^{3} (-x^3 + 4x^2 - x - 6)\, dx$$

$$= \left[\frac{x^4}{4} - \frac{4x^3}{3} + \frac{x^2}{2} + 6x\right]_{-1}^{2} + \left[-\frac{x^4}{4} + \frac{4x^3}{3} - \frac{x^2}{2} - 6x\right]_{2}^{3}$$

$$= \frac{45}{4} + \frac{7}{12} = \frac{71}{6}$$

Area between curves

In some practical problems, you may have to compute the area between two curves. Suppose $f(x)$ and $g(x)$ are functions such that $f(x) \geqslant g(x)$ on the interval $[a, b]$, as shown in the diagram. Note that we do not insist that both functions are non-negative, but we begin by showing that case for demonstration purposes.

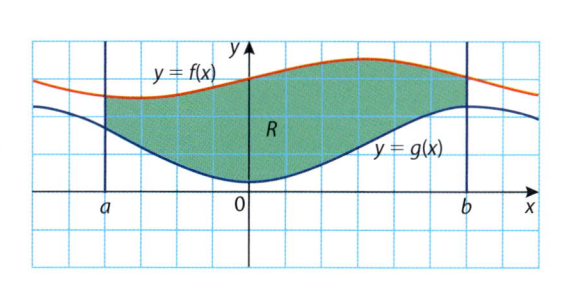

14 Integral Calculus

To find the area of the region R between the curves from $x = a$ to $x = b$, we subtract the area between the lower curve $g(x)$ and the x-axis from the area between the upper curve $f(x)$ and the x-axis; that is,

$$\text{Area of } R = \int_a^b f(x)\,dx - \int_a^b g(x)\,dx = \int_a^b [f(x) - g(x)]\,dx$$

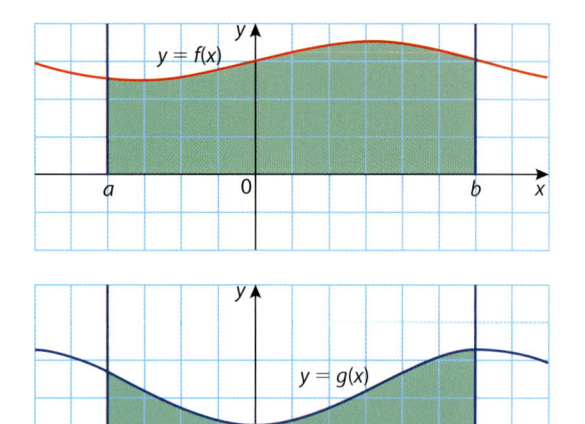

The fact just mentioned applies to all functions, not only positive functions. These facts are used to define the area between curves.

If $f(x)$ and $g(x)$ are functions such that $f(x) \geq g(x)$ on the interval $[a, b]$, then the area between the two curves is given by

$$A = \int_a^b [f(x) - g(x)]\,dx$$

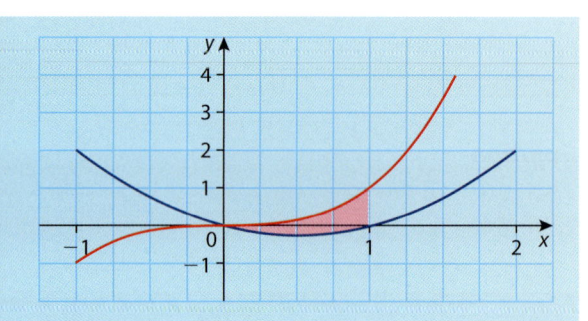

Example 11

Find the area of the region between the curves $y = x^3$ and $y = x^2 - x$ on the interval $[0, 1]$. (See diagram above.)

Solution

$y = x^3$ appears to be higher than $y = x^2 - x$ with one intersection at $x = 0$. Thus, the required area is

$$A = \int_0^1 [x^3 - (x^2 - x)]\,dx = \left[\frac{x^4}{4} - \frac{x^3}{3} + \frac{x^2}{2}\right]_0^1 = \frac{5}{12}$$

In order to take all cases into consideration, we will present here another case where you must be very careful of how you calculate the area. This is the case where the two functions in question intersect at more than one point. We will clarify this with an example.

Example 12

Find the area of the region bounded by the curves $y = x^3 + 2x^2$ and $y = x^2 + 2x$.

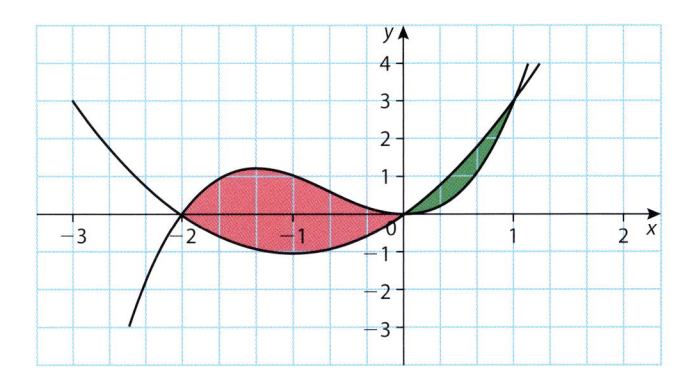

Solution

The two curves intersect when

$$x^3 + 2x^2 = x^2 + 2x \Rightarrow x^3 + x^2 - 2x = 0 \Rightarrow x(x+2)(x-1) = 0,$$

i.e. when $x = -2, 0$ or 1.

The area is equal to

$$A = \int_{-2}^{0} [(x^3 + 2x^2) - (x^2 + 2x)] \, dx + \int_{0}^{1} [(x^2 + 2x) - (x^3 + 2x^2)] \, dx$$

$$= \int_{-2}^{0} [x^3 + x^2 - 2x] \, dx + \int_{0}^{1} [-x^2 + 2x - x^3] \, dx$$

$$= \left[\frac{x^4}{4} + \frac{x^3}{3} - x^2\right]_{-2}^{0} + \left[-\frac{x^4}{4} - \frac{x^3}{3} + x^2\right]_{0}^{1}$$

$$= 0 - \left[\frac{16}{4} - \frac{8}{3} - 4\right] + \left[-\frac{1}{4} - \frac{1}{3} + 1\right] - 0 = \frac{37}{12}$$

This discussion leads us to stating the general expression you should use in evaluating areas between curves.

> If $f(x)$ and $g(x)$ are continuous functions on the interval $[a, b]$, the area between the two curves is given by
>
> $$A = \int_{a}^{b} |f(x) - g(x)| \, dx$$

The above computation can be done with your GDC as follows:

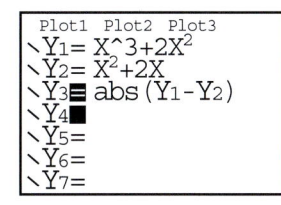

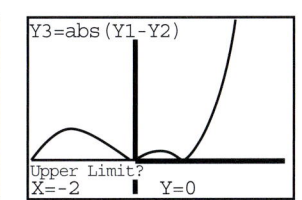

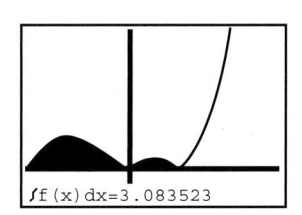

14 Integral Calculus

Exercise 14.3

Sketch the region whose area you are asked for, and then compute the required area. In each question, find the area of the region bounded by the given curves.

1. $y = x + 1, y = 7 - x^2$
2. $y = \cos x, y = x - \frac{\pi}{2}, x = -\pi$
3. $y = 2x, y = x^2 - 2$
4. $y = x^3, y = x^2 - 2, x = 1$
5. $y = x^6, y = x^2$
6. $y = 5x - x^2, y = x^2$
7. $y = 2x - x^3, y = x - x^2$
8. $y = \sin x, y = 2 - \sin x$ (one period)
9. $y = \frac{x}{2}, y = \sqrt{x}, x = 9$
10. $y = \frac{x^4}{10}, y = 3x - x^3$
11. $y = \frac{1}{x}, y = \frac{1}{x^3}, x = 8$
12. $y = 2\sin x, y = \sqrt{3}\tan x, -\frac{\pi}{4} \leq x \leq \frac{\pi}{4}$

14.4 Volumes with integrals

• **Hint:** This is an introductory section that will not be examined. It is only used to give you an idea of why we use integrals to find volumes.

Recall that the underlying principle for finding the area of a plane region is to divide the region into thin strips, approximate the area of each strip by the area of a rectangle, and then add the approximations and take the limit of the sum to produce an integral for the area. The same strategy can be used to find the volume of a solid.

The idea is to divide the solid into thin slabs, approximate the volume of each slab, add the approximations and take the limit of the sum to produce an integral of the volume.

Given a solid whose volume is to be computed, we start by taking cross sections perpendicular to the x-axis as shown in Figure 14.9. Each slab will be approximated by a cylindrical solid whose volume will be equal to the product of its base times its height.

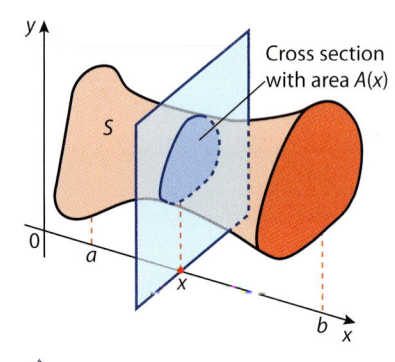

Figure 14.9

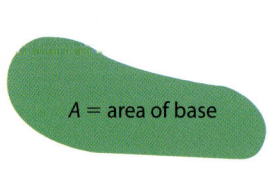

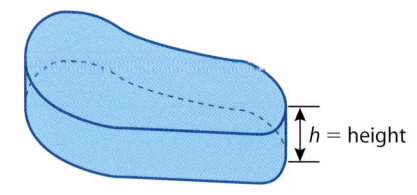

If we call the volume of the slab v_i and the area of its base $A(x)$, then

$$v_i = A(x_i) \cdot h = A(x_i) \cdot \Delta x_i$$

Using this approximation, the volume of the whole solid can be found by

$$V \approx \sum_{i=1}^{n} A(x_i) \Delta x_i$$

Taking the limit as n increases and the widths of the sub-intervals approach zero yields the definite integral

$$V = \lim_{n \to \infty} \sum_{i=1}^{n} A(x_i) \Delta x_i = \int_a^b A(x)\, dx$$

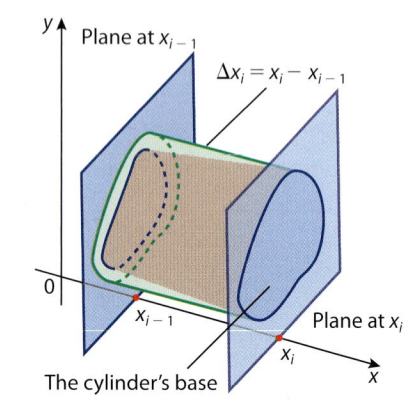

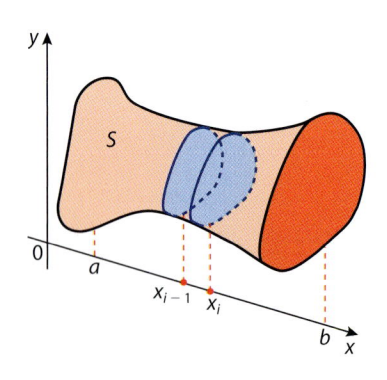

Example 13

Find the volume of the solid formed when the graph of the parabola $y = \sqrt{2x}$ over $[0, 4]$ is rotated around the x-axis through an angle of 2π radians, as shown in the diagram.

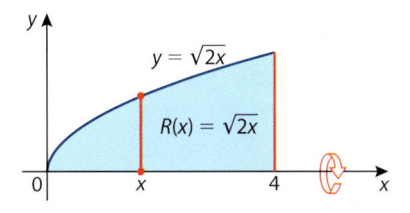

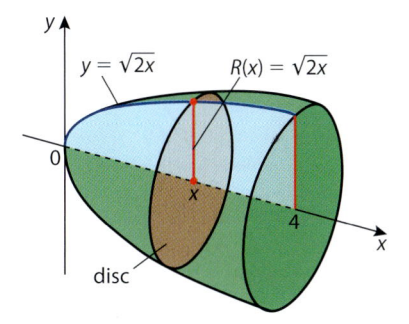

Solution

The cross section here is a circular disc whose radius is $y = \sqrt{2x}$. Therefore,
$$A(x) = \pi R^2 = \pi(\sqrt{2x})^2 = 2\pi x$$

The volume is then
$$V = \int_0^4 A(x)\, dx = \int_0^4 2\pi x\, dx = \left[2\pi \frac{x^2}{2}\right]_0^4 = 16\pi \text{ cubic units.}$$

Example 13 above is a special case of the general process for finding volumes of the so-called 'solids of revolution'.

If a region is bounded by a closed interval $[a, b]$ on the x-axis and a function $f(x)$ is rotated about the x-axis, the volume of the resulting solid of revolution is given by
$$V = \int_a^b \pi (f(x))^2\, dx$$

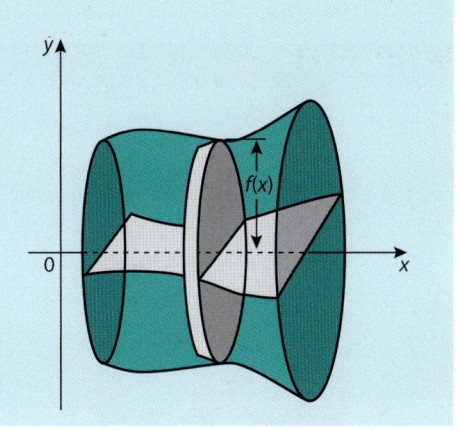

Integral Calculus

Example 14

Find the volume of a sphere with radius $R = a$.

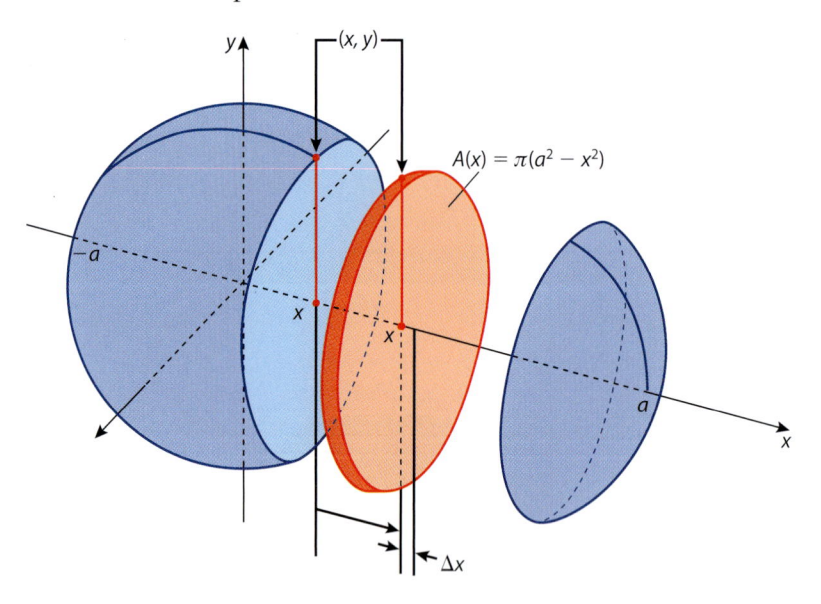

Solution

If we place the sphere with its centre at the origin, the equation of the circle will be

$$x^2 + y^2 = a^2 \Rightarrow y = \pm\sqrt{a^2 - x^2}$$

The cross section of the sphere, perpendicular to the x-axis, is a circular disc with radius y, so the area is

$$A(x) = \pi R^2 = \pi y^2 = \pi(\sqrt{a^2 - x^2})^2 = \pi(a^2 - x^2)$$

So, the volume of the sphere is

$$V = \int_{-a}^{a} \pi(a^2 - x^2)\, dx = \pi\left[a^2 x - \frac{x^3}{3}\right]_{-a}^{a}$$

$$= \pi\left(a^3 - \frac{a^3}{3}\right) - \pi\left(-a^3 - \frac{a^3}{3}\right)$$

$$= \pi\left(2a^3 - 2\frac{a^3}{3}\right) = \frac{4\pi a^3}{3}$$

Exercise 14.4

Find the volume of the solid obtained by rotating the region bounded by the given curves about the x-axis. Sketch the region, the solid and a typical disc.

1 $y = 3 - \frac{x}{3}, y = 0, x = 2, x = 3$

2 $y = 2 - x^2, y = 0$

3 $y = \sqrt{16 - x^2}, y = 0, x = 1, x = 3$

4 $y = \frac{3}{x}, y = 0, x = 1, x = 3$

5 $y = 3 - x, y = 0, x = 0$

6 $y = \sqrt{\sin x}, y = 0, 0 \leqslant x \leqslant \pi$

7 $y = \sqrt{\cos x}, y = 0, -\frac{\pi}{2} \leqslant x \leqslant \frac{\pi}{3}$

8 $y = 4 - x^2, y = 0$

9 $y = x^3 + 2x + 1, y = 0, x = 1$

10 $y = -4x - x^2, y = x^2$

14.5 Modelling linear motion

In previous sections of this text, we have examined problems involving displacement, velocity and acceleration of a moving object. In different sections of Chapter 11, we applied the fact that a derivative is a rate of change to express velocity and acceleration as derivatives. Even though our earlier work on motion problems involved an object moving in one, two or even three dimensions, our mathematical models considered the object's motion occurring only along a straight line. For example, projectile motion (e.g. a ball being thrown) is often modelled by a position function that simply gives the height (displacement) of the object. In that way, we are modelling the motion as if it were restricted to a vertical line.

In this section, we will again analyze the motion of an object as if its motion takes place along a straight line in space. This can only make sense if the mass (and thus, size) of the object is not taken into account. Hence, the object is modelled by a particle whose mass is considered to be zero. This study of motion, without reference either to the forces that cause it or to the mass of the object, is known as **kinematics**.

Displacement and total distance travelled

Recall from Chapter 11 that given time t, displacement s, velocity v and acceleration a, we have the following:

$$v = \frac{ds}{dt}, \; a = \frac{dv}{dt}, \text{ and } a = \frac{d}{dt}\left(\frac{ds}{dt}\right) = \frac{d^2s}{dt^2}$$

Let's review some of the essential terms we use to describe an object's motion.

> **Position, distance and displacement**
> - The **position** s of a particle, with respect to a chosen axis, is a measure of how far it is from a fixed point (usually the origin) *and* of its direction relative to the fixed point.
> - The **distance** $|s|$ of a particle is a measure of how far it is from a fixed point (usually the origin) and does *not* indicate direction. Thus, distance is the magnitude of position and is always positive.
> - The **displacement** is the *change* in position. The displacement of an object may be positive, negative or zero, depending on its motion.

It is important to understand the difference between displacement and distance travelled. Consider a couple of simple examples of an object moving along the *x*-axis.

1. In this first example, assume that the object does not change direction during the interval $0 \leq t \leq 5$. In other words, its velocity does not change from positive to negative or from negative to positive. If the position of the object at $t = 0$ is $x = 2$ and then the object moves so that at $t = 5$ its position is $x = -3$, its displacement, or change in position, is -5 because the object changed its position by 5 units in the negative

direction. This can be calculated by (final position) − (initial position) = −3 − 2 = −5. However, the distance travelled would be the absolute value of displacement, calculated by |final position − initial position| = |−3 − 2| = +5.

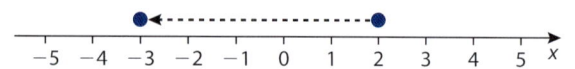

2. In this example, the object's initial and final positions are the same as in the first example – that is, at $t = 0$ its position is $x = 2$ and at $t = 5$ its position is $x = -3$. However, the object changed direction in that it first travelled to the left (negative velocity) from $x = 2$ to $x = -5$ during the interval $0 \leq t \leq 3$, and then travelled to the right (positive velocity) from $x = -5$ to $x = -3$. The object's displacement is -5 – the same as in the first example because its net change in position is just the difference between final and initial positions. However, it's clear that the object has travelled further than in the first example. But we cannot calculate it in the same way as we did in the first example. We will have to make a separate calculation for each interval where the direction changed. Hence, total distance travelled $= |-5 - 2| + |-3 - (-5)| = 7 + 2 = 9$.

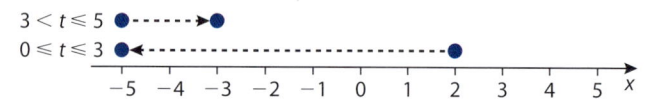

> There is no separate word to describe the magnitude of acceleration, $|a|$.

Velocity and speed
- The **velocity** $v = \dfrac{ds}{dt}$ of a particle is a measure of how fast it is moving *and* of its direction of motion relative to a fixed point.
- The **speed** $|v|$ of a particle is a measure of how fast it is moving and does *not* indicate direction. Thus, speed is the magnitude of velocity and is always positive.

Acceleration
- The **acceleration** $a = \dfrac{dv}{dt}$ of a particle is a measure of how fast its velocity is changing.

> The definite integral is a mathematical tool that can be used in applications to calculate net change of a quantity (e.g. Δ position → displacement) and total accumulation (e.g. Σ area → volume).

Example 15

The displacement s of a particle on the x-axis, relative to the origin, is given by the position function $s(t) = -t^2 + 6t$, where s in centimetres and t is in seconds.

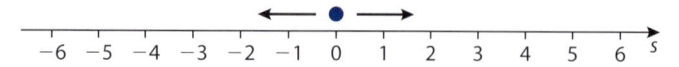

a) Find a function for the particle's velocity $v(t)$ in terms of t. Graph the functions $s(t)$ and $v(t)$ on separate axes.

b) Find the particle's position at the following times: $t = 0, 1, 3$ and 6 seconds.

c) Find the particle's displacement for the following intervals: $0 \leq t \leq 1$, $1 \leq t \leq 3$, $3 \leq t \leq 6$ and $0 \leq t \leq 6$.

d) Find the particle's total distance travelled for the following intervals: $0 \leq t \leq 1$, $1 \leq t \leq 3$, $3 \leq t \leq 6$ and $0 \leq t \leq 6$.

Solution

Position function: $s(t) = -t^2 + 6t$

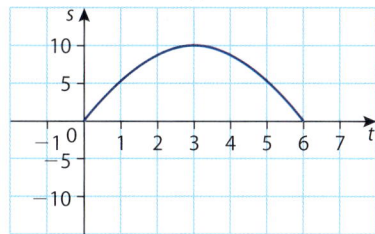

Velocity function: $v(t) = s'(t) = -2t + 6$

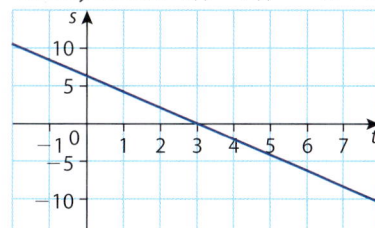

a) $v(t) = \dfrac{d}{dt}(-t^2 + 6t) = -2t + 6$

b) The particle's position at:
- $t = 0$ is $s(0) = -(0)^2 + 6(0) = 0$ cm
- $t = 1$ is $s(1) = -(1)^2 + 6(1) = 5$ cm
- $t = 3$ is $s(3) = -(3)^2 + 6(3) = 9$ cm
- $t = 6$ is $s(6) = -(6)^2 + 6(6) = 0$ cm

c) The particle's displacement for the interval:
- $0 \leq t \leq 1$ is Δ position $= s(1) - s(0) = 5 - 0 = 5$ cm
- $1 \leq t \leq 3$ is Δ position $= s(3) - s(1) = 9 - 5 = 4$ cm
- $3 \leq t \leq 6$ is Δ position $= s(6) - s(3) = 0 - 9 = -9$ cm
- $0 \leq t \leq 6$ is Δ position $= s(6) - s(0) = 0 - 0 = 0$ cm

This last result makes sense considering the particle moved to the right 9 cm then at $t = 3$ turned around and moved to the left 9 cm, ending where it started – thus, no change in net position.

d) The particle's total distance travelled for the interval:
- $0 \leq t \leq 1$ is $|s(1) - s(0)| = |5 - 0| = 5$ cm
- $1 \leq t \leq 3$ is $|s(3) - s(1)| = |9 - 5| = 4$ cm
- $3 \leq t \leq 6$ is $|s(6) - s(3)| = |0 - 9| = |-9| = 9$ cm
- $0 \leq t \leq 6$: The object's motion changed direction (velocity $= 0$) at $t = 3$, so total distance is $|s(3) - s(0)| + |s(6) - s(3)|$
$= |9 - 0| + |0 - 9| = 9 + 9 = 18$ cm

Since differentiation of the position function gives the velocity function $\left(\text{i.e. } v = \dfrac{ds}{dt}\right)$, we expect that the inverse of differentiation, integration, will lead us in the reverse direction – that is, from velocity to position. When velocity is constant, we can find the displacement with the formula:

$$\text{displacement} = \text{velocity} \times \Delta \text{ in time}$$

If we drove a car at a constant velocity of 50 km/h for 3 hours, our displacement (same as distance travelled in this case) is 150 km. If a particle travelled to the left on the x-axis at a constant rate of -4 units/sec for 5 seconds, the particle's displacement is -20 units.

The velocity–time graph on the next page depicts an object's motion with a constant velocity of 5 cm/s for $0 \leq t \leq 3$. Clearly, the object's displacement is 5 cm/s $\times$ 3 sec $= 15$ cm for this interval.

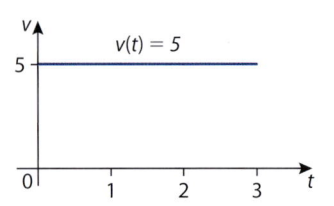

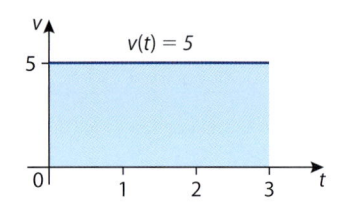

The rectangular area (3 × 5 = 15) under the velocity curve is equal to the object's displacement.

Looking back at Example 15, consider the area under the graph of $v(t)$ from $t = 0$ to $t = 3$.

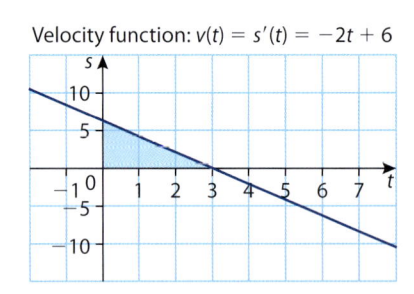

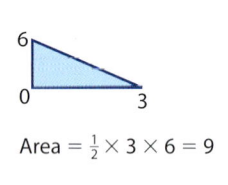

Velocity function: $v(t) = s'(t) = -2t + 6$

Area $= \frac{1}{2} \times 3 \times 6 = 9$

Given the discussion above, we should not be surprised to see that the area under the velocity curve for a certain interval is equal to the displacement for that interval. We can argue that just as the total area can be found by summing the areas of narrow rectangular strips, the displacement can be found by summing small displacements ($v \cdot \Delta t$). Consider:

$$\text{displacement} = \text{velocity} \times \Delta \text{ in time} \Rightarrow s = v \cdot \Delta t \Rightarrow s = v \cdot dt$$

We learned earlier in this chapter that if $f(x) \geq 0$ then the definite integral $\int_a^b f(x)\, dx$ gives the area between $y = f(x)$ and the x-axis from $x = a$ to $x = b$. And if $f(x) \leq 0$ then $\int_a^b f(x)\, dx$ gives a number that is the opposite of the area between $y = f(x)$ and the x-axis from a to b.

> **Using integration to find displacement and total distance travelled**
> Given that $v(t)$ is the velocity function for a particle moving along a line, then:
> $\int_a^b v(t)\, dt$ gives the displacement from $t = a$ to $t = b$
> $\left| \int_a^b v(t)\, dt \right|$ gives the total distance travelled from $t = a$ to $t = b$ if the particle does not change direction during the interval $a < t < b$.
> If a particle changes direction at some $t = c$ for $a < c < b$, the total distance travelled for the particle is given by $\left| \int_a^c v(t)\, dt \right| + \left| \int_c^b v(t)\, dt \right|$.

Let's apply integration to find the displacement and distance travelled for the two intervals $3 \leq t \leq 6$ and $0 \leq t \leq 6$ in Example 15.

- For $3 \leq t \leq 6$:

$$\text{Displacement} = \int_3^6 (-2t + 6)\, dt = \left[-t^2 + 6t \right]_3^6$$

$$= [-(6)^2 + 6(6)] - [-(3)^2 + 6(3)] = 0 - 9 = -9$$

Distance travelled $= \left|\int_3^6 (-2t + 6)\, dt\right| = \left|[-t^2 + 6t]_3^6\right|$

$= |[-(6)^2 + 6(6)] - [-(3)^2 + 6(3)]| = |0 - 9| = 9$

- For $0 \leq t \leq 6$:

Displacement $= \int_0^6 (-2t + 6)\, dt = [-t^2 + 6t]_0^6$

$= [-(6)^2 + 6(6)] - [0] = 0$

Distance travelled $= \left|\int_0^3 (-2t + 6)\, dt\right| + \left|\int_3^6 (-2t + 6)\, dt\right|$ particle changed direction at $t = 3$

$= \left|[-t^2 + 6t]_0^3\right| + \left|[-t^2 + 6t]_3^6\right|$

$= |(-9 + 18) - 0| + |0 - (-9 + 18)|$

$= |9| + |-9| = 9 + 9 = 18$

Example 16

The function $v(t) = \sin(\pi t)$ gives the velocity in m/s of a particle moving along the x-axis.

a) Determine when the particle is moving to the right, to the left, and stopped. At any time it stops, determine if it changes direction at that time.
b) Find the particle's displacement for the time interval $0 \leq t \leq 3$.
c) Find the particle's total distance travelled for the time interval $0 \leq t \leq 3$.

Solution

a) $v(t) = \sin(\pi t) = 0 \Rightarrow \sin(k \cdot \pi) = 0$ for $k \in \mathbb{Z} \Rightarrow \pi t = k\pi \Rightarrow t = k$, $k \in \mathbb{Z}$ for $0 \leq t \leq 3$, $t = 0, 1, 2, 3$. Therefore, the particle is stopped at $t = 0, 1, 2, 3$.

Since $t = 0$ and $t = 3$ are endpoints of the interval, the particle can only change direction at $t = 1$ or $t = 2$.

$v(\tfrac{1}{2}) = \sin(\pi \cdot \tfrac{1}{2}) = 1$; $v(\tfrac{3}{2}) = \sin(\pi \cdot \tfrac{3}{2}) = -1 \Rightarrow$ direction changes at $t = 1$

$v(\tfrac{3}{2}) = \sin(\pi \cdot \tfrac{3}{2}) = -1$; $v(\tfrac{5}{2}) = \sin(\pi \cdot \tfrac{5}{2}) = 1 \Rightarrow$ direction changes again at $t = 2$

b) Displacement $= \int_0^3 \sin(\pi t)\, dt = \left[-\tfrac{1}{\pi}\cos(\pi t)\right]_0^3$

$= -\tfrac{1}{\pi}\cos(3\pi) - \left(-\tfrac{1}{\pi}\cos(0)\right) = -\tfrac{1}{\pi}(-1) + \tfrac{1}{\pi}(1) = \tfrac{2}{\pi} \approx 0.637$ metres

c) Total distance travelled $= \left|\int_0^1 \sin(\pi t)\, dt\right| + \left|\int_1^2 \sin(\pi t)\, dt\right|$

$+ \left|\int_2^3 \sin(\pi t)\, dt\right| = \left|\left[-\tfrac{1}{\pi}\cos(\pi t)\right]_0^1\right|$

$+ \left|\left[-\tfrac{1}{\pi}\cos(\pi t)\right]_1^2\right| + \left|\left[-\tfrac{1}{\pi}\cos(\pi t)\right]_2^3\right|$

$= \left|\tfrac{2}{\pi}\right| + \left|-\tfrac{2}{\pi}\right| + \left|\tfrac{2}{\pi}\right| = \tfrac{6}{\pi} \approx 1.91$ metres

Note that, in Example 16, the position function is not known precisely. The position function can be obtained by finding the anti-derivative of the velocity function.

$$s(t) = \int v(t)\, dt = \int \sin(\pi t)\, dt = -\frac{1}{\pi} \cos(\pi t) + C$$

We can only determine the constant of integration C if we know the particle's initial position (or position at any other specific time). However, the particle's initial position will not affect displacement or distance travelled for any interval.

Position and velocity from acceleration

If we can obtain position from velocity by applying integration then we can also obtain velocity from acceleration by integrating. Consider the following example.

Example 17

The motion of a falling parachutist is modelled as linear motion by considering that the parachutist is a particle moving along a line whose positive direction is vertically downwards. The parachute is opened at $t = 0$ at which time the parachutist's position is $s = 0$. According to the model, the acceleration function for the parachutist's motion for $t > 0$ is given by:

$$a(t) = -54e^{-1.5t}$$

a) At the moment the parachute opens, the parachutist has a velocity of 42 m/s. Find the velocity function of the parachutist for $t > 0$. What does the model say about the parachutist's velocity as $t \to \infty$?

b) Find the position function of the parachutist for $t > 0$.

Solution

a) $v(t) = \int a(t)\, dt = \int (-54e^{-1.5t})\, dt$

$= -54\left(\dfrac{1}{-1.5}\right)e^{-1.5t} + C$

$= 36e^{-1.5t} + C$

Since $v = 42$ when $t = 0$, then $42 = 36e^0 + C \Rightarrow 42 = 36 + C \Rightarrow C = 6$

Therefore, after the parachute opens ($t > 0$) the velocity function is $v(t) = 36e^{-1.5t} + 6$.

Since $\lim\limits_{t \to \infty} e^{-1.5t} = \lim\limits_{t \to \infty} \dfrac{1}{e^{1.5t}} = 0$, then as $t \to \infty$, $\lim\limits_{t \to \infty} v(t) = 6$ m/sec.

b) $s(t) = \int v(t)\, dt = \int (36e^{-1.5t} + 6)\, dt$

$= 36\left(\dfrac{1}{-1.5}\right)e^{-1.5t} + 6t + C$

$= -24e^{-1.5t} + 6t + C$

Since $s = 0$ when $t = 0$, then $0 = -24e^0 + 6(0) + C$
$\Rightarrow 0 = -24 + C \Rightarrow C = 24$

Therefore, after the parachute opens ($t > 0$) the position function is $s(t) = -24e^{-1.5t} + 6t + 24$.

The limit of the velocity as $t \to \infty$, for a falling object, is called the **terminal velocity** of the object. While the limit $t \to \infty$ is never attained (the parachutist eventually lands on the ground), the velocity gets close to the terminal velocity very quickly. For example, after just 8 seconds, the velocity is $v(8) = 36e^{-1.5(8)} + 6 \approx 6.0002$ m/s.

Exercise 14.5

In questions 1–6, the velocity of a particle along a rectilinear path is given by the equation $v(t)$ in m/s. Find both the net distance and the total distance it travels between the times $t = a$ and $t = b$.

1. $v(t) = t^2 - 11t + 24, a = 0, b = 10$
2. $v(t) = t - \frac{1}{t^2}, a = 0.1, b = 1$
3. $v(t) = \sin 2t, a = 0, b = \frac{\pi}{2}$
4. $v(t) = \sin t + \cos t, a = 0, b = \pi$
5. $v(t) = t^3 - 8t^2 + 15t, a = 0, b = 6$
6. $v(t) = \sin\left(\frac{\pi t}{2}\right) + \cos\left(\frac{\pi t}{2}\right), a = 0, b = 1$

In questions 7–11, the acceleration of a particle along a rectilinear path is given by the equation $a(t)$ in m/s², and the initial velocity v_0 m/s is also given. Find the velocity of the particle as a function of t, and both the net distance and the total distance it travels between the times $t = a$ and $t = b$.

7. $a(t) = 3, v_0 = 0, a = 0, b = 2$
8. $a(t) = 2t - 4, v_0 = 3, a = 0, b = 3$
9. $a(t) = \sin t, v_0 = 0, a = 0, b = \frac{3\pi}{2}$
10. $a(t) = \frac{-1}{\sqrt{t+1}}, v_0 = 2, a = 0, b = 4$
11. $a(t) = 6t - \frac{1}{(t+1)^3}, v_0 = 2, a = 0, b = 2$

Practice questions

1. The graph represents the function
   $f: x \mapsto p \cos x, p \in \mathbb{N}$.

   Find
   a) the value of p
   b) the area of the shaded region.

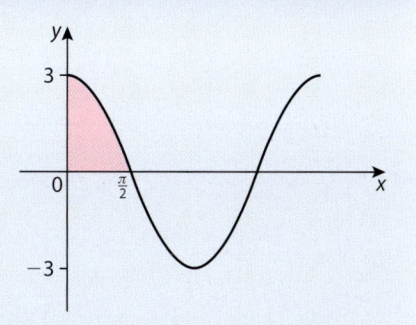

2. The diagram shows part of the graph of $y = e^{\frac{x}{2}}$.
   a) Find the coordinates of the point P, where the graph meets the y-axis.
   The shaded region between the graph and the x-axis, bounded by $x = 0$ and $x = \ln 2$, is rotated through 360° about the x-axis.
   b) Write down an integral that represents the volume of the solid obtained.
   c) Show that this volume is π cubic units.

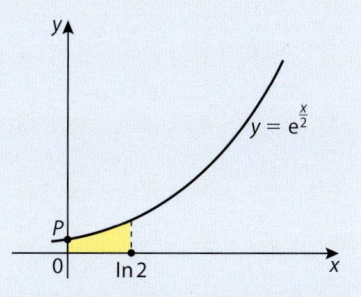

3 The diagram shows part of the graph of $y = \frac{1}{x}$. The area of the shaded region is 2 units.

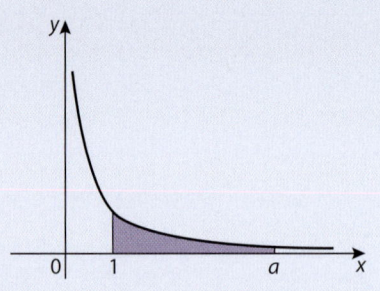

Find the exact value of a.

4 a) Find the equation of the tangent line to the curve $y = \ln x$ at the point $(e, 1)$, and verify that the origin is on this line.

  b) Show that $(x \ln x - x)' = \ln x$.

  c) The diagram shows the region enclosed by the curve $y = \ln x$, the tangent line in part **a)**, and the line $y = 0$.

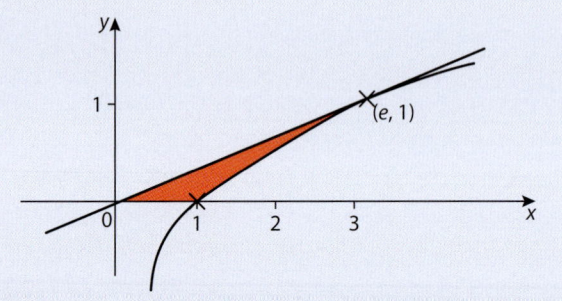

Use the result of part **b)** to show that the area of this region is $\frac{1}{2}e - 1$.

5 The main runway at Concordville airport is 2 km long. An aeroplane, landing at Concordville, touches down at point T, and immediately starts to slow down. The point A is at the southern end of the runway. A marker is located at point P on the runway.

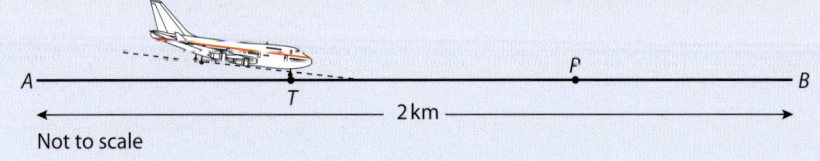

Not to scale

As the aeroplane slows down, its distance, s, from A, is given by

$$s = c + 100t - 4t^2$$

where t is the time in seconds after touchdown and c metres is the distance of T from A.

  a) The aeroplane touches down 800 m from A (i.e. $c = 800$).

   (i) Find the distance travelled by the aeroplane in the first 5 seconds after touchdown.

   (ii) Write down an expression for the velocity of the aeroplane at time t seconds after touchdown, and hence find the velocity after 5 seconds.

  The aeroplane passes the marker at P with a velocity of 36 m s^{-1}. Find

   (iii) how many seconds after touchdown it passes the marker

   (iv) the distance from P to A.

  b) Show that if the aeroplane touches down before reaching the point P, it can stop before reaching the northern end, B, of the runway.

6 a) Sketch the graph of $y = \pi \sin x - x$, $-3 \leqslant x \leqslant 3$, on millimetre square paper, using a scale of 2 cm per unit on each axis.
   Label and number both axes and indicate clearly the approximate positions of the x-intercepts and the local maximum and minimum points.
   b) Find the solution of the equation $\pi \sin x - x = 0$, $x > 0$.
   c) Find the indefinite integral
   $$\int (\pi \sin x - x)\, dx$$
   and hence, or otherwise, calculate the area of the region enclosed by the graph, the x-axis and the line $x = 1$.

7 The diagram shows the graph of the function $y = 1 + \frac{1}{x}$, $0 < x \leqslant 3$. Find the **exact** value of the area of the shaded region.

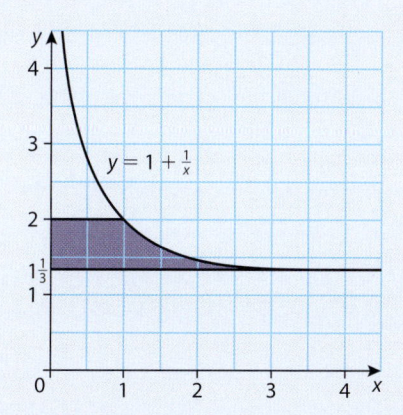

8 Note: Radians are used throughout this question.
   a) (i) Sketch the graph of $y = x^2 \cos x$, for $0 \leqslant x \leqslant 2$, making clear the approximate positions of the positive intercept, the maximum point and the endpoints.
   (ii) Write down the **approximate** coordinates of the positive x-intercept, the maximum point and the endpoints.
   b) Find the **exact value** of the positive x-intercept for $0 \leqslant x \leqslant 2$.
   Let R be the region in the first quadrant enclosed by the graph and the x-axis.
   c) (i) Shade R on your diagram.
   (ii) Write down an integral that represents the area of R.
   d) Evaluate the integral in part **c) (ii)**, either by using a graphic display calculator, or by using the following information.
   $$\frac{d}{dx}(x^2 \sin x + 2x \cos x - 2 \sin x) = x^2 \cos x$$

9 Note: Radians are used throughout this question.
   The function f is given by
   $$f(x) = (\sin x)^2 \cos x$$
   The diagram shows part of the graph of $y = f(x)$.
   The point A is a maximum point, the point B lies on the x-axis, and the point C is a point of inflexion.
   a) Give the period of f.
   b) From consideration of the graph of $y = f(x)$, find, **to an accuracy of 1 significant figure**, the range of f.

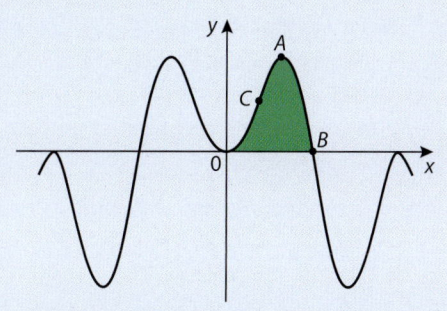

c) (i) Find $f'(x)$.
   (ii) Hence, show that, at the point A, $\cos x = \sqrt{\frac{1}{3}}$.
   (iii) Find the exact maximum value.
d) Find the exact value of the x-coordinate at the point B.
e) (i) Find $\int f(x)\, dx$.
   (ii) Find the area of the shaded region in the diagram.
f) Given that $f''(x) = 9(\cos x)^3 - 7\cos x$, find the x-coordinate at the point C.

10 **Note: Radians are used throughout this question.**
   a) Draw the graph of $y = \pi + x\cos x$, $0 \leq x \leq 5$, on millimetre square paper, using a scale of 2 cm per unit. Make clear
      (i) the integer values of x and y on each axis
      (ii) the approximate positions of the x-intercepts and the turning points.
   b) **Without the use of a calculator**, show that π is a solution of the equation
      $\pi + x\cos x = 0$.
   c) Find another solution of the equation $\pi + x\cos x = 0$ for $0 \leq x \leq 5$, giving your answer to 6 significant figures.
   d) Let R be the region enclosed by the graph and the axes for $0 \leq x \leq \pi$. Shade R on your diagram, and write down an integral which represents the area of R.
   e) Evaluate the integral in part **d)** to an accuracy of **6** significant figures. (If you consider it necessary, you can make use of the result $\frac{d}{dx}(x\sin x + \cos x) = x\cos x$.

11 The diagram right shows the graphs of $f(x) = 1 + e^{2x}$ and $g(x) = 10x + 2$, $0 \leq x \leq 1.5$.
   a) (i) Write down an expression for the vertical distance p between the graphs of f and g.
      (ii) Given that p has a maximum value for $0 \leq x \leq 1.5$, find the value of x at which this occurs.

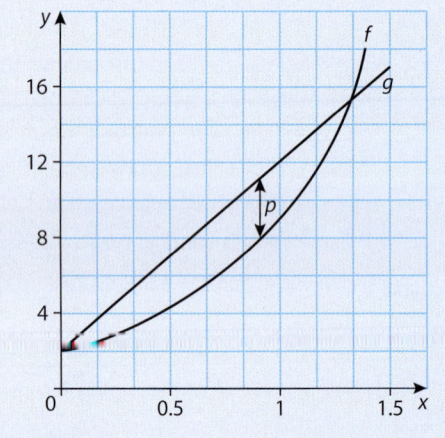

The graph of $y = f(x)$ only is shown in the diagram right. When $x = a$, $y = 5$.
   b) (i) Find $f^{-1}(x)$.
      (ii) **Hence**, show that $a = \ln 2$.
   c) The region shaded in the diagram is rotated through 360° about the x-axis. Write down an expression for the volume obtained.

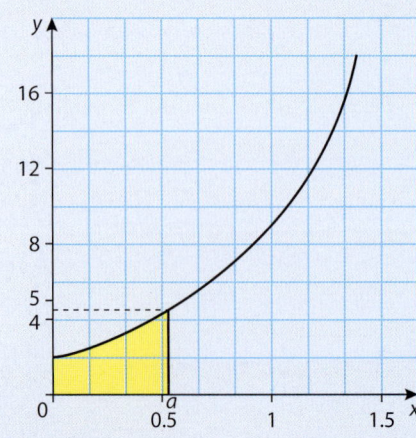

Questions 1–11: © International Baccalaureate Organization

15 Probability Distributions

Assessment statements
6.9 Concept of discrete random variables and their probability distributions.
Expected value (mean), E(x) for discrete data.
6.10 Binomial distribution.
Mean of the binomial distribution.
6.11 Normal distribution.
Properties of the normal distribution.
Standardization of normal variables.

Introduction

Investing in securities, calculating premiums for insurance policies or overbooking policies used in the airline industry are only a few of the many applications of probability and statistics. Actuaries, for example, calculate the expected 'loss' or 'gain' that an insurance company will incur and decide on how high the premiums should be. These applications depend mainly on what we call probability distributions. A probability distribution describes the behaviour of a population in the sense that it lists the distribution of possible outcomes to an event, along with the probability of each potential outcome. This can be done by a table of values with their corresponding probabilities or by using a mathematical model.

In this chapter, you will get an understanding of the basic ideas of distributions and will study two specific ones: the binomial and normal distributions.

15.1 Random variables

In Chapter 9, **variables** were defined as characteristics that change or vary over time and/or for different objects under consideration. A numerically valued variable x will vary or change depending on the outcome of the experiment we are performing. For example, suppose you are counting the number of mobile phones families in a certain city own. The variable of interest, x, can take any of the values 0, 1, 2, 3, etc. depending on the *random* outcome of the experiment. For this reason, we call the variable x a **random variable**.

> **Random variable**
> A **random variable** is a variable that takes on numerical values determined by the outcome of a random experiment.

When a probability experiment is performed, often we are not interested in all the details of the outcomes, but rather in the value of some numerical quantity determined by the result. For instance, in tossing two dice (used

15 Probability Distributions

in plenty of games), often we care about their sum and not the values on the individual dice. Consider this specific experiment: A sample space for which the points are equally likely is given in Table 15.1 below. It consists of 36 ordered pairs (a, b) where a is the number on the first die and b is the number on the second die. For each sample point, we can let the *random variable* x stand for the sum of the numbers. The resulting values of x are also presented in Table 15.1.

Table 15.1 Sample space and the values of the random variable x in the two-dice experiment.

(1, 1); $x = 2$	(2, 1); $x = 3$	(3, 1); $x = 4$	(4, 1); $x = 5$	(5, 1); $x = 6$	(6, 1); $x = 7$
(1, 2); $x = 3$	(2, 2); $x = 4$	(3, 2); $x = 5$	(4, 2); $x = 6$	(5, 2); $x = 7$	(6, 2); $x = 8$
(1, 3); $x = 4$	(2, 3); $x = 5$	(3, 3); $x = 6$	(4, 3); $x = 7$	(5, 3); $x = 8$	(6, 3); $x = 9$
(1, 4); $x = 5$	(2, 4); $x = 6$	(3, 4); $x = 7$	(4, 4); $x = 8$	(5, 4); $x = 9$	(6, 4); $x = 10$
(1, 5); $x = 6$	(2, 5); $x = 7$	(3, 5); $x = 8$	(4, 5); $x = 9$	(5, 5); $x = 10$	(6, 5); $x = 11$
(1, 6); $x = 7$	(2, 6); $x = 8$	(3, 6); $x = 9$	(4, 6); $x = 10$	(5, 6); $x = 11$	(6, 6); $x = 12$

Notice that events can be more accurately and concisely defined in terms of the random variable x; for example, the event of tossing a sum at least equal to 5 but less than 9 can be replaced by $5 \leq x < 9$.

We can think of many examples of random variables:

- X = the number of calls received by a household on a Friday night.
- X = the number of free beds available at hotels in a large city.
- X = the number of customers a sales person contacts on a working day.
- X = the length of a metal bar produced by a certain machine.
- X = the weight of newborn babies in a large hospital.

As you have seen in Chapter 9, these variables are classified as **discrete** or **continuous**, according to the values that x *can* assume. In the examples above, the first three are discrete and the last two are continuous. The random variable is discrete if its set of *possible* values is isolated points on the number line, i.e. there is a *countable* number of possible values for the variable. The variable is continuous if its set of *possible* values is an entire interval on the number line, i.e. it can take any value in an interval. Consider the number of times you toss a coin until the head side appears. The possible values are $x = 1, 2, 3, \ldots$. This is a discrete variable, even though the number of times may be infinite! On the other hand, consider the time it takes a student at your school to eat/have his/her lunch. This can be anywhere between zero and 50 minutes (given that the lunch period at your school is 50 minutes).

discrete   continuous

Example 1

State whether each of the following is a discrete or a continuous random variable.
1. The number of hairs on a Scottish Terrier
2. The height of a building

3. The amount of fat in a steak
4. A high school student's grade on a maths test
5. The number of fish in the Atlantic Ocean
6. The temperature of a wooden stove

Solution
1. Even though the number of hairs is 'almost' infinite, it is countable. So, it is a discrete random variable.
2. This can be any real number. Even when you say this building is 15 m high, the number could be 15.1 or 15.02, etc. Hence, it is continuous.
3. This is continuous, as the amount of fat could be zero or anything up to the maximum amount of fat that can be held in one piece.
4. Grades are discrete. No matter how detailed a score the teacher gives, the grades are isolated points on a scale.
5. This is almost infinite, but countable, hence discrete.
6. This is continuous, as the temperature can take any value from room temperature to 100 degrees.

Probability distribution

In Chapter 9, you learned how to work with the frequency distribution and relative or percentage frequency distribution for a set of numerical measurements on a variable x. The distribution gave the following information about x:
- The value of x that occurred.
- How often each value occurred.

You also learned how to use the mean and standard deviation to measure the centre and variability of the data set.

Here is an example of the frequency distribution of 25 families in Lower Austria that were polled in a marketing survey to list the number of litres of milk consumed during a particular week, reproduced below. As you will observe, the table lists the number of litres consumed along with the relative frequency with which that number is observed. As you recall from Chapter 10, one of the interpretations of probability is that it is understood to be the long-term relative frequency of the event.

Number of litres	Relative frequency
0	0.08
1	0.20
2	0.36
3	0.20
4	0.12
5	0.04

◀ Table 15.2

A table like this, where we replace the relative frequency with probability, is called a **probability distribution** of the random variable.

15 Probability Distributions

> The **probability distribution** for a discrete random variable is a table, graph or formula that gives the possible values of x, and the probability P(x) associated with each value of x.

Letting x be the number of litres of milk consumed by a family above, the **probability distribution** of x would be as follows:

Table 15.3

x	0	1	2	3	4	5
P(x)	0.08	0.20	0.36	0.20	0.12	0.04

The other form of representing the probability distribution is with a histogram, as shown below. Every column corresponds to the probability of the associated value of x. The values of x naturally represent mutually exclusive events. Summing P(x) over all values of x is equivalent to adding all probabilities of all simple events in the sample space, and hence the total is 1.

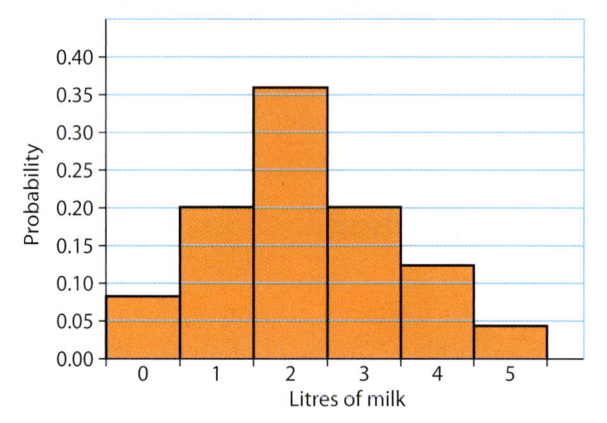

The result above can be generalized for all probability distributions:

Required properties of probability distribution functions of discrete random variables

Let x be a discrete random variable with probability distribution function, P(x). Then:
- $0 \leq P(x) \leq 1$, for any value x.
- The individual probabilities sum to 1; that is, $\sum_{x} P(x) = 1$ where the notation indicates summation over all possible values x.

Example 2

Radon is a major cause of lung cancer after smoking. It is a radioactive gas produced by the natural decay of radium in the ground. Studies in areas rich with radium revealed that one-third of houses in these areas have dangerous levels of this gas. Suppose that two houses are randomly selected and we define the random variable x to be the number of houses with dangerous levels. Find the probability distribution of x by a table, a graph and a formula.

Solution

Since two houses are selected, the possible values of x are 0, 1 or 2. To find their probabilities, we utilize what we learned in Chapter 10. The assumption here is that we are choosing the houses randomly and independently of each other!

$P(x = 2) = P(2) = P(\text{1st house with gas } and \text{ 2nd house with gas})$
$= P(\text{1st house with gas}) \times P(\text{2nd house with gas}) = \frac{1}{3} \times \frac{1}{3} = \frac{1}{9}$

$P(x = 0) = P(0) = P(\text{1st house without gas } and \text{ 2nd house without gas})$
$= P(\text{1st house without gas}) \times P(\text{2nd house without gas})$
$= \frac{2}{3} \times \frac{2}{3} = \frac{4}{9}$

$P(x = 1) = 1 - [P(0) + P(2)] = 1 - \left[\frac{4}{9} + \frac{1}{9}\right] = \frac{4}{9}$

Table

x	0	1	2
$P(x)$	$\frac{4}{9}$	$\frac{4}{9}$	$\frac{1}{9}$

Graph

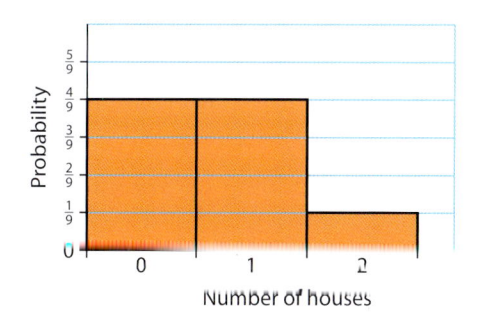

Any type of graph can be used to give the probability distribution, as long as it shows the possible values of x and the corresponding probabilities. The probability here is graphically displayed as the height of a rectangle. Moreover, the rectangle corresponding to each value of x has an area equal to the probability $P(x)$. The histogram is the preferred tool due to its connection to the continuous distributions discussed later in the chapter.

Formula/rule

The probability distribution of x can also be given by the following rule. Don't be concerned now with how we came up with this formula, as we will discuss it later in the chapter. The only reason we are looking at it now is to illustrate the fact that a formula/rule can sometimes be used to give the probability distribution.

$$P(x) = \binom{2}{x} \cdot \left(\frac{1}{3}\right)^x \cdot \left(\frac{2}{3}\right)^{2-x}, \text{ where } \binom{2}{x} \text{ represents the binomial coefficient you saw in Chapter 3.}$$

Notice that when x is replaced by 0, 1 or 2 we obtain the results we are looking for:

$$P(0) = \binom{2}{0} \cdot \left(\frac{1}{3}\right)^0 \cdot \left(\frac{2}{3}\right)^{2-0} = 1 \cdot 1 \cdot \frac{4}{9} = \frac{4}{9}$$

$$P(1) = \binom{2}{1} \cdot \left(\frac{1}{3}\right)^1 \cdot \left(\frac{2}{3}\right)^{2-1} = 2 \cdot \frac{1}{3} \cdot \frac{2}{3} = \frac{4}{9}$$

$$P(2) = \binom{2}{2} \cdot \left(\frac{1}{3}\right)^2 \cdot \left(\frac{2}{3}\right)^{2-2} = 1 \cdot \frac{1}{9} \cdot 1 = \frac{1}{9}$$

15 Probability Distributions

Example 3
Many universities have the policy of posting the grade distributions for their courses. Several of the universities have a grade-point average that codes the grades in the following manner: A = 4, B = 3, C = 2, D = 1 and F = 0. During the spring term at a certain large university, 13% of the students in an introductory statistics course received A's, 37% B's, 45% C's, 4% received D's and 1% received F's. The experiment here is to choose a student at random and mark down his/her grade. The student's grade on the 4-point scale is a random variable x.

Here is the probability distribution of x:

x	0	1	2	3	4
P(x)	0.01	0.04	0.45	0.37	0.13

Is this a probability distribution?

Solution
Yes, it is. Each probability is between 0 and 1, and the sum of all probabilities is 1.

What is the probability that a randomly chosen student receives a B or better?

$$P(x \geqslant 3) = P(x = 3) + P(x = 4) = 0.37 + 0.13 = 0.40$$

Example 4
In the codes example in Chapter 10, we saw the probability with which people choose the first digits for the codes for their cellphones. The probability distribution is copied below for reference.

First digit	0	1	2	3	4	5	6	7	8	9
Probability	0.009	0.300	0.174	0.122	0.096	0.078	0.067	0.058	0.051	0.045

Here, x is the first digit chosen.

What is the probability that you pick a first digit and it is more than 5? Show a probability histogram for the distribution.

Solution

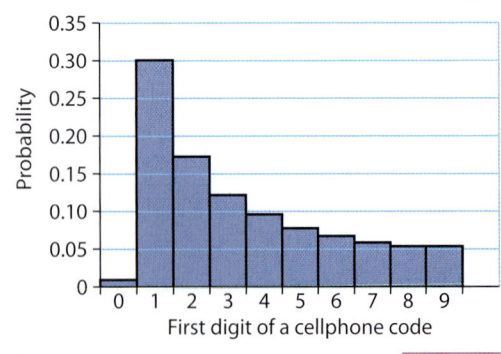

$$P(x > 5) = P(x = 6) + P(x = 7) + P(x = 8) + P(x = 9) = 0.221$$

Note that the height of each bar shows the probability of the outcome at its base. The heights add up to 1, of course. The bars in this histogram have the same width, namely 1. So, the areas also display the probability assignments of the outcomes. Think of such histograms (probability histograms) as idealized pictures of the results of very many repeated trials.

500

Expected values

The probability distribution for a random variable looks very similar to the relative frequency distribution discussed in Chapter 9. The difference is that the relative frequency distribution describes a *sample* of measurements, whereas the probability distribution is constructed as a *model* for the *entire population*. Just as the mean and standard deviation gave you measures for the centre and spread of the sample data, you can calculate similar measures to describe the centre and spread of the population.

The population mean, which measures the average value of x in the population, is also called the **expected value** of the random variable x. It is the value that you would *expect* to observe *on average* if you repeat the experiment an infinite number of times. The formula we use to determine the expected value can be simply understood with an example.

Let's revisit the milk consumption example. Let x be the number of litres consumed. Here is the table of probabilities again:

x	0	1	2	3	4	5
P(x)	0.08	0.20	0.36	0.20	0.12	0.04

Suppose we choose a large number of families, say 100 000. Intuitively, using the relative frequency concept of probability, you would expect to observe 8000 families consuming no milk, 20 000 consuming 1 litre, and the rest similarly done: 36 000, 20 000, 12 000 and 4000.

The average (mean) value of x, as defined in Chapter 9, would then be equal to

$$\frac{\text{sum of all measurements}}{n}$$

$$= \frac{0 \cdot 8000 + 1 \cdot 20\,000 + 2 \cdot 36\,000 + 3 \cdot 20\,000 + 4 \cdot 12\,000 + 5 \cdot 4000}{100\,000}$$

$$= \frac{0 \cdot 8000}{100\,000} + \frac{1 \cdot 20\,000}{100\,000} + \frac{2 \cdot 36\,000}{100\,000} + \frac{3 \cdot 20\,000}{100\,000} + \frac{4 \cdot 12\,000}{100\,000} + \frac{5 \cdot 4000}{100\,000}$$

$$= 0 \cdot 0.08 + 1 \cdot 0.20 + 2 \cdot 0.36 + 3 \cdot 0.20 + 4 \cdot 0.12 + 5 \cdot 0.04$$

$$= 0 \cdot P(0) + 1 \cdot P(1) + 2 \cdot P(2) + 3 \cdot P(3) + 4 \cdot P(4) + 5 \cdot P(5) = 2.2$$

That is, we expect to see families, *on average*, consuming 2.2 litres of milk! This does not mean that we know what a family *will* consume, but we can say what we *expect* to happen.

> Let x be a discrete random variable with probability distribution P(x). The mean or **expected value** of x is given by
>
> $$\mu = E(x) = \sum x \cdot P(x)$$

Insurance companies make extensive use of expected value calculations. Here is a simplified example.

An insurance company offers a policy that pays you €10 000 when you totally damage your car or €5000 for major damages (50%). They charge you €50 per year for this service. The question is, how can they make profit?

15 Probability Distributions

To understand how they can afford this, suppose that the 'total damage' car accident rate, in any year, is 1 out of every 1000 cars, and that another 2 out of 1000 will have serious damages. Then we can display the probability model for this policy in a table like this:

Type of accident	Amount paid x	Probability $P(X = x)$
Total damage	10 000	$\frac{1}{1000}$
Major damage	5000	$\frac{2}{1000}$
Minor or no damage	0	$\frac{997}{1000}$

The expected amount the insurance company pays is given by

$$\mu = E(X) = \sum xP(x) = €10\,000\left(\frac{1}{1000}\right) + €5000\left(\frac{2}{1000}\right)$$
$$+ €0\left(\frac{997}{1000}\right) = €20$$

This means that the insurance company *expects* to pay, on average, an amount of €20 per insured car. Since it is charging people €50 for the policy, the company *expects* to make a profit of €30 per car. Thinking about the problem in a different perspective, suppose they insure 1000 cars, then the company would expect to pay €10 000 for 1 car and €5000 to each of two cars with major damage. This is a total of €20 000 for all cars, or an average of $\frac{20\,000}{1000} = €20$ per car.

Of course, this expected value is not what actually happens to any *particular* policy. No individual policy actually costs the insurance company €20. We are dealing with random events, so a few car owners may require a payment of €10 000 or €5000, many others receive **nothing**! Because of the need to anticipate such variability, the insurance company needs to know a measure of this variability, which is nothing but the **standard deviation.**

Variance and standard deviation

- **Hint:** This part is a very important and helpful concept in the discussion of random variables; however, it is optional in the sense that it is not included in the IB syllabus.

For data in Chapter 9, we calculated the variance by computing the deviation from the mean, $x - \mu$, and then squaring it. We do that with random variables as well.

We can use similar arguments to justify the formulae for the population variance σ^2 and, consequently, the population standard deviation σ. These measures describe the spread of the values of the random variable around the centre. We similarly use the idea of the 'average' or 'expected' value of the squared deviations of the x-values from the mean μ or $E(x)$.

Let x be a discrete random variable with probability distribution $P(x)$ and mean μ. The **variance of x** is given by

$$\sigma^2 = E((x - \mu)^2) = \sum(x - \mu)^2 \cdot P(x)$$

(This is sometimes called Var(x).)

Note: It can also be shown, similar to what you saw in Chapter 9, that you have another 'computation' formula for the variance:

$$\sigma^2 = \sum(x-\mu)^2 \cdot P(x) = \sum x^2 \cdot P(x) - \mu^2 = \sum x^2 \cdot P(x) - [E(x)]^2$$
$$= \sum x^2 \cdot P(x) - \left[\sum xP(x)\right]^2$$

> The **standard deviation** σ of a random variable x is equal to the positive square root of its variance.

Let us go back to the milk consumption example. Recall that we calculated the expected value, mean, to be 2.2 litres. In order to calculate the variance, we can tabulate our work to make the manual calculation simple.

x	$P(x)$	Deviation $(x-\mu)$	Squared deviation $(x-\mu)^2$	$(x-\mu)^2 \cdot P(x)$
0	0.08	−2.2	4.84	0.3872
1	0.20	−1.2	1.44	0.2880
2	0.36	−0.2	0.04	0.0144
3	0.20	0.8	0.64	0.1280
4	0.12	1.8	3.24	0.3888
5	0.04	2.8	7.84	0.3136
		Total	$\sum(x-\mu)^2 \cdot P(x)$	1.52

So, the variance of the milk consumption is 1.52 litres², or the standard deviation is 1.233 litres.

GDC notes

The above calculations, along with the expected value calculation, can be easily done using your GDC.

First, store x and $P(x)$ into L1 and L2.

```
L1      L2      L3    2
0       .08
1       .2
2       .36
3       .2
4       .12
5       .04
L2(1)=.08
```

Then, to find $x P(x)$, we multiply L1 and L2 and store the result in L3.

```
L1*L2 →L3
(0 .2 .72 .6 .4...
```

To find the expected value, you simply get the sum of the entries in L3, since they correspond to $\sum x \cdot P(x)$.

```
L1*L2 →L3
(0 .2 .72 .6 .4...
sum(L3)
              2.2
```

15 Probability Distributions

```
L1-2.2→L4
(-2.2 -1.2 -.2…
(L4)²*L2→L5
(.3872 .228 .01…
sum(L5)
            1.52
```

To find the variance, we need to find the deviations from the mean; so we make L4 that deviation, i.e. we store L1 − 2.2 into L4. Then, to get the squared deviations multiplied by the corresponding probability, we set up L5 to be L4 squared multiplied by L2, the probability. Now, to find the variance, just add the terms of L5.

Software note

In the comfort of home/class, the above calculation can be performed on a computer with a simple spreadsheet like the following one:

x	P(x)	xP(x)	x − μ	(x − μ)²	(x − μ)²P(x)	
0	0.08	0	−2.2	4.84	0.3872	
1	0.2	0.2	−1.2	1.44	0.288	
2	0.36	0.72	−0.2	0.04	0.0144	
3	0.2	0.6	0.8	0.64	0.128	
4	0.12	0.48	1.8	3.24	0.3888	E4*B4
5	0.04	0.2	2.8	7.84	0.3136	
Totals	1	2.2			1.52	
				A3 − 2.2	E6^2	
		A2*B2		SUM(C2:C7)		

Example 5

A computer store sells a particular type of laptop. The daily demand for the laptops is given in the table below. x is the number of laptops in demand. They have only 4 laptops left in stock and would like to know how well they are prepared for all eventualities. Work out the expected value of the demand and the standard deviation.

x	0	1	2	3	4	5
P(x)	0.08	0.40	0.24	0.15	0.08	0.05

Solution

$$E(x) = \sum xP(x) = 0 \times 0.08 + 1 \times 0.40 + 2 \times 0.24 + 3 \times 0.15 \\ + 4 \times 0.08 + 5 \times 0.05 = 1.90$$

$$\text{Var}(x) = \sigma^2 = \sum (x - \mu)^2 P(x)$$
$$= (0 - 1.9)^2 \cdot 0.08 + (1 - 1.9)^2 \cdot 0.40 + (2 - 1.9)^2 \cdot 0.24$$
$$+ (3 - 1.9)^2 \cdot 0.15 + (4 - 1.9)^2 \cdot 0.08 + (5 - 1.9)^2 \cdot 0.05 = 1.63$$

$$\sigma = 1.28$$

Spreadsheet output is also given.

x	P(x)	xP(x)	x − μ	(x − μ)²	(x − μ)²P(x)
0	0.08	0	−1.9	3.61	0.2888
1	0.4	0.4	−0.9	0.81	0.324
2	0.24	0.48	0.1	0.01	0.0024
3	0.15	0.45	1.1	1.21	0.1815
4	0.08	0.32	2.1	4.41	0.3528
5	0.05	0.25	3.1	9.61	0.4805
Totals	1	1.9			1.63

The graph of the probability distribution is given below.

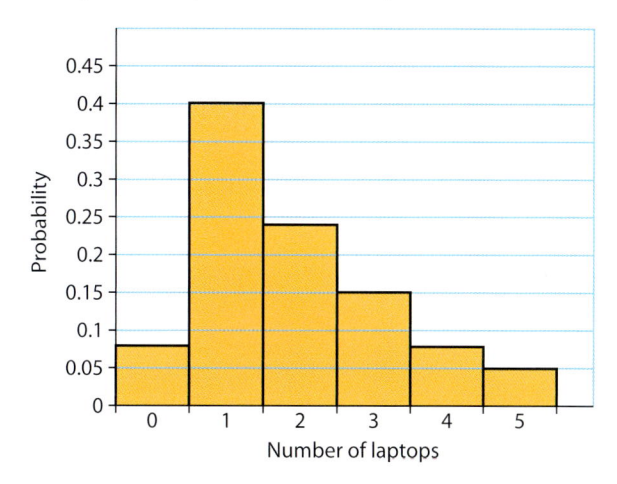

As an approximation, we can use the *empirical rule* to see where most of the demand is expected to be. Recall that the empirical rule tells us that about 95% of the values would lie within 2 standard deviations from the mean. In this case $\mu \pm 2\sigma = 1.9 \pm 2 \times 1.28 \Rightarrow (-0.66, 4.46)$. This interval does not contain the 5 units of demand. We can say that it is unlikely that 5 or more customers of this shop will want to buy a laptop today.

GDC

After entering the demand in L1 and the probabilities in L2, it is enough to find the sum of their product.

For the variance, we follow the same procedure as described in the previous example, see right.

Notice here that we combined several steps in one.

```
sum(L1*L2)
              1.9
```

```
L1*L2→L3
(0 .4 .48 .45 ....
(L1-1.9)²*L2→L5
(.2888 .324 .00...
sum(L5)
              1.63
```

Exercise 15.1

(Problems marked with (*) are optional.)

1 Classify each of the following as discrete or continuous random variables.
   a) The number of words spelled correctly by a student on a spelling test.
   b) The amount of water flowing through the Niagara Falls per year.
   c) The length of time a student is late to class.
   d) The number of bacteria per cc of drinking water in Geneva.
   e) The amount of CO produced per litre of unleaded gas.
   f) The amount of a flu vaccine in a syringe.
   g) The heart rate of a lab mouse.
   h) The barometric pressure at Mount Everest.
   i) The distance travelled by a taxi driver per day.
   j) Total score of football teams in national leagues.
   k) Height of ocean tides on the shores of Portugal.
   l) Tensile breaking strength (in Newtons per square metre) of a 5 cm diameter steel cable.
   m) Number of overdue books in a public library.

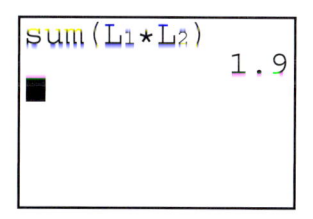

15 Probability Distributions

2 A random variable y has this probability distribution:

y	0	1	2	3	4	5
$P(y)$	0.1	0.3		0.1	0.05	0.05

a) Find P(2).
b) Construct a probability histogram for this distribution.
c) Find μ and σ.(*)
d) Locate the interval $\mu \pm \sigma$ as well as $\mu \pm 2\sigma$ on the histogram.(*)

3 A discrete random variable x can assume five possible values: 12, 13, 15, 18, and 20. Its probability distribution is shown below.

x	12	13	15	18	20
$P(x)$	0.14	0.11		0.26	0.23

a) What is P(15)?
b) What is the probability that x equals 12 or 20?
c) What is $P(x \leq 18)$?
d) Find $E(x)$.
e) Find $V(x)$.(*)

4 Medical research has shown that a certain type of chemotherapy is successful 70% of the time when used to treat skin cancer. In a study to check the validity of such a claim, researchers chose different treatment centres and chose five of their patients at random. Here is the probability distribution of the number of successful treatments for groups of five:

x	0	1	2	3	4	5
$P(x)$	0.002	0.029	0.132	0.309	0.360	0.168

a) Find the probability that at least two patients would benefit from the treatment.
b) Find the probability that the majority of the group does not benefit from the treatment.
c) Find $E(x)$ and interpret the result.
d) Show that $\sigma(x) = 1.02$.(*)
e) Graph P(x). Locate μ, $\mu \pm \sigma$ and $\mu \pm 2\sigma$ on the graph. Use the empirical rule to approximate the probability that x falls in this interval. Compare this with the actual probability.

5 The probability function of a discrete random variable X is given by
$$P(X = x) = \frac{kx}{2}, \text{ for } x = 12, 14, 16, 18.$$
Set up the table showing the probability distribution and find the value of k.

6 x has probability distribution as shown in the table.

x	5	10	15	20	25
$P(x)$	$\frac{3}{20}$	$\frac{7}{30}$	k	$\frac{3}{10}$	$\frac{13}{60}$

a) Find the value of k.
b) Find $P(x > 10)$.
c) Find $P(5 < x \leq 20)$.
d) Find the expected value and the standard deviation.

7 The discrete random variable Y has a probability density function

$$P(Y = y) = k(16 - y^2), \text{ for } y = 0, 1, 2, 3, 4.$$

a) Find the value of the constant k.
b) Draw a histogram to illustrate the distribution.
c) Find $P(1 \leqslant y \leqslant 3)$.
d) Find the mean and variance.

8 The probability distribution of students categorized by age that visit a certain movie house on weekends is given below. The probabilities for 18- and 19-year-olds are missing. We know that

$$P(x = 18) = 2P(x = 19)$$

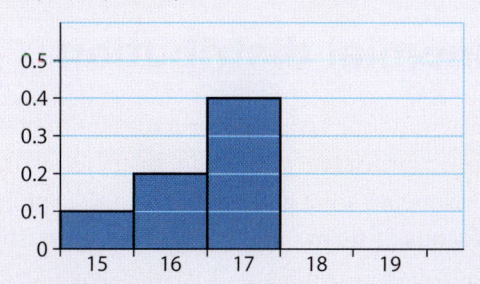

a) Complete the histogram and describe the distribution.
b) Find the expected value and the variance.

9 In a small town, a computer store sells laptops to the local residents. However, due to low demand, they like to keep their stock at a manageable level. The data they have indicate that the weekly demand for the laptops they sell follows the distribution given in the table below.

X: number of laptops bought	0	1	2	3	4	5
P(X = x)	0.10	0.40	0.20	0.15	0.10	0.05

a) Find the mean and standard deviation of this distribution.
b) Use the empirical rule to find the approximate number of laptops that is sold about 95% of the time.
c) Is it likely that 5 or more customers buy a laptop in any week?

10 The discrete random variable x has probability function given by

$$P(x) = \begin{cases} \left(\frac{1}{4}\right)^{x-1} & x = 2, 3, 4, 5, 6 \\ k & x = 7 \\ 0 & \text{otherwise} \end{cases}$$

where k is a constant. Determine the value of k and the expected value of x.

11 The following is a probability distribution for a random variable y.

y	0	1	2	3
P(Y = y)	0.1	0.11	k	$(k - 1)^2$

a) Find the value of k.
b) Find the expected value.

507

12 A closed box contains eight red balls and four white ones. A ball is taken out at random, its colour noted, and then returned. This is done three times. Let X represent the number of red balls drawn.
   a) Set up a table to show the probability distribution of X.
   b) What is the expected number of red balls in this experiment?

13 A discrete random variable Y has the following probability distribution function
$$P(Y = y) = k(4 - y), \text{ for } y = 0, 1, 2, 3 \text{ and } 4.$$
   a) Find the value of k.
   b) Find $P(1 \leq y < 3)$.

15.2 The binomial distribution

Examples of discrete random variables are abundant in everyday situations. However, there are a few discrete probability distributions that are widely applied and serve as *models* for a great number of the applications. In this book, we will study one of them only: the **binomial distribution**.

We will start our discussion of the binomial distribution with an example.

Suppose a cereal company puts miniature figures in boxes of cornflakes to make them attractive for children and thus boost sales. The manufacturer claims that 20% of the boxes contain a figure. You buy three boxes of this cereal. What is the probability that you'll get exactly three figures?

To get three figures means that the first box contains a figure (0.20 chance), as does the second (also 0.20), and the third (0.20). You want three figures; therefore, this is the intersection of three events and the probability is simply $0.20^3 = 0.008$.

If you want to calculate the probability of getting exactly two figures, the situation becomes more complicated. A tree diagram can help you visualize it better.

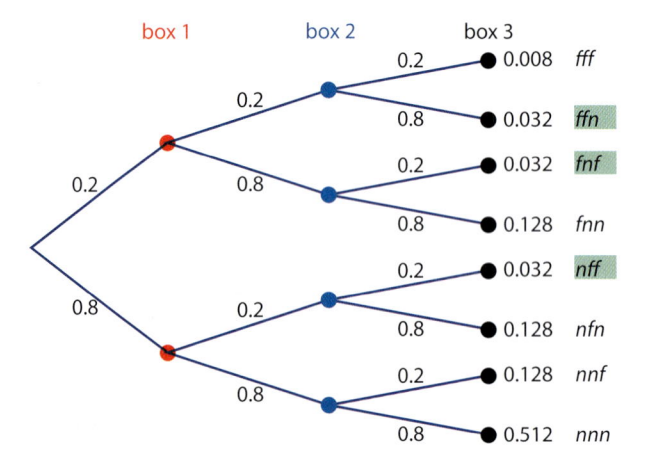

Let f stand for figure and n for no figure. There are three events of interest to us. Since we are interested in two figures, we want to see *ffn*, which has a probability of $0.2 \times 0.2 \times 0.8 = 0.2^2 \times 0.8 = 0.032$, and the other events of interest are *fnf* and *nff*, with probabilities 0.2 × 0.8 × 0.2 = 0.032 and 0.8 × 0.2 × 0.2 = 0.032.

Since the order of multiplication is not important, you see that three probabilities are the same. These three events are disjoint, as can be clearly seen from the tree diagram, and hence the probability of exactly two figures is the sum of the three numbers: 0.032 + 0.032 + 0.032. Of course, you may realize by now that it would be much simpler if you wrote 3(0.032), since there are three events with the same probability.

What if you have five boxes?

The situation is similar, of course. However, a tree diagram would not be useful in this case, as there is too much information to assemble to see the solution. As you have seen above, no matter how you succeed in finding a figure, whether it is in the first box, the second or the third, it has the same probability, 0.2. So, to have two successes (finding figures) in the five boxes, you need the other three to be failures (no figures), with a probability of 0.8 for each failure. Therefore, the chance of having a case like *ffnnn* is $0.2^2 \times 0.8^3$. However, this can happen in several disjoint ways. How many? If you count them, you will find 10! This means the probability of having exactly two figures in five boxes is $10 \times 0.2^2 \times 0.8^3 = 0.2048$.
(Here are the 10 possibilities: *ffnnn, fnfnn, fnnfn, fnnnf, nffnn, nnffn, nnnff, nfnfn, nnfnf, nfnnf*.)

The number 10 is nothing but the *binomial* coefficient (Pascal's entry) you saw in Chapter 3. This is also the 'combination' of three events out of five. (The proof of this result is beyond the scope of this book.)

The previous result can be written as $\binom{5}{2} 0.2^2 \cdot 0.8^3$, where $\binom{5}{2}$ is the binomial coefficient.

You can find experiments like this one in many situations. Coin-tossing is only a simple example of this. Another very common example is opinion polls which are conducted before elections and used to predict voter preferences. Each sampled person can be compared to a coin – but a biased coin! A voter you sample in favour of your candidate can correspond to either a 'head' or a 'tail' on a coin. Such experiments all exhibit the typical characteristics of the **binomial experiment**.

A **binomial experiment** is one that has the following five characteristics:
1. The experiment consists of n identical trials.
2. Each trial has one of two outcomes. We call one of them success, S, and the other failure, F.
3. The probability of success on a single trial, p, is constant throughout the whole experiment. The probability of failure is $1 - p$, which is sometimes denoted by q. That is, $p + q = 1$.
4. The trials are independent.
5. We are interested in the number of successes x that are possible during the n trials. That is, $x = 0, 1, 2, \ldots, n$.

15 Probability Distributions

In the cereal company's example above, we started with $n = 3$ and $p = 0.2$ and asked for the probability of two successes, i.e. $x = 2$. In the second part, we have $n = 5$.

Let us imagine repeating a binomial experiment n times. If the probability of success is p, the probability of having x successes is $pppp\ldots$, x times (p^x), because the order is not important, as we saw before. However, in order to have exactly x successes, the rest, $(n - x)$ trials, must be failures, that is, with probability of $qqqq\ldots$, $(n - x)$ times (q^{n-x}). This is only one order (combination) where the successes happen the first x times and the rest are failures. In order to cater for 'all orders', we have to count the number of orders (combinations) possible. This is given by the binomial coefficient $\binom{n}{x}$.

We will state the following result without proof.

> **The binomial distribution**
> Suppose that a random experiment can result in two possible mutually exclusive and collectively exhaustive outcomes, 'success' and 'failure', and that p is the probability of a success resulting in a single trial. If n independent trials are carried out, the distribution of the number of successes 'x' resulting is called the **binomial distribution**. Its probability distribution function for the binomial random variable x is:
>
> P(x successes in n independent trials)
> $= P(x) = \binom{n}{x}p^x(1-p)^{n-x} = \binom{n}{x}p^x q^{n-x}$, for $x = 0, 1, 2, \ldots n$.
>
> **Notation:**
> The notation used to indicate that a variable has a binomial probability distribution with n trials and success probability of p is: $x \sim B(n, p)$.

Example 6

The computer shop orders its notebooks from a supplier, which like many suppliers has a rate of defective items of 10%. The shop usually takes a sample of 10 computers and checks them for defects. If they find two computers defective, they return the shipment. What is the probability that their random sample will contain two defective computers?

Solution

We will consider this to be a random sample and the shipment large enough to render the trials independent of each other. The probability of finding two defective computers in a sample of 10 is given by

$$P(x = 2) = \binom{10}{2}0.1^2 0.9^{10-2} = 45 \times 0.01 \times 0.43047 = 0.194$$

Of course, it is a daunting task to do all the calculations by hand. A GDC can do this calculation for you in two different ways.

The first possibility is to let the calculator do all the calculations in the formula above: Go to the math menu, then choose PRB, then go to #3.

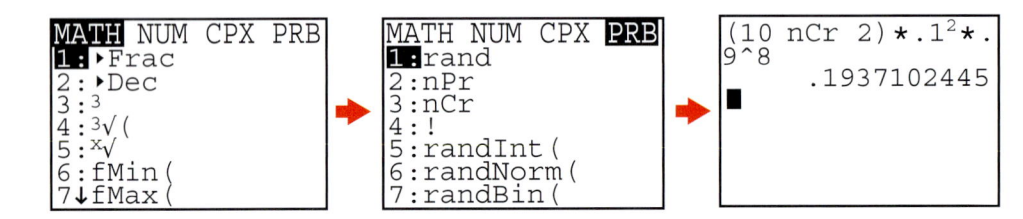

The second one is direct. We go to the 'DISTR' button, then scroll down to 'binompdf' and write down the two parameters followed by the number of successes:

```
DISTR DRAW          binompdf(10,.1,2
1:normalpdf(        )
2:normalcdf(              .1937102445
3:invNorm(
4:invT(
5:tpdf(
6:tcdf(
7↓X²pdf(
```

Using a spreadsheet, you can also produce this result or even a set of probabilities covering all the possible values. The command used here for Excel is (BINOMDIST(B1:G1,10,0.1,FALSE)) which produced the table below:

x	0.00	1.00	2.00	3.00	4.00	5.00	6.00	7.00	8.00	9.00	10.00
$P(x)$	0.349	0.387	0.194	0.057	0.011	0.001	0.000	0.000	0.000	0.000	0.000

Similarly, the GDC can also give you a list of the probabilities:

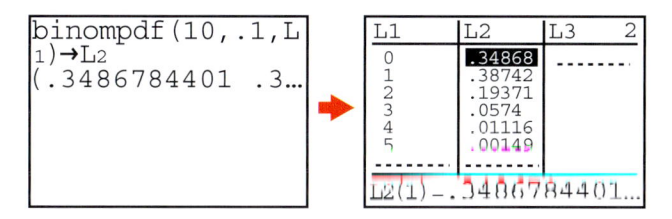

```
binompdf(10,.1,L        L1    L2      L3    2
1)→L2                   0   .34868    ------
(.3486784401 .3...      1   .38742
                        2   .19371
                        3   .0574
                        4   .01116
                        5   .00149
                        ----  ----
                        L2(1)=.3486784401...
```

Like other distributions, when you look at the binomial distribution, you want to look at its expected value and standard deviation.

Using the formula we developed for the expected value, $\sum xP(x)$, we can of course add $xP(x)$ for all the values involved in the experiment. The process would be long and tedious for something we can intuitively know. For example, in the defective items sample, if we know that the defective rate of the computer manufacturer is 10%, it is natural to *expect* to have $10 \times 0.1 = 1$ defective computer! If we have 100 computers with a defective rate of 10%, how many would you expect to be defective? Can you think of a reason why it would not be 10?

This is so simple that many people would not even consider it. The expected value of the successes in the binomial is actually nothing but the number of trials n multiplied by the probability of success, i.e. np!

The binomial probability model

n = number of trials

p = probability of success, $q = 1 - p$ probability of failure

x = number of successes in n trials

$P(x) = \binom{n}{x}p^x(1-p)^{n-x} = \binom{n}{x}p^x q^{n-x}$, for $x = 0, 1, 2, \ldots n$

Expected value = $\mu = np$

Variance = $\sigma^2 = npq$, $\sigma = \sqrt{npq}$

So, in the defective notebooks case, the expected number of defective items in the sample of 10 is $np = 10 \times 0.1 = 1$!

And the standard deviation is $\sigma = \sqrt{npq} = \sqrt{10 \times 0.1 \times 0.9} = 0.949$.

Example 7

Among the studies carried out to examine the effectiveness of advertising methods, a study reported that 4 out of 10 web surfers remember advertisement banners after they have seen them.

a) If 20 web surfers are chosen at random and shown an ad, what is the expected number of surfers that would remember the ad?

b) What is the chance that 5 of those 20 will remember the ad?

c) What is the probability that at most 1 surfer would remember the ad?

d) What is the chance that at least two surfers would remember the ad?

Solution

a) $x \sim (20, 0.4)$. The expected number is simply $20 \times 0.4 = 8$. We expect 8 of the surfers to remember the ad. Notice on the histogram below that the area in red corresponds to the expected value 8.

```
binompdf(20,.4,5
)
          .0746470195
```

b) $P(5) = \binom{20}{5} 0.4^5 (0.6)^{15} = 0.0746$, or see the output from the GDC to the right. Graphically, this area is shown on the histogram as the green area.

```
binompdf(20,.4,0
)
          3.65615844E-5
binompdf(20,.4,1
)
          4.87487792E-4
          4.87487792E-4
```

c) $P(x \leq 1) = P(x = 0) + P(x = 1) = 0.000\,524$

d) $P(x \geq 2) = 1 - P(x \leq 1)$
$ = 1 - 0.000\,524 = 0.999\,475$

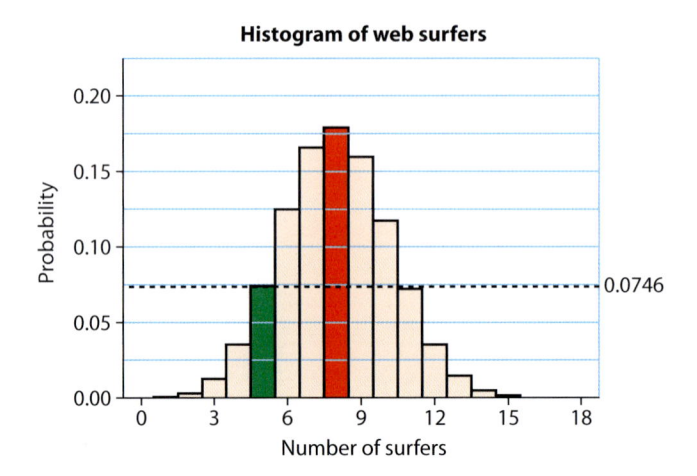

Histogram of web surfers

Exercise 15.2

(Problems marked with (*) are optional.)

1. Consider the following binomial distribution
$$P(x) = \binom{5}{x}(0.6)^x(0.4)^{5-x}, x = 0, 1, \ldots, 5$$
   a) Make a table for this distribution.
   b) Graph this distribution.
   c) Find the mean and standard deviation in two ways:
      (i) by formula
      (ii) by using the table of values you created in part a).
   d) Locate the mean μ and the two intervals $\mu \pm \sigma$ and $\mu \pm 2\sigma$ on the graph.
   e) Find the actual probabilities for x to lie within each of the intervals $\mu \pm \sigma$ and $\mu \pm 2\sigma$ and compare them to the empirical rule.

2. A poll of 20 adults is taken in a large city. The purpose is to determine whether they support banning smoking in restaurants. It is known that approximately 60% of the population supports the decision. Let x represent the number of respondents in favour of the decision.
   a) What is the probability that 5 respondents support the decision?
   b) What is the probability that none of the 20 supports the decision?
   c) What is the probability that at least 1 respondent supports the decision?
   d) What is the probability that at least two respondents support the decision?
   e) Find the mean and standard deviation of the distribution.

3. Consider the binomial random variable with $n = 6$ and $p = 0.3$.
   a) Fill in the probabilities below.

k	0	1	2	3	4	5	6
$P(x \leq k)$							

   b) Fill in the table below. Some cells have been filled for you to guide you.

Number of successes x	List the values of x	Write the probability statement	Explain it, if needed	Find the required probability
At most 3				
At least 3				
More than 3	4, 5, 6	$P(x > 3)$	$1 - P(x \leq 3)$	0.070 47
Fewer than 3				
Between 3 and 5 (inclusive)				
Exactly 3				

4. Repeat question 3 with $n = 7$ and $p = 0.4$.

5. A box contains 8 balls: 5 are green and 3 are white, red and yellow. Three balls are chosen at random without replacement and the number of green balls y is recorded.
   a) Explain why y is not a binomial random variable.
   b) Explain why, when we repeat the experiment with replacement, then y is a binomial.
   c) Give the values of n and p and display the probability distribution in tabular form.
   d) What is the probability that at most 2 green balls are drawn?
   e) What is the expected number of green balls drawn?
   f) What is the variance of the number of balls drawn?
   g) What is the probability that some green balls will be drawn?

6 On a multiple choice test, there are 10 questions, each with 5 possible answers, one of which is correct. Nick is unaware of the content of the material and guesses on all questions.
   a) Find the probability that Nick does not answer any question correctly.
   b) Find the probability that Nick answers at most half of the questions correctly.
   c) Find the probability that Nick answers at least one question correctly.
   d) How many questions should Nick expect to answer correctly?

7 Houses in a large city are equipped with alarm systems to protect them from burglary. A company claims their system to be 98% reliable. That is, it will trigger an alarm in 98% of the cases. In a certain neighbourhood, 10 houses equipped with this system experience an attempted burglary.
   a) Find the probability that all the alarms work properly.
   b) Find the probability that at least half of the houses trigger an alarm.
   c) Find the probability that at most 8 alarms will work properly.

8 Harry Potter books are purchased by readers of all ages! 40% of Harry Potter books were purchased by readers 30 years of age or older! 15 readers are chosen at random. Find the probability that
   a) at least 10 of them are 30 or older
   b) 10 of them are 30 or older
   c) at most 10 of them are younger than 30.

9 A factory makes computer hard disks. Over a long period, 1.5% of them are found to be defective. A random sample of 50 hard disks is tested.
   a) Write down the expected number of defective hard disks in the sample.
   b) Find the probability that three hard disks are defective.
   c) Find the probability that more than one hard disk is defective.

10 Car colour preferences change over time and according to the area the customer lives in and the car model he/she is interested in. In a certain city, a large dealer of BMW cars noticed that 10% of the cars he sells are 'metallic grey'. Twenty of his customers are selected at random, and their car orders are checked for colour. Find the probability that
   a) at least five cars are 'metallic grey'
   b) at most 6 cars are 'metallic grey'
   c) more than 5 are 'metallic grey'
   d) between 4 and 6 are 'metallic grey'
   e) more than 15 are not 'metallic grey'.
   In a sample of 100 customer records, find
   f) the expected number of 'metallic grey' car orders
   g) the standard deviation of 'metallic grey' car orders.(*)
   According to the empirical rule, 95% of the 'metallic grey' orders are between a and b.
   h) Find a and b.(*)

11 Dogs have health insurance too! Owners of dogs in many countries buy health insurance for their dogs. 3% of all dogs have health insurance. In a random sample of 100 dogs in a large city, find
   a) the expected number of dogs with health insurance
   b) the probability that 5 of the dogs have health insurance
   c) the probability that more than 10 dogs have health insurance.

12 A balanced coin is tossed 5 times. Let x be the number of heads observed.
   a) Using a table, construct the probability distribution of x.
   b) What is the probability that no heads are observed?
   c) What is the probability that all tosses are heads?
   d) What is the probability that at least one head is observed?
   e) What is the probability that at least one tail is observed?
   f) Given that the coin is unbalanced in such a way that it shows 2 heads in every 10 tosses, answer the same questions above.

15.3 The normal distribution

Continuous random variables

When a random variable x is discrete, you assign a positive probability to each value that x can take and get the probability distribution for x. The sum of all the probabilities associated with the different values of x is 1.

You have seen, in the discrete variable case, that we graphically represent the probabilities corresponding to the different values of the random variable x with a probability histogram (relative frequency histogram), where the area of each bar corresponds to the probability of the specific value it represents.

Consider now a continuous random variable x, such as height and weight, and length of life of a particular product – a TV set for example. Because it is continuous, the possible values of x are over an interval. Moreover, there are an infinite number of possible values of x. Hence, we cannot find a probability distribution function for x by listing all the possible values of x along with their probabilities, as you see in the histogram below. If we try to assign probabilities to each of these uncountable values, the probabilities will no longer sum to 1, as is the case with discrete variables. Therefore, you must use a different approach to generate the probability distribution for such random variables.

Suppose that you have a set of measurements on a continuous random variable, and you create a relative frequency histogram to describe their distribution. For a small number of measurements, you can use a small number of classes, but as more and more measurements are collected, you can use more classes and reduce the **class width**.

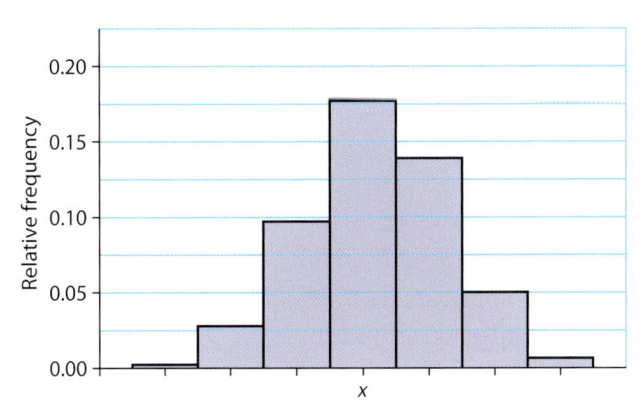

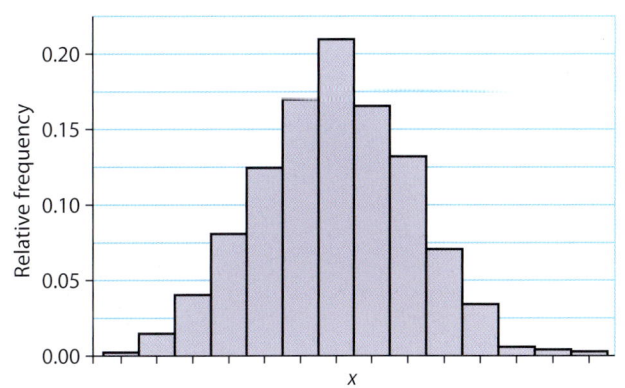

The histogram will slightly change as the class width becomes smaller and smaller, as shown in the diagrams on the next page. As the number of measurements becomes very large and the class width becomes very narrow, the relative frequency histogram appears more and more like the smooth curve you see below. This is what happens in the continuous case, and the smooth curve describing the probability distribution of the continuous random variable becomes the **PDF** (**probability density**

function) of *x*, represented by a curve $y = f(x)$. This curve is such that the entire area under the curve is 1 and the area between any two points is the probability that *x* falls between those two points.

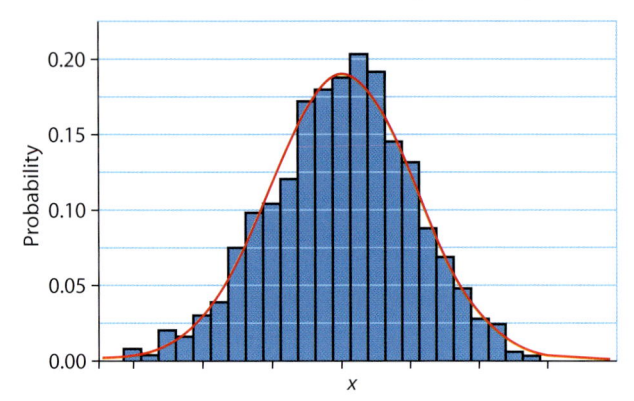

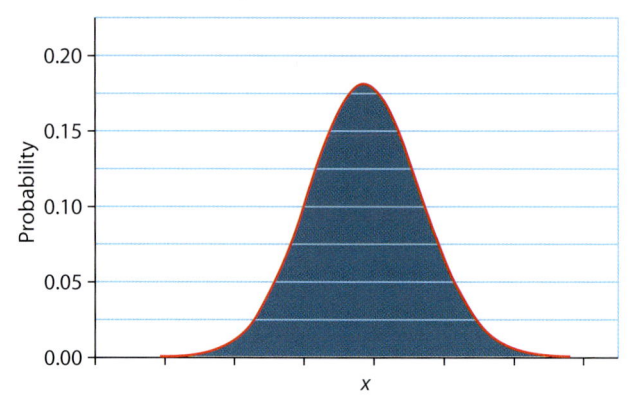

Probability density function

Let *x* be a continuous random variable. The probability density function, $f(x)$, of the random variable is a function with the following properties:
1. $f(x) > 0$ for all values of *x*.
2. The area under the probability density function $f(x)$ over all values of the random variable *x* is equal to 1.0.
3. Suppose this density function is graphed. Let *a* and *b* be two possible values of the random variable *x*, with $a < b$. Then the probability that *x* lies between *a* and *b* [$P(a < x < b)$] is the area under the density function between these points.

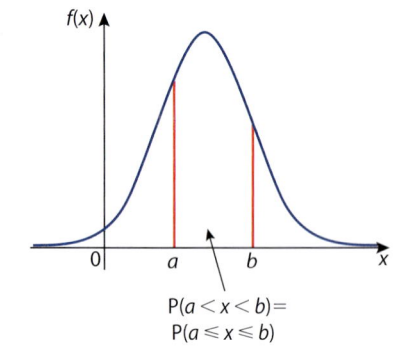

Notice that, based on this definition, the probability that *x* equals any point *a* is 0. This is so because the area above a value, say *a*, is a rectangle whose width is 0. So, for the continuous case, regardless of whether the endpoints *a* and *b* are themselves included, the area included between *a* and *b* is the same.

$$P(a < x < b) = P(a \leq x \leq b) = P(a \leq x < b) = P(a < x \leq b)$$

Continuous probability distributions can assume a variety of shapes. However, for reasons of staying within (with some extensions) the boundaries of the IB syllabus, we will focus on one distribution. In fact, a large number of random variables observed in our surroundings possess a frequency distribution that is approximately bell-shaped. We call that distribution the **normal probability distribution**.

The normal distribution

The most important type of continuous random variable is the *normal* random variable. The probability density function of a normal random variable *x* is determined by two parameters: the mean or expected value μ and the standard deviation σ of the variable.

The normal probability density function is a bell-shaped density curve that is symmetric about the mean μ. Its variability is measured by σ. The larger the value of σ the more variability there is in the curve. That is, the higher the probability of finding values of the random variable further away from the mean. Figure 15.1 represents three different normal density functions with the same mean but different standard deviations. Note how the curves 'flatten' as σ increases. This is so because the area under the curve has to stay equal to 1.

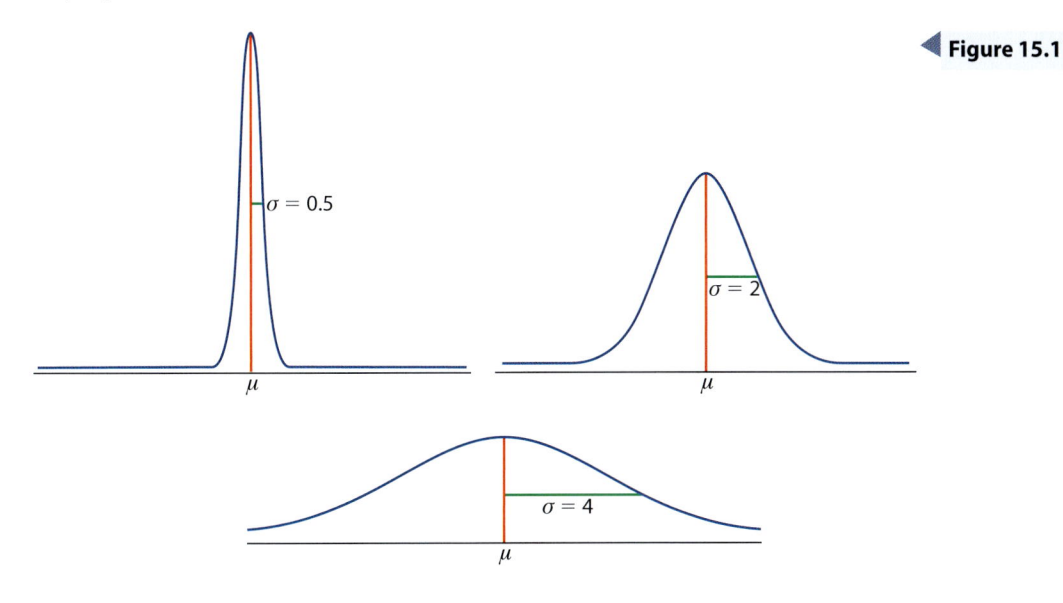

◀ Figure 15.1

Probability density function of the normal distribution

The probability density function for a normally distributed random variable x is

$$f(x) = \frac{1}{\sigma\sqrt{2\pi}}e^{-\frac{(x-\mu)^2}{2\sigma^2}} = \frac{1}{\sigma\sqrt{2\pi}}e^{-\frac{1}{2}\left(\frac{x-\mu}{\sigma}\right)^2} \text{ for } -\infty < x < \infty$$

where μ and σ^2 are any number such that $-\infty < \mu < \infty$ and $0 < \sigma^2 < \infty$, and where e and π are the well-known constants $e = 2.718\,28\ldots$ and $\pi = 3.141\,59\ldots$.

Notation:
When a variable is normally distributed, we write $X \sim N(\mu, \sigma^2)$.

Although we will not make direct use of the formula above, it is interesting to note its properties, because they help us understand how the normal distribution works. Notice that the equation is completely determined by the mean μ and the standard deviation σ.

The graph of a normal probability distribution is shown in Figure 15.2. As you notice, the mean or expected value locates the centre of the distribution, and the distribution is symmetric about this mean. Since the total area under the curve is 1, the symmetry of the curve implies that the area to the right of the mean and the area to the left are both equal to 0.5. The shape, or how 'flat' it is, is determined by σ, as we have seen in Figure 15.1. Large values of σ tend to reduce the height of the curve and increase the spread, and small values of σ increase the height to compensate for the narrowness of the distribution.

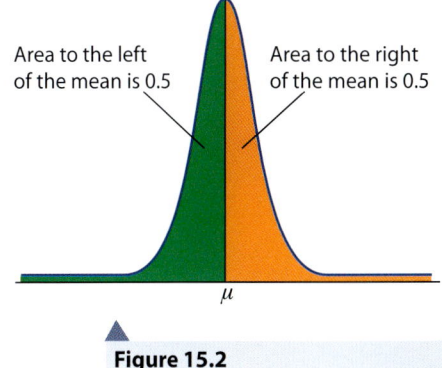

Area to the left of the mean is 0.5

Area to the right of the mean is 0.5

▲ Figure 15.2

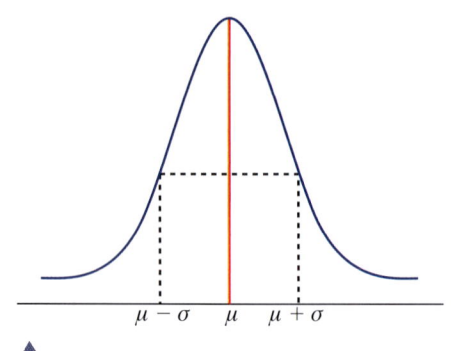

Figure 15.3

So, the normal distribution is fully determined by its mean, μ, and its standard deviation, σ. Changing μ without changing σ moves the normal curve along the horizontal axis without changing its spread. As you have seen above, the standard deviation σ controls the spread of the curve. You can also locate the standard deviation by eye on the curve. One σ to the right or left of the mean μ marks the point where the curvature of the curve changes. That is, as you move right from the mean, at the point where $x = \mu + \sigma$, the curve changes its curvature from downwards to upwards. Similarly, as you move one σ to the left from the mean the curve changes its curvature from downwards to upwards.

Although there are many normal curves, they all have common properties. Here is one important one that you have seen in Chapter 9:

> **The empirical rule – restated**
> In the normal distribution with mean μ and standard deviation σ:
> - Approximately 68% of the observations fall within σ of the mean μ.
> - Approximately 95% of the observations fall within 2σ of the mean μ.
> - Approximately 99.7% of the observations fall within 3σ of the mean μ.

Figure 15.4 illustrates this rule. Later in this section, you will learn how to find these areas from a table or from your GDC.

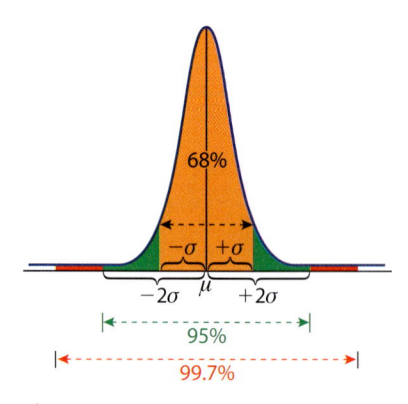

Figure 15.4

Example 8

Heights of young German men between 18 and 19 years of age follow a distribution that is approximately normal, with a mean of 181 cm and a standard deviation of 8 cm (approximately). Describe this population of young men.

Solution
According to the empirical rule, we find that approximately 68% of those young men have a height between 173 cm and 189 cm, 95% of them between 165 cm and 197 cm, and 99.7% between 157 cm and 205 cm. Looking further, you can say that only 0.15% are taller than 205 cm, or shorter than 157 cm.

As the empirical rule suggests, all normal distributions are the same if we measure in units of size σ about the mean μ as centre. Changing to these units is called *standardizing*. To standardize a value, measure how far it is from the mean and express that distance in terms of σ. This is how the calculation can be done:

> **Standardizing**
> If x is a normal random variable, with mean μ and standard deviation σ, the standardized value of x is
> $$z = \frac{x - \mu}{\sigma}$$
> A standardized value is also called the **z-score**.

The quantity $x - \mu$ tells us how far our value is from the mean; dividing by σ then tells us how many standard deviations that distance is equal to.

The standardizing process, as you notice, is a transformation of the normal curve. For discussion purposes, assume the mean μ to be positive. The transformation $x - \mu$ shifts the graph back μ units. So, the new centre is shifted from μ back μ units. That is, the new centre is 0! Dividing by σ is going to 'scale' the distances from the mean and express everything in terms of σ. So, a point that is one standard deviation from the mean is going to be 1 unit above the new mean, i.e. it will be represented by +1. Now, if you look at the empirical rule we discussed earlier, points that are within one standard deviation from the mean will be within a distance of 1 in the new distribution. Instead of being at $\mu + \sigma$ and $\mu - \sigma$, they will be at $0 + 1$ and $0 - 1$ respectively, i.e. -1 and $+1$. (See Figure 15.6.)

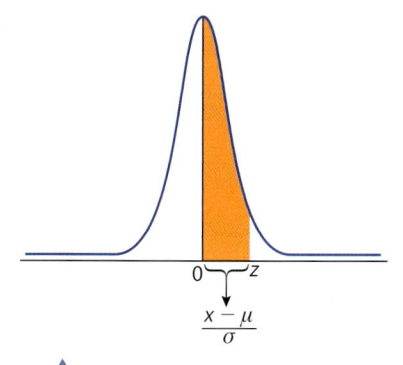

Figure 15.5

The new distribution we created by this transformation is called the **standard normal distribution**. It has a mean of 0 and a standard deviation of 1. It is a very helpful distribution because it will enable us to read the areas under any normal distribution through the standardization process, as will be demonstrated in the examples that follow.

> **Probability density function of the standard normal distribution**
> The probability density function for standard normal distribution is
> $$f(z) = \frac{1}{\sqrt{2\pi}} e^{-\frac{1}{2}(z)^2} \text{ for } -\infty < z < \infty$$

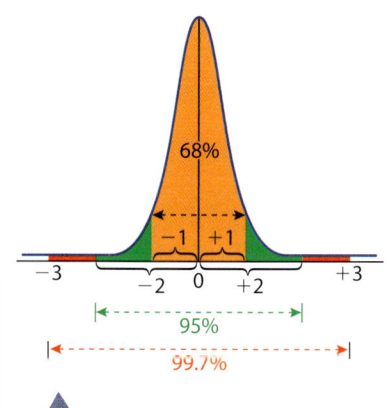

Figure 15.6

Since linear transformations can transform all normal functions to standard, this becomes a very convenient and efficient way of finding the area under any normal distribution.

Let us look at an example.

A young German man with a height of 192 cm has a z-score of

$$z = \frac{x - \mu}{\sigma} = \frac{192 - 181}{8} = 1.375$$

or 1.375 standard deviations above the mean. Similarly, a young man with a height of 175 cm is

$$z = \frac{x - \mu}{\sigma} = \frac{175 - 181}{8} = -0.75$$

or 0.75 standard deviations below the mean.

To find the probability that a normal variable x lies in the interval a to b, we need to find the area under the normal curve $N(\mu, \sigma^2)$ between the points a and b. However, there is an infinitely large number of normal curves – one for each mean and standard deviation. (See Figure 15.7.)

A separate table of areas for each of these curves is obviously not practical. Instead, we use one table for the standard normal distribution, which gives us the required areas. When you standardize a and b, you get two standard numbers z_1 and z_2 such that the area between z_1 and z_2 is the same as the area we need.

15 Probability Distributions

Figure 15.7

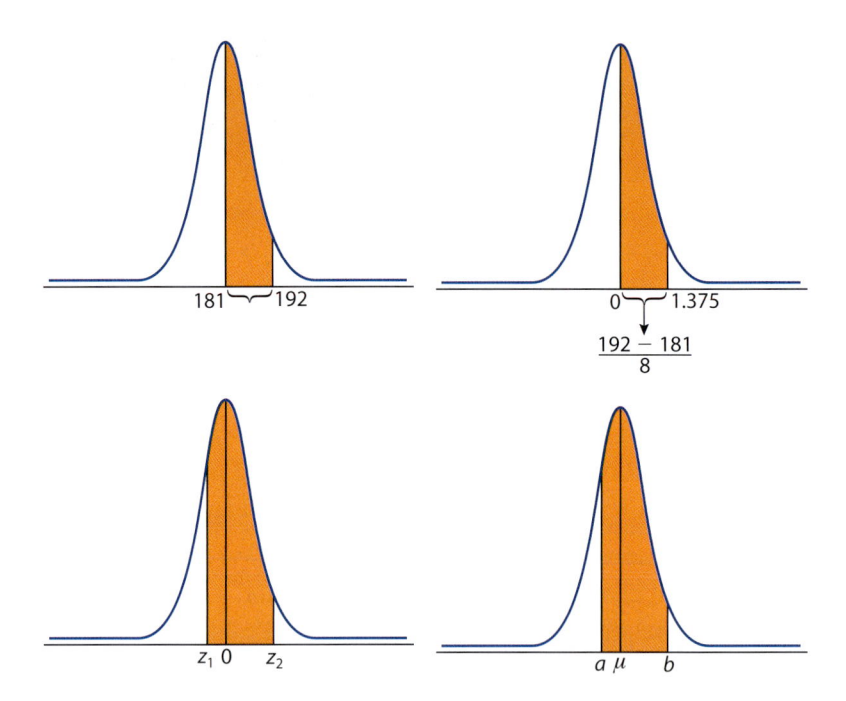

For a complete normal distribution table visit www.heinemann.co.uk/hotlinks, enter the express code 4235P and click on weblink 4.

In the example above, if we are interested in the proportion of young German men whose height is between 175 cm and 192 cm, we calculate the z-scores for these numbers and then read the area from the table. Here is an abbreviated version of the table and instructions on how to use it. (There are many tables of the areas under normal distributions. We will use a table constructed in a similar way to the one used on IB examinations.)

z	0.00	0.01	→	0.05	0.06	0.07	0.08	0.09
0.0	0.5000	0.5040		0.5199	0.5239	0.5279	0.5319	0.5359
0.1	0.5398	0.5438		0.5596	0.5636	0.5675	0.5714	0.5753
0.2	0.5793	0.5832		0.5987	0.6026	0.6064	0.6103	0.6141
1.2	0.8849	0.8869		0.8944	0.8962	0.8980	0.8997	0.9015
↓			→			↓		
↓			→			↓		
1.3	0.9032	0.9049		0.9115	0.9131	**0.9147**	**0.9162**	0.9177
1.4	0.9192	0.9207		0.9265	0.9279	0.9292	0.9306	0.9319
1.5	0.9332	0.9345		0.9394	0.9406	0.9418	0.9429	0.9441
1.6	0.9452	0.9463		0.9505	0.9515	0.9525	0.9535	0.9545

$p = P(Z \leq z)$

Figure 15.8

The table, as constructed, gives you the areas under the normal distribution to the left of some value z, as you see in Figure 15.8.

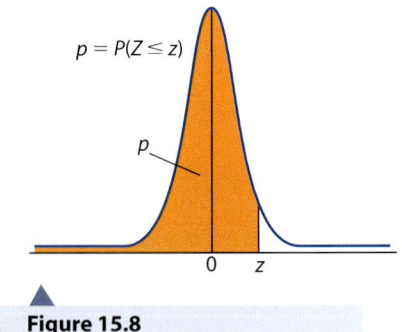

The table starts at 0, and gives the areas till $z = 3.9$. To read an area to the left of a number z, say 1.37, you read the first column to find the first two digits of z. So, in the first column, we stop at the cell containing 1.3. To get the area for 1.37, we look at the first row and choose the column corresponding to 0.07. Where the row at 1.3 meets the column at 0.07 is the area under the normal distribution corresponding to 1.37, namely, 0.9147. That is, the probability of at most a height with $z = 1.37$ is 0.9147. Since the table does not go to 4 decimal places, our answers will not be very precise. So, to find the probability corresponding to a height of 192 cm, we need a z of 1.375, which is not in the table. We can use 1.37, 1.38, or take an average. If we want an average, we read the neighbouring area of 0.9162, and get the average to be 0.915 45.

Unfortunately, due to limitations of space, this type of table does not cater for negative values of z. The good news is that, due to the symmetry of the distribution, the area to the left of a negative value of z is the same as the area to the right of its absolute value. So, if we are interested in the area to the left of -0.75, we look for the area to the right of 0.75, i.e. $1 - P(z < 0.75) = 1 - 0.7734 = 0.2266$ (see Figure 15.10). So, in the example above, if we want to know the probability of a young German man, chosen at random, having a height between 175 cm and 192 cm, we look up the corresponding area under the standard normal distribution between -0.75 and 1.375. Since these two areas are cumulative, we need to subtract them, i.e. the required area is $0.915\,45 - 0.2266 = 0.6885$.

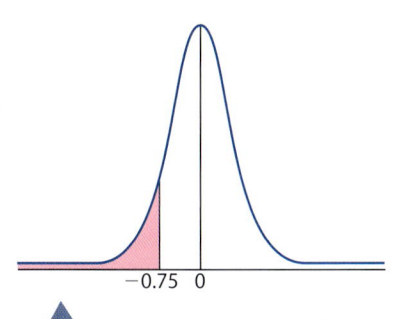

Figure 15.9

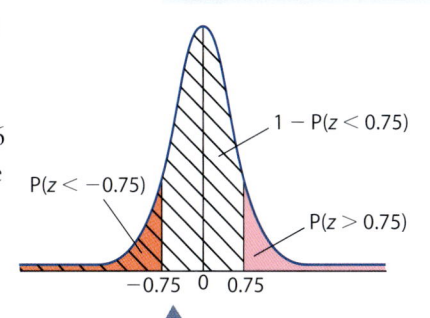

Figure 15.10 $P(z > 0.75) = 1 - P(z < 0.75)$.

Figure 15.11

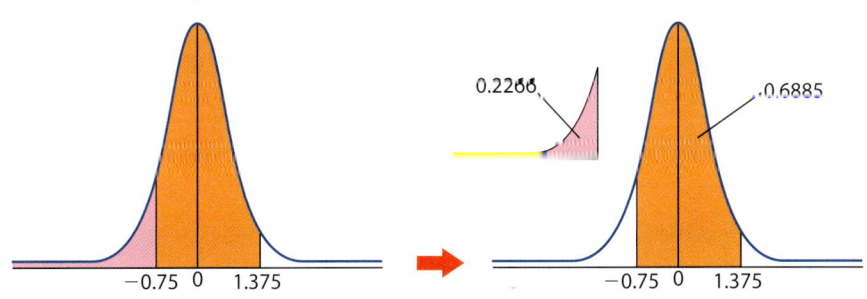

What is the chance that a young German man is taller than 175 cm?

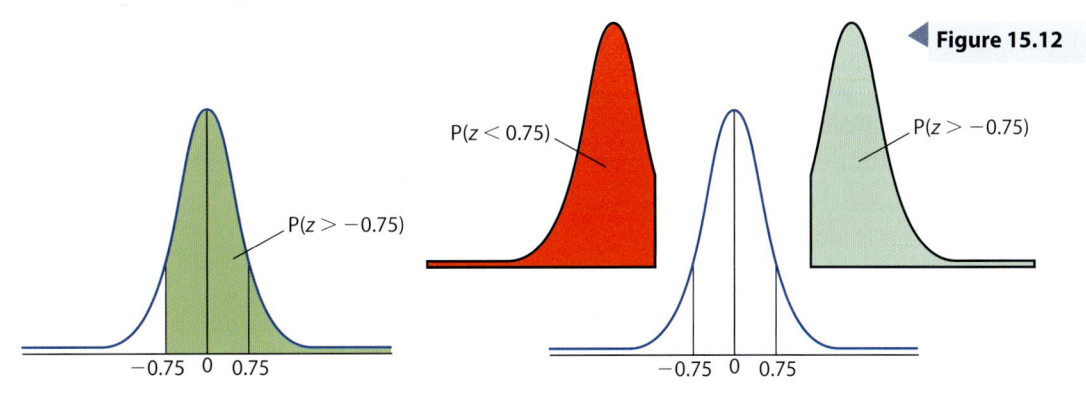

Figure 15.12

This means that we have to look at the area above -0.75. Due to symmetry, the area in question, which is to the right of -0.75, is equal to the area below 0.75, which in turn can be read directly from the table as 0.7734.

These calculations are much easier to calculate using a GDC, of course. Also, with the GDC, you do not need to standardize your variables either. However, because there are cases where you need to understand standardization and other cases where you are *required to use a table*, you need to know both methods.

Here is how your GDC can give you your answers.

You first go to the 'Distribution' menu and choose 'normalcdf'. Then you enter the numbers in the following order: *Lower limit, upper limit, mean, and standard deviation*. The result will be the area you need. See the screen images below.

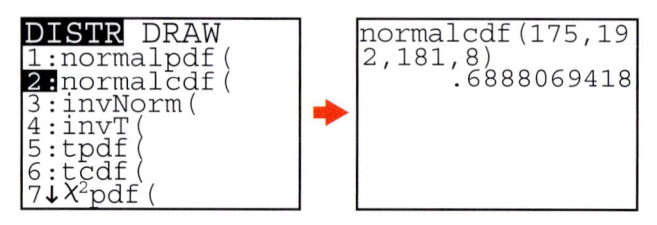

```
DISTR DRAW
1:normalpdf(
2:normalcdf(
3:invNorm(
4:invT(
5:tpdf(
6:tcdf(
7↓X²pdf(
```

```
normalcdf(175,19
2,181,8)
          .6888069418
```

```
normalcdf(-.75,1
.375)
          .6888069418
```

If you want to use the standard normal, your commands will be the same, but you do not need to include the mean and standard deviation. They are the default.

If you need the probability that a young man is taller than 175 cm, you can also read it either by looking at the distribution with the original data or by standardizing.

Example 9

The age of graduate students in engineering programmes throughout the US is normally distributed with mean $\mu = 24.5$ and standard deviation $\sigma = 2.5$.

If a student is chosen at random,
a) what is the probability he/she is younger than 26 years old?
b) what proportion of students is older than 23.7 years?
c) what percentage of students is between 22 and 28 years old?
d) what percentage of the ages falls within 1 standard deviation of the mean? 2 standard deviations? 3 standard deviations?

Solution
If we let X = age of students, then $X \sim N(\mu = 24.5, \sigma^2 = 6.25)$.
a) To answer this, we can either standardize and then read the table for the area left of 0.6:

$$P\left(z < \frac{26 - 24.5}{2.5}\right) = P(z < 0.6) = 0.7257, \text{ or use a GDC}$$

```
normalcdf(0,26,2
4.5,2.5)
          .7257469354
```

Notice here that we put 0 as a lower limit. You can put a number as a lower limit far enough from the mean to make sure you are receiving the correct cumulative distribution.

b) This can be done similarly:
$$P(x > 23.7) = P\left(z > \frac{23.7 - 24.5}{2.5} = -0.32\right)$$
So, by symmetry we know that
$$P(z > -0.32) = P(z < 0.32) = 0.6255$$

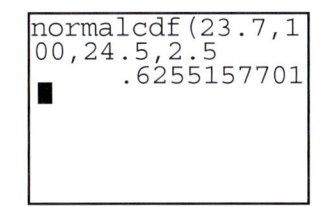

With a GDC:
```
normalcdf(23.7,1
00,24.5,2.5
          .6255157701
```

Also, notice here that we wrote 100 as an upper limit, which is an arbitrary number far enough to the right to be sure we include the whole population.

c) $P(22 < x < 28) = P\left(\frac{22 - 24.5}{2.5} < z < \frac{28 - 24.5}{2.5}\right) = P(-1 < z < 1.4)$

We find the area to the left of 1.4 and to the left of -1 and subtract them
$$= 0.9192 - 0.1587 = 0.7606 = 76.06\%$$

With a GDC:
```
normalcdf(22,28,
24.5,2.5)
          .7605880293
```

d) This, as you know, is the empirical rule we talked about before. Let us see what percentage of the approximately normal data will lie within 1, 2 or 3 standard deviations.
We start with the traditional table:
$$P(-1 \leq z \leq 1) = P(z \leq 1) - P(z \leq -1) = 0.8413 - 0.1587$$
$$= \mathbf{0.6826}$$
This is the exact value corresponding to the empirical rule's 68%!
$$P(-2 \leq z \leq 2) = P(z \leq 2) - P(z \leq -2) = 0.9772 - 0.0228$$
$$= \mathbf{0.9544}$$
Again, this is the exact value corresponding to the empirical rule's 95%!
$$P(-3 \leq z \leq 3) = P(z \leq 3) - P(z \leq -3) = 0.9987 - 0.0013$$
$$= \mathbf{0.9973}$$
And again, this is the exact value corresponding to the empirical rule's 99.7%!

The inverse normal distribution

Another type of problem arises in situations similar to the one above when we are given a cumulative probability and would like to find the value in our data that has this cumulative probability. For example, what age

15 Probability Distributions

marks the 95th percentile? That is, what age is higher or equal to 95% of the population? To answer this question, we need to reverse our steps. So far, we are given a value and then we look for the area corresponding to it. Now, we are given the area and we have to look for the number. That is why this is called the **inverse normal distribution**. Again, the approach is to find the *standard inverse normal number and then to 'de-standardize' it.* That is, to find the value from the original data that corresponds to the z-value at hand.

There is an inverse normal table available online (see box on page 520). We will produce a part of the inverse normal table here for explanation.

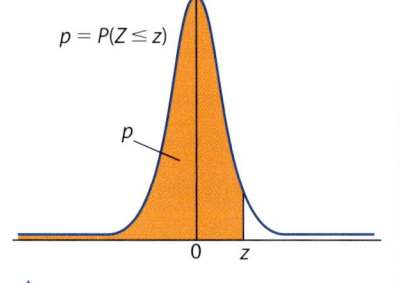

Figure 15.13

$p = P(Z \leq z)$

p	0.000	0.001	→	0.005	0.006	0.007	0.008	0.009
0.50	0.0000	0.0025		0.0125	0.0150	0.0175	0.0201	0.0226
0.51	0.0251	0.0276		0.0376	0.0401	0.0426	0.0451	0.0476
0.52	0.0502	0.0527		0.0627	0.0652	0.0677	0.0702	0.0728
0.53	0.0753	0.0778		0.0878	0.0904	0.0929	0.0954	0.0979
↓			→					
↓			→					
0.74	0.6433	0.6464		0.6588	0.6620	0.6651	0.6682	0.6713
0.75	0.6745	0.6776		0.6903	0.6935	0.6967	0.6999	0.7031
0.76	0.7063	0.7095		0.7225	0.7257	0.7290	0.7323	0.7356
0.77	0.7388	0.7421		0.7554	0.7588	0.7621	0.7655	0.7688
0.78	0.7722	0.7756		0.7892	0.7926	0.7961	0.7995	0.8030
0.79	0.8064	0.8099		0.8239	0.8274	0.8310	0.8345	0.8381

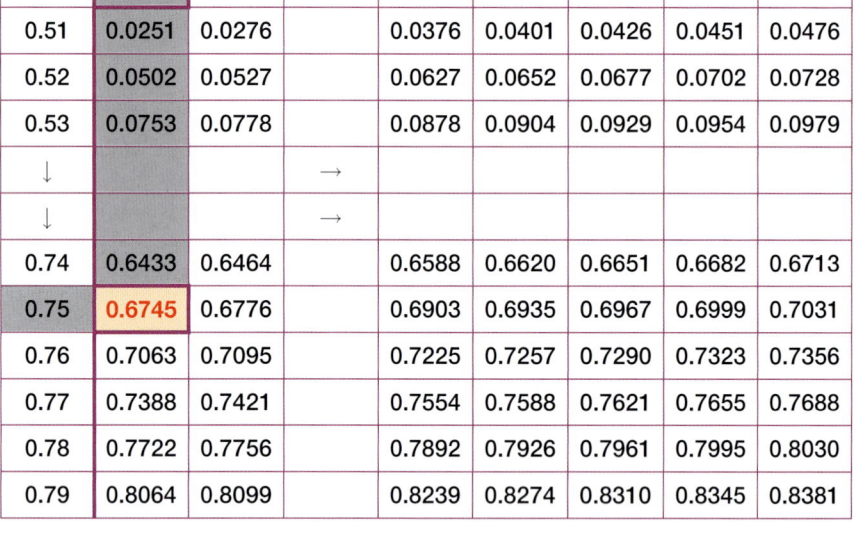

The table gives a selection of probabilities above the mean and the body of the table gives the z-value corresponding to that area. You know that 0 has a cumulative probability of 0.5. Look at the table and observe the intersection of the 0.50 row and the 0.000 column. It is 0, the mean of the standard normal distribution.

If we need to know what z-score the third quartile Q_3 is, for example, we need to look up 0.75. The z-score corresponding to Q_3 is 0.6745 as you see.

Suppose you want to find the z-score that leaves an area of 0.915 below it.

p	0.000	→	0.005
0.50	0.0000		0.0125
0.51	0.0251		0.0376
↓			
0.91	1.3408		1.3722

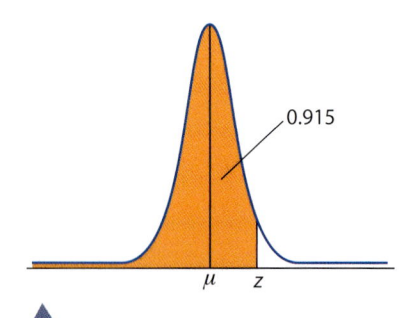

Figure 15.14

In the first column, we choose 0.91, then at the intersection of the row at 0.91 and the column at 0.005 the z-score corresponds to 0.915. So,

$$P(z < 1.3722) = 0.9151$$

The GDC can also be used in this case. The process is identical to the normal calculation. The difference is in choosing 'invNorm' instead.

```
invNorm(.5)
                    0
invNorm(.915)
           1.37220381
invNorm(.75)
            .6744897495
```

In the young German men example, we would like to find what height leaves 95% of the population below it.

In this case, we look up the z-score corresponding to 0.95 and we find that it is $z = 1.6449$.

Now $z = 1.6449 = \dfrac{x - 181}{8} \Rightarrow x - 181 = 8 \times 1.6449$

$\Rightarrow x = 181 + 8 \times 1.6449 = 194.16$.

So, 95% of the young German men are shorter than 194.16 cm.

The GDC gives you this number with less effort:

```
invNorm(.95,181,8)
            194.158829
```

Example 10

Since November 2007, the average time it takes fast trains (Eurostar) to travel between London and Paris is 2 hours 15 minutes, with a standard deviation of 4 minutes. Assume a normal distribution.
a) What is the probability that a randomly chosen trip will take longer than 2 hours and 20 minutes?
b) What is the probability that a randomly chosen trip will take less than 2 hours and 10 minutes?
c) What is the IQR of a trip on these trains?

Solution

We will do each problem using a table and a GDC to acquaint you with both methods.
a) The mean $\mu = 2.25$ and $\sigma = 0.067$.
2 hours 20 minutes = 2.33

$$P(x > 2.33) = P\left(z > \dfrac{2.33 - 2.25}{0.067}\right) = P(z > 1.25)$$

From the table: $P(z > 1.25) = 1 - P(z < 1.25) = 1 - 0.8944 = 0.1056$
Using your GDC:

```
normalcdf(2.3333
,100,2.25,.06667
)
            .10575261
```

The number 100 is arbitrary!

b) 2 hours 10 minutes = 2.167

$$P(x < 2.167) = P\left(z > \frac{2.167 - 2.25}{0.067}\right) = P(z < -1.25)$$

From the table, and by symmetry, this is the same as $P(z > 1.25)$, which we found in part a) above.

GDC

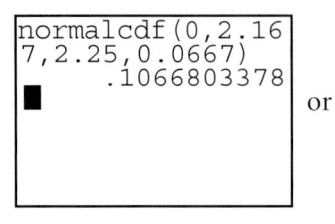

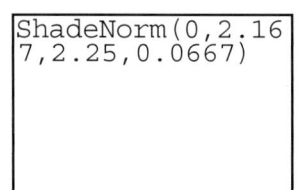

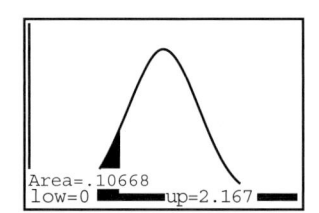

```
normalcdf(0,2.16
7,2.25,0.0667)
      .1066803378
■
```
or
```
ShadeNorm(0,2.16
7,2.25,0.0667)
```

Area=.10668
low=0   up=2.167

c) To find the IQR, we need to find Q_1 and Q_3.

Q_1 is the number that leaves 25% of the data before it. So, we need to find the inverse normal variable that has an area of 0.25 before it.

From the table we can only do so using symmetry. So, we find the z-score that corresponds to 0.25 by finding its symmetrical number, which is the z-score with 0.75. So, we only need to find $z(0.75)$. The table of standard inverse normal gives us $z = 0.6745$.

So, Q_1 corresponds to -0.6745.

$$z = -0.6745 = \frac{x - 2.25}{0.067} \Rightarrow x - 2.25 = 0.067 \times (-0.6745)$$
$$\Rightarrow x = 2.25 - 0.045 = 2.205$$

Q_3 corresponds to 0.6745.

$$z = 0.6745 = \frac{x - 2.25}{0.067} \Rightarrow x - 2.25 = 0.067 \times (0.6745)$$
$$\Rightarrow x = 2.25 + 0.045 = 2.295$$

IQR = 2.295 − 2.205 = 0.090 of an hour, i.e. 5.4 minutes.

Example 11

The age at which babies develop the ability to walk can be described by a normal model. It is known that 5% of babies learn how to walk by the age of 10 months and 25% need more than 13 months. Find the mean and standard deviation of the distribution.

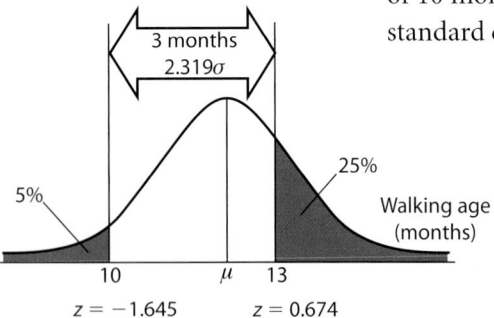

Solution

Looking at the diagram left will help you visualize the solution. There are several approaches to this problem. Here is one:

Look at the distance between 10 and 13 months in two different ways. First, 10 and 13 months are 3 months apart. When standardized, the respective z-scores are −1.645 and

3 months
2.319σ

5%
25%
Walking age (months)

10   μ 13
$z = -1.645$ $z = 0.674$

0.674. The z-scores are 2.319 standard deviations apart. So, 3 months must be the same as 2.319 standard deviations. Here is the calculation:

To find the z-score for the lowest 5% and the highest 25%, we can use the inverse table or a GDC:

```
invNorm(.05)
        -1.644853626
invNorm(.75)
         .6744897495
```

$2.319\sigma = 3$, $\sigma = 1.294$, and

$$z = \frac{x - \mu}{\sigma} \Rightarrow 0.674 = \frac{13 - \mu}{1.294} \Rightarrow \mu = 12.128 \text{ or, alternatively,}$$

$$-1.645 = \frac{10 - \mu}{\sigma} \Rightarrow \mu - 1.645\sigma = 10,$$

$$0.674 = \frac{13 - \mu}{\sigma} \Rightarrow \mu + 0.674\sigma = 13$$

Solving the system of two equations in μ and σ will give the same result as above.

Exercise 15.3

1. The time it takes to change the batteries of your GDC is approximately normal with mean 50 hours and standard deviation of 7.5 hours.
   Find the probability that your newly equipped GDC will last
   a) at least 50 hours
   b) between 50 and 75 hours
   c) less than 42.5 hours
   d) between 42.5 and 57.5 hours
   e) more than 65 hours
   f) 47.5 hours

2. Find each of the following probabilities.
   a) $P(|z| < 1.2)$
   b) $P(|z| > 1.4)$
   c) $P(x < 3.7)$, where $x \sim N(3, 3)$
   d) $P(x > -3.7)$, where $x \sim N(3, 3)$

3. A car manufacturer introduces a new model that has an in-city mileage of 11.4 litres/100 kilometres. Tests show that this model has a standard deviation of 1.26. The distribution is assumed to be normal.
   A car is chosen at random from this model.
   a) What is the probability that it will have a consumption less than 8.4 litres/100 kilometres?
   b) What is the probability that the consumption is between 8.4 and 14.4 litres/100 kilometres?

4. Find the value of z that will be exceeded only 10% of the time.

5. Find the value of $z = z_0$ such that 95% of the values of z lie between $-z_0$ and $+z_0$.

6 The scores on a public schools examination are normally distributed with a mean of 550 and a standard deviation of 100.
   a) What is the probability that a randomly chosen student from this population scores below 400?
   b) What is the probability that a student will score between 450 and 650?
   c) What score should you have in order to be in the 90th percentile?
   d) Find the IQR of this distribution.

7 A company producing and packaging sugar for home consumption put labels on their sugar bags noting the weight to be 500 g. Their machines are known to fill the bags with weights that are normally distributed with a standard deviation of 5.7 g. A bag that contains less than 500 g is considered to be underweight and is not appreciated by consumers.
   a) If the company decides to set their machines to fill the bags with a mean of 512 g, what fraction will be underweight?
   b) If they wish the percentage of underweight bags to be at most 4%, what mean setting must they have?
   c) If they do not want to set the mean as high as 512, but instead at 510, what standard deviation gives them at most 4% underweight bags?

8 In a large school, heights of students who are 13 years old are normally distributed with a mean of 151 cm and a standard deviation of 8 cm. Find the probability that a randomly chosen child is
   a) shorter than 166 cm   b) within 6 cm of the average.

9 The time it takes Kevin to get to school every day is normally distributed with a mean of 12 minutes and standard deviation of 2 minutes. Estimate the number of days when he takes
   a) longer than 17 minutes
   b) less than 10 minutes
   c) between 9 and 13 minutes.
   There are 180 school days in Kevin's school.

10 X has a normal distribution with mean 16. Given that the probability that X is less than 16.56 is 64%, find the standard deviation, σ, of this distribution.

11 X has a normal distribution with variance of 9. Given that the probability that X is more than 36.5 is 2.9%, find the mean, μ, of this distribution.

12 X has a normal distribution such that the probability that X is larger than 14.6 is 93.5% and $P(x > 29.6) = 2.2\%$. Find the mean, μ, and the standard deviation, σ, of this distribution.

13 $X \sim N(\mu, \sigma^2)$, $P(x > 19.6) = 0.16$ and $P(X < 17.6) = 0.012$. Find μ and σ.

14 Bottles of mineral water sold by a company are advertised to contain 1 litre of water. To guarantee customer satisfaction the company actually adjusts its filling process to fill the bottles with an average of 1012 ml. The process follows a normal distribution with standard deviation of 5 ml.
   a) Find the probability that a randomly chosen bottle contains more than 1010 ml.
   b) Find the probability that a bottle contains less than the advertised volume.
   c) In a shipment of 10 000 bottles, what is the expected number of 'underfilled' bottles?

15 Cholesterol plays a major role in a person's heart health. High blood cholesterol is a major risk factor for coronary heart disease and stroke. The level of cholesterol in the blood is measured in milligrams per decilitre of blood (mg/dl). According to the WHO, in general, less than 200 mg/dl is a desirable level, 200 to 239 is borderline high, and above 240 is a high risk level and a person with this level

has more than twice the risk of heart disease as a person with less than a 200 level.

In a certain country, it is known that the average cholesterol level of their adult population is 184 mg/dl with a standard deviation of 22 mg/dl. It can be modelled by a normal distribution.
a) What percentage do you expect to be borderline high?
b) What percentage do you consider are high risk?
c) Estimate the interquartile range of the cholesterol levels in this country.
d) Above what value are the highest 2% of adults' cholesterol levels in this country?

16 A manufacturer of car tyres claims that the treadlife of its winter tyres can be described by a normal model with an average life of 52 000 km and a standard deviation of 4000 km.
   a) You buy a set of tyres from this manufacturer. Is it reasonable for you to hope they last more than 64 000 km?
   b) What fraction of these tyres do you expect to last less than 48 000 km?
   c) What fraction of these tyres do you expect to last between 48 000 km and 56 000 km?
   d) What is the IQR of the treadlife of this type of tyre?
   e) The company wants to guarantee a minimum life for these tyres. That is, they will refund customers whose tyres last less than a specific distance. What should their minimum life guarantee be so that they do not end up refunding more than 2% of their customers?

17 Chicken eggs are graded by size for the purpose of sales. In Europe, modern egg sizes are defined as follows: very large has a mass of 73 g or more, large is between 63 and 73 g, medium is between 53 and 63 g, and small is less than 53 g. The small size is usually considered as undesirable by consumers.
   a) Mature hens (older than 1 year) produce eggs with an average mass of 67 g. 98% of the eggs produced by mature hens are above the minimum desirable weight. What is the standard deviation if the egg production can be modelled by a normal distribution?
   b) Young hens produce eggs with a mean masss of 51 g. Only 28% of their eggs exceed the desired minimum. What is the standard deviation?
   c) A farmer finds that 7% of his farm's eggs are 'underweight', and 12% are very large. Estimate the mean and standard deviation of this farmer's eggs.

Practice questions

1 Residents of a small town have savings which are normally distributed with a mean of $3000 and a standard deviation of $500.
   a) What percentage of townspeople have savings greater than $3200?
   b) Two townspeople are chosen at random. What is the probability that both of them have savings between $2300 and $3300?
   c) The percentage of townspeople with savings less than d dollars is 74.22%. Find the value of d.

2 A box contains 35 red discs and 5 black discs. A disc is selected at random and its colour noted. The disc is then replaced in the box.
   a) In eight such selections, what is the probability that a black disc is selected
      (i) exactly once?
      (ii) at least once?
   b) The process of selecting and replacing is carried out 400 times. What is the expected number of black discs that would be drawn?

3 The graph shows a normal curve for the random variable X, with mean μ and standard deviation σ.

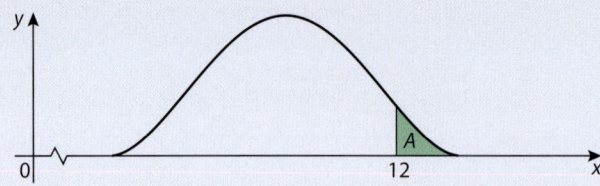

It is known that $P(X \geq 12) = 0.1$.
 a) The shaded region A is the region under the curve where $x \geq 12$. Write down the area of the shaded region A.
It is also known that $P(X \leq 8) = 0.1$.
 b) Find the value of μ, explaining your method in full.
 c) Show that $\sigma = 1.56$ to an accuracy of 3 significant figures.
 d) Find $P(X \leq 11)$.

4 A fair coin is tossed eight times. Calculate
 a) the probability of obtaining exactly 4 heads
 b) the probability of obtaining exactly 3 heads
 c) the probability of obtaining 3, 4 or 5 heads.

5 The lifespan of a particular species of insect is normally distributed with a mean of 57 hours and a standard deviation of 4.4 hours.
The probability that the lifespan of an insect of this species lies between 55 and 60 hours is represented by the shaded area in the following diagram. This diagram represents the standard normal curve.
 a) Write down the values of a and b.
 b) Find the probability that the lifespan of an insect of this species is
  (i) more than 55 hours
  (ii) between 55 and 60 hours.
90% of the insects die after t hours.
 c) (i) Represent this information on a standard normal curve diagram, similar to the one shown, indicating clearly the area representing 90%.
  (ii) Find the value of t.

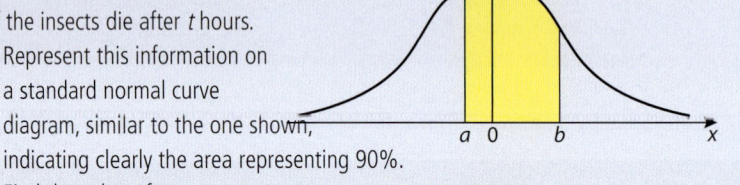

6 An urban highway has a speed limit of 50 km h^{-1}. It is known that the speeds of vehicles travelling on the highway are normally distributed, with a standard deviation of 10 km h^{-1}, and that 30% of the vehicles using the highway exceed the speed limit.
 a) Show that the mean speed of the vehicles is approximately 44.8 km h^{-1}.
(*The following part is optional and is currently not in the syllabus of Maths SL.*)
The police conduct a 'Safer Driving' campaign intended to encourage slower driving, and want to know whether the campaign has been effective. It is found that a sample of 25 vehicles has a mean speed of 41.3 km h^{-1}.
 b) Given that the null hypothesis is
    H0: the mean speed has been unaffected by the campaign state
    state H1, the alternative hypothesis.
 c) State whether a one-tailed or two-tailed test is appropriate for these hypotheses, and explain why.
 d) Has the campaign had significant effect at the 5% level?

7 Intelligence quotient (IQ) in a certain population is normally distributed with a mean of 100 and a standard deviation of 15.
  a) What percentage of the population has an IQ between 90 and 125?
  b) If two persons are chosen at random from the population, what is the probability that both have an IQ greater than 125?

(*The following part is optional and is currently not in the syllabus of Maths SL.*)
  c) The mean IQ of a random group of 25 persons suffering from a certain brain disorder was found to be 95.2. Is this sufficient evidence, at the 0.05 level of significance, that people suffering from the disorder have, on average, a lower IQ than the entire population? State your null hypothesis and your alternative hypothesis, and explain your reasoning.

8 Bags of cement are labelled 25 kg. The bags are filled by machine and the actual weights are normally distributed with mean 25.7 kg and standard deviation 0.50 kg.
  a) What is the probability a bag selected at random will weigh less than 25.0 kg?
  In order to reduce the number of underweight bags (bags weighing less than 25 kg) to 2.5% of the total, the mean is increased without changing the standard deviation.
  b) Show that the increased mean is 26.0 kg.
  It is decided to purchase a more accurate machine for filling the bags. The requirements for this machine are that only 2.5% of bags be under 25 kg and that only 2.5% of bags be over 26 kg.
  c) Calculate the mean and standard deviation that satisfy these requirements.
  The cost of the new machine is $5000. Cement sells for $0.80 per kg.
  d) Compared to the cost of operating with a 26 kg mean, how many bags must be filled in order to recover the cost of the new equipment?

9 The mass of packets of a breakfast cereal is normally distributed with a mean of 750 g and standard deviation of 25 g.
  a) Find the probability that a packet chosen at random has mass
    (i) less than 740 g
    (ii) at least 780 g
    (iii) between 740 g and 780 g.
  b) Two packets are chosen at random. What is the probability that both packets have a mass that is less than 740 g?
  c) The mass of 70% of the packets is more than x grams. Find the value of x.

10 In a country called Tallopia, the height of adults is normally distributed with a mean of 187.5 cm and a standard deviation of 9.5 cm.
  a) What percentage of adults in Tallopia have a height greater than 197 cm?
  b) A standard doorway in Tallopia is designed so that 99% of adults have a space of at least 17 cm over their heads when going through a doorway. Find the height of a standard doorway in Tallopia. Give your answer to the nearest cm.

11 It is claimed that the masses of a population of lions are normally distributed with a mean mass of 310 kg and a standard deviation of 30 kg.
  a) Calculate the probability that a lion selected at random will have a mass of 350 kg or more.
  b) The probability that the mass of a lion lies between a and b is 0.95, where a and b are symmetric about the mean. Find the values of a and b.

12 Reaction times of human beings are normally distributed with a mean of 0.76 seconds and a standard deviation of 0.06 seconds.

The graph below is that of the standard normal curve. The shaded area represents the probability that the reaction time of a person chosen at random is between 0.70 and 0.79 seconds.

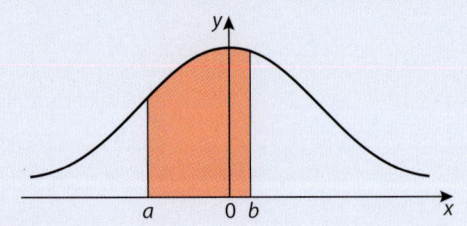

a) Write down the values of a and b.
b) Calculate the probability that the reaction time of a person chosen at random is
   (i) greater than 0.70 seconds
   (ii) between 0.70 and 0.79 seconds.

Three per cent (3%) of the population have a reaction time less than c seconds.
c) (i) Represent this information on a diagram similar to the one above. Indicate clearly the area representing 3%.
   (ii) Find c.

13 A factory makes calculators. Over a long period, 2% of them are found to be faulty. A random sample of 100 calculators is tested.
a) Write down the expected number of faulty calculators in the sample.
b) Find the probability that three calculators are faulty.
c) Find the probability that more than one calculator is faulty.

14 The speeds of cars at a certain point on a straight road are normally distributed with mean μ and standard deviation σ. 15% of the cars travelled at speeds greater than 90 km h^{-1} and 12% of them at speeds less than 40 km h^{-1}. Find μ and σ.

15 Bag A contains 2 red balls and 3 green balls. Two balls are chosen at random from the bag without replacement. Let X denote the number of red balls chosen. The following table shows the probability distribution for X.

x	0	1	2
$P(X = x)$	$\frac{3}{10}$	$\frac{6}{10}$	$\frac{1}{10}$

a) Calculate $E(X)$, the mean number of red balls chosen.

Bag B contains 4 red balls and 2 green balls. Two balls are chosen at random from bag B.
b) (i) Draw a tree diagram to represent the above information, including the probability of each event.
   (ii) Hence, find the probability distribution for Y, where Y is the number of red balls chosen.

A standard die with six faces is rolled. If a 1 or 6 is obtained, two balls are chosen from bag A, otherwise two balls are chosen from bag B.
c) Calculate the probability that two red balls are chosen.
d) Given that two red balls are obtained, find the conditional probability that a 1 or 6 was rolled on the die.

16 Ball bearings are used in engines in large quantities. A car manufacturer buys these bearings from a factory. They agree on the following terms: The car company chooses a sample of 50 ball bearings from the shipment. If they find more than 2 defective bearings, the shipment is rejected. It is a fact that the factory produces 4% defective bearings.
   a) What is the probability that the sample is clear of defects?
   b) What is the probability that the shipment is accepted?
   c) What is the expected number of defective bearings in the sample of 50?

17 Each CD produced by a certain company is guaranteed to function properly with a probability of 98%. The company sells these CDs in packages of 10 and offers a money-back guarantee that all the CDs in a package will function.
   a) What is the probability that a package is returned?
   b) You buy three packages. What is the probability that exactly 1 of them must be returned?

18 The table below shows the probability distribution of a random variable X.

x	0	1	2	3
$P(X = x)$	$2k$	$2k^2$	$k^2 + k$	$2k^2 + k$

   a) Calculate the value of k.
   b) Find $E(X)$.

19 It is estimated that 2.3% of the cherry tomato fruits produced on a certain farm are considered to be small and cannot be sold for commercial purposes. The farmers have to separate such fruits and use them for domestic consumption instead.
   a) 12 tomatoes are randomly selected from the produce. Calculate
      (i) the probability that three are not fit for selling
      (ii) the probability that at least four are not fit for selling.
   b) It is known that the sizes of such tomatoes are normally distributed with a mean of 3 cm and a standard deviation of 0.5 cm. Tomatoes that are categorized as large will have to be larger than 2.5 cm. What proportion of the produce is large?

Questions 1–15: © International Baccalaureate Organization

16 Internal Assessment – Portfolio Tasks

Section I – Preparation for portfolio tasks: *Problem solvers*

These four introductory '*problem solvers*' are intended to prepare you for successfully completing a set of portfolio tasks, in order to satisfy the internal assessment component of the Mathematics SL course. Each of the problem solvers will challenge you to think mathematically and, importantly, to clearly document your thinking and working that leads to your solution.

In the process of documenting your solution, it is important to address the following four components of a comprehensive solution.

1. Conceptual understanding – the '*what*'
   - Indicate important information, assumptions and mathematical concepts you apply.
2. Processes and strategies – the '*how*'
   - Clearly show and explain the thinking and procedural steps that lead to your answer.
3. Confirmation – the '*check*'
   - Verify your solution by looking at further examples or finding an alternative method.
4. Communication – the '*connecting path*'
   - The solution needs to be thorough and complete, with all of the parts (concepts, strategies and confirmation) fitting together using tables, diagrams, pictures, graphs and/or words.

Problem solver 1: Finish line dilemma

An automobile race is arranged between two drivers – Steady Eddy and Rapid Robert. Steady Eddy – living up to his reputation – manages to average exactly 150 kilometres per hour (kph) for the race. Rapid Robert has some mechanical problems early in the race and averages 140 kph for the first half of the length of the race. Rapid Robert recovers during the later stages of the race and averages 160 kph for the second half of the length of the race. What is the outcome of the race?

Problem solver 2: Modelling sunrise, sunset and hours of daylight

The following graph shows the times for sunrise (dashed curve) and sunset (solid curve) on the 21st of each month, from January to December, for the city of Aberdeen, Scotland (latitude 58° 12′ N).

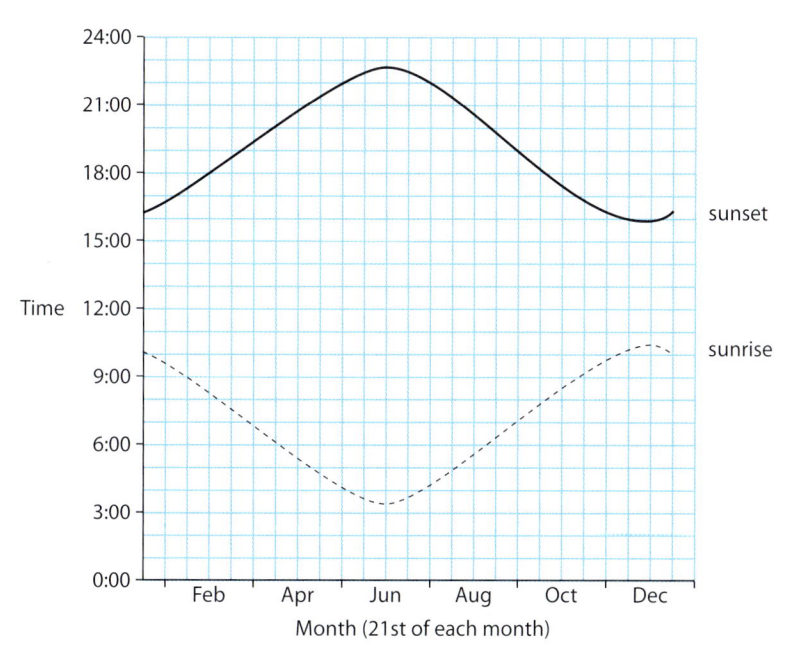

From the shape of the two curves, we can conjecture that each can be modelled by a sinusoidal function of the form $y = a\sin(b(x + c)) + d$. The independent variable x represents the time in months. To simplify assigning values for x, assume that each month has 30 days and denote the 21st day of successive months by 0 to 11 — that is, January 21 is 0, February 21 is 1, March 21 is 2, and so on. Estimate the value for each of the parameters a, b, c and d for both the sunset curve and the sunrise curve to conjecture sinusoidal functions that accurately model both curves. Use technology to graph your two functions. Comment on how well they model the given curves for sunset and sunrise times.

Use your two functions for sunset and sunrise times to compute the hours of daylight for the 21st day of each month. Display your results in a table. Find another sinusoidal function that accurately models the relationship between the time in months (x) and the amount of daylight.

An *equinox* – a day in which the hours of daylight and darkness are nearly equal – occurs each spring (March) and autumn (September) when the Sun shines directly at the equator. The word *equinox* derives from the Latin words *aequus* (equal) and *nox* (night). A *solstice* is a day in which the hours of daylight is nearly a maximum or minimum. The name is derived from the Latin *sol* (Sun) and *sistere* (to stand still). The summer (June) solstice – the longest day of the year in the northern hemisphere – occurs when the Earth tilts the northern hemisphere closest to the Sun, and the winter (December) solstice – the shortest day of the year in the northern hemisphere – occurs when the southern hemisphere is closest to the Sun.

Use your sinusoidal function that gives the amount of daylight as a function of time in months to predict the two days of the year when the hours of daylight and darkness are equal. How do the predictions from your model compare to the actual dates for the March and September equinoxes? Again, use your sinusoidal function that gives the amount of daylight as a function of time in months to predict the days of the year when the hours of daylight is a minimum and a maximum. How do the predictions from your model compare to the actual dates for the June and December solstices?

Problem solver 3: Happy Birthday to you … two … at least

What is the minimum number of people in a group so that the probability of at least two people in the group having the same birthday is 100%? State any important facts or assumptions that need to be considered.

Having answered the previous question, now tackle this one. What is the minimum number of people in a group so that the probability of at least two people in the group having the same birthday is over 50%?

Problem solver 4: Towers of Hanoi … extended

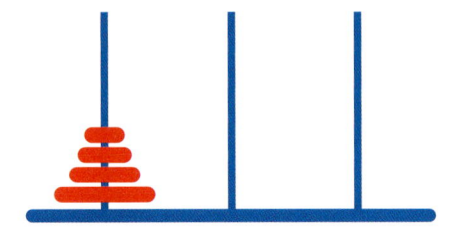

In the Towers of Hanoi puzzle, the aim is to move an 'ordered tower' of discs (stacked up from largest to smallest) from one peg to one of the two other pegs, subject to two rules. First, a 'move' consists of only one disc changing position. Secondly, a larger disc may not be placed on top of a smaller disc. The challenge of the puzzle is to determine the least number of moves necessary for a tower of discs to be transferred from one peg to another.

The Towers of Hanoi puzzle was invented in 1883 by Edouard Lucas, a French mathematician. The traditional Towers of Hanoi puzzle has three pegs. Mathematicians have verified that in the case of the traditional puzzle there is an optimal strategy (or procedure) that will give the least number of moves for any number of discs. For three pegs, find the least number of moves for the following number of discs: 1, 2, 3, 4 and 5. Why is the least number of moves always odd? Representing the number of discs with n and the least number of moves with m, find a formula that expresses m in terms of n for three pegs.

Surprisingly, mathematicians have not found an optimal strategy for determining the least number of moves for a Towers of Hanoi puzzle with more than three pegs. Consider a Towers of Hanoi puzzle with four pegs. How do you think the minimum moves necessary for transferring a tower of n discs from one peg to another will compare to the minimum moves for the same number of discs when there are three pegs? Try to find the minimum moves with four pegs for the following number of discs: 1, 2, 3, 4 and 5. Can you find any patterns?

Section II – Understanding expectations and assessment of portfolio tasks

Expectations

Every student must produce a portfolio containing two pieces of work completed during the course. Each piece of work in the portfolio is internally assessed by the teacher using an established set of criteria (shown at the end of this section). A sample of student portfolios from each school is then externally moderated to ensure uniformity of standards. The portfolio is worth 20% of the total score for the Mathematics SL course.

There are two types of tasks: mathematical investigation (Type I) and mathematical modelling (Type II). Each final portfolio must contain one Type I task and one Type II task. In a mathematical investigation task (Type I), you are expected to generate data that allows you to search for patterns, leading you to pose, test and confirm a general statement. In a mathematical modelling task (Type II), you are expected to define variables and manipulate data, leading to the construction and evaluation of a mathematical model.

You must be aware that any portfolio work submitted for assessment must be entirely your own work – you will be required to sign a document to verify this. You are not allowed to collaborate with any other students – in or out of school. However, you should discuss with your teacher any questions or concerns that you may have while completing a portfolio task.

The need for proper mathematical notation and terminology, as opposed to calculator or computer notation, is important to recognize – as well as adequate documentation of your application of technology to the development of each portfolio task. You are expected and required to reflect on the mathematical processes and algorithms the technology is performing, and communicate them clearly and succinctly.

It is quite likely that you have limited experience in reflecting and writing about mathematical problems. You are expected to write 'correct' mathematics, but the most important aspect of completing a portfolio task is **not** about 'right' or 'wrong' answers – it is much more about your own thoughts, reflections, and insights and, ultimately, what you can write clearly about mathematical processes and documenting results in a coherent and readable manner.

The portfolio tasks aim to
- reward you for mathematics carried out without the time limitations and pressure associated with written examinations
- increase your understanding of mathematical concepts and processes
- develop your personal insights into the nature of mathematics, and to develop your ability to ask questions about mathematics
- allow you to experience the satisfaction of applying mathematical processes on your own

- enable you to discover, use and appreciate the power of a calculator or computer as a tool for doing mathematics
- develop your qualities of patience and persistence, and enable you to reflect on the significance of results obtained
- allow you to show, with confidence, what you know and what you can do.

Above all, it is hoped that you will find the portfolio tasks both stimulating and rewarding.

Essential skills

Type I – Mathematical investigation
- Produce a strategy
- Generate data
- Recognize patterns or structures
- Search for further cases
- Form a general statement
- Test a general statement
- Justify a general statement
- Appropriate use of technology

Type II – Mathematical modelling
- Identify problem variables
- Construct relationships between variables
- Manipulate data relevant to the problem
- Estimate values of model parameters not measured or calculated from the data
- Evaluate the usefulness of the model
- Communicate the entire process
- Appropriate use of technology

Assessment

The five criteria for assessing each of your portfolio tasks – with descriptors for each achievement level – are printed here along with brief explanatory notes for each criterion. Each task has a maximum of 20 marks. Note that the descriptors for criteria C and D are different for Type I and Type II tasks.

Criterion A: Use of notation and terminology

Achievement level	Descriptor
0	The student does **not use** appropriate notation and terminology.
1	The student **uses some** appropriate notation and/or terminology.
2	The student **uses** appropriate notation and terminology in a **consistent manner** and does so throughout the work.

Explanatory notes:
The key idea behind this criterion is to assess how well your use of terminology describes the context and addresses appropriate use of mathematical symbols (for example, use of '$\approx$' instead of '$=$'). Word processing a document does not increase the level of achievement for this criterion, or for criterion B. You should take care to write (either handwritten or using word-processing software) appropriate mathematical symbols (especially if your word-processing software does not supply them). Calculator/computer notation should **not** be used. For example writing x^2 instead of x^2 or ABS(x) instead of $|x|$ is considered inappropriate and will be penalised.

Criterion B: Communication

Achievement level	Descriptor
0	The student **neither provides** explanations **nor uses** appropriate forms of representation (for example, symbols, tables, graphs and/or diagrams).
1	The student **attempts** to provide explanations or uses **some** appropriate forms of representation (for example, symbols, tables, graphs and/or diagrams).
2	The student provides **adequate** explanations or arguments, and communicates them using appropriate forms of representation (for example, symbols, tables, graphs, and/or diagrams).
3	The student provides **complete**, **coherent** explanations or arguments, and communicates them **clearly** using appropriate forms of representation (for example, symbols, tables, graphs, and/or diagrams).

Explanatory notes:
The work can achieve a good mark if the reader does not need to refer to the wording used to set the task. Level 2 **cannot** be achieved if you only write down mathematical computations without explanation. Graphs, tables and diagrams should accompany your work in the appropriate place and not be attached to the end of the document. Graphs must be correctly labelled and neatly drawn on graph paper. Graphs generated by a computer program or a calculator 'screen dump' are acceptable, providing that all items are correctly labelled, even if the labels are written in by hand. Graphs generated by calculator or computer should present the variables and labels appropriate to the task. Colour keying the graphs can increase clarity of communication.

Criterion C: Mathematical process
Type I – Mathematical investigation: searching for patterns

Achievement level	Descriptor
0	The student does **not** attempt to use a mathematical strategy.
1	The student **uses** a mathematical strategy to produce data.
2	The student **organizes** the data generated.
3	The student **attempts to analyse** data to enable the formulation of a general statement.
4	The student **successfully analyses** the **correct** data to enable the formulation of a general statement.
5	The student **tests the validity** of the general statement by considering further examples.

Explanatory notes:
This criterion assesses your *process* in obtaining a general statement. The *correctness* of your statement is assessed in criterion D. Level 3 can **only** be achieved if the amount of data generated is sufficient to warrant an analysis.

Type II – Mathematical modelling: developing a model

Achievement level	Descriptor
0	The student does **not define** variables, parameters or constraints of the task.
1	The student **defines some** variables, parameters or constraints of the task.
2	The student **defines** variables, parameters and constraints of the task and **attempts** to create a mathematical model.
3	The student **correctly analyses** variables, parameters and constraints of the task to **enable the formulation** of a mathematical model that is relevant to the task and consistent with the level of the course.
4	The student **considers how well** the model fits the data.
5	The student **applies** the model to other situations.

Explanatory notes:
Any form of definition of variables, parameters or constraints is acceptable. For example, labelling a graph or table, or stating the domain and range of a function. Level 4 may be obtained by use of an appropriate qualitative and/or quantitative analysis. At achievement level 5, applying the model to other situations could include, for example, a change of parameter or more data.

Criterion D: Results

Type I – Mathematical investigation: generalization

Achievement level	Descriptor
0	The student does **not** produce any general statement consistent with the patterns and/or structures generated.
1	The student **attempts** to produce a general statement that is consistent with the patterns and/or structures generated.
2	The student **correctly** produces a general statement that is consistent with the patterns and/or structures generated.
3	The student expresses the **correct** general statement in **appropriate** mathematical terminology.
4	The student **correctly states** the scope or limitations of the general statement.
5	The student gives a correct informal justification of the general statement.

Explanatory notes:
A correct justification of the general statement that does not take into account scope or limitations can only achieve, at most, level 4. It is important to note the difference between '**a** (i.e. any) general statement' in level 2 and '**the** general statement' in level 3.

Type II – Mathematical modelling: interpretation

Achievement level	Descriptor
0	The student has **not** arrived at any results.
1	The student **has** arrived at **some** results.
2	The student **has not interpreted** the reasonableness of the results of the model in the context of the task.
3	The student **has attempted to interpret** the reasonableness of the results of the model in the context of the task, to the appropriate degree of accuracy.
4	The student **has correctly interpreted** the reasonableness of the results of the model in the context of the task, to the appropriate degree of accuracy.
5	The student has **correctly** and **critically** interpreted the reasonableness of the results of the model in the context of the task, **including** possible limitations and modifications of the results, to the appropriate degree of accuracy.

Explanatory notes:
'… appropriate degree of accuracy' means appropriate in the context of the task.

Criterion E: Use of technology

Achievement level	Descriptor
0	The student uses a calculator or computer for **only routine** calculations.
1	The student **attempts** to use a calculator or computer in a manner **that could enhance** the development of the task.
2	The student makes **limited** use of a calculator or computer in a manner **that enhances** the development of the task.
3	The student makes **full and resourceful** use of a calculator or computer in a manner that **significantly enhances** the development of the task.

Explanatory notes:
The emphasis in this criterion is on the contribution of the technology to the mathematical development of the task rather than to the presentation/communication. Using a computer and/or a GDC to generate graphs or tables may not necessarily enhance the development of the task, and, therefore, may not merit more than a level 1.

Criterion F: Quality of work

Achievement level	Descriptor
0	The student has shown a **poor** quality of work.
1	The student has shown a **satisfactory** quality of work.
2	The student has shown an **outstanding** quality of work.

Explanatory notes:
Making a satisfactory attempt to address all of the requirements of the task should achieve level 1. In order to achieve level 2, work must show precision, insight and a sophisticated level of mathematical understanding.

Section III – Sample portfolio tasks
Sample portfolio task 1
Investigation: *Rates of change for exponential functions*

You have learned to find formulae for the instantaneous rate of change for a few elementary functions. You know that the instantaneous rate of change for functions of the type $f(x) = x^a$ (that is, a power function) can be found by using the formula $f'(x) = ax^{a-1}$. This is, of course, for constant values of a. In this portfolio task, you will investigate rates of change for functions of the type $f(x) = a^x$. This is an exponential function where the base a is a constant greater than zero and the exponent is the variable x.

First, examine the exponential function $f(x) = 2^x$. Find the *approximate* rate of change of f at several different values of x. There are several ways to do this, so you will need to describe your method. By manipulating your inputs, you should be able to increase the accuracy of your approximation to any reasonable number of significant digits. **Use this evidence to show that the following statement is false: If $f(x) = 2^x$, then $f(x) = x \cdot 2^{x-1}$.**

Next, begin comparing your approximations for the instantaneous rate of change for $f(x) = 2^x$ at various x-coordinates to the y-coordinates of $f(x) = 2^x$ at those locations. Answer the following question: How does the *rate* value compare to the *function* value at any chosen x-coordinate? Please note that you are *not* asked to find a formula for the rate in terms of the x-coordinate, but instead to compare rates to y-coordinates. You should explore a few possibilities before you pick one answer. For example, is there a constant difference between the rate and the y-coordinate?

Continue your investigation by examining the function $g(x) = 3^x$. Do the same comparison of rate to function value that you did for $f(x) = 2^x$.

Complete your investigation by examining an exponential function where the base is between zero and one. Using all of your results, conjecture a rule for finding the derivative of an exponential function $f(x) = a^x$ where a is a constant greater than zero.

Sample portfolio task 2
Modelling: *Half-life*

A substance whose quantity is decreasing over time (decay) is often described in terms of its half-life, which is the time required for half the material to decay. The rates of decay for radioactive materials and other substances are often given in terms of their half-lives. The half-life of carbon-14, a radioactive isotope of carbon, is used by scientists to date organic materials up to 60 000 years old. The half-life of anti-cancer drugs is used to determine the amount and frequency of chemotherapy doses. For example, one of the first chemotherapy drugs, mechlorethamine, has a half-life of less than one minute – i.e. it takes less than one minute for half of the drug to be eliminated from the body. In contrast, lead that gets into human bone tissue takes ten years for half of it to be eliminated.

The decay of a substance, and describing its half-life, can be simulated by the following experiment. Collect as many six-sided dice as you can – at least twenty. Using a large cup, or similar container, roll all of the dice simultaneously and remove any of the dice that shows a six. Carefully record the roll number and the number of dice remaining. The number of dice for roll number 'zero' will be the number of dice you have at the start of the experiment. After removing the dice showing a six, replace the remaining dice in the container and roll all of them. Again, remove any dice showing a six. Continue to roll in this manner until just one die remains.

It is important to collect a sufficient amount of data – in other words, repeating the experiment (trial) a sufficient number of times. Consider using a computer or calculator program to simulate the experiment and to collect data for a large number of trials.

State at least one function that models the relationship between roll number and the number of dice remaining. Using technology, find at least one other function that fits your data. Which of your functions best models your data? Explain your choice.

Using your functions, predict on which roll the number of dice will be half of the original number. Compare these theoretical results with your experimental data. Determine what you consider to be the best value for the half-life (in terms of the number of rolls) for the quantity of dice.

Figure 16.1

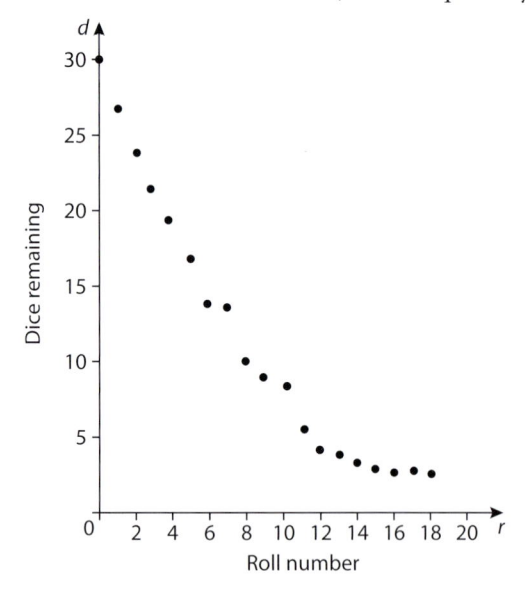

Figure 16.2

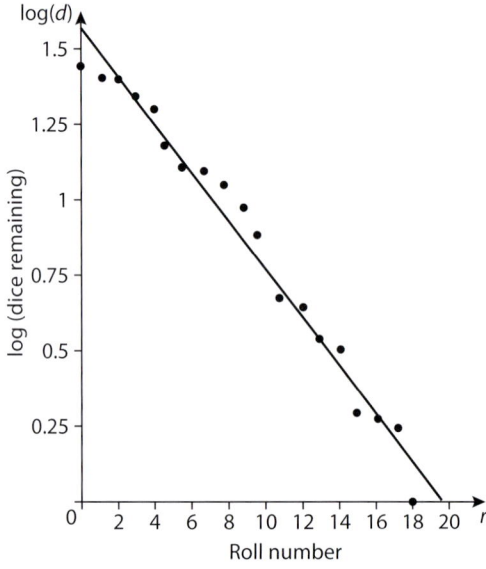

A student performed this half-life simulation and obtained data that is graphed in Figure 16.1. The student then plotted roll number versus the common logarithm (base 10) of the dice remaining, as shown in Figure 16.2.

The student used a calculator to find a line of best fit (regression line) for the data in Figure 16.2. The equation of the regression line is $y = -0.0804x + 1.577$, where x represents the roll number (r) and y represents the common logarithm of the number of dice remaining ($\log_{10} d$). Hence, $\log_{10} d = -0.0804r + 1.577$. The student then used this linear equation and laws of logarithms to find the following exponential equation for the original data set.

$$d = 37.76(0.831)^r$$

Find an exponential model for your data using the same process as outlined above. Find the half-life that this function predicts. Compare this function with the previous functions that you found to model your data.

Sample portfolio task 3
Investigation: *Crossed lines*

a) Using suitable technology, create a graph that links each point $(-n, n^2)$ for $2 \leq n \leq 16$ to each point (m, m^2) for $2 \leq m \leq 16$, where n and m are integers.

When sufficiently expanded, the lower middle part of your graph should look something like the one below.

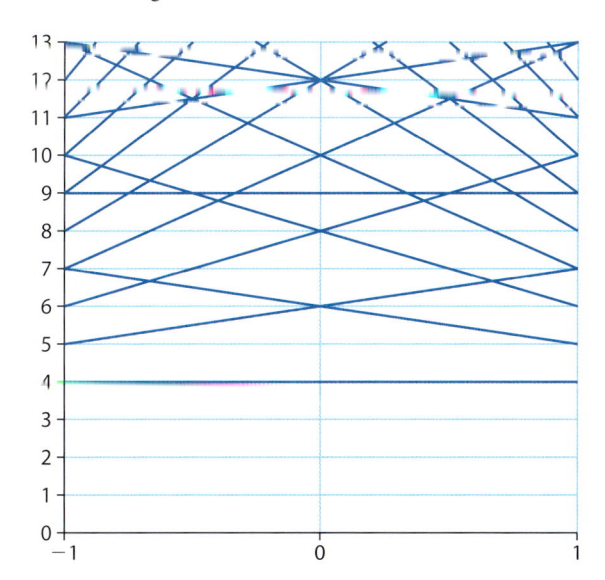

b) Note that some integer values on the y-axis have lines going through them and some do not. Make a list of all the values starting from $y = 2$ that do not have lines through them.
c) What do the smaller y-values in your list have in common?
d) At what y-value does the common rule break down? What could you do to continue the rule further?
e) Find a general equation in terms of n and m for the line joining $(-n, n^2)$ to (m, m^2) and hence find a general expression for the y-intercept of the line. How does this help explain the rule you found in part c)?
f) Now consider all the points on the y-axis for which there are lines passing through.

How many lines go through each point? (For example, you should see that 4 lines pass through (0, 12).)
What is the numerical connection between the number of lines and the *y*-coordinate of the point?
Is the relationship true for all points or only as far as a certain limit?
Give an explanation for the relationship.

Sample portfolio task 4
Modelling: *Hanging objects*

The object of this task is to compare mathematical models for a curve that results when a chain, rope or cable is suspended between two points at equal heights. Your first task is to gather data for such an object that you choose and to 'fit' the resulting curve with an equation. Your second task is to see if this best-fit model can be applied to structural arches.

Part 1: Gathering data
First, you will need to choose an object to suspend. There are many good choices, including an appropriate necklace, a chain, a typical thick rope, or an extremely loose, flexible cable of some sort. The object must be suspended between two posts of equal height so that it hangs between them. The resulting curve must be smooth, with no bumps or irregularities.

Once you have suspended your object, you need to establish a list of data points. This can be done in a number of ways, but all must adhere to the following conditions:

- You must report the length of your (unsuspended) object.
- You must clearly establish a square coordinate system for the object as it hangs.
- You must use units on the coordinate grid that match 'real' units with which to measure the hanging object.
- You must gather enough data points (at least 6) so that a reasonable model can be tested.
- You must provide evidence (such as a photograph) so that your method is clear. This should include evidence of how you established a scale.

Part 2: A first model
The curve that traces the arc of such a smooth hanging object may be reasonably modelled by a parabola. Use your data points to determine a quadratic function that fits your curve. To assess the quality of this model, perform the following three tests:

- Use your model to predict the height of the suspended object at a displacement from its vertex that is *different* from one of the displacements in your data set. Check the accuracy of your prediction with a measurement. Do this for more than one point.

- On the picture or diagram of your object, carefully draw a line tangent to the curve at some point. Find the slope of that line. Use a derivative to see if your model matches the actual slope of your curve at that point.

- Use the arc-length formula $\left(\int_a^b \sqrt{1 + (f'(x))^2}\, dx\right)$ to estimate the length of your object according to the model, and compare this answer to its actual length. You should use technology to evaluate the integral.

Part 3: A second model
There are many functions other than quadratics that appear to be 'u-shaped'. Find a second *polynomial* that also may fit your curve. As in Part 2, use your data set to determine the equation for such a polynomial. Perform the same three tests you performed in Part 2 to see if this model is more or less accurate than your first.

Part 4: Beyond polynomials
Consider other 'u-shaped' curves this time. In particular, consider the curve formed by the sum of two exponential functions that are reflections of each other across the y-axis. Finding the coefficients to make this curve match your data may require a different technique from the one you used in previous parts of this problem. You may wish to note that any function of the form $f(x) = a^x$ can be written in the form $f(x) = e^{kx}$, which may make the calculus application easier. Once you have a reasonable model, perform the same three tests on your model that you performed in Parts 2 and 3. Again, compare its accuracy to that of your other models.

Part 5: An extension
Many 'real' curves in architecture and engineering may appear to be parabolic. Find an arch of a bridge, a power line, a bridge cable, or another such curve that you can photograph. When you do this, remember to take a level, 'head-on' picture so that you can eliminate errors resulting from awkward angles. Find a mathematical model that describes the arch in your photograph and explain why you chose that model. For this part, you will NOT need to establish scale or determine actual arc length. Finally, consider whether arches that are merely decorative are different from those that are used for architectural support.

Sample portfolio task 5
Investigation: *Surface area and volume of a cone*

A cone is made by removing the smaller sector ABC from a circle of radius r and then joining side AB to side CB. Angle ABC – labelled θ – is measured in radians and must be less than π so that sector ABC is less than half of the circle. The radius of the original circle is labelled R. After the edges, AB and CB, of the larger remaining sector are joined together to make the lateral surface of a cone, a circle of radius r is constructed to be the base of the cone.

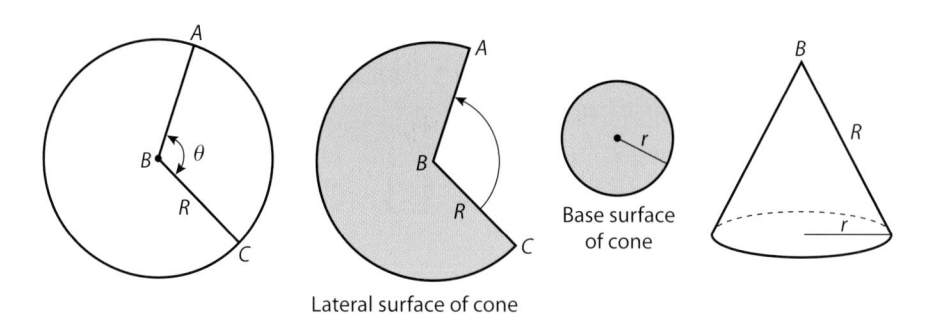

Find an expression for the total surface area (lateral plus base) of the cone, S, in terms of the angle θ (in radians) and the radius R (in centimetres) of the original circle. State and explain any constraints on the values of θ and R. By setting $R = 1$, express S as a function of θ. Graph this function. Describe the behaviour of the function $S(\theta)$ and explain this behaviour in connection with the physical construction of the cone.

Find an expression for the volume of the cone, V, in terms of the angle θ (in radians) and the radius R (in centimetres) of the original circle. By setting $R = 1$, express V as a function of θ. Graph this function. Use technology to find the value of θ that gives a maximum volume for the cone. Explore the maximum volume of cones made from a circle with radius other than $R = 1$. What do you notice about the value of θ for the maximum volume of the different size cones? Explain this result.

The volume of the cone is measured in units of cubic centimetres and the surface area of the cone is measured in units of square centimetres. For many applications involving three-dimensional objects or organisms, it is important to know the ratio of volume to surface area – that is, the number of cubic centimetres of volume per one square centimetre of surface area. The volume to surface area ratio is an important factor in determining heat loss or gain, and also for the manufacture of efficient containers.

With $R = 1$, graph the ratio of volume to surface area versus θ. Use technology to find the value of θ that gives a maximum ratio for volume to surface area. Compare this to the ratio of volume to surface area for the value of θ that produced a maximum volume. Does the ratio of volume to surface area occur at the same value of θ for other size cones made from circles with a radius other than $R = 1$. Comment on your results.

Acknowledgments:
We are grateful for contributions to this chapter from the following two teachers:
Peter Ashbourne, United World College Mostar
Ronald Sellke, American International School of Vienna

17 Sample Examination Papers

Paper 1 – Non-GDC paper

Full marks are not necessarily awarded for a correct answer with no working. Answers must be supported by working and/or explanations. Where an answer is incorrect, some marks may be given for a correct method, provided this is shown by written working. You are therefore advised to show all working.

Sample paper 1 - A

Section A

1 [*Maximum mark:* 5]
   From January till July, the mean number of visitors per month to a museum was 1200. From August till December, the mean number of visitors to the museum was 1500.
   What was the mean number of visitors per month for the whole year?

2 [*Maximum marks* 6]
   In an arithmetic sequence, the first term is -3 and the fifth term is 13.
   a) Find the common difference d.
   b) If the nth term is 101, find the value of n.

3 [*Maximum mark:* 10]
   Let $f(x) = \ln x$ and $g(x) = \dfrac{x+3}{2x}, x \neq 0$.
   Find
   a) $(g \circ f)(e)$, where e is the base of the natural logarithm
   b) $g^{-1}(1)$
   c) the domain of $(f \circ g)(x)$.

4 [*Maximum mark:* 6]
   The diagram shows a circle with radius 10 cm. The minor arc subtends an angle of 2 radians at the centre.
   Find
   a) the length of the minor arc
   b) the area of the shaded region.

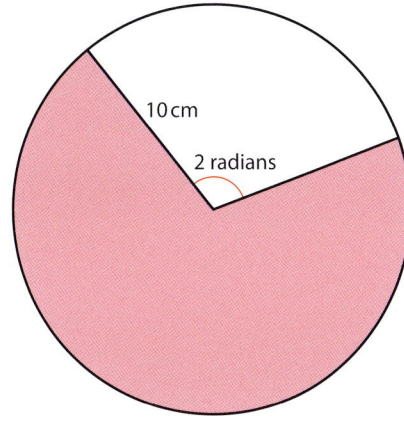

5 [*Maximum mark: 6*]
A vector equation of a line is
$$\begin{pmatrix} x \\ y \end{pmatrix} = \begin{pmatrix} -2 \\ 3 \end{pmatrix} + t\begin{pmatrix} 3 \\ 5 \end{pmatrix}, t \in \mathbb{R}.$$
Find the equation of the line in the form $ax + by = c$, where a, b and $c \in \mathbb{N}$.

6 [*Maximum mark: 6*]
Consider the expansion of $\left(2x - \dfrac{3}{x^2}\right)^9$. Find

a) the number of terms in the expansion
b) the value of the term independent of x in this expansion.

7 [*Maximum mark: 6*]
A particle is moving eastward along a straight line with the speed, at any time $t > 0$, given by $v(t) = \dfrac{3}{2}\sqrt{t + 4} - 1$.
Find the displacement $s(t)$ if $s(0) = 2$.

Section B

8 [*Maximum mark: 18*]
A parallelogram $ABCD$ is given with consecutive vertices $A(-3, -4)$, $B(4, -2)$ and $C(5, 3)$.
a) Find the vector $\overrightarrow{AB}$ and hence the coordinates of D.
b) Find a vector equation of the line (DC), i.e. $\mathbf{r} = \mathbf{r}_0 + t\mathbf{v}$.
c) Show that the point $E(12, 5)$ is on the line (DC) and find the corresponding value of t for this point.

A point T with coordinates $(x, 10)$ is given.
d) If $|\overrightarrow{TE}| = 14$, find the possible values of x.
e) In each case above, find the cosine of the angle TED.

9 [*Maximum mark: 14*]
Consider the function $f(x) = x^3 + \dfrac{3}{2}x^2 - 4x$.
a) Show that the x-coordinates of the two points where this function has a tangent parallel to the line with equation $y = 2x - 7$ are $x = -2$ and $x = 1$.
b) Find the area of the shaded region which is bounded by the curve of the function $f(x)$, the x-axis and the two lines $x = -2$ and $x = 1$.

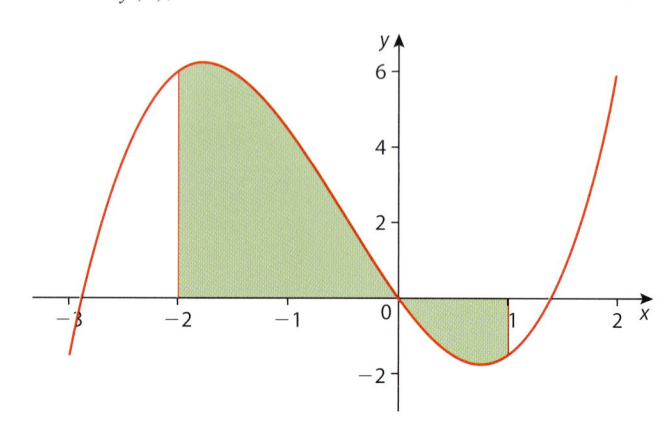

10 [*Maximum mark: 13*]

The heights of students in an IB maths class at a large school are normally distributed with mean height of 174 cm. It is known that the proportion of the class whose height is less than 185 cm is 97.5%.
  a) A student is selected at random. What is the probability that he/she is taller than 185 cm?
  b) The probability that a student is taller than k is 97.5%. Find the value of k.
  c) The class has 220 students. What is the expected number of students with heights between 163 and 185 cm?
  d) Three students from this class are chosen at random. What is the probability that at least two of them are taller than 185 cm?

Sample Paper 1 - B
Section A

1 [*Maximum mark: 6*]
A box contains 15 red marbles and 7 white ones. Three marbles are selected at random, one after the other, without replacement.
  a) What is the probability that exactly two of the marbles are red?
  b) The first marble is white. What is the probability that the second and third marbles are red?

2 [*Maximum mark: 5*]
The diagram shows part of the graph of $y = a(x - h)^2 + k$.

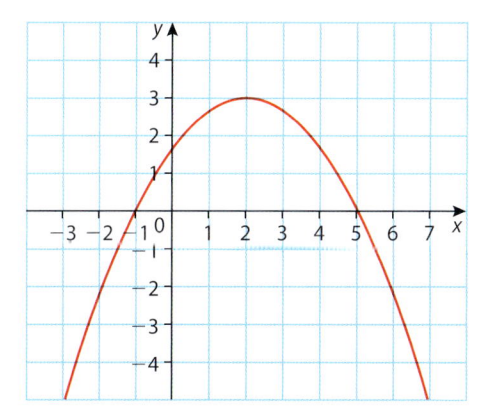

The graph has its vertex at the point (2, 3) and crosses the x-axis at the point (5, 0).
Find the value of:
  a) h b) k c) a

3 [*Maximum mark: 7*]
A sequence is defined by
$$v_n = 5 + \tfrac{1}{2}(n - 1).$$
  a) Write down the value of v_1 and v_{30}.
  b) Find $\displaystyle\sum_{n=1}^{30}\left(\frac{9+n}{2}\right)$.

4 [*Maximum mark: 6*]
The diagram shows the graph of the function
$f(x) = 2x^2 - 20x + 51$.

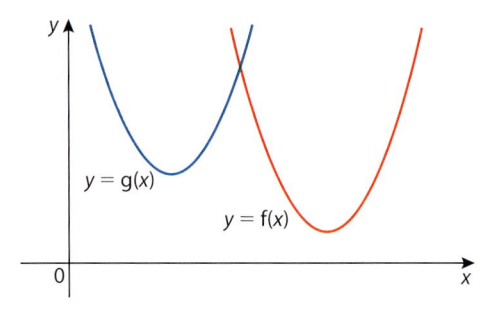

a) Write $f(x)$ in the form $f(x) = a(x - h)^2 + k$.
b) $g(x)$ is a translation of $f(x)$ by a vector $\begin{pmatrix} -3 \\ 2 \end{pmatrix}$. Write the equation of the function $g(x)$.

5 [*Maximum mark: 7*]
Consider the angle θ such that $\pi \leq \theta \leq \frac{3\pi}{2}$ and $\sin\theta = -\frac{20}{29}$. Find
a) $\cos\theta$
b) $\cos 2\theta$
c) $\cos\left(\frac{\pi}{2} - \theta\right)$

6 [*Maximum mark: 7*]
In triangle ABC, $AB = 6$ cm, $AC = 11$ cm, and the angle at A has $\cos A = \frac{3}{4}$.
Find a) $\sin A$
b) the area of triangle ABC
c) BC

7 [*Maximum mark: 7*]

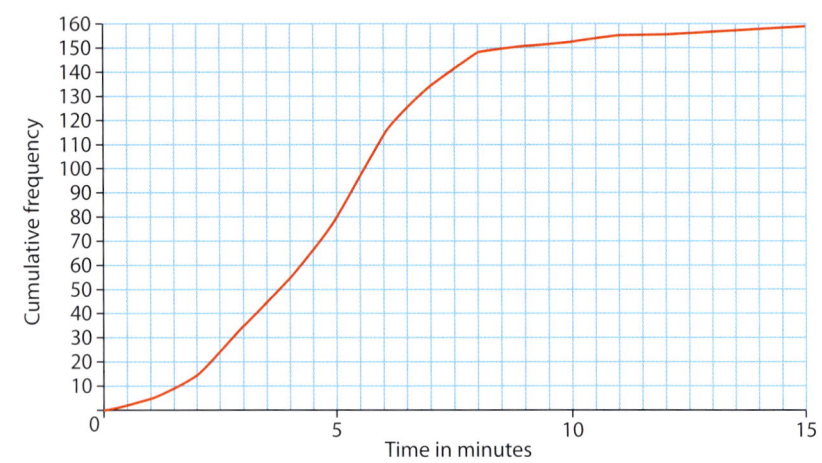

Marco recorded the time he has to wait for the school bus every morning for a period of 160 days. The cumulative frequency graph for his waiting times is shown in the diagram.
a) Use the graph to find
  (i) the median
  (ii) the interquartile range.
b) For 65% of the time, he had to wait more than k minutes. Find k.

Section B

8 [*Maximum mark: 16*]

Two lines L_1 and L_2 are described by their respective vector equations

$$r_1 = \begin{pmatrix} -2 \\ 3 \\ 1 \end{pmatrix} + \lambda \begin{pmatrix} 1 \\ 2 \\ 1 \end{pmatrix}; r_2 = \begin{pmatrix} -4 \\ 9 \\ 4 \end{pmatrix} + \mu \begin{pmatrix} -3 \\ 4 \\ 2 \end{pmatrix}$$

a) Find the cosine of the acute angle formed by these two lines.

b) (i) A is a point on L_1 corresponding to $\lambda = 1$. Find the coordinates of the point A.

   (ii) Show that A is also on L_2.

c) A third line L_3 has the equation $r_3 = \begin{pmatrix} 2 \\ 1 \\ 0 \end{pmatrix} + m \begin{pmatrix} x \\ y \\ 10 \end{pmatrix}$. Find x and y if L_3 is perpendicular to both L_1 and L_2.

9 [*Maximum mark: 16*]

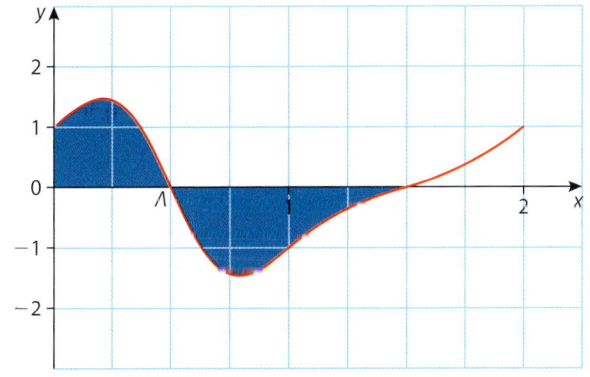

A part of the graph of the function $f(x) = e^{\sin x} \cos x$ is given in the diagram.

a) Find the coordinates of the point A.

b) At what other point does the graph cross the horizontal axis on the interval $[0, 2\pi]$?

c) Find the derivative of the function, and hence show that the maximum value is at the point with x-coordinate $\arcsin\left(\dfrac{a + \sqrt{b}}{2}\right)$, i.e. find the values of a and b.

d) Find the area of the shaded region.

10 [*Maximum mark: 13*]

a) Show that $\dfrac{1}{6}\begin{pmatrix} -5 & 2 & 3 \\ 2 & -2 & 0 \\ 14 & -2 & -6 \end{pmatrix}$ is the inverse of $\begin{pmatrix} 2 & 1 & 1 \\ 2 & -2 & 1 \\ 4 & 3 & 1 \end{pmatrix}$.

b) Hence, find the solution of the system

$$\begin{cases} 2x + y + z = -3 \\ 2x - 2y + z = -9 \\ 4x + 3y + z = -1 \end{cases}$$

c) Find the value of k so that the determinant of $A = -3$.

$$A = \begin{pmatrix} k & 1 & 1 \\ k & -2 & 1 \\ 4 & 3 & 1 \end{pmatrix}$$

Paper 2 – GDC paper

Full marks are not necessarily awarded for a correct answer with no working. Answers must be supported by working and/or explanations. Where an answer is incorrect, some marks may be given for a correct method, provided this is shown by written working. You are therefore advised to show all working.

Sample paper 2 - A

Section A

1 [*Maximum mark: 7*]
   Two ships A and B start from the same port. Ship A moves west at a constant speed of 25 km/h and ship B moves in a direction N 50° E (clockwise from north) at a constant speed of 30 km/h. How far apart are the two ships after two hours?

2 [*Maximum mark: 6*]
   $F(x)$ is an anti-derivative of $f(x) = \dfrac{2}{x+e} + 2\cos x$, $x \neq -e$. $F(0) = 3$. Find an expression for $F(x)$.

3 [*Maximum mark: 6*]
   Consider the arithmetic series: $3 + 7 + 11 + \ldots$
   a) Find an expression for S_n, the sum of the first n terms.
   b) Find the value of n such that $S_n = 3916$.

4 [*Maximum mark: 7*]
   A particle moves in a straight line path with speed, in metres per second, given by
   $$v = \frac{2t}{3 + t^2}$$
   a) Find the distance travelled in the first 3 seconds.
   b) Find the acceleration after 3 seconds.

5 [*Maximum mark: 6*]
   Find the angle between the two vectors **u** = 2**i** − 3**j** + 2**k** and **v** = **i** + 2**j** − 3**k**

6 [*Maximum mark: 6*]
   Find the value of n such that the determinant of the matrix A is zero.
   $$A = \begin{pmatrix} 1 & 2 & k \\ k & 1 & -1 \\ 4 & 3 & 2 \end{pmatrix}$$

7 [*Maximum mark: 7*]
   Antonio plays darts and has a probability of hitting the bullseye 40% of the time. He throws 6 darts at the target.
   a) Find the probability that he hits the bullseye exactly 3 times.
   b) Find the probability that he hits the bullseye at least 3 times.

Section B

8 [*Maximum mark:* 10]

The figure right shows part of the graph of the function
$f(x) = x^3 - 7x^2 + 14x - 6$.

a) Find the x-coordinate of the points where the curve crosses the x-axis.

b) Find the area of the shaded region.

c) Find the volume of the solid resulting from rotating the shaded region around the x-axis through an angle of 360°.

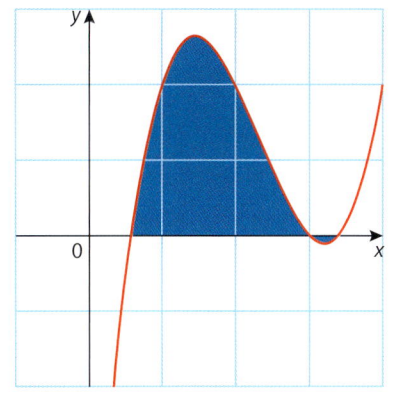

9 [*Maximum mark:* 18]

The masses of bags of flour filled by a machine are normally distributed with a mean of 500 g and standard deviation of 6 g.

If the weight of a bag is less than a, it is considered 'underweight' and has to be refilled. If the weight of the bag is more than b, it is considered 'overweight' and has to be refilled.

a) If 3% of the bags are underweight, show that the value of $a = 488.72$ g.

b) If 2% of the bags are overweight, show that the value of $b = 512.32$ g.

c) We need to adjust the mean so that the machine will give us less underweight bags. (Keep the standard deviation at 6.) What should the new mean be so that the underweight bags will only be 2% of the output?

d) We keep the mean at 500 g, but adjust the standard deviation. How large should the standard deviation be so that we receive only 2% underweight bags?

e) If we want the underweight bags to be 2% and the overweight bags 1%, what should the mean and standard deviations be to achieve this?

10 [*Maximum mark:* 17]

The three vertices of a triangle PQR lie on the following two lines:

$$L_1: r = \begin{pmatrix} 0 \\ 3 \\ 1 \end{pmatrix} + \lambda \begin{pmatrix} 1 \\ 1 \\ -2 \end{pmatrix}; \quad L_2: r = \begin{pmatrix} 2 \\ -1 \\ -2 \end{pmatrix} + \mu \begin{pmatrix} 3 \\ -3 \\ -5 \end{pmatrix}$$

a) L_1 and L_2 intersect at the vertex P. Show that P has the coordinates $(-1, 2, 3)$.

b) Q lies on line L_1 and corresponds to $\lambda = 2$, while R lies on L_2 and corresponds to $\mu = 1$. Find the lengths of the sides of triangle PQR.

c) Find the measure of the vertex angle at P.

d) Find the area of the triangle.

e) Find a vector equation of the line (QR).

f) Find a point S on [QR] such that $\overrightarrow{QS} = 2\overrightarrow{SR}$.

Sample paper 2 - B

Section A

1. [*Maximum mark: 7*]
   The diagram shows two concentric circles with radii of 10 and 7 cm. The shaded region is 53.41 cm². Find the measure of the central angle α, giving your answer to the nearest degree.

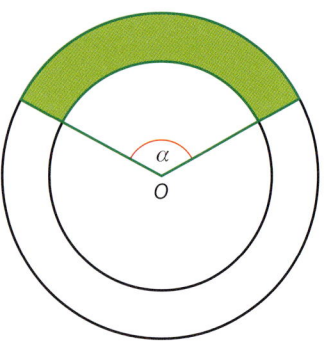

2. [*Maximum mark: 6*]
   Let $f(x) = \cos(2x - 1), 0 \leq x \leq \pi$.
   a) Sketch the graph of this function, starting with $y = \cos x$ and then applying all necessary transformations.
   b) Find the exact locations of the extreme values.

3. [*Maximum mark: 7*]
   In triangle *IBO*, $IB = 8$ cm, $OB = 11$ cm, and the area of the triangle is 38.105 cm². Find the two possible values of angle *IBO*.

4. [*Maximum mark: 6*]
   The two lines L_1 and L_2 intersect at a point *A*.
   L_1: $\mathbf{r} = \mathbf{i} + 6\mathbf{j} + t(-\mathbf{i} + 3\mathbf{j})$
   L_2: $\mathbf{r} = 3\mathbf{i} + 2\mathbf{j} + s(\mathbf{i} - \mathbf{j})$
   a) Find the coordinates of the point *A*.
   b) Find the angle between the two lines.
   c) Find the area of triangle *ABC*, where *B* is a point on L_1 that corresponds to $t = 3$ and *C* is a point on L_2 corresponding to $s = 3$.

5. [*Maximum mark: 6*]
   The speed of an object moving in a rectilinear fashion is given by
   $$v(t) = 2t - \sin t,$$
   where *t* represents time after 12:00. The displacement of the object at $t = 0$ is 2 cm. Find an expression for the displacement of the object in terms of *t*.

6. [*Maximum mark: 7*]
   The sum of the first two terms of a geometric sequence is 9. The sum to infinity of the corresponding geometric series is 12. Find all possible values of the first term and the common ratio.

7. [*Maximum mark: 6*]
   A supermarket packages tomatoes in bags with masses that are normally distributed with a mean of 2.5 kg and standard deviation of 0.3 kg. 20% of the bags weigh more than *x* kg. Find the value of *x*.

Section B

8. [*Maximum mark: 17*]
   Consider the function $f(x) = \dfrac{x^2 - 4}{e^x}$ defined over the set of real numbers.
   a) Find the derivative of this function.

b) Find the zeros of the function.
c) Identify all asymptotes.
d) Find the extreme values and where they happen, and graph the function.
e) Find the area of the region bounded by the function and the *x*-axis.
f) Set up an integral to give the volume of the solid generated as we rotate the region in e) about the *x*-axis through 360°, giving your answer to 3 d.p.

9 [*Maximum mark:* 11]

In a small country there is a wave of new 'start-up' technical companies. The number of companies is given by
$$N = 2840 + 100t$$
where *t* is the number of years since 1992.
a) (i) How many companies were there at the start of 2006?
   (ii) In what year did the number of companies first reach 4000?

In 1986 there were 120 000 people working in that industry. After *t* years, the size of the workforce in this industry is described by
$$W = 120\,000(1.03)^t.$$
b) (i) Find the number of workers at the beginning of 2006.
   (ii) In what year will the number of workers first exceed 220 000?

10 [*Maximum mark:* 17]

An IB group (year 1 and year 2) in a large school has 310 students. The three best subjects are given in the table below.

	Year 1	Year 2
English	60	45
Maths	75	55
Humanities	20	55

a) A student is chosen at random.
   (i) Find the probability that the student's best subject is maths.
   (ii) Find the probability that a student is in year 2 with maths as a best subject.
   (iii) Are the events year 2 and maths independent? Justify.
b) (i) A year 1 student is chosen. What is the probability that he/she has maths as a best subject?
   (ii) A maths student is selected. What is the probability that he/she is in year 2?
c) Assume this population is representative of all IB candidates around the world and that the probability that a student chooses maths as a best subject is independent of other students' decisions.
   (i) From a group of 20 students, what is the probability that maths is a best subject for 4 of them?
   (ii) From a group of 20 students, what is the probability that maths is a best subject for at most 4 of them?

18 Theory of Knowledge

What is TOK?

Theory of knowledge is concerned with how we know what we claim to know. As an IB diploma student you take classes in a number of areas of knowledge corresponding to the IB hexagon. While we call what we learn in each of these subjects 'knowledge', each seems to go about the process of getting this knowledge in a different way. Theory of knowledge examines these different ways of knowing and asks a number of questions about what sort of things can be considered facts, knowledge, good evidence and truth in each of the IB subjects.

Mathematics is rather puzzling as an area of knowledge. Most other subjects that we study in the IB base their knowledge claims upon observations of the world. Mathematics does not. Yet mathematics has profoundly practical applications in the world. How can this be? Knowledge claims in the sciences – while often fairly secure – are nonetheless provisional in some sense. Science allows the possibility that it is wrong – that some new observation or discovery will overturn previously held beliefs. The statements of mathematics, on the other hand, are certain. $1 + 1 = 2$ is not just probably true. It is certain. **It cannot be otherwise.** This is because $1 + 1 = 2$ can be proved. These features give mathematics a special place in TOK.

Explain why probability theory is certain even though it deals in probabilities.

Think of your favourite topic in the SL course. In this topic, can you identify (1) a mathematical transformation and (2) an invariant under this transformation? If you get stuck, ask your maths teacher. (Hint: when studying a function $f(x)$ defined on the real numbers, the function itself is a transformation of the whole real number line, and the set of points that are unmoved by the function, i.e. for which $f(x) = x$ are the invariants. This set is the fixed point set of the function. These points would be represented graphically as the points where the graph of $y = f(x)$ intersected the 45 degree line $y = x$.)

What is mathematics?

It is remarkably difficult to pin down exactly what mathematics is about. A first attempt might be: 'mathematics is the study of numbers'. Certainly, much of our school mathematics is concerned with operations on numbers and the relations between them. This is what is called **arithmetic**. But there is much more to mathematics than numbers, and mathematicians do not take kindly to being thought of as simply good at adding up the bill in the restaurant (actually many of them are not). One of the oldest fields of mathematical thought is **geometry**. When we study geometrical objects such as points, lines, planes, triangles, circles and ellipses we are not studying numbers as such. Rather we are studying the structure of space itself – in particular those aspects that stay unchanged under various types of geometrical transformation. These aspects we call **invariants**. Modern mathematics takes this idea further and studies structures, which are far removed from numbers or even our everyday intuitions about space and time. We could do far worse than define mathematics as the study of transformations and invariants.

What are the foundations of mathematics?

Sets

Modern mathematicians build up the raw materials of their subject from quite humble beginnings. Let us look at how they do this. They start off with some basic concepts about sets. A set, as you know, is just a collection of elements placed inside curly brackets. For example, we could consider a set $A = \{1, 2, 3, 4\}$. We can say that 1 belongs to A: $1 \in A$, but that 5 does not belong to A: $5 \notin A$. The notions of what it is to be a set and to belong to a set are **primitive.** This means that they cannot be explained in terms of more simple notions. If you keep on asking the question 'why?' (as some small children do), the questions stop when you get to a primitive concept (you find yourself answering: 'it just is'). Aristotle was aware of this when he stated that any explanation has to end somewhere. We can now answer him that explanations end in primitive concepts.

Think about an explanation in one of your IB subjects. Keep on asking the question 'why?' until you can go no further. What you are left with is a primitive notion. Are the primitive notions in physics different from those in history?

Mappings

We also need the idea of a mapping between sets. A mapping from A to B is a rule that assigns an element of B to each element of A. The functions that you study in your Standard Level Mathematics course are examples of mappings between the set of real numbers and itself.

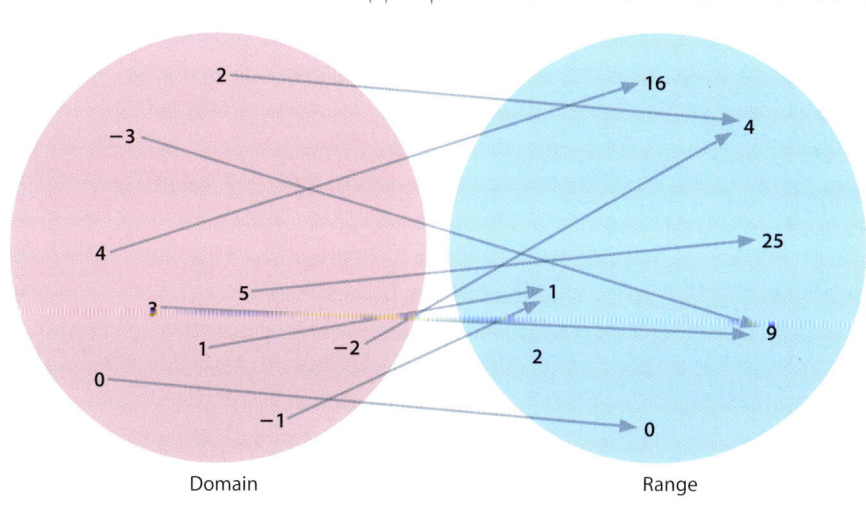

Notice that for a mapping to be well defined, every member of the domain set has to have an arrow (and only one arrow) pointing from it. But some members of the range set can have more than one arrow pointing to them (and the number 2 has no arrow pointing to it). This is an example of a many-to-one mapping. What is the mapping represented here?

▲ An example of a mapping between two sets.

Bertrand Russell and A. N. Whitehead, in their monumental book *Principia Mathematica* (1913), reduced the whole of mathematics to these simple notions. With a bit of work and a great deal of care and patience we can establish basic truths of arithmetic, such as $1 + 1 = 2$. (Proving this takes about four pages of quite sophisticated mathematical argument; this is a surprise to many students who think that 2 is *defined* to be $1 + 1$.)

Because we can build the whole of mathematics out of these primitive ideas of sets and mappings, does this mean that this is what mathematics is about?

559

THEORY OF KNOWLEDGE

Russell's paradox[1]

In constructing mathematics from set theory, we must be careful that we do not allow sets to be members of themselves. Consider the collection $D = \{d: d$ is a set **and** d contains more than 1 element$\}$. By this definition, D actually belongs to itself, since D contains more than one element. There is something rather strange about this, which might make us suspicious. The self-reference involved in thinking about sets that are members of themselves leads to a famous paradox discovered by the English philosopher and mathematician Bertrand Russell. He considered the set that is defined as follows: $S = \{s: s$ is a set **and** s does not belong to itself$\}$. The question he then asked was: Does S belong to itself or not? If the answer is yes – S does belong to itself, then, by the definition of S, S does not belong to itself. If S does not belong to itself, then, by the definition of S, S does belong to itself. Either way we get a contradiction. Russell realized that certain large collections (such as that of all sets) were actually too big to be a set. A collection like this is called a **proper class**.

[1] Bertrand Russell *The Principles of Mathematics* (1903) Cambridge

The barber of Seville

Russell's paradox is similar to the story of the barber of Seville. There was a man who lived in Seville who was a barber. He had a monopoly on the shaving industry in Seville. He shaved every man in the town who did not shave himself. What is contradictory about this?

Mathematics and the real world

1 + 1 = 2?

The objects of mathematics, such as the number systems that we use, are built up from elementary ideas about sets. In this sense, mathematics can be seen as a rather elaborate abstract game, which seems to be about nothing in particular. Bertrand Russell wrote: 'Mathematics is a subject where we do not know what we are talking about, nor whether what we are saying is true.' A possible response to this could be: 'We don't need to establish formally that 1 + 1 = 2. It is easy to prove. Here I have one apple and there another apple. I put them together and I have two apples!' What is wrong with this approach? Think carefully about what abstractions we are making from the real world in order that this argument works. Does it still work with two glasses of water poured together, or two piles of sand pushed together, or two rabbits (male and female) left together for a suitable length of time? These are all examples of the rather curious and sometimes awkward connection between the world of mathematics and the real world.

The Platonist view of mathematics

Plato was aware of the tension between the world of perfect geometrical objects – points with no area, lines with no width, perfect circles and triangles – and the messy physical world. There are no perfectly thin lines, infinitely small points and perfect circles in the real world. But he thought there was a world of perfect mathematical objects underlying the imperfect physical world of our everyday experiences. This mathematical world existed independently of human beings. There would still be nine planets in the solar system long after human beings have ceased to exist on Earth (well, eight actually!). Plato's thinking can help explain the usefulness of mathematics. After all, mathematics is often described as the language of the natural sciences – it is almost impossible to do biology, chemistry or physics without it. But increasingly, mathematics is becoming the *lingua franca* of the social sciences. For example, cutting-edge research in economics is highly mathematical. Governments use highly complex mathematical models to make predictions about future inflation, unemployment and growth rates. This makes a lot of sense, if we grant that mathematics is 'out there' as part of the structure of our physical and social world, as Plato thought it was. That would explain why mathematical methods are so effective in solving real world problems. This is called the **Platonist** view of mathematics.

Formalist and constructivist mathematics

There are two responses to Plato's view that mathematics is 'out there'. One emphasizes the game-like nature of mathematics and the other the fact that it is played by human beings. The **formalist** approach treats mathematics as an abstract formal game. The game proceeds using an agreed set of rules from agreed starting points, or **axioms**. The individual symbolic statements of mathematics mean nothing outside the game, just as 'checkmate' is meaningless outside chess and 'fifteen-love' is meaningless outside the game of tennis.

> The formalist must concede that any use mathematics has in the outside world is largely a coincidence. Is this a point against the formalist view of mathematics?

The **constructivist** sees mathematics as a human activity. To this way of thinking, when there are no more humans there will be no more mathematics. Mathematics is produced by individuals or societies in much the same way as literature and other cultural artefacts. Again, the constructivist has the problem of explaining the success of mathematics in describing, understanding and predicting the outside world. How can a man-made system fit the non-human world so well?

There is another problem with the constructivist view of mathematics. It seems that we are accountable to the truths of mathematics. Mathematicians speak of mathematics as having an independent existence – maths is there to be discovered rather than being man-made. It is certainly true, as we shall see later, that maths can throw up quite unexpected results. Is this compatible with the description of mathematics as being built up out of a few basic and abstract raw materials? In order to try to answer this question, we need to look a little more closely at what constitutes mathematical truth.

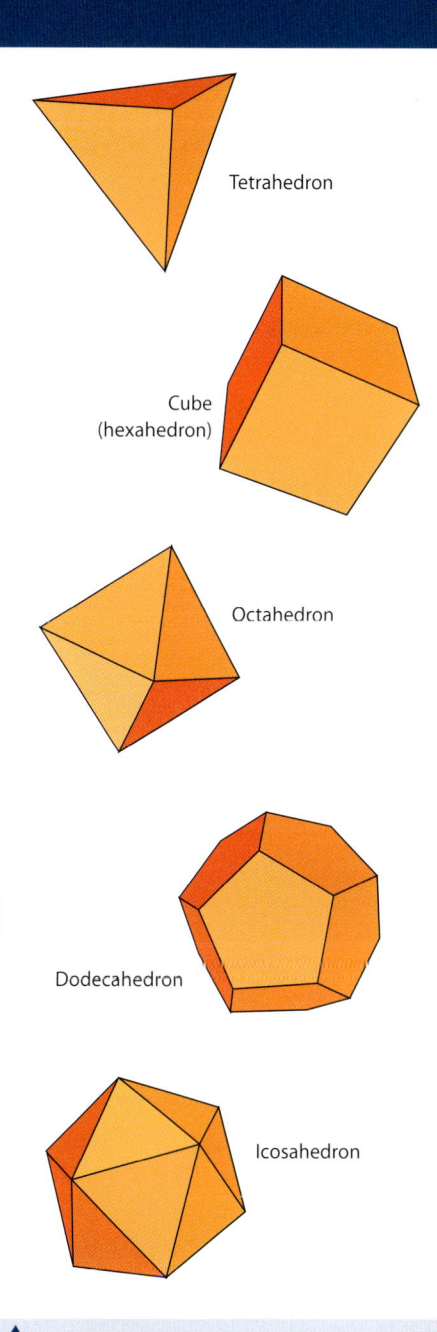

Tetrahedron

Cube (hexahedron)

Octahedron

Dodecahedron

Icosahedron

▲ Plato thought that underlying the messy real world was the perfect world of mathematics. The five regular polyhedra shown are often called the Platonic solids.

> Think about the question of whether mathematics is 'out there' in the world or whether it is an invention of human beings. Does this question occur in other areas of knowledge? Does it make sense to ask if English literature is out there in the world? Does it make sense to ask if chemistry is invented by humans?

What is truth in mathematics?

Let us look again at what we mean by **mathematical truth**. Mathematical statements are true if they can be proved. Before it is proved, a mathematical statement is called a **conjecture**. Once it is proved it is called a **theorem**. So, theorems are mathematical truths.

The idea of proof in mathematics is very old. In around 300 BC, the geometer Euclid of Alexandria formalized the notion of proof in his book *The Elements*. He proved a number of truths about geometrical figures. A proof is a list of statements. Each statement is derived from the preceding statement in the list using only the rules of logic. This is called **chain reasoning**. But what starts the chain in the first place? The first statement in the chain must be, in some sense, either true by definition or self-evident in some way. These self-evident truths are called **axioms**. They are considered to be basic or primitive mathematical truths. By definition, they cannot be proved. A mathematical proof builds a chain of reasoning from the axioms to final mathematical results – theorems.

They are very special from a TOK perspective because it seems that a theorem is an example of knowledge that is certain. A mathematical theorem is not just probably true. It is true in the sense that, given the definitions of the terms it uses and the axiom system used to prove it, **it cannot be otherwise**. In TOK, we rarely meet truths that are certain in this absolute sense.

Theorem, theory and proof

Be careful that you do not confuse the word 'theorem' with the similar-sounding 'theory'. In mathematics, a theory is an established piece of mathematical work that might contain many theorems. In other words, mathematical theories are pieces of true mathematics. In science, the word is more problematic. It might apply to an established piece of science that has been tested and found to yield accurate predictions and to give good explanations of phenomena in the physical world. But the term can also refer to a more tentative idea that has not yet been thoroughly tested. It is a common mistake in TOK essays to make a statement such as: 'The theory of evolution is only a theory so it cannot be considered knowledge'. Evolution theory belongs to the first type of theory – it is as well supported by evidence as the fact that water is H_2O – but the essay treats it as belonging to the second.

A word of warning is also needed about the word **proof**. Strictly speaking, proof is the mathematical process outlined above – where a mathematical statement is derived from axioms in a step-by-step manner. Proof implies absolute certainty. Be careful applying this word outside mathematics.

Part of a manuscript by the French mathematician Evariste Galois.

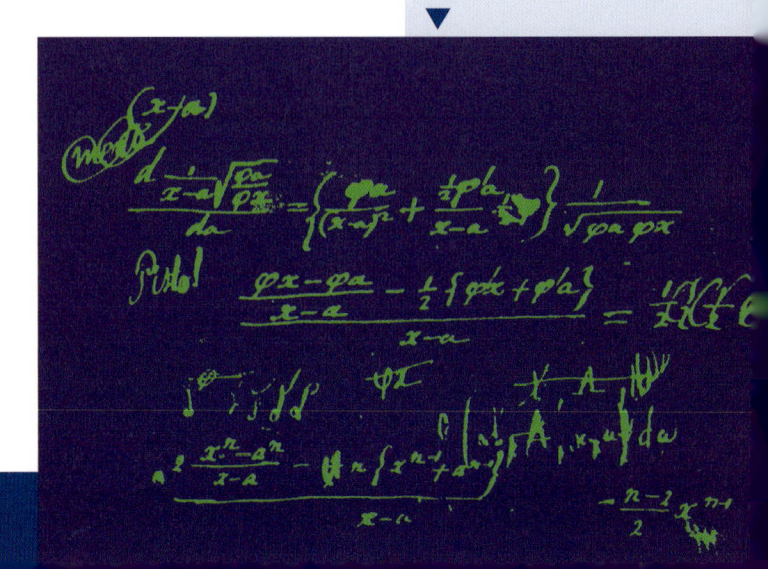

Absolute certainty is generally not achievable in science for a number of reasons that you may have discovered in your TOK course. Scientific results are not proved in this strict sense, it is better to describe them as being 'secure' or 'well supported by the evidence'.

To see how mathematical proof works, let us prove a simple theorem.

Theorem: Let x and y be odd integers. Then $x + y$ is an even integer.

Proof: The definition of an odd number is that it is an even number plus 1. An even number is a number in the 2× table.

So, write $x = 2m + 1$ for some integer m. In a similar fashion, $y = 2n + 1$ for some integer n.

$x + y = (2m + 1) + (2n + 1) = 2m + 2n + 2$

We can take out the common factor of 2 to give: $x + y = 2(m + n + 1)$

Since m, n and 1 are integers, it follows that $m + n + 1$ is also an integer.[2]

Hence, $x + y$ is 2× an integer and so must be even. QED

We write QED (*Quad Erat Demonstrandum* – meaning 'which was to be shown') at the end, to show that the proof is finished.

Moser's circle problem illustrates the difference between an experimental approach to mathematics – a semantic method (trying out a conjecture to see if it works) and proving it – a syntactic method.

Let us take a closer look at some features of this method of proof. First notice that we have in effect proved an infinite sequence of statements including: 3 + 5 is even, 3 + 7 is even, 5 + 7 is even, and so on. We could have attempted to do a sort of mathematical experiment by checking whether the result holds for some randomly chosen odd numbers: 3 + 5 = 8, which is even; 3 + 7 = 10, which is even; 5 + 7 = 12, which is even; and so on. But this is not a proof. There is always an infinite number of examples that we have not tried and for which the result might not hold. This is what mathematicians call a **semantic** method. But, as you have probably learned from studying the natural sciences in TOK, it takes a single counter-example to disprove a conjecture. The same is true in mathematics. Why not try Moser's circle problem (shown right) to see what we mean?

A proof is a **syntactic** method. It does not look at particular examples of odd numbers but rather depends on features that all odd numbers have in common (namely their oddness!). We have been able to do this by using algebra. We have substituted letters for numbers to allow us to talk generally about odd numbers rather than specific examples. This is typical of a mathematical

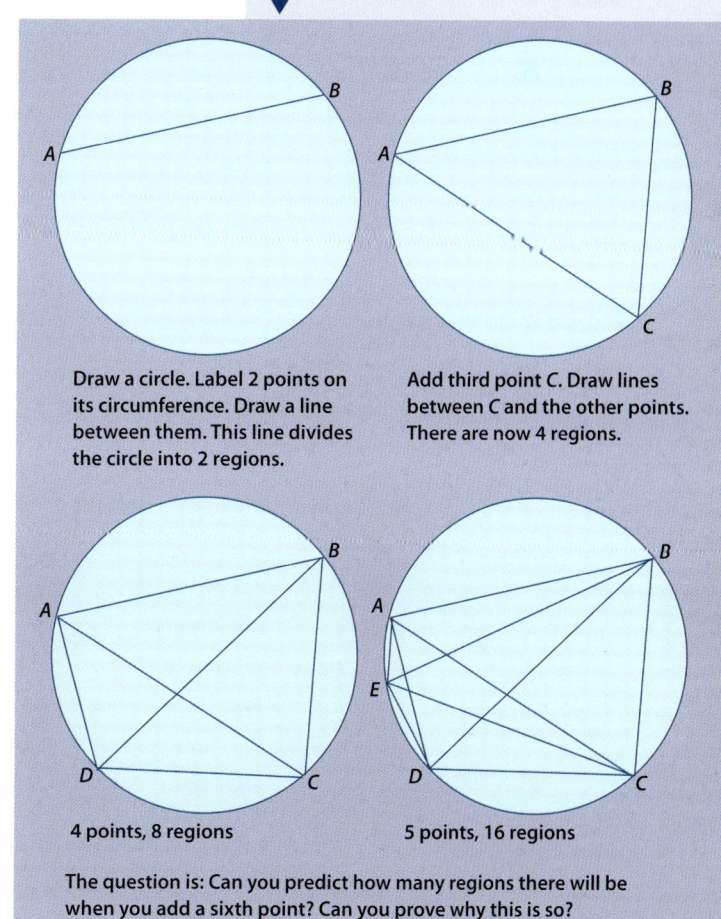

Draw a circle. Label 2 points on its circumference. Draw a line between them. This line divides the circle into 2 regions.

Add third point C. Draw lines between C and the other points. There are now 4 regions.

4 points, 8 regions

5 points, 16 regions

The question is: Can you predict how many regions there will be when you add a sixth point? Can you prove why this is so?

[2]There is a further subtlety here in the statement that $n + m + 1$ is an integer because m and n are. This is because the integers are closed under addition because $\mathbb{Z}^+$ is a group – closure is a property of groups. Groups are structures that underlie most mathematics.

563

THEORY OF KNOWLEDGE

proof. Once the conjecture is proved we can state categorically that it is true, now and for all time. It does not depend on culture, nationality, personal points of view, language or gender. It does not matter who proved it. It could be a university professor of mathematics or it could be an eight-year-old. It simply does not matter. In mathematics, proof means truth and that is the end of the story.

Axioms

A mathematical statement is true if it could be derived from the basic axioms of set theory by using only the rules of logic. In the SL course, the rules of logic are packaged in a convenient way to help us solve problems. We call this package 'the rules of algebra'. These are rules such as: You can add the same number to both sides of an equality and it remains equal (if $y = x$ then $y + 5 = x + 5$).

We can use these rules of algebra to solve mathematical problems. Each problem we solve is a little theorem. An example is: If $x + 5 = 10$ then $x = 5$. This is rather a simple theorem, but it is a theorem nevertheless. If you write any of your standard maths problems in the form '**If** … (problem to be solved) **then** (solution)' you get a theorem. (This assumes that you have solved the problem correctly!) But there is one additional set of assumptions that we do not explicitly mention when we solve these problems (or prove these theorems). That is, the assumptions that the axioms of set theory on which we base all our mathematics (and without which none of our mathematics would mean anything) are true. But how do we choose which axioms to use? How do we know that we have chosen a good set of axioms? These questions are not easy to answer. We shall examine them using a concrete example.

Euclid's postulates

What axioms did Euclid propose for doing plane geometry?
Here are Euclid's axioms. He called them 'postulates'.

> **Euclid's postulates**
> 1. A straight line segment can be drawn joining any two points.
> 2. Any straight line segment can be extended indefinitely in a straight line.
> 3. Given any straight line segment, a circle can be drawn having the given line segment as radius and one endpoint as centre.
> 4. All right angles are congruent.
> 5. If two lines are drawn, which intersect a third in such a way that the sum of the inner angles on one side is less than two right angles, then the two lines inevitably must intersect each other on that side, if extended far enough.

In some sense, Euclid's axioms express mathematical intuitions about the nature of geometrical objects. What is clear in any case is that they are not established using observation of the external world. Objects such as points, lines, circles and planes do not exist in the real world with the perfect qualities they possess in mathematics.

Euclidian geometry

Let us try to use Euclid's postulates to do some geometry (see right).

Let us now examine the construction and see which postulates were used.

Step 1: drawing the arcs is allowed by postulate 3 (twice).

Step 2: drawing the line segments AC and BC is allowed by postulate 1 (twice).

It follows from step 1 that the line segments AC and BC are both equal to AB. Therefore, they must be equal (this is sometimes quoted as a separate axiom – that two line segments equal to the same line segment must be of equal length).

Non-Euclidian geometry

Take a look at Euclid's postulate 5. This cannot be proved as a theorem from the other axioms (that this is impossible can itself be proved!) although many people have attempted this. Euclid himself only used the first four axioms in the first 28 propositions of the *Elements*, but he was forced to use the fifth axiom, so-called 'parallel postulate', in the 29th proposition. The independence of the parallel postulate means that we can choose whether we accept it or not. If we accept it, parallel lines do not meet. The geometry we get is the familiar geometry of the plane. This is the geometry that we can use to construct buildings and other physical objects. In 1823, Janos Bolyai and Nicolai Lobachevsky independently realized that entirely self-consistent non-Euclidean geometries could be envisaged in which the fifth axiom did not hold. There are two quite different geometries in this case: those in which parallel lines meet at some point – elliptical geometry – and those in which parallel lines diverge – hyperbolic geometry. An example of elliptical geometry is the geometry of long distance travel on the Earth's surface. The shortest path between two points (say the most efficient route of a jet airliner) is a curve called a great circle.[3] The parallel lines of longitude are great circles that meet at the poles. If parallel lines diverge, we get so-called hyperbolic geometry. An example of doing hyperbolic geometry would be to draw lines on a saddle.

Problem: To construct an equilateral triangle on a given line segment using Euclid's axioms.

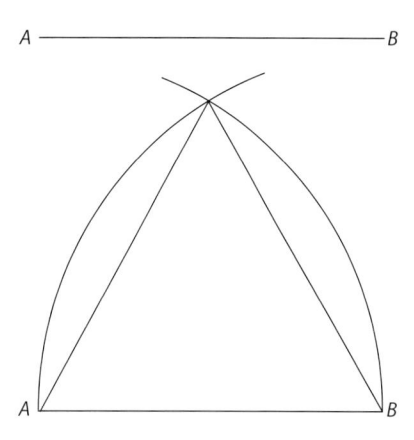

Why not try out a construction yourself, using the postulates of Euclidean geometry? Extend the arcs below the segment AB to meet again at D. Join CD with a line segment. The task is to prove that CD is the perpendicular bisector of AB using Euclid's postulates. (The perpendicular bisector of AB is a line segment CD that cuts AB exactly in half and the angle it makes with AB is exactly a right angle.)

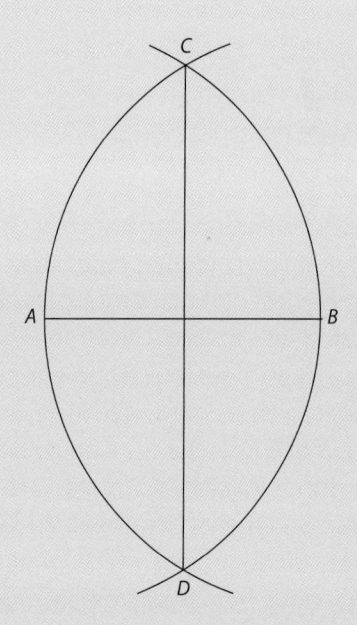

[3] A 'great circle' is the largest circle that can be drawn on a sphere, and is the intersection between the surface of a sphere and a plane passing through the centre of the sphere. The shortest path between two points on a sphere follows a great circle.

THEORY OF KNOWLEDGE

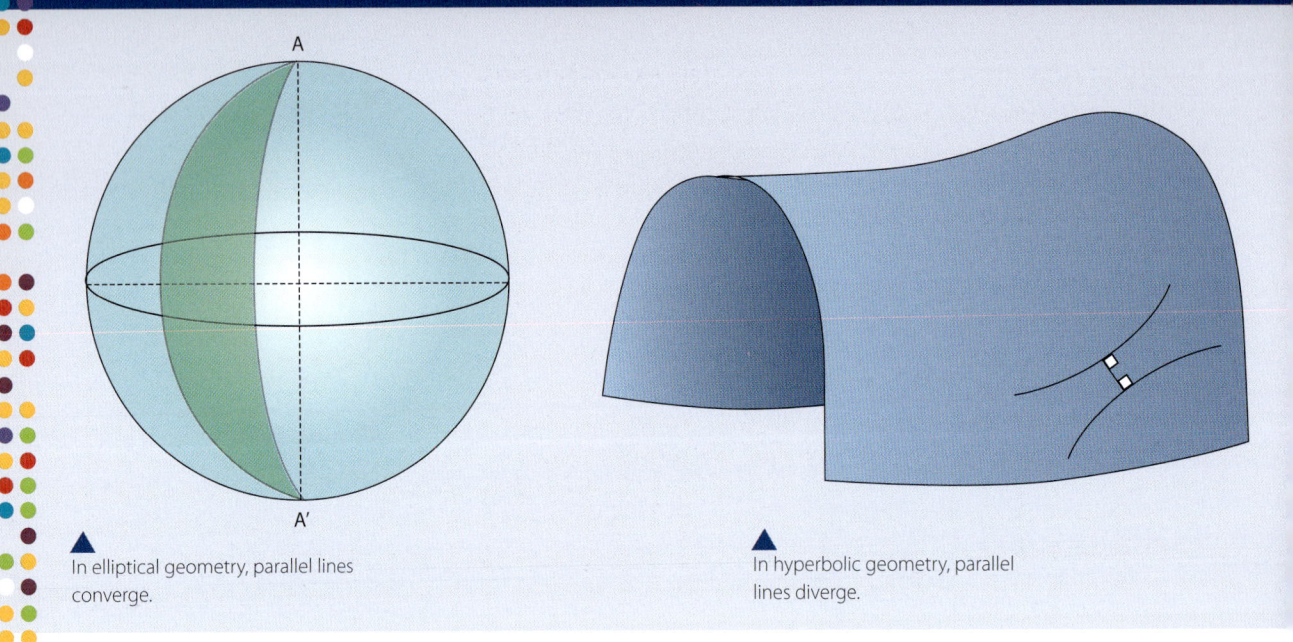

▲ In elliptical geometry, parallel lines converge.

▲ In hyperbolic geometry, parallel lines diverge.

Consistency and completeness

We saw that postulate 5 is independent of the other four - that it could not be derived from them. More generally, there are two questions that can be asked of any set of axioms:

(1) Is the set **consistent**? In other words, is it impossible to derive a contradiction from them (to derive both the statements 'P is true' and 'not P is true')?

(2) Is the set **complete**? That is, any (semantically) true statement can be derived from them.

Are Euclid's axioms complete? Surprisingly, the German mathematician David Hilbert[4] showed that Euclid needs another 15 axioms to have a complete set to do what we now call Euclidian geometry.

In 1931, the Austrian logician Kurt Godel[5] proved the devastating result that you could not prove the consistency and the completeness of the axioms for set theory that were used in *Principia Mathematica*. This famous incompleteness theorem proves, by an ingenious argument, that consistency and completeness is unprovable in any system rich enough to include the laws of arithmetic. So, it could be that mathematics is based upon rather shaky foundations. This might mean that there is a true statement of mathematics lurking somewhere in the recesses of the subject, which is not provable within the system. More serious, from a mathematical point of view, is the possibility that we can derive a contradiction of the form 'P is true' and 'not P is true' from the axioms using the rules of logic. Producing a contradiction means instant death for any area of knowledge. If you believe that 'P is true' and that 'not P is true' then one of your beliefs has to be false. This makes the combined belief 'P is

[4] David Hilbert *Foundations of Geometry* (1902) Gottingen

[5] Über formal unentscheidbare Sätze der Principia Mathematica und verwandter Systeme, I. *Monatshefte für Mathematik und Physik* 38, 173-98 (1931)

> Do you hold any contradictory beliefs? If so, what are the implications for what you consider to be knowledge?

true and not P is true' false under all circumstances. So, if an area of knowledge throws up a contradiction, it simply cannot ever be true. It is condemned to being false whatever the actual state of the world. Since knowledge can be thought of (at least as a first approximation) as justified true belief, a statement that is forever false cannot be knowledge.

Beautiful equations

Einstein suggested that the most incomprehensible thing about the universe was that it was comprehensible. From a TOK point of view, the most incomprehensible thing about the universe is that it is comprehensible in the language of mathematics. Galileo wrote: 'Philosophy is written in this grand book, the universe … It is written in the language of mathematics, and its characters are triangles, circles, and other geometric figures….'[6]

What is perhaps most puzzling is not just that we can describe the universe in mathematical terms, but the mathematics we need to do this is mostly simple, elegant and even beautiful.

To illustrate this, let us look at some of the famous equations of physics. Most of you will be familiar with at least some of the following.

> To what extent is mathematics really a language?

> [6]Galileo, Il Saggiatore (1623) Rome

Relation between force and acceleration: $F = ma$ (more generally this is $F = \frac{d}{dt}(mv)$)

Gravitational force between two bodies: $F = \frac{Gm_1m_2}{r^2}$

Energy of rest mass: $E = mc^2$

Kinetic energy of a moving body: $E = \frac{1}{2}mv^2$

Electrostatic force between two charges: $F = \frac{kq_1q_2}{r^2}$

Maxwell's equations: $\nabla \times \mathbf{B} - \frac{d\mathbf{E}}{dt} = 4\pi \mathbf{J}$ $\nabla \times \mathbf{E} + \frac{d\mathbf{B}}{dt} = 0$ $\nabla \cdot \mathbf{B} = 0$ $\nabla \cdot \mathbf{E} = 4\pi \rho$

Einstein's field equation for general relativity: $R_{\mu\nu} - \frac{1}{2}g_{\mu\nu} = 8\pi T_{\mu\nu}$

> Similar equations can be found in the other natural sciences. Can you think of any?
>
> Is there a sense in which these equations are elegant or beautiful?

I must admit that I find it perplexing that the whole crazy complex universe can be described by such simple, elegant and even beautiful equations. It seems that our mathematics fits the universe rather well. It is difficult to believe that maths is just a mind game that we humans have invented.

But the argument for simplicity and beauty goes further. Symmetry in the underlying algebra led mathematical physicists to propose the existence of new fundamental particles, which were subsequently discovered. In some cases, beauty and elegance of the mathematical description have even been used as evidence of its truth. The physicist Paul Dirac said: 'It seems that if one is working from the point of view of getting beauty in one's equations, and if one has really a sound insight, one is on a sure line of progress.'

THEORY OF KNOWLEDGE

'God used beautiful mathematics in creating the world.'
Paul Dirac

Dirac's own equation for the electron must qualify for being one of the most profoundly beautiful of all. Its beauty lies in the extraordinary neatness of the underlying mathematics – it all seems to fit so perfectly together:

$$\left(\beta mc^2 + \sum_{k=1}^{3} \alpha_k p_k c\right) \psi(x, t) = ih\frac{d\psi}{dt}(x, t)$$

The physicist and mathematician Palle Jorgensen[7] has written: '[Dirac] … liked to use his equation for the electron as an example, stressing that he was led to it by paying attention to the beauty of the math, more than to the physics experiments.'

I shall leave the last word on this subject to Dirac himself, writing in *Scientific American* in 1963:

'I think that there is a moral to this story, namely that it is more important to have beauty in one's equations than to have them fit experiment.'

By any standards, this is an extraordinary statement for a mathematical physicist to make.

[7] Palle Jorgensen *Operator Commutation Relations* (1984) New York

How good are your mathematical intuitions?

Mathematics can sometimes surprise us. Our mathematical intuitions can sometimes let us down, badly. In this section, we shall try out two basic scenarios upon our unsuspecting intuitions and see how they fare.

Scenario 1: The rare genetic disease

Consider the following. There is a very rare genetic disease amongst the population. Very few people have the disease. As a precaution, a test has been developed to check in a particular case whether a person has the disease or not. Although the test is quite good, it is not perfect – it is only 99% accurate. A person X takes the test and it shows positive. The question for your intuition: What is the probability that X actually has the disease?

Think about this for a moment before we go on with the analysis.

Many of the students and teachers that I have worked with in the past have given the same answer: The probability that X actually has the disease, given a positive test result, is around 99%. Did you say the same?

If you did, your mathematical intuition let you down – very badly.

Let us put some numbers into this problem. For the sake of simplicity, let us assume that the country in which the test takes place has a population of 10 million. We are told that the disease is very rare. Let us assume that only 100 people have it in the whole country.

We are told that the test is 99% accurate so that of the 100 cases of the disease the test would show positive in 99 cases and negative in 1 case. So far, so good.

Now let us look at the 9 999 900 people who do not have the disease. In 99% of these cases the test does its job and records a negative result. But in 1% of these cases the test records a positive result. 1% of 9 999 900 is 99 999. This means that if the whole population were tested 99 999 + 99 = 100 098 test results would be positive. Of these, only 99 people have the disease. Therefore, the probability of having the disease, given a positive test result, is 99/100 098 = 0.0989% – in other words, about a tenth of a per cent or one in one thousand. This is a bit different from the 99% that most people guess. How well did you do?

What went wrong with the intuition here?

The important factor in this problem is not just the accuracy of the test *but the accuracy of the test relative to the incidence of the disease*. Because the disease is so rare, the actual number of people with the disease is overwhelmed by the false positive results of the test – the 1% or so of the population who do not have the disease, but the test shows positive anyway. If more people had the disease and if the test was more accurate, the test scoring positive would be a better predictor of X actually having the disease.

Try this problem out with some other numbers to check how the test could be made more useful.

Scenario 2: The Monty Hall game

The second scenario is also based on probability theory. The problem refers to a TV game show, which is loosely based on the actual show *Let's Make a Deal*[8]. A contestant in the show is shown three doors and told (truthfully), by the game show host Monty Hall, that behind one of the doors is a luxury sports car and behind the other two doors are goats. The contestant is told that she must pick a door. She will be allowed to take home whatever is behind the door she picks. We shall assume at this stage that she prefers to win the car. So she goes ahead and picks a door. At this point, Monty Hall opens another door to reveal a goat. (Whenever this game is played, Monty Hall chooses a door concealing a goat.) He then asks the competitor: 'Do you want to switch to the other closed door?'

What does your intuition tell you? Should the contestant switch or should she stick to her original choice?

[8] A widely known statement of the problem was published in Marilyn vos Savant's *Ask Marilyn* column in *Parade* (1990).

◀ The Monty Hall problem: should the contestant switch?

Take a little time to think this through. You might like to try this game with a friend to see experimentally what the best strategy is.

Clearly, because there is one car and two goats, the probability of picking the car if the competitor does not switch doors is 1/3.

If she does switch, what is the probability of winning the car? Let us ask a related question. If she does switch, under what circumstances can she lose the car? Clearly, the only way she can lose the car is if her original choice was right. In other words, she has a 1/3 probability of losing. This must mean that by switching, her probability of winning the car is 2/3.

In other words, by switching she doubles her probability of winning.

Does this make sense? Even after this explanation many of the students and teachers that attend my workshops are not convinced. They argue that they cannot see how an asymmetry has been introduced into the situation.

The crucial point is that Monty Hall knows where the car is. He always opens a door to reveal a goat. It is this act that produces the required asymmetry.

Consider an extreme version of the Monty Hall problem. Imagine 100 closed doors containing 1 car and 99 goats. Let us suppose, for the sake of the argument, that our contestant chooses door number 1. Monty Hall then opens 98 doors to reveal goats. The contestant would be foolish not to switch to the one remaining door (and multiply her probability of winning by a factor of 99).

Try this problem out on your friends and relatives. Are their mathematical intuitions letting them down?

Is the fact that mathematics can surprise us and go against our intuitions evidence that mathematics exists independently of us?

[9] John R. Searle *The Construction of Social Reality* (1995) London

What is a social fact?

The philosopher John Searle[9] points out that many of the facts in our lives are actually socially constructed. He uses money as his central example. Money is money because we believe it to be money. There is something rather strange about this. Normally speaking, when we define a term X, we do not expect the definition to refer to X. Did we not learn in TOK that it was bad to define X in terms of itself? Was this not the reason why our TOK teacher advised us to keep clear of dictionary definitions: 'knowledge – that which is known'. Searle thinks that this sort of circularity is characteristic of what he calls a **socially constructed fact**. He asserts that the social agreements that we make collectively that something should be money makes it such. So 'X functions as money in society S' is a socially constructed fact. As such, statements about it are objective and capable of being evaluated as true or false. Our socially constructed reality includes the concepts of wife, girlfriend, driving licence, bank account, traffic lights, rules of etiquette, nationality, legality, country, nationality, debt, honour, and so on. Many of the physical objects around us are defined in terms of their function, and hence in terms of our intentions, and hence are socially constructed. The concept of a chair or a knife is socially constructed. This is what makes them so difficult to define without using the words 'function' or 'intention'.

Try to define a chair without making reference to human intentions.

Is mathematics a social fact?

Reuben Hersch[10] argues that numbers (and any other mathematical entities) are social constructions. If we acknowledge that they are not just out there in the world independent of human beings and they are not just thoughts in people's heads (our intuitions can be wrong after all) then what are they? There is a third possibility. Mathematics is a construction of human society.

Hersch proposes that mathematics is itself a whole interconnected web of socially constructed reality. Here he is in an interview with John Brockman on the Edge website:[11]

[10] Reuben Hersch *What Is Mathematics, Really?* (1997) Oxford

[11] http://www.edge.org/3rd_culture/hersh/hersh_p1.html (accessed Feb 2008)

THEORY OF KNOWLEDGE

'Mathematics is neither physical nor mental, it's social. It's part of culture, it's part of history, it's like law, like religion, like money, like all those very real things, which are real only as part of collective human consciousness. Being part of society and culture, it's both internal and external. Internal to society and culture as a whole, external to the individual, who has to learn it from books and in school. That's what math is.'

When asked what he called his theory of mathematics, Hersch replied that he calls it humanism 'because it's saying that math is something human. There's no math without people. Many people think that ellipses and numbers and so on are there whether or not any people know about them; I think that's a confusion.'

Hersch points out that we do use numbers to describe physical reality and that this seems to contradict the idea that numbers are a social construction. It is important to note here that we use numbers in two distinct ways: as nouns and adjectives. When we say nine apples, nine is an adjective. 'If it's an objective fact that there are nine apples on the table, that's just as objective as the fact that the apples are red, or that they're ripe, or anything else about them, that's a fact'. The problem occurs when we make a subconscious switch to 'nine' as an abstract noun in the sort of problems we deal with in maths class. Hersch thinks that this is not really the same nine. They are connected, but the number nine is an abstract object as part of a number system. It is a human creation.

Politics and maths learning

Hersch sees both a political and a pedagogic dimension to his thinking about mathematics. He thinks that a humanistic vision of mathematics chimes in with more progressive politics. How can politics enter mathematics? As soon as we think of mathematics as a social construction then the exact arrangements by which this construction comes about – the institutions that build and maintain it – become important. These arrangements are political. Particularly interesting for us here is how a different view of maths can bring about changes in maths teaching and learning. Let us return to Hersch:

'Let me state three possible philosophical attitudes towards mathematics. Platonism says mathematics is about some abstract entities, which are independent of humanity. Formalism says mathematics is nothing but calculations. There's no meaning to it at all. You just come out with the right answer by following the rules. Humanism sees mathematics as part of human culture and human history. It's hard to come to rigorous conclusions about this kind of thing, but I feel it's almost obvious that Platonism and Formalism are anti-educational, and interfere with understanding, and Humanism at least doesn't hurt and could be beneficial. Formalism is connected with rote, the traditional method, which is still common in many parts of the world. Here's an algorithm; practise it for a while; now here's another one. That's certainly what makes a lot of people hate mathematics. (I don't mean that mathematicians who are formalists advocate teaching by rote. But the formalist conception of mathematics fits naturally with the rote method of instruction.) There are various kinds of Platonists. Some are good teachers, some are bad. But the Platonist idea, that, as my friend Phil Davis puts it, Pi is in the sky, helps to make mathematics intimidating and remote. It can be an excuse for a pupil's failure to learn, or for a teacher's saying, "Some people just don't get it." The humanistic philosophy brings mathematics down to earth, makes it accessible psychologically, and increases the likelihood that someone can learn it, because it's just one of the things that people do.'

Do you agree with Reuben Hersch's humanist picture of mathematics – that mathematics is a social construction? Do you think he is right in his association of formalism with rote learning of maths and Platonism with the idea of maths being something remote that some people simply 'do not get'?

Are you really only intelligent if you can do maths?

There is a possibility that the arguments explored in this section might cast light on an aspect of mathematics learning which has seemed puzzling – why it is that mathematical ability is seen to be closely correlated with a certain type of intelligence. Mathematics has, moreover, seemed to polarize society into two distinct groups: those that can do it and those that cannot. Those that cannot do it often feel the stigma of failure. Is Hersch right in attributing this to a formalistic or platonic view? Is he right to suggest that if maths is just a meaningless set of formal exercises, then it will not be valued in the main by society? If maths is out there to be discovered, it does seem reasonable to imagine that a particular individual who does not make the discovery might experience a sense of failure. The interesting question in this case is: What practical consequences in the classroom would follow from a humanist view of mathematics?

The golden ratio

There are some intriguing links between mathematics and the arts. One link that seems to fascinate many students of mathematics is the ancient idea of the **golden ratio**. Consider a line segment AB. The Greek mathematicians were interested in dividing AB by placing a point X in such a way that the ratio of the smaller piece to the longer piece was equal to the ratio of the longer piece to the whole line.

In other words: $XB/AX = AX/AB$

```
A               X           B
|---------------|-----------|
```

Let us rescale our units so that $AB = 1$ unit. Let $AX = x$. Then $XB = 1 - x$.

The equation above gives us: $\dfrac{1-x}{x} = \dfrac{x}{1}$

Rearranging gives us: $1 - x = x^2$

This gives the quadratic equation: $x^2 + x - 1 = 0$

Solving this equation using the quadratic formula gives:

$x = \dfrac{-1 + \sqrt{5}}{2}$ and $x = \dfrac{-1 - \sqrt{5}}{2}$ or $x = 0.618\,033\,988\,75\ldots$ or $-1.618\,033\,988\,75\ldots$

THEORY OF KNOWLEDGE

The first of these solutions is known as the **golden ratio**. Because of the special symmetry of the relationship between the different parts of the line segment above, this ratio was thought to be special or perfect in some way. Rectangles in which the ratio of the shorter to the longer side is equal to the golden ratio were thought to be especially beautiful. Try this out yourself in the rectangle beauty contest. Choose the rectangle that is most pleasing to you. Measure the sides and calculate the ratio between the shorter and the longer side. How close are you to the golden ratio?

A4 paper has dimensions of 210 mm × 297 mm. 210/297 = 0.707, which is a little high. A4 paper is a little too 'fat' to be a golden rectangle.

Measure some rectangles in your school or home environment – for example, credit cards, postcards, books, tables. How close are they to golden rectangles?

Rectangle beauty contest

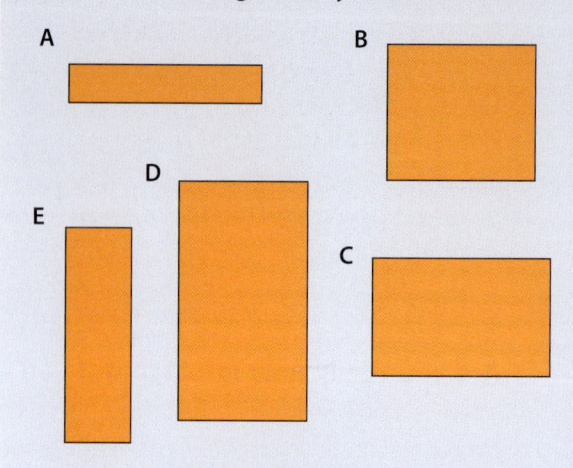

Which rectangle do you find the most pleasing?

A2, A3 and A5 paper are also all a little too fat to be golden rectangles. Why is this?

The golden ratio and the arts

There are many studies of the occurrence of the golden ratio in the natural and human worlds. It occurs in nature in connection with spirals and the Fibonacci sequence. The golden ratio has also been exploited by human beings in art, architecture and music. For example, the golden ratio was exploited by the ancient Greeks in their designs for temples and other buildings. The Parthenon in Athens is constructed using the golden section at key points.

The Greek letter ϕ is often used for the golden section.

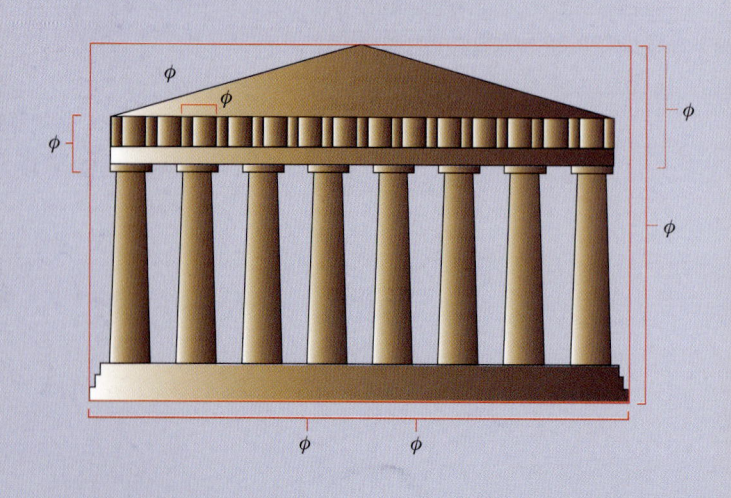

Golden ratios have been consciously used in the structure of some musical compositions. The French composer Debussy is known to have used this ratio in his orchestral piece *La Mer*, for example. The 55 bar introduction to 'Dialogue du vent et de la mer' breaks down into five sections of 21, 8, 8, 5 and 13 bars in length, which are numbers in the Fibonacci sequence. The golden ratio point of bar 34 in this passage is signalled by the entry of the trombones and percussion. More generally, we can ask ourselves how many pieces of music (or films or plays or dance performances) have some sort of structurally significant event roughly two-thirds of the way through the piece?

> Think of a film you have seen recently. At what point in the film did the moment of highest tension occur? How far into the film did this happen? Calculate this as a proportion. Is it close to 62%?

The Fibonacci sequence

The golden ratio is linked closely to the Fibonacci sequence:

1, 1, 2, 3, 5, 8, 13, 21, 34, 55, 89, …

What are the next two terms in the sequence?

If we divide successive terms in the sequence:

$\frac{1}{1} = 1$, $\frac{1}{2} = 0.5$, $\frac{2}{3} = 0.6667$, $\frac{3}{5} = 0.6$, $\frac{5}{8} = 0.625$, $\frac{8}{13} = 0.6154$, $\frac{13}{21} = 0.6190$, $\frac{21}{34} = 0.6176$, …

What is going on here?

Much has been written about how this sequence occurs in nature. It is naturally associated with certain types of growth. Ian Stewart, in his book *Nature's Numbers,* describes how these numbers are naturally associated with the spiral growth of many types of shell, for example. There is nothing mystical about this link. But it is tempting to think again about the Platonists and their view of mathematics as somehow embedded in the outside world.

The golden ratio suggests a strong link between mathematics and the arts. In theory of knowledge, it also raises a set of interesting questions about the nature of beauty. If we find certain rectangles pleasing because of the golden ratio, we might also find certain faces beautiful because of the ratios between the features, and find certain paintings or pieces of music beautiful, because of their proportions. Beauty would not be entirely in the eye of the beholder – it would be in the mathematics.

Answers

Chapter 1
Exercise 1.1
1. distance = $\frac{17}{4}$
2. distance = 9
3. distance = 7.4
4. distance = $\frac{26}{3}$
5. distance = $\frac{11\pi}{3}$
6. distance = $\frac{17}{12}$
7. $-5 \leq x \leq 3$ closed interval, bounded
8. $-10 < x \leq -2$ half-open interval, bounded
9. $x \geq 1$ half-open interval, unbounded
10. $x < 4$ open interval, unbounded
11. $0 \leq x < 2\pi$ half-open interval, bounded
12. $a \leq x \leq b$ closed interval, bounded
13. $]6, \infty[$
14. $]-\infty, -8]$
15. $]2, 9[$
16. $[0, 12[$
17. $]-5, \infty[$
18. $[-3, 3]$
19. $x \geq 6$ $[6, \infty[$
20. $4 \leq x < 10$ $[4, 10[$
21. $x < 0$ $]-\infty, 0[$
22. $0 < x < 25$ $]0, 25[$
23. $\{1, 3, 5, 7\}$
24. $\{1, 2, 3, 4, 5, 6, 7, 8, 9\}$
25. $\emptyset$
26. $\{1, 2, 3, 4, 5, 6, 7, 8\}$
27. $\{2, 4, 6\}$
28. $\{1, 2, 3, 4, 5, 6, 7, 8, 9\}$
29. $\mathbb{Z} \subset \mathbb{R}$
30. $\mathbb{N} \subset \mathbb{Q}$
31. $\mathbb{N} \subset \mathbb{Z}$
32. $\mathbb{Z} \subset \mathbb{Q}$
33. $|x| < 6$
34. $|x| \geq 4$
35. $|x| \leq \pi$
36. $|x| > 1$
37. 13
38. 4
39. -25
40. -5
41. $3 - \sqrt{3}$
42. -1
43. $x = -5, 5$
44. $x = -1, 7$
45. $x = -4, 16$
46. $x = -2, -\frac{4}{3}$

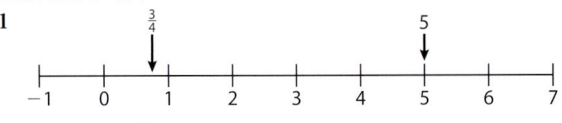

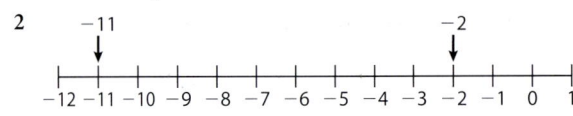

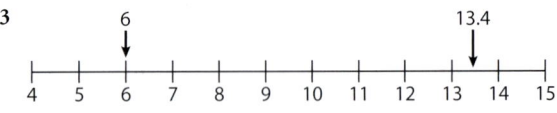

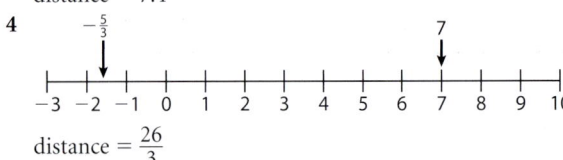

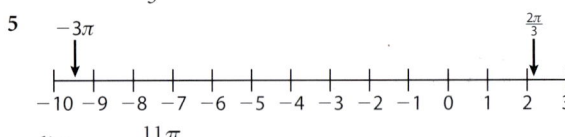

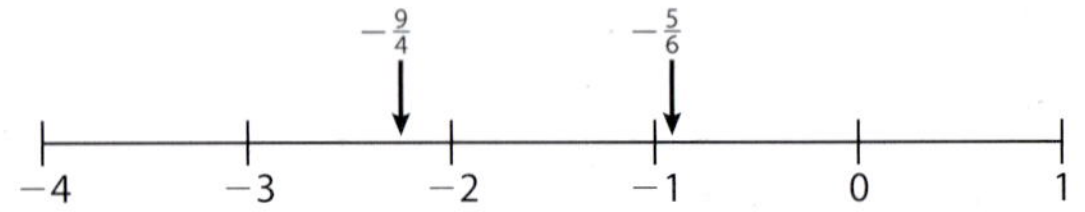

Exercise 1.2
1. $2\sqrt{2}$
2. 2
3. 6
4. 3
5. 2
6. $\frac{\sqrt{3}}{2}$
7. $5\sqrt{2}$
8. $3\sqrt{7}$
9. $12\sqrt{2}$
10. $4\sqrt{2}$
11. $10\sqrt{3}$
12. $10\sqrt{3} - \sqrt{2}$
13. $4\sqrt{6} + \sqrt{3}$
14. $\frac{\sqrt{2}}{2}$
15. $\frac{3\sqrt{5}}{5}$
16. $\frac{2\sqrt{21}}{7}$
17. $\frac{\sqrt{3}}{9}$
18. $\frac{4\sqrt{2}}{3}$
19. $\frac{\sqrt{6}}{3}$

Exercise 1.3
1. 2
2. 27
3. 16
4. 16
5. 8
6. 8
7. $\frac{4}{9}$
8. $\frac{3}{4}$
9. $\frac{125}{8}$
10. $\frac{1}{9}$
11. 1
12. $\frac{16}{3}$
13. $\frac{-64}{27}$
14. $3a^2 b^4$
15. $-3a^3 b^6$
16. $9a^2 b^4$
17. $10x^5 y^3$
18. $\frac{4}{3w}$
19. $\frac{3m^6}{4n^4}$
20. $\frac{m^6}{8n^6}$
21. 3^{2m+n}
22. $\frac{y^2}{x^2}$
23. $\frac{b}{4a^5}$
24. x
25. $\frac{4(a+b)^2}{3}$
26. $\frac{(x+4y)^{\frac{3}{2}}}{2}$
27. $\sqrt{p^2 + q^2}$
28. 2^{6n+2m}

Exercise 1.4
1. 2.54×10^2
2. 7.81×10^{-3}
3. 7.41×10^6
4. 1.04×10^{-6}
5. 4.98
6. 1.99×10^{-3}
7. 1.49×10^8
8. 8.99×10^{-5}
9. 0.0027
10. $50\,000\,000$
11. $0.000\,000\,090\,35$
12. $4\,180\,000\,000\,000$
13. 2.5×10^3
14. 2×10^4
15. 8.2×10^{-5}
16. 5.6×10^{18}

Exercise 1.5
1. $n^2 - n - 20$
2. $10y^2 - 9y - 9$
3. $x^2 - 49$
4. $25m^2 + 20m + 4$
5. $x^3 - 3x^2 + 3x - 1$
6. $1 - a$
7. $a^2 + a - b^2 + b$
8. $4x^2 + 12x + 9 - y^2$
9. $a^3 + 3a^2 b + 3ab^2 + b^3$
10. $a^2 x^2 + 2abx + b^2$
11. -4
12. $4x^3 - 8x^2 + 13x - 5$
13. $12(x+2)(x-2)$
14. $x^2(x-6)$
15. $(x+4)(x-3)$
16. $-(m-1)(m+7)$
17. $(x-8)(x-2)$
18. $(y+1)(y+6)$
19. $3(n-5)(n-2)$
20. $2x(x+1)(x+9)$
21. $(a+4)(a-4)$
22. $(3y+1)(y-5)$
23. $(5n^2 + 2)(5n^2 - 2)$
24. $a(x+3)^2$
25. $(m+1)^2(2n-1)$
26. $(x+1)(x-1)(x^2+1)$
27. $y(6-y)$
28. $2y^2(2y^2 - 5y - 48)$
29. $(2x-5)^2$
30. $(2x+3)^{-3}(4x+3) = \dfrac{4x+3}{(2x+3)^3}$

31 $\dfrac{1}{x+1}$ **32** $\dfrac{1}{2n}$ **33** $\dfrac{a+b}{5}$

34 $x+2$ **35** -1 **36** $4x+h$

37 $\dfrac{-2x+5}{15}$ **38** $\dfrac{b-a}{ab}$ **39** $\dfrac{-8x+6}{2x-1}$

40 $\dfrac{x^2+x+3}{x^2+3x}$ **41** $\dfrac{2x}{x^2-y^2}$ **42** $\dfrac{-2}{x-2}$

43 6 **44** $\dfrac{3y-10}{y^2-3y-10}$ **45** $\dfrac{1}{ab-b^2}$

46 $\dfrac{-5x^2-5x}{2}$ **47** $\dfrac{3+\sqrt{2}}{7}$ **48** $10-5\sqrt{3}$

49 $\dfrac{11+4\sqrt{6}}{5}$ **50** $\dfrac{7-\sqrt{5}}{44}$

Exercise 1.6

1 $x = h - \dfrac{n}{m}$ **2** $a = \dfrac{v^2+t}{b}$

3 $b_1 = \dfrac{2A}{h} - b_2$ **4** $r = \pm\sqrt{\dfrac{2A}{\theta}}$

5 $k = \dfrac{gh}{f}$ **6** $t = \dfrac{x}{a+b}$

7 $r = \sqrt[3]{\dfrac{3V}{\pi h}}$ **8** $k = \dfrac{g}{F(m_1+m_2)}$

9 $y = -\tfrac{2}{3}x - 5$ **10** $y = -4$

11 $y = \tfrac{5}{4}x + 6$ **12** $x = \tfrac{7}{3}$

13 $y = -4x + 11$ **14** $y = -\tfrac{5}{2}x - 7$

15 a) 17 b) $\left(0, \tfrac{5}{2}\right)$

16 a) $\sqrt{40}$ b) $(2, 3)$

17 a) $\dfrac{\sqrt{82}}{3}$ b) $\left(-1, \tfrac{7}{6}\right)$

18 a) $\sqrt{533}$ b) $\left(1, \tfrac{11}{2}\right)$

19 $k = 1$ or 9 **20** $k = -11$ or -3

21 $(\sqrt{5})^2 + (\sqrt{45})^2 = (\sqrt{50})^2$

22 Sides are: $\sqrt{29}, \sqrt{29}, \sqrt{58}$

23 Sides are: $\sqrt{45}, \sqrt{10}, \sqrt{45}, \sqrt{10}$

24 $(5, 1)$ **25** $\left(4, \tfrac{1}{2}\right)$ **26** $(3, -4)$ **27** $(3.8, -1.6)$

28 No solution   **29** $(-1, 2)$ **30** $(-1, 3)$ **31** $(-3, -8)$

32 Lines are coincident; solution set is all points on the line $y = -\tfrac{1}{4}x - \tfrac{3}{4}$

33 $\left(\tfrac{20}{3}, \tfrac{40}{3}\right)$ **34** $\left(\tfrac{1}{2}, 3\right)$ **35** $(-5, 10)$

36 $(5, -3)$ **37** $(14.1, 10.4)$ **38** $\left(\tfrac{11}{19}, -\tfrac{18}{19}\right)$

Chapter 2

Exercise 2.1

1 a) G   b) Function   **2** a) L   b) Function
3 a) H   b) Function   **4** a) K   b) Not function
5 a) J   b) Function   **6** a) C   b) Function
7 a) A   b) Function   **8** a) I   b) Function
9 a) F   b) Function

10 $A = \dfrac{C^2}{4\pi}$ **11** $A = \dfrac{l^2\sqrt{3}}{4}$ **12** $x \in \mathbb{R}$

13 $x \in \mathbb{R}$ **14** $t \leq 3$ **15** $t \in \mathbb{R}$ **16** $r \geq 0$

17 $x \in \mathbb{R}, x \neq \pm 3$

18 No, a vertical line does not represent a function.

19 Domain: $x \in \mathbb{R}, x \neq 5$, range: $y \in \mathbb{R}, y \neq 0$

20 a) (i) $\sqrt{17}$ (ii) 7 (iii) 0
b) $x < 4$ c) Domain: $x \geq 4$, range: $y \geq 0$

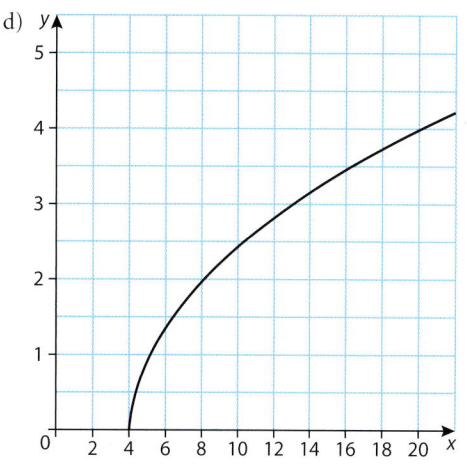

d)

21 Domain: $x \in (-\infty, -3) \cup (3, \infty)$; range: $y \in (0, \infty)$

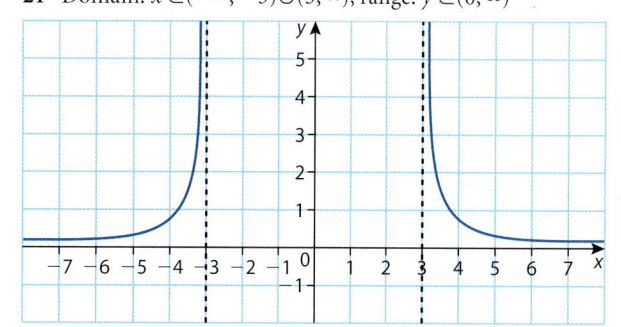

Exercise 2.2

1 a) $(f \circ g)(5) = 1, (g \circ f)(5) = \tfrac{1}{7}$
b) $(f \circ g)(x) = \dfrac{2}{x-3}, (g \circ f)(x) = \dfrac{1}{2x-3}$

2 a) 1 b) -7 c) 7
d) -47 e) -1 f) -79
g) $1 - 2x^2$ h) $-4x^2 + 12x - 7$
i) $4x - 9$ j) $-x^4 + 4x^2 - 2$

3 $(f \circ g)(x) = 12x + 7$, domain: $x \in \mathbb{R}$;
$(g \circ f)(x) = 12x - 1$, domain: $x \in \mathbb{R}$

4 $(f \circ g)(x) = 4x^2 + 1$, domain: $x \in \mathbb{R}$;
$(g \circ f)(x) = -2x^2 - 2$, domain: $x \in \mathbb{R}$

5 $(f \circ g)(x) = \sqrt{x^2 + 2}$, domain: $x \in \mathbb{R}$;
$(g \circ f)(x) = x + 2$, domain: $x \geq -1$

6 $(f \circ g)(x) = \dfrac{2}{x+3}$, domain: $x \in \mathbb{R}, x \neq -3$;
$(g \circ f)(x) = -\dfrac{x+2}{x+4}$, domain: $x \in \mathbb{R}, x \neq -4$

7 $(f \circ g)(x) = x$, domain: $x \in \mathbb{R}$; $(g \circ f)(x) = x$, domain: $x \in \mathbb{R}$

8 a) $(g \circ h)(x) = \sqrt{9 - x^2}$, domain: $-3 \leq x \leq 3$, range: $y \geq 0$
b) $(h \circ g)(x) = -x + 11$, domain: $x \geq 1$, range: $y \leq 10$

9 $h(x) = x + 3, g(x) = x^2$

10 $h(x) = x - 5, g(x) = \sqrt{x}$

11 $h(x) = \sqrt{x}, g(x) = 7 - x$

12 $h(x) = x + 3, g(x) = \dfrac{1}{x}$

13 $h(x) = x + 1, g(x) = 10^x$

14 $h(x) = x - 9, g(x) = \sqrt[3]{x}$

15 a) Domain of f: $x \geq 0$ b) Domain of g: $x \in \mathbb{R}$
c) $(f \circ g)(x) = \sqrt{x^2 + 1}$, domain of $(f \circ g)$: $x \in \mathbb{R}$

16 a) $D(f) = \{x \in \square : x \neq 0\}$ b) $D(g) = \square$
c) $f(g(x)) = f(x+3) = \dfrac{1}{x+3} \Rightarrow D(f \circ g) = \{x \in \square : x \neq -3\}$

Answers

17 a) Domain of $f: x \neq \pm 1$ b) Domain of $g: x \in \mathbb{R}$
c) $(f \circ g)(x) = \dfrac{3}{x^2 + 2x}$, domain of $(f \circ g): x \neq 0, -2$

18 a) Domain of $f: x \in \mathbb{R}$ b) Domain of $g: x \in \mathbb{R}$
c) $(f \circ g)(x) = x + 3$, domain of $(f \circ g): x \in \mathbb{R}$

Exercise 2.3

1 a) 2   b) 6
2 a) −1   b) b
3 4
4 6
5 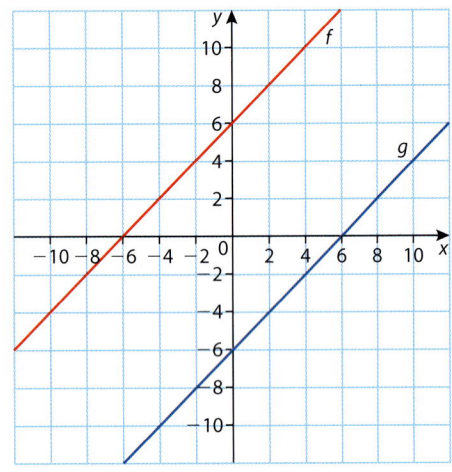
6 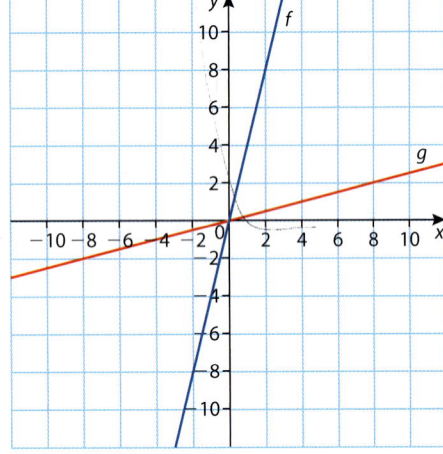
7 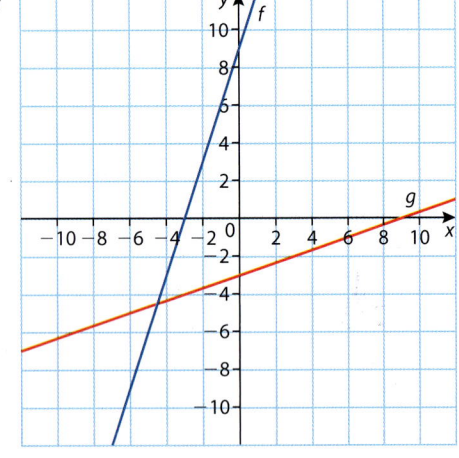
8 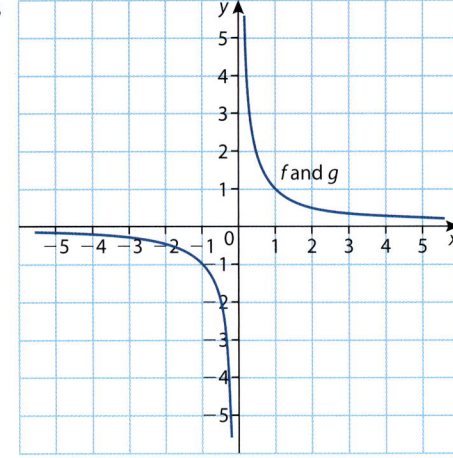
9 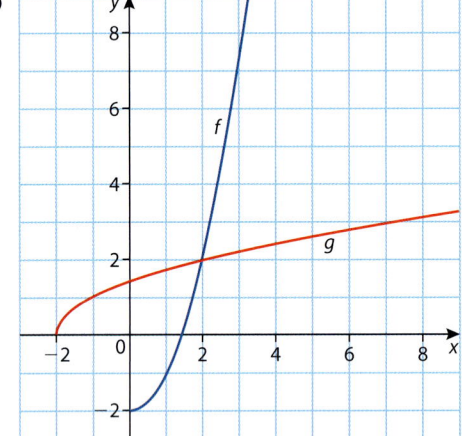
10

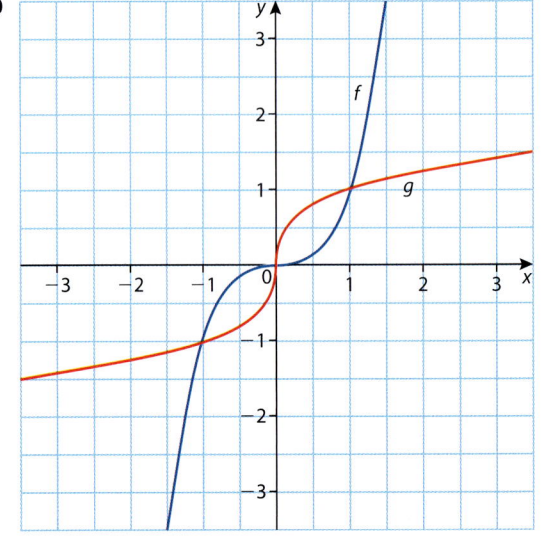

578

11

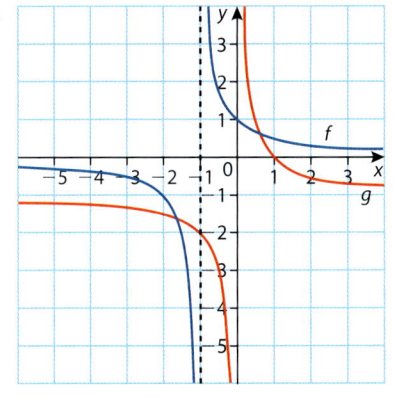

12

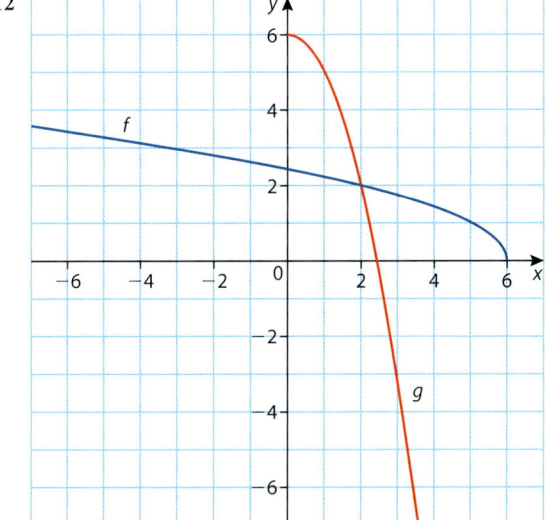

13 $f^{-1}(x) = \frac{1}{2}x + \frac{3}{2}, x \in \mathbb{R}$

14 $f^{-1}(x) = 4x - 7, x \in \mathbb{R}$

15 $f^{-1}(x) = x^2, x \geq 0$

16 $f^{-1}(x) = \frac{1}{x} - 2, x \in \mathbb{R}, x \neq 0$

17 $f^{-1}(x) = \sqrt{4 - x}, x \leq 4$

18 $f^{-1}(x) = x^2 + 5, x \geq 0$

19 $f^{-1}(x) = \frac{1}{a}x - \frac{b}{a}, x \in \mathbb{R}$

20 $f^{-1}(x) = \sqrt{x + 1} - 1, x \geq -1$

21 $\frac{3}{2}$

22 5

23 -4

24 $\frac{7}{2}$

25 $g^{-1} \circ h^{-1} = \frac{1}{2}x - 1$

26 $h^{-1} \circ g^{-1} = \frac{1}{2}x + \frac{1}{2}$

27 $(g \circ h)^{-1} = \frac{1}{2}x + \frac{1}{2}$

28 $(h \circ g)^{-1} = \frac{1}{2}x - 1$

29 $f(f(x)) = f\left(\frac{a}{x+b} - b\right) = \frac{a}{\frac{a}{x+b} - b + b} - b = \frac{a}{\frac{a}{x+b}} - b$
$= \frac{a}{1} \cdot \frac{x+b}{a} - b = x + b - b = x$

Since $f(f(x)) = x$, then the function f is its own inverse.

Exercise 2.4

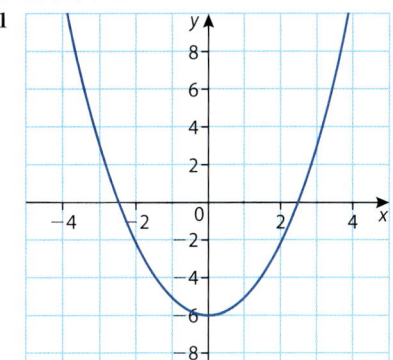

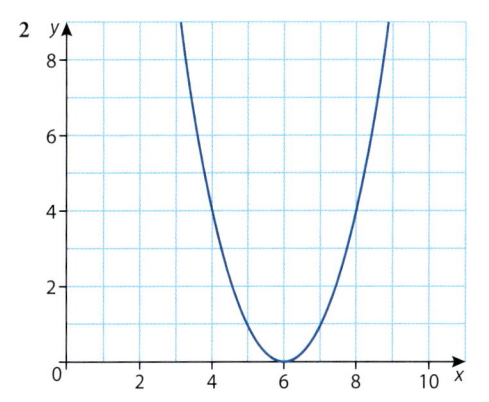

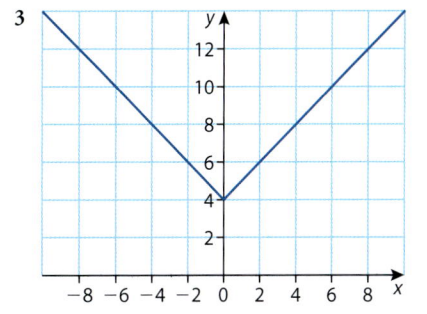

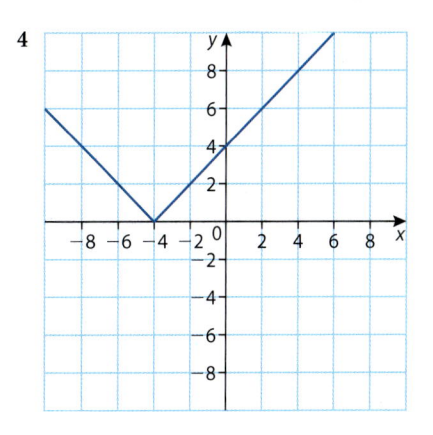

1

2

3

4

Answers

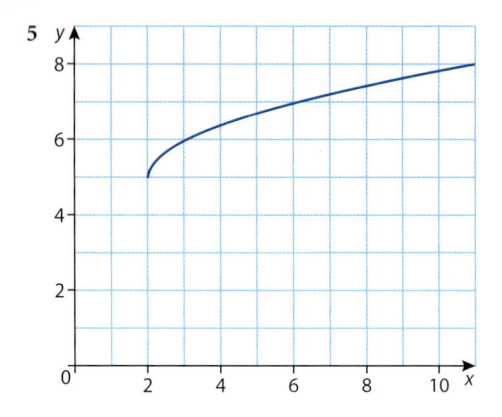

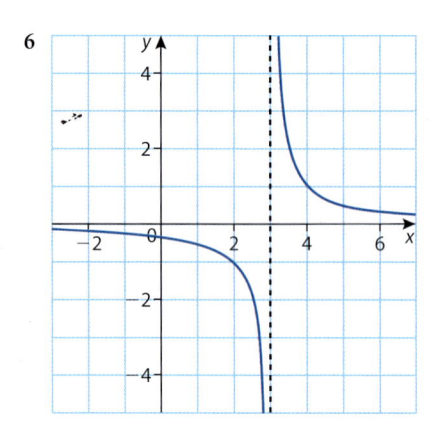

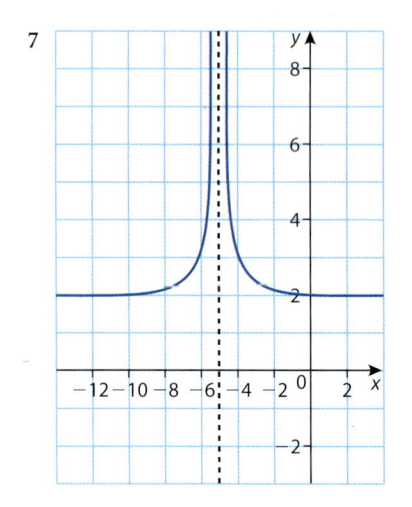

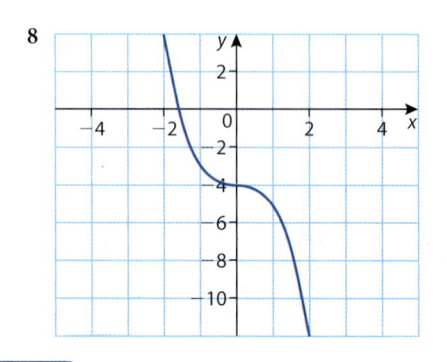

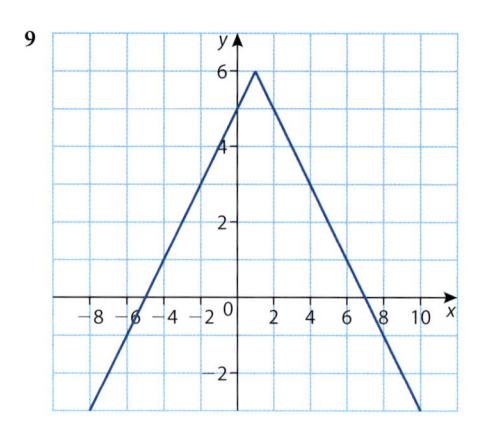

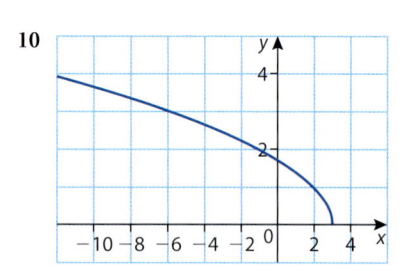

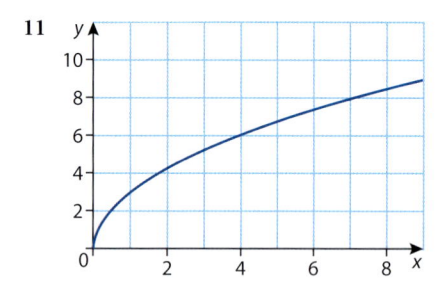

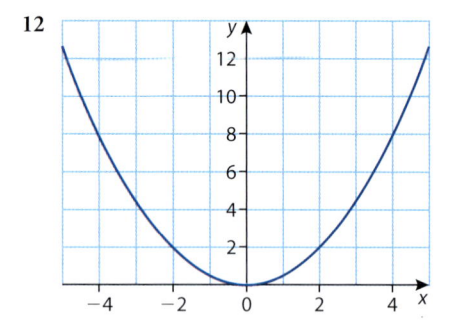

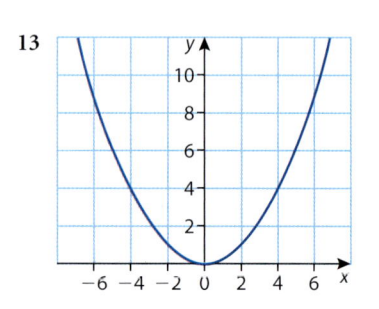

5–13. (Graphs)

580

14

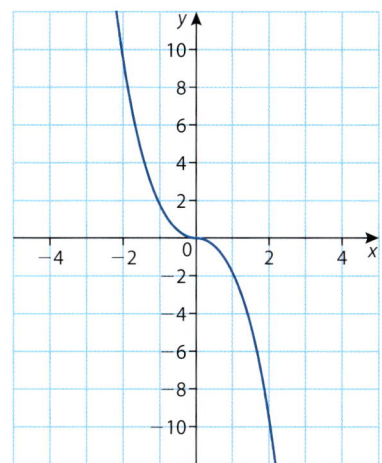

15 $y = -x^2 + 5$
16 $y = \sqrt{-x}$
17 $y = -|x + 1|$
18 $y = \dfrac{1}{x-2} - 3$

19 a) 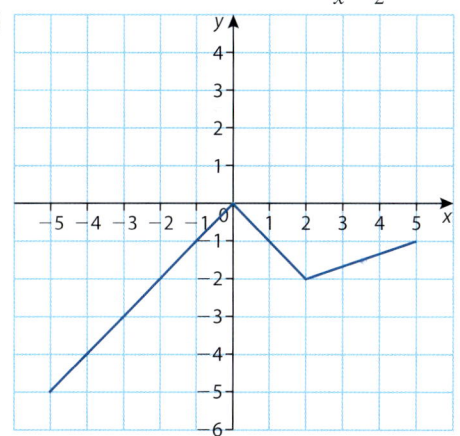
    b) 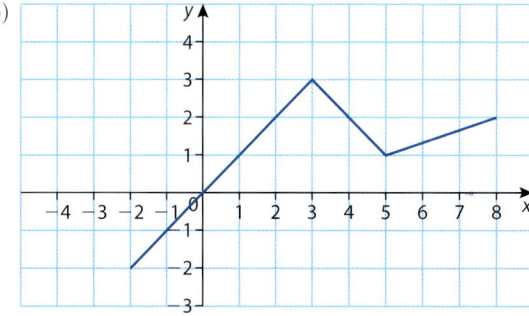
    c) 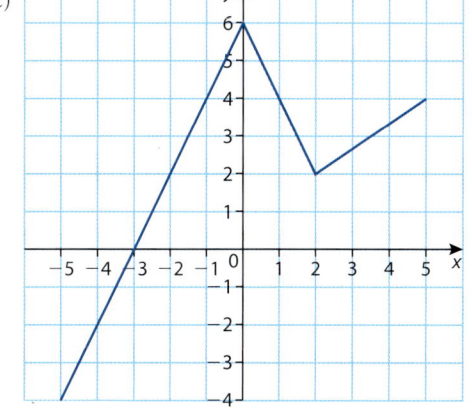
    d) 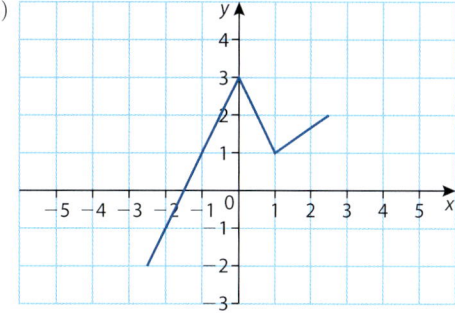
    e) 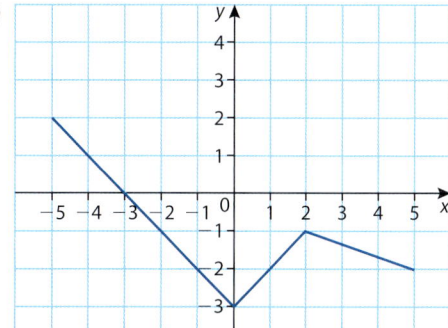
    f) 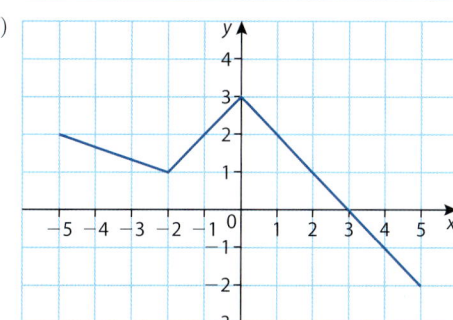
    g)

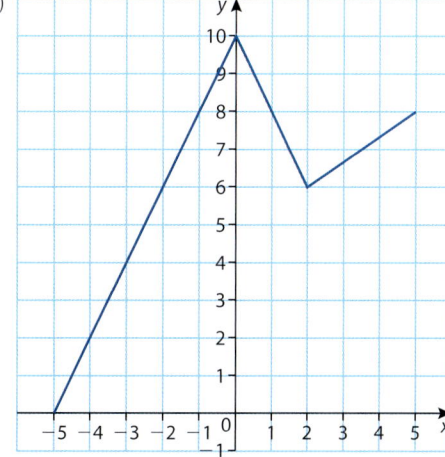

20 Horizontal translation 3 units right; vertical translation 5 units up (or reverse order).
21 Reflect over the x-axis; vertical translation 2 units up (or reverse order).
22 Horizontal translation 4 units left; vertical shrink by factor $\tfrac{1}{2}$ (or reverse order).
23 Horizontal translation 1 unit right; horizontal shrink by factor $\tfrac{1}{3}$; vertical translation 6 units down.

581

Answers

24 a) $f(x) = (x+3)^2 - 7$ **b)** vertex $(-3, -7)$
25 a) $f(x) = (x-1)^2 + 3$ **b)** vertex $(1, 3)$
26 a) $f(x) = 4\left(x - \frac{1}{2}\right)^2 - 2$ **b)** vertex $\left(\frac{1}{2}, -2\right)$

Exercise 2.5

1 a) $f(x) = (x-5)^2 + 7$; axis of symmetry: $x = 5$, vertex $(5, 7)$
   b) Horizontal translation 5 units right; vertical translation 7 units up.
   c) Minimum: 7
2 a) $f(x) = (x+3)^2 - 1$; axis of symmetry: $x = -3$, vertex $(-3, -1)$
   b) Horizontal translation 3 units left; vertical translation 1 unit down.
   c) Minimum: -1
3 a) $f(x) = -2(x+1)^2 + 12$; axis of symmetry: $x = -1$, vertex $(-1, 12)$
   b) Horizontal translation 1 unit left; reflection over x-axis; vertical stretch by factor 2; vertical translation 12 units up.
   c) Maximum: 12
4 a) $f(x) = 4\left(x - \frac{1}{2}\right)^2 + 8$; axis of symmetry: $x = \frac{1}{2}$, vertex $\left(\frac{1}{2}, 8\right)$
   b) Horizontal translation $\frac{1}{2}$ unit right; vertical stretch by factor 4; vertical translation 8 units up.
   c) Maximum: 8
5 a) $f(x) = \frac{1}{2}(x+7)^2 + \frac{3}{2}$; axis of symmetry: $x = -7$, vertex $\left(-7, \frac{3}{2}\right)$
   b) Horizontal translation 7 units left; vertical shrink by factor $\frac{1}{2}$; vertical translation $\frac{3}{2}$ unit up.
   c) Minimum: $\frac{3}{2}$
6 $x = 2, x = -4$
7 $x = 5, x = -2$
8 $x = \frac{3}{2}, x = 0$
9 $x = 6, x = -1$
10 $x = 3$
11 $x = \frac{1}{3}, x = -4$
12 $x = 3, x = 2$
13 $x = 2, x = \frac{1}{4}$
14 $x = -2 \pm \sqrt{7}$
15 $x = 5, x = -1$
16 No real solution
17 $x = -4 \pm \sqrt{13}$
18 $x = 2, x = -4$
19 $x = \dfrac{2 \pm \sqrt{22}}{2}$
20 a) $x = 2 \pm \sqrt{5}$
   b) axis of symmetry: $x = 2$
   c) minimum value of f is -5
21 Two real solutions   **22** No real solutions
23 Two real solutions   **24** No real solutions
25 $p = \pm 2\sqrt{2}$ **26** $k < 4$
27 $k < -1, k > 1$ **28** $m < -3, m > 3$

Practice questions

1 a) $a = -3, b = 1$ **b)** range: $y \geqslant 0$
2 a) 5   **b)** -9
3 a) $g^{-1}(x) = -3x + 4$ **b)** $x = \frac{2}{3}$
4 a) $(g \circ h)(x) = 2x - 3$ **b)** 24
5 a) $f(x) = (x+4)^2 - 5$ **b)** $f^{-1}(x) = -4 + \sqrt{x+5}$
   c) domain: $x \geqslant 5$

6 a)

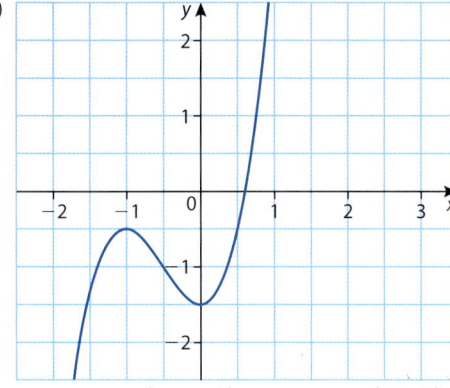

b) Maximum at $\left(-1, -\frac{1}{2}\right)$; minimum at $\left(0, -\frac{3}{2}\right)$
7 a) $k = \frac{1}{2}$ **b)** $p = -5$ **c)** $q = 3$
8 a)

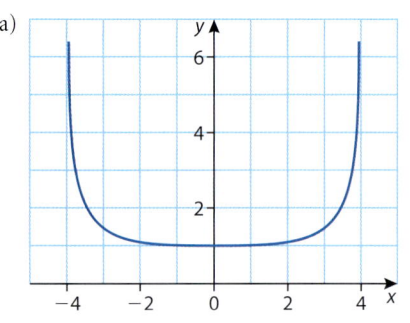

b) $x = 4, x = -4$ **c)** range: $y \geqslant 1$

9 a)

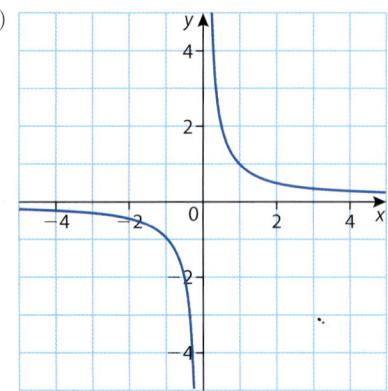

b) $h(x) = \dfrac{1}{x+4} - 2$
c) (i) x-intercept: $\left(-\frac{7}{2}, 0\right)$; y-intercept: $\left(0, -\frac{7}{4}\right)$
     (ii) Vertical asymptote: $x = -4$; horizontal asymptote: $y = -2$
     (iii)

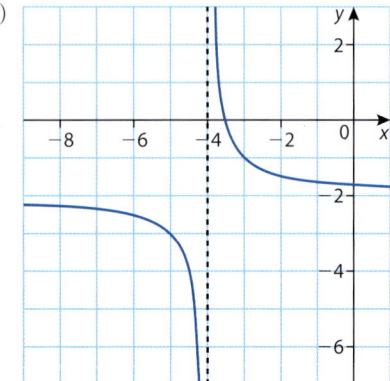

582

10 a) (i) $\sqrt{11}$ (ii) 7   (iii) 0
    b) $x < -3$ c) $g(f(x)) = x - 2, x \geq -3$
11 a) 4   b) $(g^{-1} \circ h)(x) = 2x^2 + 6$ c) $x = \pm 2\sqrt{2}$
12 a) $f^{-1}(x) = \frac{1}{3}x + \frac{1}{3}$
    b) $(f \circ g)(x) = \frac{12}{x} - 1$
    c) $(f \circ g)^{-1}(x) = \frac{12}{x+1}$
    d) $(g \circ g)(x) = x$
13 a) $f(x) = 2(x + 2)^2 + 9$
    b) $g(x) = 2(x - 3)^2 + 11$
14 a) $g(x) = 3(x - 1)^2 - 7$
    b) Vertex: $(1, -7)$ c) $x = 1$
    d) y-intercept: $(0, -4)$ e) $p = 3, q = 21, r = 3$
15 a) (i) $a = 8$ (ii) $b = -3$
    b) Reflection over x-axis
16 a)

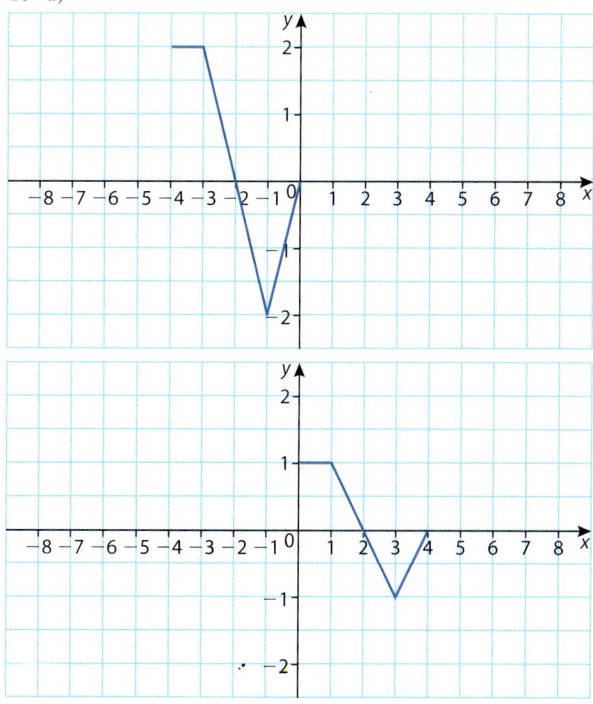

    b) $A'(-3, -2)$
17 a) $p = -3, q = \frac{1}{3}$ b) $x = -\frac{4}{3}$
    c) $f(x) = x^2 + \frac{8}{3}x - 1$
18 $A(-5, 0), B\left(-\frac{3}{2}, \frac{49}{4}\right), C(2, 0)$

Chapter 3

Exercise 3.1
1   $-1, 1, 3, 5, 7, 97$
2   $2, 6, 18, 54, 162, 4.786 \times 10^{23}$
3   $\frac{2}{3}, -\frac{2}{3}, \frac{6}{11}, -\frac{4}{9}, \frac{10}{27}, -\frac{50}{1251}$
4   $1, 2, 9, 64, 625, 1.776 \times 10^{83}$
5   $3, 11, 27, 59, 123, 4.50 \times 10^{15}$
6   $0, 3, \frac{3}{7}, \frac{21}{13}, \frac{39}{55}$, approx. 1
7   $2, 6, 18, 54, 162, 4.786 \times 10^{23}$
8   $-1, 1, 3, 5, 7, 97$

Exercise 3.2
1   $3, \frac{19}{5}, \frac{23}{5}, \frac{27}{5}, \frac{31}{5}, 7$
2 a) Arithmetic, $d = 2, a_{50} = 97$
    b) Arithmetic, $d = 1, a_{50} = 52$
    c) Arithmetic, $d = 2, a_{50} = 97$
    d) Not arithmetic, *no common difference*
    e) Not arithmetic, *no common difference*
    f) Arithmetic, $d = -7, a_{50} = -341$
3 a) 26
    b) $a_n = -2 + 4(n - 1)$
    c) $a_1 = -2, a_n = a_{n-1} + 4$ for $n > 1$
4 a) 1
    b) $a_n = 29 - 4(n - 1)$
    c) $a_1 = 29, a_n = a_{n-1} - 4$ for $n > 1$
5 a) 57
    b) $a_n = -6 + 9(n - 1)$
    c) $a_1 = -6, a_n = a_{n-1} + 9$ for $n > 1$
6 a) 9.23
    b) $a_n = 10.07 - 0.12(n - 1)$
    c) $a_1 = 10.07, a_n = a_{n-1} - 0.12$ for $n > 1$
7 a) 79
    b) $a_n = 100 - 3(n - 1)$
    c) $a_1 = 100, a_n = a_{n-1} - 3$ for $n > 1$
8 a) $-\frac{27}{4}$
    b) $a_n = 2 - \frac{5}{4}(n - 1)$
    c) $a_1 = 2, a_n = a_{n-1} - \frac{5}{4}$ for $n > 1$
9   $13, 7, 1, -5, -11, -17, -23$
10   $299, 299\frac{1}{4}, 299\frac{1}{2}, 299\frac{3}{4}, 300$
11   $a_n = -10 + 4(n - 1) = 4n - 14$
12   $a_n = -\frac{142}{3} + \frac{11}{3}(n - 1) = -51 + \frac{11}{3}n$

Exercise 3.3
1   $3, 6, 12, 24, 48, 96$
2 a) Arithmetic, $d = 3, a_{10} = 27$
    b) Geometric, $r = 2, b_{10} = 4096$
    c) Neither, $c_{10} = -1534$
    d) Geometric, $r = 3, u_{10} = 78\,732$
    e) Geometric, $r = 2.5, a_{10} = 7629.39453125$
    f) Geometric, $r = -2.5, a_{10} = -7629.39453125$
    g) Arithmetic, $d = 0.75, a_{10} = 8.75$
    h) Geometric, $r = -\frac{2}{3}, a_{10} = -\frac{1024}{2187}$
3 a) $\frac{2187}{64}$ b) $a_n = -2\left(-\frac{3}{2}\right)^{n-1}$
    c) $a_1 = -2, a_n = -\frac{3}{2}a_{n-1}, n > 1$
4 a) $\frac{390\,625}{117\,649}$ b) $a_n = 35\left(\frac{5}{7}\right)^{n-1}$
    c) $a_1 = 35, a_n = \frac{5}{7}a_{n-1}, n > 1$
5 a) $-\frac{3}{64}$ b) $a_n = -6\left(\frac{1}{2}\right)^{n-1}$
    c) $a_n = -6, a_n = \frac{1}{2}a_{n-1}, n > 1$
6 a) 1216   b) $9.5 \times 2^{n-1}$
    c) $a_1 = 9.5, a_n = 2a_{n-1}, n > 1$
7 a) $69.833\,729\,609\,375 = \frac{893\,871\,739}{12\,800\,000}$
    b) $a_n = 100\left(\frac{19}{20}\right)^{n-1}$
    c) $a_1 = 100, a_n = \frac{19}{20}a_{n-1}, n > 1$

Answers

8 a) $0.00208568573 = \dfrac{2187}{1048576}$
   b) $a_n = 2\left(\dfrac{3}{8}\right)^{n-1}$ c) $a_1 = 2, a_n = \dfrac{3}{8}a_{n-1}, n > 1$
9 7, 35, 175, 875, 4375   10 36
11 1.5, $a_n = 24\left(\dfrac{1}{2}\right)^{n-1}$ 12 $\dfrac{49}{3}$
13 10th term   14 Yes, 10th term
15 €2228.92   16 £945.23
17 €2968.79   18 7745 thousands

Exercise 3.4
1 11 280   2 $-\dfrac{105469}{1024}$ 3 0.7
4 $\dfrac{10}{7}$ 5 $\dfrac{16 + 4\sqrt{3}}{39}$
6 a) $\dfrac{52}{99}$ b) $\dfrac{449}{990}$ c) $\dfrac{7459}{2475}$
7 13 026.135 (£13 026.14)

Exercise 3.5
1 a) $x^5 + 10x^4y + 40x^3y^2 + 80x^2y^3 + 80xy^4 + 32y^5$
   b) $a^4 - 4a^3b + 6a^2b^2 - 4ab^3 + b^4$
   c) $x^6 - 18x^5 + 135x^4 - 540x^3 + 1215x^2 - 1458x + 729$
   d) $16 - 32x^3 + 24x^6 - 8x^9 + x^{12}$
   e) $x^7 - 21bx^6 + 189b^2x^5 - 945b^3x^4 + 2835b^4x^3 - 5103b^5x^2 + 5103b^6x - 2187b^7$
   f) $64n^6 + 192n^3 + 240 + \dfrac{160}{n^3} + \dfrac{60}{n^6} + \dfrac{12}{n^9} + \dfrac{1}{n^{12}}$
   g) $\dfrac{81}{x^4} - \dfrac{216}{x^2\sqrt{x}} + \dfrac{216}{x} - 96\sqrt{x} + 16x^2$
2 a) 56   b) 0   c) 1225   d) 32   e) 0
3 a) $x^7 + 14x^6y + 84x^5y^2 + 280x^4y^3 + 560x^3y^4 + 672x^2y^5 + 448xy^6 + 128y^7$
   b) $a^6 - 6a^5b + 15a^4b^2 - 20a^3b^3 + 15a^2b^4 - 6ab^5 + b^6$
   c) $x^5 - 15x^4 + 90x^3 - 270x^2 + 405x - 243$
   d) $x^{18} - 12x^{15} + 60x^{12} - 160x^9 + 240x^6 - 192x^3 + 64$
   e) $x^7 - 21bx^6 + 189b^2x^5 - 945b^3x^4 + 2835b^4x^3 - 5103b^5x^2 + 5103b^6x - 2187b^7$
   f) $64n^6 + 192n^3 + 240 + \dfrac{160}{n^3} + \dfrac{60}{n^6} + \dfrac{12}{n^9} + \dfrac{1}{n^{12}}$
   g) $\dfrac{81}{x^4} - \dfrac{216}{x^2\sqrt{x}} + \dfrac{216}{x} - 96\sqrt{x} + 16x^2$
   h) 112   i) $1792\sqrt{3}$
   j) 16   k) $-23 + 10i\sqrt{2}$
4 a) $x^{45} - 90x^{43} + 3960x^{41}$
   b) Does not exist as the powers of x decrease by 2's starting at 45. There is no chance for any expression to have zero exponent.
   c) $\binom{45}{43}x^2\left(\dfrac{-2}{x}\right)^{43} + \binom{45}{44}x\left(\dfrac{-2}{x}\right)^{44} + \left(\dfrac{-2}{x}\right)^{45} = -\binom{45}{43}\dfrac{2^{43}}{x^{41}} + \binom{45}{44}\dfrac{2^{44}}{x^{43}} - \dfrac{2^{45}}{x^{45}}$
   d) $\binom{45}{21}x^{24}\left(\dfrac{-2}{x}\right)^{21} = -\binom{45}{21}\cdot 2^{21}x^3$
5 $\binom{n}{k} = \dfrac{n!}{k!(n-k)!} = \dfrac{n!}{(n-k)!k!} = \dfrac{n!}{(n-k)!(n-(n-k))!} = \binom{n}{n-k}$
6 $(1+1)^n = \binom{n}{0} + \binom{n}{1} + \binom{n}{2} \cdots + \binom{n}{n}$
   $2^n = 1 + \binom{n}{1} + \binom{n}{2} \cdots + \binom{n}{n} \Rightarrow 2^n - 1 = \binom{n}{1} + \binom{n}{2} \cdots \binom{n}{n}$
7 Answers vary   8 $\left(\dfrac{1}{3} + \dfrac{2}{3}\right)^6 = 1$
9 $\left(\dfrac{2}{5} + \dfrac{3}{5}\right)^8 = 1$ 10 $\left(\dfrac{1}{7} + \dfrac{6}{7}\right)^n = 1$

Practice questions
1 $-1, 1, 3, 5, 7$ 2 $-1, 1, 5, 13, 29$
3 $\dfrac{3}{2}, \dfrac{3}{4}, \dfrac{3}{8}, \dfrac{3}{16}, \dfrac{3}{32}$ 4 5, 8, 11, 14, 17
5 $1, 7, -5, 19, -29$ 6 3, 7, 13, 21, 31
7 Arithmetic, $d = 3$ 8 Geometric, $r = -3$
9 Geometric, $r = 2$ 10 Neither
11 Neither   12 Arithmetic, $d = 1.3$
13 a) 32   b) $-3 + 5(n-1)$
    c) $a_1 = -3, a_n = a_{n-1} + 5$ for $n > 1$
14 a) -9 b) $19 - 4(n-1)$
    c) $a_1 = 19, a_n = a_{n-1} - 4$ for $n > 1$
15 a) 69   b) $-8 + 11(n-1)$
    c) $a_1 = -8, a_n = a_{n-1} + 11$ for $n > 1$
16 a) 9.35   b) $10.05 - 0.1(n-1)$
    c) $a_1 = 10.05, a_n = a_{n-1} - 0.1$ for $n > 1$
17 a) 93   b) $100 - (n-1)$
    c) $a_1 = 100, a_n = a_{n-1} - 1$ for $n > 1$
18 a) $-\dfrac{17}{2}$ b) $2 - 1.5(n-1)$
    c) $a_1 = 2, a_n = a_{n-1} - 1.5$ for $n > 1$
19 a) 384   b) $3 \times 2^{n-1}$
    c) $a_1 = 3, a_n = 2a_{n-1}$ for $n > 1$
20 a) 8748   b) $4 \times 3^{n-1}$
    c) $a_1 = 4, a_n = 3a_{n-1}$ for $n > 1$
21 a) -5 b) $5 \times (-1)^{n-1}$
    c) $a_1 = 5, a_n = -a_{n-1}$ for $n > 1$
22 a) -384 b) $3 \times (-2)^{n-1}$
    c) $a_1 = 3, a_n = -2a_{n-1}$ for $n > 1$
23 a) $-\dfrac{4}{9}$ b) $972 \times \left(-\dfrac{1}{3}\right)^{n-1}$
    c) $a_1 = 972, a_n = \left(-\dfrac{1}{3}\right)a_{n-1}$ for $n > 1$
24 $15, 9, 3, -3, -9, -15, -21$
25 $99, 99.25, 99.5, 99.75, 100$
26 $a_n = 4n - 1$
27 $a_n = -86 + \left(\dfrac{19}{3}\right)(n-1)$
28 7, 21, 63, 189, 567, 1701
29 ± 24
30 $a_4 = \pm 3, r = \pm\left(\dfrac{1}{2}\right), a_n = 24\left(\dfrac{1}{2}\right)^{n-1}$ or $a_n = 24\left(-\dfrac{1}{2}\right)^{n-1}$
31 $\dfrac{98}{9}$ 32 10th term   33 Yes, 10th term
34 €3714.87   35 £2921.16   36 €2098.63
37 11 400   38 $\dfrac{210938}{177147} \approx 1.191$ 39 49.2
40 $\dfrac{6}{5}$ 41 $\dfrac{3 + \sqrt{6}}{2}$
42 a) $\dfrac{7}{9}$ b) $\dfrac{38}{110}$ c) $\dfrac{31808}{9900}$
43 -145152 44 $35a^3$ 45 96 096
46 $243n^5 - 810n^4m + 1080n^3m^2 - 720n^2m^3 + 240nm^4 - 32m^5$
47 7 838 208
48 $d = 5, n = 20$
49 a) Nick: 20
       Charlotte: 17.6
    b) Nick: 390
       Charlotte: 381.3
    c) Charlotte will exceed the 40 hours during week 14.
    d) In week 12 Charlotte will catch up with Nick and exceed him.
50 a) Loss for the second month = 1060 g
       Loss for the third month = 1123.6 g
    b) Plan A loss = 1880 g
       Plan B loss = 1898.3 g
    c) (i) Loss due to plan A in all 12 months = 17 280 g
       (ii) Loss due to Plan B in all 12 months = 16 869.9 g

584

51 a) €895.42 b) €6985.82
52 a) 142.5
   b) 19 003.5
53 a) On the 37th day
   b) 407 km
54 a) 1.5
   b) 207 595
   c) 2009
   d) 619 583
   e) Market saturation
55 $-4, 3006$
56 a) $\sqrt{\frac{1}{4} + \frac{1}{4}} = \frac{\sqrt{2}}{2}$ b) $\frac{1}{2}$
   c) (i) $\frac{1}{4}$ (ii) $\frac{1}{2}$ d) (i) $\frac{1}{512}$ (ii) 2
57 a) 1220 b) 36 920
58 a) Area A = 1, Area B = $\frac{1}{9}$
   b) $\frac{1}{81}$
   c) $1 + \frac{8}{9}, 1 + \frac{8}{9} + \left(\frac{8}{9}\right)^2$
   d) 0
59 a) Neither, geometric converging, arithmetic, geometric diverging
   b) 6
60 a) (i) Kell: 18 400, 18 800; YBO: 18 190, 19 463.3
       (ii) Kell: 198 000; YBO: 234 879.62
       (iii) Kell: 21 600; YBO: 31 253.81
   b) (i) After the second year
       (ii) 4th year
61 a) 62 b) 936
62 a) $7000(1 + 0.0525)^t$
   b) 7 years
   c) Yes, since 10 084.7 > 10 015.0
63 a) 11 b) 2 c) 15

Chapter 4
Exercise 4.1 and 4.2

1

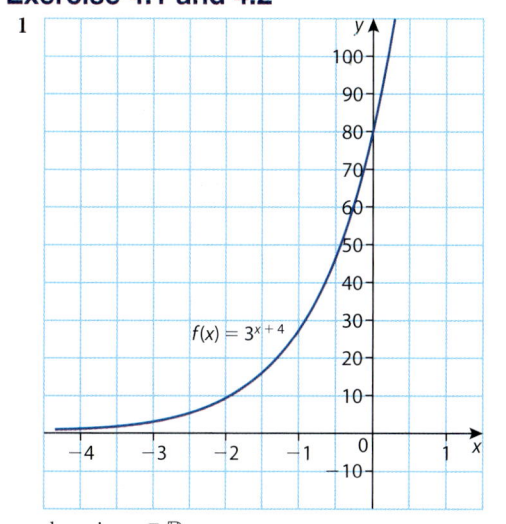

domain: $x \in \mathbb{R}$
range: $y > 0$
y-intercept: (0, 81)
horizontal asymptote: $y = 0$ (x-axis)

2

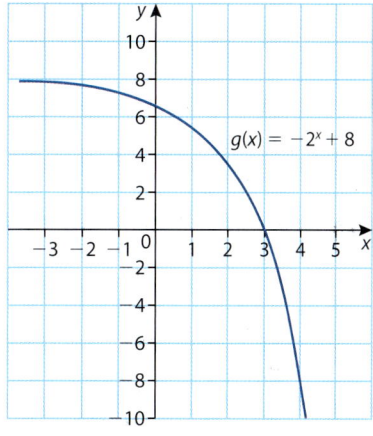

domain: $x \in \mathbb{R}$
range: $y < 8$
y-intercept: (0, 7)
horizontal asymptote: $y = 8$

3

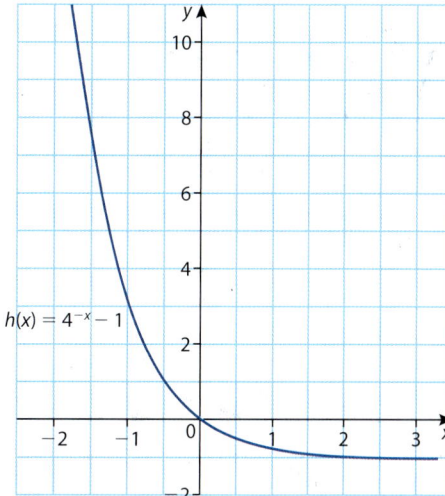

domain: $x \in \mathbb{R}$
range: $y > -1$
y-intercept: (0, 0)
horizontal asymptote: $y = -1$

4 Domain: $x \in \mathbb{R}$
range: if $a < 0 \Rightarrow y > d$, if $a < 0 \Rightarrow y < d$
y-intercept: $(0, ab^{-c} + d)$
horizontal asymptote: $y = d$

5

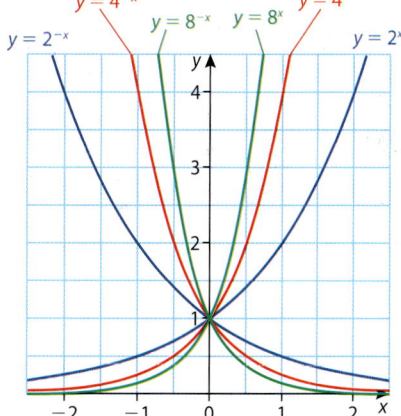

585

Answers

6 a) $y = \left(\frac{1}{2}\right)^x$ b) $y = \left(\frac{1}{4}\right)^x$ c) $\left(\frac{1}{8}\right)^x$

7 $y = b^x$ is steeper

8 $P(t) = 100\,000(3)^{\frac{t}{25}}$, where t is number of years
   a) 900 000   b) 2 167 402   c) 8 100 000

9 $N(t) = 10^4(2)^{\frac{t}{3}}$
   a) 20 000   b) 80 000
   c) 5 120 000   d) 10 485 760 000

10 a) $17 204.28   b) $29 598.74   c) $50 922.51

11 a) $A(t) = 5000\left(1 + \frac{0.09}{12}\right)^{12t}$

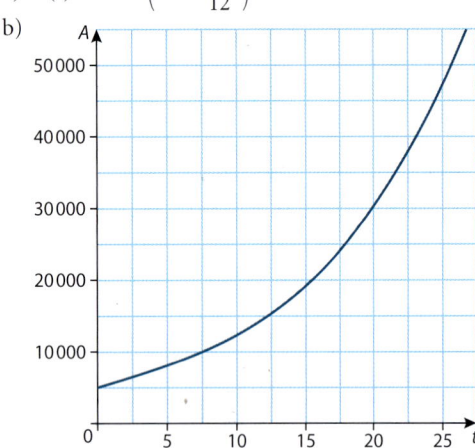

b) [graph showing exponential curve from ~5000 at t=0 rising to ~50000 at t=25]

c) Minimum number of years is 16.

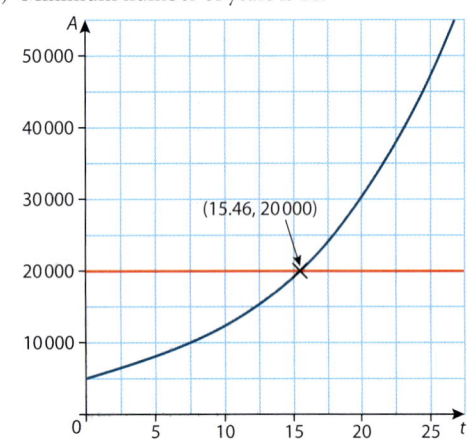

[graph showing exponential curve intersecting horizontal line at (15.46, 20 000)]

12 a) $16 850.58   b) $17 289.16
    c) $17 331.09   d) $17 332.47

13 a) $2   b) $2.61   c) $2.71   d) $2.72   e) $2.72

14 a) 240 310   b) 192 759

15 8.90%

16 $0.0992A_0$ (or 9.92% of A_0 remains)

17 $b > 0$ because if $b = 0$ then the result is always zero, and if $b < 0$ then b^x gives a positive result when x is an even integer and a negative result when x is an odd integer.

18 Payment plan II gives the largest salary. You will get paid $10 737 418.23 after 30 days.

Exercise 4.3

1 As $x \to \infty$, $\left(1 + \frac{1}{x}\right)^x \to e \approx 2.718\,281\,828\ldots$; $y = \left(x + \frac{1}{x}\right)^x$ will never intersect $y = 2.72$

2 Bank A: earn 113.71 euros in interest.
   Bank B: earn 114.07 euros in interest.
   Bank B account earns 0.36 euros more in interest.

3 Blue Star has the greater total of $1358.42, which is $11.93 more than the Red Star.

4 a) 0.976 kg   b) 0.787 kg   c) 0.0916 kg
   d) 0.002 54 kg

5 a) 5 kg   b) 70.7%

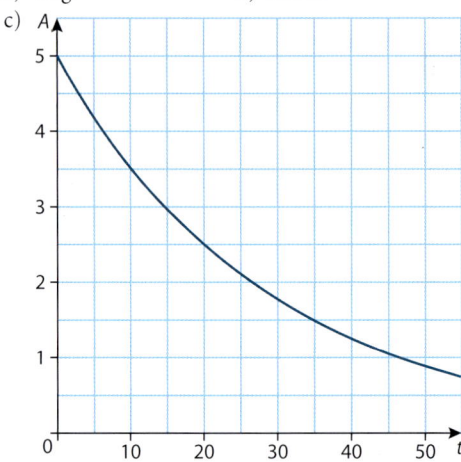

c) [graph showing decreasing exponential curve from 5 at t=0 to ~0.75 at t=50]

d) 20 days

6 $8\frac{1}{2}$% compounded semi-annually is the better investment.

Exercise 4.4

1 $2^4 = 16$ 2 $e^0 = 1$ 3 $10^2 = 100$
4 $10^{-2} = 0.01$ 5 $7^3 = 343$ 6 $e^{-1} = \frac{1}{e}$
7 $10^y = 50$ 8 $e^{12} = x$ 9 $e^3 = x + 2$
10 $\log_2 1024 = 10$ 11 $\log_{10} 0.0001 = -4$
12 $\log_4\left(\frac{1}{2}\right) = -\frac{1}{2}$ 13 $\log_3 81 = 4$ 14 $\log_{10} 1 = 0$
15 $\ln 5 = x$ 16 $\log_2 0.125 = -3$ 17 $\ln y = 4$
18 $\log_{10} y = x + 1$ 19 6   20 3
21 -3 22 5   23 0
24 6   25 -3 26 $\sqrt{2}$
27 3   28 $\frac{1}{2}$ 29 -2
30 -3 31 $\frac{1}{2}$ 32 18
33 $\frac{1}{3}$ 34 π 35 1.6990
36 0.2386   37 3.912   38 0.5493
39 1.398   40 0.2090   41 4.605
42 13.82   43 $x > 2$ 44 $x \in \Box^*$.
45 $x > 0$ 46 $f(x) = \log_4 x$ 47 $f(x) = \log_2 x$
48 $f(x) = \log_{10} x$ 49 $f(x) = \log_3 x$
50 $\log_2 2 + \log_2 m = 1 + \log_2 m$ 51 $\log 9 - \log x$
52 $\frac{1}{5}\ln x$ 53 $\log a + 3\log b$
54 $\log 10x + \log(1 + r)^t = \log 10 + \log x + t\log(1 + r)$
55 $3\ln m - \ln n$ 56 $\log x$ 57 $\log_3 72$
58 $\ln\left(\frac{y^4}{4}\right)$ 59 $\log_b 4$ 60 $\log\left(\frac{p}{qr}\right)$
61 $\ln\left(\frac{36}{e}\right)$ 62 9.97   63 -5.32
64 2.06   65 -0.179 66 4.32
67 1.86   68 $\log_b a = \frac{\log_a a}{\log_a b} = \frac{1}{\log_a b}$
69 $\log e = \frac{\ln e}{\ln 10} = \frac{1}{\ln 10}$
70 $dB = 10\log\left(\frac{I}{10^{-16}}\right) = 10(\log I - \log 10^{-16}) = 10(\log I + 16)$
    $= 10\log I + 160$
    $10\log 10^{-4} + 160 = 10(-4) + 160 = 120$ decibels

Exercise 4.5
1. 0.699
2. 2.5
3. 7.99
4. 3.64
5. −1.92
6. 2.71
7. 0.434
8. 2.12
9. 4.42
10. 0.225
11. 0.642
12. 22.0
13. a) $6248.58 b) $9\frac{1}{4}$ years
14. 12.9 years
15. 20 hours (≈ 19.93)
16. a) 24 years (≈ 23.45) b) 12 years (≈ 11.9)
    c) 9 years (≈ 8.04)
17. 6 years
18. a) 99.7% b) 139 000 years
19. a) 37 dogs b) 9 years
20. a) 458 litres b) 8.89 minutes ≈ 8 minutes 53 seconds
    c) 39 minutes
21. a) 5 kg b) 17.7 days
22. $x = \frac{20}{3}$
23. $x = 104$
24. $x = \frac{1}{e^3}$
25. $x = 4$
26. $x = 98$
27. $x = \pm e^8$
28. $x = 2$ or $x = 4$
29. $x = 9$
30. $x = \frac{13}{5}$
31. $x = 3$
32. $x = 1$ or $x = 100$

Practice questions
1. a) $x = 2$ b) $x = 3$ c) $x = \frac{1}{2}$ d) $x = 3$
2. a) $x \approx 2.58$ b) $x \approx 1.17$ c) $x = 2$ d) $x \approx 0.304$
3. a) $\log_2(9x)$ b) $\ln\left(\frac{3\sqrt{x-4}}{x}\right)$
4. a) 1.89 b) 4.85
5. a) 2597 euros b) 11 years c) 7.18%
6. $2x − 2y − 6z$
7. a) $1474.47 b) 5.7%
8. a) 1 b) $\frac{3}{2}$ c) 36
9. a) 604 b) 13 years
10. 95.8%
11. a) 88% b) $11 610 c) 2011
12. a) Domain: $x \in \mathbb{R}$, range: $y > 0$
    b) y-intercept: $\left(0, \frac{1}{e^2}\right)$; asymptote: $y = 0$ (x-axis)
    c) $f^{-1}(x) = 2 + \ln x$
    d) domain: $x > 0$, range: $y \in \mathbb{R}$
13. a) 631 b) 1270
    c) (i) $A_0 = 500$ (ii) $b = 1.06$
    d) $k = \ln 1.06 \approx 0.058\,27$
14. a) Domain: $x < 0, x > 2$ b) domain: $x > 2$
    c) $x = -\frac{2}{99}$ d) no solution
15. a) $C = 5000, k \approx 0.0556$ b) 140 753

Chapter 5
Exercise 5.1 and 5.2
1. a) (i) $\begin{pmatrix} x-1 & x-3 \\ y+3 & y+1 \end{pmatrix}$ (ii) $\begin{pmatrix} -x-7 & 3x+3 \\ 3y-7 & 11-y \end{pmatrix}$
   b) $x = -3, y = 5$ c) $x = 3, y = -3$
   d) $AB = \begin{pmatrix} 2x-2 & xy-2x+6 \\ xy-x+y+11 & -3 \end{pmatrix}$;

   $BA = \begin{pmatrix} -2x-3y+1 & x^2+x-9 \\ y^2-3y-6 & 4x+3y-6 \end{pmatrix}$
2. a) $x = 2, y = -10$
   b) $p = 2, q = -4$
3. a) $\begin{pmatrix} 0 & 1 & 0 & 0 & 1 & 2 & 0 \\ 1 & 0 & 1 & 1 & 1 & 1 & 0 \\ 0 & 1 & 0 & 2 & 0 & 0 & 2 \\ 0 & 1 & 2 & 0 & 1 & 0 & 0 \\ 1 & 1 & 0 & 1 & 0 & 1 & 0 \\ 2 & 1 & 0 & 0 & 1 & 0 & 0 \\ 0 & 0 & 2 & 0 & 0 & 0 & 0 \end{pmatrix}$ b) $\begin{pmatrix} 6 & 3 & 1 & 2 & 3 & 2 & 0 \\ 3 & 5 & 2 & 3 & 3 & 3 & 2 \\ 1 & 2 & 9 & 1 & 3 & 1 & 0 \\ 2 & 3 & 1 & 6 & 1 & 2 & 4 \\ 3 & 3 & 3 & 1 & 4 & 3 & 0 \\ 2 & 3 & 1 & 2 & 3 & 6 & 0 \\ 0 & 2 & 0 & 4 & 0 & 0 & 4 \end{pmatrix}$

Matrix signifies the number of routes between each pair that go via one other city.

4. a) $A + C = \begin{pmatrix} x+1 & 10 & y+1 \\ 0 & -x-3 & y+3 \\ 2x+y+7 & x-3y & -x+2y-1 \end{pmatrix}$
   b) $\begin{pmatrix} 17m+2 & -6 \\ 4-9m & 9 \\ 7m-2 & -17 \end{pmatrix}$
   c) Not possible d) $x = 3, y = 1$
   e) Not possible f) $m = 3$
5. $a = -3, b = 3, c = 2$
6. $x = 4, y = -3$
7. $m = 2, n = 3$
8. Shop A: €18.77
9. a) $\begin{pmatrix} 2 & 4 \\ -4 & 12 \end{pmatrix}$ b) associative
   c) $\begin{pmatrix} -18 & 16 \\ 42 & -7 \end{pmatrix}$ d) associative
10. $AB = [88\,142]$, which represents total profit.
11. There is no unique solution for r.
12. a) (i) $\begin{pmatrix} 1 & 2 \\ 0 & 1 \end{pmatrix}$ (ii) $\begin{pmatrix} 1 & 3 \\ 0 & 1 \end{pmatrix}$
    (iii) $\begin{pmatrix} 1 & 4 \\ 0 & 1 \end{pmatrix}$ (iv) $\begin{pmatrix} 1 & n \\ 0 & 1 \end{pmatrix}$
    b) (i) $\begin{pmatrix} 9 & 18 \\ 0 & 9 \end{pmatrix}$ (ii) $\begin{pmatrix} 27 & 81 \\ 0 & 27 \end{pmatrix}$
    (iii) $\begin{pmatrix} 81 & 324 \\ 0 & 81 \end{pmatrix}$ (iv) $B^n = \begin{pmatrix} 3^n & n \cdot 3^n \\ 0 & 3^n \end{pmatrix}$

Exercise 5.3
1. a) $\begin{pmatrix} -9 & -7 \\ 4 & 3 \end{pmatrix}$ b) $M = \begin{pmatrix} -9 & -7 \\ 4 & 3 \end{pmatrix}\begin{pmatrix} 2 & 1 \\ 3 & 5 \end{pmatrix}$
   c) $\begin{pmatrix} -39 & -44 \\ 17 & 19 \end{pmatrix}$
   d) (i) $N = \begin{pmatrix} 2 & 1 \\ 3 & 5 \end{pmatrix}\begin{pmatrix} -9 & -7 \\ 4 & 3 \end{pmatrix}$ (ii) $N = \begin{pmatrix} -14 & -11 \\ -7 & -6 \end{pmatrix}$
   e) If $AB = C$ then $B = A^{-1}C$, while if $BA = C$, then $B = CA^{-1}$. Also, $A^{-1}C \neq CA^{-1}$.
2. $\begin{pmatrix} 1 & -\frac{3}{5} \\ 0 & 1 \end{pmatrix}$
3. a) $|A| = -5 \neq 0$ b) $\begin{pmatrix} \frac{9}{5} & \frac{11}{5} & -\frac{8}{5} \\ \frac{6}{5} & \frac{9}{5} & -\frac{7}{5} \\ 1 & 1 & -1 \end{pmatrix}$ c) $\begin{pmatrix} \frac{1}{2} \\ -1 \\ \frac{1}{5} \end{pmatrix}$
4. a) $\begin{pmatrix} \frac{\sqrt{3}}{2} & \frac{1}{2} \\ -\frac{1}{2} & \frac{\sqrt{3}}{2} \end{pmatrix}$ b) $\begin{pmatrix} \frac{3}{a}+1 & -1 \\ -a-2 & a \end{pmatrix}$
5. $x = 2$ or $x = 3$
6. $n = 0.5$
7. a) $X = \begin{pmatrix} \frac{1}{2} & 0 \\ \frac{3}{4} & -\frac{7}{6} \end{pmatrix}$ b) $Y = \begin{pmatrix} 1 & \frac{13}{12} \\ -1 & -\frac{5}{3} \end{pmatrix}$
   c) $X \neq Y$ – not commutative

Answers

8 a) $PQ = \begin{pmatrix} 5 & -4 & 3 \\ 33 & 5 & -1 \\ 2 & -3 & 2 \end{pmatrix}, QP = \begin{pmatrix} 4 & -5 & -8 \\ 8 & 0 & -4 \\ 7 & 10 & 8 \end{pmatrix};$

b) $P^{-1} = \begin{pmatrix} 1 & 0 & -1 \\ -\frac{7}{5} & \frac{1}{5} & \frac{11}{5} \\ 1 & 0 & -2 \end{pmatrix}, Q^{-1} = \begin{pmatrix} 0 & \frac{1}{4} & 0 \\ 1 & -1 & 1 \\ 2 & -\frac{7}{4} & 1 \end{pmatrix}$

$P^{-1}Q^{-1} = \begin{pmatrix} -2 & 2 & -1 \\ \frac{23}{5} & -\frac{22}{5} & \frac{12}{5} \\ -4 & \frac{15}{4} & -2 \end{pmatrix}$

$Q^{-1}P^{-1} = \begin{pmatrix} -\frac{7}{20} & \frac{1}{20} & \frac{11}{20} \\ \frac{17}{5} & -\frac{1}{5} & -\frac{26}{5} \\ \frac{109}{20} & -\frac{7}{20} & -\frac{157}{20} \end{pmatrix}$

$(PQ)^{-1} = \begin{pmatrix} -\frac{7}{20} & \frac{1}{20} & \frac{11}{20} \\ \frac{17}{5} & -\frac{1}{5} & -\frac{26}{5} \\ \frac{109}{20} & -\frac{7}{20} & -\frac{157}{20} \end{pmatrix}$

$(QP)^{-1} = \begin{pmatrix} -2 & 2 & -1 \\ \frac{23}{5} & -\frac{22}{5} & \frac{12}{5} \\ -4 & \frac{15}{4} & -2 \end{pmatrix}$

Practice questions

1 a) $(-1, 4, 0)$ b) $(1, 1, 2)$

2 a) $\begin{pmatrix} \frac{1}{2a} & -\frac{1}{2a} \\ \frac{1}{2a} & \frac{1}{2a} \end{pmatrix}$ b) singular for $a = 0$, $\begin{pmatrix} 1 & -\frac{1}{a} \\ -\frac{2}{a} & \frac{3}{a^2} \end{pmatrix}$

c) singular for $a = 0$, $\frac{1}{e^{4a}-1}\begin{pmatrix} e^a & -e^{2a} \\ -e^{-2a} & e^{3a} \end{pmatrix}$

d) $\begin{pmatrix} \sin a & \cos a \\ -\cos a & \sin a \end{pmatrix}$

3 $x = 2$ or $x = -\frac{1}{3}$

4 a) $\begin{pmatrix} k^2+9 & 3k-3 \\ 3k-3 & 10 \end{pmatrix}$ b) $k = 2$ c) $x = 2, y = 3$

5 $N = \begin{pmatrix} 1 & 1 \\ 1 & 1 \end{pmatrix}$

6 No real solutions for k.

7 a) $m = 1, n = -1$ b) $(x, y, z) = (1, -1, 2)$

8 $m = 1$, or $m = 2$

9 $m = 1, n = 6$

10 $x = 0$, or $x = -\frac{1}{2}$

11 $m = \frac{13}{8}; n = \frac{37}{8}; p = -\frac{23}{8}; q = -\frac{39}{8}$

12 a) $\begin{pmatrix} 1 & -2 \\ -3a & 6a+1 \end{pmatrix}$ b) $\begin{pmatrix} 15 & -36 \\ -8 & 20 \end{pmatrix}$

13 a) $a = -1$ or $a = \frac{1}{2}$

b) $\begin{pmatrix} -\frac{6}{7} & \frac{4}{7} & \frac{5}{7} \\ \frac{5}{7} & -\frac{1}{7} & -\frac{3}{7} \\ \frac{2}{7} & \frac{1}{7} & -\frac{4}{7} \end{pmatrix}, \begin{pmatrix} -\frac{6}{7} & 1 & \frac{2}{7} \\ \frac{1}{2} & -\frac{1}{4} & 0 \\ \frac{5}{7} & -\frac{1}{2} & -\frac{4}{7} \end{pmatrix}$ c) $(24, -6, -1)$

14 $x = -7$, or $x = 1$

15 a) $\begin{pmatrix} a^4+4 & 2a-2 \\ 2a-2 & 5 \end{pmatrix}$

b) $a = -1; \begin{pmatrix} x \\ y \end{pmatrix} = \begin{pmatrix} 1 \\ -1 \end{pmatrix}$

16 $B = \begin{pmatrix} 1 & 3 \\ 4 & 12 \end{pmatrix}$

17 $a = \frac{28}{33}; b = \frac{59}{33}; c = \frac{20}{33}; d = \frac{28}{33}$

18 a) $A^{-1} = \begin{pmatrix} \frac{1}{19} & \frac{2}{19} \\ \frac{-7}{19} & \frac{5}{19} \end{pmatrix}$

b) (i) $X = (C-B)A^{-1}$ (ii) $X = \begin{pmatrix} 2 & -3 \\ -4 & 1 \end{pmatrix}$

19 a) $A+B = \begin{pmatrix} a+1 & b+2 \\ c+d & 1+c \end{pmatrix}$

b) $AB = \begin{pmatrix} a+bd & 2a+bc \\ c+d & 3c \end{pmatrix}$

20 a) $\begin{pmatrix} 0.1 & 0.4 & 0.1 \\ -0.7 & 0.2 & 0.3 \\ -1.2 & 0.2 & 0.8 \end{pmatrix}$

b) $x = 1.2, y = 0.6, z = 1.6$

21 a) $Q = \begin{pmatrix} -3 & 2 \\ 1 & 14-a \\ & 3 \end{pmatrix}$

b) $CD = \begin{pmatrix} -14 & -4+4a \\ -2 & 2+7a \end{pmatrix}$

c) $D^{-1} = \frac{1}{5a+2}\begin{pmatrix} a & -2 \\ 1 & 5 \end{pmatrix}$

Chapter 6

Exercise 6.1

1 $\frac{\pi}{3}$ **2** $\frac{5\pi}{6}$ **3** $-\frac{3\pi}{2}$ **4** $\frac{\pi}{5}$

5 $\frac{3\pi}{4}$ **6** $\frac{5\pi}{18}$ **7** $-\frac{\pi}{4}$ **8** $\frac{20\pi}{9}$

9 $-\frac{8\pi}{3}$ **10** $135°$ **11** $-630°$ **12** $115°$

13 $210°$ **14** $-143°$ **15** $300°$ **16** $15°$

17 $89.95° \approx 90°$ **18** $480°$ **19** $390°, -330°$

20 $\frac{7\pi}{2}, -\frac{\pi}{2}$ **21** $535°, -185°$ **22** $\frac{11\pi}{6}, -\frac{13\pi}{6}$

23 $\frac{11\pi}{3}, -\frac{\pi}{3}$ **24** $3.25 + 2\pi \approx 9.5, 3.25 - 2\pi \approx -3.03$

25 12.6 cm **26** 14.7 cm

27 1.5 radians, or approx. 85.9° **28** $r \approx 7.16$

29 area $\approx 13.96 \approx 14.0$ cm² **30** area ≈ 131 cm²

31 $\alpha = 3$ (radian measure), or $\alpha = 172°$

32 32 cm

33 6.77 cm

Exercise 6.2

1 a) I b) $\left(\frac{\sqrt{3}}{2}, \frac{1}{2}\right)$

2 a) IV b) $\left(\frac{1}{2}, -\frac{\sqrt{3}}{2}\right)$

3 a) IV b) $\left(\frac{\sqrt{2}}{2}, -\frac{\sqrt{2}}{2}\right)$

4 a) Negative x-axis b) $(0, -1)$

5 a) II b) $(-0.416, 0.909)$

6 a) I b) $\left(\frac{\sqrt{2}}{2}, \frac{\sqrt{2}}{2}\right)$

7 a) IV b) $(0.540, -0.841)$

588

8 a) II b) $\left(-\frac{\sqrt{2}}{2}, \frac{\sqrt{2}}{2}\right)$

9 a) III b) $(-0.929, -0.369)$

10 $\sin\frac{\pi}{3} = \frac{\sqrt{3}}{2}, \cos\frac{\pi}{3} = \frac{1}{2}, \tan\frac{\pi}{3} = \sqrt{3}$

11 $\sin\frac{5\pi}{6} = \frac{1}{2}, \cos\frac{5\pi}{6} = -\frac{\sqrt{3}}{2}, \tan\frac{5\pi}{6} = -\frac{\sqrt{3}}{3}$

12 $\sin\left(-\frac{3\pi}{4}\right) = -\frac{\sqrt{2}}{2}, \cos\left(-\frac{3\pi}{4}\right) = -\frac{\sqrt{2}}{2}, \tan\left(-\frac{3\pi}{4}\right) = 1$

13 $\sin\frac{\pi}{2} = 1, \cos\frac{\pi}{2} = 0, \tan\frac{\pi}{2}$ is undefined

14 $\sin\left(-\frac{4\pi}{3}\right) = \frac{\sqrt{3}}{2}, \cos\left(-\frac{4\pi}{3}\right) = -\frac{1}{2}, \tan\left(-\frac{4\pi}{3}\right) = -\sqrt{3}$

15 $\sin 3\pi = 0, \cos 3\pi = -1, \tan 3\pi = 0$

16 $\sin\frac{3\pi}{2} = -1, \cos\frac{3\pi}{2} = 0, \tan\frac{3\pi}{2}$ is undefined

17 $\sin\left(-\frac{7\pi}{6}\right) = \frac{1}{2}, \cos\left(-\frac{7\pi}{6}\right) = -\frac{\sqrt{3}}{2}, \tan\left(-\frac{7\pi}{6}\right) = -\frac{\sqrt{3}}{3}$

18 $\sin(1.25\pi) = -\frac{\sqrt{2}}{2}, \cos(1.25\pi) = -\frac{\sqrt{2}}{2}, \tan(1.25\pi) = 1$

19 $\sin\frac{13\pi}{6} = \sin\frac{\pi}{6} = \frac{1}{2}; \cos\frac{13\pi}{6} = \cos\frac{\pi}{6} = \frac{\sqrt{3}}{2}$

20 $\sin\frac{10\pi}{3} = \sin\frac{4\pi}{3} = -\frac{\sqrt{3}}{2}; \cos\frac{10\pi}{3} = \cos\frac{4\pi}{3} = -\frac{1}{2}$

21 $\sin\frac{15\pi}{4} = \sin\frac{7\pi}{4} = -\frac{\sqrt{2}}{2}; \cos\frac{15\pi}{4} = \cos\frac{7\pi}{4} = -\frac{\sqrt{2}}{2}$

22 $\sin\frac{17\pi}{6} = \sin\frac{5\pi}{6} = \frac{1}{2}; \cos\frac{17\pi}{6} = \cos\frac{5\pi}{6} = -\frac{\sqrt{3}}{2}$

Exercise 6.3

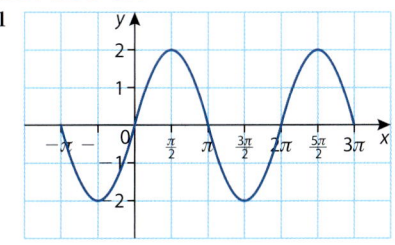

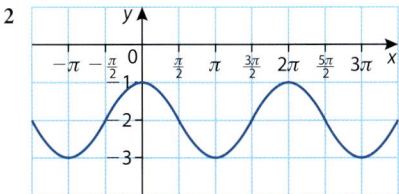

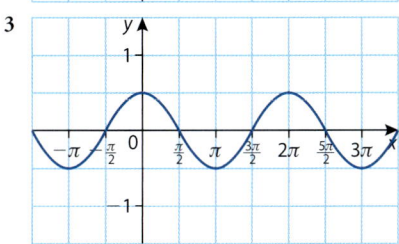

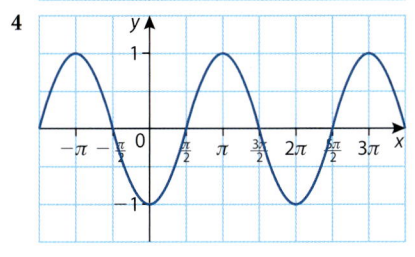

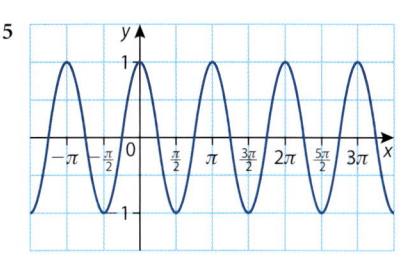

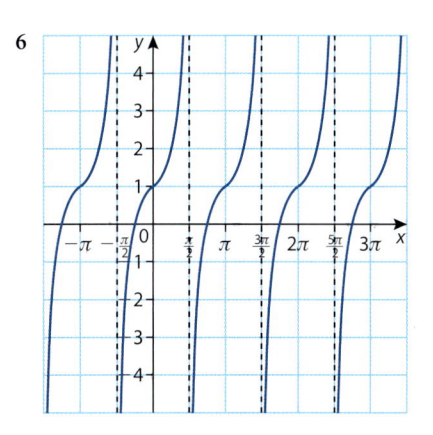

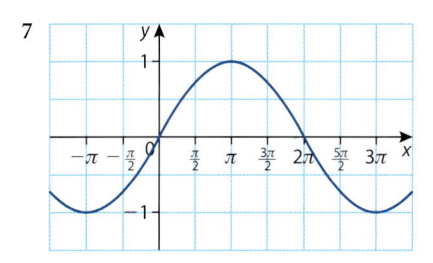

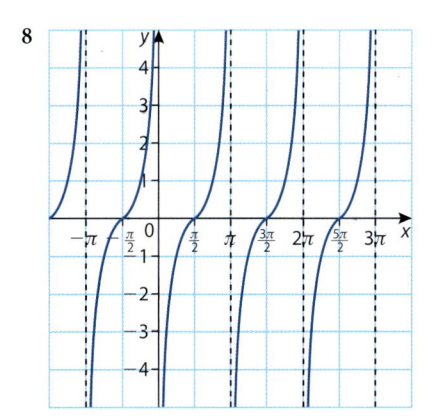

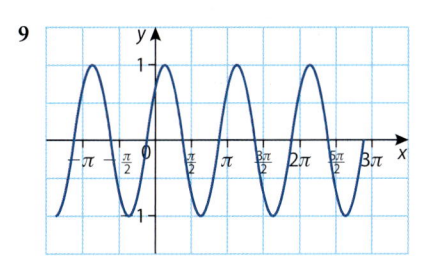

Answers

10. a)

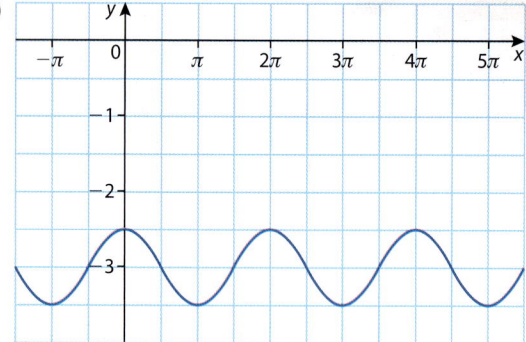

amplitude = $\frac{1}{2}$, period = 2π

b) Domain: $x \in \mathbb{R}$, range: $-3.5 \leq y \leq -2.5$

11. a)

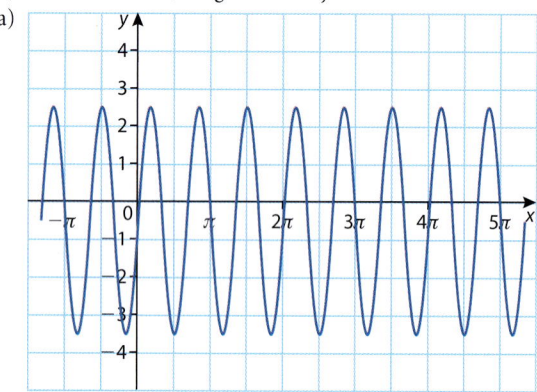

amplitude = 3, period = $\frac{2\pi}{3}$

b) Domain: $x \in \mathbb{R}$, range: $-3.5 \leq y \leq 2.5$

12. a)

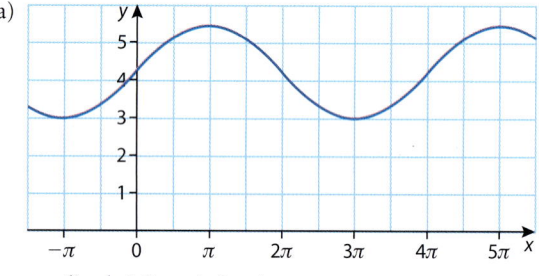

amplitude 1.2, period = 4π

b) Domain: $x \in \mathbb{R}$, range: $3.1 \leq y \leq 5.5$

13. $A = 3, B = 7$
14. $A = 2.7, B = 5.9$
15. $A = 1.9, B = 4.3$
16. a) $p = 8$ b) $q = 6$

Exercise 6.4

1. $x = \frac{\pi}{3}, \frac{5\pi}{3}$
2. $x = \frac{7\pi}{6}, \frac{11\pi}{6}$
3. $x = \frac{\pi}{4}, \frac{5\pi}{4}$
4. $x = \frac{\pi}{3}, \frac{2\pi}{3}$
5. $x = \frac{\pi}{4}, \frac{3\pi}{4}, \frac{5\pi}{4}, \frac{7\pi}{4}$
6. $x = \frac{\pi}{6}, \frac{5\pi}{6}, \frac{7\pi}{6}, \frac{11\pi}{6}$
7. $x = \frac{\pi}{4}, \frac{3\pi}{4}, \frac{5\pi}{4}, \frac{7\pi}{4}$
8. $x = \frac{\pi}{3}, \frac{2\pi}{3}, \frac{4\pi}{3}, \frac{5\pi}{3}$
9. $x = 0, \frac{3\pi}{4}, \pi, \frac{7\pi}{4}, 2\pi$
10. $x = 0, \frac{\pi}{2}, \pi, \frac{3\pi}{2}, 2\pi$
11. $x \approx 0.412, 2.73$
12. $x \approx 1.91, 4.37$
13. $x \approx 1.11, 4.25$
14. $x \approx 0.508, 1.06, 3.65, 4.20$
15. $x \approx 0.961, 3.32$
16. $x \approx 1.28, 4.42$
17. $-\frac{5\pi}{2}, -\frac{3\pi}{2}, -\frac{\pi}{2}, \frac{\pi}{2}, \frac{3\pi}{2}, \frac{5\pi}{2}$
18. $-\frac{11\pi}{6}, \frac{\pi}{6}$
19. $\frac{7\pi}{12}, \frac{19\pi}{12}$
20. $0, \frac{\pi}{4}, \frac{\pi}{2}, \frac{3\pi}{4}, \pi, \frac{5\pi}{4}, \frac{3\pi}{2}, \frac{7\pi}{4}, 2\pi$
21. $x = \frac{5\pi}{6}, \frac{3\pi}{2}$
22. $x = \frac{\pi}{4}, \frac{5\pi}{4}$
23. $x = \frac{\pi}{6}, \frac{\pi}{3}, \frac{7\pi}{6}, \frac{4\pi}{3}$
24. $x = \frac{\pi}{6}, \frac{5\pi}{6}, \frac{7\pi}{6}, \frac{11\pi}{6}$
25. $t \approx 1.5$ hours
26. a) 80th day (March 21) and approximately 263rd day (September 20)
    b) 105th day (April 15) and approximately 238th day (August 26)
    c) 94 days – from 125th day to 218th day
27. $x = \frac{\pi}{2}, \frac{2\pi}{3}, \frac{4\pi}{3}, \frac{3\pi}{2}$
28. $x = \frac{\pi}{2}, \frac{7\pi}{6}, \frac{11\pi}{6}$
29. $x = \frac{\pi}{2}, -\frac{\pi}{2}$
30. $x \approx 0.375, 2.77$
31. $x \approx -0.785, 1.11$
32. $x = \frac{\pi}{4}, \frac{3\pi}{4}$
33. $x = 0, \frac{\pi}{3}, \frac{5\pi}{3}, 2\pi$
34. $x \approx 0.983, 4.12$
35. a) $\cos x = \frac{4}{5}$ b) $\cos 2x = \frac{7}{25}$ c) $\sin 2x = \frac{24}{25}$
36. a) $\sin x = \frac{\sqrt{5}}{3}$ b) $\sin 2x = -\frac{4\sqrt{5}}{9}$ c) $\cos 2x = -\frac{1}{9}$

Practice questions

1. a) 135 cm   b) 85 cm
   c) $t = 0.5$ sec.   d) 1 sec.
2. $x = \frac{\pi}{3}, \frac{\pi}{2}, \frac{5\pi}{3}$
3. $\theta \approx 2.28$ (radian measure)
4. a) (i) -1 (ii) 4π
   b) Four
5. a) $p = 35$ b) $q = 29$ c) $m = \frac{1}{2}$
6. $x \approx 0.483, 0.571, 2.42, 2.86$
7. a) $x = \frac{2\pi}{3}, \frac{4\pi}{3}$ b) $x = \frac{\pi}{6}, \frac{\pi}{2}, \frac{5\pi}{6}, \frac{3\pi}{2}$
8. a) $\sin x = \frac{1}{3}$ b) $\cos 2x = \frac{7}{9}$ c) $\sin 2x = -\frac{4\sqrt{2}}{9}$
9. a) $1.6 \sin\left(\frac{2\pi}{11}\left(x - \frac{9}{4}\right)\right) + 4.2$
   b) Approximately 3.15 metres
   c) Approximately 12:27 p.m. to 7:33 p.m.
10. $x \approx 0.785, 1.89$
11. a) 15 cm   b) area ≈ 239 cm^2
12. $k > 2.5, k < -2.5$
13. $k = 1, a = -2$

Chapter 7

Exercise 7.1

1. $\frac{5}{\sqrt{89}} = \frac{5\sqrt{89}}{89}$
2. $\frac{8}{\sqrt{89}} = \frac{8\sqrt{89}}{89}$
3. $\frac{5}{8}$
4. $\frac{8}{\sqrt{89}} = \frac{8\sqrt{89}}{89}$
5. $\frac{5}{\sqrt{89}} = \frac{5\sqrt{89}}{89}$
6. $\frac{8}{5}$
7. $B\hat{A}C \approx 32.0°, A\hat{B}C \approx 58.0°$
8. $\cos\theta = \frac{4}{5}, \tan\theta = \frac{3}{4}$
9. $\sin\theta = \frac{\sqrt{39}}{8}, \tan\theta = \frac{\sqrt{39}}{5}$
10. $\sin\theta = \frac{2}{\sqrt{5}} = \frac{2\sqrt{5}}{5}, \cos\theta = \frac{1}{\sqrt{5}} = \frac{\sqrt{5}}{5}$
11. $\sin\theta = \frac{\sqrt{51}}{10}, \tan\theta = \frac{\sqrt{51}}{7}$
12. $\sin\theta = \frac{1}{\sqrt{10}} = \frac{\sqrt{10}}{10}, \cos\theta = \frac{3}{\sqrt{10}} = \frac{3\sqrt{10}}{10}$
13. $\sin\theta = \frac{3}{4}, \tan\theta = \frac{3}{\sqrt{7}} = \frac{3\sqrt{7}}{7}$
14. $\frac{\sqrt{2}}{2}$
15. $\frac{\sqrt{3}}{2}$
16. 1
17. $\frac{\sqrt{3}}{2}$
18. $\frac{\sqrt{3}}{3}$
19. $\frac{1}{2}$
20. $60°, \frac{\pi}{3}$
21. $45°, \frac{\pi}{4}$
22. $60°, \frac{\pi}{3}$
23. $60°, \frac{\pi}{3}$
24. $45°, \frac{\pi}{4}$
25. $30°, \frac{\pi}{6}$
26. $46.5°, 0.812$
27. $43.5°, 0.759$
28. $52.3°, 0.913$
29. $80.6°, 1.41$
30. $28.2°, 0.492$
31. $33.1°, 0.577$
32. $x \approx 86.6$
33. $x \approx 8.60$
34. $x \approx 20.6$
35. $x \approx 374$
36. $x = 18$
37. $x = 200$
38. $Q\hat{P}R = 75°, r \approx 5.36, q \approx 20.7$
39. $B\hat{A}C \approx 22.6°, A\hat{B}C \approx 67.4°$
40. 114 metres
41. $67.4°$
42. 4.05 metres
43. 5.71 metres
44. $44.0°$
45. 572 metres

Exercise 7.2

1. $\sin\theta = \frac{3}{5}, \cos\theta = \frac{4}{5}, \tan\theta = \frac{3}{4}$
2. $\sin\theta = \frac{5}{13}, \cos\theta = -\frac{12}{13}, \tan\theta = -\frac{5}{12}$
3. $\sin\theta = -\frac{1}{\sqrt{2}} = -\frac{\sqrt{2}}{2}, \cos\theta = \frac{1}{\sqrt{2}} = \frac{\sqrt{2}}{2}, \tan\theta = -1$
4. $\sin\theta = -\frac{1}{2}, \cos\theta = -\frac{\sqrt{3}}{2}, \tan\theta = \frac{1}{\sqrt{3}} = \frac{\sqrt{3}}{3}$
5. $\sin\theta = \frac{3}{\sqrt{10}} = \frac{3\sqrt{10}}{10}, \cos\theta = \frac{1}{\sqrt{10}} = \frac{\sqrt{10}}{10}, \tan\theta = 3$
6. $\sin\theta = -\frac{1}{\sqrt{2}} = -\frac{\sqrt{2}}{2}, \cos\theta = -\frac{1}{\sqrt{2}} = -\frac{\sqrt{2}}{2}, \tan\theta = 1$
7. a) $\sin 120° = \frac{\sqrt{3}}{2}, \cos 120° = -\frac{1}{2}, \tan 120° = -\sqrt{3}$
   b) $\sin 135° = \frac{\sqrt{2}}{2}, \cos 135° = -\frac{\sqrt{2}}{2}, \tan 135° = -1$
   c) $\sin 150° = \frac{1}{2}, \cos 150° = -\frac{\sqrt{3}}{2}, \tan 120° = -\frac{\sqrt{3}}{3}$
8. a) $\sin 225° = -\frac{\sqrt{2}}{2}, \cos 225° = -\frac{\sqrt{2}}{2}, \tan 225° = 1$
   b) $\sin 330° = -\frac{1}{2}, \cos 330° = \frac{\sqrt{3}}{2}, \tan 330° = -\frac{\sqrt{3}}{3}$
   c) $\sin\frac{7\pi}{6} = -\frac{1}{2}, \cos\frac{7\pi}{6} = -\frac{\sqrt{3}}{2}, \tan\frac{7\pi}{6} = \frac{\sqrt{3}}{3}$
   d) $\sin(-60°) = -\frac{\sqrt{3}}{2}, \cos(-60°) = \frac{1}{2}, \tan(-60°) = -\sqrt{3}$
   e) $\sin 270° = -1, \cos 270° = 0, \tan 270°$ is undefined
   f) $\sin\frac{5\pi}{3} = -\frac{\sqrt{3}}{2}, \cos\frac{5\pi}{3} = \frac{1}{2}, \tan\frac{5\pi}{3} = -\sqrt{3}$
   g) $\sin(-120°) = -\frac{\sqrt{3}}{2}, \cos(-120°) = -\frac{1}{2}, \tan(-120°) = \sqrt{3}$
   h) $\sin\left(-\frac{\pi}{4}\right) = -\frac{\sqrt{2}}{2}, \cos\left(-\frac{\pi}{4}\right) = \frac{\sqrt{2}}{2}, \tan\left(-\frac{\pi}{4}\right) = -1$
   i) $\sin\pi = 0, \cos\pi = -1, \tan\pi = 0$
9. $\sin\theta = -\frac{4}{5}, \tan\theta = -\frac{4}{3}$
10. $\cos\theta = -\frac{15}{17}, \tan\theta = -\frac{8}{15}$
11. $\sin\theta = -\frac{12}{13}, \cos\theta = \frac{5}{13}$
12. $\cos\theta = -1, \tan\theta = 0$
13. a) (i) $30°$ (ii) $85°$
    b) (i) $45°$ (ii) $7°$
    c) (i) $60°$ (ii) $20°$
14. a) $6\sqrt{3}$ units^2 b) 88.9 units^2 c) $675\sqrt{2}$ units^2
15. a) $75\pi \approx 236$ cm^2 b) $\frac{225\sqrt{3}}{4} \approx 97.4$ cm^2
16. a) $\frac{50\pi}{3} - 25\sqrt{3} \approx 9.06$ cm^2 b) $54\pi - 36\sqrt{2} \approx 119$ cm^2
17. 121.4 cm^2

Exercise 7.3 and 7.4

1. Infinite, not one triangle
2. One triangle
3. One triangle
4. One triangle
5. Two triangles
6. One triangle
7. $BC \approx 17.9, AC \approx 27.0, A\hat{C}B = 115°$
8. $AB \approx 18.1, BC \approx 22.5, B\hat{A}C = 65°$
9. $AB \approx 3.91, BC \approx 1.56, A\hat{B}C = 111°$
10. $AB \approx 326, AC \approx 149, B\hat{A}C = 43°$
11. $AB \approx 74.1, B\hat{A}C \approx 60.2°, A\hat{B}C \approx 48.8°$
12. $B\hat{A}C \approx 75.5°, A\hat{B}C \approx 57.9°, A\hat{C}B \approx 46.6°$
13. $B\hat{A}C \approx 81.6°, A\hat{B}C \approx 60.6°, A\hat{C}B \approx 37.8°$
14. Two possible triangles:
    (1) $B\hat{A}C \approx 55.9°, A\hat{C}B \approx 81.1°, AB \approx 40.6$
    (2) $B\hat{A}C \approx 124.1°, A\hat{C}B \approx 12.9°, AB \approx 9.17$
15. Two possible triangles:
    (1) $A\hat{B}C \approx 72.2°, A\hat{C}B \approx 45.8°, AB \approx 0.414$
    (2) $A\hat{B}C \approx 107.8°, A\hat{C}B \approx 10.2°, AB \approx 0.102$
16. 10.8 cm and 30.4 cm
17. $51.3°, 51.3°, 77.4°$
18. 25.8 metres
19. $71.6°$ or $22.4°$
20. Distance ≈ 743 metres
21. $20.7°$
22. Area ≈ 151.2 cm^2
23. a) $BC = 5\sin 36°$ or $BC \geq 5$
    b) $5\sin 36° < BC < 5$
    c) $BC < 5\sin 36°$
24. a) $BC = 5\sqrt{3}$ or $BC \geq 10$
    b) $5\sqrt{3} < BC < 10$
    c) $BC < 5\sqrt{3}$
25. $x \approx 64.9$ m, $y \approx 56.9$ m

Answers

Exercise 7.5
1. a) $\tan 70° \approx 2.75$ b) $y = x\tan 70°$
2. a) $\tan(-20°) \approx -0.364$ b) $y = x\tan(-20°)$
3. a) -1 b) $y = -x + 2$
4. a) $\tan 22° \approx 0.404$ b) $y = x\tan 22° - \frac{3}{2}$
5. $45°$
6. $33.7°$
7. $60.3°$
8. $71.6°$
9. $45°$
10. a) $y = \frac{\sqrt{3}}{3}x$ b) $56.6°$
11. $AB \approx 35.0$ cm
12. $P\hat{R}O \approx 71.8°$, $S\hat{R}O \approx 51.3°$
13. 4104 metres
14. 406.1 metres
15. 2.70 metres
16. a) 1291.8 km b) $12.8°$
17. 59.5 cm
18. $\triangle ABC = 72$ cm², $\triangle ABD = 24\sqrt{3} \approx 41.6$ cm², $\triangle BCD \approx 34.6$ cm², $\triangle ACD \approx 84.8$ cm²
19. $D\hat{E}F \approx 41.9°$
20. 43.0 metres

Practice questions
1. $\sin A\hat{O}B = \frac{24}{25}$
2. $\sin 2\theta = \frac{21}{29}$, $\cos 2\theta = \frac{20}{29}$
3. $101.5°$
4. $\sin 2A = -\frac{120}{169}$
5. a) 29.1 m b) 41.9 m
6. $C\hat{A}B \approx 86.4°$
7. a) $38.2°$ b) 17.3 cm²
8. a) $A\hat{C}B \approx 116°$ b) 155 cm²
9. 78.5 km
10. $J\hat{K}L \approx 31°$
11. a) 3.26 cm b) 7.07 cm²
12. $70.5°$
13. a) 91 m b) $1690\sqrt{3}$
    c) (ii) $A_2 = 26x$ (iii) $x = 40\sqrt{3}$
    d) (i) Supplementary angles have equal sines.

Chapter 8

Exercise 8.1 and 8.2
1.

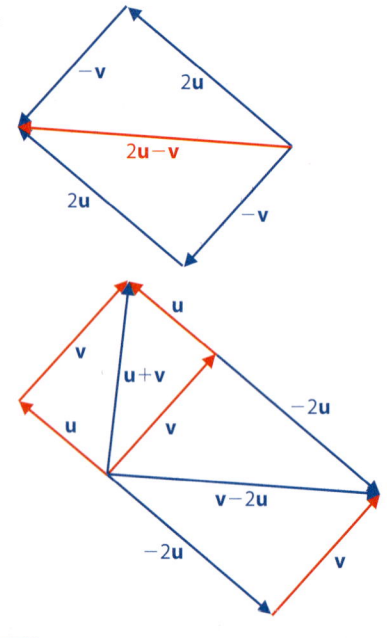

2. a) $\sqrt{41}$ b) $\mathbf{u} = (4, -5)$
   c) $\mathbf{v} = \left(\frac{4}{\sqrt{41}}, \frac{-5}{\sqrt{41}}\right)$ d) 1
3. a) $\sqrt{53}$ b) $\mathbf{u} = (7, -2)$
   c) $\mathbf{v} = \left(\frac{7}{\sqrt{53}}, \frac{-2}{\sqrt{53}}\right)$ d) 1
4. a) $\vec{PQ} = (5, -6)$ b) $\sqrt{61}$ d) $(4, -5)$
5. a) $\vec{PQ} = (4, 6)$ b) $2\sqrt{13}$ d) $(3, 7)$
6. a) $\vec{PQ} = (5, 5)$ b) $5\sqrt{2}$ d) $(4, 6)$
7. a) $\vec{PQ} = (4, 6)$ b) $2\sqrt{13}$ d) $(3, 7)$
8. $(1, -1)$
9. $(8, -1)$
10. $(4, 8)$
11. $(-5, -5)$
12. a) $\mathbf{u} + \mathbf{v} = 2\mathbf{i} + 2\mathbf{j}$, $\mathbf{u} - \mathbf{v} = 4\mathbf{i} - 4\mathbf{j}$, $2\mathbf{u} + 3\mathbf{v} = 3\mathbf{i} + 7\mathbf{j}$, $2\mathbf{u} - 3\mathbf{v} = 9\mathbf{i} - 11\mathbf{j}$
    b) $|\mathbf{u} + \mathbf{v}| = 2\sqrt{2}$, $|\mathbf{u} - \mathbf{v}| = 4\sqrt{2}$, $|\mathbf{u}| + |\mathbf{v}| = 2\sqrt{10}$, $|\mathbf{u}| - |\mathbf{v}| = 0$
    c) $|2\mathbf{u} + 3\mathbf{v}| = \sqrt{58}$, $|2\mathbf{u} - 3\mathbf{v}| = \sqrt{202}$, $2|\mathbf{u}| + 3|\mathbf{v}| = 5\sqrt{10}$, $2|\mathbf{u}| - 3|\mathbf{v}| = -\sqrt{10}$
13. $\left(\frac{11}{8}, -\frac{1}{4}\right)$
14. $\mathbf{u} = \frac{8}{5}\mathbf{i} - \frac{7}{5}\mathbf{j}$; $\mathbf{v} = -\frac{1}{5}\mathbf{i} + \frac{4}{5}\mathbf{j}$
15. $\sqrt{13}, \sqrt{17}$
16. a) $\mathbf{v} + \mathbf{u}$ b) $\mathbf{v} + 0.5\mathbf{u}$ c) $\mathbf{v} - \mathbf{u}$ d) $0.5(\mathbf{v} - \mathbf{u})$

Exercise 8.3
1. a) $0°$ b) $90°$ c) $180°$ d) $56.31°$ e) $135°$
2. a) $\sqrt{13}, 33.69°$ b) $\sqrt{13}, 213.69°$
   c) $2\sqrt{13}, 33.69°$ d) $3\sqrt{13}, 213.69°$
   e) $5\sqrt{13}, 213.69°$ f) $\sqrt{13}, 33.69°$
3. a) $(145.54, 273.71)$ b) $(40.70, 14.49)$
4. a) $\left(\frac{3}{5}, \frac{4}{5}\right)$ b) $\frac{2}{\sqrt{29}}\mathbf{i} - \frac{5}{\sqrt{29}}\mathbf{j}$
5. $\frac{21}{5}\mathbf{i} - \frac{28}{5}\mathbf{j}$
6. a) $\vec{P} = (840\cos 80°, -840\sin 80°)$; $\vec{W} = (60\cos 30°, -60\sin 30°)$
   b) $\vec{V} = (840\cos 80° + 60\cos 30°, -840\sin 80° - 60\sin 30°)$ $= (197.83, -857.24)$
   c) Speed = 879.77 km/h, bearing $167°$
7. a) $\vec{P} = (520\cos 110)\vec{i} + (520\sin 110)\vec{j}$ $= -177.85\vec{i} + 488.64\vec{j}$
   $\vec{W} = (64\cos 160)\vec{i} + (64\sin 160)\vec{j} = -60.14\vec{i} + 21.89\vec{j}$
   b) $\vec{V} = (-177.85 - 60.14)\vec{i} + (488.64 + 21.89)\vec{j}$ $= -237.99\vec{i} + 510.53\vec{j}$
   c) Speed = 563.28 km/h, bearing $335.01°$
8. 24.15, 6.47
9. 200 m east of the initial point.
10. Force = 8176.158 N at an angle of $-10.85°$ to the x-axis.
11. Water = 12.36, boat = 38.04
12. $T = 35.89$, $S = 41.57$
13. 35.4 km/h at N $12.88°$ W
14. At N $11.54°$ W

Exercise 8.4
1. a) $0, 90°$ b) $13, 54°$
   c) $11, 42°$ d) $2\sqrt{3}, 30°$
2. a) -1 b) -1 c) $(57, -38)$
   d) $(-12, -15)$ e) -6 f) 3
   g) Scalar multiplication is distributive over addition of vectors. Multiplication is not associative.

3 a) 2000 b) 6450
4 a) 26.6, 63.4, 90 b) 41.4, 74.5, 64.1
   c) 41.6, 116.6, 21.8
5 a) $(5t, -3t)$ b) $(3t, 2t)$
6 a) $(x - 1)(x - 3) + (y - 2)(y - 4) = 0$
   b) $(x - 3)(x + 1) + (y - 4)(y + 7) = 0$
7 No
8 $t = \dfrac{21}{5}$
9 $b = \sqrt{6}$ or $b = -\sqrt{6}$
10 $\left(\dfrac{4\sqrt{3} + 3}{10}, \dfrac{4 - 3\sqrt{3}}{10}\right)$
11 $t = 0$
12 Sides of rhombus: $\vec{a}$ and $\vec{b}$ with $|\vec{a}| = |\vec{b}|$, diagonals are $\vec{a} + \vec{b}$ and $\vec{a} - \vec{b} \Rightarrow (\vec{a} + \vec{b})(\vec{a} - \vec{b}) = (\vec{a})^2 - \vec{a}\vec{b} + \vec{a}\vec{b} - (\vec{b})^2 = 0$

Practice questions
1 a) $\mathbf{v} - \mathbf{u}$
   b) $(\tfrac{1}{2})(\mathbf{v}-\mathbf{u})$
   c) $(\tfrac{1}{2})(\mathbf{u}+\mathbf{v})$
   d) $(\tfrac{3}{2})\mathbf{v} - (\tfrac{1}{2})\mathbf{u}$
2 a) $(6, -1)$ b) $\dfrac{6}{\sqrt{37}}(6, -1)$
3 a) $OR = 15$ b) $\begin{pmatrix}-5\\5\sqrt{5}\end{pmatrix}$ c) $\dfrac{1}{\sqrt{6}}$ d) $75\sqrt{5}$
4 a) $\overrightarrow{MR} = \begin{pmatrix}11\\4\end{pmatrix}$ b) $\overrightarrow{AC} = \begin{pmatrix}-3\\6\end{pmatrix}$
   c) 83.4°; $\mathbf{u} = \tfrac{1}{2}\overrightarrow{MR}$, $\mathbf{v} = -\tfrac{1}{2}\overrightarrow{MR} \Rightarrow \mathbf{u} \parallel \mathbf{v}$ and $|\mathbf{u}| = |\mathbf{v}|$
5 $m = \dfrac{63}{46}, n = \dfrac{37}{46}$
6 a) 15 km/h, 19.7 km/h
   b) $\begin{pmatrix}4.5\\6\end{pmatrix}; \begin{pmatrix}9\\-4\end{pmatrix}$
   c) 11.4 km
   d) At 8 a.m.
   e) 12.2 km
   f) 54 minutes
7 a)

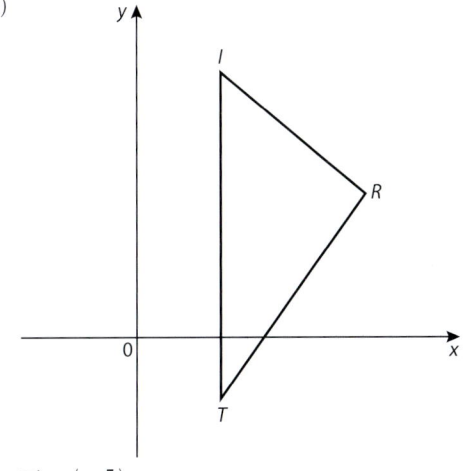

   b) $\overrightarrow{IR} = \begin{pmatrix}5\\-\tfrac{25}{6}\end{pmatrix}$
8 a) $\begin{pmatrix}745\\1000\end{pmatrix}$ b) 600 km/h c) at 1.5 hrs
   d) $\begin{pmatrix}325\\940\end{pmatrix}$ e) 451 km
9 $2n^2 - n + 12 = 0$ does not have real solutions, so it is not possible.

Chapter 9
Exercise 9.1
1 a) Student, all students in a community, random sample of few students, qualitative
   b) Exam, 10th-grade students in a country, a sample from a few schools, quantitative
   c) Newborns, heights of newborns in a city, sample from a few hospitals, quantitative
   d) Children, eye colour of children in a city, sample of children at schools, qualitative
   e) Working persons, commuters in a city, sample of few districts, quantitative
   f) Country leaders, sample of few presidents, qualitative
   g) Students, origin countries of a group of international school students, qualitative

2 Answers are not unique!
   a) Skewed to the right as few players score very high
   b) Symmetric
   c) Skewed to the right
   d) Unimodal, or bi-modal, symmetric or skewed, etc.

3 a) b) Quantitative
   c) d) Qualitative

4 a) Discrete b) Continuous
   c) Continuous d) Discrete
   e) Continuous f) Discrete (debatable!)

5

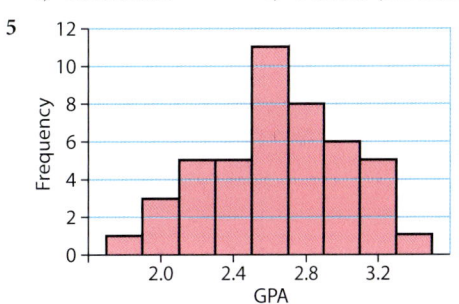

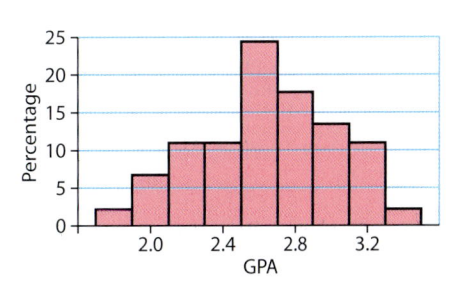

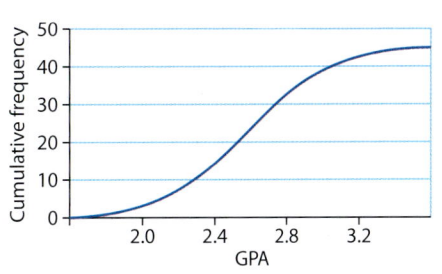

Relatively symmetric. No outliers.

Answers

6

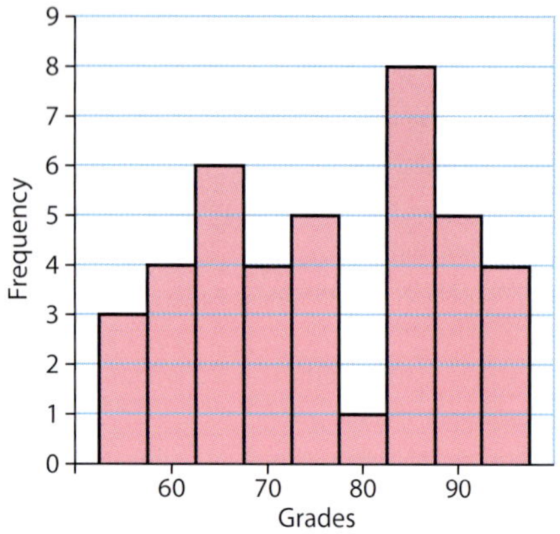

The grades appear to be divided into two groups, one with mode around 65 and the other around 85. No outliers are detected.

7 a)

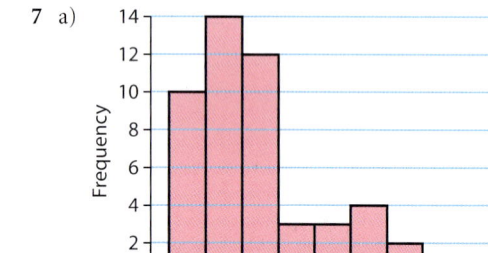

b) The data is skewed to the right.

c)

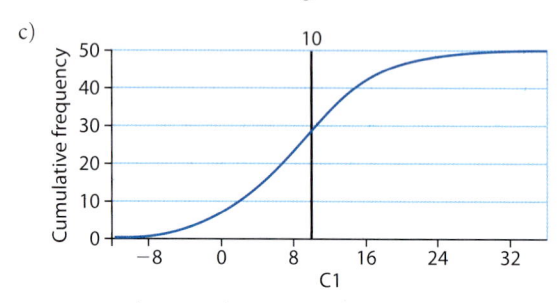

Apparently, more than 35 out of the 50 will lose the licence, about 70%.

8 a)

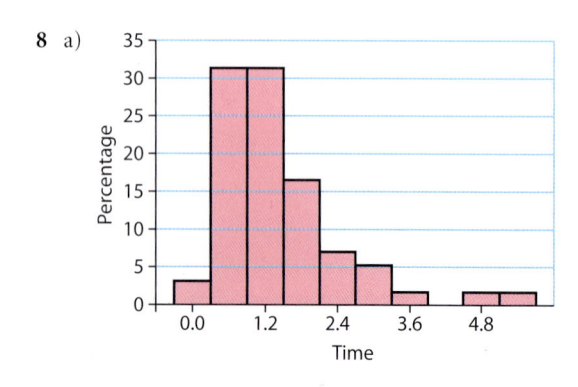

b)

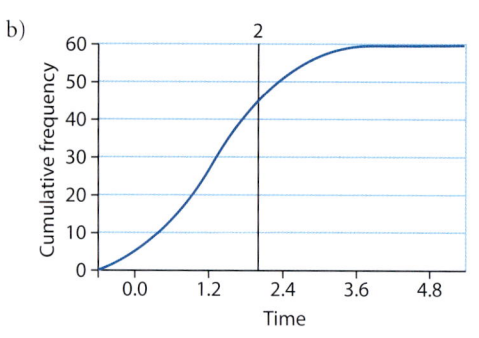

Apparently, about 10 customers have to wait more than 2 minutes.

9 a) Skewed to the right, there is a mode at about 7 days stay, and a few extremes that stayed more than 20 days. A good proportion stayed for about 3 days.

b)

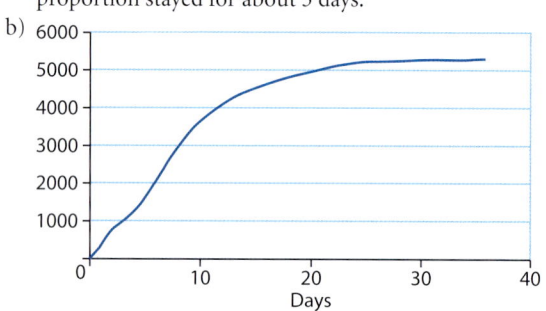

c) Approximately 35% of the patients

10 a) 40 minutes
b) Approximately 30%
c)

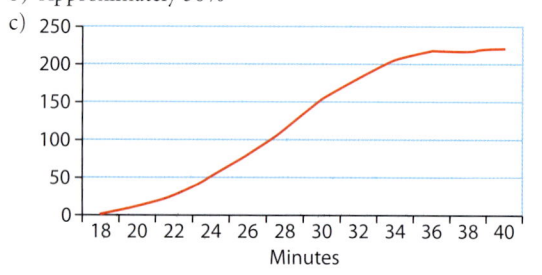

Exercise 9.2 and 9.3

1 a) 6   **b)** 6
c) It appears to be symmetric as the mean and median are the same. A histogram supports this view.

2 a) 7.8   **b)** 7.5   **c)** 7 or 8

3 Average = 1.16, median = 1. Median is more appropriate as the data is skewed to the right.

4 Mean = 7494.7, median = 837.5. There are extreme values and hence the median is more appropriate.

5 Mean = median = 430. It appears to be symmetric and hence either measure would be fine.

6 a) 49.56   **b)** 49.93

7 a)

$x \leq 10$	$x \leq 20$	$x \leq 30$	$x \leq 40$	$x \leq 50$
15	65	165	335	595

$x \leq 60$	$x \leq 70$	$x \leq 80$	$x \leq 90$	$x \leq 100$
815	905	950	980	1000

594

b)

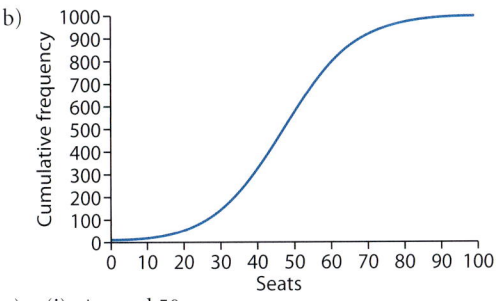

c) (i) Around 50
   (ii) Q1 = 40, Q3 = 60, IQR = 20
   (iii) About 170 days
   (iv) Approximately 70 seats

8 2.05

9 a) Q1 = 165.1, median = 167.64, Q3 = 177.8, minimum = 152, maximum = 193
   b) 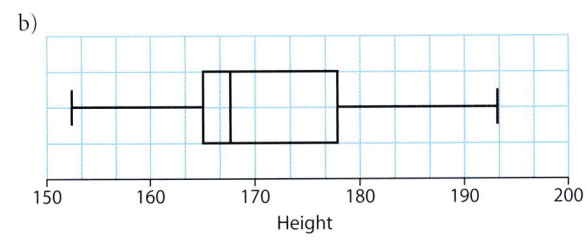 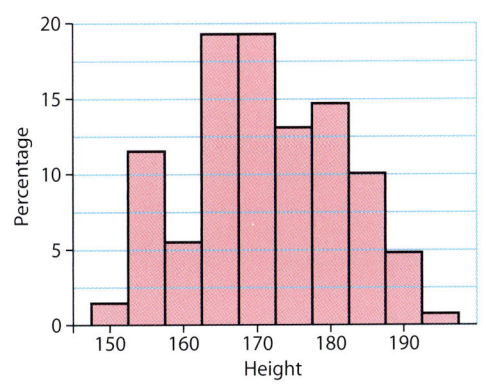

   c) Mean = 170.5, standard deviation = 9.61
   d) The heights are widely spread from very short to very tall players. Heights are slightly skewed to the right, bimodal at 165 and 170, no apparent outliers. The heights between the first quartile and the median are closer together than the rest of the data.
   e) 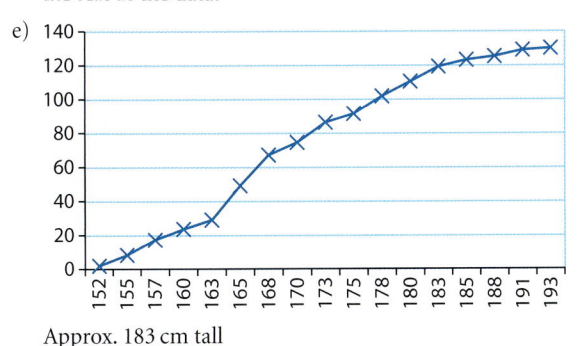

   Approx. 183 cm tall
   f) 171.3

Practice questions

1 a) 12   b) $\sqrt{30.83}$
2 4
3 a)

Time	1.6	2.1	2.6	3.1	3.6	4.1	4.6	5.1	5.6	6.1	6.6
Frequency	2	2	6	4	11	10	5	5	3	2	0

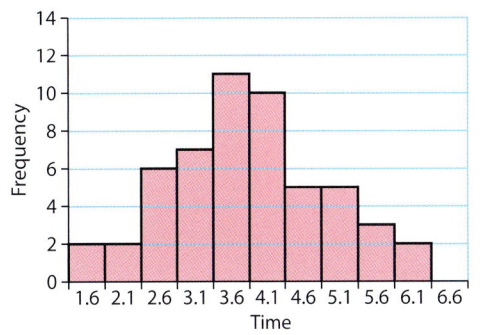

   b) 86%   c) approx. 4   d) 3.86, 1.1
   e)

   f) Minimum = 1.6, Q1 = 3, median = 4, Q3 = 4.5, maximum = 6.2 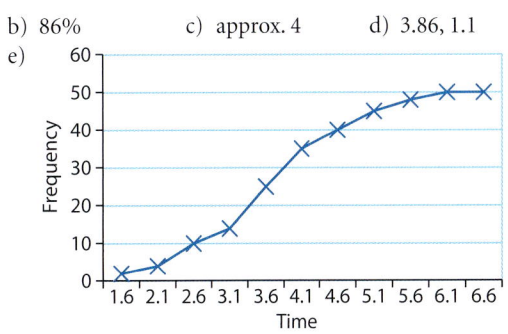

4 a) Median and IQR as the data is skewed with outliers.
   b) Mean = 682.6, standard deviation = 876.2
   c) 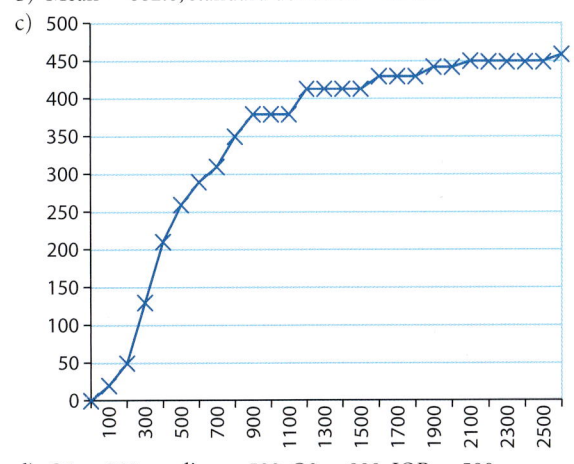

   d) Q1 = 300, median = 500, Q3 = 800, IQR = 500
   e) There are a few outliers on the right side. Outliers lie above Q3 + 1.5IQR = 1550.
   f) Data is skewed to the right, with several outliers from 1600 onwards. It is bimodal at 300–400.

5 a) Spain, Spain   b) France
   c) On average, it appears that France produces the more expensive wines as 50% of its wines are more expensive than most of the wines from the other countries. Italy's prices seem to be symmetric while France's prices are skewed to the left. Spain has the widest range of prices.

Answers

6 a) Mean = 52.65, standard deviation = 7.66
b) Median = 51.34, IQR = 2.65
c) Apparently, the data is skewed to the right with a clear outlier of 112.72! This outlier pulled the value of the mean to the right and increased the spread of the data. The median and IQR are not influenced by the extreme value.

7 a) The distribution does not appear to be symmetric as the mean is less than the median, the lower whisker is longer than the upper one and the distance between Q1 and the median is larger than the distance between the median and Q3. Left skewed.
b) There are no outliers as Q1 − 1.5IQR = 37 < 42 and Q3 + 1.5IQR = 99 > 86.
c) [box plot: 42, 60.25, 70, 75.75, 86]
d) See a)

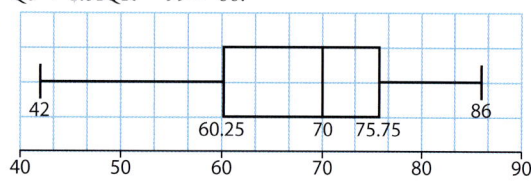

8 a) 225
b) Q1 = 205, Q3 = 255, 90th percentile = 300, 10th percentile = 190
c) IQR = 50, since Q1 − 1.5IQR = 130 > minimum and Q3 + 1.5IQR = 330 < 400 then there are outliers on both sides.
d) [box plot with median 227.5, outliers on both sides]
e) The distribution has many outliers. Apparently skewed to the right with more outliers there. The middle 50% seem to be very close together while the whiskers appear to be quite spread.

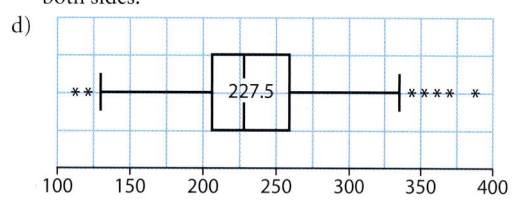

9 a)

Speed	Frequency
26–30	8
31–34	15
35–38	31
39–42	24
43–46	10
47–50	10
51–54	2

b) [histogram]

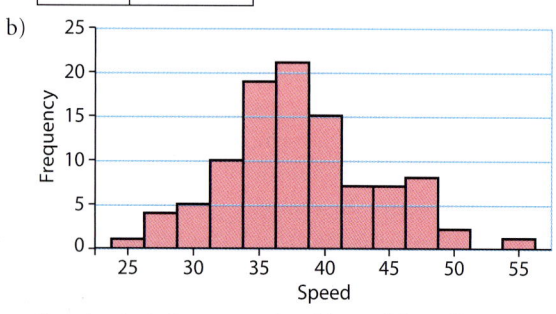

Data is relatively symmetric with possible outlier at 55. The mode is approximately 37.

Histogram created from table: [histogram with bins 28.5, 32.5, 36.5, 40.5, 44.5, 48.5, 52.5]

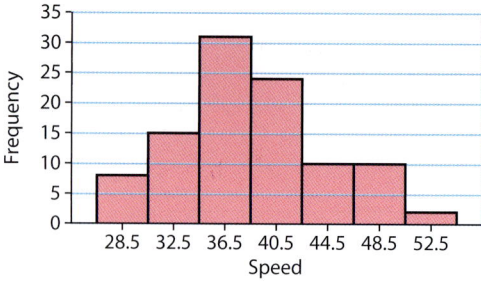

c) Mean = 38.2, standard deviation = 5.7
d)

Speed	Cu. frequency
26–30	8
31–34	23
35–38	54
39–42	78
43–46	88
47–50	98
51–54	100

e) Median = 37.6, Q1 = 34.5, Q3 = 41.3, IQR = 6.8
f) There are outliers on the right since Q3 + 1.5IQR = 51.5 < maximum = 54.

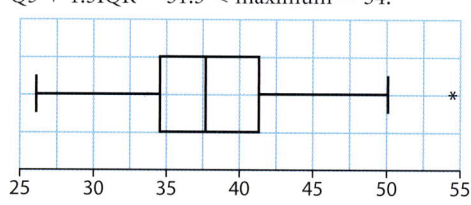

[box plot]

10 a) Mean = 1846.9, median = 1898.6, standard deviation = 233.8, Q1 = 1711.8, Q3 = 2031.3, IQR = 319.5
b) Q1 − 1.5IQR = 1232.55 > minimum, so there is an outlier on the left.
c) [box plot]
d)]1613, 2081[
e) The mean and standard deviation will get larger. The rest will not change much.

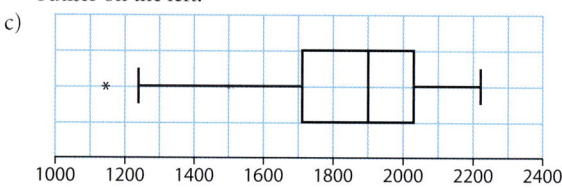

11 a) 49.6 minutes b) 48.9 minutes

12 a)

≤10	≤20	≤30	≤40	≤50	≤60	≤70	≤80	≤90	≤100
30	130	330	670	1190	1630	1810	1900	1960	2000

b) [cumulative frequency graph]
c) (i) 47 (ii) About 500 (iii) Above 60

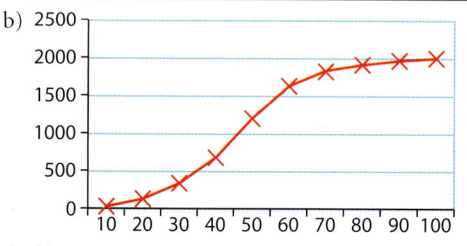

13 1.74

14 a) $m = 12$ b) Standard deviation = 5
15 a) 97.2
b)
30	60	90	120	150	180	210	240
5	20	53	74	85	92	97	100

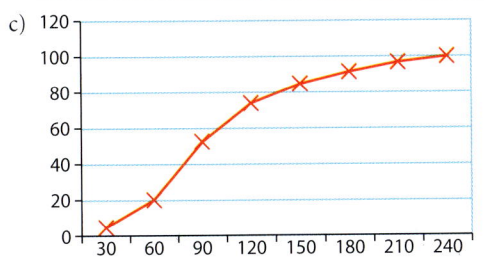

c)

d) Median = 88
Q1 = 66
Q3 = 124

16 a) (i) 10   (ii) 24
b) Mean = 63, standard deviation = 20.5
c) Skew to the left   d) 65

17 a) 7.41
b)
Weight	Number of packets
$w \leq 85$	5
$w \leq 90$	15
$w \leq 95$	30
$w \leq 100$	56
$w \leq 105$	69
$w \leq 110$	76
$w \leq 115$	80

c) (i) Median = 97   (ii) Q3 = 101   d) 0   e) 0.282

18 a) 98.2
b) (i) $a = 165$, $b = 275$
(ii)

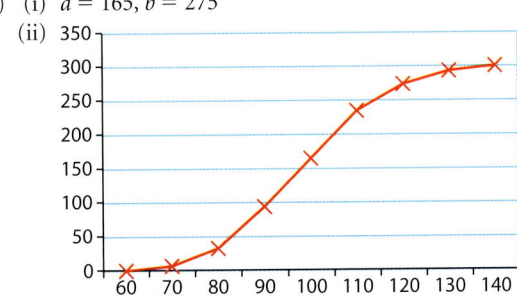

c) (i) 34%   (ii) 115

19 a) (i) 24   (ii) 158   b) 40   c) 7%
20 $a = 3$
21 a)

b) IQR = 110   c) $a = 7$, $b = 6$
d) 199   e) (i) 9   (ii) $\frac{15}{28}$

22 a) (i) 20   (ii) 24   b) 10
23 a)
Mark	[0, 20[	[20, 40[	[40, 60[	[60, 80[	[80, 100[
Number of students	22	50	66	42	20

b) Pass mark = 43%
24 a) 183   b) 14
25 $a = 3$, $b = 7$, $c = 11$, $d = 11$
26 a) 100   b) $a = 55$, $b = 75$

Chapter 10

Exercise 10.1 and 10.2

1 a) {left-handed, right-handed}
b) All real numbers from (say) 50 cm to 210 cm.
c) All real numbers from 0 to 720 (say).
2 {(1, h), (2, h), …, (1, t), …, (6, t)}
3 a) {(1, hearts), …, (king, hearts), (1, spades), …}
b) {[(1, hearts), (king, diamonds)], …,[(1, spades), (10, diamonds)],…}
c) a: 52, b: 1326
4 a) 0.47   b) Anywhere from 0 to 20
c) 10 000
5 a) {(1, 1), (1, 2), …, (4, 4)}   b) {3, 4, …, 9}
6 a) {(b, b), (b, g), (b, y), (g, b), (g, g), (g, y), (y, b), (y, g), (y, y)}
b) {(y, y), (y, b), (y, g)}
c) {(b, b), (g, g), (y, y)}
7 a) {(b, g), (b, y), (g, b), (g, y), (y, b), (y, g)}
b) {(y, b), (y, g)}   c) ∅
8 a) {(t, t, t), (t, t, h), (t, h, t), (h, t, t), (h, t, h), (h, h, t), (t, h, h), (h, h, h)}
b) {(h, t, h), (h, h, t), (t, h, h), (h, h, h)}
9 {(I, fly), (I, dr), (I, tr), (H, dr), (H, b)}
{(I, fly)}
10 a) {(1, g), (1, f), …, (0, c)}
b) {(0, c), (0, s)}
c) {(1, g), (1, f), (0, g), (0, f)}
d) {(1, g), (1, f), (1, s), (1, c)}

Exercise 10.3

1 a) $\frac{3}{10}$ b) $\frac{3}{4}$
2 a) 0.63   b) 1
3 a) (i) $\frac{1}{52}$ (ii) $\frac{7}{26}$ (iii) $\frac{4}{13}$ (iv) $\frac{10}{13}$
b) (i) $\frac{1}{51}$ (ii) $\frac{13}{17}$
c) (i) $\frac{1}{52}$ (ii) $\frac{10}{13}$
4 a) $\frac{4}{5}$ b) $\frac{11}{30}$ c) 1
5 a) $\frac{1}{2}$ b) $\frac{1}{12}$
6 a) $\frac{1}{7}$ b) $\frac{4}{7}$
7 a) (i) {(1, 1), (1, 2), …, (6, 6)}
(ii) $\frac{1}{6}$ (iii) $\frac{2}{9}$ (iv) $\frac{5}{6}$
b) (i) 0   (ii) $\frac{1}{9}$ (iii) $\frac{5}{36}$ (iv) 0
8 a) 0.04   b) 0.55   c) 0.1548
d) 0.060 372   e) 0.104 022
9 a) Yes   b) no   c) no
10 a) 0.06   b) 0.42   c) 0.3364   d) 0.412
11 a) 0.183   b) 0.69

Answers

Exercise 10.4

1. $\frac{7}{20}$

2. a) $\frac{5}{10}$ b) $\frac{4}{10}$ c) $\frac{2}{10}$ d) $\frac{1}{10}$ e) $\frac{2}{3}$

3. a) 92%
   b) (i) 0.64%   (ii) 15.36%   (iii) 14.72%
   c) 48.68%

4. a) 10 000   b) $\frac{9}{10}$ c) 0.3439   d) $\frac{1000}{3439}$

5. a) $\frac{15}{16}$ b) $\frac{4}{5}$ c) $\frac{1}{5}$

6. a) $\{(1, 1), (1, 2), \ldots, (6, 6)\}$
   b)
x	2	3	4	5	6	7	8	9	10	11	12
P(x)	$\frac{1}{36}$	$\frac{1}{18}$	$\frac{1}{12}$	$\frac{1}{9}$	$\frac{5}{36}$	$\frac{1}{6}$	$\frac{5}{36}$	$\frac{1}{9}$	$\frac{1}{12}$	$\frac{1}{18}$	$\frac{1}{36}$

   c) $\frac{11}{36}$ d) $\frac{11}{12}$ e) $\frac{1}{3}$ f) $\frac{2}{3}$

7. a) $\frac{7}{15}$ b) $\frac{11}{75}$ c) $\frac{9}{35}$
   d) $\frac{46}{75}$ e) $\frac{11}{20}$
   f) No: P(female) ≠ P(female/grade 12) − for example

8. a) (i) 0.56   (ii) 0.15
   b) $\frac{15}{56}$ c) no

9.
P(A)	P(B)	Conditions for events A and B	P(A∩B)	P(A ∪ B)	P(A|B)
0.3	0.4	Mutually exclusive	0.00	0.7	0.00
0.3	0.4	Independent	0.12	0.58	0.30
0.1	0.5	Mutually exclusive	0.00	0.60	0.00
0.2	0.5	Independent	0.10	0.60	0.20

10. a) 0.30   b) yes
11. a) 65%   b) 35%   c) 52%

Practice questions

1. a) 0.30   b) 0.72   c) 0.70
2. a) 0.0004   b) 0.9996   c) 0.0004
3. 0.999 98
4. a) (i) 0.85   (ii) 0.80   (iii) 0.15   b) 0.083
5. a) (i) 0.3405   (ii) 0.0108   (iii) 0.9622   (iv) 0.30
   b) Yes
6. a) 0.63   b) 0.971
7. a) 0.60   b) yes, $P(B|A) = P(B) = 0.60$ c) 0.42
8. a)
	Boys	Girls
Passed the ski test	32	16
Failed the ski test	14	12
Training, but did not take the test yet	20	16
Too young to take the test	4	6

   b) (i) 0.6167   (ii) 0.56   (iii) 0.1463

9. a) $\frac{3}{32}$ b) $\frac{3}{4}$ c) $\frac{5}{32}$
10. a) 0.02   b) 0.64
11. a) 0.4   b) 0.6
12. a) 0.38   b) 0.283
13. b) $\frac{11}{36}$

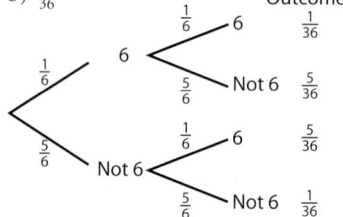

14. a) [Venn diagram: A ∪ B complement shaded]
    b) (i) 2   (ii) $\frac{1}{18}$ c) $n(A \cap B) \neq 0$

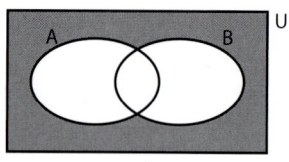

15. a)
	Male	Female	Total
Unemployed	20	40	60
Employed	90	50	140
Total	110	90	200

    b) (i) $\frac{1}{5}$ (ii) $\frac{9}{14}$

16. $\frac{44}{65}$

17. a) [Venn diagram: B only shaded]
    b) 35   c) 0.35

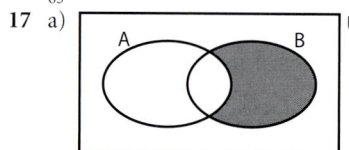

18. a) [Tree diagram: C (0.4) → B (0.6), B' (0.4); C' (0.6) → B (0.5), B' (0.5)]
    b) 0.54   c) 0.444

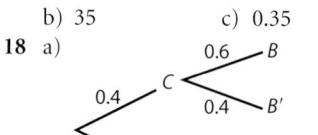

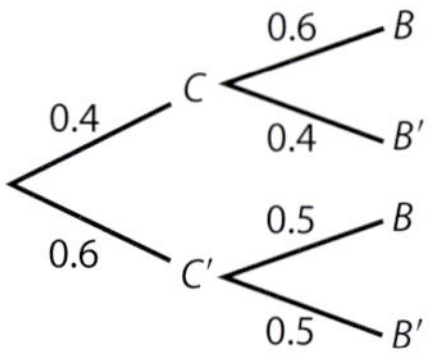

19. a) $\frac{7}{12}$ b) $\frac{11}{36}$ c) $\frac{1}{3}$

20. a) $\frac{1}{11}$ b) $\frac{12}{121}$

21. a) Independent   b) M   c) N

22. a) $a = 21, b = 11, c = 17$
    b) (i) $\frac{1}{8}$ (ii) $\frac{21}{32}$
    c) (i) 0.253   (ii) 0.747

23. $\frac{31}{66}$

24. a) [Tree diagram: W ($\frac{7}{8}$) → L ($\frac{1}{4}$), L' ($\frac{3}{4}$); W' ($\frac{1}{8}$) → L ($\frac{3}{5}$), L' ($\frac{2}{3}$)]
    b) $\frac{47}{160}$ c) $\frac{35}{47}$

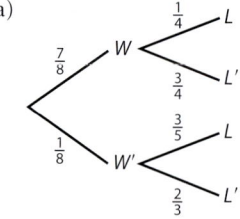

25. a) $\frac{1}{3}$ b) $\frac{7}{12}$ c) $\frac{3}{7}$

26. a) [Tree diagram: Red (0.4) → Grows (0.9), Does not Grow (0.1); Yellow (0.6) → Grows (0.8), Does not Grow (0.2)]
    b) (i) 0.36   (ii) 0.84   (iii) 0.429

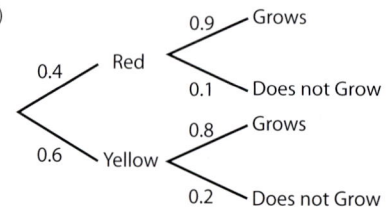

27. a) $\frac{1}{6}$ b) $\frac{1}{12}$ c) $\frac{2}{9}$

28. a) (i) $\frac{8}{21}$ (ii) $\frac{1}{6}$ (iii) no, $P(A \cap B) \neq P(A)P(B)$
    b) $\frac{10}{17}$ c) $\frac{200}{399}$

[Tree diagram for Q13: 6 ($\frac{1}{6}$) → 6 ($\frac{1}{6}$) outcome $\frac{1}{36}$, Not 6 ($\frac{5}{6}$) outcome $\frac{5}{36}$; Not 6 ($\frac{5}{6}$) → 6 ($\frac{1}{6}$) outcome $\frac{5}{36}$, Not 6 ($\frac{5}{6}$) outcome $\frac{1}{36}$]

Chapter 11

Exercise 11.1

1. 4
2. $3x^2$
3. $2x$
4. 6
5. 0
6. $\frac{5}{2}$
7. d.n.e. (increases without bound)
8. $\lim_{c \to \infty} \left(1 + \frac{1}{c}\right)^c = e$
9. $\lim_{x \to \infty} f(x) = \lim_{x \to -\infty} f(x) = 3$
10. As $x \to a$, $g(x) \to +\infty$
11. a) Horizontal: $y = 3$; vertical: $x = -1$
    b) Horizontal: $y = 0$ (x-axis); vertical: $x = 2$
    c) Horizontal: $y = b$; vertical: $x = a$

Exercise 11.2

1. $f'(x) = -2x$
2. $g'(x) = 3x^2$
3. $h'(x) = \frac{1}{2\sqrt{x}}$
4. $r'(x) = -\frac{2}{x^3}$
5. (i) slope $= -2$
   (ii) slope $= 3$
   (iii) slope $= \frac{1}{2}$
   (iv) slope $= -2$
6. a) $y' = 6x - 4$ b) -4
7. a) $y' = -2x - 6$ b) 0
8. a) $y' = -\frac{6}{x^4}$ b) -6
9. a) $y' = 5x^4 - 3x^2 - 1$ b) 1
10. a) $y' = 2x - 4$ b) 0
11. a) $y' = 2 - \frac{1}{x^2} + \frac{9}{x^4}$ b) 10
12. a) $y' = 1 - \frac{2}{x^3}$ b) 3
13. $a = -5$, $b = 2$
14. $(0, 0)$
15. $(2, 8)$ and $(-2, -8)$
16. $\left(\frac{5}{2}, -\frac{21}{4}\right)$
17. $(1, -2)$
18. a) Between A and B
    b) Rate of change is positive at A, B and F. Rate of change is negative at D and E. Rate of change is zero at C.
    c) Pair B and D, and pair E and F.
19. $a = 1$, $b = 5$
20. $a = 1$
21. $(3, 6)$
22. a) 12.61   b) 12
23. $f'(x) = 2ax + b$
24. a) $4.\overline{6}$ degrees Celsius per hour
    b) $C'(t) = 3\sqrt{t}$
    c) $t = \frac{196}{81} \approx 2.42$ hours

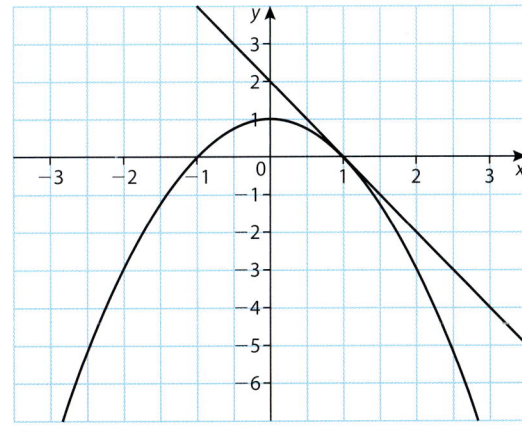

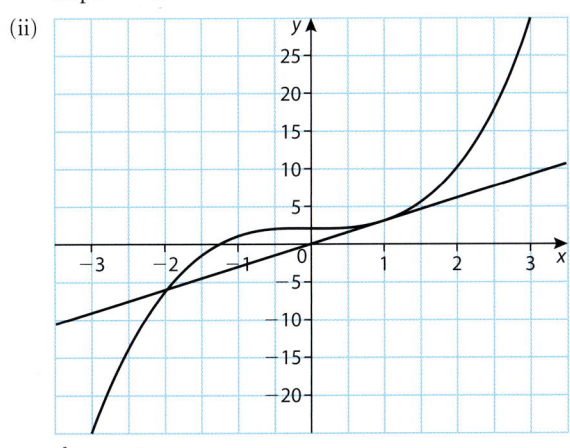

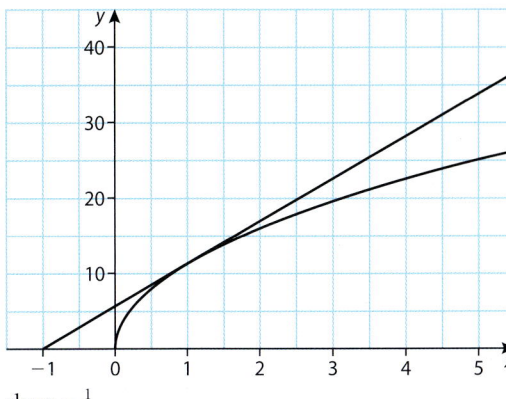

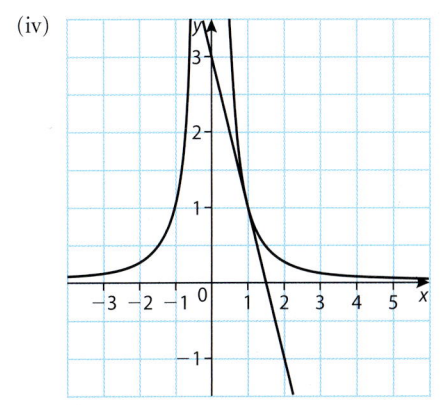

599

Answers

Exercise 11.3

1. $(1, -7)$
2. $\left(-\frac{3}{2}, 8\right)$
3. $(3, 2)$
4. a) $y' = 2x - 5$
   b) increasing for $x > \frac{5}{2}$
   c) decreasing for $x < \frac{5}{2}$
5. a) $y' = -6x - 4$
   b) increasing for $x < -\frac{2}{3}$
   c) decreasing for $x > -\frac{2}{3}$
6. a) $y' = x^2 - 1$
   b) increasing for $x > 1, x < -1$
   c) decreasing for $-1 < x < 1$
7. a) $y' = 4x^3 - 12x^2$
   b) increasing for $x > 3$
   c) decreasing for $x < 0, 0 < x < 3$
8. a) $(3, -130), (-4, 213)$
   b) $(3, -130)$ minimum because 2nd derivative is positive at $x = 3$
   $(-4, 213)$ maximum because 2nd derivative is negative at $x = -4$
   c) [graph showing stationary points at $(-4, 213)$ and $(3, -130)$]
9. a) $(0, -5)$
   b) Stationary point is neither a maximum nor minimum because 1st derivative is always positive.
   c) [graph showing stationary point at $(0, -5)$]
10. a) $(1, 4), (3, 0)$
    b) $(1, 4)$ maximum because 2nd derivative is negative at $x = 1$
    $(3, 0)$ minimum because 2nd derivative is positive at $x = 3$
    c) [graph showing stationary points at $(1, 4)$ and $(3, 0)$]
11. a) $(-1, 4), (0, 6), \left(\frac{5}{2}, -\frac{279}{16}\right)$
    b) $(-1, 4)$ minimum because 2nd derivative is positive at $x = -1$
    $(0, 6)$ maximum because 2nd derivative is negative at $x = 0$
    $\left(\frac{5}{2}, -\frac{279}{16}\right)$ minimum because 2nd derivative is positive at $x = \frac{5}{2}$
    c) [graph showing stationary points at $(-1, 4)$, $(0, 6)$, and $\left(\frac{5}{2}, -\frac{279}{16}\right)$]
12. a) $(-1, 14), \left(\frac{7}{3}, -\frac{122}{27}\right)$
    b) $(-1, 14)$ maximum because 2nd derivative is negative at $x = -1$
    $\left(\frac{7}{3}, -\frac{122}{27}\right)$ minimum because 2nd derivative is positive at $x = \frac{7}{3}$
    c) [graph showing stationary points at $(-1, 14)$ and $\left(\frac{7}{3}, -\frac{122}{27}\right)$]
13. a) $\left(\frac{1}{4}, -\frac{1}{4}\right)$
    b) $\left(\frac{1}{4}, -\frac{1}{4}\right)$ minimum because 2nd derivative is positive at $x = \frac{1}{4}$

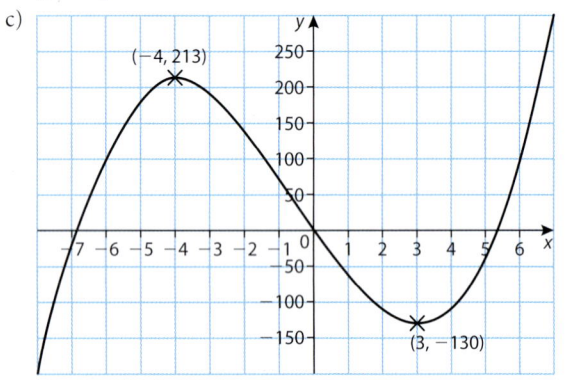

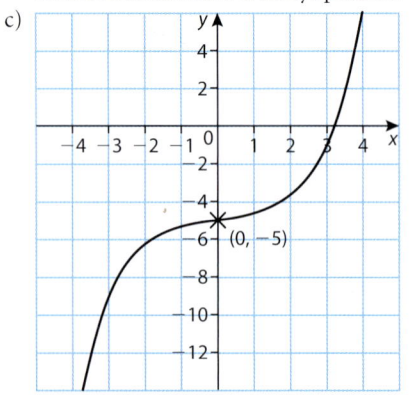

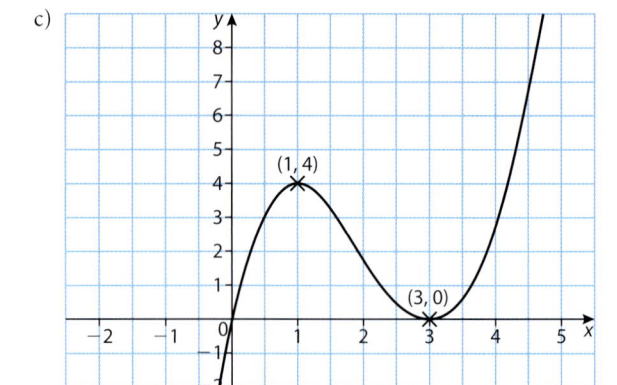

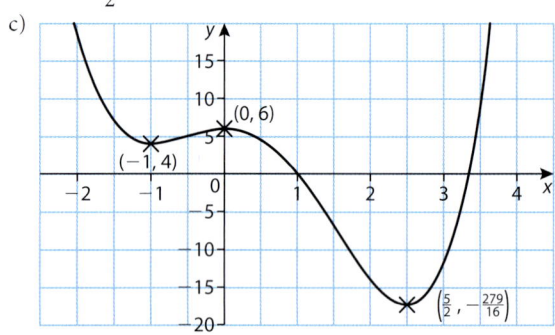

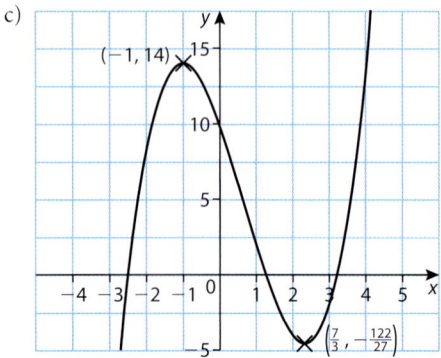

c)

[Graph showing curve with point $(\frac{1}{4}, -\frac{1}{4})$ marked]

14 a) $v(t) = 3t^2 - 8t + 1$; $a(t) = 6t - 8$

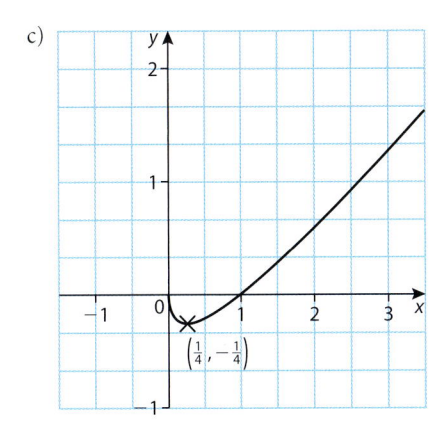

b) [Three graphs: Displacement function $s(t) = t^3 - 4t^2 + t$; Velocity function $v(t) = 3t^2 - 8t + 1$; Acceleration function $a(t) = 6t - 8$]

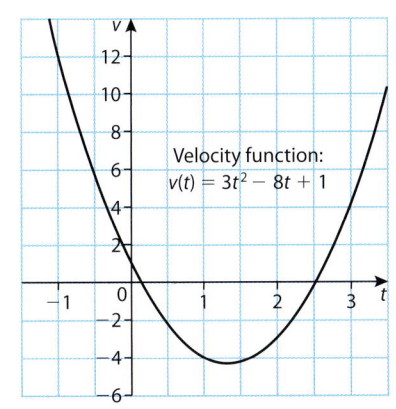

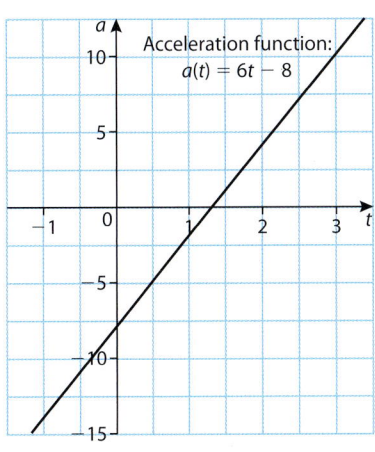

c) $t \approx 0.131$, displacement ≈ 0.0646
d) $t = 1.\overline{3}$, velocity $= -4.\overline{3}$

e) Object moves right at a decreasing velocity, then turns left with increasing velocity, then slowing down and turning right with increasing velocity.

15 Relative maximum at $(-2, 16)$; relative minimum at $(2, 16)$; inflexion point at $(0, 0)$.

16 Absolute minima at $(-2, -4)$ and $(2, -4)$; relative maximum at $(0, 0)$; inflexion points at $\left(-\frac{2\sqrt{3}}{3}, -\frac{20}{9}\right)$ and $\left(\frac{2\sqrt{3}}{3}, -\frac{20}{9}\right)$.

17 Relative maximum at $(-2, -4)$; relative minimum at $(2, 4)$; no inflexion points.

18 Relative minimum at $\left(-\frac{\sqrt[3]{4}}{2}, \frac{3\sqrt[3]{2}}{2}\right)$; inflexion point at $(1, 0)$.

19 Relative minimum at $(-1, -2)$; relative maximum at $(1, 2)$; inflexion points at $\left(-\frac{\sqrt{2}}{2}, -\frac{7\sqrt{2}}{8}\right)$, $(0, 0)$ and $\left(\frac{\sqrt{2}}{2}, \frac{7\sqrt{2}}{8}\right)$.

20 Relative minimum at $(-1, 0)$; absolute minimum at $(2, -27)$; relative maximum at $(0, 5)$; inflexion points at $(1.22, -13.4)$ and $(-0.549, 2.32)$.

21 a) $v(0) = 27 \text{ m s}^{-1}$, $a(0) = -66 \text{ m s}^{-2}$
b) $v(3) = 45 \text{ m s}^{-1}$, $a(3) = 78 \text{ m s}^{-2}$
c) $t = \frac{1}{2}$ and $t = 2\frac{1}{4}$; where displacement has a relative maximum or minimum
d) $t = \frac{11}{8} = 1.375$; where acceleration is zero

22 $x \approx 5.77$ tonnes; $D \approx 34.6$ ($\$34\,600$); this cost is a minimum because cost decreases to this value then increases.

23 $a - 3, b = 4, c = -2$

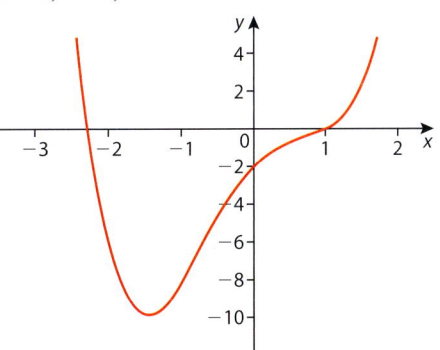

[Graph of curve]

24 Relative maximum at $\left(-2, -\frac{15}{4}\right)$, stationary inflexion point at $(1, 3)$.
$f(x) \to x$ as $x \to \pm\infty$

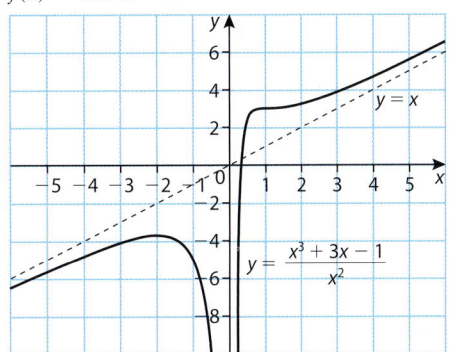

[Graph of $y = \frac{x^3 + 3x - 1}{x^2}$ with asymptote $y = x$]

Exercise 11.4

1 a) $y = -4x - 8$ b) $y = \frac{4}{27}$
  c) $y = -x + 1$ d) $y = -2x + 4$

2 a) $y = \frac{1}{4}x + \frac{19}{4}$ b) $x = -\frac{2}{3}$
  c) $y = x + 1$ d) $y = \frac{1}{2}x + \frac{11}{4}$

Answers

3. At $(0, 0)$; $y = 2x$; at $(1, 0)$; $y = -x + 1$; at $(2, 0)$; $y = 2x - 4$
4. $y = -2x$
5. a) $x = 1$
   b) For $y = x^2 - 6x + 20$, tangent is $y = -4x + 19$
      For $y = x^3 - 3x^2 - x$, tangent is $y = -4x + 1$
6. Normal: $y = \frac{1}{2}x - \frac{7}{2}$; intersection pt: $\left(-\frac{1}{2}, -\frac{15}{4}\right)$
7. Tangent: $y = -3x + 3$; normal: $y = \frac{1}{3}x - \frac{1}{3}$
8. $a = -4, b = 1$
9. a) $y = 2x + \frac{5}{2}$ b) $\left(\frac{2}{3}, \frac{41}{27}\right)$
10. Tangent: $y = -\frac{3}{4}x + 1$; normal: $y = \frac{4}{3}x - \frac{22}{3}$

Practice questions

1. a) Gradient $= 3$ b) $y = 3x - \frac{9}{4}$
   c) [graph]
   d) $Q\left(\frac{3}{4}, 0\right), R\left(0, -\frac{9}{4}\right)$
   f) $y = 2ax - a^2$
   g) $T\left(\frac{a}{2}, 0\right), U(0, -a^2)$
   h) x-coord.: $\frac{a+0}{2} = \frac{a}{2}$; y-coord.: $\frac{a^2 - a^2}{2} = 0$
2. $A = 1, B = 2, C = 1$
3. a) $4x - 15x^4$ b) $-\frac{1}{x^2}$
4. a) $x = 2$ or -2; $f'(1) = -6 < 0$ (decreasing) and $f'(3) = \frac{10}{9} > 0$ (increasing) $\therefore f(2)$ is a turning point
   b) Vertical asymptote: $x = 0$ (y-axis); oblique asymptote: $y = 2x$
5. $\left(\frac{1}{2}, 3\right)$
6. $a = 1$
7. a) $y = 5x - 7$ b) $y = -\frac{1}{5}x + \frac{17}{5}$
8. a) $x = 1$ b) $-3 < x < -2, 1 < x < 3$
   c) $x = -\frac{1}{2}$
   d) [graph: inflexion pt at $x = -\frac{1}{2}$, maximum at $x = 1$, minimum at $x = -2$]
9. $b = 2, c = 3$

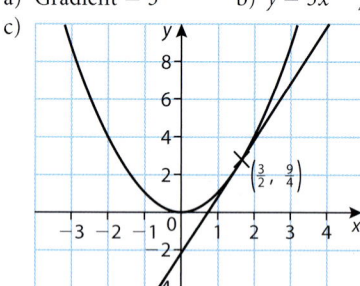

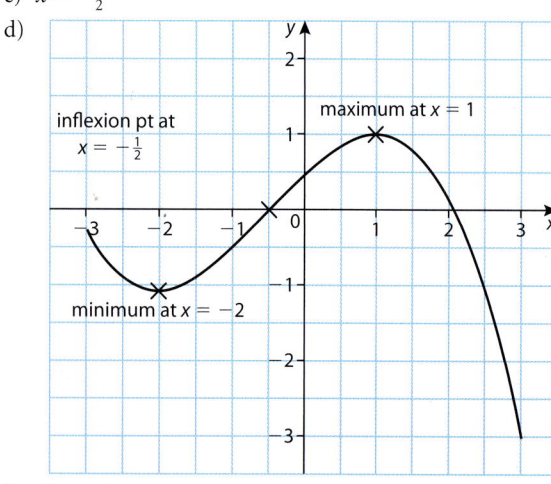

10.
function	diagram
f_1	d
f_2	e
f_3	b
f_4	a

11. a) (i) $x = 0$ (y-axis)   (ii) $y = 3$
    b) $\frac{dy}{dx} = \frac{2}{x^2}$
    c) Increasing on $x < 0, x > 0$; nowhere decreasing
    d) None; $\frac{dy}{dx} \neq 0$
12. Maximum at $(-1, 1)$, minimum at $(0, 0)$, maximum at $(1, 1)$
13. $a = \frac{8}{3}, b = \frac{16}{5}$
14. a) 10 m s^{-1} b) 10 seconds   c) 50 metres
15. a) $v = 14 - 9.8t, a = -9.8$
    b) $t \approx 1.43$ seconds, $h = 10$
    c) Velocity $= 0$, acceleration $= -9.8 \text{ m s}^{-2}$

Chapter 12

Exercise 12.1

1. a) $\left(\frac{5}{2}, -2, 0\right)$
   b) $(3, 2\sqrt{3}, 0)$
   c) $(-1, 2, -2)$
   d) $(a, -4a, -a)$
2. a) $Q\left(-\frac{1}{2}, -3, 2\right)$
   b) $P\left(\frac{5}{2}, -2, 0\right)$
   c) $Q(0, -4a, 3a)$
3. a) $(x, y, z) = (t, t, 5 - 5t)$, or $(x, y, z) = (1 + t, 1 + t, -5t)$
   b) $(x, y, z) = (-1 + 4t, 5t, 1 - 3t)$
   c) $(x, y, z) = (2 - 4t, 3 - 6t, 4 + t)$
4. a) $C(7, -8, -1)$
   b) $C\left(-1, \frac{11}{2}, \frac{29}{3}\right)$
   c) $C(2 - a, 4 - 2a, -b - 2)$
5. a) $\left(-\frac{1}{3}, 1, \frac{1}{3}\right)$
   b) $\left(1, -\frac{5}{3}, -1\right)$
   c) $\left(\frac{a+b+c}{3}, \frac{2a+2b+2c}{3}, a+b+c\right)$
6. a) $D(-1, 1, -6)$
   b) $D(-2\sqrt{2}, 2\sqrt{3}, 1 - 4\sqrt{5})$
   c) $D\left(\frac{5}{2}, -\frac{2}{3}, -4\right)$
7. $m = 5, n = 1$
8. a) $\mathbf{v} = \frac{2}{3}\mathbf{i} + \frac{2}{3}\mathbf{j} - \frac{1}{3}\mathbf{k}$
   b) $\mathbf{v} = \frac{3}{\sqrt{14}}\mathbf{i} - \frac{2}{\sqrt{14}}\mathbf{j} + \frac{1}{\sqrt{14}}\mathbf{k}$
   c) $\mathbf{v} = \frac{2}{3}\mathbf{i} - \frac{1}{3}\mathbf{j} - \frac{2}{3}\mathbf{k}$
9. a) $|\mathbf{u} + \mathbf{v}| = \sqrt{29}$
   b) $|\mathbf{u}| + |\mathbf{v}| = \sqrt{14} + \sqrt{5}$
   c) $|-3\mathbf{u}| + |3\mathbf{v}| = 3\sqrt{14} + 3\sqrt{5}$
   d) $\frac{1}{|\mathbf{u}|}\mathbf{u} = \frac{\mathbf{i}}{\sqrt{14}} + \frac{3\mathbf{j}}{\sqrt{14}} - \frac{2\mathbf{k}}{\sqrt{14}}$
   e) $\left|\frac{1}{|\mathbf{u}|}\mathbf{u}\right| = 1$
10. a) $(3, 4, -5)$ b) $(0, -2, 5)$
11. a) $\left(1, -\frac{4}{3}\right)$ b) $\sqrt{6}(4\mathbf{i} + 2\mathbf{j} - 2\mathbf{k})$ c) $-\frac{2}{3}\mathbf{i} + \frac{8}{3}\mathbf{j} - 2\mathbf{k}$
12. 0

Exercise 12.2

1. a) $-16, 117.65°$ b) $-20, 64.68°$ c) $13, 40.24°$
   d) $-15, 151.74°$ e) $6, 60°$ f) $-6, 120°$
2. a) Orthogonal b) acute c) orthogonal
3. a) $\mathbf{v} \cdot \mathbf{u} = 0 = \mathbf{wu}$ b) $\frac{3}{\sqrt{13}}\mathbf{i} + \frac{2}{\sqrt{13}}\mathbf{j}, \frac{-3}{\sqrt{13}}\mathbf{i} - \frac{2}{\sqrt{13}}\mathbf{j}$
4. a) (i) $\cos\alpha = \frac{2}{\sqrt{14}}, \cos\beta = \frac{-3}{\sqrt{14}}, \cos\gamma = \frac{1}{\sqrt{14}}$
      (ii) $\cos^2\alpha + \cos^2\beta + \cos^2\gamma = \frac{2^2}{14} + \frac{(-3)^2}{14} + \frac{1^2}{14} = 1$
      (iii) $\alpha \approx 58°, \beta \approx 143°, \gamma \approx 74°$
   b) (i) $\cos\alpha = \frac{1}{\sqrt{6}}, \cos\beta = \frac{-2}{\sqrt{6}}, \cos\gamma = \frac{1}{\sqrt{6}}$
      (ii) $\cos^2\alpha + \cos^2\beta + \cos^2\gamma = \frac{1^2}{6} + \frac{2^2}{6} + \frac{1^2}{6} = 1$
      (iii) $\alpha \approx 66°, \beta \approx 145°, \gamma \approx 66°$
   c) (i) $\cos\alpha = \frac{3}{\sqrt{14}}, \cos\beta = \frac{-2}{\sqrt{14}}, \cos\gamma = \frac{1}{\sqrt{14}}$
      (ii) $\cos^2\alpha + \cos^2\beta + \cos^2\gamma = \frac{3^2}{14} + \frac{(-2)^2}{14} + \frac{1^2}{14} = 1$
      (iii) $\alpha \approx 37°, \beta \approx 122°, \gamma \approx 74°$
   d) (i) $\cos\alpha = \frac{3}{5}, \cos\beta = 0, \cos\gamma = \frac{-4}{5}$
      (ii) $\cos^2\alpha + \cos^2\beta + \cos^2\gamma = \frac{3^2}{25} + \frac{0^2}{25} + \frac{4^2}{25} = 1$
      (iii) $\alpha \approx 53°, \beta \approx 90°, \gamma \approx 143°$
5. a) $m = -\frac{9}{8}$ b) $m = 1$ or $-\frac{1}{4}$
6. $m = -14$
7. a) $127°$ b) $63°$ c) $73°$
8. a) $m = \frac{1}{3}$ b) $m = -\frac{1}{4}$
9. $m_A: \mathbf{r} = (4, -2, -1) + m(-1, 0, \frac{3}{2}); m_B: \mathbf{r} = (3, -5, -1) + n(\frac{1}{2}, \frac{9}{2}, \frac{3}{2})$
   $m_C: \mathbf{r} = (3, 1, 2) + k(\frac{1}{2}, -\frac{9}{2}, -3)$; centroid $(\frac{10}{3}, -2, 0)$
10. $90, 90, 82, 74, 60, 54, 53, 52, 47, 43, 38, 37$
11. 68.22
12. $103.3°, 133.5°, 46.5°$
13. 0
14. $k = 2$
15. $x = -20, y = -14$
16. $x = 5$
17. $117°, \overrightarrow{AC} = \begin{pmatrix} 0 \\ 6 \\ 3 \end{pmatrix}, 33°$
18. a) $b = -\frac{1}{2}$
    b) $b = \frac{1}{2}$
20. $(-140.8, 140.8, 18)$

Exercise 12.3

1. a) $\mathbf{r} = \begin{pmatrix} -1 \\ 0 \\ 2 \end{pmatrix} + t\begin{pmatrix} 1 \\ 5 \\ -4 \end{pmatrix}$ $\begin{pmatrix} x \\ y \\ z \end{pmatrix} = \begin{pmatrix} -1 + t \\ 5t \\ 2 - 4t \end{pmatrix}$
   b) $\mathbf{r} = \begin{pmatrix} 3 \\ -1 \\ 2 \end{pmatrix} + t\begin{pmatrix} 2 \\ 5 \\ -1 \end{pmatrix}$ $\begin{pmatrix} x \\ y \\ z \end{pmatrix} = \begin{pmatrix} 3 + 2t \\ -1 + 5t \\ 2 - t \end{pmatrix}$
   c) $\mathbf{r} = \begin{pmatrix} 1 \\ -2 \\ 6 \end{pmatrix} + t\begin{pmatrix} 3 \\ 5 \\ -11 \end{pmatrix}$ $\begin{pmatrix} x \\ y \\ z \end{pmatrix} = \begin{pmatrix} 1 + 3t \\ -2 + 5t \\ 6 - 11t \end{pmatrix}$

2. a) $\mathbf{r} = \begin{pmatrix} -1 \\ 4 \\ 2 \end{pmatrix} + t\begin{pmatrix} 8 \\ 1 \\ -2 \end{pmatrix}$ b) $\mathbf{r} = \begin{pmatrix} 4 \\ 2 \\ -3 \end{pmatrix} + t\begin{pmatrix} -4 \\ -4 \\ 4 \end{pmatrix}$
   c) $\mathbf{r} = \begin{pmatrix} 1 \\ 3 \\ -3 \end{pmatrix} + t\begin{pmatrix} 4 \\ -2 \\ 5 \end{pmatrix}$

3. a) $\mathbf{r} = \begin{pmatrix} 3 \\ -2 \end{pmatrix} + t\begin{pmatrix} 2 \\ 3 \end{pmatrix}$ b) $\mathbf{r} = \begin{pmatrix} 0 \\ -2 \end{pmatrix} + t\begin{pmatrix} 5 \\ 2 \end{pmatrix}$
4. $2x + 3y = 7$
5. $\mathbf{r} = 2\mathbf{i} - 3\mathbf{j} + \lambda(4\mathbf{i} - 3\mathbf{j})$
6. $\mathbf{r} = (-2, 1, 4) + t(3, -4, 7)$
7. a) $(1, -1, 2)$ b) $(-17, -1, 1)$
   c) No d) No
8. a) $\mathbf{r} = (2, -1) + t(1, 3)$ $\begin{pmatrix} x \\ y \end{pmatrix} = \begin{pmatrix} 2 + t \\ -1 + 3t \end{pmatrix}$
   b) $\mathbf{r} = (2, -1) + t(-3, 7)$ $\begin{pmatrix} x \\ y \end{pmatrix} = \begin{pmatrix} 2 - 3t \\ -1 + 7t \end{pmatrix}$
   c) $\mathbf{r} = (2, -1) + t(7, 3)$ $\begin{pmatrix} x \\ y \end{pmatrix} = \begin{pmatrix} 2 + 7t \\ -1 + 3t \end{pmatrix}$
   d) $\mathbf{r} = (0, 2) + t(2, -4)$ $\begin{pmatrix} x \\ y \end{pmatrix} = \begin{pmatrix} 2t \\ 2 - 4t \end{pmatrix}$
9. a) $t = \frac{3}{2}$ b) no c) $m = \frac{7}{2}$
10. a) (i) $(3, -4)$ (ii) $(7, 24)$ (iii) 25
    b) (i) $(-3, 1)$ (ii) $(5, -12)$ (iii) 13
    c) (i) $(5, -2)$ (ii) $(24, -7)$ (iii) 25
11. a) $(-96, 128)$ b) $\left(\frac{2040}{13}, -\frac{850}{13}\right)$
12. a) $(24, 18)$
    b) $\mathbf{r} = (3, 2) + t(24, 18)$
    c) In 10 minutes
13. a) $a = -3, b = -5$
    b) $-\frac{\sqrt{21}}{6}$
    c) $\frac{\sqrt{15}}{6}, \frac{\sqrt{35}}{2}$
14. a) $146.8°$ b) 3.87
    c) (i) $L_1: \mathbf{r} = (2, -1, 0) + t(0, 1, 2); L_2: \mathbf{r} = (-1, 1, 1) + t(1, -3, -2)$
15. a) $(x, y, z) = (1 + t, 3 - 2t, -17 + 5t)$
    b) $(4, -3, -2)$

Practice questions

1. a) $\overrightarrow{OD} - \overrightarrow{OC}$ b) $\frac{1}{2}(\overrightarrow{OD} - \overrightarrow{OC})$ c) $\frac{1}{2}(\overrightarrow{OD} + \overrightarrow{OC})$
2. a) $5\mathbf{i} + 12\mathbf{j}$ b) $10\mathbf{i} + 24\mathbf{j}$
3. a) $|\overrightarrow{OA}| = |\overrightarrow{OB}| = |\overrightarrow{OC}| = 6$
   b) $\overrightarrow{AC} = \begin{pmatrix} -1 \\ \sqrt{11} \end{pmatrix}$ c) $\frac{1}{\sqrt{12}}$ d) $6\sqrt{11}$
4. a) $(10, 5)$ b) $(-3, 6); 90°$
5. $a = 2, b = 8$
6. $\mathbf{r} = (3, -1) + t(4, -5)$
7. a) 39.4 b) (i) $(9, 12), (18, -8)$ (ii) $\sqrt{481}$
   c) 7 a.m. d) 24.4 km e) 54 minutes
8. $\mathbf{r} = t(2\mathbf{i} + 3\mathbf{j})$
9. b) $(2, 3.25)$
10. c) $90°$
    d) (i) $12x - 5y = 301$ (ii) $(28, 7)$

Answers

11 117°
12 $2x + 3y = 5$
13 a) (6, 20) b) (i) (6, −8) (ii) 10
    c) $4x + 3y = 84$ d) collide at 15:00
    f) 26 km
14 72°
15 a) 3.94 m b) 1.22 m/s
    c) $x − 0.7y = 2$ d) $\left(\frac{170}{29}, \frac{160}{29}\right)$
    e) speed = 1.24 m/s
16 $\binom{x}{y} = \binom{1}{3} + t\binom{5}{2}$
17 $2x^2 + 7x − 15 = 0, x = \frac{3}{2}, x = −5$
18 a) (ii) (288, 84) (iii) 50 minutes b) 20.6°
    c) (i) (99, 168) (iii) $XY = 75$ d) 180 km
19 $3x + 2y = 7$
20 a) $\vec{ST} = \binom{9}{9}$, $V(−4, 6)$ b) $\mathbf{r} = (−4, 6) + \lambda(1, 1)$
    c) $\lambda = 5$ d) (i) $a = 5$ (ii) 157°
21 81.9°
22 a) 13 b) $\frac{1}{5}(3\mathbf{i} + 4\mathbf{j})$ c) $\frac{56}{65}$
23 (2, 3)
24 a) (3, −2) c) (iii) 23 square units
25 a) $\vec{OB} = \binom{-1}{7}$; $\vec{OC} = \binom{8}{9}$ b) $d = 11$
    c) $\vec{BD} = \binom{12}{-3}$ d) (i) $\binom{x}{y} = \binom{-1}{7} + t\binom{12}{-3}$ (ii) $t = 0$
26 a) (i) $\vec{AB} = \binom{-5}{1}$ (ii) $AB = \sqrt{26}$ b) $\vec{AD} = \binom{d-2}{25}$
    c) (ii) $\vec{OD} = \binom{7}{23}$ d) $\vec{OC} = \binom{2}{24}$ e) 130
27 a) (i) $\vec{BC} = −6\mathbf{i} − 2\mathbf{j}$ (ii) $\vec{OD} = −2\mathbf{i}$ b) 82.9°
    c) $\mathbf{r} = \mathbf{i} − 3\mathbf{j} + t(2\mathbf{i} + 7\mathbf{j})$ d) $15\mathbf{i} + 46\mathbf{j}$
28 a) (5, 5, −5) b) (−5, 0, 5) c) (5, 5, −5)
29 b) (i) (49, 32, 0) (ii) 54 km/h
    c) (i) $\frac{5}{6}$ hours (ii) (9, 12, 5)
30 a) (i) $\vec{AB} = \binom{800}{600}$
    b) (ii) $\binom{-400}{-50}$ (iii) at 16:00 hours
    c) 27.8 km

Chapter 13

Exercise 13.1

1 a) $y' = \cos x + \sin x$ b) $y' = −e^x$
   c) $y' = 1 + \frac{1}{x}$ [or $\frac{x+1}{x}$] d) $y' = \frac{2e^x}{5}$
   e) $y' = 3x^2 − 2\sin x$ f) $y' = \frac{2e}{x}$
2 a) $y = \frac{1}{2}x + \frac{3\sqrt{3} - \pi}{6}$ b) $y = 2x + 1$
   c) $y = \frac{x}{2e}$
3 a) $x = \frac{\pi}{6}, x = \frac{5\pi}{6}$
   b) Maximum at $\frac{\pi}{6}$, minimum at $\frac{5\pi}{6}$; $g''\left(\frac{\pi}{6}\right) < 0, g''\left(\frac{5\pi}{6}\right) > 0$
4 (0, −1) is an absolute maximum; $y''(0) < 0$

5 $\frac{d^2y}{dx^2} = \frac{1}{x^2} \neq 0 \therefore$ no points of inflexion
6 $x = \frac{\pi}{2}$
7 a) $f'(x) = e^x − 3x^2$; $f''(x) = e^x − 6x$
   b) $x \approx 3.73$ or $x \approx 0.910$ or $x \approx −0.459$
   c) Decreasing on $(−\infty, −0.459)$ and $(0.910, 3.73)$
      Increasing on $(−0.459, 0.910)$ and $(3.73, \infty)$
   d) $x \approx −0.459$ (minimum); $x \approx 0.910$ (maximum);
      $x \approx 3.73$ (minimum)
   e) $x \approx 0.204$ or $x \approx 2.83$
   f) Concave up on $(−\infty, 0.204)$ and $(2.83, \infty)$; concave down on $(0.204, 2.83)$
8 $\frac{1}{e}$
9 $\frac{d}{dx}[\log_b x] = \frac{1}{x \ln b}$

Exercise 13.2

1 a) $y' = 12(3x − 8)^3$ b) $y' = −\frac{1}{2\sqrt{1-x}}$
   c) $y' = \frac{2}{x}$ d) $y' = \cos\left(\frac{x}{2}\right)$
   e) $y' = −\frac{4x}{(x^2 + 4)^3}$ f) $y' = −3e^{-3x}$
   g) $y' = −\frac{1}{2\sqrt{(x+2)^3}}$ [or $−\frac{1}{(2x+4)\sqrt{x+2}}$]
   h) $y' = −2\sin x \cos x$ i) $y' = 2xe^{x^2}$
   j) $y' = \frac{-6x + 5}{(3x^2 - 5x + 7)^2}$
   k) $y' = \frac{2}{3\sqrt[3]{(2x+5)^2}}$ l) $y' = \frac{2x}{x^2 - 9}$
2 a) $y = −12x − 11$ b) $y = \frac{9}{5}x − \frac{2}{5}$
   c) $y = 2x − 2\pi$
3 a) $v(t) = −2t \sin(t^2 − 1)$ b) velocity = 0
   c) $t = \sqrt{\pi + 1} \approx 2.04, t = 1$
   d) Accelerating to the right then slowing down, turning around, accelerating to the left, slowing down, turning around again, and then accelerating to the right.
4 a) $\frac{dy}{dx} = −1$ for $x < −1$, $\frac{dy}{dx} = 1$ for $x > −1$
   b) $\frac{dy}{dx} = −\frac{\cos x}{\sin^2 x}$ c) $\frac{dy}{dx} = (1 + \sqrt{x})^2 \left[3x + \frac{3\sqrt{x}}{2}\right]$
   d) $\frac{dy}{dx} = (−\sin x)e^{\cos x}$ e) $\frac{dy}{dx} = \frac{2 \ln x}{x}$
   f) $\frac{dy}{dx} = −\frac{3}{\sqrt{(2x+1)^3}}$ or $−\frac{3}{(2x+1)\sqrt{2x+1}}$
5 a) $y = −12x + 38$
   b) $y = \frac{1}{12}x + \frac{7}{4}$
6 a) $y = \frac{2}{3}x + \frac{5}{3}$
   b) $y = −\frac{3}{2}x + 6$
7 a) $y = 4x − 4$
   b) $y = −\frac{1}{4}x + \frac{1}{4}$
8 a) $f'(x) = (\ln 2)2^x$ or $2^x \ln 2$
   b) $y = x \ln 2 + 1$
   c) $f'(x) = 2^x \ln 2 \neq 0$ for any x
9 a) $\frac{dy}{dx} = 2\cos\left(2x − \frac{\pi}{2}\right)$; $\frac{d^2y}{dx^2} = −4\sin\left(2x − \frac{\pi}{2}\right)$
   b) $\left(\frac{\pi}{4}, 0\right)$ and $\left(\frac{3\pi}{4}, 0\right)$

Exercise 13.3

1. a) $y' = x^2 e^x + 2xe^x$
   b) $y' = \sqrt{1-x} - \dfrac{x}{2\sqrt{1-x}}$ or $\dfrac{2-3x}{2\sqrt{1-x}}$
   c) $y' = 1 + \ln x$
   d) $y' = \cos^2 x - \sin^2 x$ or $2\cos^2 x - 1$
   e) $\dfrac{y' = xe^x - e^x}{x^2}$
   f) $y' = -\dfrac{2}{(x-1)^2}$
   g) $y' = 2(2x-1)^2(7x^4 - 2x^3 + 3)$
   h) $y' = \dfrac{x\cos x - \sin x}{x^2}$
   i) $y' = \dfrac{-xe^x + e^x - 1}{(e^x - 1)^2}$
   j) $y' = \dfrac{33}{(3x+2)^2}$
   k) $y' = 2x\ln(3x) + \dfrac{x^2 - 1}{x}$ or $\dfrac{2x^2\ln(3x) + x^2 - 1}{x}$
   l) $y' = 0$

2. a) $y = -\tfrac{1}{2}x + 2$ b) $y = \tfrac{1}{2}x + \tfrac{1}{2}$ c) $y = 5x - 3$

3. a) $(-1, -2e)$ and $\left(3, \dfrac{6}{e^3}\right)$
   b) $(-1, -2e)$ is a minimum, $\left(3, \dfrac{6}{e^3}\right)$ is a maximum
   c) (i) $h(x) \to 0$ as $x \to 0$ (ii) $h(x) \to \infty$ as $x \to -\infty$
   d) Horizontal asymptote: $y = 0$ (x-axis)
   e)

4. $\dfrac{d}{dx}(c \cdot f(x)) = \dfrac{d}{dx}(c) \cdot f(x) + c \cdot \dfrac{d}{dx}(f(x)) = 0 \cdot f(x) + c \cdot \dfrac{d}{dx}(f(x))$
   $= c \cdot \dfrac{d}{dx}(f(x))$

5. a) (i) $(0, 0)$ and $(4, 0)$ (ii) $\left(\dfrac{4}{3}, \dfrac{256}{27}\right)$ (iii) $\left(\dfrac{8}{3}, \dfrac{128}{27}\right)$
   b)

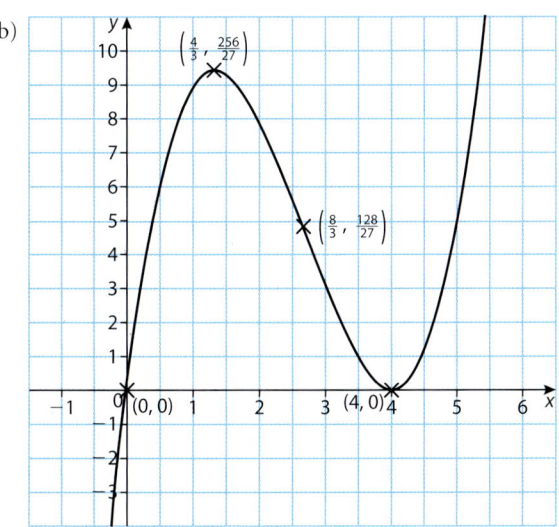

6. c) $g''(3.8) = 0$ and $g''(3) = \dfrac{1}{3} > 0$, $g''(4) = -\dfrac{2}{625} < 0$, therefore graph of g changes concavity from up to down at $x = 3.8$, verifying that graph of g does have an inflexion point at $x = 3.8$

Exercise 13.4

1. $\sqrt{2}$ by $\dfrac{\sqrt{2}}{2}$
2. $13\tfrac{1}{3}$ cm by $6\tfrac{2}{3}$ cm
3. b) $S = 4x^2 + \dfrac{3000}{x}$ c) $7.21\text{ cm} \times 14.4\text{ cm} \times 9.61\text{ cm}$
4. $x = 5\sqrt{2\pi} \approx 12.5$ cm
5. $x \approx 3.62$ m
6. Longest ladder ≈ 7.02 m
7. $d \approx 2.64$ km

Exercise 13.5

1. $y' = 7(x-1)^6$
2. $y' = -\dfrac{1}{5x^2}$
3. $y' = 2(3x+4)^3(15x+4)$
4. $y' = \dfrac{\sin x + \sin x \cos^2 x}{\cos^2 x}$
5. $y' = 6e^{6x}$
6. $y' = 3x^2 \ln x + x^2$
7. $y' = 4\cos\left(4x - \dfrac{\pi}{4}\right)$
8. $y' = -2xe^{-x^2}$
9. $y' = \dfrac{3(\ln x)^2}{x}$
10. $y' = \dfrac{1 - 2x^2}{\sqrt{1-x^2}}$
11. $y' = 3^x \ln 3$
12. $y' = \dfrac{2x^2 + 6x + 2}{(2x+3)^2}$
13. $y' = \dfrac{2x\sin x \cos x + x^2}{\cos^2 x}$
14. $y' = \dfrac{e^x \cos x + e^x \sin x}{\cos^2 x}$
15. $y' = \dfrac{2x + 2}{3\sqrt[3]{(x^2 + 2x - 4)}}$
16. $y' = -2\sin(2x)\cos(\cos(2x))$
17. $y' = -2e^{1-2x}$
18. $y' = e^x \ln(x^2) + \dfrac{2e^x}{x}$
19. $y' = \dfrac{e^x - 2}{e^x - 2x}$
20. $y' = \tfrac{3}{2}(\ln 2)8^x$
21. a)

   b) $1°\text{C/hr}$
   c) $C'(t) = \dfrac{\pi}{2}\sin\left(\dfrac{\pi}{12}t - \dfrac{\pi}{6}\right)$
   d) $\dfrac{\pi\sqrt{2}}{4} \approx 1.11°\text{C/hr}$
   e) $t = 14$ hrs, $C = 31°$

605

Answers

22 $\left(\frac{\sqrt{6}}{2}, \frac{5}{2}\right)$ and $\left(-\frac{\sqrt{6}}{2}, -\frac{5}{2}\right)$

23 $y = -\frac{1}{2}x - \frac{3}{2}$; $P(-3, 0)$, $Q\left(0, -\frac{3}{2}\right)$

24 34.6 km/hr

25 a) $a = 16$ b) $a = -54$

c) If $x > 0$, then for $f'(x) = 0$, $a > 0$ and if both $x > 0$ and $a > 0$ then $\frac{d^2y}{dx^2} = \frac{2a}{x^3} + 2$ is always positive; hence, graph of f is always concave up – and no maximum is possible. If $x < 0$, then for $f'(x) = 0$, $a < 0$ and if both $x < 0$ and $a < 0$ then $\frac{d^2y}{dx^2} = \frac{2a}{x^3} + 2$ is always positive – and again no maximum is possible.

26 $y = -\frac{2}{3}x + 4$

27 a) Stationary pt: $\left(0, \frac{1}{\sqrt{2\pi}}\right)$, inflexion pts: $\left(1, \frac{1}{\sqrt{2e\pi}}\right)$ and $\left(-1, \frac{1}{\sqrt{2e\pi}}\right)$

b) As $x \to \pm\infty$, $f(x) \to 0$; horizontal asymptote: $y = 0$ (x-axis)

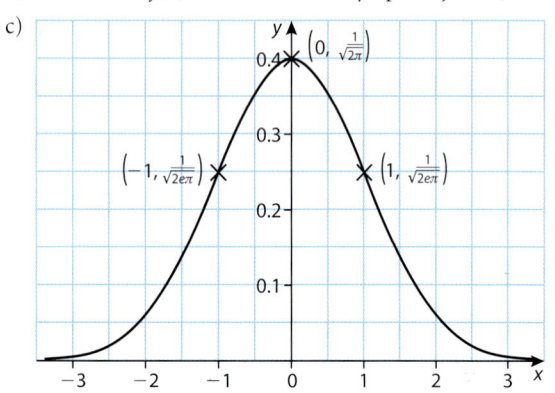

c)

28 Tangent: $y = \left(\frac{\pi + 2}{2}\right)x - \frac{\pi^2}{8}$; normal: $y = \left(-\frac{2}{\pi + 2}\right)x + \frac{\pi^2 + 4\pi}{4\pi + 8}$

29 $\frac{8}{\pi + 4}$ metres by $\frac{4}{\pi + 4}$ metres (or 1.26 m by 0.56 m)

30 a) 4 metres

b) $t \approx 0.644$ seconds, $s = 5$ m

c) $t \approx 3.79$ seconds, $s = -5$ m

d) $v(t) = 3\cos t - 4\sin t$, $a(t) = -3\sin t - 4\cos t$

e) $t \approx 5.36$ seconds, $v = 5 \text{ ms}^{-1}$

f) $t \approx 2.21$ seconds, $s = 0$ m and $v = -5 \text{ ms}^{-1}$

Practice questions

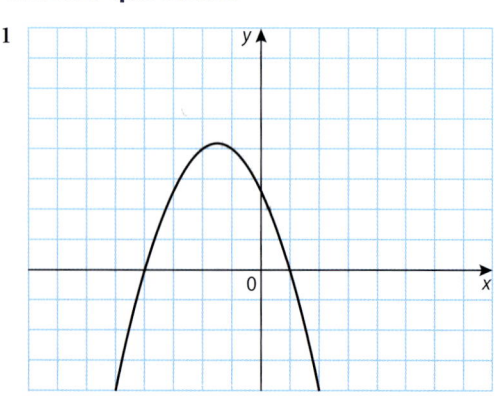

1

2 a) (i) $(-5, 0)$ and $(0, 0)$ (ii) $\left(-\frac{5}{3}, \frac{500}{27}\right)$ (iii) $\left(-\frac{10}{3}, \frac{250}{27}\right)$

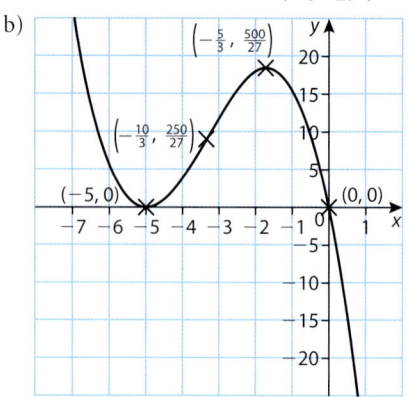

b)

3 $\left(\frac{1}{3}, 1\right)$

4 $y = 3x + 1$

5 a) (i) $a = -4$ (ii) $b = 2$

b) (i) $f'(x) = -3x^2 - 4x + 8$

(ii) $\frac{-2 + 2\sqrt{7}}{3}, \frac{-2 - 2\sqrt{7}}{3}$ (iii) $f(1) = 5.05$

c) (i) $y = 8x$ (ii) $x = -2$

6 a) (i) $v(0) = 0$ (ii) $v(10) \approx 51.3$

b) (i) $a(t) = 9.9e^{-0.15t}$ (ii) $a(0) = 9.9$

c) (i) 66   (ii) 0

(iii) as object falls it approaches terminal velocity

7 a) $\left(-\frac{2}{3}, -\frac{149}{27}\right)$ is a minimum, $(-4, 13)$ is a maximum

b) $\left(-\frac{7}{3}, \frac{101}{27}\right)$ is an inflexion point

8 a) (i) $g'(x) = -\frac{3}{e^{3x}}$

(ii) $e^{3x} > 0$ for all x; hence, $-\frac{3}{e^{3x}}$ for all x – therefore, $f(x)$ is decreasing for all x

b) (i) $e + 2$ (ii) $g'\left(-\frac{1}{3}\right) = -3e$

c) $y = -3ex + 2$

9 b) $f'(3) = 0$ and $f''(3) > 0 \Rightarrow$ stationary point at $x = 3$ and graph of f is concave up at $x = 3$, so $f(3)$ is a minimum

c) $(4, 0)$

10 a) $-\frac{4}{(2x + 3)^3}$ b) $5\cos(5x)e^{\sin(5x)}$

11 $A = 1, B = 2, C = 1$

12 a) (i) $\frac{1}{x}$ (ii) $-\frac{1}{x^2}$

b) (ii) When $x = e$, $\frac{dy}{dx} = 0$, $\frac{d^2y}{dx^2} = -\frac{1}{e^3} < 0$ and $f(e) = \frac{1}{e}$; therefore, the curve has a maximum value of $\frac{1}{e}$ at $x = e$

c) $x = 4$ or $x = 2$

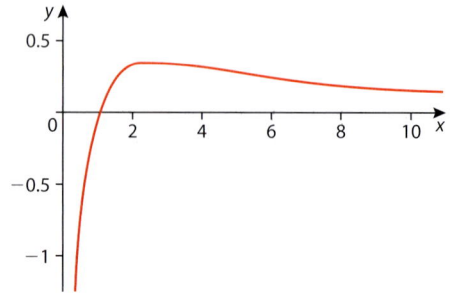

13 a) $\frac{x^4 + 3x^2}{(x^2 + 1)^2}$ b) $2e^x\cos(2x) + e^x\sin(2x)$

14 $a = 1$

15 a) $x = 3$; sign of $h''(x)$ changes from negative (concave down) to positive (concave up) at $x = 3$
   b) $x = 1$; $h'(x)$ changes from positive (h increasing) to negative (h decreasing) at $x = 1$
   c)

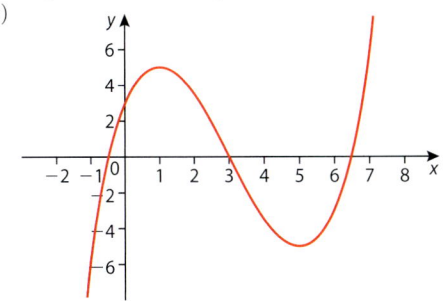

16 $y = 2ex - e$

17 $h = 8$ cm, $r = 4$ cm

18 Maximum area is 32 square units; dimensions are 4 by 8

19 a) E b) A c) C

20 $y = -\frac{1}{5}x + \frac{32}{5}$

Chapter 14

Exercise 14.1

1 $\frac{x^2}{2} + 2x + c$
2 $t^3 + t + c$
3 $\frac{x}{3} - \frac{x^4}{14} + c$
4 $\frac{2t^3}{3} + \frac{t^2}{2} - 3t + c$
5 $\frac{5u^{\frac{7}{5}}}{7} - u^4 + c$
6 $\frac{4x\sqrt{x}}{3} - 3\sqrt{x} + c$
7 $-3\cos\theta + 4\sin\theta + c$
8 $t^3 + 2\cos t + c$
9 $\frac{4x^2\sqrt{x}}{5} - \frac{10x\sqrt{x}}{3} + c$
10 $3\sin\theta - 2\tan\theta + c$
11 $\frac{1}{3}e^{3t-1} + c$
12 $2\ln|t| + c$
13 $\frac{1}{6}\ln(3t^2 + 5) + c$
14 $e^{\sin\theta} + c$
15 $\frac{(2x+3)^3}{6} + c$
16 $-\frac{5x^4}{4} + \frac{2x^3}{3} + cx + k$
17 $-\frac{x^5}{5} + \frac{x^4}{4} + \frac{x^2}{2} + 2x - \frac{11}{20}$
18 $\frac{4t^3}{3} + \sin t + ct + k$
19 $3x^4 - 4x^2 + 7x + 3$
20 $2\sin\theta + \frac{1}{2}\cos 2\theta + c$

Exercise 14.2

1 24
2 40
3 $\frac{24}{25}$
4 0
5 $\frac{176\sqrt{7} - 44}{5}$
6 0
7 2
8 -268
9 $\frac{64}{3}$
10 2
11 $\ln\left(\frac{11}{3}\right)$
12 $\frac{44}{3} - 8\sqrt{3}$
13 3
14 $\sqrt{\pi} + 1$
15 a) 6 b) 6 c) 12

Exercise 14.3

1 $\frac{125}{6}$
2 $\frac{9\pi^2}{8} + 1$
3 $4\sqrt{3}$
4 $\frac{10}{3}$
5 $\frac{8}{21}$
6 $\frac{125}{24}$
7 $\frac{13}{12}$
8 4π
9 $\frac{59}{12}$
10 Approx. 361.95 (4 points of intersection!)
11 $3\ln 2 - \frac{63}{128}$
12 Between $-\frac{\pi}{6}$ and $\frac{\pi}{6}$, $\sqrt{3}\ln\left(\frac{3}{4}\right) - 2\sqrt{3} + 4$

Exercise 14.4

1 $\frac{127\pi}{27}$
2 $\frac{64\sqrt{2}\pi}{15}$
3 $\frac{70\pi}{3}$
4 6π
5 9π
6 2π
7 $\left(\frac{\sqrt{3}}{2} + 1\right)\pi$
8 $\frac{512\pi}{15}$
9 Approx. 5.937π
10 $\frac{32\pi}{3}$

Exercise 14.5

1 $\frac{70}{3}$ m, 65 m
2 8.5 m to the left, 8.5 m
3 1 m, 1 m
4 2 m, $2\sqrt{2}$ m
5 18 m, 28.67 m
6 $\frac{4}{\pi}$ m, $\frac{4}{\pi}$ m
7 $3t$, 6 m, 6 m
8 $t^2 - 4t + 3$, 0, 2.67 m
9 $1 - \cos t$, $\left(\frac{3\pi}{2} + 1\right)$ m, $\left(\frac{3\pi}{2} + 1\right)$ m
10 $4 - 2\sqrt{t+1}$, 2.43 m, 2.91 m
11 $3t^2 + \frac{1}{2(1+t)^2} + \frac{3}{2}$, 11.3 m, 11.3 m

Practice questions

1 a) $p = 3$ b) 3 square units
2 a) $(0, 1)$ b) $V = \int_0^{\ln 2}\left(e^{\frac{x}{2}}\right)^2 dx$
3 $a = e^2$
4 a) $y = \frac{x}{e}$
5 a) (i) 400 m (ii) $v = 100 - 8t$, 60 m/s
   (iii) 8 s (iv) 1344 m
   b) Distance needed 625
6 b) 2.31 c) $-\pi\cos x - \frac{x^2}{2} + c$, 0.944
7 $\ln 3$
8 a) (ii) $(1.57, 0)$; $(1.1, 0.55)$; $(0, 0)$, $(2, -1.66)$
   b) $x = \frac{\pi}{2}$ c) (ii) $\int_0^{\frac{\pi}{2}} x^2 \cos x \, dx$ d) $\frac{\pi^2}{2} - 2$
9 a) 2π
   b) Range: $\{y \mid -0.4 < y < 0.4\}$
   c) (i) $-3\sin^3 x + 2\sin x$ (iii) $\frac{2\sqrt{3}}{9}$
   d) $\frac{\pi}{2}$
   e) (i) $\frac{1}{3}\sin^3 x + c$ (ii) $\frac{1}{3}$
   f) $\arccos\frac{\sqrt{7}}{3} \approx 0.491$
10 c) 3.69672 d) $\int_0^\pi (\pi + x\cos x)dx$ e) $\pi^2 - 2 \approx 7.86960$
11 a) (i) $10x + 1 - e^{2x}$ (ii) $\frac{\ln 5}{2} \approx 0.805$
   b) (i) $f^{-1}(x) = \frac{\ln(x-1)}{2}$ c) $v = \pi\int_0^{\ln 2}(1 + e^{2x})^2 \, dx$

Answers

Chapter 15
Exercise 15.1
1. a) Discrete b) Continuous
   c) Continuous d) Discrete
   e) Continuous f) Continuous
   g) Discrete h) Continuous
   i) Continuous j) Discrete
   k) Continuous l) Continuous
   m) Discrete
2. a) 0.4
   b)

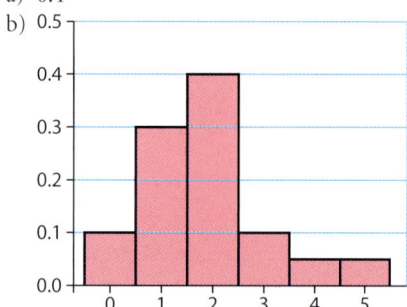

   c) 1.85, 1.19
3. a) 0.26 b) 0.37
   c) 0.77 d) 16.29
   e) 8.126
4. a) 0.969 b) 0.163
   c) 3.5
5. $k = \frac{1}{30}$

x	12	14	16	18
$P(X = x)$	$6k$	$7k$	$8k$	$9k$

6. a) $k = \frac{1}{10}$ b) $\frac{37}{60}$ c) $\frac{19}{30}$
   d) $E(x) = 16$, SD = 7
7. a) $\frac{1}{50}$
   b)

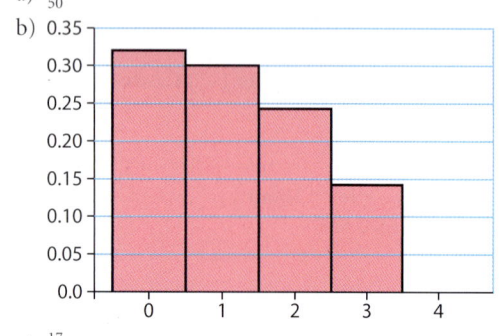

   c) $\frac{17}{25}$
   d) $\mu = 1.2$; var. = 1.08
8. a) $P(x = 18) = 0.2$, $P(x = 19) = 0.1$, symmetric distribution
   b) $\mu = 17$, var. = 1.2
9. a) $\mu = 1.9$, SD = 1.34
   b) between 0 and 5
10. $k = 0.667$, $E(x) = 5.44$
11. a) $k = 0.3$ or 0.7
    b) For $k = 0.3$: $E(x) = 2.18$; for $k = 0.7$: $E(x) = 1.78$
12. a)

y	0	1	2	3
$P(Y = y)$	$\frac{1}{27}$	$\frac{2}{9}$	$\frac{4}{9}$	$\frac{8}{27}$

   b) 2
13. a) $k = \frac{1}{10}$ b) $\frac{1}{2}$

Exercise 15.2
1. a)

x	0	1	2	3	4	5
$P(X = x)$	0.010 24	0.0768	0.2304	0.3456	0.2592	0.077 76

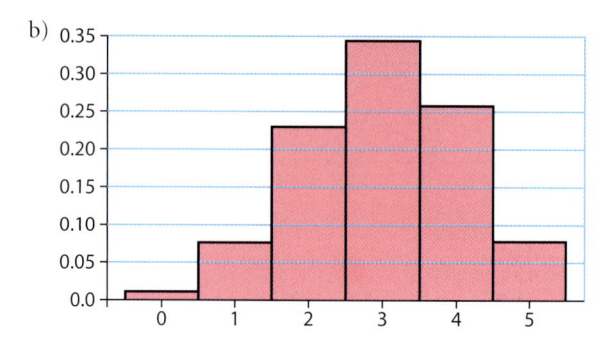

   b)

   c) (i) Mean = 3, SD = 1.095
      (ii) Mean = 3, SD = 1.095
   d) Between 2 and 4, and between 1 and 5
   e) 0.8352, 0.990. Slightly more than the empirical rule.

2. a) 0.001 294 494
   b) 0.000 000 011
   c) 0.999 999 99
   d) 0.999 999 66
   e) Mean = 12, SD = 2.19

3. a)

k	$P(x \leq k)$
0	0.117 65
1	0.420 17
2	0.744 31
3	0.929 53
4	0.989 07
5	0.999 27
6	1

   b)

Number of successes x	List the values of x	Write the probability statement	Explain it, if needed	Find the required probability
At most 3	0, 1, 2, 3	$P(x \leq 3)$	$P(x \leq 3)$	0.929 53
At least 3	3, 4, 5, 6	$P(x \geq 3)$	$1 - P(x \leq 2)$	0.255 69
More than 3	4, 5, 6	$P(x > 3)$	$1 - P(x \leq 3)$	0.070 47
Fewer than 3	0, 1, 2	$P(x \leq 2)$	$P(x \leq 2)$	0.744 31
Between 3 and 5 (inclusive)	3, 4, 5	$P(3 \leq x \leq 5)$	$P(x \leq 5) - P(x \leq 2)$	0.254 96
Exactly 3	3	$P(x = 3)$	$P(x = 3)$	0.185 22

4 a)

k	P(x ≤ k)
0	0.027 99
1	0.158 63
2	0.419 90
3	0.710 21
4	0.903 74
5	0.981 16
6	0.998 36
7	1

b)

Number of successes x	List the values of x	Write the probability statement	Explain it, if needed	Find the required probability
At most 3	0, 1, 2, 3	P(x ≤ 3)	P(x ≤ 3)	0.710 21
At least 3	3, 4, 5, 6, 7	P(x ≥ 3)	1 − P(x ≤ 2)	0.580 10
More than 3	4, 5, 6, 7	P(x > 3)	1 − P(x ≤ 3)	0.289 79
Fewer than 3	0, 1, 2	P(x ≤ 2)	P(x ≤ 2)	0.419 90
Between 3 and 5 (inclusive)	3, 4, 5	P(3 ≤ x ≤ 5)	P(x ≤ 5) − P(x ≤ 2)	0.561 26
Exactly 3	3	P(x = 3)	P(x = 3)	0.290 304

5 a) p is not constant, trials are not independent
  b) p becomes constant
  c) $n = 3$, $p = \frac{5}{8}$

y	0	1	2	3
P(Y = y)	0.052 73	0.263 672	0.439 453	0.244 141

  d) 0.755 86   e) 1.875   f) 0.703 125   g) 0.947 27
6 a) 0.107 374 b) 0.993 63 c) 0.892 63   d) 2
7 a) 0.817 073 b) 1   c) 0.016 1776
8 a) 0.033 833 b) 0.024 486 c) 0.782 722
9 a) 0.75   b) 0.032 5112 c) 0.172 678
10 a) 0.043 1745 b) 0.997 614 c) 0.011 2531
   d) 0.130 567   e) 0.956 826 f) 10
   g) 3   h) 4, 16
11 a) 3   b) 0.101 308   c) 0.000 214 925
12 a)

x	0	1	2	3	4	5
P(x)	0.031 25	0.156 25	0.312 50	0.312 50	0.156 25	0.031 25

  b) 0.031 25   c) 0.031 25
  d) 0.968 75   e) 0.968 75
  f) a)

x	0	1	2	3	4	5
P(x)	0.327 68	0.409 60	0.204 80	0.051 20	0.006 40	0.000 32

  b) 0.327 68   c) 0.000 32
  d) 0.672 32   e) 0.999 68

Exercise 15.3

(Some answers are rounded)
1 a) 0.5   b) 0.499 571   c) 0.158 655
  d) 0.682 690   e) 0.022 750   f) 0
2 a) 0.769 86   b) 0.161 514   c) 0.656 947
  d) 0.343 053
3 a) 0.008 634   b) 0.982 732
4 1.28
5 1.96
6 a) 0.066 807   b) 0.682 69   c) 678.16
  d) 134.898
7 a) 1.8%   b) 509.975   c) 5.71
8 a) 0.9696   b) 0.546 746
9 a) 1 day   b) 29 days   c) 112 days
10 1.56
11 30.81
12 $\mu = 21.037$, $\sigma = 4.252$
13 $\mu = 18.988$, $\sigma = 0.615$
14 a) 0.655 422 b) 0.008 198   c) 82 bottles
15 a) 0.227 319 b) 0.55%   c) 29.678
   d) 229.182
16 a) Not likely: chance is 0.14%   b) 15.87%
   c) 68.27%   d) 5396   e) 43 785
17 a) 6.817   b) 3.4315
   c) $\mu = 64.14$, $\sigma = 7.545$

Practice questions

1 a) 34.5%   b) 0.416   c) 3325
2 a) (i) 0.393   (ii) 0.656
  b) 50
3 a) 0.1   b) 10   d) 0.739
4 a) $\frac{35}{128}$ b) $\frac{7}{32}$ c) $\frac{91}{128}$
5 a) $a = -0.455$, $b = 0.682$
  b) (i) 0.675   (ii) 0.428
  c)

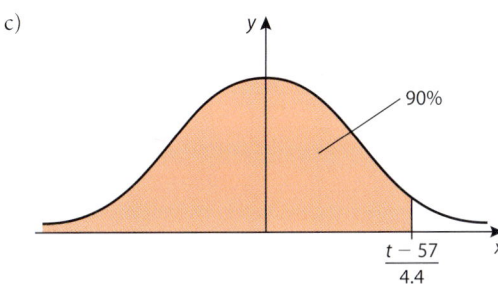

  (ii) $t = 62.6$
6 a) $\mu = 50 - 10(0.522\,44) \approx 44.8$
  b) H1: the mean speed has been affected by the campaign.
  c) One-tailed test, as we are interested in a decrease in the mean only (not also an increase).
7 a) 70.1%   b) 0.002 26   c) p-value = 5.48%
8 a) 0.0808
  c) $\mu = 25.5$, $\sigma = 0.255$
  d) 12 500
9 a) (i) 0.345   (ii) 0.115   (iii) 0.540
  b) 0.119   c) 737
10 a) 15.9%   b) 227 cm
11 a) 0.0912   b) $a = 251$, $b = 369$

Answers

12 a) $a = -1, b = 0.5$
b) (i) 0.841   (ii) 0.533
c) (i)

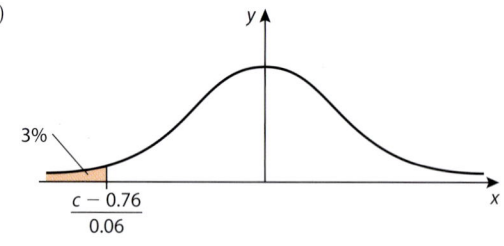

[Normal distribution curve with 3% shaded in left tail at $\frac{c - 0.76}{0.06}$]

(ii) 0.647

13 a) 2
b) 0.182
c) 0.597

14 $\mu = 66.6, \sigma = 22.6$

15 a) 0.8
b) (i)

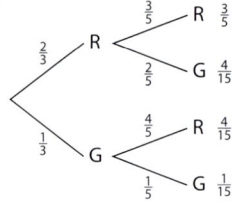

[Tree diagram:
- $\frac{2}{3}$ R → $\frac{3}{5}$ R $\frac{3}{5}$; $\frac{2}{5}$ G $\frac{4}{15}$
- $\frac{1}{3}$ G → $\frac{4}{5}$ R $\frac{4}{15}$; $\frac{1}{5}$ G $\frac{1}{15}$]

(ii)

X	0	1	2
$P(X = x)$	$\frac{1}{15}$	$\frac{8}{15}$	$\frac{2}{5}$

c) $\frac{3}{10}$ d) $\frac{1}{9}$

16 a) 0.129 886   b) 0.676 714   c) 2

17 a) 0.1829   b) 0.3664

18 a) $\frac{1}{5}$ b) $\frac{7}{5}$

19 a) (i) 0.217%   (ii) 0.012%
b) 84.13%

Index

Page numbers in italics refer to information boxes and hint boxes.

A

absolute values 4–5
accelerations 375–8, 486, 490
actuaries 495
additive inverses 5
al-Khowarizmi 1
algebra 1
algebraic expressions 14–20
algebraic fractions 18–20
amplitudes 181
angles 161, *204*
  bearings *265*
  complementary angles 208
  degrees 161, *162*, 163–5
  depression 208–11
  elevation 208–11
  incidence 453
  radians 162–5, *349*, 422, 455
  refraction 453
  vectors 268
anti-derivatives 463–5, 467, 474
arcs 162, 165–6
areas
  integration 470–4, 477–81
  sectors 221
  triangles 218–20
Aristotle 559
arithmetic 558
arithmetic means (*see* means)
arithmetic operations 5
arithmetic sequences 80–2
arithmetic series 90–1
asymptotes 38–9, 111, *349*, *350*
average velocities 373
axes of symmetry, parabolas 66
axioms 561, 562, 564, 565, 566
  completeness 566
  consistency 566

B

Babylonians *162*
base vectors 256
bases (exponents) 106
bearings *265*
Bernoulli, Bernoulli & Bernoulli *117*
binomial coefficients 98–100
binomial distribution 508–12
binomial experiments 509–10
binomial probability model 511
binomial theorem 100–2
binomials *14*, 96–7
Bolyai, Janos 565
bounded intervals 3
box plots 293–4
Briggs, Henry *122*
Buffon, Count *317*

C

calculus 347, 360
Cartesian coordinate system 33
categorical (qualitative) data 276
Cayley, Arthur *141*
Celsius scale *92*
chain reasoning 562
chain rule 431–6, 437, 467, 468
change of base formula 127–8
circular functions (*see* trigonometric functions)
closed intervals 2
column vectors 142
common logarithmic function 124
commutative property 146, 258
complementary angles 208
complements (events) 323
completeness, axioms 566
completing the square 67–8
components, vectors 253–4, 256, 257, 393
composite functions 41–44, 46–7
  derivative 431–6
  domains 44
  ranges 44
compound fractions 19–20
compound interest 85–6, 115–16, 118–19
concave graphs 378–9, 380
conditional probabilities 331–5
conjectures 562
conjugates 15
consistency, axioms 566
constants of integration 465
constructivism 561
continuous compound interest formula 119
continuous functions 368, *369*
continuous random variables 496, 515–16
coordinate planes 23, 390
coordinates 1
cosine function 169–76, 189, 196–7, 205–8, 214–18
  derivative 422–4
  domain 172
  graph 178–85
cosine rule (law of cosines) 229–33
critical points 369
critical values (optimization) 448
cube roots 7
cumulative frequency distributions 279–81, 300

D

data sets 275
definite integrals 472–6, *486*, 488
degrees (angles) 161, *162*, 163–5
degrees (polynomials) 65
denominators 8–9, 20
density functions (probability) 515–16
dependent variables 33
depression (angles) 208–11
derivatives 356–65
  chain rule 431–6, 437, 467, 468
  composite functions 431–6
  cosine function 422–4
  exponential functions 425–7, 437
  first derivative test 371–2, 455–6
  natural logarithmic function 427–30
  product rule 439–42
  quotient rule 442–6
  second derivative test 380–1, 456
  second derivatives 374–80
  sine function 421–2
  tangent function 424, 443–4
Descartes, René 33, *383*
descriptive statistics 275
determinants 153
diagonal matrices 142
difference of two squares 15
difference quotients 354
differentiable functions *369*
Dirac, Paul 567–8
directed line segments 251
direction angles, vectors 260–2, 399–400
discrete random variables 496, 498, 502, 508
disjoint events 323, *330*
displacements 373, 375–7, 485–90
distance formula 25–6, 253, 259
distances *376*, 485–90
distributive property 15, 258
domains 34, 35–6, 37, *38*
  composite functions 44
  exponential functions 111
  inverse functions 46, 47, 49–50
  logarithmic functions 122
  one-to-one functions 48–9
  trigonometric functions 172, 177
dot (scalar) products 266–70, 398–9
double angle identities 197

E

e (number) 117–120
Einstein 567
elevation (angles) 208–11
elimination 28–9
elliptical geometry 565
empty set 3
equally likely outcomes 325
equations 22
equinoxes 535
etymologies
  algebra 1
  equinox 535

Index

quadratic 65
secant 355
solstice 535
tangent 355
trigonometry 160
Euclid 562, 564
Euclid's postulates 564–5
Euler, Leonard 38, *119*
evaluation theorem 475
even functions 179
events 319–20, 328–31
expansion (algebra) 15–16
expected values 501–2
explicit definitions, sequences 79
exponential equations 130–3
exponential functions 106–13, 424–5, 542–3
   derivative 425–7, 437
   domain 111
   e (number) 117–120
   graph 111–13
   inverse function 122
   natural exponential function 120
   range 111
exponential models 113–16, *427*
exponents 9–12, 106
expressions (algebra) 14–20
extremes (functions) *370*

F

factorization (algebra) 16–18
   quadratic functions 72
Fahrenheit scale *92*
Fermat, Pierre *22*, 447
Fibonacci sequence 575
finite sequences 78
first derivative test 371–2, 455–6
five number summaries 293
formalism 561, 572
formulae 22
frequency distribution tables 277–8
functions 34–40
   composite functions 41–44, 46–7
   continuous functions 368, *369*
   differentiable functions *369*
   domains 34, 35–6, 37, *38*
   graphs 53–64
   identity function 48
   inverse functions 45–7, 48–52
   limits 350
   one-to-one functions 48–9
   ranges 34, 35–6, 37, *38*
fundamental theorem of calculus 474–5

G

Galileo 567
geometric sequences 82–7
geometric series 91–6
geometry 558
Gibbs, J. Williard *251*
Godel, Kurt 566
golden ratio 573–5
gradients (slopes) 23, 25, 235–7, 352–6
graphs 22–6, 53–64

areas 470–4, 477–81
cosine function 178–85
exponential functions 111–3
logarithmic functions 123–4, 125
normal lines 385–6
quadratic functions 66–8, 72–3
reflections 57–82
secant lines 354, 364
sine function 177–85
slopes 23, 25, 235–7, 352–6
stretches 59–64
tangent function 185–7
tangent lines 353–4, 382–5
translations 54–7
grouped data 296–7

H

half-lives 114, 543–5
half-open intervals 2
haphazard events 318
health-adjusted life expectancies 273–4
Hersch, Reuben 571–3
Hilbert, David 566
histograms 278–9
horizontal stretches 60–3
horizontal translations 55–6
humanism 572
hyperbolic geometry 565
hypotenuses 204

I

identities 22
identity function 48
identity matrices 145–6
incidence (angles) 453
indefinite integrals *465*
independent events 330, 336–7
independent variables 33
indices (exponents) 7
infinite sequences 78
infinite sums (*see* series)
infinity symbol (∞) *38*
inflexion points 378–80
instantaneous velocities 373
integers 2
integrands *465*
integration 466–70
   areas 470–4, 477–81
   definite integrals 472–6, *486*, 488
   substitution 468–70
   volumes 482–4
interest 84–6
interquartile ranges (IQRs) 292–3, 294
intersections (events) 329
intersections (lines) 405
intersections (sets) 3
intuition 569–71
intervals 2–3
invariants 558
inverse functions 45–52
   domains 46, 47, 48–52
   exponential functions 122
   ranges 46, 47
inverse matrices 151–5
inverse normal distribution 524–7

inverse operations (*see also* additive inverses, multiplicative inverses) 46
irrational numbers 2

J

Jorgensen, Palle 568

K

kinematics 485

L

Laplace, Pierre-Simon *122*
law of cosines 229–33
law of sines 223–9, 238, 240
Leibniz, Gottfried 360, *439*
Leibniz notation 367, 374
limits 348–51, 354–5
line segments *204*
linear equations (*see also* systems of equations) 23, 149–50
linear functions 65
lines
   three-dimensional space 402–9
   two-dimensional space 23–5, 235–7
Lobachevsky, Nicolai 565
logarithmic functions 121–8
   common logarithmic function 124
   graph 123–4, 125
   natural logarithmic function 124

M

magnitudes (vectors) 251, 259, 391
mappings (*see also* functions) 559
matrices 140–7, 151–5
   addition 143
   determinants 153
   identity matrices 145–6
   multiplication 144–7
   multiplicative inverses 151–5
   subtraction 143
maxima 66, 369–70, 371–2, 380–1, 447
mean value theorem 464–5
means 284, 285, 296, 298
medians 284, 285–6
midpoints 26
minima 66, 369–70, 371–2, 380–1, 447, 450
modes 284, 286, 287
monomials *14*
Monty Hall game 570–1
multiplication rule (probabilities) 330, 333
multiplicative inverses 5, *54*
   matrices 151–5
mutually exclusive events 323, 329

N

Napier, John *122*
natural exponential function 120
   derivative 425–7
natural logarithmic function 124
   derivative 427–30

natural numbers 2
negatively skewed distributions 288
Newton, Isaac 360
Newton notation 367, 374
Newton's law of cooling 427
non-Euclidian geometries 565
non-rigid transformations 59
normal distribution 516–23
    inverse normal distribution 524–7
    standard normal distribution 519
normal lines 385–6
numerical (quantitative) data 276

O

odd functions 179
ogives 279, 294–5
one-to-one correspondences 1
one-to-one functions 48–9
open intervals 2
optimization 447–53
ordered pairs 23, 33
Oughtred, William 122
outliers 293, 298

P

parabolas 66, 368
parallel lines 25, 404
parallel vectors 255, 267, 393, 402
parameters (algebra) 22, 150
parameters (statistics) 284, 285
parametric equations 402–3, 406
partial sums 88
Pascal's triangle 97–8, 100
Pearson, Karl 317
percentiles 291–2
periodic functions 169, 421
perpendicular lines 25, 385
perpendicular vectors 267, 398
phase shifts 183
Plato 560–1
Platonism 561, 572, 573
polynomials 14–18
population growth 86–7, 427
populations (statistics) 274, 289
position vectors 253
positively skewed distributions 288
power functions 359
powers (see exponents)
primitive notions 559
principal square roots 7, 39
principle of least time 447, 453
probabilities 316–7, 323–5
    conditional probabilities 331–5
probability distributions 497–500
    binomial distribution 508–12
    density functions 515–16
    expected values 501–2
    normal distribution 516–27
    standard deviations 503–5
probability models 318
probability tables 326
product rule (derivatives) 439–42
projections 267
proofs 562–3
proper classes 560

proper subsets 2
Pythagoras' theorem 22, 25, 205, 231, 253
Pythagorean identities 197

Q

quadratic equations 69–71
quadratic formula 68, 70
quadratic functions 65–73, 367
    factorization 72
    graph 66–8, 72–3
qualitative data 276
quantitative data 276
quartiles 292, 295, 524
quotient rule (derivatives) 442–6

R

radians 162–5, 349, 422, 455
radicals 7–9
radioactive materials 114, 427, 498
random events 318
random variables 495–6
    continuous random variables 496, 515–16
    discrete random variables 496, 498, 502, 508
ranges (data) 288–9, 293
ranges (functions) 34, 35–6, 37, 38
    composite functions 44
    exponential functions 111
    logarithmic functions 122
    inverse functions 46, 47
    one-to-one functions 48, 49
    trigonometric functions 172
rational numbers 2
real number line 1, 168
real numbers 1–5
recursive definitions, sequences 79, 83
reflections 57–82
refraction (angles) 453
relations 33, 36
relative cumulative frequency distributions 279
relative frequency theory 322
resolving, vectors 262
right triangles 204–11, 239
rigid transformations 59
roots (see also solutions) 7
row vectors 142
Russell, Bertrand 559, 560
Russell's paradox 560

S

sample spaces 318–19
samples 274, 289
scalar products 266–70, 398–9
scalars 144, 251, 254–5
scale factors 59
scientific notation 12–14
Searle, John 571
secant lines 354, 364
second derivative test 380–1, 456
second derivatives 374–80
sectors 166–7

self-inverse functions 48
semantic method 563
sequences 78–9
    arithmetic sequences 80–2
    geometric sequences 82–7
series 88–96
    arithmetic series 90–1
    geometric series 91–6
    sigma notation 88–90
sets 2–3, 319, 559, 560
sexagesimal number system 162
shrinks (see stretches)
sigma notation 88–90
significant figures 13
similar triangles 205
simultaneous equations (see also systems of equations) 27–30
sine function 169–76, 189, 196–7, 205–8, 214–18
    derivative 421–2
    domain 172, 177
    graph 177–85
sine rule (law of sines) 223–9, 238, 240
singular matrices 153
sinusoidal functions 535–6
skew lines 403, 404
skewed distributions 287–8
slide rules 122
slopes 23, 25, 235–7, 352–6
Snell's law 453
socially constructed facts 571
solids of revolution 483
solstices 535–6
solution sets 22, 23, 149
speeds 376, 486
square roots 7
standard deviations
    data 289–91, 296, 297, 298
    random variables 503–5
standard form (scientific notation) 12–14
standard normal distribution 519
stationary points 368–9, 370–1, 380–1
statistics 284, 285
stretches 59–64
submatrices 142
subsets 2
substitution (integration) 468–70
substitution (simultaneous equations) 29–30
surds (radicals) 7–9
surface areas 548
symmetric distributions 287
syntactic method 563
systems of equations (see also simultaneous equations) 27–9, 154–5

T

tangent function 169–76, 189, 205–7, 215–17
    derivative 424, 443–4
    domain 172
    graph 185–7
tangent lines 353–4, 382–5

Index

terminal velocities *490*
Thales of Miletus *205*
theorems 562
three-dimensional trigonometry 241–3
total change 475
Towers of Hanoi 536
transformations (*see* reflections, stretches and translations)
translations 54–7
tree diagrams 325–6
triangular matrices 143
triangular rule 392
trigonometric equations 189–96
trigonometric functions 169–76, 205–7, 214–18
  domain 172
  graphs 177–87
  range 172
trigonometric identities 196–200, *217*
trigonometry 160–1
trinomials *14*

U

unbounded intervals 3
unions (events) 329
unions (sets) 3
unit circle 163
unit vectors 256, 259, 394–6

V

variables (statistics) 275
variances
  data 289, 291, 296, 300
  random variables 502–4
vector equations 406–7
vectors 142, 250–8, 391–6
  addition 255–6
  base vectors 256
  components 253–4, 256, 257, 393
  direction angles 260–2, 399–400
  parallel vectors 255, 267, 393, 402
  perpendicular vectors 267, 398
  scalar products 266–70, 398–9
  unit vectors 256, 259, 394–6
velocities 375–8, 407–8, 486–8, 490
Venn diagrams *2*, 319
vertexes (graphs) 66, 367–8
vertical stretches 59–60, 64
vertical translations 54, 56, 64
volumes 482–4, 548

W

Whitehead, A.N. 559
Wiles, Andrew 22
work *270*
World Health Organization 273–4
wrapping functions 169

Z

z-scores 518, 524
zero matrices 142
zeros, functions 69